The Noise Manual
Revised
Fifth Edition

The Noise Manual
Revised
Fifth Edition

Editors

Elliott H. Berger, MS, INCE Bd. Cert.
Senior Scientist, Auditory Research
E•A•R/Aearo Company
Indianapolis, IN 46268-1657
317-692-3031
eberger@compuserve.com

Larry H. Royster, PhD
Professor Emeritus
Department of Mechanical and
  Aerospace Engineering
North Carolina State University
Raleigh, NC 27695-7910
919-515-5225
royster@eos.ncsu.edu

Dennis P. Driscoll, MS, PE
President
Associates in Acoustics, Inc.
Evergreen, CO 80439
303-670-9270
noisecontrol@compuserve.com

Julia Doswell Royster, PhD, CCC-A/SLP
President
Environmental Noise Consultants, Inc.
Raleigh, NC 27622-0698
919-782-1624
effective_hcps@compuserve.com

Martha Layne
President
Marty Layne Audiology
Portland, ME 04103-6004
207-878-5540
mlayne@gwi.net

Available from:
American Industrial Hygiene Association
2700 Prosperity Avenue, Suite 250
Fairfax, VA 22031
www.aiha.org

ISBN 0-931504-02-4

Stock #619-BP-03

Printed in the United States of America

The inside pages are of paper crafted with 20% post-consumer fiber to meet the
executive order on recycled papers and are acid free to provide archival quality that
ensures the longevity of the printed piece. The entire book has been printed with
alcohol-free soybean inks.

Authors

Robert Anderson
Principal
James Anderson and Assoc., Inc.
Okemos, MI 48864
517-349-8066
RobertAnderson@jaa-inc.com

Elliott H. Berger, MS, INCE Bd. Cert.
Senior Scientist, Auditory Research
E•A•R/Aearo Company
Indianapolis, IN 46268-1657
317-692-3031
eberger@compuserve.com

Warren Blazier, MS
President
Warren Blazier Associates, Inc.
San Francisco, CA 94123
415-474-5673
warrenblaz@aol.com.

John G. Casali, PhD, CPE
Chaired Professor
Grado Dept. of Industrial and Systems
  Engineering
Director, Auditory Systems Laboratory
Virginia Tech
Blacksburg, VA 24061
540-231-5073
jcasali@vt.edu

Robert Dobie, MD
Clinical Professor
Dept. of Otolaryngology
UC-Davis Medical Center
Sacramento, CA 95817
916-734-2704
radobie@ucdavis.edu

Dennis P. Driscoll, MS, PE
President
Associates in Acoustics, Inc.
Evergreen, CO 80439
303-670-9270
noisecontrol@compuserve.com

John J. Earshen, MS
President
Angevine Acoustical Consultants
Elma, NY 14059
716-652-0282
angevineac@aol.com

Susan Cooper Megerson, MA
Intercampus Program in Communicative
  Disorders
University of Kansas
Kansas City, KS 66160
913-962-1759
scmegerson@mindspring.com

Paul Ostergaard, SM, PE
Retired
Fairview, PA 16415
814-474-4634
leadoxide@alum.mit.edu

Gary Robinson, PhD
Research Associate Professor
Auditory Systems Laboratory
Industrial and Systems Engineering
Virginia Tech, Blacksburg, VA 24061
540-231-2680
grobins@vt.edu

(continued next page)

Authors (cont.)

Julia D. Royster, PhD, CCC-A/SLP
President
Environmental Noise Consultants, Inc.
Raleigh, NC 27622-0698
919-782-1624
effective_hcps@compuserve.com

Larry H. Royster, PhD
Professor Emeritus
Department of Mechanical and
  Aerospace Engineering
North Carolina State University
Raleigh, NC 27695-7910
919-515-5225
royster@eos.ncsu.edu

Andrew P. Stewart, MA
Occupational Audiologist
Durham, NC 27701
919-530-1334
andystewart@mindspring.com

Noral D. Stewart, PhD
President
Stewart Acoustical Consultants
Raleigh, NC 27622
919-781-8824
noral@sacnc.com

Alice H. Suter, PhD
Alice Suter and Associates
Ashland, OR 97520
541-488-8077
ahsuter@charter.net

W. Dixon Ward, PhD
Professor Emeritus
Departments of Communication
  Disorders and Otolaryngology
University of Minnesota
Minneapolis, MN 55455-3007
Deceased

Foreword

The American Industrial Hygiene Association's fifth edition of *The Noise Manual* represents the most current information available on the subject of noise and hearing conservation. It is a valuable reference not only for those readers with a technical interest in noise, but also for those who desire an improved general understanding of the subject. Besides addressing noise-related issues within the workplace, this edition has been expanded to also include noise-related problems arising in the community as well.

Many people have worked diligently to develop this manual. The American Industrial Hygiene Association is pleased to acknowledge their contributions in making this new edition possible. We are confident that all readers of this manual will benefit through a better knowledge of noise, and methods for its control.

James R. Thornton
President
American Industrial Hygiene Association

This work is dedicated

to W. Dixon Ward (1924 – 1996) for his pioneering research into the effects of noise on hearing, and for his contributions to this text and to the preceding 4th Edition,

to Ross Gardner, Jr. (1933 – 2000) for changing the face of hearing conservation by his invention of the slow-recovery foam earplug,

and to all of those individuals who have suffered hearing loss on and off the job as a result of harmful exposure to noise.

Preface

We, the Editors, are pleased to bring you the revised and expanded fifth edition of the AIHA *Noise Manual*. Many of us were involved with the fourth edition and welcomed the opportunity to update, and perfect an already highly acclaimed and often-quoted reference handbook.

Our intention for this fifth edition of *The Noise Manual* was to be comprehensive in scope and practical in content. The result is a thoroughly referenced and indexed text that will appeal to the novice industrial hygienist, while still providing a detailed reference source for experienced practitioners and for those involved in the regulatory arenas. At the same time, due to the multidisciplinary nature of noise and hearing conservation (hygiene, safety, acoustics, audiology, occupational medicine and nursing, engineering, behavioral, and legal), we feel the Manual should prove valuable to those beyond the industrial hygiene community. It is our hope that it will also continue to serve as a principal or supplemental textbook for courses and seminars in hearing conservation, occupational audiology, and the effects of noise on people.

The fifth edition includes completely new materials in seven chapters (1, 6, and 13–17) and three appendices (II–IV), new authors and hence different treatments in three chapters (9, 11, and 18), complete revisions of the materials in two additional chapters (2 and 10), and revisions and updates to the remaining chapters (3–5, 7, 8, and 12). An original introductory chapter was specifically designed to help the professional promote the value of hearing and the worth of hearing conservation, and an overview chapter was developed to provide the framework for managing an effective hearing conservation program. New areas of great interest to the hygiene and acoustical communities have been incorporated: room noise criteria, speech and signal detection, community noise, prediction of hearing loss due to noise exposure, and standards and regulations. A dramatically expanded and useful table of symbols and abbreviations may also be found at the front of this edition.

The history of the Manual began with the first edition in 1958, and continued through subsequent editions in 1966, 1975, and the fourth edition in 1986. Due to the continuing interest in noise, the excellent sales of the fourth edition (approximately 13,000 copies sold in 8 printings), and the changes in hearing conservation over a period of about a decade, the AIHA Noise Committee began considering preparation of a fifth edition in 1994. Subsequent to initial planning and agreements, work began in earnest in 1997 with the manuscript being completed in the spring of 2000.

Through the five editions, both the contents and the title of this text have changed, as the stature and importance of the book in the noise community continued to grow. For the first three editions the text was called *The Industrial Noise Manual*, changing to *The Noise & Hearing Conservation Manual* for the fourth edition, and in the current version, *The Noise Manual*. The name changes have reflected the broadening scope of the text from one primarily concerned with industrial/occupational noise, its measurement, effects, and control, to a book that provided a more expansive view of hearing conservation (fourth edition), and finally today, a text that also examines annoyance, the acoustic habitability of spaces, and communication issues.

Each chapter in this edition was primarily written by the author(s) with whose name(s) it is associated. Prior to editing, the chapters received critical peer reviews from the editors as well as the following readers, whose comments are gratefully acknowledged. Subsequently, the editors wheedled, cajoled, persuaded, and nit-picked to strive for accuracy, clarity, completeness, and consistency (both within and among chapters).

Jim Banach	Donald Gasaway	Paul Ostergaard
William Cavanaugh	Lee Hager	Charles Pavlovic
Allen Cudworth	Daniel Johnson	Dylan Romo
Robert Dobie	Thomas Lloyd	Paul Schomer
John Earshen	Mary McDaniel	Andrew Stewart
John Franks	Susan Megerson	Noral Stewart
Ross Gardner, Jr.	Deanna Meinke	Laurie Kastner-Wells

We extend our sincerest appreciation to the AIHA Noise Committee for the initiation and support of this project, and to Fran Kuecker, Manager, Publishing, and her staff of the American Industrial Hygiene Association for their extensive and dedicated efforts in the preparation of this manuscript for publication. Additionally, we wish to acknowledge the administrative assistance and research efforts of Cyd Kladden, Technical Secretary, E•A•R.

May, 2000

Elliott Berger
Larry Royster
Julie Royster
Dennis Driscoll
Marty Layne

Contents

(continued next page)

Contents (cont.)

Symbols and Abbreviations[1]

a	acceleration. Unit: m/s^2, in/s^2
$\mathbf{a_o}$	reference acceleration (10×10^{-6} m/s^2, 394×10^{-6} in/s^2)
a	absorption. Unit: m^2, ft^2
$\mathbf{A_{total}}$	attenuation at each octave band for outdoor sound propagation, composed of A_{div} (geometrical divergence), A_{air} (air absorption), A_{env} (environmental effects) and A_{misc} (miscellaneous other factors). Unit: dB
AAO-HNS	American Academy of Otolaryngology—Head and Neck Surgery
AAOO	American Academy of Ophthalmology and Otolaryngology
ACFM	actual cubic feet per minute. Unit: ft^3/min
ACGIH	American Conference of Governmental Industrial Hygienists
ACOEM	American College of Occupational and Environmental Medicine
ACOM	American College of Occupational Medicine
ADA	Americans with Disabilities Act
ADBA	audiometric database analysis
AI	articulation index. Unit: dimensionless
AIHA	American Industrial Hygiene Association
AMA	American Medical Association
ANSI	American National Standards Institute
ARI	Air-Conditioning and Refrigeration Institute
ASA	Acoustical Society of America. Also, until 1966, American Standards Association
ASHA	American Speech-Language-Hearing Association
ASHRAE	American Society of Heating, Refrigerating and Air-Conditioning Engineers
ASTM	American Society for Testing and Materials
BC	bone and tissue conduction
BI	binaural impairment of hearing. Unit: percent
BW	bandwidth as in effective filter bandwidth
c	speed of sound [344 m/s, 1128 ft/s @ 21°C]. Unit: m/s, ft/s

---

[1] Certain terms in this table may appear both as an abbreviation that can be used in a sentence and as a letter symbol that can be used in an equation. In such cases the letter symbols that are associated with the abbreviations will always appear as a leading capital letter with appended subscripts. Examples are DNL and L_{dn}, and PWL and L_W.

$C_1..C_2..C_N$ OSHA designator for total time of exposure at a stated A-weighted noise level corresponding to allowable exposures T_N. Running subscripts refer to particular noise levels for a range of values from 1 to N. Unit: min, h (also see $T_1..T_2..T_N$)

CAOHC Council for Accreditation in Occupational Hearing Conservation

CEC Council of the European Communities

CF Correction Factor for metric or English units in equations for sound pressure level. Unit: dB

CFR Code of Federal Regulations

CHABA Committee on Hearing, Bioacoustics, and Biomechanics

CNEL community noise equivalent level. Unit: dBA

cps cycles per second (also see hertz)

D noise dose as a percentage of maximum permitted daily noise dose

dB decibel

dBA decibel measured using A frequency-weighting (also see L_A)

dBC decibel measured using C frequency-weighting (also see L_c)

df degrees of freedom

DI directivity index. Unit: dB

DIS draft international standard

DNL day-night average sound level (A-weighting implicit). Unit: dBA (also see L_{dn})

DOD Department of Defense

DOL Department of Labor

DR dose rate accumulation. Unit: %/min

DRC damage-risk criteria

e napierian base = 2.71828 ... (natural logarithms)

E dynamic modulus. Unit: N/m^2

E_A A-weighted sound exposure computed with 3-dB exchange rate. Unit: Pa^2h

E_{AT} A-weighted sound exposure with measurement time period, T. E_{CT} denotes C-weighting. (Note: reporting of T is optional). Unit: Pa^2h

E_C C-weighted sound exposure. Unit: Pa^2h

ENT ear, nose, and throat

EPA Environmental Protection Agency

ER exchange rate. Unit: dB per time doubling

f frequency (cycles per second). Unit: hertz (Hz)

f_c center frequency. Unit: Hz

f_n natural frequency, also called resonant frequency. Unit: Hz

FDA Food and Drug Administration

FECA Federal Employees' Compensation Act

FFT	fast Fourier transform
g	acceleration of gravity [9.81 m/s^2, 32.2 ft/s^2]. Unit: m/s^2, ft/s^2
h	hour
H*	actual duration of workshift. Unit: min, h
HCA	Hearing Conservation Amendment (OSHA CFR 1910.95, paragraphs c – p)
HCP	hearing conservation program
HH	hearing handicap
HL	hearing level. Unit: dB
HPD	hearing protection device
HRTF	head-related transfer function
HTL	hearing threshold level. Unit: dB
HTLA	hearing threshold level associated with age. Unit: dB
HTLAN	hearing threshold level associated with age and noise. Unit: dB
HVAC	heating, ventilation, and air conditioning
Hz	hertz (cycles/second; also see cps)
HML	High, Medium, Low, hearing protector rating. Unit: dB
I	sound intensity. Unit: watts/m^2
I$_o$	reference sound intensity (10^{-12} w/m^2)
IEC	International Electrotechnical Commission
I-INCE	International Institute of Noise Control Engineering
IHC	inner hair cells
IL	insertion loss. Unit: dB
INCE	Institute of Noise Control Engineering
INEP	industrial-noise-exposed population
ISO	International Organization for Standardization
k	dynamic stiffness, as in a spring constant. Unit: N/m, lb/ft
k	Constant used in various equations relating sound power and sound pressure. The value differs for metric and English units, and is specific to each equation.
K	values used for determining tolerance limits as listed in Table 7.11. Unit: dimensionless
kg	kilogram
L$_a$	acceleration level (vibration). Unit: dB
L$_A$	A-weighted sound level. Unit: dBA
L$_{A8hn}$	A-weighted average sound level with 3-dB exchange rate, normalized to 8 hours. Same as L$_{EX,8h}$ as defined in ISO 1999. Contrast with L$_{eq,T}$ which is a non-normalized quantity. Unit: dBA
L$_{AE}$	sound exposure level with A-weighting. Unit: dBA (also see SEL)
L$_{Aeq,T}$	see L$_{eq,T}$

$L_{AF}(t)$ A-weighted and fast response sound level as a function of time. Use of S instead of F denotes slow response. Unit: dBA

L_{Apk} peak sound level with A-weighting. L_{Cpk} denotes C-weighting, and lack of either an A or a C denotes no frequency weighting. Unit: dBA

L_{AS} sound level with A-weighting and slow response. Similarly L_{AF} denotes fast response. Unit: dBA

L_{ASmax} maximum sound level with A-weighting and slow response. Unit: dBA

L_{ASX} level exceeded X% of the time, measured using A-weighting and slow response. L_{AFX} denotes fast response. The total measurement duration must be stated. Unit: dBA

$L_{A}(t)$ A-weighted sound level as a function of time. Unit: dBA

L_{Awkn} A-weighted weekly average sound level, computed by energy averaging (3-dB exchange) the daily TWAs (5-dB exchange), normalized to a 5-day exposure. Unit: dBA

L_{Ayrn} A-weighted yearly average sound level, computed by energy averaging (3-dB exchange) either the weekly TWAs or values of L_{Awkn}, normalized to a 48- or 50-week exposure, as appropriate. Unit: dBA

L_{c} criterion level (A-weighted). Exposure at L_{c} for the criterion time generates 100% noise dose. Unit: dBA

L_{C} C-weighted sound level. Unit: dBC

L_{Cdn} day-night average level, C-weighting used. Unit: dBC (also see L_{dn})

L_{CE} sound exposure level with C-weighting. Unit: dBC (also see SEL)

L_{CS} sound level with C-weighting and slow response. L_{CF} denotes fast response. Unit: dBC

L_{dn} day-night average sound level (A-weighting implicit), use of L_{Adn} is optional. Unit: dBA (also see DNL).

$L_{eq,T}$ equivalent-continuous sound level, also called average sound level, during time period T, using a 3-dB exchange rate. Weighting must be specified separately as in $L_{Aeq,T}$. Contrast to L_{A8hn} for normalized average sound level, but note that for an 8-hr measurement, $L_{A8hn} = L_{Aeq,8h}$. Unit: dB, dBA, or dBC.

L_{F} sound levels or sound pressure levels measured with fast time constant; 125-ms exponential weighted time average instrument response, often called "fast response." (also see L_{S}.)

L_{mt} masked threshold. Unit: dB

$L_{OSHA,T}$ A-weighted average sound level, during the time period T, with 5-dB exchange rate and slow meter response, applied in OSHA/MSHA practice. The time period over which the average is computed should be stated, but in most contemporary instruments and in common usage the averaging time is reported separately, hence the "T" in the subscript does not appear. The equivalent metric based on a 3-dB exchange rate is $L_{Aeq,T}$; the 8-hr normalized quantity, with a 5-dB exchange rate, is TWA. Unit: dBA

L_p	sound pressure level. Unit: dB (also see SPL)
L_{pb}	band pressure level, i.e., the sound pressure level for sound contained within a restricted band of frequencies (e.g., octave, 1/3 octave). Unit: dB
L_{ps}	spectrum sound pressure level. Unit: dB/Hz
L_S	sound levels or sound pressure levels measured with slow time constant; 1-s exponential weighted time average instrument response, often called "slow response." (also see L_F)
L_W	sound power level. Used with A or C subscript (e.g., L_{WA}) denotes use of A- or C-weighting. Unit: dB (also see PWL)
$L_1..L_2..L_N$	OSHA designator for a specific A-weighted noise level at which there is a permissible time of exposure (also called reference duration). Running subscripts refer to particular noise levels for a range of values from 1 to N. Unit: dBA
LCL	lower confidence limit. Unit: dimensionless
log	logarithm. When no subscript appears, base 10 is assumed.
m	meter
m	mass. Unit, kg, slugs or lb(m).
MAF	minimum audible field. Unit: dB
MAP	minimum audible pressure. Unit: dB
ms	1/1000 s (millisecond)
MSHA	Mine Safety and Health Administration
N	newton (Kg-m/s^2)
N	rotational speed. Unit: rpm
NASED	National Association of Special Equipment Distributors
NBS	National Bureau of Standards. Current name has been changed to NIST
NC	noise criterion level. Unit: dB
NCB	balanced noise criteria. Unit: dB
NCPF	noise control priority factor.
NHCA	National Hearing Conservation Association
NIHL	noise-induced hearing loss. Unit: dB
NINEP	non–industrial-noise-exposed population
NIOSH	National Institute for Occupational Safety and Health
NIPTS	noise-induced permanent threshold shift. Unit: dB
NIST	National Institute of Standards and Technology
NR	noise reduction. Unit: dB (in Chapter 13 NR is used to mean Noise Rating)
NRC	noise reduction coefficient. Unit: dimensionless
NRR	Noise Reduction Rating. Often a trailing subscript, as in NRR_{84}, is used to indicate the percentage of the population that is protected. Unit: dB

NRR(SF)	Noise Reduction Rating (Subject Fit). Unit: dB
NSC	National Safety Council
NVLAP	National Voluntary Laboratory Accreditation Program
OAE	otoacoustic emissions
OB	octave band. Often preceded by 1/1 or 1/3 to indicate octave or one-third octave band.
OHC	occupational hearing conservationist (in Chapter 4 also used for outer hair cells)
ONIPTS	occupational-noise-induced permanent threshold shift
OPM	Office of Personnel Management
OSHA	Occupational Safety and Health Administration
p	sound pressure. Unit: Pa
p_o	reference sound pressure (20μPa)
p_{rms}	root-mean-square sound pressure. Unit: Pa
Pa	pascal
PEL	permissible exposure limit. A-weighted sound level at which exposure for a stated time, typically 8 hrs., accumulates 100% noise dose. Unit: dBA
PNC	preferred noise criteria. Unit: dB
PNR	predicted noise level reduction. Unit: dB
PSIL	preferred speech interference level. Unit: dB
PTS	permanent threshold shift. Unit: dB
PU	polyurethane
PVC	polyvinyl chloride
PWL	sound power level. Unit: dB (also see L_W)
$\%B_b$	percent of employees showing a shift of 15 dB or more toward better hearing at any test frequency (500 – 6000 Hz) in either ear relative to their baseline audiogram. Unit: %
$\%W_b$	percent of employees showing a shift of 15 dB or more toward worse hearing at any test frequency (500 – 6000 Hz) in either ear relative to their baseline audiogram. Unit: %
$\%BW_s$	percent of employees showing a shift of 15 dB or more toward either better or worse hearing at any test frequency (500 – 6000 Hz) in either ear between two sequential audiograms. Unit: %
$\%W_s$	percent of employees showing a shift of 15 dB or more toward worse hearing at any test frequency (500 – 6000 Hz) in either ear between two sequential audiograms. Unit: %
q	nondimensional parameter that determines the exchange rate in computing noise dose. For 5-dB exchange rate, q = 16.61; for 3-dB exchange rate, q = 10.
Q	directivity factor. Unit: dimensionless
QAI	quality assessment index, used in conjunction with RC Mark II method for rating a noise spectrum. Unit: dB

r	radius, effective radius, or distance from source. Unit: m, ft
rad	radians
R	room constant. Unit: m^2, ft^2 (also called metric sabins or sabins)
R	mechanical vibration isolation. Unit: dimensionless
RC	room criterion level. Unit: dB
REAT	real-ear attenuation at threshold. Unit: dB
REL	recommended exposure limit. NIOSH recommended 8-hr time-weighted average limit (85 dB), using a 3-dB exchange rate, which if equaled or exceeded, indicates a hazardous exposure. Unit: dBA
rms	root-mean-square
s	second
S	surface area of room. Unit: m^2, ft^2
SCFM	standard cubic feet per minute. Unit: ft^3/min
SD	standard deviation. Unit: same units as measured quantity
SEL	sound exposure level. Unit: dB (also see L_{AE})
SII	speech intelligibility index. Unit: dimensionless
SIL	speech interference level. Unit: dB
SLC	Sound Level Conversion. Often a trailing subscript, as in SLC_{80}, is used to indicate the percentage of the population that is protected. Unit: dB
SLM	sound level meter
SN	serial number
S/N	speech-to-noise or signal-to-noise ratio. Unit: dB
SNR	Single Number Rating. Unit: dB
SPL	sound pressure level. Unit: dB (also see L_p)
SRT	speech recognition threshold (previously called speech reception threshold). Unit: dB
STC	sound transmission class. Unit: dB
STS	standard threshold shift in hearing, as defined in OSHA HCA. Unit: dB
t	time. Unit: s, min, h
t	refers to values of Student's t-distribution as found in Table 7.10. Unit: dimensionless
T_c	criterion sound duration. In OSHA/MSHA practice, $T_c = 8$ h. Unit: h
T_p	Permissible time of exposure, with any defined exchange rate, at a stated A-weighted sound level which accumulates 100% noise dose. Compare to OSHA permissible time, T_N. Unit: min, h

$T_1..T_2..T_N$	OSHA designator for permissible time of exposure (also called reference duration) with a 5-dB exchange rate, at a stated A-weighted sound level which accumulates 100% noise dose. Running subscripts refer to particular noise levels for a range of values from 1 to N. Unit: min, h (compare to T_p for permissible time with other than 5-dB exchange rate, and also see $C_1..C_2..C_N$)
T_{60}	reverberation time. Unit: s
T^*	Actual measurement time for worker exposure to noise. Unit: min, h
TL	transmission loss. Unit: dB
TLV®	threshold limit value.
TR	transmissibility, also called transmission ratio. Unit: dimensionless
TTS	temporary threshold shift. Unit: dB
TTS_2	temporary threshold shift measured 2 minutes post-exposure. Unit: dB
TWA	A-weighted average sound level with 5-dB exchange rate and slow meter response, applied in OSHA/MSHA practice. The TWA is normalized to 8 hours. Contrast with L_{OSHA} for a non-normalized quantity. Unit: dBA
UCL	upper confidence limit. Unit: dimensionless
v	velocity. Unit: m/s, in/s
w	watt
W	sound power. Unit: watts
W_o	reference sound power (10^{-12} acoustic watts)
WC	workers' compensation
WHO	World Health Organization
Z	standard scores for the normal distribution as listed in Table 7.8. Unit: dimensionless
α	sound absorption coefficient. Unit: dimensionless
α'	air attenuation coefficient. Unit: dB/km
δ	static deflection. Unit: cm, in
η	loss factor. Unit: dimensionless
λ	wavelength. Unit: m, ft
μPa	micropascal ($10^{-6}Pa$)
ρ	density of air. Unit: kg/m³
τ	transmission coefficient. Unit: dimensionless
ω	angular frequency = $2\pi f$. Unit: rad/s
ω_n	angular natural frequency. Unit: rad/s
ζ	ratio of viscous damping constant to critical damping value. Unit: dimensionless
%B, %W	see definitions alphabetized under "p" for percent

Quiet places
are the think tank of the soul.

Gordon Hempton
The Sound Tracker

The Noise Manual, revised 5th edition, edited by E.H. Berger,
L.H. Royster, J.D. Royster, D.P. Driscoll, and M. Layne
©2003 American Industrial Hygiene Association

1

Noise Control and Hearing Conservation: Why Do It?

Elliott H. Berger

Contents

To my daughter Jessica
who brightens my life in countless ways.

"At our company hearing conservation is more than a priority; it is a value. That is an important distinction, because although priorities may change, values never do."

— Charles Robinson, Safety and Health Systems Manager
NORPAC (Weyerhaeuser), Washington

Introduction

It will come as no great surprise to readers of *The Noise Manual* that excessive exposure to noise causes hearing loss. Neither should they be surprised to be reminded that noise-induced hearing loss (NIHL) can be avoided by reducing the level or duration of the exposure or by the use of hearing protection. And many should also already be aware that beyond the potential loss of hearing due to noise, other issues might argue for the value of hearing conservation. But what the reader may fail to perceive is the true extent of noise in our society, the prevalence of NIHL, the importance of sound in the enjoyment, quality, and productivity of our daily existence, or the full justification for the existence of occupational hearing conservation programs (HCPs).

Therefore the purpose of this chapter is to frame the significance of noise and its effects in a compelling manner for those new to the field. Moreover, another goal is to provide concepts and resources that may strike a chord for all hearing conservationists to enable them to better accomplish their job of educating and motivating others to measure, assess, and control noise, and to protect the hearing of those who are exposed.

The Extent of the Noise Problem

Noise is virtually everywhere. In fact, silence, by which is meant in this context, complete absolute quiet, is so rare that for those few who experience it, they can likely cite the time and place — perhaps in a well-isolated acoustical test

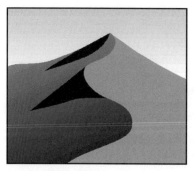

chamber during an auditory experiment or maybe in the deserts of the U.S. southwest in the moments between aircraft overflights. The opportunities to savor such an experience are few and far between. For example, nature recordist and sound tracker Gordon Hempton reported that of the 20 locations he surveyed in Washington state that had "noise-free" intervals in excess of 15 minutes in 1984, only three remained as pristine

5 years later (Grossmann, 1995). Even when one is willing to forgo total silence, quiet is elusive, especially as population density increases. In the European community 40% of the population is exposed to transportation noise with an A-weighted equivalent continuous sound level $(L_{Aeq,T})$ exceeding 55 dB daytime and 20% exceeding 65 dB, levels that are intrusive or annoying to many (WHO, 1995). Throughout our lives we are faced with noise from transportation, construction and public works, our neighbors and their pets, home appliances, and shop tools. Additionally, we willingly expose ourselves to noise from leisure activities such as target shooting and hunting, motor sports, snowmobiling, speed boating, attendance at public sporting events, concerts and movie theaters, and of course home stereos and Walkman-type devices. It seems as though we rarely can or do give our ears a rest.

The occupational scene is even more daunting. Noise is arguably the most pervasive hazardous agent in the workplace with estimates suggesting upwards of 5 million and perhaps as many as 30 million U.S. workers[1] are exposed to hazardous occupational noise levels (regular exposure above 85 dBA), with millions more at risk from other ototraumatic agents (NIOSH, 1996). Translating exposures to actual cases of NIHL is not straightforward, but a 1990s consensus conference concluded that hearing loss afflicts 28 million Americans, 10 million cases of which are "at least partially attributable to NIHL" (NIH, 1990).

The extent of the noise problem was demonstrated by the identification of NIHL in the 1980s as one of the 10 leading work-related diseases and injuries (NIOSH, 1988). This was reinforced a decade later when hearing loss due to noise exposure was listed as one of the eight most critical occupational diseases and injuries requiring research and development activities within the framework of the National Occupational Research Agenda (NIOSH, 1996). The experience in other countries is similar. For example in British Columbia where extensive workers' compensation records are maintained for all industry in that province, data on disabilities and fatalities from the years 1987 through 1996 indicate that claims for occupational NIHL[2] exceeded claims for each of the other physical agents for which data were recorded (with the exception of repetitive-motion injuries) by a factor of approximately three, and also exceeded the combined claims for all chemical agents (WCB, 1998). Similarly in Chile, the number of employers contending with noise in the workplace surpasses the next most common hazard (use of solvents) by a factor of five, and exceeds others such as pesticides, dust, and chemicals by even greater amounts (Dummer, 1997).

---

[1] Exact numbers are difficult to specify. NIOSH has estimated 30 million exposed workers (NIOSH, 1996), but their most recent publication provides a figure of 5 million (1998). Although based on more elderly data (Simpson and Bruce, 1981), the estimate of 9 million found in Table 16.1 may be one of the most reliable. The 1996 NIOSH estimate may be higher because it is based on sound levels and not 8-hr. exposures, because it includes additional occupational sectors not covered by Table 16.1, and because of other procedural differences.

[2] Accepted claims are those with a pensionable degree of hearing loss, defined as the average of 500, 1000, and 2000 Hz greater than 28 dB.

Brief Historical Overview and Perspective

The problem of hearing loss from occupational exposure to noise dates from at least the Middle Ages where workers in professions such as blacksmithing, mining, and church bell ringing were known to suffer such impairments (WHO, 1995). As early as 1831 "blacksmiths' deafness" with the concomitant feature of tinnitus (referred to as "ringing and noise in the ears") was cited in the medical literature (Fosboke, 1831). One-half century later another medical article referred to it as "boiler-maker's deafness," since the author at that time based his findings upon examination of 40 men from the steam-boiler shops in Portland, Maine (Holt, 1882). The effect, namely loss of hearing, was clearly identified, but the mechanism was poorly understood; Holt ascribed it to constant agitation of the joints of the ossicles, thereby causing ankylosis (stiffening due to the growth of a fibrous or bony union), especially of the stapes. Mechanisms of prevention were also not well known at the time, with Holt reporting that men tried stopping their ears with cotton wool and pads but derived no benefit therefrom; he had no alternative suggestions. At the same time in Scotland it was reported that men (also boilermakers) were prejudiced against use of cotton earplugs because it would predispose them to catching a cold when the plugs were removed at night (Barr, 1896). Other objections included interference with hearing and itchiness, bona fide complaints that are still voiced today.

In large part, serious and sustained interest in hearing conservation developed as a result of World War II, subsequent to which untold numbers of soldiers returned home with hearing loss. In fact, one of the earliest regulations dealing with hearing conservation was Air Force Regulation 160-3, issued in 1948 (Dept. of the Air Force, 1948). Industrial HCPs began to appear in the late 1940s and early 1950s with some of the first reported programs established in the aviation and metals industries (Bolger, 1956; Haluska, 1964; Hatton, 1956; Wilkins, 1956). Government noise regulations followed in the late 1960s (U.S. DOL, 1969) and became more prominent and widely enforced with the enactment of the Occupational Safety and Health Act (OSHA) of 1970 and promulgation of the noise standard in 1971 (OSHA, 1971). It took an additional decade for OSHA to produce the hearing conservation amendment (OSHA, 1981 and 1983), which specified the details of an occupational hearing conservation program that were only hinted at in the original 1971 standard. See Driscoll (1991) for a more complete historical overview, Appendix I for the complete OSHA noise standard, and Chapter 16 for a more extensive discussion of noise standards and regulations.

Although today we have a much improved understanding of the mechanisms of NIHL (damage to the hair cells of the inner ear as discussed in Chapter 5), a wealth of successful and innovative engineering noise control measures at our disposal (see Chapter 9), and a broad panoply of hearing protection devices (HPDs; see Chapter 10), the affliction is still prevalent, and some would say that hearing conservation has been unsuccessful. This is unfortunate indeed since long-term occupational NIHL is a completely preventable injury. As was stated over a decade ago in *The Development of a National Noise Strategy* (Berger and Royster, 1987, p. 40), and is still true today,

> "In large part what is needed is *not* the development of new solutions, but rather the broad dissemination of existing techniques plus the education and motivation of management and labor alike to speed the implementation of effective programs."

Unfortunately this hasn't yet occurred. According to a late 1980s estimate based on a U.S. national occupational exposure survey of approximately 4500 establishments, compliance with the law is sketchy at best, as shown in Table 1.1 (Franks and Burks, 1998). It would appear that in small industry hearing conservation merely consists of providing hearing protection devices. It would probably be safe to presume that this equates solely to making HPDs available without the substantial commitment to education, motivation, training, and enforcement that is necessary to make them work. Regrettably, few approach hearing conservation with the zeal expressed by the quote at the beginning of this chapter.

TABLE 1.1

Percent of companies surveyed, by size, with sound levels above 85 dBA, providing elements of a hearing conservation program (from Franks and Burks, 1998).

Company Size	Monitoring	Audiometry	Hearing Protection
Small	0	0	16
Medium	4	5	38
Large	29	19	84

A more focused set of data is available from the state of Michigan, which was involved from 1994–1997 in a special emphasis program for NIHL (Rosenman et al., 1998). Follow-up of approximately 1800 reported cases of hearing loss indicated that 46% of the noisy companies where the persons worked did not have an HCP at the time of employment of the individual. Although they found that over the 4 years studied the number of companies not providing regular audiometric evaluations had decreased among manufacturing companies with more than 100 employees, this was not the case in smaller companies or in the construction and farming sectors. Remember, *these data do not even examine the potential effectiveness of current HCPs, but simply whether or not the required components of the programs exist.*

Much remains to be accomplished. Whether we call our efforts a "hearing conservation program," as has been popular since the 1950s, and is the term used in *The Noise Manual*, or a "hearing loss prevention program," as defined by NIOSH

(Franks et al., 1996; NIOSH, 1998), or a "noise management program" as it is called in Australia and New Zealand (SA/SNZ, 1998), it is apparent that management, workers, regulators, in short, society in general, must be galvanized to act. The ideal would be to design new equipment and retrofit existing installations to reduce noise to safe levels, but the costs, difficulties, and maintenance aspects of addressing many noise problems through engineering controls suggest that relying solely on such an approach is doomed to failure. Similarly, directing one's efforts toward simply assessing the noise, or measuring hearing thresholds, or dispensing hearing protection, cannot adequately resolve the problem. A concerted multi-faceted approach is required. An overview is provided in Chapter 6.

Value of Hearing Conservation

Quality of Life: The Value of Good Hearing

Some have argued that of all our senses, hearing is the most vital (Gasaway, 1985); such a contention may be debated. Regardless, it is clear that hearing is fundamental to language, communication, and socialization. Language is so overwhelmingly oral that of the many thousands of languages spoken in the course of human history only around 100 have ever been committed to writing to a degree sufficient to have produced a literature, and most have never been written at all (Ong, 1982). And, whether or not a language is committed to text, as poet and author Maya Angelou has observed,

> "Words mean more than what is set down on paper. It takes the human voice to infuse them with shades of deeper meaning."

Moreover, sound by its nature is evanescent — it ceases to exist even as it is produced; by the time the last syllable of a word is uttered, its initial sounds have faded. Another unique characteristic of sound as compared to the other principal sense, vision, is that sounds pour into the hearer's ears whereas sight places the observer on the outside looking at, or looking in. Vision comes to the viewer from one direction whereas sound confronts us from all directions and places us in the center of an auditory space (Ong, 1982; Schafer, 1993), enveloping us and hence often greatly impacting attitudes and emotions.

The ability to hear is undeniably a key quality-of-life issue, from communication with coworkers, family, friends, and loved ones, to times of relaxation or appreciation, to hearing warning sounds and other signals. Many of life's joys involve activities and social interaction. That interplay is generally acoustic and oral in nature — conversing over a meal, playing at the beach, or listening to one's mate, or child, or someone special, whisper "I love you." The impact of hearing loss is often felt as much by the family of the impaired person, as by the person him or herself. In particular, spouses have reported that the hearing loss limits companionship and intimate communications, with only the most serious matters discussed because more casual conversation can take too much effort

(Hetu et al., 1995). Hearing loss can also create tension between the partners and make the children feel uneasy (Hetu and Getty, 1991).

Alone time can also be listening time, and this too can be diminished by hearing loss. Whether it be a restful afternoon at home tuned in to a favorite musical recording, daydreaming to the evocative sounds of sleet on a cold and frosted window pane, or chuckling at the rapid sniffing of one's new puppy as it investigates a dirty sock, those joys can be lost to the hearing impaired. For a worker, good hearing can mean the ability to identify the ping of a small spring-loaded part flying off the work bench so that it can be more easily located as it lands on the

floor, or the detection of changes in a machine spectrum distinctive of poor production quality or a failure mode, or the ability to distinguish a warning sound indicating the need for immediate action.

Because NIHL is a cumulative effect that often takes many years to fully develop, it is revealing to examine its impact on the elderly. Bess et al. (1989) studied the relationship between hearing loss and functional disturbance for 153 patients over age 65 who were seeing primary care internists for conditions such as diabetes, hypertension, and osteoarthritis. Using a standardized questionnaire that assessed physical and psychosocial function in a behavioral context they found an association between degree of hearing impairment and functional disability. Comparing the scores they reported to other data in the literature for unimpaired adults, patients one year after a heart transplant, those with chronic pulmonary disease, and terminally ill cancer and stroke patients, they concluded the hearing impairment was associated with a clinically significant level of functional impairment having a lasting degrading impact on the quality of life.

In a similar vein Sixt and Rosenhall (1997) posited that hearing impairment in the elderly results in social inactivity and isolation, which can lead to a reduction in life span. This was based on their finding that hearing impairment was correlated with factors indicative of poor health and increased mortality in a group of approximately 1600 elderly with ages ranging from 70 to 88.

Because of the terrible personal toll of hearing loss, one might hope, or suppose that the impairment could be ameliorated by the use of hearing aids, much like eyeglasses can restore normal vision. Such is not the case. Eyeglasses correct for the inability of the lens of the eye to properly focus light on the sensory cells in the retina, whereas the hearing aid, although it can amplify and filter sound, cannot rectify a situation in which the sensory cells affected by noise exposure are absent altogether. Regardless of the auditory information presented to the inner ear, the nerve cells specifically designed to respond to certain sounds are absent and others must fill in. Although audition is improved with amplification, it is not restored to its pre-noise pristine state.

And finally, consider the stigma of hearing loss. Those with such impairments are subject to stereotyping and prejudice; they are often presumed to be rude or stupid or both (Noble, 1996). For example the common retort, "Are you *deaf!*" rarely is an actual inquiry into the hearing ability of the listener. Instead it conveys the demeaning message, "Are you so *socially inept* as to be unable to respond appropriately to me or even to be able to respond to me at all?" Even worse, among coworkers, even when it is apparent that fellow workers' hearing loss is due to job-related conditions to which they are all exposed, it is often found that hearing impairment is best concealed. Otherwise, the impaired are made fun of, taken advantage of, experience restrictions on job advancement, and also have less job security (Hetu, 1996).

It is no small wonder with such barriers facing the hearing impaired that use of a hearing aid, which is a visible indication and reminder of their impairment, is frequently shunned. Eyeglasses are often a fashion accessory if not a fashion statement, whereas the "best" hearing aid is one that is so tiny as to not be seen at all.

Quality of Life: Quiet Ears

Almost as important as the ability to hear sound is the ability to hear "quiet." For many this is not possible. It is estimated that as many as 36 million Americans suffer from tinnitus, sounds heard within the head in the absence of actual sounds in the environment. For about 7 million the tinnitus is severe (Shulman, 1991). Tinnitus can be experienced in many forms, such as ringing, hissing, whistling, buzzing, or clicking. It can be disabling, dramatically affecting and diminishing the quality of life. An apt optical analogy would be a bright red dot in the middle of one's visual field, a dot that never, ever goes away. The sound of tinnitus can be equally as intrusive and disconcerting.

Although there are many causes of tinnitus, major epidemiologic studies of tinnitus in the adult population have revealed that both age and hearing loss are significant, with hearing loss the more significant. Thus, another important reason to preserve good hearing and avoid NIHL is to retain the ability to appreciate ears that make no sounds of their own. Also to be considered is that in some states tinnitus is compensable under the workers' compensation system (see Chapter 18).

On-the-Job Communication

Besides the far-reaching long-term effects on the quality of life due to hearing loss caused by noise, there are also the immediate communications problems that noise creates — the masking of sounds that must be heard (see Chapter 14). This affects those with normal hearing, and to an even greater extent, those already hearing impaired. Face-to-face, telephone, or even amplified communications can be difficult or impossible, and messages can be lost or misunderstood. The costs and difficulty of specifying, purchasing, and maintaining electronic communications systems to overcome these problems in critical environments must also be considered.

Extra-Auditory Effects: Productivity, Lost-Time Accidents, and Related Issues

Unprotected workers in high-noise environments have more lost-time accidents, are less productive, and in general experience more problems than do those with lower noise exposures. Over 60 years ago Weston and Adams (1935) studied the effects of wearing hearing protection on English weavers with a noise exposure of 96 dBA. In their initial experiment one group of 10 weavers spent alternate weeks either protected or unprotected for a period of 6 months. In a second study, 10 controls with no hearing protection and 10 experimental subjects wearing hearing protection were studied for a period of 1 year. The initial study found a 12% increase in personal efficiency and the latter one a 7½% increase, an effect of considerable magnitude in a purely manual process and one that would certainly cost justify an HCP.

Subsequent behavioral studies of weavers in India and Egypt have also examined the effects of noise. The Indian study included 100 weavers who were tested with coordination and dexterity tasks in their work environment (approximately 103 dBA). Those wearing hearing protection were found to perform significantly better than those without (Bhattacharya et al., 1985). The Egyptian study covered 2458 workers exposed to average noise levels from 80 to 99 dBA, so comparisons were actually between different departments rather than for one department under differing conditions of hearing protection (Noweir, 1984). Nevertheless the results indicated that workers in lower noise had less disciplinary actions and absenteeism, and statistically significant greater productivity than those with higher exposures, but the productivity gains (about 1%) were not as great as reported by Weston and Adams.

Another study, this time of boiler plant workers with exposures of approximately 95 dBA, examined issues beyond simple personal efficiency or productivity (Cohen, 1976). Data were compared for 2-year periods, before and after the advent of an HCP involving the use of hearing protectors. Results indicated fewer job injuries, medical problems, and absences in the post-HCP period, as typified by the results in Figure 1.1. Since a control population of low-noise-exposed workers exhibited no pre/post-HCP reduction in absenteeism, but the high-noise group did, it is likely that reduced noise exposure as a result of HPD usage was the controlling variable. Another significant finding was that comparisons of injury data before and after the advent of the HCP suggested that use of HPDs reduced rather than increased the number of mishaps. This provides evidence to counter the notion that wearing HPDs increases the likelihood of accidents by interfering with the ability to detect warning sounds and other acoustic cues in a background of noise. (See Chapters 10 and 14 for additional related discussions.)

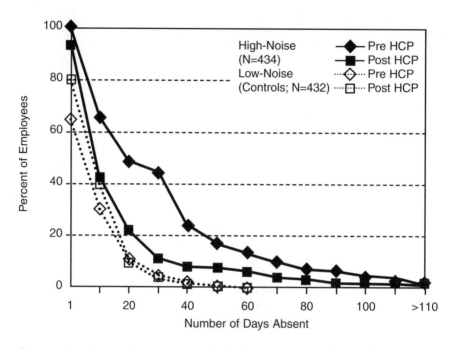

Figure 1.1 — Cumulative frequency distribution of workers with specifiable number of days absent, separated into high- and low-noise groups. Each point represents percentage of workers having had as many or more days absent as read from the abscissa. After Cohen (1976).

Schmidt et al. (1982) conducted a study similar to Cohen's, in which they examined industrial injury data for 5 years preceding and 5 years following the implementation of an HCP at a cotton yarn manufacturing plant. They too found a significant reduction in reported injuries for the approximately 150 workers who were studied. Of equal interest was the finding, based on audiometric records, that females wore their HPDs more effectively and received better protection, and it was they who showed a greater reduction in industrial injuries, thus demonstrating a direct link between HPD usage and reduced injury rates.

Although the use of hearing protection does not generally increase the likelihood of accidents, hearing loss well might. In one case-controlled study of 300 manual workers in a Dutch shipyard who experienced injuries, vs. 300 who did not, the odds ratio for occupational injuries (ratio of the odds that workers with hearing loss experience injuries to the odds that workers without hearing loss experience injuries) was found to be 1.9 for hearing loss of greater than 20 dB at 4 kHz (van Charante and Mulder, 1990). In a much larger study based on data for about 450,000 workers aged 18 to 65 who were participants in a U.S. National Health Interview Survey, hearing impairment (defined by self report during an interview) was found to be associated with a 55% increased risk of occupational injury (odds ratio of 1.55 with a 95% confidence interval of 1.29 –

1.87), which was equal to that found for epilepsy (Zwerling et al., 1997). With the exception of blindness and deafness, which had odds ratios of 3.21 and 2.19, respectively, hearing impairment also exceeded all other risk factors (e.g., visual impairment, extremity impairment, arthritis, etc.). It would appear that there are sound reasons to maintain good hearing.

And finally, a very positive study on the benefits of engineering noise controls was reported by Her Majesty's Chief Inspector of Factories in England (Staples, 1981). A ball bearing grinder generating noise levels from 103 to 114 dBA was fitted with enclosures that cost about $20,000 and netted approximately a 20-dB noise reduction. Besides the obvious and valuable protection of hearing that was provided, absenteeism in the department, which had been excessive, was reduced to a level no higher than elsewhere in the factory, and productivity of the machine operations increased by about 20% — a valuable payback indeed.

Extra-Auditory Effects: Health

Although extra-auditory effects of noise exposure have been frequently hypothesized and reported widely, especially in the lay press, there is disagreement as to the validity and interpretation of the supporting data. Clearly the principal consequence of noise is upon the organ designed to respond to it, namely the ear. The one health effect that has been identified most often is that prolonged equivalent daily exposures of at least 85 dBA "may contribute to increased blood pressure and hypertension" (Smoorenburg et al., 1996).

Recently, two independent studies have examined the presence of cortisol and other stress hormones in noise-exposed workers, as well as subjective assessments of fatigue and postwork irritability (Melamed and Bruhis, 1996; Sudo et al., 1996), and have confirmed the findings of an earlier study (Ising et al., 1979). In the best-controlled of the three studies (Melamed and Bruhis, 1996), 35 textile workers who were exposed to noise levels of 85–95 dBA were examined one day before and on the last day of a 7-day working period, during which they wore earmuffs. Decreased cortisol levels and reduced fatigue and postwork irritability were observed in the hearing-protected condition, suggesting that chronic noise exposure increases stress and reduces the quality of life for those exposed. A concomitant observation of Melamed and Bruhis was that reduction of noise at the ear by HPDs, as opposed to reduction of the noise in which the person is immersed, was sufficient to influence the extra-auditory health effects of the noise exposure. This supports the prevailing knowledge that potential stress effects of noise are mediated by the auditory system (Suter, 1989) and thus it is necessary to hear the noise for adverse effects to occur.

Worker Attitude

As can be inferred from the preceding section on extra-auditory effects of noise on productivity etc., reducing noise exposures via either noise controls or hearing protection can improve worker attitudes, with consequent benefits derived therefrom. Just as hard hats, respirators, safety shoes, and good workplace design are valued and implemented in modern business, so too should an HCP be considered an integral aspect of the overall safety program. Not only can this improve employee morale on the job, but if the educational program is well-designed employees will be alerted to the need and value of taking safe hearing practices home to their private lives, thus providing enrichment for all concerned. (See Chapter 8 for additional details.)

Some have argued that hearing health should be viewed as a "wellness" issue in the same context as other health-related company benefits (James, 1998). They contend that hearing relates directly to job safety and performance and as such it must be accounted for if we expect workers to function in teams and communicate effectively. Companies should include hearing health as part of overall health campaigns, even inviting employees' families to participate, thus expanding the hearing conservation message to spouses and children. The potential for audiograms to detect other medical pathologies, besides NIHL, is a benefit to all employees.

Personalizing Hearing Conservation

In talking to employees about the prominence of our sense of hearing and the poignancy of its loss, it often has greater impact if one imparts a personal story that might also directly relate to lives of those listening, perhaps about an incident or time when the vital nature of hearing was brought into sharp focus. In one recent article an audiologist shared four "true-life" incidents that highlighted the importance of her hearing, and hopefully, in so doing, reached her intended audience at a gut level. The events related to everyday occurrences in life such as listening to an audiocassette of her daughter at age 2, enjoying the myriad instrumental details in a bluegrass music recording, spending a few hours in "girl talk" with a best friend, and listening to her congregation intone a favorite hymn (Watkins, 1997). One story that I have shared relates to my love of the wilderness.

> Over Thanksgiving I had the thrill of joining a close friend for a hike up Mount Moosilauke in the White Mountains of New Hampshire. The morning was cool as we started our climb, just above freezing, the breeze slight, and the mountain remote and devoid of people. Just past the trailhead we hiked along a small stream, stopping to listen to the water flowing over rocks and through and beneath the encrusted ice. That gentle soothing sound was our companion until we split from the stream and headed sharply up slope in a few inches of snow.

> As the creek was lost from view, so too was its acoustic presence. We were then set free in a soft, almost soundless winter snowscape. Soon we stopped for a break and a "listen" — a peaceful moment to use our ears to their fullest. I wanted to hear what you can't in a town, or near neigh-

bors or airports or highways or anywhere in the proximity of civilization. I listened to a pristine silence, punctuated by the barely audible and deliciously delicate crinkling sound of a nearly frozen brook trickling just under the snow, an occasional twittering bird, and the whispering of a sporadic breeze through fir and stands of leafless birch. ... AND THEN, AND NOT UNTIL THEN, I noticed the view ... what a breathtaking occasion, sound and sight merged into one incredible experience!

Both my friend and I are blessed with good hearing — hearing that we have protected. Had we been careless and let noise rob us of our hearing ability, the adventure on Mt. Moosilauke would have lost much of its fullness and joy; if the loss had been compounded by noise-induced tinnitus our experience would have also been marred by noises in our own ears that would have imprinted an ugly blemish on the sound pictures.

The joy in telling this story is seeing the expression on some of the faces in the audience indicating that the tale also touched a part of their experience. Direct and immediate feedback after the lecture confirmed its meaningfulness. Especially gratifying were the remarks I received the following year from a couple of the employees who had been in attendance. They related how, when in the woods some months later, they recalled my message concerning their hearing, and of its splendor and the importance of preserving it.

Another key to personalizing hearing conservation is to remind listeners of how many of their exciting, moving, and joyful life experiences rely heavily on sound. Film is one of those experiences. In a review of a riveting 1994 thriller *Blink*, starring Madeline Stowe as a blind woman recovering from eye surgery who witnesses a murder, or thinks she does, David Ansen (1994) wrote:

> "But the secret ingredient of this adrenaline-pumper is the sound mix, supervised by Chris Newman, who also happened to work on *The Exorcist*, *The French Connection*, and *The Silence of the Lambs*, tense movies all. We're rarely conscious of it, but what really frightens us in the movies is often not what we see but what we hear. Not the guy with the knife but the man at the dials, splicing in an electronic 'boo'!"

Concluding Remarks

Occupational hearing loss is often overlooked because it usually occurs insidiously, without dramatic consequences such as bleeding, deformity, or death.

Nevertheless, NIHL has a terrible impact on the quality of life and human interaction. Those who suffer its debilitating effects say things like,
- "My daughter no longer seems to speak clearly."
- "I always miss the punch lines and my friends get tired of repeating themselves."
- "To get the sound from things, now you have to see them."
- "I miss the birds. I miss the whispers. I miss all the good sounds."
- "You always have this ringing in your ears."
- "It kind of puts you away from the outside world, kind of leaves you out by yourself. I feel lost. I feel I'm in a place by myself. There are people around me. I can see them, but I can't hear them."

And years later, those who ameliorate the impact of NIHL by purchasing and using hearing aids may likely exclaim, "I had no idea (or I had forgotten) what all I was missing!"

For a particularly telling viewpoint we turn to the words of Helen Keller, a woman who had neither her hearing nor her vision. When asked to compare her loss of vision with her loss of hearing, she is reputed to have said (Walker, 1986),

> "Blindness cuts people off from things; deafness cuts people off from people."

An alternative and fanciful perspective that demonstrates the spiritual aspect of sound can be gleaned from the Native American culture, wherein the following creation myth is told by the Hopi of the southwest (Schafer, 1993).

> "Palongawhoya, traveling throughout the earth, sounded out his call as he was bidden. All the vibrating centers along the earth's axis from pole to pole resounded to his call. The whole earth trembled; the universe quivered in tone. Thus, he made the whole world an instrument of sound, and sound an instrument of carrying messages, resounding praise to the creator of all."

The ability to hear and to hear well — clearly, deeply, and keenly, is an important aspect of life and living. The ability to experience and hear a range of sounds from those near the threshold of hearing to sounds that are full, bright, boisterous, and moving, adds depth and beauty to life. The value of preserving that ability should be self-evident; nevertheless, it has been argued at length in the preceding paragraphs. Hearing conservation is the key, both occupationally where millions of people are noise exposed and programs are often regulated, and in nonoccupational settings where it usually requires personal motivation and the perseverance of individual action.

In industry, programs can be nominally compliant and accomplish little except perhaps avoidance of OSHA citations, or they can be meaningful and prevent hearing loss. The cost differential is often trivial, and effective HCPs as noted above, deliver a supplementary return on investment in terms of benefits such as enhanced employee attitudes, improved productivity, and a better company image. It's the employer's choice whether to implement the additional efforts required to achieve a truly effective HCP. If you want to, this book will tell you how.

References

Ansen, D. (1994). Movie review, *Newsweek*, January 31.

Barr, T. (1896). *Manual of Diseases of the Ear*, James Maclehose and Sons, Glasgow.

Berger, E. H., and Royster, J. D. (1987). "The Development of a National Noise Strategy," *Sound and Vibration 21*(1), 40–44.

Bess, F. H., Lichtenstein, J. J., Logan, S. A., Burger, M. C., and Nelson, E. (1989). "Hearing Impairment as a Determinant of Function in the Elderly," *J. Am. Geriatrics Soc. 37*, 123–128.

Bhattacharya, S. K., Roy, A., Tripathi, S. R., and Chatterjee, S. K. (1985). "Behavioural Measurements in Textile Weavers Wearing Hearing Protectors," *Indian J. Med. Res. 82*(July), 55–64.

Bolger, A. (1956). "Ryan's Hearing Conservation Program," *Ind. Hyg. Qtrly. 17*(March), 52–54.

Cohen, A. (1976). "The Influence of a Company Hearing Conservation Program on Extra-Auditory Problems in Workers," *J. Saf. Res. 8*(4), 146–162.

Dept. of the Air Force (1948). "Precautionary Measures Against Noise Hazards," AFR 160-3, Washington, DC.

Driscoll, D. P. (1991). "Historical Overview of Hearing Loss Compensation," *Spectrum Suppl. 1*(8), 20.

Dummer, W. (1997). "Occupational Health and Workman's Compensation in Chile," *Appl. Occup. Environ. Hyg. 12*(12), 805–812.

Fosboke, J. (1831). "Practical Observations on the Pathology and Treatment of Deafness, No. II," *Lancet VI*, 645–648.

Franks, J., and Burks, A. (1998). "Engineering Noise Controls and Personal Protective Equipment," in *Control of Workplace Hazards for the 21st Century — Setting the Research Agenda*, National Institute for Occupational Safety and Health, Cincinnati, OH.

Franks, J. R., Stephenson, M. R., and Merry, C. J. (1996). "Preventing Occupational Hearing Loss — A Practical Guide," U.S. Dept. of HHS (NIOSH), Pub. No. 96-110, Cincinnati, OH.

Gasaway, D. C. (1985). *Hearing Conservation — A Practical Manual and Guide*, Prentice-Hall, Inc., Englewood Cliffs, NJ.

Grossmann, J. (1995). "The Sound of Silence," *American Way* (April), 75–78 and 114–116.

Haluska, F. P. (1964). "Hearing Conservation in a Metal Stamping and Forging Plant," *National Safety Congress Transactions*, Vol. 3, 11–15.

Hatton, J. F. (1956). "Lockheed's Ear Protection Program," *Ind. Hyg. Qtrly. 17*(March), 48–49.

Hetu, R. (1996). "The Stigma Attached to Hearing Impairment," *Scand. Audiol. Suppl 43*, 12–24.

Hetu, R., and Getty, L. (1991). "The Nature of the Handicaps Associated with Occupational Hearing Loss: Implications for Prevention," in *Proceedings, National Seminar Series on Occupational Noise-Induced Hearing Loss, Prevention and Rehabilitation*, edited by L. Getty, R. Hetu, W. G. Noble, and R. Waugh, National Occupational Health and Safety Comm., Sydney, Australia, 64–85.

Hetu, R., Getty, L., and Quoc, H. T. (1995). "Impact of Occupational Hearing Loss on the Lives of Workers," in *Occupational Medicine: State of the Art Reviews — Vol. 10, No. 3*, edited by T. C. Morata and D. E. Dunn, Hanley & Belfus, Inc., Philadelphia, PA, 495–512.

Holt, E. E. (1882). "Boiler-Maker's Deafness and Hearing in a Noise," *Trans. Am. Otol. Soc.* *3*, 34–44.

Ising, H., Gunter, T., Havestadt, C., Krause, C., Markert, B., Melchert, H. U., Schoknecht, G., Thefeld, W., and Tietze, K. W. (1979). "Study on the Quantification of Risk for the Heart and Circulatory System Associated with Noise Workers," EPA translation TR-79-0857, Office of Noise Abatement and Control, Washington, DC.

James, R. (1998). "Workshop #3 — Marketing Hearing Conservation Programs," *Spectrum Suppl. 1*(15), 18.

Melamed, S., and Bruhis, S. (1996). "The Effects of Chronic Industrial Noise Exposure on Urinary Cortisol, Fatigue, and Irritability," *J. Occup. Environ. Med. 38*(3), 252–256.

NIH (1990). "Consensus Conference: Noise and Hearing Loss," *J. Am. Med. Assoc. 263*(23), 3185–3190.

NIOSH (1988). "A Proposed National Strategy for the Prevention of Noise-Induced Hearing Loss," in *Proposed National Strategies for the Prevention of Leading Work-Related Diseases, Part 2*, Association of Schools of Public Health, 51–63.

NIOSH (1996). "National Occupational Research Agenda," National Institute for Occupational Safety and Health, DHHS (NIOSH) Pub. No. 96-115, Cincinnati, OH.

NIOSH (1998). "Criteria for a Recommended Standard: Occupational Noise Exposure, Revised Criteria 1998," National Institute for Occupational Safety and Health, DHHS (NIOSH) Pub. No. 98-126, Cincinnati, OH.

Noble, W. (1996). "What Is a Psychosocial Approach to Hearing Loss?" *Scand. Audiol. Suppl. 43*, 6–11.

Noweir, M. H. (1984). "Noise Exposure as Related to Productivity, Disciplinary Actions, Absenteeism, and Accidents Among Textile Workers," *J. Saf. Res. 15*(4), 163–174.

Ong, W. J. (1982). *Orality & Literacy, The Technologizing of the Word*, Routledge, New York, NY.

OSHA (1971). "Occupational Noise Exposure," Occupational Safety and Health Administration, 29CFR1910.95 *Fed. Regist. 36*(105), 10518.

OSHA (1981). "Occupational Noise Exposure; Hearing Conservation Amendment," Occupational Safety and Health Administration, 29CFR1910.95 *Fed. Regist. 46*(11), 4078 – 4181.

OSHA (1983). "Occupational Noise Exposure; Hearing Conservation Amendment; Final Rule," Occupational Safety and Health Administration, 29CFR1910.95 *Fed. Regist. 48*(46), 9738–9785.

Rosenman, K. D., Reilly, M. J., Deliefde, B., and Kalinowski, D. J. (1998). "1997 Annual Report on Occupational Noise Induced Hearing Loss in Michigan," Michigan State Univ., Lansing, MI.

SA/SNZ (1998). "Occupational Noise Management Part 0: Overview," Standards Australia and Standards New Zealand, AS/NZS 1269.0:1998, Homebush, Australia.

Schafer, R. M. (1993). *Voices of Tyranny, Temples of Silence*, Arcana Editions, Ontario, Canada.

Schmidt, J. W., Royster, L. H., and Pearson, R. G. (1982). "Impact of an Industrial Hearing Conservation Program on Occupational Injuries," *Sound and Vibration 16*(5), 16–20.

Shulman, A. (1991). "Epidemiology of Tinnitus," in *Tinnitus, Diagnosis / Treatment*, edited by A. Shulman, J-M. Aran, J. Tonndorf, H. Feldmann, and J. A. Vernon, Lea & Febiger, Philadelphia, PA, 237–247.

Simpson, M., and Bruce, R. (1981). "Noise in America: The Extent of the Problem," U.S. Environmental Protection Agency, EPA Rept. No. 550/9-81-101, Washington, DC.

Sixt, E., and Rosenhall, U. (1997). "Presbycusis Related to Socioeconomic Factors and State of Health," *Scand. Audiol. 26*(3), 133–140.

Smoorenburg, G. F., Axelsson, A., Babisch, W., Diamond, I. G., Ising, H., Marth, E., Miedema, H. M. E., Ohrstrom, E., Rice, C. G., Abbing, E. W. R., van de Wiel, J. A. G., Passchier-Vermeer, W. (1996). "Effects of Noise on Health," *Noise News Int. 4*(3), 137–150.

Staples, N. (1981). "Hearing Conservation — Is Management Short Changing Those at Risk?," *Noise and Vib. Control Worldwide 12*(6), 236–238.

Sudo, A., Luong, N. A., Jonai, H., Matsuda, S., Villaneuva, M. B. G., Sotoyama, M., Cong, N. T., Trinh, L. V., Hien, H. M. Trong, N. D., and Sy, N. (1996). "Effects of Earplugs on Catecholamine and Cortisol Excretion in Noise-Exposed Textile Workers," *Ind. Health 34*(32), 279–286.

Suter, A. (1989). "The Effects of Noise on Performance," U.S. Army Human Eng. Lab., Tech. Memo 3-89, Aberdeen Proving Ground, MD.

U.S. DOL (1969). "Occupational Noise Exposure," *Fed. Regist. 34*, 7946ff.

van Charante, A. W. M., and Mulder, P. G. H. (1990). "Perceptual Acuity and the Risk of Industrial Accidents," *Am. J. Epidemiology 131*(4), 652–663.

Walker, L. (1986). *A Loss for Words*, Harper and Row, New York, NY, 20.

Watkins, D. (1997). "So What's the Big Deal?" *Spectrum 14*(4), 11.

WCB (1998). "Occupational Diseases in British Columbia, 1979 – 1996," Workers' Compensation Board of British Columbia, Vancouver, Canada.

Weston, H. C., and Adams, S. (1935). "The Performance of Weavers Under Varying Conditions of Noise," Med. Res. Council Ind. Health Res. Board, Report No. 70, London, England.

WHO (1995). "Community Noise," *Archives of the Center for Sensory Research 2*(1), edited by B. Berglund and T. Lindvall, prepared for World Health Organization by Stockholm University, Sweden.

Wilkins, R. (1956). "One Approach to Hearing Conservation," *Ind. Hyg. Qtrly. 17*(March), 54–55.

Zwerling, C., Whitten, P. S., Davis, C. S., and Sprince, N. L. (1997). "Occupational Injuries Among Workers with Disabilities, The National Health Interview Survey, 1985-1994," *J. Am. Med. Assoc. 278*(24), 2163–2166.

The Noise Manual, revised 5th edition, edited by E.H. Berger,
L.H. Royster, J.D. Royster, D.P. Driscoll, and M. Layne
©2003 American Industrial Hygiene Association

2 Physics of Sound and Vibration

Paul B. Ostergaard

Contents

Basic Characteristics of Sound

In air, sound amplitude is usually described in terms of a variation of pressure (p) above and below ambient atmospheric pressure. This pressure oscillation, commonly known as sound pressure, can be generated in many ways including a vibrating surface or string, the impact of two bodies, or by fluctuations in pressure in turbulent fluid flow. *Sound* is formally defined as the fluctuations in pressure above and below the ambient pressure of a medium that has elasticity and viscosity. Sound is also defined as the auditory sensation evoked by the oscillations in pressure described above (ANSI S1.1-1994). Oscillations that occur in solids are referred to as vibration.

The *amplitude* of a pressure oscillation is the amount by which the instantaneous pressure varies from the ambient or atmospheric pressure in airborne sound. In a solid the amplitude is the displacement of particles of the solid from their at-rest position within the solid.

The *speed* (c) with which sound travels through air is determined by the density and pressure of the medium, characteristics that in turn depend on temperature. In most situations the pressure can be ignored and only temperature need be considered, but for most practical purposes, the speed of sound is essentially constant. In air, the speed of sound is approximately 344 meters/second (m/s) (1130 feet/second [ft/s]) for standard atmospheric conditions (20 °C [68 °F], and 1.013×10^5 Pa [14.7 psi]). In liquids and solids the speed of sound is higher due to higher density, that is to say, about 1500 m/sec (4900 ft/sec) in water, and about 6100 m/sec (20,000 ft/sec) in steel. See Appendix III for typical properties of waves in different materials.

If conditions warrant, near standard atmospheric temperature, 20° C (68° F) and pressure 1.013×10^5 Pa (14.7 psi), the speed of sound can be computed from

$$c = 331.5 + 0.58 \text{ °C} \text{ m/s} \tag{2.1a}$$

where °C is the temperature in degrees Celsius. In English units,

$$c = 1054 + 1.07 \text{ °F} \text{ ft/s} \tag{2.1b}$$

where °F is temperature in degrees Fahrenheit. For a temperature change of 20 or 30 degrees from the standard atmospheric temperature, and/or ambient pressures much different from sea level (e.g., Denver), see Beranek and Ver (1992).

The distance traveled by a sound or vibration wave during one pressure cycle is called the *wavelength* (λ) which is measured in meters (m) or feet (ft). *Frequency* (f), the rate of oscillation of the wave at a fixed position in space or in a solid medium, is expressed in hertz (Hz). One hertz is equivalent to one cycle per second (cps). While frequency is an objective measurement of the physical number of oscillations in a wave, *pitch* is the corresponding subjective response of a listener. Pitch usually depends on frequency, with higher frequencies corresponding to higher pitch, but it also depends to a small extent on the magnitude of the sound pressure as well.

The speed at which sound travels (c) is related to frequency and wavelength as shown in Equation 2.2:

$$c = f \lambda \quad \text{m/s (ft/s)} \tag{2.2}$$

Using the speed of sound in air calculated at standard atmospheric conditions (Equation 2.1a or 2.1b), the wavelength of a sound wave at 1000 Hz is found to be approximately 0.344 m (1.13 ft). Looking at Equation 2.2, it can be seen that the wavelength is inversely proportional to the frequency. From Equations 2.1a and 2.1b it also can be seen that the speed of sound changes about 0.5 m/s for a one degree C (1 ft/s for one degree F) change in temperature. This temperature change will change the wavelength about 0.3% at 100 Hz (0.1% for one degree F). The change is even less at higher frequencies. Hence in most practical situations a change of temperature of 10 degrees or more is needed to make any significant impact in any calculations where the wavelength is important.

Bandwidth

The acoustic energy of sound and vibration sources normally is widely distributed over the frequency spectrum. In addition, the acoustic or vibration response of structures is frequency dependent. Because the frequency range in acoustics is too broad to handle as a single unit for noise control, or when it is necessary to analyze a sound or meet some criterion, the frequency range must be broken into smaller units. The most common bandwidth (BW) or range of frequencies used for noise measurements is the *octave band* (OB). A frequency band is said to be an octave wide when its upper band-edge frequency, f_2, is twice the lower band-edge frequency, f_1:

$$f_2 = 2\, f_1 \quad \text{Hz} \tag{2.3}$$

Octave-band measurements often are used for noise-control work because they provide a useful and often essential amount of information with a reasonable number of measurements.

When more detailed characteristics of a noise are required, as might be the case for pinpointing a particular noise source in a background of other sources, it is necessary to use frequency bands narrower than octave bands. *One-third octave bands* are used for this purpose. A 1/3 octave band is defined as a frequency band whose upper band-edge frequency, f_2, is the cube root of two times the lower band-edge frequency, f_1:

$$f_2 = \sqrt[3]{2}\, f_1 \quad \text{Hz} \tag{2.4}$$

There are three 1/3 octave bands in each octave band.

The *center frequency*, f_c, of any of these bands is the geometric mean of the upper and lower band-edge frequencies:

$$f_c = \sqrt{f_1 f_2} \ \text{Hz} \qquad (2.5)$$

where f_1 and f_2 are the lower and upper frequency band limits in Hz, respectively. Further discussion of standardized octave and 1/3 octave band frequencies, and upper and lower band-edge frequencies (ANSI S1.11-1986), may be found in Chapter 3, Table 3.3.

Both octave and 1/3 octave band analyzers are constant percentage bandwidth filters. As a result, the width of each of these frequency bands increases as the center frequency increases. The bandwidth of an octave band is 70.7% of the center frequency, and for the 1/3 octave band 23.1% of the center frequency. When fine details of the frequency content of a sound are required a constant bandwidth filter analyzer is used (see Chapter 3, *Accessories and Other Instruments, Frequency or Spectrum Analyzers*).

Power and Pressure

The *sound power* (W) of a source is the total acoustic output that it produces in watts (w). For most practical situations the sound power of a source is constant regardless of its location in different environments.

The amplitude of the *sound pressure* disturbance can be related to the displacement amplitude of the vibrating sound source. Pressure is expressed as force per unit area. The preferred unit of pressure is the pascal (Pa) or one newton per square meter (N/m^2).

In acoustics, as in electricity and other fields, the maximum value of the fluctuating or oscillating quantity is not a good measure of its effectiveness, except in some instances for impulse or impact sounds. Since sound fluctuations would have an average value of zero, an ordinary average is not useful. A more useful descriptor of the effective sound pressure is the *root-mean-square* (rms) value of the time-varying sound pressure. It is defined as the square root of the time-averaged, squared sound pressure, or

$$p_{rms} = \sqrt{\frac{1}{T} \int_0^T p(t)^2 \, dt} \ \ \text{Pa} \qquad (2.6)$$

where p_{rms} is the root-mean-square (or effective) sound pressure, $p(t)$ is the time-varying instantaneous sound pressure, and T is the measurement period. The term inside the square root sign simply denotes the averaging of p^2.

Units and Sound Radiation

Decibels

In acoustics, decibel notation is utilized for most quantities. The *decibel* (dB) is a dimensionless quantity based on the logarithm of the ratio of two power-like quantities and is defined as follows:

$$L = 10 \log \left(\frac{A}{B}\right) \ dB \qquad\qquad (2.7)$$

where L is the level, and A and B are quantities related to power. Any time a "level" is referred to in acoustics, decibel notation is implied. In acoustics all levels are defined as the logarithm of the ratio of two quantities with the denominator as the reference quantity, term B in Equation 2.7. Keep in mind that decibels can be negative if the measured value is less than the reference quantity, or can be equal to zero if the measured value is equal to the reference quantity.

Sound Power Level and Intensity

For convenience, the sound power output of a source is expressed in terms of its *sound power level* (abbreviated PWL and symbolized L_W) and defined as

$$L_W \text{ or PWL} = 10 \log \left(\frac{W}{W_0}\right) \ dB \text{ re } 10^{-12} \text{ watt} \qquad (2.8)$$

where W is the sound power of the source in watts, and W_0 is the reference sound power defined as 10^{-12} watt. Figure 2.1 shows the relationship between sound power in watts and PWL in dB re 10^{-12} watt. Note that every factor of 10 in increase in sound power in watts results in an increase of 10 dB in PWL.

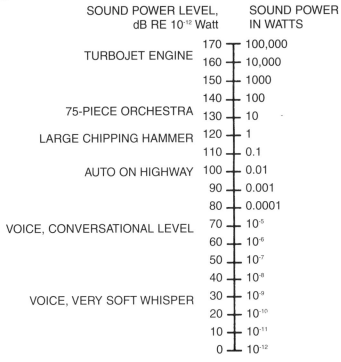

Figure 2.1 — Relationship between sound power level (L_W) and sound power (W).

When sound power is radiated from a *point source* (one that radiates uniformly in all directions) into free space, the acoustic power is evenly distributed over a sphere. The resulting sound power per unit area of the sphere is designated as *intensity* and has the units of watts/m². Although intensity diminishes as distance from a sound source increases, the power radiated, which is the product of the intensity and the area over which it is spread, remains constant. The decrease in intensity with distance is 6 dB with each doubling of the distance from the source.

There is no practical way to measure sound power directly. Modern instruments make it possible to measure intensity, but the instruments are expensive and must be used carefully. Under most conditions of sound radiation, sound intensity is proportional to the square of the sound pressure. Since sound pressure can be measured more easily, instruments are normally built to measure sound pressure level in decibels (ANSI S1.4-1983; ANSI S1.4A-1985). Using appropriate methods for measuring sound pressure level, it is possible to then calculate the sound power radiated using the known source intensity.

Directivity

The *directivity factor* (Q) is a dimensionless quantity that is a measure of the degree to which sound emitted by a source is concentrated in particular directions rather than radiated uniformly in a spherical pattern as it would be from a point source. Directivity factors for radiation patterns associated with various surfaces surrounding a sound source are shown in Figure 2.2. Each radiation pattern shown is actually a portion of a sphere with the source at the center of the sphere.

DIRECTIVITY FACTOR (Q), SIMPLIFIED RELATIONSHIPS

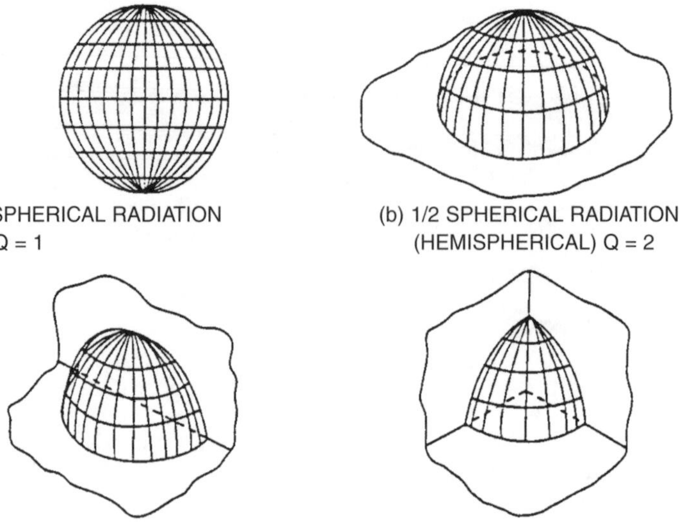

(a) SPHERICAL RADIATION
Q = 1

(b) 1/2 SPHERICAL RADIATION
(HEMISPHERICAL) Q = 2

(c) 1/4 SPHERICAL RADIATION
Q = 4

(d) 1/8 SPHERICAL RADIATION
Q = 8

Figure 2.2 — Directivity factor (Q) for varying boundary conditions.

In the following examples of the effects of directivity, the sound source and distance to an observer, r, remain constant and there are no nearby objects. For a sound source located in free space, the radiation pattern is illustrated in Figure 2.2a. This type of radiation is nondirectional and Q = 1. For hemispherical radiation (Figure 2.2b), where the sound source is on a floor in the center of a large room or at ground level outdoors, the sound intensity at the distance r would be twice as great as that for spherical radiation because the surface area over which the sound is spread has been reduced by a factor of 2, that is to say, one-half the surface area of a sphere. If the sound source is near the intersection of the floor and a wall of a room such that the area into which the source radiates is one-quarter of a sphere (Figure 2.2c), Q would be 4 since the spherical area has been reduced by a factor of 4. In like manner, if the sound source is near the intersection of the floor and two walls (Figure 2.2d), the sound source radiates into one-eighth of a sphere so that Q would equal 8. Note that the surface area into which the source radiates is the surface area of the sphere divided by the directivity factor.

In the discussion above, the directivity of the noise source is only related to the source location in relation to the reflecting surfaces. If the noise source itself exhibits directional characteristics, both influences must be considered. In real-world environments noise sources typically have directionality due to both the source location and the directional characteristics of the source itself. The directional characteristics are frequency dependent and typically increase with frequency.

For a given sound source, Q is the ratio of the sound pressure, measured at some distance r, to the sound pressure expected at the same distance r, but under conditions in which the source radiates the same total acoustic power in a uniform spherical pattern. Knowing the directivity factor of the sound source, one can modify the directivity factor of the source position in the room to obtain a combined directivity factor accurate enough for most applications of industrial noise control (Beranek and Ver, 1992; Harris, 1979).

A term related to the directivity factor, Q, is the *directivity index* (DI). The directivity index in decibels is defined as:

$$DI = 10 \log Q \quad dB \tag{2.9}$$

From the previous discussion on directivity and Equation 2.9, it can be seen that for hemispherical radiation Q = 2, and the equivalent DI = 3 dB, the sound pressure level (see below) measured on the hemisphere for a source of given sound power output will be 3 dB higher than for spherical radiation.

Sound Pressure Level

The readings obtained using a sound level meter represent *sound pressure levels* (abbreviated SPL and symbolized L_p), in decibels referred to as reference sound pressure, which for measurements in air is 20 micropascals (20 µPa, also expressed as 20 µN/m²). This reference was selected to approximately equal the threshold of normal human hearing at 1000 Hz. The equation for sound pressure level is

$$L_p \text{ or SPL} = 20 \log \left(\frac{p}{p_0}\right) \text{ dB re 20 } \mu\text{Pa} \qquad (2.10)$$

where p is the measured root-mean-square (rms) sound pressure and p_0 is the reference rms sound pressure. Note that the multiplier is 20 and not 10 as with the PWL (Equation 2.8). This is because sound power and intensity are proportional to the square of the pressure under normal conditions and $10 \log p^2 = 20 \log p$.

Figure 2.3 shows the equivalence between sound pressure in Pa and the SPL in dB re 20 μPa. It illustrates the great advantage of using the decibel notation rather than the corresponding wide range of sound pressure. Note that any range over which the sound pressure is doubled is equivalent to 6-dB change in SPL, and a range over which sound pressure increases by a factor of 10 results in a 20-dB change in SPL, whether at low or high sound levels. For example, a change in pressure from 20 to 40 μPa or from 10 to 20 Pa both represent a change of 6 dB in SPL.

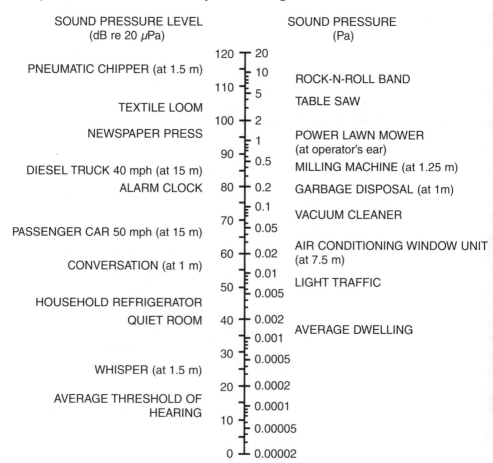

Figure 2.3 — Relationship between sound pressure level in decibels (dB) and sound pressure in Pa (N/m²).

Weighted Sound Levels

The *A-weighted sound level* (dBA) is used in many noise regulations. It is a reading obtained from a sound level meter using the A-weighting frequency network that characterizes the sound with a single number as opposed to the many readings that would be obtained using octave or 1/3 octave band readings. This weighting network reduces the influence of the lower frequencies (below about 500 Hz) (see Chapter 3, Sound Level Meters, Weighting Filters).

The A-weighted sound level, in decibels, can be measured directly or may be computed from unweighted octave or 1/3 octave-band SPL measurements using the frequency weightings given in Chapter 3, Table 3.1. These frequency weightings represent arithmetic adjustments applied to the measured octave-band SPLs in each frequency band (see Example 2.2). Sometimes A-weighting also is applied to octave-band sound power levels to obtain an A-weighted sound power level.

The *C-weighted sound level* uses another frequency weighting network in a sound level meter. It does not have as substantial a low-frequency roll-off as the A-weighting network (Table 3.1). As with A-weighting, it also provides a single number sound level. Measurements taken with the C-weighting differ little from those taken with no frequency weighting unless the sound being measured has strong frequency content below 25 Hz or above 10,000 Hz. The use of no frequency weighting is referred to as a *flat-* or *linear-frequency response*. When measurements are made in octave or 1/3 octave bands, they can be either weighted or linear as appropriate to the needs of the analysis.

Decibel Manipulations

At times it is necessary to use decibel addition or subtraction with sound levels. An example is the addition of frequency-band sound pressure levels to obtain the overall or total sound level. Another example is the estimation of the total SPL due to the addition of a new machine of known sound output to an existing noise environment of known characteristics. Addition can also be accomplished using A-weighted or C-weighted sound levels. Subtraction may be necessary in the correction for background sound that contaminates the desired measurement. Again subtraction can be done with all types of bandwidth and frequency weightings.

The equation for the addition of the sound levels of N random, uncorrelated sounds is

$$L_p = 10 \log \left[\sum_{i=1}^{N} 10^{\frac{L_{pi}}{10}} \right] \text{ dB} \qquad (2.11)$$

where L_p is the total sound pressure level in decibels generated by N sources and L_{pi} represents the individual SPLs to be added.

TABLE 2.1
Table for combining in decibels, levels of uncorrelated sounds.

Numerical difference between levels L_{p1} and L_{p2} (dB)	L_{p3}: Amount to be added to the higher of L_{p1} or L_{p2} (dB)
0.0 to 0.6	3.0
0.7 to 1.6	2.5
1.7 to 3.1	2.0
3.2 to 4.7	1.5
4.8 to 7.2	1.0
7.3 to 13.9	0.5
13.0 to ∞	0.0

Step 1: Determine the difference between the two levels to be added (L_{p1} minus L_{p2}).
Step 2: Find the number (L_{p3}) corresponding to this difference in the table.
Step 3: Add the number (L_{p3}) to the higher of L_{p1} and L_{p2} to obtain the resultant level (L_p).

Table 2.1 also can be used to add random, uncorrelated sound levels. The arithmetic difference between the levels is used to find the amount, in decibels, to be added arithmetically to the larger of L_{p1} or L_{p2} to obtain the sum of the two levels. If more than two are to be added, the sum of the first two must be added to the third, the resultant of the three levels to the fourth, and so forth, until all levels have been added.

To use Equation 2.11 for addition of the octave-band SPLs with a hand-held calculator or computer spreadsheet, begin by keying in the first SPL; divide by 10 and raise 10 to the result, that is to say, take the antilog of $0.1 \times L_{p1}$. Key in the second SPL, divide by 10 and raise 10 to the result. Add the two numbers. This is now the sum of the antilogs of one-tenth of the two SPLs. Continue to add the antilogs of one-tenth of the remaining SPLs until all have been added. Then, take the logarithm (to the base 10) of the result and multiply it by 10. This will be the total of all of the levels added. This procedure can also be expressed as

$$L_p = 10 \log \left[10^{\frac{L_{p1}}{10}} + 10^{\frac{L_{p2}}{10}} + ... + 10^{\frac{L_{pn}}{10}} \right] \text{ dB} \qquad (2.12)$$

Note that the term $10^{\frac{L_{pi}}{10}}$ in Equations 2.11 and 2.12 is equivalent to power.

EXAMPLE 2.1, Addition of Sound Levels

The overall sound pressure level produced by a random noise source can be calculated by adding the sound pressure levels measured in individual octave bands. As an example, consider the data presented in Table 2.2.

TABLE 2.2
Octave-band sound pressure levels utilized in Example 2.1.

Octave-band center frequency (Hz)	31.5	63	125	250	500	1000	2000	4000	8000
Sound pressure level (dB re 20 µPa)	85	88	94	94	95	100	97	90	88

When adding a series of levels using Table 2.1, begin with the highest levels so that calculations may be stopped when addition of lower values to the sum does not contribute significantly to the total. In this example, the levels of 100 dB and 97 dB have a difference of 3 dB, which corresponds to a 2.0 dB addition as found in Table 2.1. Thus, the combination of 100 dB and 97 dB will result in 100 + 2 = 102 dB. Combining 102 dB with 95 dB, the next higher level, gives 102 + 1 = 103 dB, which is the total of the highest three bands. This procedure is continued, adding one band at a time, and the overall SPL will be found to be 104.1 dB.

Table 2.3 illustrates why larger differences in sound levels can be ignored when adding two levels. For example, when adding two levels that differ by 10 dB, the combined level increases by only 0.4 dB.

TABLE 2.3
Illustration of sound level addition significance (all values in dB).

Lower level	70	80	85	86	87	88	89	90
Higher level	90	90	90	90	90	90	90	90
Sum	90.0	90.4	91.2	91.5	91.8	92.1	92.5	93.0

Subtraction of sound levels is done in a manner analogous to addition. In this case Equation 2.12 must be changed to

$$L_{p1} = 10 \log \left[10^{\frac{L_p}{10}} - 10^{\frac{L_{p2}}{10}} \right] dB \qquad (2.13)$$

where L_p is the total sound pressure level of L_{p1} and L_{p2}.

Remember that in most noise control problems it is not necessary to carry decibel calculations to more than one decimal place with the result rounded to integer values. Rarely can an accuracy of 0.5 dB be achieved in practical noise measurements. However, in using Equations 2.11, 2.12, and 2.13 the significant figures must be carried through to the last calculation to avoid significant errors. In using Equations 2.11, 2.12, and 2.13 the noise sources are assumed to be uncorrelated, random sources. Not all actual sources exhibit these characteristics. However, the equations give reliable results. An example is a mechanical room in which there are several machines of the same type and size.

Noise Control Terminology

Noise reduction (NR) is the arithmetic difference between the levels measured at two locations, one on either side of a noise control device (i.e., an enclosure or a barrier):

$$NR = L_{p1} - L_{p2} \quad dB \qquad (2.14)$$

where L_{p1} is the level at location 1 and L_{p2} is the level at location 2. Note in this case and in the one that follows, the levels are manipulated arithmetically rather than logarithmically. These levels as well as the ones that follow can be weighted or linear, overall, or measured in octave or 1/3 octave bands.

Insertion loss (IL) is the difference in levels at the same fixed measuring location before and after a noise control method has been applied to the noise source:

$$IL = L_p - L_{p'} \text{ dB} \tag{2.15}$$

where IL is the insertion loss in decibels, L_p is the level before noise control, and $L_{p'}$ is the level at the same location after noise control.

Attenuation is the reduction in SPL or PWL, in decibels, measured at increasing distances from a noise source (i.e., outdoors or down an air-conditioning duct system).

In working with noise sources in architectural spaces an important consideration is the amount of *sound absorption* that is present. Sound absorption is a function of frequency. A measure of the sound energy absorbed by a material is the dimensionless quantity, the *sound absorption coefficient* (α). It represents the fraction of the sound incident upon the material that is absorbed, that is to say, if $\alpha = 0.75$, 75% of the incident sound is absorbed. All materials absorb some sound, but to be "sound absorbing" a material should absorb a good fraction (i.e., the normally accepted value is at least one-half of the incident energy at the higher frequencies) of the sound energy incident upon it. The sound absorption of architectural materials is normally measured in the laboratory using a diffuse sound field (see next section below) (ASTM C424 - 1985; Hedeen, 1980; ISO 364:1985). Sound absorption is useful in controlling noise in a reverberant sound field (see below and Chapter 9 for a fuller discussion).

The sound isolation (or insulation) provided by an impervious barrier between two spaces is quantified by its sound *transmission loss* (abbreviated TL). The TL of a panel in decibels is 10 times the logarithm to the base 10 of the ratio of the sound power incident upon the panel to the sound power transmitted through the panel to the adjoining space (ASTM E90-1990; Hedeen, 1980; ISO 140-3:1990). The TL of a material or structure is governed by the physical properties of the material and type of construction used. The TL of a wall generally increases 5 to 6 dB for each doubling of the wall weight per unit of surface area in the mass-controlled region and for each doubling of frequency. See Chapter 9 for a fuller discussion of sound transmission loss.

Sound Fields, Sound Pressure, and Sound Power

Many noise-control problems require a practical knowledge of the relationship between PWL and SPL. This relationship allows the calculation of the approximate SPL that a particular machine will produce in a specified environment at a specified location, based upon a known PWL and directivity of the machine. Before this relationship can be presented several terms need to be defined.

A *free field* exists when sound radiates into space from a source and there is nothing to impede the sound energy as it flows from the source. The free field can also exist above a reflecting plane, such as the ground.

Consider a small, nondirectional sound source that is radiating sound equally in all directions, that is to say, Q=1. In a free field, since there are no reflections of sound, the SPL will be the same at any point equidistant from the source, that is, any point on the surface of a sphere centered on the source. The sound intensity will diminish inversely as the square of the distance, r, from the source since the sound energy is spreading over the increasing area of the sphere ($4\pi r^2$). Thus in a true free field, the SPL decreases 6 dB with each doubling of distance from the source. However, under most actual conditions, particularly indoors, reflection of sound (for example, from the floor or ground) will reduce the attenuation due to distance. Doubling the distance will produce, instead of a decrease of 6 dB, a drop of possibly only 4–5 dB.

A *direct sound field* is one in which sound travels directly to a receiver, without reflection from any surfaces and is the principal means by which sound gets to the receiver. However, many industrial noise problems are complicated by the fact that the noise is confined in a room. Reflections from the walls, floor, ceiling, and equipment in the room create a *reverberant sound field* which exists in conjunction with the sound radiated directly from the source to the receiver (the direct sound field). The reverberant field alters the sound field characteristics from those described above for a free field. If reverberant-field SPLs are uniform throughout a room and sound waves travel in all directions (due to reflections) with equal probability, the sound field is said to be *diffuse*. In actual practice, perfectly direct, reverberant, or diffuse sound fields rarely exist; rather, in most cases, the sound fields are usually something in between.

When the operator of a machine is close to its dominant noise source, he or she may be in the acoustical *near field* where there is no simple relationship between sound intensity and sound pressure. The extent of the near field depends on the frequency, on a characteristic source dimension, and on the phases of the radiating parts of the surface, so it is difficult to establish general limits for the near field. Many of the equations for predicting sound fields around machines, such as those following, assume *far field* conditions where sound behaves in a predictable manner. Therefore it is sometimes necessary to be aware that errors may arise when working in proximity to machines. (See Chapter 9, Noise Control Engineering, for additional guidance.)

The *room constant* (R) is a variable that is used in calculations relating the PWL to the SPL in a room. It has the units of m^2 (ft^2). The value of R is an indication of the average amount of sound absorption in a room. The room constant R can be calculated at any frequency from the following equation:

$$R = \frac{S_t \bar{\alpha}}{1 - \bar{\alpha}} \quad m^2 \ (ft)^2 \qquad (2.16)$$

where R is the room constant, S_t is the total surface area of the room in m^2 (ft^2) and $\bar{\alpha}$ is the average sound absorption coefficient of the room surfaces computed using

$$\overline{\alpha} = \frac{\sum\limits_{i=1}^{N} S_i \alpha_i}{\sum\limits_{i=1}^{N} S_i} \qquad (2.17)$$

where S_i is the area of the individual N room surfaces in m² (ft²) and α_i is the corresponding sound absorption coefficient for the ith surface. The room constant R can be estimated from Figure 2.4.

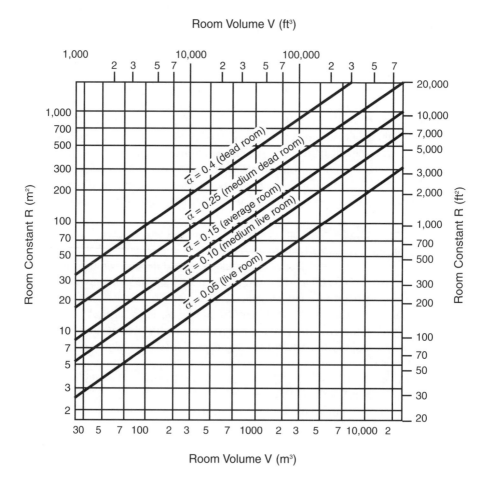

Figure 2.4 — Room constant for typical rooms. Value of the room constant R as a function of room volume for rooms with proportions of about 1:1.5:2. The parameter is the average sound-absorption coefficient for the room. From Beranek (1954).

One simple relationship between SPL and PWL is for a free-field, nondirectional sound source. The relation is given by the following equation:

$$L_p = L_W - 20 \log r - k + CF \quad \text{dB re 20 } \mu\text{Pa} \tag{2.18}$$

where L_p is the SPL, L_W is the PWL in decibels re 10^{-12} watt, r is the distance from the source in m (ft), k is a constant (11.0 dB for metric units [0.5 dB for English units]), and CF is a correction factor, in decibels, for atmospheric temperature and pressure. Since most industrial noise problems are concerned with air at or near standard conditions, that is, 20 °C (68 °F) and 1.013×10^5 Pa (14.7 psi), CF is usually negligible. For extreme conditions the correction factor can be calculated from

$$CF = 10 \log \left[\left(\frac{273 + °C}{293} \right)^{1/2} \left(\frac{1013}{B} \right) \right] \text{dB} \tag{2.19a}$$

where °C is temperature and B is the barometric pressure in millibars, or in English units,

$$CF = 10 \log \left[\left(\frac{460 + °F}{528} \right)^{1/2} \left(\frac{30}{B} \right) \right] \text{dB} \tag{2.19b}$$

where °F is the temperature and B is the barometric pressure in inches of mercury.

If the sound is directional in a free field, the relationship between sound power level and sound pressure level becomes

$$L_p = L_W - 20 \log r + DI - k + CF \quad \text{dB re 20 } \mu\text{Pa} \tag{2.20}$$

where k has the same value as in Equation 2.18.

In an enclosed space, the relation between PWL and SPL, which includes the effects of reverberation, is as follows:

$$L_p = L_W + 10 \log \left[\frac{Q}{4\pi r^2} + \frac{4}{R} \right] + k + CF \quad \text{dB re 20 } \mu\text{Pa} \tag{2.21}$$

where Q is the directivity factor, R is the room constant, k is 0 dB for metric units (10.5 dB for English units), and CF is defined in Equation 2.19a or 2.19b. (**Note that although the constant used in Equations 2.18 and 2.21 is denoted by the same symbol, k, it has different values in each equation.**) The relationship described by Equation 2.21 is shown graphically in Figure 2.5 without temperature and pressure corrections. It can be seen from the figure that for a given distance and directivity factor, the reverberant field sound level will decrease with an increase in sound absorption.

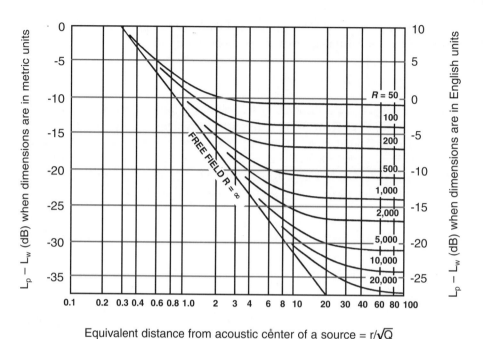

Equivalent distance from acoustic center of a source = $r/\sqrt{Q}$

Figure 2.5 — Curves for determining the sound pressure level in a large room relative to the sound power level as a function of the directivity factor Q, distance r, and
room constant R. The ordinate is calculated from Equation 2.21. When all dimensions are in English units with length in feet, use right-hand ordinate; for metric
dimensions with length in meters, use left-hand ordinate. From Beranek (1988).

Figure 2.5 is very useful in noise control work because it combines the basic characteristics of sound fields in a manner that is easy to use. The horizontal portions of
the room constant (R) curves illustrate the reverberant field with its uniform sound
levels as distance from the source increases and, at a given sound source distance, the
decreasing sound level of a reverberant field as the amount of sound absorption is
increased. The downward slanting line (6 dB per doubling of distance) illustrates the
direct sound field decrease in sound level with increasing distance. The curved portions of the room constant curves show the transition from a direct sound field to a
reverberant field.

The PWL of a source is generally independent of environment. However, the SPL
at some distance, r, from the source, is dependent on that distance, and on the source
directionality and sound-absorbing characteristics of the environment, both of which
are frequency dependent. Therefore, it is important to state under what conditions the
SPL from a sound source at a particular place was measured or is calculated.

EXAMPLE 2.2, Addition of Machine Sound Levels to Determine A-Weighted Sound Level

A vendor is proposing a small machine for installation in a room where it would be the only noise source of consequence. The PWL octave-band spectrum of this machine is shown by line 1 of Table 2.4. The machine is to be installed on the floor in the center of a live room, that is to say, a room with little sound absorption, having dimensions of $18.3 \times 42.7 \times 6.1$ m (4767 m³). If the operator is positioned 1 m from the machine during the work shift, would the noise level at the operator's position exceed 90 dBA? (If the machine is quite large, the operator will be in the near field, and the calculated predictions must be questioned.)

From Figure 2.5, for a live room with a volume of 4767 m³, R is approximately 90 m². The operator position dictates that r = 1 m. Since the machine is to be installed on the floor near the center of this large room, hemispherical radiation is assumed, Q = 2. Referring to Figure 2.5, compute $r/\sqrt{Q}$, which equals 0.71, and proceed vertically from 0.7 to R = 90 m² and horizontally to the left to find that the SPLs are approximately 7 dB below the PWLs of line 1 of Table 2.4. Hence, the SPLs shown on line 2 are obtained. To obtain the A-weighted sound levels from the octave-band levels, select the appropriate correction factors from Table 3.1 as reproduced on line 3 in Table 2.4. By arithmetically subtracting the correction factors of line 3 from the SPLs of line 2, obtain the octave-band, A-weighted sound levels shown on line 4. Now, add the A-weighted OB levels of line 4 using Equation 2.12 to obtain an A-weighted sound level of 85 dBA. Thus the noise level at the operator's position does not exceed 90 dBA.

TABLE 2.4
Computations described in Example 2.2.

Line No.	Description	Octave Band Center Frequency (Hz)							
		63	125	250	500	1000	2000	4000	8000
1	L_W (given), dB re 10^{-12} watt	81	95	90	89	88	84	79	65
2	$L_p = (L_W -7)$, dB re 20 µPa	74	88	83	82	81	77	72	58
3	A-weighting, dB (see Table 3.1)	-26	-16	-9	-3	0	+1	+1	-1
4	L_p (A-weighted)	48	72	74	79	81	78	73	57

EXAMPLE 2.3, Estimation of Source Sound Power Level to Meet Octave-Band Criteria

A machine is to be installed in the corner of a room with low R, a reverberant room. The dimensions of the room are $6.1 \times 15.2 \times 30.5$ m (2828 m³). From Figure 2.5, R is 65 m². Since the machine is to be installed in one corner of the room, Q = 8. The operator has to be as close as 1 m to the machine, so r = 1 m. (Again the operator may be in the near field.) The octave-band SPL design criterion is shown by line 1 of Table 2.5. What would the maximum allowable octave-band sound power levels be for this new machine?

TABLE 2.5
Computations described in Example 2.3.

Line No.	Description	Octave Band Center Frequency (Hz)							
		63	*125*	*250*	*500*	*1000*	*2000*	*4000*	*8000*
1	Criteria, L_p (dB re 20 µPa)	106	100	94	90	90	90	90	91
2	Allowance for multiple noise sources, dB	6	6	6	6	6	6	6	6
3	Single machine criteria, L_p (dB re 20 µPa)	100	94	88	84	84	84	84	85
4	Single machine criteria, L_W (dB re 10^{-12} watt)	101	95	89	85	85	85	85	86

Line 1 provides the criterion octave-band sound pressure levels that have been established. The design goal octave-band sound pressure levels, line 3, are 6 dB below the criteria to allow for noise generated by other possible sources in the room. If an allowance were not made for multiple sources, it would never be possible to add other sources without exceeding the criteria. The single-machine criterion is obtained by subtraction as shown by line 3. But this is the SPL and the requirement is for PWL. Referring to Figure 2.5, the relationship between SPL and PWL can be found. Starting from $r/\sqrt{Q} = 0.35$ and proceeding upward to R = 65 m², and horizontally to the left, the graph indicates that the SPL is 1 dB less than the PWL. Hence, add 1 dB to line 3 to obtain the allowable maximum octave-band PWLs as shown on line 4. This can then be used as a machine purchase specification.

Basic Characteristics of Mechanical Vibration

The study of vibration uses some of the same concepts and procedures found in prior discussions relating to airborne sound. Vibration, however, deals with the motion of structures such as strings, beams, plates, and solid bodies. These structures are set into motion and vibrate, causing airborne or structure-borne sound problems. When a structure is moved by some force the structure is accelerated and this acceleration is called *vibration acceleration* measured in m/s² (ft/s² or in/s²).

A *resonance frequency* (f_n), also known as a *natural frequency* of vibration of a system or structure, is a frequency at which it will oscillate freely if excited by an external force or displacement or a moving part of the structure. There can be a single dominant frequency or there can be many of them. If the system is excited by a steady-state excitation at a frequency that is the same as or close to the resonance frequency, the motion of a structure can be extreme. All structures have many frequencies of vibration, but it is usually the first few lower ones that create problems.

When a normal spring is compressed or extended its displacement is often assumed to be a linear function of the force applied. The ratio of force to displacement is known as the *spring constant* (k) in N/m (lb/in or lb/ft). To compute the resonance frequency for a system, f_n, the following equation is used:

$$f_n = \frac{1}{2\pi} \sqrt{\frac{k}{m}} \text{ Hz} \qquad (2.22a)$$

where k is the stiffness of the spring in N/m and m is mass of the supported body in kg. In English units

$$f_n = \frac{1}{2\pi} \sqrt{\frac{kg}{m}} \text{ Hz} \qquad (2.22b)$$

where k is the spring stiffness in lb/ft, g is the acceleration of gravity, 32.2 ft/sec² and m is the weight in pounds.

Additionally, for a single degree of freedom system, f_n can be predicted if the static deflection of the isolator, under load, is measured. When the mass is placed on the spring, it compresses it by an amount equal to mg/k, which is called the *static deflection* (δ). In this case the natural frequency can be easily computed as:

$$f_n = \frac{1}{2\pi} \sqrt{\frac{g}{\delta}} = \frac{4.98}{\sqrt{\delta_1}} = \frac{3.13}{\sqrt{\delta_2}} \text{ Hz} \qquad (2.23)$$

where δ_1 is the static deflection in centimeters and δ_2 is the static deflection in inches.

EXAMPLE 2.4, Calculation of Resonance Frequency

A small air compressor, mass equals 1000 kg, needs to be mounted on four springs so that its natural frequency is 20 Hz. Substituting 20 Hz for f_n and 1000 for m in Equation 2.22a and solving the equations, the spring constant, k, is found to be 1.579 × 10⁷ N/m. Assuming equal loading on each spring means that each of the four springs must have a spring constant of 3.94 × 10⁶ N/m. Alternately, one may find it convenient to solve for the static deflection of the spring. By inverting Equation 2.23 and plugging in the value of 20 Hz for f_n, the deflection is found to be 0.062 cm.

In the usual situations no vibrating system has motion in only one direction. Any system is free to move in many directions, and the number of independent directions or types of motion it can have is called the *degrees of freedom* (df). For example, a rigid body supported on springs usually exhibits six degrees of freedom: three linear and three rotational.

Other terms in vibration (and sound) are *impulse* and *impact*. Impulses are time dependent in that they are the product of force and the time during which the force is applied, that is to say, a short steam blow-off. An impact on the other hand is the single collision of one mass in motion with a second mass, which may be in motion or at rest, for example, a hammer striking a nail.

Units of Vibration

The measurement of vibration is described in terms of levels. The measurement quantity used to describe vibration is *vibration acceleration level*, L_a, and

$$L_a = 20 \log \left(\frac{a}{a_0} \right) \text{ dB re } 10 \text{ } \mu\text{m/s}^2 \text{ } (394 \text{ } \mu\text{in/s}^2) \qquad (2.24)$$

where L_a is the vibration acceleration level, a is the rms vibration acceleration and a_0 is the reference acceleration, $10 \text{ } \mu\text{m/s}^2$ ($394 \text{ } \mu\text{in/s}^2$). This reference acceleration is very close to 10^{-6} of the acceleration due to gravity, g (9.81 m/s^2) (32.2 ft/s^2). An rms vibration level of 1g is 120 dB re 10 μm/s^2. At times the acceleration is also expressed as some fraction of g. Seldom is the acceleration in vibration of machinery expressed in m/s^2.

Types of Vibration

Vibration can be classified as either single frequency or multiple-frequency, steady-state, random, or transient. *Single frequency vibration* or *multiple frequency vibration* occurs when a limited number of frequencies excite a structure. An example would be a floor-mounted electric motor driving a compressor. This type of vibration is often associated with an imbalanced force in an electric motor with a gear that can produce several harmonically related frequencies. *Random frequency vibration* occurs when many unrelated frequencies are involved in the vibration excitation, and it is not possible to specify the instantaneous vibration at any given instant in time except in statistical terms. *Transient vibration* is that which occurs when a normally stable body is excited by either an impulse or an impact. The body generally comes back to rest due to losses (damping) in the system during oscillation at its resonance frequency.

References

ANSI (1983). "Specification for Sound Level Meters," American National Standards Institute, Sl.4-1983 (R 1990), New York, NY.

ANSI (1985). "Amendment to Sl.4-1983," American National Standards Institute, S1.4A-1985, New York, NY.

ANSI (1986). "Specification for Octave-Band and Fractional-Octave-Band Analog and Digital Filters," American National Standards Institute, Sl.1 1-1986 (R1993), New York, NY.

ANSI (1994). "Acoustical Terminology," American National Standards Institute, Sl.1-1994, New York, NY.

ASTM (1985). "Recommended Practice for Measurement of Sound Absorption, C 424-85," American Society for Testing Materials, Philadelphia, PA.

ASTM (1990). "Recommended Practice for Laboratory Measurement of Airborne Sound Transmission Loss of Building Floors and Walls, E90-90," American Society for Testing Materials, Philadelphia, PA.

Beranek, L. L. (1954). *Acoustics,* McGraw-Hill, New York, NY.

Beranek, L. L. (1988). *Noise and Vibration Control,* Institute of Noise Control Engineering, Poughkeepsie, NY.

Beranek, L. L., and Ver, I. L. (1992). *Vibration and Noise Control Engineering,* Wiley, New York, NY.

Harris, C. M. (1979). *Handbook of Noise Control, 2nd Edition,* McGraw-Hill, New York, NY.

Hedeen, R.A. (1980). *Compendium of Materials for Noise Control,* U.S. Dept. of HEW, NIOSH Pub. No. 80-116, Cincinnati, OH.

ISO (1985). "Acoustics — Measurements of Sound Absorption in a Reverberation Room," International Organization for Standardization, ISO 364:1985, Switzerland.

ISO (1990). "Acoustics — Measurement of Sound Insulation in Buildings and of Building Elements — Part 3: Laboratory Measurements of Airborne Sound Insulation of Building Elements," International Organization for Standardization, ISO 140-3:1990, Switzerland.

The Noise Manual, revised 5th edition, edited by E.H. Berger,
L.H. Royster, J.D. Royster, D.P. Driscoll, and M. Layne
©2003 American Industrial Hygiene Association

3

Sound Measurement: Instrumentation and Noise Descriptors

John J. Earshen

Contents

Introduction

In industrial hygiene practice, noise exposure is measured to assess its potential detrimental effects on humans. While emphasis is placed on hearing damage, subjective effects such as speech interference and annoyance are also of interest. Two categories of measurement are necessary to meet the objectives of reducing hazards and assessing subjective effects. One provides data for predicting the impact of sound on humans; the other provides data for engineering purposes directed to mitigating or ameliorating impacts.

During the past 25 years, microprocessors and computers have produced a revolutionary change in measuring instruments, and in data processing and retention. Solutions to problems previously considered impractical or uneconomical are now possible within the scope of industrial hygiene practice. Choices among instruments, and data processing algorithms and programs, are large and tend to obscure how problems can be addressed. To develop a structured understanding of measurement, assessment, and mitigation, the salient factors of the overall process have to be identified.

Sound is a dynamic pressure fluctuation superimposed on existing atmospheric pressure. It varies in both time and space, and is propagated as compressional waves. If simultaneous measurements could be made at all points of a defined space and recorded as continuous functions of time, the sound pressure field would be uniquely determined for reference in determining noise exposures. Clearly, this is impossible. Practical measurements are limited to observations at a finite number of points, each of which may exhibit unique time variability. The basic measurement problem involves choosing the number and location of observation points, and deciding whether to record detailed time histories of pressure variations or to perform on-site averaging or other data-compression operations (for example, the basic sound level meter [SLM] measures a short-duration moving average of the square of acoustic pressure).

Curiously, in the domain of measuring physical acoustics variables, rarely are the descriptors stated in pressure or other physical units. Instead, the decibel notation is used because of historical precedents in psychoacoustics and circuit analysis. Such notation does not infer that such measured values have direct application to human factors without further conversion. However since human hearing spans an enormous dynamic range, the compression produced by a logarithmic scale facilitates graphical presentation of time-varying sound.

Nearly all instruments used for noise analysis have evolved from the basic SLM, which senses acoustic pressure and indicates sound levels. Specialized instruments have been developed to overcome SLM limitations and to automatically compute various noise measures. This chapter describes the generic instruments, and identifies noise measures that are derived from them and are applicable to identifying hearing hazards and subjective response. Additionally, guidelines for instrument selection and utilization are presented. An annually updated guide to commercially available instruments and their manufacturers is published in *Sound and Vibration* magazine (Anon., 1998).

Specialized noise measures are essential for quantifying hazards to hearing, and are pivotal in establishing regulatory protection against such hazards. Other specialized measures are necessary for quantifying subjective effects, and are required for use in regulatory practice as well as in performance specifications for equipment and devices. It is vital to separate measurement confidence limits on quantifying physical variables from those defined for predicting effects on humans. Too often this boundary is blurred, leading to erroneous conclusions. It must be remembered that all empirically derived and defined measures have limitations of their own on predicting results. It is incorrect to conclude that significant improvements in accuracy and precision of measuring instruments will necessarily result in improved accuracy and precision in the prediction and assessment of effects on humans.

In evaluating the reliability of exposure assessments, three separate factors have to be considered:

1. The nature of the noise exposure environment characterized by its variability over time and space. This will determine requirements for the types of survey instruments to use and their needed accuracy and precision. Accuracy refers to the closeness of a measurement to the quantity intended to be determined; precision refers to the way in which repeated measurements of a fixed quantity conform to themselves.

2. The measurement protocol and procedure for deriving the defined measures. Some procedures use statistically sampled data, others use continuous data. Futhermore, some are based on measurements made on or in the immediate vicinity of a human subject; others require that the subject be removed from the region of interest during measurement.

3. The fundamental validity of the measure itself. For example, selection of decibel exchange rates for averaging, criterion levels, and base periods for exposures have different adherents and are vigorously debated.

When identical instruments and procedures are employed to make successive measurements of exposure to noise presumed to have constant properties, the end results may be highly variable. Such variability can be caused by subtle but significant changes in the noise sources, the position and orientation of human subjects, and an inability to replicate instrument performance and measurement procedures. Fortunately, many of the measurements made in industry can be replicated with a high degree of confidence. Nevertheless, when initiating a measurement program and evaluating results, careful attention should be paid to observing and assessing causes and limits of variability.

The industrial hygienist must be aware of the types of problems that may be encountered in a measurement and assessment program. An understanding must be acquired of the characteristics and limitations of noise measures, the properties and applications of measuring instruments, the nature of sound fields, and the measurement protocols. (See also Chapters 2 and 7.)

Generic Sound Level Analyzers

A substantial portion of the metrics used to quantify noise-induced threats to hearing and subjective response was derived and based on measurements ranging as far back as 50 years with progenitor sound level meters. Such instruments had dynamic responses and performance capabilities constrained by then contemporary technology. Measurements made with current, more advanced instruments must account for, and incorporate where necessary, the unique properties of the early instruments. This is necessary to produce results compatible with bodies of data assembled in the past.

By definition, sound pressure level is 10 times the logarithm to the base 10 of the mean square sound pressure expressed as a ratio to the square of the reference pressure of 20 micropascals (see Equation 2.10). All sound level measuring instruments of interest to the industrial hygienist incorporate this function. Although it is now common for an instrument to have capabilities to generate many of the derived measures, understanding the generic types required for particular measures is essential. Figure 3.1 illustrates functional relationships among generic instruments. Initially, acoustic pressure sensed by the microphone produces an electrical signal input to a preamplifier. The conditioned signal is then processed through frequency-selective or weighting filters. (Commonly provided choices are listed in the functional block and discussed later.) Other frequency-

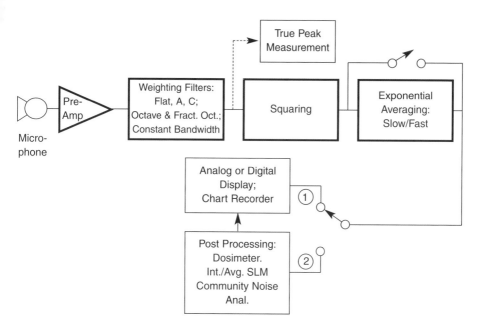

Figure 3.1 — Generic sound level analyzers (blocks shown with bold borders represent basic sound level meter components).

weighting filters may be included for obtaining particular noise measures. The next block indicates a squaring operation required because sound pressure level (SPL) is a function of pressure squared. Next is an exponential averaging filter; this generates a form of a moving averaging window in time. Functionally, this block characterizes the instrument's dynamic response characteristics (SLOW or FAST). Some integrating/averaging sound level meters eliminate the exponential averaging and substitute true averaging of the pressure squared (ANSI S1.43-1997; IEC, 1985). Note the bypass selector switch which shows that exponential averaging may be eliminated in certain instruments.

Following the basic processing, additional functional blocks are shown as switch selectable for illustration. Actual instruments may allow selection of functions or may be totally dedicated to particular operations. The functions depicted in the post-processing block may be performed in real time or off-line by subsequent processing of stored time history data.

Basic Sound Level Meter (SLM)

The component blocks indicated with bold borders in Figure 3.1 are representative of basic SLMs. Historically, instruments used moving needle meters, the dynamics of which influenced the definition and computation of many derived noise measures. Contemporary instruments can have displays that use moving needle meters, digital indicators, analog plots of sound levels, or combinations thereof.

To gain understanding of the operation of an SLM, assume that the microphone is subjected to sound having frequencies spanning the entire range of hearing. Examination of the signal from the microphone will show waveforms having frequency components up to approximately 20 kHz. (Special microphones and analyzers may be needed for measurement of higher frequencies.) Should a sensitive set of high-quality earphones be connected here, the sound heard would be a good replication of the sound sensed by the microphone.

After amplification and filtering, the noise signal has been frequency weighted, and if auditioned, would appear to have been modified by a tone control. For example, if an A-weighting filter is used, the sound in the earphones will have its low-frequency components diminished, and frequencies above 500 Hz will dominate (see Figure 3.3). At this point in the circuit, actual instruments may have an output for connecting earphones or a sound recorder. It is useful to listen to such a signal. If the monitored sound is distorted or otherwise grossly different from that heard by the surveyor in the actual environment, instrument failure or overloading should be suspected and investigated.

The next block introduces a drastic change to the signal by the squaring process, which generates a waveform that is, in principle, proportional to the instantaneous power of the acoustic signal. Through the earphones, there would no longer be a recognizable association with the microphone input since squaring is a nonlinear operation. If this waveform drove a moving needle indicator capable of following perfectly its instantaneous variations, the observer would only see a blur. For the general case, when the input waveform contains frequencies up to 20 kHz, the output of the squaring circuit would contain frequencies up to 40 kHz. Due to limitations of human vision, even if a display or needle movement responded that rapidly, it could not be perceived and would be impractical.

To make the output visually usable, some means of reducing the rate of indicator variations is necessary. One logical approach would be to divide the pressure squared waveform into sequential segments. Each segment could be averaged, and the running output would represent a moving average. The duration of each segment should be sufficiently long so that updated average readings would be presented at a rate that is easily perceived by human vision. Such an instrument is a time-history version of an averaging SLM, and is realizable through analog or digital signal processing. However, the concept of the SLM was introduced over 60 years ago, and the available technology of the early years undoubtedly influenced the choice of a dynamic averaging process based upon an exponential averager. Similar problems existed at the time in other areas in the monitoring of rapidly varying waveforms with moving needle meters (e.g., modulation monitors in transmitters, recording level indicators, etc.). Comparable solutions were applied to these problems. From a pragmatic viewpoint, a means to "slow down the meter" was absolutely mandatory to make it readable. By combining the inertial response of a moving meter with simple electrical circuits, acceptable performances were achieved.

Two instrument dynamic characteristics were established in the early days of development and have persisted to the present. They were designated SLOW and FAST. These dynamics are each described by specific time constants. To illustrate the properties of a time constant, consider what the hypothetical response of the meter movement of an SLM would be for a suddenly applied constant sound pressure. If the meter has instantaneous response, the needle would instantly rise to conform to the input. As previously discussed, such response is not desirable. By control of the physical parameters determining the needle response, it is possible to obtain a sluggish response that tends to follow the short-time average value of the applied waveform. If it has such dynamics, the meter needle rises asymptotically toward the value of the suddenly applied constant sound amplitude. Because such a response curve theoretically only reaches the maximum value at infinite time, a convention has been adopted to characterize the rising response curve in terms of the time delay it takes to reach 63 percent of its maximum value. This is called the time constant. The response obtained with such dynamics is defined as exponential averaging (Yeager and Marsh, 1991).

The two SLM time constants are respectively 1 second (SLOW) and 0.125 second (FAST). Why these specific numerical time constraints were chosen rather than others of similar magnitude is difficult to ascertain. In practice, the SLOW constant is used when one attempts to determine the average or slowly changing average value of observed sound. In contrast, the FAST time constant is used to estimate the variability in the observed sound where only the limits are desired. In Figure 3.1, the basic SLM is implemented by moving the selector to position 1, thus connecting a visual display of sound levels.

The literature contains a sparsity of discussions developing rationales for the given time constants based on psychoacoustic and physiological models of the hearing process. Furthermore, the particular values selected are not substantially supported or universally accepted. One can reasonably speculate that instrument designers originally chose time constants and dynamics attainable with the available technology that were "comfortable" to the instrument user. Subsequent investigations recorded measurements made with such instrument dynamics and built up bodies of data that were and are empirically matched to physiological and psychoacoustic references.

Instrument dynamics associated with basic SLMs significantly affect how important noise measures are defined and computed. Though modern computer-based instruments can produce measurements in a mathematically elegant manner, extensive collections of data on effects of sound on humans are based on the traditional dynamics. Accordingly, various noise measures and regulations require that computing instruments incorporate SLOW, FAST, or both. For example, there has been extensive debate regarding SLM dynamics to be incorporated in measuring and computing noise dose when there is exposure to impulsive noise (Kundert, 1982; Rockwell, 1981; Suter, 1983). Nevertheless, the Occupational Safety and Health Administration (OSHA) guidelines (Appendix I) mandate that SLOW response be used. For detailed analyses of how selected dynamic response affects measured results, see Earshen, 1980 and 1985; Kamperman, 1980; and Kundert, 1982.

Graphic Level Recorder

When the selector is at position 1, the signal can be stored on a chart recorder as a function of time. Such plots can be useful when there is a need to observe trends or to store the information for further graphical analyses. One such application is to determine decay rates in order to compute reverberation times for implementing engineering controls. Such applications are obsolescent. Considering the flexibility and ease with which such records can be stored and reproduced from computer files, recording data initially on only hard-copy charts is not an attractive option.

Dosimeter

Various measures exist that purport to quantify the threat to the hearing of humans exposed to excessive sound. As discussed later, such measures are derived by processing sound-level measurements obtained in the immediate vicinity of a person's ear (hearing zone) with an SLM. A dosimeter is a body-worn instrument that performs two functions. The microphone senses sound pressure in the "hearing zone" while the remainder of the instrument automatically computes the desired noise measures. If the selector in Figure 3.1 is in position 2, a basic dosimeter can be configured.

Dosimeters are battery-powered and have evolved from simple devices computing single-number exposure measures to highly sophisticated monitors that compute and store comprehensive data. It is important to recognize that a dosimeter derives directly from an SLM, and was developed to simplify measurement and computational procedures. If measures derived from SLMs and dosimeters are to be commensurate, the dosimeter must correctly duplicate the dynamic characteristics of the SLM. In the U.S., specific requirements prescribing such characteristics are set forth in the pertinent ANSI standard on personal dosimeters (ANSI S1.25-1991) and in OSHA regulations (Appendix I). In other countries and jurisdictions, other dynamic responses are often specified.

Integrating/Averaging SLM

The integrating/averaging SLM has properties similar to those of a basic dosimeter. However, there is an important difference. Once again referring to Figure 3.1, this instrument configuration is attained by placing the selector at position 2. Note that the symbolic bypassing switch is also closed around the exponential averaging block. This eliminates the defined dynamic response of SLOW or FAST. Consequently integrating or averaging takes place on the frequency-weighted but otherwise unmodified pressure-squared signal. Averaging under these constraints (ANSI S1.43-1997; IEC, 1985) is only defined for 3-dB exchange rates (discussed later). Such measurements are not applicable to OSHA exposure metrics.

True Peak Measuring SLM

There are applications that require measurement of true peak sound pressure levels (e.g., gunshots, punching or stamping machines, etc.). Measurements should be unencumbered by conventional SLM dynamic response characteristics. To illustrate the processing path followed, Figure 3.1 shows a separate signal path from the microphone output to the true peak processor. The signal may have frequency weighting such as A or C, or may have FLAT (i.e., uniform) response over a specified band of frequencies. The processing may retain the highest peak observed during an observation period. Another version may record the highest peaks observed in each of a series of contiguous time intervals having selectable durations.

Community Noise Analyzer

Yet another member of the generic group is the community noise analyzer, represented in Figure 3.1 by position 2 of the selector. It automates the computation of special noise measures (discussed later) and collection of data to identify and regulate emission of environmental noise. Such instruments have very broad capabilities for processing sound pressure data, employing multiple frequency weightings such as A and C, and filtering options such as octave band and narrow band. Processing may selectively employ (sequentially or simultaneously) multiple frequency weightings or dynamic response settings such as SLOW, FAST, or IMPULSE (see *Impulse Dynamics*). The computed measures include ones based on integration, averaging, and statistical processing.

Typically community noise analyzers are enclosed in weatherproof containers and used with special all-weather outdoor microphones. This makes it possible to conduct continuous observations outdoors over many days or on a permanent basis. In addition, these capabilities permit measurement of industrial noise for a hearing conservation program under ambient conditions that are so adverse as to preclude use of conventional SLMs or dosimeters.

Forms of Implementation

The generic instruments have been implemented entirely with analog circuits and devices, or in combination with digital processors. The analog/digital combinations are most common. For properly designed instruments, the implementation scheme employed does not affect the overall performance. Without additional information, the internal implementation of the instrument cannot be deduced by examining only input and output data. Many current instruments employ analog circuits in the blocks up to the selector switch (indicated in Figure 3.1) and digital processors beyond that point. Significant advancements in digital signal processing chips and capabilities of personal computers (PCs) have now made possible the entire digital processing of the signal after the input preamplifier.

Sound Level Meters

Types

Contemporary electronic technology has produced large-scale integrated circuits on solid-state chips, each containing astronomical numbers of analog and/or digital components. The number of individual chips used in instruments has been dramatically reduced. It is now feasible to integrate all functions on a single solid-state chip. As a consequence, SLMs have become smaller, more rugged, more reliable, stable and accurate, less demanding of power, and, most significant of all, capable of being integrated into comprehensive instruments that overcome operator limitations on making readings, and on processing and recording data. Increasingly, SLMs are being integrated into instruments performing multiple functions. The current SLM standard (ANSI S1.4-1983) specifies four types of SLMs, none of which incorporates integration or averaging. It provides for three grades of instruments — types 0, 1, and 2 — and for a special-purpose type S, each specifying accuracies applicable to a particular use:

- Type 0: Laboratory Standard — It is intended for use in the laboratory as a high-precision reference standard, and is not required to satisfy environmental requirements for a field instrument.
- Type 1: Precision — An instrument intended for measurements in the field and laboratory. (As a rough guide, measurements with a type 1 SLM will have errors not exceeding 1 dB.)
- Type 2: General Purpose — An instrument with more lenient design tolerances than type 1, intended for general field use, particularly in applications where high-frequency (over 10 kHz) sound components do not dominate. (Use

of this type of instrument is specified by OSHA and Mine Safety and Health Administration (MSHA) for determining compliance. For a type 2 SLM, estimated errors are not to exceed 2 dB.)

- Type S: Special Purpose — May have design tolerances associated with any of the three grades, but is not required to contain all of the functions stipulated for a numbered type.

Inexpensive sound level meters intended for use by hobbyists do not conform to required ANSI standards and are not suitable for compliance measurements. They can however serve an educational function permitting individuals to estimate noise levels of commonly encountered sources such as lawnmowers, power tools, and rock bands.

Impulse Dynamics

For types 0, 1, and 2, the weighting filters A and C and two exponential-time-averaging characteristics, SLOW and FAST, are required. An additional meter response characteristic, IMPULSE, is defined but not required. This characteristic specifies an exponential rise time constant of 35 milliseconds, and an asymmetric decay time constant of 1.5 seconds. It is mainly used in certain community noise measures and in rating the emissions of business machines. In North America, IMPULSE response is not commonly used to rate industrial noise. Historically, it was defined to enable SLMs to respond to a short burst of sound, and to hold the maximum value long enough for a human operator to perceive it before it decays. *This dynamic response characteristic is not suitable for measuring true peak sound level.*

True Peak SLM

Various regulatory agencies and professional practice require that accurate measurements be made of true peak SPL (e.g., OSHA requires that unprotected peak exposures be kept below 140 dB for impulsive sound). For such applications, an optional peak measuring mode is defined. A type 0 true peak SLM must adequately measure a 50-microsecond pulse (ANSI S1.4-1983). For types 1 and 2 instruments, the true peak reading SLM must adequately measure a pulse of 100 – microseconds duration.

Pertinent terminology used in current practice should be clarified. Impact and impulse noise are used interchangeably as descriptors of short transient sounds. In industrial hygiene practice, an impulse (impact) sound is commonly defined as a transient having less than 1-second duration, which may repeat after a delay of more than 1 second. This definition is ambiguous, and operation of instruments stipulated as making such measurements should be carefully reviewed and understood to prevent serious departures from true values.

The term "peak sound level" is loosely used in some instrument specifications when in fact reference is actually to maximum levels measured with SLOW or FAST (or even IMPULSE) response. The reader is cautioned to specifically establish what is meant by such designation for a particular response. The reader is warned *not to use* "IMPULSE" response as defined in ANSI S1.4 to measure true peak SPL.

For many years, measurement of true peak sound levels was understood to mean measurement of peak sound pressure levels, i.e., measurement without any frequency weighting such as A or C. The interpretation has been that measurements should be made with FLAT weighting (i.e., uniform frequency response). This presents the possibility to obtain inconsistent readings from different instruments measuring the same transient. In real instruments, uniform frequency response is bounded by upper and lower frequency limits. The SLM standards do not specify such limits explicitly. In contrast, C weighting is acceptably flat and has specified upper and lower frequency roll-off properties. To promote compatibility of measurements, contemporary professional guidelines recommend C weighting for measuring transients (ACGIH, 1998).

Many instruments presently in use may employ "flat" or A frequency weighting for measurement of peak levels. Readings obtained with such weightings in many instances are not likely to produce radically different readings compared to each other or to C-weighted results. However, differences are possible when significant components of sound exist at very low frequencies. For particular impulsive waveforms, potential differences may be assessed by making simultaneous measurements with instruments having different weightings. Conclusions may also be drawn by analyzing the detailed spectral content of given impulses to detect presence of significant low-frequency components.

Weighting Filters

As noted earlier, various acoustical measuring instruments employ frequency-selective weighting filters. These derive their characteristics from the perception of loudness of pure tones by human hearing. Others have bandpass filters (such as octave-band) to analyze the spectral content of sound waveforms. Functionally, the weighting filters (A and C) can be thought of as "tone controls." This can be demonstrated with SLMs, which have provision for connecting earphones in the amplifier chain following the filters. For a monitored sound source having broad spectral content, a distinctive change in perceived tonal quality will be observed as various weighting filters are switched in. However, when instead bandpass filters are switched in, interesting observations can be made regarding the influence of various frequency components in the perceived sound (e.g., speech with high-frequency components filtered out). The effect is particularly striking when octave-band or fractional octave-band filters are inserted.

The rationale for using band-selective filters (octave, fractional octave, narrow, etc.) to analyze spectral content of sound waveforms is self-evident. In contrast, it should be noted that the A- and C-weighting filters currently in use were historically derived from the Fletcher–Munson curves originally published in the 1930s (Fletcher and Munson, 1933). They characterized equal loudness perception by humans for pure tones of variable frequency as is shown in Figure 3.2. (An updated version of these curves (ISO, 1987) appears as Figure 4.6). They are interpret-

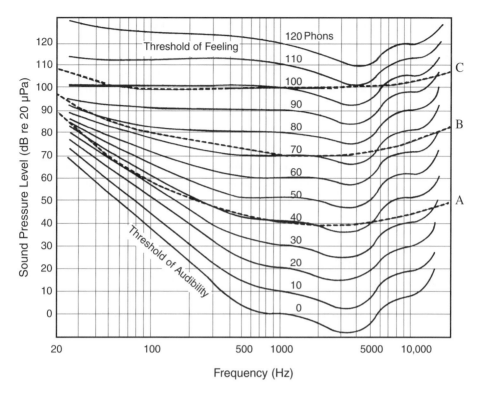

Figure 3.2 — Fletcher-Munson curves with weighting filter response overlay.
See Figure 4.6 for an updated version.

ed as follows: Given a starting reference pure tone at a fixed SPL at 1 kHz, each one of the family of solid line curves shows how the SPL of a pure tone must be modified when its frequency is different from the reference frequency in order to produce a sensation of loudness equal to that produced by the reference tone. Observe that the curves change substantially as the reference level is varied from 100 dB down to 0 dB. This demonstrates that human hearing response to pure tones across frequencies is very much a function of the absolute level of the tone.

After the Fletcher–Munson curves were developed and published, proposals were set forth for design of instruments to measure the loudness of complex sounds. Agreement was reached to standardize three frequency weighting curves that would attempt to approximate the equal loudness response of human hearing at low, medium, and high SPLs. The first was characterized by the 40-phon equal-loudness contour; its approximation by an electrically achievable filter was designated the A filter. Similarly, the 70-phon curve was designated for approximation by the B filter, and the 100-phon contour was designated for representation by the C filter. The resulting A-, B-, and C-filter curves relative to their values at 1 kHz are shown in Figure 3.3, and have been superposed as dashed lines over the 40-, 70-, and 100-phon contours in Figure 3.2. Note that the A-, B-, and C-filter curves do not appear to conform very closely to the corresponding 40-, 70-, and 100-phon equal loudness curves. The original Fletcher–Munson curves had much closer correspondence to these weighting filter responses. Although subsequent research has produced modified equal loudness curves, the originally established weighting filter responses have been unchanged (ANSI S1.4-1983).

The curvature of the filters seems to slope the wrong way in the superposed sketches in Figure 3.2. However, this is a perfectly valid representation. Note that in actual practice (see Figure 3.3), the filter curves are "turned upside down" to modify the otherwise flat (ideal) frequency response of an unweighted instrument in order to make it approximate the chosen properties of hearing. Increasingly through the years, B weighting has found few applications and is presently not provided in new instruments.

Various texts tend to mislead the reader with statements such as "the A-weighting curve approximates frequency response of human hearing and thus is incorporated in measuring instruments." A more informative, correct statement is that the A-weighting curve is an approximation of equal loudness perception characteristics of human hearing for pure tones relative to a reference of 40 dB SPL at 1 kHz (i.e., the 40-phon curve). Its application to the measurement of noise exposure for hearing protection and other purposes is only remotely, if at all, related to equal loudness perception. As a result of investigations in which a variety of weighting filters have been compared, it has been concluded that empirically derived measures using A weighting give a better estimation of the threat to hearing by given noise waveforms than do the other weightings. Because of simplicity and substantiated results, A weighting has continued to receive wide acceptance.

A weighting is also often used in determination of annoyance measures. Again, this use is based on empirical data derived from studies with human subjects. A weighting is not used to the exclusion of other descriptors in annoyance determination as extensively as it is in the area of hearing risk due to noise exposure. Nonetheless, it has very wide acceptance.

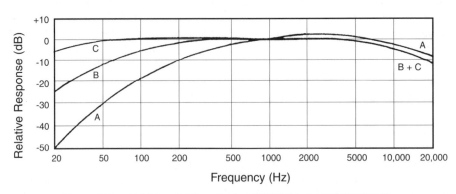

Figure 3.3 — SLM weighting curves [ANSI S1.4-1983 (R1977)]

The C filter has acceptance in applications where blast-type waveforms are encountered. Some additional use is made of C weighting by comparing successive readings made with A and C weightings in order to determine approximately whether or not a sound has significant low-frequency components. If it does, the C reading will be higher than the A; if it does not, the two readings will be very similar. (See Figure 3.3 and note that between 500 and 20,000 Hz, the A and C weightings differ only slightly.) Such C-A techniques are useful, for example, in evaluation of the adequacy of hearing protection devices (see Chapter 10). For more accurate analyses, instruments such as octave, fractional octave, and narrow band analyzers are commonly available.

For reasons previously discussed (see *True Peak SLM*), C weighting is strongly recommended in processing measurements of true impulse sound. Impulses having significant frequency components only above several hundred hertz will not produce greatly different readings between A and C weightings. However, OSHA regulations require unweighted measurements for impulse sounds. This can cause significant variations in readings between instruments.

Table 3.1 lists in detail the relative response levels for the A and C curves. These data are particularly useful when calculations are made to combine octave- or 1/3 octave-band levels for the purpose of determining equivalent A- or C-weighted levels. The A- or C-weighting filters are not used to determine loudness of complex sound waveforms. Instruments for such purposes do exist and are used to evaluate product sound characteristics, among other uses. Such instruments use specialized complex filtering and signal processing. Thus, use of C (or possibly A) weighting would have to be properly qualified as based on professional judgment.

SLM Measurement Tolerances

Table 3.2 shows the allowed tolerance for relative response levels for sound arriving at random incidence to the instrument microphone for the three types of SLMs. Careful examination of the table shows why it is not possible to give a simple answer to the question of the accuracy of a given type of SLM. It is evident that at low and again at high frequencies, the tolerance limit "opens up." Thus, the accuracies must be stated in terms of envelope limits that are functions of frequency. These characteristics are principally imposed by the limitation of practical microphones. (Note that the envelope specifies the frequency response of any given instrument of a particular type. It is not a measure of repeatability or precision for measurements with a particular instrument.)

To illustrate the point, consider two simple measurements made with a type 2 instrument. If a measurement is made of machine noise having frequencies ranging between 100 and 1250 Hz, Table 3.2 indicates that deviations of up to ± 1.5 dB will be permitted. In contrast, if measurements are made of noise having frequencies ranging between 4000 and 8000 Hz, deviations of up to ± 5 dB may exist. It is evident that attainable measurement accuracies are substantially affected by the predominant frequency content of the noise.

TABLE 3.1
Relative response for A and C weighting (ANSI S1.4-1983 R1997).

Nominal Frequency (Hz)	A Weighting (dB)	C Weighting (dB)
10	-70.4	-14.3
12.5	-63.4	-11.2
16	-56.7	-8.5
20	-50.5	-6.2
25	-44.7	-4.4
31.5	-39.4	-3.0
40	-34.6	-2.0
50	-30.2	-1.3
63	-26.2	-0.8
80	-22.5	-0.5
100	-19.1	-0.3
125	-16.1	-0.2
160	-13.4	-0.1
200	-10.9	0
250	-8.6	0
315	-6.6	0
400	-4.8	0
500	-3.2	0
630	-1.9	0
800	-0.8	0
1000	0	0
1250	0.6	0
1600	1.0	-0.1
2000	1.2	-0.2
2500	1.3	-0.3
3150	1.2	-0.5
4000	1.0	-0.8
5000	0.5	-1.3
6300	-0.1	-2.0
8000	-1.1	-3.0
10000	-2.5	-4.4
12500	-4.3	-6.2
16000	-6.6	-8.5
20000	-9.3	-11.2

TABLE 3.2
Tolerances for random-incidence frequency response of SLMs
(ANSI S1.4-1983 R1997).

Nominal Frequency (Hz)	Type 0 (dB)	Type 1 (dB)	Type 2 (dB)
10	+ 2,-5	± 4	+ 5, -∞
12.5	+ 2,-4	± 3.5	+ 5, -∞
16	+ 2,-3	± 3	+ 5, -∞
20	± 2	± 2.5	± 3
25	± 1.5	± 2	± 3
31.5	± 1	± 1.5	± 3
40	± 1	± 1.5	± 2
50	± 1	± 1	± 2
63	± 1	± 1	± 2
80	± 1	± 1	± 2
100	± 0.7	± 1	± 1.5
125	± 0.7	± 1	± 1.5
160	± 0.7	± 1	± 1.5
200	± 0.7	± 1	± 1.5
250	± 0.7	± 1	± 1.5
315	± 0.7	± 1	± 1.5
400	± 0.7	± 1	± 1.5
500	± 0.7	± 1	± 1.5
630	± 0.7	± 1	± 1.5
800	± 0.7	± 1	± 1.5
1000	± 0.7	± 1	± 1.5
1250	± 0.7	± 1	± 1.5
1600	± 0.7	± 1	± 2
2000	± 0.7	± 1	± 2
2500	± 0.7	± 1	± 2.5
3150	± 0.7	± 1	± 2.5
4000	± 0.7	± 1	± 3
5000	± 1	± 1.5	± 3.5
6300	+ 1,-1.5	+ 1.5,-2	± 4.5
8000	+ 1,-2	+ 1.5,-3	+ 5
10000	+ 2,-3	+ 2,-4	+ 5, -∞
12500	+ 2,-3	+ 3,-6	+ 5, -∞
16000	+ 2,-3	+ 3, -∞	+ 5, -∞
20000	+ 2,-3	+ 3, -∞	+ 5, -∞

For type 2, a sharp cutoff at 10 kHz is allowed and used in most instruments.
The given tolerances include both microphone and circuit performances.

In industry, ambient noise commonly contains dominant components at frequencies ranging between 100 and 3000 Hz. It is evident from Table 3.2 that the envelope has modest width. Above 3 kHz and up to 10 kHz, the envelope widens out to ± 5 dB. Since there are some sources that have significant components in that range (e.g., steam or air jets, or high-speed mechanical components), their spectral content should be examined when assessing accuracy of measurements. Use of type 1 instruments may be indicated.

Root Mean Square Value (rms)

Throughout this text, reference is made to rms values of sound pressure. The use of this measure stems from the fact that sound is a dynamic pressure fluctuation whose arithmetic average value is zero. Therefore ordinary arithmetic averaging cannot provide useful measures. An alternate approach recognizes that decibel notation and many other acoustical measures are based on sound power, which is proportional to the square of sound pressure. It is possible to define an effective sound pressure, p_{rms}, for a given averaging time, T, which produces the same average power as the actual time-varying sound pressure. Thus,

$$p_{rms} = \left[\frac{1}{T} \int_0^T p^2 dt \right]^{1/2} \tag{3.1}$$

The term inside the brackets simply denotes the averaging of p^2.

Crest Factor and Pulse Range

One of the instrumentation performance measures is *crest factor*. It is used to specify the ability of a sound-measuring instrument to correctly process waveforms that have peak values substantially higher than their rms average. Crest factor is rigorously defined only for steady-state repetitive waveforms, and is equal to the ratio of peak acoustic pressure to the rms value of the waveform. Alternatively, crest factor, stated in decibels, is equal to 20 times the logarithm of the ratio of peak to rms pressures of the waveform. Historically, the concept of crest factor probably had its origin in electrical utility system measurements and in sound recording. In such applications, it is common to encounter periodic waveforms that depart substantially from a sinusoidal form and have high peak excursions. The crest-factor metric was adopted for rating electrical and recording meters for their capability to correctly measure the average value of repetitive waveforms having high peaks. It was appropriately extended to rating sound-measuring instruments.

It is important to examine the functional properties of crest factor to gain appreciation for numerical magnitudes, and to recognize limitations of its utility for characterizing waveforms and specifying instrument capabilities. For a steady-state sine wave, it is well known that the rms value is 0.707 times the peak value. Accordingly, the inverse of that ratio, the crest factor, is 1.41 or 3 dB.

Consider next a repetitive train of sound pulses shown in Figure 3.4. They have constant amplitude, P, repetition periods, T, and individual durations, ΔT. Through application of Equation 3.1 and the definition of crest factor the following relationship can be developed:

$$\frac{p_{peak}}{p_{rms}} = \frac{P}{\left[\dfrac{P^2 \Delta T}{T}\right]^{1/2}} = \left[\frac{T}{\Delta T}\right]^{1/2} \tag{3.2}$$

Crest factor for the pulse train is determined by the ratio of the repetition period divided by the pulse duration. As an example, for 0.1-second pulses occurring once per second, the crest factor equals 20 log of Equation 3.2, which is 20 log $(1/0.1)^{1/2} = 10$ dB.

In the past, SLMs and dosimeters used to determine worker exposure to noise were required to have a minimum crest factor of 10 dB (ANSI S1.25-1978), which corresponds to a pressure ratio of 3.16. Equation 3.2 shows that this corresponds to a ratio of 10:1 between repetition period and pulse duration. Although this might be viewed to be inadequate, much of the noise encountered

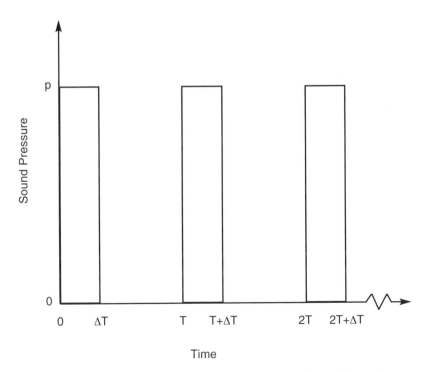

Figure 3.4 — Crest factor determination for a repetitive pulse train.

in industry can be successfully measured with such instruments. This is surprising when sources such as impact wrenches, chippers, continuous riveters, and impact drills are considered, since the subjective impression is that they all produce waveforms with very high crest factors. The requirement that pulse duration be no shorter than one-tenth the repetition period permits measurement of a large number of sources. Nevertheless, there are sources which generate higher crest factor waveforms that would not be adequately measured. For the illustrative pulse train, consider how a large crest factor may be manifested. For a fixed repetition period, shortening the duration period increases the crest factor. Alternately, for a fixed pulse duration, increasing the repetition period increases the crest factor. Taken to the extreme, a single pulse can be considered to have an infinite repetition period and thus an infinite crest factor! Though extending the analysis is not rigorously justified, it is evident that isolated impulses can be characterized as having very large crest factors. This leads to a curious conclusion. Consider a chipping gun that produces a pulse of fixed duration every time a working surface is impacted. Single or widely spaced blows would exhibit a very large crest factor, yet a continuous rapidly repetitive chipping burst would exhibit a much lower crest factor.

For these and other reasons related to the properties of exponential averaging, crest factor rating is not suitable for rating the capability of an instrument to correctly measure the average value of isolated impulses (e.g., punching, stamping, forging, etc.) (Earshen, 1985; Kundert, 1982).

A new measure has been adopted to rate the ability of an instrument to measure either the integral or the average value of pulsed transients. This measure is called *pulse range* (IEC, 1985), and stipulates tolerances to be met in determining averages for isolated pulses of given duration and amplitude. The referenced document stipulates measurement tolerances for level excursions up to 73 dB for pulses as short as one millisecond. Another instrument standard incorporating pulse range rating is ANSI S1.25-1991, "American National Standard Specification for Personal Dosimeters." This measure is adequate for identifying the capability of an instrument to correctly measure *both* isolated and repetitive impulses and high crest factor waveforms.

Note that the crest factor and pulse range only rate the ability of an instrument to average or integrate sound waveforms accurately. Neither rates the capability of an instrument to measure true instantaneous peak values. In fact, many instruments having high crest factor and pulse range capabilities have no provisions at all for measuring peak values separately.

Dynamic Range

An important rating parameter descriptive of instrument performance is *dynamic range*. It defines the range over which the signal can be processed or amplified without exceeding specified limits on linearity and corruption by self-generated noise. Every instrument or amplifying device has a maximum not-to-exceed value of signal, if specified linearity is to be maintained. With no input, each has internally generated noise that is not inherently distinguishable from an actual signal. This minimum value is called the *noise floor*. The range of magnitudes between the maximum signal level and the noise floor across the range of frequencies establishes the dynamic range of the device. Dynamic range is specified in decibels, and is equal to 20 times the logarithm to the base 10 of the ratio of the maximum value to the noise floor.

Microphones

Characteristics

Microphones are transducers that sense acoustic pressure and convert it into an electrical signal for subsequent processing. Ideally, conversion should be linear over a wide range of amplitudes and should be independent of frequency for the range of application. It is also desirable that the output of microphones not be susceptible to excessive mechanical vibration or induced interference from electromagnetic fields. Microphone response is specified in terms of conversion sensitivity as a function of frequency and the angle of incidence of impinging sound waves. The referenced sound pressure amplitude at the point of measurement is that of the plane wave undisturbed by the presence of the microphone. Such performance specifications are called the free-field sensitivity response levels.

A microphone having infinitesimal dimensions would have no change in sensitivity versus frequency as a function of angle of arrival. It is the finite size and geometry of a real microphone relative to the wavelength of the incoming sound that determine directional sensitivity. Directional sensitivity is manifested when microphone dimensions approach or exceed about one-fourth of the wavelengths of sound being sensed. The internal transducer mechanism that ultimately converts the pressure to an electrical signal does not control these properties. However, the internal transducer mechanism independently has its own frequency response characteristic.

All real microphones have variable sensitivity versus angle of arrival of plane waves (Johnson et al., 1991). Two classes are commonly used with measuring instruments — free-field and diffuse-field microphones. Other terms describing them respectively are perpendicular-incidence and random-incidence microphones. The perpendicular incidence or free-field microphone is constructed to have the flattest (i.e., frequency independent) response when pointed toward a source of plane waves (free-field conditions). In other directions, it has increased variations versus frequency. In contrast, the random-incidence microphone is configured to achieve a flat frequency response when sound arrives from all directions simultaneously as is the case in a diffuse field. Its response at some orientation angles will be higher than the true pressure at the point of measurement, while at others, it will be lower. The variations in response are symmetrically distributed over angles of orientation. When there are simultaneous arrivals from all directions, compensation occurs and the response is flattened out in frequency.

When sound is not diffuse but is composed of a dominant plane wave, the manufacturer specifies a preferred direction of orientation for measurement with a random-incidence microphone. At such orientation, the response is nearly independent of frequency (i.e., flat). Typically such angles are between 70° and 80° from the direction of plane wave arrival (Hassall, 1991).

Most microphones in instruments used in industrial hygiene practice have diameters of one-half inch or less. Sensitivity of such microphones is nearly independent of angle of arrival for sound at frequencies below approximately 6000 Hz. In this range either class can be considered omni-directional. At higher frequencies, angle of arrival becomes significant and the random-incidence microphone becomes preferred for diffuse-field conditions. On the other hand, if free-field conditions exist at such frequencies, the random-incidence microphone should be oriented at the manufacturer's recommended direction (70° – 80° from normal incidence). Under the same conditions, a free-field microphone should point toward the source.

The primary types of measurement microphones currently in use are listed by transducer mechanism in the following sections.

Capacitor (Condenser)

The transducer mechanism consists of variation in electrical capacitance caused by sound pressure to produce an electrical signal. The name, condenser, is a holdover from the earlier term for capacitor.

A thin, stretched diaphragm forms one plate of an electrical capacitor. The diaphragm is displaced by pressure fluctuation of incident sound waves. A second capacitor plate is stationary, and is mounted in the body of the microphone behind the diaphragm. The capacitance of the two-plate configuration varies proportionally to incident sound pressure. A constant voltage, defined as the polarizing potential, is applied across the capacitor to establish a constant static charge on the plates. Variation in capacitance causes voltage variation about the polarizing potential. These voltage variations alone are transmitted as output from the microphone. Such microphones have excellent linearity and frequency response and are available in all quality grades. They have low sensitivity to vibration; however, they are relatively delicate and can be damaged by rough handling.

Since the polarizing potential is applied across the very small separation distance of the capacitor, high electric field gradients are produced. Should contaminants (such as dust or more particularly moisture) invade the space between the plates, electrical discharges and short circuits can occur. Such microphones are quite susceptible to condensation of atmospheric moisture. Even if shielded from rain, a threat is still present should the temperature drop below the dew point. Specialized desiccant containers installed coaxially with the microphone can provide some protection. For prolonged use in adverse conditions, special protective enclosures and provisions can be used. If shorting or discharges occur, they will be manifested by erratic readings, and can be readily detected if the operator monitors the audio signal.

Electret

Electrets are similar to capacitor types in construction and performance, but do not require an external polarization voltage. Instead, they have an electret built into either the diaphragm or the stationary condenser plate. An electret is an electric charge analog of a permanent magnet. It is typically made of special waxes or plastics that have the ability to retain fixed electrical charge distribution (i.e., polarization). Such microphones are available in all grades and are somewhat less stable and more affected by temperatures in comparison to capacitor types. Like those, they have low sensitivity to vibration but are similarly delicate and can be damaged by rough handling. Because there is no polarizing potential, sensitivity to moisture is lower than that of capacitor types.

Piezoelectric

The piezoelectric microphones employ crystalline elements such as quartz or tourmaline, which have the property of generating an output potential when they are strained by an applied force. Furthermore, the polarity of the output is related to the direction of the strain. Compression produces an output of one polarity while extension produces an opposite polarity. Because the conversion coefficients are not equal for tension and compression, linearization is achieved by prestressing the element. Although force on the piezoelectric element can be applied directly by sound pressure, increased sensitivity is obtained by using a diaphragm that is mechanically connected to the element.

Piezoelectric microphones are highly linear and have superior ability to handle high sound pressure levels over extremely wide frequency ranges (such as those generated by sonic booms or explosive blasting) without distortion. When used in the cited applications, they are sometimes referred to as blast gauges. They are very rugged and have low susceptibility to moisture, relatively low sensitivity, and high noise floors (approximately 50 dBA). They are sensitive to vibration and have relatively high changes in sensitivity with changes in ambient temperature especially below 50° F (10° C).

Ceramic

Ceramic microphones are sometimes loosely classified as piezoelectric. Their transducer mechanism uses a special polarized ceramic such as barium titanate, which generates an electrical potential when it is strained. This conversion is similar to piezoelectric effect, however, the mechanism differs in that *both* compressive and extensive strains produce the *same* polarity of potential. If untreated, such a transducer would produce a highly distorted output. By a process called polarization (not identical to that for the capacitor type) the equivalent of an off-set strain is introduced in the material. This off-set linearizes the output. Retention of polarization is temperature dependent. At a critical elevated temperature polarization is lost.

The stressing force is derived by linkage to a diaphragm that moves in response to sound pressure. Such microphones are more rugged than capacitor or electret microphones. Even when used in dosimeters, they can survive rough treatment. They are also significantly less susceptible to impairment by condensation of atmospheric moisture.

Ceramic microphones have greater susceptibility to excitation by vibration, have higher noise floors, and have a more limited high-frequency response when compared to capacitor and electret types. Though they can be made in higher grades, they commonly are only available for type 2 instruments. These limitations are acceptable for dosimetry, and they are used in such applications because of their other desirable properties.

Accessories and Other Instruments

Acoustical Calibrators

Sound measuring instruments use many forms of analog and digital electronic circuits to process the electrical output from microphones. Confirmation of processing accuracies and calibration of equipment functions other than the microphone are best done by insertion of an appropriate electrical signal representative of all sound waveforms that the instruments are designed to measure. Modern equipment is quite stable unless physically mistreated. Comprehensive verification of specifications and calibration only needs to be done infrequently, either following repairs or to comply with company or regulatory agency requirements for periodic recalibration. The user, however, must not become complacent and should learn how to detect equipment malfunction.

The weakest link in the measurement chain is the microphone. Sensitivity and frequency response of this component can change as a result of either varying ambient conditions or physical damage. For legal purposes, calibration with an acoustical calibrator should be documented before and after a test.

The pertinent standard for acoustical calibrators ANSI Sl.40-1984 (R1997) states:

"[Calibrators] . . . normally include a sound source which generates a known sound pressure level in a coupler into which a microphone is inserted. A diaphragm or piston inside the coupler is driven sinusoidally and generates a specified SPL and frequency within the coupler. The calibrator presents to the inserted microphone of a sound level meter or other sound measuring system a reference or known acoustic signal so one can verify the system sensitivity or set the system to indicate the correct SPL at some frequency."

Some acoustical calibrators provide two or more nominal SPLs and operate at two or more frequencies. Multiple levels and multiple frequencies in some instances may be useful for checking the linearity and, in a limited way, the frequency response of a measuring system. The latter is useful for gross checks of microphones and weighting filters. Additional signals such as tone bursts may be provided for use in checking some important electroacoustic characteristics of SLMs and other acoustical instruments. Such waveforms can also be useful in checking performance of instruments used to measure sound exposure levels or average sound levels over time periods of variable duration.

The ANSI calibrator standard specifies tolerances for SPLs produced by calibrators. These range from ±0.3 dB for calibration of microphones used with types 0 and 1 SLM instruments to ±0.4 dB for calibration of type 2 instruments, including dosimeters.

The frequency range over which coupler-type acoustical calibrators can be designed to operate is limited by certain physical constraints. It is not practical to develop such calibrators to span the entire operating frequency range of acoustical instruments. Limitations are encountered when the wavelengths of sound frequencies generated become small compared to the dimensions of the microphone and calibrator cavity.

There are less common calibrators that produce:

- controlled outputs—step-attenuated for testing instrument linearity,
- repeated tone burst to test crest factor (limited to low crest factor), and
- transient signals to test instrument dynamic response (i.e., SLOW or FAST).

Such calibrators are not widely used because they are inadequate for the comprehensive calibrations provided by calibration laboratories, and too cumbersome for use in the field.

Calibrators may be configured to accept only one type of microphone, or they may accommodate adapters permitting calibration of many types and sizes. Coupler-type calibrators should only be used with microphones for which they are designed. Instructions supplied by the manufacturer on use and corrections for barometric pressure and temperature should be carefully followed. Corrections are uniquely related to the design of specific calibrators and are not universally applicable. For example, some require only very small corrections over a large range of altitudes and resulting pressures.

Another method exists for spot calibration of microphones during measurement surveys. It uses an electrostatic actuator that is based on the principle that application of an electrical potential across parallel capacitor plates produces attraction between the plates. In a capacitor or electret microphone, the sensing diaphragm is actually one plate of a capacitor. Sound pressure normally displaces the diaphragm, but displacement can also take place if an external plate is positioned close to the diaphragm and an electrical potential is applied across the pair. In this method of calibration, sound is not produced. The calibrator capacitor plate is actually a metallic grid that permits unencumbered passage of sound under normal operation. Because a very close, stable separation is required, such a calibrator is normally semi-permanently attached to a microphone, and is not routinely removed.

Frequency or Spectrum Analyzers

To support development and implementation of engineering controls, to select hearing protectors (in critical situations), and to evaluate noise annoyance, the need sometimes exists to determine spectral properties with higher resolution than is provided by A or C filtering.

Some instruments having high-resolution frequency-selective filtering are generically related to SLMs; others are similar but have distinctive properties. For the generic group, the frequency-selective filters are substituted for the A and C filters (see Figure 3.1). The output may be a meter indication or a graphic plot, although instruments capable of storing the measurement data as computer files are widely available and used. The significant similarity to the SLM is the inclusion of an exponential averaging function. The output of such analyzers presents observed sound pressure or weighted levels as functions of frequency.

The most common spectrum analyzer is the octave-band instrument, which uses selectable filters having center frequencies related by multiples of 2. The passband of such filters is bounded by f_2, the upper band-edge frequency, and f_1, the lower band-edge frequency. The center frequency for the bands, f_c, is the geometric mean of the band-edge frequencies (see Chapter 2, *Bandwidth*).

$$f_c = \sqrt{f_1 f_2} \ \text{Hz} \qquad (3.3)$$

The band-edge frequencies define the points where the response curves of adjacent filters cross over and are each attenuated by 3 dB below the passband level. This assures that pure-tone components at or near the cross-over frequencies will be correctly measured in the two adjacent filters. If the measured levels in each filter are added, the true level will be obtained.

A simple way to recall the mean frequencies of octave filters is to reference the commonly used calibrator frequency of 1000 Hz. The higher frequency filters can then be identified by repeatedly multiplying by 2 (e.g., 1000, 2000, 4000, etc.), and the lower frequency filters can be obtained by successively dividing by 2 (e.g., 1000, 500, 250, etc.). The band-edge frequencies of adjacent filters are also related harmonically by a multiplier or divisor of 2. Accordingly, octave-band filters have progressively wider bandwidths as the mean frequency is increased, and each successive filter has double the bandwidth of the lower adjacent filter. They are classified as constant percentage filters. The bandwidth of each is 70.7% of the center frequency. Should narrower band filtering be desired, 1/3 octave filters are available. The bandwidth of these is 23.1% of the center frequency. Established nomenclature designates the octave-band types "1/1 octave" and 1/3 octave types "1/3 octave." Still narrower fractional octave filters are provided in some instruments (e.g., 1/10, 1/12, etc.).

Table 3.3 lists nominal center frequencies for octave and 1/3 octave-band filters. It also shows the approximate octave and 1/3 octave passbands. The table is derived from ANSI S1.11-1986, "Specification for Octave-Band and Fractional-Octave Band Analog and Digital Filters," and from ANSI S1.6-1984, "Preferred Frequencies, Frequency Levels, and Band Numbers for Acoustical Measurements." The defining formulas for center frequencies yield some values that are not integral numbers or depart somewhat from being harmonically related. A rationalized progression of center frequencies has been adopted and has international concurrence. It is tabulated in ANSI S1.6-1984 and the rationalized values are designated nominal frequencies. The band edge frequencies are computed from the relation given in Equation 3.3. Small adjustments have to be

TABLE 3.3
Nominal center and approximate band edge frequencies for contiguous octave and 1/3 octave bands (values in Hz).

Band	1/1 Octave Bands			1/3 Octave Bands		
	Lower	Center	Upper	Lower	Center	Upper
10				9.2	10	10.9
11				10.9	12.5	14.3
12	11	16	22.4	14.3	16	17.9
13				17.9	20	22.4
14				22.4	25	28
15	22.4	31.5	45	28	31.5	35.5
16				35.5	40	45
17				45	50	56
18	45	63	90	56	63	71
19				71	80	90
20				90	100	112
21	90	125	180	112	125	140
22				140	160	180
23				180	200	224
24	180	250	355	224	250	280
25				280	315	355
26				355	400	450
27	355	500	710	450	500	560
28				560	630	710
29				710	800	900
30	710	1000	1400	900	1000	1120
31				1120	1250	1400
32				1400	1600	1800
33	1400	2000	2800	1800	2000	2240
34				2240	2500	2800
35				2800	3150	3550
36	2800	4000	5600	3550	4000	4500
37				4500	5000	5600
38				5600	6300	7100
39	5600	8000	11200	7100	8000	9000
40				9000	10000	11200
41				11200	12500	14000
42	11200	16000	22400	14000	16000	18000
43				18000	20000	22400

made in these cross-over frequencies, depending on the specific formulation used to implement the bandpass filters.

The constant percentage bandwidth of such filters makes their use questionable when the objective is to identify spectral content with a fixed degree of resolution. The origin of octave-band filters relates to measurements pertinent to psychoacoustics. Many rating schemes based on such filters have evolved and are in wide use (e.g., properties of acoustical materials, community noise criteria, sound reinforcement balancing, etc.).

In applying various types of spectrum analyzers, it is vital to thoroughly understand the relations between time functions and frequency functions. The fundamental presumption that a unique relation exists between the two is only rigorously defined for functions of infinite duration and invariant statistical properties. Analyses of *all* real-world waveforms involve varying degrees of approximations and assumptions. Use of spectrum analyzers and application of results should be approached with caution, and guidelines provided with instruments and contained in reference texts should be observed (Randall, 1991).

Data Recording

Despite the versatility and broad capabilities of modern instruments not only to sense sound, but to also process or compute online many different specialized measures used to interpret its effects, circumstances do exist where it is very useful to have a high-fidelity transcription of the sound under consideration. With most derived measures, once the base data have been processed, it is completely impossible to back track in order to perform alternate computations. For transcriptions to have utility, they must be essentially free of distortion and artifacts. In considering the option of using a recording versus on-line processing, one important determining factor is potential degradation in accuracy by the overall process as compared to the on-line approach. To support such examination, the capabilities of the recording means should be compared to those of the processing instrument or computer. Salient parameters include dynamic range, frequency and phase response, linearity, and stability of the recording medium (Bies and Hansen, 1996).

Ordinary direct magnetic tape recording is not well suited for use in instrumentation. Its principal shortcomings include limited dynamic range (40–65 dB), amplitude, phase, and frequency distortion, and poor low-frequency response. (Good phase response is critical to accurate recording of transients such as impulses.)

Most attractive for use in industrial hygiene and development of engineering controls applications are digital recorders, including digital audio tape recorders (DAT) and computers. Digital audio tape recorders have superior dynamic range (70–80 dB), high fidelity of recording, and substantial freedom from phase and frequency distortion. Their capabilities equal those of most analyzers and processors and, thus, off-line processing is broadly possible. Recorders made for audio reproduction by audiophiles may be used, but care must be exercised in matching inputs and outputs. Furthermore, reproducing attenuation settings pertinent to signal calibrations is problematic.

Another digital recording option is to use portable PCs. Interfaces for digitizing and signal conditioning are available. Configurations can be set up to provide capabilities compatible with off-line analyzers. Storage in high-capacity hard drives is possible for lengthy period records, but the available working space on the hard drives can be reduced especially if multiple long records must be retained. An attractive alternate is to use recordable compact disks (CDs). Recorders to produce disks for storage and transfer are available as PC accessories.

New, high-performance stand-alone digital processing and recording components are continuously being introduced as a result of the exponential growth of the underlying technology. Though the prospect is attractive, the end user should exercise extreme caution regarding interfacing such components with acoustical measuring and analytical instruments. Hardware and software compatibility issues must be correctly addressed to assure proper operation. For such ventures, competent professional assistance is advised. In contrast, when such integration is performed by instrument and analyzer manufacturers, unpleasant surprises are avoided. Regrettably, the market potential for advanced capability acoustical instruments is limited; thus, there is a notable lag between the introduction of improved components and incorporation in new instruments.

Noise Measures

Single discrete SPLs are rarely of value to measure in and of themselves. Rather, a number of specific noise measures, which are derived by averaging, integrating, or otherwise manipulating the basic measurements, are generally employed for evaluating and regulating noise hazard and annoyance. Many of these quantities are computed online by instruments themselves, but it is also possible to generate some of these measures by postprocessing of stored data. Such would be the case, for example, if SLM records were later analyzed to compute noise dose.

The measures discussed here are derived from time–function variables. Of the large number of noise measures that have been defined (Bishop and Schomer, 1991; Schultz, 1972), the ones most applied in industrial hygiene practice are based on A-weighted sound levels. Subsequent discussion defines and explains application of two groups: average levels and accumulated exposure or dose measures.

Average Levels
Equivalent Continuous Sound Level ($L_{eq,T}$, $L_{Aeq,T}$)
The average is perhaps the most obvious measure for classifying a time-varying sound level for a given observation period. Since sound levels are expressed in decibels, which are logarithmic, they cannot be averaged arithmetically. Therefore, it is the underlying acoustic pressure squared that is averaged arithmetically, and the result expressed in decibels. Note that the restriction on averaging pertains only to processing of the fundamental pressure squared variable.

Statistical processing of sampled derived measures, such as an ensemble of $L_{eq,T}$ values, is not similarly restricted and is discussed in Chapter 7.

The defining function is:

$$L_{eq,T} = 10 \log \left[\frac{1}{T} \int_0^T 10^{L_p(t)/10} \, dt \right] = 10 \log \left[\frac{1}{T} \int_0^T \left[\frac{p(t)}{po} \right]^2 dt \right] \quad \textbf{(3.4a)}$$

where $L_p(t)$ is sound pressure level as a function of time, T is the observation or averaging time, p(t) is sound pressure as a function of time, and p_o is the reference pressure of 20 micropascals (µPa). Adoption of the equivalent sound level terminology was likely motivated to forestall erroneous use of arithmetic averaging. Nevertheless, the term *average sound level* is also in wide use and is functionally identical.

The relationship remains valid if instead a frequency-weighted sound level is averaged to obtain an equivalent weighted sound level. For example:

$$L_{Aeq,T} = 10 \log \left[\frac{1}{T} \int_0^T 10^{L_A(t)/10} \, dt \right] \quad \textbf{(3.4b)}$$

where $L_A(t)$ is A-weighted sound level as a function of time.

The SLM dynamics are normally not explicitly stated in the defining relation, and in many applications are not relevant. Other forms of the function used by regulatory agencies, such as L_{OSHA} (see Equation 3.13), require that sound levels be measured with specific dynamics (e.g., SLOW). In such applications, appropriate forms of the variables are included.

For sound levels stated for discrete time increments, the overall equivalent level can be obtained by the following combining relationship. The levels for each discrete time period may be constants or may themselves be equivalent levels. Thus, the combined equivalent level is (A-weighted levels used in the example):

$$L_{Aeq,T} = 10 \log \left[\frac{1}{T} \sum_{i=1}^N t_i 10^{L_{Ai}/10} \right] \quad \textbf{(3.5)}$$

where i is the i_{th} increment, N is the total number of increments, t_i is the duration of the i_{th} increment, L_{Ai} is the A-weighted sound level for the increment, and T is the sum of the individual time increments. The comments concerning meter dynamics that follow Equation 3.4b apply here as well.

The equivalent sound level measure has many possible applications in regulations governing hearing protection, measurements of the average noise level emitted by a source, and community noise regulations. Currently, equivalent sound level is referenced by a number of regulatory bodies for hearing protection in North America. In recent times, the U.S. Air Force and the

American Conference of Governmental Industrial Hygienists (ACGIH, 1998) have adopted this metric. The currently published National Institute for Occupational Safety and Health (NIOSH) criterion document (1998) and the International Standards Organization (ISO, 1990) also reference this measure. Regulations commonly stipulate that unprotected exposure must be limited to a fixed maximum level of noise averaged over a shift duration of 8 hours during a calendar day. This defines a limit that must not be exceeded, but does not of itself indicate how an actual noise level that varies over time must be determined.

Consider that a fixed decibel level as stated in the regulatory limit corresponds to a fixed magnitude of pressure squared. Acoustic power is proportional to sound pressure squared, and power is defined as energy transfer per unit time. If the limit stated as pressure squared is multiplied by the shift duration time, the product can be identified dimensionally as energy. Consider:

$$\text{Proportionality constant} \times (\text{pressure})^2 \times \text{time} = \text{power} \times \text{time} \qquad \textbf{(3.6)}$$

$$\text{Power} \times \text{time} = \frac{\text{energy}}{\text{time}} \times \text{time} = \text{energy} \qquad \textbf{(3.7)}$$

It is evident that the exposure limit stated by a fixed decibel level for an 8-hour period corresponds functionally to an accumulation of sound energy not to be exceeded. In actual circumstances where sound levels are continuously varying, the equivalent sound level equals a constant sound level that appropriately integrated over the averaging time would result in the same energy as integrating the corresponding variable sound level over the same time.

Given the above, the regulatory limit may be restated to require that the instantaneous pressure squared, integrated over the exposure time, not exceed the product of the limiting pressure squared times 8 hours. The same total energy can be accumulated for an infinite set of combinations of power amplitude and exposure duration. The requirement is that the product of the two must remain constant and under the regulatory limit (i.e., equal energy must be maintained). It further follows that doubling the power amplitude requires a halving of exposure duration if equal energy is to be maintained. Accordingly, a 3-dB exchange rule applies to sound level and exposure duration. (A doubling of power corresponds to a 3-dB increase in level; a halving corresponds to a 3-dB decrease.)

Next, consider the concept of noise dose. A daily limit on accumulated sound energy was described above. Although the limit could be stated in joules, regulatory agencies normalize the allowed limit by stating it as percentage dose, where a dose of 100% corresponds to an exposure at the *criterion level* (L_c) for a duration equal to the *criterion time* (T_c), normally specified as 8 hours. Note that the criterion time is a specified quantity that is distinctly different from, and not dependent on, the actual workshift duration during which exposure is accumulated. In some instances, both can be equal to 8 hours.

There are applications where it is required that accumulated exposure be stated in terms of other measures. One of these is the A-weighted average sound level with 3-dB exchange rate, normalized to 8 hours. It is given by:

$$L_{A8hn} = 10 \log \left[\frac{1}{8} \int_0^T 10^{\,L_A(t)/10} \, dt \right] = L_{Aeq,T} + 10 \log \left(\frac{T}{8} \right) \qquad (3.8)$$

The analogous 5-dB 8-hour normalized quantity is the time weighted average (TWA, see Equation 3.14). It is the level at which exposure for 8 hours will produce the same accumulated exposure as obtained during the arbitrary duration T.

Another metric that quantifies accumulated exposure to noise is *sound exposure level* (SEL). It is expressed by:

$$L_{AE} = L_{Aeq,T} + 10 \log \left[\frac{T}{T_o} \right] \qquad (3.9)$$

where T is the exposure duration in seconds, and T_o is a reference duration of 1 second. The SEL shown here is for A weighting. For C weighting, the symbol is L_{CE} and $L_{Ceq,T}$ must be used in the computation.

Yet another measure is *sound exposure* (E_A). For A-weighted sound pressure, it is defined by:

$$E_A = \int_0^T p^2(t) \, dt \qquad (3.10)$$

The units are pascal squared hours (although pascal squared seconds are also used). This measure expresses exposure to sound in absolute physical units simplifying interpretation as compared to normalized measures (i.e., ratios) such as dose and TWA. Many current instruments are capable of generating this measure.

To gain appreciation about the numerical magnitudes of two prominent measures denoting exactly the same exposure to noise, consider that exposure to an equivalent level of 85 dBA for 8 hours corresponds to a sound exposure, E_A, of 1 pascal squared-hour (Pa^2h). If the sound exposure is doubled to 2 Pa^2h, the equivalent level for 8 hours increases by 3 dBA. (The relationship may be further examined by applying Equations 3.4b and 3.10, where time is expressed in hours.)

Equivalent Continuous Levels Based on 4- and 5-dB Exchange Rates—$L_{eq(4)}$, L_{OSHA}, TWA

Additional metrics, defined to serve regulatory purposes, are similar in concept to the equivalent continuous level based on the 3-dB exchange rate, as described above. However, since these alternative metrics require that the product of time with pressure squared raised to a power other than 1 be constant, they employ exchange rates other than 3 dB (compare Equations 3.4a, 3.11, and 3.13). These alternative functional relationships have no intrinsic meaning; they are simply approximations that have been selected to correspond to agreed-upon damage risk criteria.

The *exchange rate* (ER) is the trade-off relationship between an increase (or decrease) in sound level and the corresponding change in allowed exposure duration (Botsford, 1967); when the sound level increases by the decibel value of the ER, the allowed duration is halved. The U.S. Department of Defense chose the 4-dB exchange rate in the 1970s, although subsequently, the Army and Air Force switched to the 3-dB ER, leaving only the Navy with a 4-dB ER. OSHA selected the 5-dB ER.

In the personal dosimeter standard (ANSI S1.25-1991) a normalized dose based on the concept discussed is defined and presented in a parametric Equation. By appropriate choice of parameters, it is possible to define dose and to derive functions for computing equivalent continuous sound levels based on 3-, 4-, or 5-dB ERs. Among practitioners and instrument manufacturers, there have been many letter symbols used to represent the various measures. The symbols $L_{av(4)}$, $L_{eq(4)}$, and L_{DOD} have been used interchangeably in North America as have $L_{eq(5)}$, $L_{av(5)}$, and L_{OSHA}.

Though it is obsolescent, where applied, the equivalent continuous sound level using 4-dB exchange rate is:

$$L_{eq(4)} = 13.29 \log \left[\frac{1}{T} \int_0^T 10^{L_{AS(t)/13.29}} \, dt \right] \tag{3.11}$$

where $L_{AS}(t)$ is A-weighted sound level measured with SLOW exponential averaging. The term 13.29 is an evaluation of the parameter q related to the 4-dB ER by:

$$q = \frac{ER}{\log 2} = \frac{4}{\log 2} = 13.29 \tag{3.12}$$

As applied by OSHA, the equivalent continuous sound level using 5-dB exchange rate is:

$$L_{OSHA} = 16.61 \log \left[\frac{1}{T} \int_0^T 10^{L_{AS(t)/16.61}} \, dt \right] \tag{3.13}$$

where the parameter q of Equation 3.12 becomes $\frac{5}{\log 2} = 16.61$.

In applying Equation 3.13, attention is called to certain constraints on computation that have been established by administrative decisions in OSHA. To evaluate compliance under separate regulatory clauses, OSHA practice is to preprocess and edit the A-weighted sound level measured with SLOW response contained in Equation 3.13. Two cut-off or threshold levels are referenced — 90 (for engineering controls) and 80 dBA (for hearing conservation). Depending on the application, one of the two is used. In the processing, all values of L_{AS} smaller than the referenced cut-off are replaced by 0. (This is only true in principle. Because of the logarithmic nature of decibels, a small replacement decibel level is actually used in computations.) This in effect discards those contributions.

The concept of a cut-off or threshold level rests on tenuous grounds, but is well entrenched, and its implications should be understood. Use of a threshold implies that exposure to sound below the threshold is nonhazardous. Another rationale for its use is to make compatible measurements made with instruments having different noise floors. A basis for the 90-dB cut-off is presumed to be the result of arbitrary administrative interpretation of regulations in the early days of OSHA. As a practical matter, use of the 90-dB cut-off can seriously bias results, whereas an 80-dB cut-off is of minor consequence (Seiler and Giardino, 1994).

Data represented as a percent dose can also be reported in terms of a normalized average sound level designated *time-weighted average* (TWA). Contrary to common misinterpretation, TWA *does not* represent sound level averaged over any time period, but rather is normalized to 8 hours. It is defined for OSHA applications as:

$$\text{TWA} = 16.61 \log \left[\frac{1}{8} \int_0^T 10^{L_{AS(t)/16.61}} dt \right] = L_{OSHA} + 16.61 \log \left(\frac{T}{8} \right) \qquad (3.14)$$

Compare Equation 3.14 and Equation 3.8 to see how the choice of a 5-dB exchange rate affects the integration.

Equivalent Sound Levels and Community Noise

In community noise regulations, equivalent sound level commonly specifies maximum allowed average sound level as measured at the property line of a source or receiver. The averaging period must be specified but may be of arbitrary duration. Common durations include 1 and 24 hours.

A special version of equivalent sound level used in environmental applications is the *day–night average level*, L_{dn}. This is only defined for a 24-hour period and is obtained by combining the equivalent sound levels for two segments of the 24-hour period. The segments are 0700 to 2200, and 2200 to 0700. A 10-dB penalty increment is added to the equivalent sound level of the second segment before combining the two segments. The penalty is based on the premise that people are more annoyed by a given level of noise during nominal sleeping hours, and that nighttime background levels are approximately 10-dB lower than daytime levels. (See Chapter 15 for additional details.)

Noise Dose

Noise dose is given by the following defining relationship:

$$D = \frac{100}{T_c} \int_0^T 10^{[L_{AS(t)} - L_c]/q} \, dt \tag{3.15}$$

where T_c is the *criterion time* and L_c is *the criterion level*. Exposure to a sound level for a duration equal to T_c at a level equal to L_c will yield an exposure of 100% noise dose. [Note: The concept of a criterion level, L_c, is distinctly different from the threshold or cut-off level described in the previous section.] The parameter q is related to exchange rate and is 13.29 for a 4-dB exchange rate and 16.61 for a 5-dB exchange rate (see Equation 3.12). For OSHA at present, T_c is 8 hours and L_c is 90 dBA. T is the measurement duration, in hours.

If the sound level readings are discrete for increments of time, dose in percent is:

$$D = \frac{100}{T_c} \sum_{i=1}^{N} t_i 10^{[L_{ASi} - L_c]/q} \tag{3.16}$$

where L_{ASi} is the A-weighted SLOW sound level for the i_{th} interval, N is the total number of intervals, and t_i is the duration of the i_{th} interval.

An alternate method for computing noise dose under OSHA guidelines when sound levels for discrete time intervals are known is:

$$D = 100 \left[\frac{C_1}{T_1} + \frac{C_2}{T_2} + \dots + \frac{C_N}{T_N} \right] \tag{3.17}$$

where C_N is the actual time duration of the N_{th} interval during which a worker is exposed to a constant sound level L_{ASN}. The values for T_N are listed in Table G-16a in Appendix A of the Hearing Conservation Amendment (Appendix I of this text). These values represent the maximum permissible times during which a worker can be exposed to a corresponding sound level, L_{ASN}, without exceeding a dose of 100%.

Table G-16a is derived from the relationship:

$$T_p = T_c / 2^{[L_{AS} - L_c]/ER} \tag{3.18}$$

where ER is the exchange rate in decibels, and T_p is *maximum permissible exposure time* (also called the reference duration) at sound level L_{AS}. Note that T_p is defined for any exchange rate. T_N, which is only defined for a 5-dB exchange rate, is the same as T_p when ER = 5 dB. Alternately, Equation 3.18 can be inverted to solve for the permissible sound level for a given exposure time, as shown in Equation 3.19.

$$L_{AS} = [ER/\log(2)] \ [\log(T_c/T_p)] + L_c \tag{3.19}$$

Based on Equation 3.18, Figure 3.5 shows a series of plots relating maximum permissible exposure times for particular ERs and criterion levels. Any combination of criterion level, criterion time, and ER may be selected and curves plotted using Equations 3.18 and 3.19. Three curves are actually shown (Figure 3.5), all based on 8-hour criterion time. The 5-dB exchange, 90-dBA criterion level curve represents OSHA practice. The 3-dB exchange, 90-dBA criterion level curve rep-

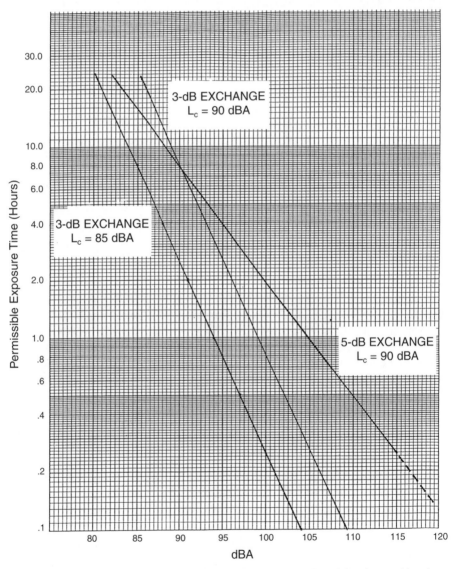

Figure 3.5 — Examples of permissible exposure time vs. A-weighted sound levels.

resents practice in many other countries. The 3-dB exchange, 85-dBA criterion level curve represents the ACGIH guidelines and is the current NIOSH (1998) criterion. The warning that special limitations may apply is based on the fact that permissible exposure time and sound level may not always be exchanged without restrictions. For example, currently OSHA does not allow any duration of unprotected exposure above 115 dBA, except for impulsive sound.

Instrument Selection and Utilization

Although noise and sound measurements are made for a wide variety of purposes, in industrial hygiene practice, the principal objectives normally are:

- evaluation of threat to worker hearing relative to statutory and/or scientific limits,
- collection of data for legal proceedings,
- collection of data to support development of engineering controls,
- test of machines or equipment components to determine compliance with particular specifications,
- collection of data on signal and speech perception and annoyance problems,
- qualifying interior levels of audiometric booths.

In addition to the above, an industrial hygienist may be called upon for advice or recommendations regarding evaluations of miscellaneous subjective and nonauditory effects of noise on humans.

In planning for measurements toward the principal objectives, the initial step is to identify the specific purpose of the measurements and which data must be collected. To determine both, it is necessary to understand:

- hearing damage criteria and related measures to be derived from noise measurements,
- psychoacoustic aspects of the human auditory system,
- properties of acoustic fields, and
- capabilities and limitations of measuring instruments.

Subsequent discussion emphasizes the use and salient operating characteristics of instruments. The objectives and procedures for conducting surveys are addressed in Chapter 7.

Use of Microphones

The primary sensor for noise measurement is the microphone. Proper application requires selection of a type matched to the properties of the field being measured. This information can be found in the instrument manufacturers' instructions and application notes. Extensive discussions are presented in several texts on acoustics and noise control (Bies and Hansen, 1996; Hassall, 1991). In addition, ANSI S1.13-1995 describes standardized procedures for use of microphones and instruments in making measurements. Also, Chapter 2 of this text contains definitions and properties of free and diffuse sound fields.

Regardless of the microphone type used, some basic facts and procedures pertain to all measurements. Whether attached to an instrument or hand held, a microphone is susceptible to errors in measurement produced by reflections. Figure 3.6 illustrates the variations in measurement that can occur due to the presence of the operator relative to an SLM or microphone (Hassall, 1991). Such effects can be minimized by the simple expedient of orienting the microphone and operator in a plane perpendicular to the propagation path from the source. The microphone itself should be pointed in the direction recommended by the manufacturer to produce the most uniform frequency response in the ambient acoustic field. (For example, for a free-field microphone in a free field, point toward the source; for a random-incidence microphone in a free field, point an angle of 70° to 80° from the direction to the source.) The worst possible place to stand is either in front of the microphone and risk casting an acoustic shadow, or behind the microphone and risk reflective interference. If a diffuse field is being measured, there is no identifiable propagation path. Then the microphone should be held as far from the operator as is practical. A suitable extension cable may be used to isolate the microphone.

Note that Figure 3.6 shows errors in measurement due to reflections that occur for pure tones at individual frequencies under the free-field conditions. This represents worst-case conditions, which may be encountered infrequently. If measurements are made over broad bands of frequencies (e.g., A weighted) and in a dif-

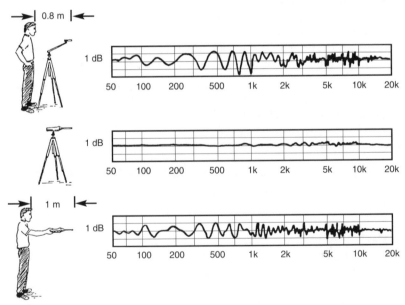

Figure 3.6 — Typical measurements showing influence of observer and meter case on measurement of sound. From: Harris, C.M. (ed.). *Handbook of Acoustical Measurements and Noise Control,* New York: McGraw-Hill, 1991, p. 9.14. Reprinted with permission of C.M. Harris, Columbia University, Educational and Professional Publishing Group, New York.

fuse field, the effects would be greatly reduced. The admonition about minimizing the influence of the observer should be respected, however, knowledge of the character of the sound field being measured may permit relaxation of the dictate.

Similarly, care should be exercised in measurements near reflecting surfaces, such as walls or large objects. "Near" is a relative term; the governing parameter is the wavelength of the sound being measured. Good practice dictates staying at least ¼ wavelength away from a reflector. (Such a restriction cannot be observed, of course, when the noise exposure of a person is being measured.) In measurement of noise containing many frequency components, it is the wavelength of the lowest frequency component that governs. For 100 Hz, a ¼ wavelength is approximately 76 cm. At 200 Hz, the corresponding distance is 38 cm. Distances can be scaled up and down to match frequencies of interest. (See Chapter 2 for additional discussions on wavelength effects.)

The preceding guideline will fail in the special situation where a standing wave exists. Such an effect occurs when a noise source radiates sound toward a reflecting surface which is uniform and does not scatter the sound. If the sound is reflected toward the source, the incident and reflected waves superpose and produce a "standing wave" by constructive and destructive interference. This condition is detected by moving the microphone along the principal propagation path and noting that a succession of maxima and minima of sound levels occurs. Because a worker in the area exhibiting such conditions may move about randomly in space and time, an estimate of the average sound level can be made. Note the maximum and minimum levels. If the difference is less than 8 dB, use the arithmetic average; if greater than 8 dB, use a dosimeter or integrating/averaging SLM. Standing-wave effects are not common in industrial environments.

The operator taking measurements should strive to minimize disturbance of an ambient sound field by appropriate microphone positioning, especially in investigating sound fields for engineering control purposes. The related question, whether to remove workers from a measuring zone when the objective is to determine their personal exposure, is not simple to answer. Logically it would appear that it is the noise actually impacting the worker's ears that matters. Unfortunately, there is no clear concurrence in the professional communities and regulatory agencies regarding the method to follow. Consider for example, if a sound-level measurement is made at the very entry of the auditory canal, should the value obtained be corrected to the value that would have existed in an undisturbed ambient field, at the center of the employee's head? This is one of a number of perplexing issues that affect choice and implementation of noise measures used to predict hearing damage. Current OSHA practice stipulates locating the microphone in the worker's hearing zone (a hypothetical sphere of approximately 30 cm radius centered in the head). MSHA more specifically requires microphone placement on the top-middle of the shoulder. (Though more restrictive, it is within the previously defined hearing zone.)

A fundamental split exists over the types of microphones to be used in measurements defined by standards of European origin versus those of North American origin (Bruel, 1981). The European approach is based on free-field microphones. In North America, random-incidence microphones are preferred and almost universally supplied with SLMs and dosimeters. The user of microphones and associated instruments should verify that the appropriate type is selected for measurements complying with applicable technical and regulatory standards (see *Microphones, Characteristics*). The manufacturer's instructions and applications engineers should be consulted for clarification.

Long exposure of any microphone to very high humidity should be avoided. This is especially true when condensation of water on the microphone may take place. Although ceramic, capacitor, and electret microphones are not basically damaged by exposure to high humidity, their operation can be seriously affected unless proper precautions are taken. Some microphones have extremely small ports that can be blocked by water from condensation. This materially affects performance. While humidity and some condensed water would not damage the microphones, should these occur in corrosive atmospheres, permanent damage is possible. Manufacturers' recommendations should be sought and followed regarding operation in such environments.

For proper operation, it is essential that very little electrical leakage resistance be developed across the microphone circuit. The exposed insulating surface of the microphone is specially treated to maintain this low leakage, even under conditions of high humidity. Despite this precaution, the leakage can be excessive under extreme conditions. It is then advisable to keep the microphone at a temperature higher than ambient to prevent condensation and thus avoid leakage. Microphone preamplifiers supplied by some manufacturers can include an electrically powered resistance heater to achieve this goal.

In climates where the humidity is normally high, it is recommended that the microphone be stored at a temperature above ambient to avoid condensation. A problem can easily develop when a microphone is stored in an air-conditioned office. If it is suddenly transferred to a manufacturing floor on a day having high humidity, condensation can occur. Note that the problem lies with the microphone being at a temperature below the dew point even though the dew point temperature has not been approached on the shop floor. The recommended procedure is to keep the microphone in a sealed container until its temperature is equalized.

Moisture problems may be detected in several ways. Should they occur, placement of a calibrator on such a microphone may produce readings that fluctuate or stabilize at levels significantly changed from the last calibration. The capacitor type, with its high polarization potential, will emit a crackling sound similar to static on an AM radio, which can be heard if the output signal is monitored with earphones. Electret microphones are less susceptible to degradation and especially noise generation under conditions of high humidity, as compared to the capacitor type that have high polarization voltage (see *Microphones, Characteristics*). An electret microphone can tolerate normal variations in temperature and humidity for long periods without a significant change in sensitivity. Nevertheless, if the humidity is abnormally high for long periods, an electret microphone should be stored in a small container with silica gel or other desiccant.

Although many industrial noise measurements are made indoors at average room temperatures, some measurements must be made at higher or lower temperatures. Under extreme conditions, it is essential to know the limitations of the equipment as specified by the manufacturer. Most microphones will withstand operating temperatures of -25° C to +55° C without damage. Preamplifiers for capacitor microphones are limited to about 80° C. SLMs usually cannot be operated below -10° C without special low temperature batteries. When electret and ceramic microphones are stored or shipped, maximum nonoperating temperatures must not be exceeded. Manufacturers' recommendations should be observed. Microphones are normally calibrated at room temperature (21° C). If a microphone is operated at other temperatures, its sensitivity may be somewhat different, and a correction may have to be applied as specified by the manufacturer.

Summation of Decibels and Corrections for Background Noise

In performing noise measurements, it is often necessary to add or subtract sound level readings obtained at a common point. Simple arithmetic addition or subtraction of the levels in decibels is invalid. Chapter 2 discusses the proper methods. Frequently, a quick estimate of summations must be obtained without resorting to tables or formulas. To do so, apply the following approximations. For equal values of the levels to be added, add 3 dB to the common level. For levels differing by 10 dB, add 0.4 dB to the higher one. It is evident that for a difference of 10 dB or more, the lower level adds little to the sum (Figure 3.7a, Equation 2.12).

In using the above, it is important that the sound levels involve the same frequency weightings. For example, if one of the levels is A weighted, then the other one must also be A weighted. It is not correct to use the given method to add or subtract an A-weighted and a C-weighted level measurement.

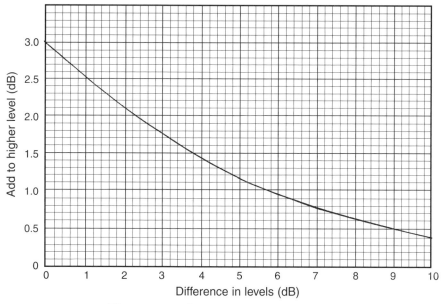

Figure 3.7a — Summation of sound levels.

When measurements are made on an individual noise source, the possibility exists that errors will be introduced by the added contribution of background noise. If measurements are obtained of the background alone, a correction may be computed through use of the nomograph of Figure 3.7b or Equation 2.13. It is evident that for a difference of 10 dB between total and background alone, the correction is only approximately 0.4 dB.

Limits on Measurement Imposed by System Noise Floor

An ideal microphone and associated amplifiers should have exactly zero output if no sound is present. Real instruments have self-generated internal noise sources; therefore, the output can never be exactly zero for a total absence of sound.

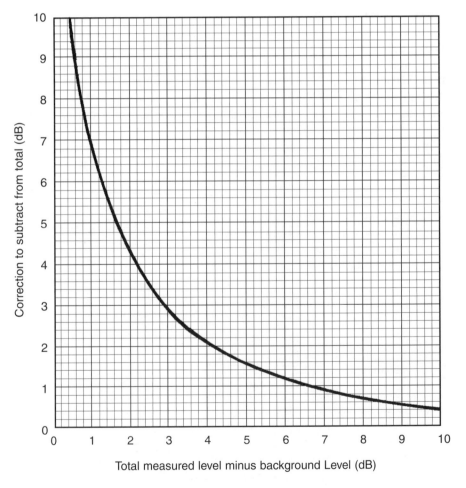

Figure 3.7b — Background noise correction.

Modern instruments have self-generated noise levels that are much below commonly encountered industrial levels, so they impose no constraints on such measurements. There are circumstances which necessitate examination of the limits imposed by system noise floors. One example involves measurement of community background noise when intrusive industrial sources are shut down. Another of particular current significance pertains to octave-band measurement of background noise in audiometric booths and rooms.

The principal sources of noise that establish a system measurement floor are the electrical circuit noise and noise generated by air turbulence at the microphone. Both noise types have individual spectral characteristics. It is therefore necessary to obtain data stipulating the noise floor levels for every weighting filter and microphone that might be selected for use with the SLM or other instrument. Such data are usually available from the instrument or microphone manufacturer.

A possible point of confusion to be avoided results from the specific manner in which noise floor data are identified. No universal nomenclature is recognized, so descriptors may vary among manufacturers. Conceptually, the floor levels may be identified as the "self noise level" or the "minimum measurable level." The two are not synonymous. Consider hypothetically a self noise level (electrical) of 25 dB being stipulated for a given filter (for example, A weighting). In the complete absence of acoustic noise at the microphone, an SLM can be expected to have an average reading of 25 dB. If a broadband noise having an A-weighted envelope and a level of 25 dB is presented to the microphone, the SLM will have an average reading of 28 dB (recall that summing two random noise signals having equal levels results in a combination having a level 3 dB higher). Strict application of this principle implies that measurements of levels equal to the "self noise level" can be made (see Equations 2.12 and 2.13 and Figures 3.7a and b). In practice, this should be avoided because validity requires that both signals be stationary. Standards of good practice dictate that the self noise floor be at least 10 dB below the level at which a highly reliable measurement can be made.

A second limit is imposed by turbulence-generated noise produced by air flow past the microphone. Such noise increases with the magnitude of the velocity of the flow. A manufacturer may supply data for wind-generated noise as sensed with particular windscreens and frequency-weighting filters. For a given velocity, noise level is greater in amplitude at lower frequencies. It is commonly observed that an A-weighted reading of turbulence noise will be substantially lower than that taken with C or flat weighting because the A-weighting filter discriminates against low frequencies.

Air-flow noise can be greatly reduced by enclosing the microphone in a windscreen. The screen may range from a small plastic foam ball to a wire-framed sphere several feet in diameter, covered with a fine nylon mesh. Such screens must be nearly transparent acoustically, and they must minimize air flow turbulence and thus noise.

Data reported for windscreens are typically stated as equivalent self noise for a given wind speed. It is the responsibility of the user to decide how much margin to use in deciding the minimum measurable level of ambient noise when a given wind speed is encountered. For example, one manufacturer stipulates a self noise of 56 dBA for a screened microphone exposed to 25-mph wind. If one chooses to use a 10-dB margin, the minimum measurable level is 66 dBA. Clearly, this may be an unacceptable level for some community noise measurements. One solution is to wait for a calm day.

A simple test to explore whether turbulence noise exists is to make the measurement with and without a windscreen. If use of the screen lowers the reading, turbulence is present. If the reading is lowered by 10 dB or more, the reading is dominated by turbulence. This may also be detected by listening to the output of the SLM with high-quality headphones. Turbulence may add a noise to the sound measurement that is not perceived in the actual sound field.

Since turbulence noise is highest at low frequencies, it can result in undetected overload in some instruments when spectral analyses are being made at frequencies different from those of the turbulence. This can easily occur with octave- and narrow-band analyzers. Although many instruments will indicate overload conditions regardless of cause, some do not. If there is a suspicion that the problem exists, care must be exercised to assure that high-level out-of-band frequency components do not overload the measuring instrument. Refer to instrument and microphone manufacturer for precautions to be observed.

In industrial environments with high sound levels, turbulence-generated noise is not commonly of concern. Should very high air velocity flows be encountered, the possibility of interference with the sound measurement should be evaluated (e.g., near the opening to a glass melting furnace, high-velocity cooling air jets may be encountered). In environments where ambient noise levels are low, use of comfort cooling fans may influence measurements and should be accounted for.

Calibration

The electronics of contemporary instruments are highly stable and do not normally require adjustment on a day-to-day basis. To a lesser degree, microphones and other electromechanical transducers are also stable, although transducers have greater susceptibility to damage and changes in environmental conditions. In spite of this, it is good practice to check calibration of instruments with an acoustical calibrator at the beginning and end of a survey to provide verification of continued reliable operation and not to make minor corrections, which are usually counterproductive (see Chapter 7).

There are two levels of instrument calibration that should be performed. The first is the routine pre- and post-test calibration with an acoustical calibrator for documentation; the second is comprehensive calibration by a laboratory facility, both of instruments and acoustical calibrators. For legal reference, certificates of calibration should be kept on file.

Spot checks of calibration are limited in scope. They are intended to detect faulty connections and defective operation indicated either by significant departure in calibration, or by manifested, long-term trends indicative of incipient failure. After an acoustical calibration is applied, there are two options. One is to note and document the instrument reading (unless a major failure is detected); the second is to note and document the difference in reading, and then adjust the instrument to conform with the calibration. The choice of documentation versus adjustment *and* documentation is influenced by the type of calibration adjustments that are provided. With mechanical trim pots, repeated readjustment is inappropriate as the pots themselves can wear out or otherwise prove unstable. However, in newer systems with software-controlled adjustments, the decision to repeatedly readjust is acceptable as long as it is done with care and documentation. Either method will generate a trail of documentation. If the choice is made to only document but not adjust the instrument reading, survey records must ultimately be corrected to compensate for the departure from calibration level.

Probably, the most important function to be performed by the calibrator is to detect gross changes in instrument performance, in particular the microphone. In condenser and electret microphones, there is always the possibility of puncturing the diaphragm, particularly because most of these types of microphones are easily disassembled. Should this happen, the low-frequency response of the microphone tends to be affected more than the high. If a multiple frequency calibrator is available, this defect can be detected. Should a deviation be noted, simply substituting another presumably undamaged microphone cannot be relied upon to correct the problem. A laboratory test should be made of the entire instrument.

Calibrators should be selected with care. It is not safe to assume that because a microphone fits snugly into a calibrator cavity that the internally generated pressure is necessarily at its nominal value. The presence of the microphone itself can affect the calibration signal. The proper approach is to consult the manufacturer of the calibrator and/or instrument to determine if a given microphone and calibrator are compatible.

Changes in atmospheric pressure affect the output levels of calibrators. The effect is instrument-specific and may be very small for some. Appropriate manufacturer's reference information should be consulted to apply necessary corrections.

Where possible, periodically perform a calibration with two or more calibrators to obtain comparisons. Successful cross checks among multiple calibrators can provide justification for extending the time intervals between laboratory recalibration of calibrators. In the short run, attribute correct levels to those produced by the majority. The excluded calibrator(s) should be sent for laboratory recalibration. Eventually, they should be compared against the retained calibrators.

There is no simple answer to the question of how often calibrators should be recalibrated. Certainly, any time a calibrator has been dropped or is otherwise suspected to have been damaged, it should be recalibrated. Regulatory agencies and user organizations sometime stipulate that laboratory calibration of instruments as well as acoustical calibrators be performed at specified intervals. To retain confidence, establish a regular schedule for recalibration for both.

There are enhanced capability calibrators that permit testing of instrument time constants, linearity, crest factor, etc. These are useful instruments but should be used in the same spirit as the basic calibrator: namely, monitoring verification of continued satisfactory field performance rather than fundamental calibration. An important additional step in using calibrators is to verify that the internal batteries are in good condition, sufficient to permit acceptable output by the calibrator. Some instruments have automatic shutdown when battery voltage falls below an acceptable level; others require that the user monitor a battery-condition indicator. Again, it is good practice to comply with all the manufacturer's recommendations.

Measuring Techniques for Fluctuating Noise

Reading a sound level meter is not difficult when the indicator reading is constant; questions arise only when it is fluctuating. Instructions of instrument manufacturers and those incorporated in some standard reference texts for interpreting variable readings (e.g., Bruce, 1988) have to be carefully assessed to determine when they are appropriate for the objectives of the measurement. To appreciate the significance of what follows, a basic understanding is required of the workings and dynamics of an SLM and the noise measures used to evaluate and regulate hazards to hearing. Difficulties may arise because guidelines presented in many references strictly apply only to measurements pertaining to engineering controls and regulatory compliance based on 3-dB exchange averaging. When measurements are used for predicting or verifying compliance with regulations based on other than 3-dB exchange, subtle, yet significant constraints arise. This is especially important for large fluctuations (over 8 dB). For fluctuations over a range up to approximately 8 dB, the central tendency (i.e., arithmetic average) may be recorded as representing the sound level. It is equally applicable for the various exchange rates in common use (i.e., 3, 4, 5 dB). For example, this procedure should provide an estimate of L_{OSHA} within approximately 1 dBA of the true value (see Chapter 7).

For larger excursions and in the presence of impulsive noise, various sampling methodologies were developed when only primitive noise measuring instruments were available. Use of contemporary state-of-the-art integrating/averaging sound level meters and dosimeters eliminates the need for uneconomical time-consuming sampling protocols, and also eliminates estimation penalties incurred with "eye-ball averaging." Such instruments factor in requisite parameters and dynamic responses and continuously process time-varying sound levels. They should be used when exposures to noise have significant variations in time and space. Note that 4- and 5-dB exchange rates and SLOW-response parameter settings are not provided in all integrating/averaging SLMs. For compliance measurements required by OSHA or other regulatory agencies, such options are necessary and instruments having them must be selected.

Dosimeter Use

Calibration and Programming

Before and after use, a dosimeter must be calibrated and a record made of the readings. With contemporary dosimeters, calibration procedure is similar to that for SLMs. As is the case for SLMs, it is recommended that observed significant (>0.5 dB) changes in calibration be recorded for subsequent reference. Unlike the case for SLMs, it is preferable that prior to use, but not after, an adjustment be made to the dosimeter setting to conform to the calibrator. This will simplify need for postsurvey adjustment of data. Otherwise, if survey data are recorded in decibels, the adjustment would simply add or subtract the calibration offset in dB. However, should data be in percentage dose, the adjustment becomes more complicated and should be avoided for simplicity. A post-test calibration that shows a change in excess of 0.5 dB should bring the test results into question.

Some older types of dosimeters, which are obsolete but still in use, require a more complicated calibration procedure. Sound from an acoustical calibrator must be supplied for timed duration periods. Should this type of dosimeter have to be used, first check that performance satisfies current instrument standards. If it does, follow the manufacturer's instructions for calibration. It is particularly important that modern calibrators satisfying ANSI S1.40-1984 be used.

The fundamental functions of measuring and computing noise dose, time-weighted average, and/or sound exposure level are common to all current dosimeters. In addition, other dosimeters have expanded capabilities including performing statistical distribution analyses, generating time-history profiles of incremental averages of sound levels, and so forth. Most dosimeters can also function as sound level meters, while others can process noise measures applicable to assessment of community noise.

In preparation for a survey, a dosimeter should have at least the following parameters set as required for a particular application:

- exchange rate (3, 4, 5, or 6 dB),
- criterion level (e.g., 84, 85, 90, or special dB),
- cut-off or threshold level(s) (80, 90, or other dB),
- criterion duration (hrs),
- exponential averaging (SLOW, FAST, or none),
- weighting (A or C).

The functions required to generate particular metrics (such as time-history recording in incremental averages and exposure in pascal-squared hours) should be selected, and function-specific parameters should also be set.

Incorporation of parameters and selection of functions can be provided by:

- factory presetting,
- dip switch setting,
- dosimeter key pad insertion,
- downloading of computer-stored files of settings.

Verification of Exchange Rate

Availability of a simple method for verifying the exchange rate setting of a dosimeter is especially useful for instruments having factory or internal settings. If the dosimeter can display average sound levels, the following procedure is recommended. After initial calibration, supply an acoustical input with the same calibrator for a period of time long enough to obtain stable average level readings. This could be as short as 30 seconds. The dosimeter is then switched to stand-by mode. The average level indicated at the end of the fixed exposure period should correspond exactly to the output level of the calibrator. Next, the calibrator is turned off but kept covering the microphone as an acoustic shield. Finally, the dosimeter is again placed in operation for a period of time identical to the first. At the end of the second period, the average level should be observed. No effective additional acoustical input was supplied, yet the averaging time was doubled. The end value of average level should decrease by an amount corresponding to the ER. Thus for an initial calibrator input of 94 dB, the final average level should be 89 dB for a 5-dB ER, 90 dB for a 4-dB ER, and 91 dB for a 3-dB ER.

Microphone Placement

Whether proper measurements to determine exposures to noise should be made with the microphone positioned on the person, or at the location of exposure with the person removed is a matter not having universal agreement. Although choice of position may be based on professional judgment, OSHA compliance directives to inspectors stipulate microphone placement in the hearing zone and, if known, on the side of the person facing the highest noise levels. The "hearing zone" is bounded by a hypothetical sphere of 30-cm radius enclosing the head. More specifically, this author recommends placement of the microphone on the top-middle of the shoulder. Note that in many instances for safety and/or access reasons, use of a dosimeter is the only feasible approach.

The extent to which a person's body can affect measurements is influenced by the nature of the sound field and its spectral content, absorption by clothing, and position of the microphone on the body. Departures from the undisturbed field between -1 and +5 dB have been measured experimentally in individual frequency bands (Kuhn, 1986; Kuhn and Guernsey, 1983; Seiler, 1982). The conclusions are that minimum disturbances occur when position is on the shoulder or collar (Marsh and Richings, 1991). For measurements of A-weighted sound levels in industry, a typical increase is estimated to be 0.5-1.0 dBA above that of the undisturbed field.

Hearing Zone versus Area Monitoring

An established method for determining noise exposure is to make measurements in the hearing zone of a worker during a work shift. Such measurements may be made by a surveyor who tracks the worker and positions the measuring microphone in the hearing zone. More commonly they are made with a dosimeter placed on the worker and the microphone mounted in the hearing zone.

Various other schemes have been proposed to minimize survey effort. Profiling or task-based assessment requires determination of noise varying over time and space in an area where a defined task is performed by workers (James, 1992). Time and motion analyses establish worker locations during task performance.

From the two sets of data, noise exposures for the task can be computed. The scheme is attractive because it eliminates the need for monitoring particular workers, and permits computing various exposures by combining different tasks. Theoretically, this is an approach that can capitalize on computer processing.

The task-based analytic procedure is critically dependent on accurate definition of the noise field at all points of the task area versus time. In addition, the location versus time of the worker during the task must be accurately described. The hearing zone method in contrast does not require correlated independent determinations of worker location versus time, and noise field amplitudes versus time at all locations.

In practice, the utility of a particular method is highly dependent on the variabilities of the noise field and worker location. As one extreme, consider noise exposure of a worker tending machines producing steady noise levels in a highly reverberant environment. An example would be a power plant having large areas where noise is nearly constant, and no worker closely approaches individual sources having levels above the general background. In such a case, it is only necessary to establish duration of time spent in the area to determine exposure. At the other extreme, the noise field can have large variations in magnitude over small changes in distance (a foot or less). Accurately establishing the noise field characteristics as well as the location versus time of the worker are problematical. To illustrate how sensitive variability in space can be, consider measurements made on a worker polishing metal parts with a hand-held flap-wheel grinder (Earshen, 1997). Two identical dosimeters were placed on the worker with each microphone placed on opposite shoulders. Noise doses measured in compliance with OSHA procedures for each shoulder position differed by 184%, and were 523% and 339% respectively.

The key issues to consider in planning measurements pertain to variability. The less that is known about the pattern of exposure, the more that hearing zone measurements are indicated. Furthermore, if a required accuracy is needed, then each work-function exposure must be statistically assessed (see Chapter 7).

Measurement Artifacts

Limiting unprotected exposure to impulsive noise to 140 dB peak is recommended by ACGIH (1998) and required by OSHA. Some dosimeters can record true peak SPL of impulsive waveforms in addition to factoring in their average values. With instruments having peak-reading capabilities, care must be exercised in interpreting readings because microphones respond to tapping and rubbing. It is possible to produce impulse readings that may approach or exceed 140 dB peak by intentionally or accidentally tapping the microphone because it and the cable attached to it are sensitive to vibration. The result is to indicate spurious exceedance of the 140-dB limit, and to possibly increase measured TWA or dose.

Investigations have been made to assess the extent of potential measurement degradation as a result of various artifacts (Royster, 1997). Experiments were conducted with dosimeters connected to their normal microphones with pink noise signals electrically added. The average value of the pink noise was set to approximately 84 and 90 dBA. The microphones were then subjected to a variety of spurious inputs at the rate of 5 per hour. The artifacts consisted of shouting and blowing into the microphones as well as thumping them. Thumping and blowing registered peak pressure levels near and above 140 dB. To guard against such artifacts, use of recording dosimeters that identify time of occurrence of impulses is particularly useful. Inspection of the record should provide leads for further investigation to determine if real sources are producing peaks.

In contrast to the findings regarding spurious peak recordings, it was observed that increases in TWA or noise dose by thumping or shouting artifacts were very small (a fraction of a decibel), for 5-dB ER with background levels of 84 and 90 dBA. Blowing into the microphone produced approximately 2-dBA increase against the 85-dBA background, and 1 dBA against the 90-dBA background. As the background increases, the added contributions by the artifacts to the TWA decrease. Procedures, findings, and conclusions of the referenced investigation are discussed in Chapter 7.

To minimize vibration effects, proper attachments of microphones and their connecting cables should be made. The microphones should be covered with small windscreens provided by the manufacturer or otherwise obtained. The windscreens will guard against rubbing and touching, and can also provide some isolation against blowing. After positioning on the person, the microphone and connecting cable should be secured in place with surgical or duct tape. Care should be exercised not to cover the microphone. A common error should be avoided by assuring that the dosimeter is not collecting data during the attachment or removal of the microphone and cable. (This is particularly serious for single-number dosimeters because corrections are not possible.) Note that the mini-windscreens normally available with dosimeters are not very effective in shielding against turbulence-produced noise by high-velocity cooling and ventilation fans. Some reports claim that their use can *increase* turbulence. Should measurements be necessary in high air-velocity environments, supplementary investigations should be made by protecting the microphones against direct exposure to the air blasts.

It is evident that spurious peak level recordings are possible. Where TWA and noise dose are concerned, the spurious events are most likely to produce minimal or small changes. Precautions in mounting will reduce susceptibility to inadvertent excitation of the microphones. Some measure of defense against blowing would be provided by the mini-windscreens. In the final analysis, critical examinations of time-history dosimetry records provide a method detecting spurious events. A concerted effort to sabotage data can succeed but would be susceptible to detection by examination of time-history records.

There are circumstances where microphone and instrument attachment are difficult (e.g., no shirt being worn or protective clothing being worn). A convenient solution is to use reflective-marker safety vests as repositories. Microphones, cables, and dosimeters can be attached and taped to such vests. Some practitioners have commissioned custom-fabricated vests to accommodate dosimeters and microphones.

Impulse Noise

Sound level meters (ANSI S1.4-1983) and early-design dosimeters (ANSI S1.25-1978) are not suitable for measuring noise that has significant components of impulsive noise because of the introduction of serious errors. For such applications modern dosimeters (ANSI S1.25-1991) and integrating/averaging sound-level meters (ANSI S1.43-1997) should be used.

Wet Environments

A microphone and instrument problem encountered in practice yet rarely addressed in manufacturers' instructions is the monitoring of a worker in an extremely wet or rainy environment. Protection of the dosimeter itself is not difficult, as many currently available dosimeters are water-resistant. For those that are not, placement inside a plastic food or sandwich bag will suffice.

Microphone protection is a different, more challenging problem because any physical protective covering can affect its frequency response. Many enclosures have been tried, including sandwich bags, balloons, plastic wrapping material, etc. In controlled laboratory evaluations, each has sometimes been found to affect frequency response and therefore, cannot be used without constraints. An effective but limited solution is to wait until it stops raining and foul weather subsides. Another is to use an all-weather microphone and enclosure, but cost, weight, and size limit this option. Furthermore, attachment to the person is not practical. One alternative is to make measurements in the work location with all-weather analysis equipment, and draw inferences regarding exposure of the worker.

As an absolutely last resort, covering the microphone with a makeshift waterproof enclosure may be a solution; however, extreme care must be used to detect and eliminate errors. *Serious errors can occur.* The thinnest and limpest covering should be employed. A flimsy sandwich bag has on occasion been used by some practitioners. It should be placed over a microphone already covered by a windscreen. The bag should be placed in position very loosely and tied securely to hold water out. It is mandatory to verify that there is minimal effect on the frequency response of the microphone. Accurate evaluation is difficult, requiring careful measurements. If a noise to be monitored can be observed from a dry place, the validity of the scheme can be tested by comparing successive measurements with the microphone wrapped and unwrapped. This is not easy to do, but can be accomplished with diligence. If the differences in average sound level do not vary by more than 1–2 dB, the scheme should be usable. *This makeshift solution should not be relied upon for compliance, but it can provide valuable survey data when no alternative exists.*

Potential Discrepancies Between SLMs and Dosimeters

Logically, sound level meters and dosimeters should produce identical measurement results since they are generically related in a prescribed manner. When comparisons are to be made, it is important that the particular instruments involved have commensurate performance capabilities. Such capabilities must also recognize human factors affecting results (i.e., SLM measurements entail visual observation and interpretation of reading). Discrepancies *are possible,* but they are principally caused by improper utilization of incompatible instruments. The manifestations of, and remedies for, such discrepancies are summarized in this section.

Microphone Placement

The microphone used with a dosimeter is placed on the person being monitored, except where area measurement is required. When placed on a person, acoustical boundary conditions at the microphone can affect readings. In contrast, the SLM microphone is commonly located away from nearby objects during measurements. In such an eventuality, differences may be produced. Note, however, that OSHA requires the measuring microphone to be in "the hearing zone" of the person being monitored. If results are to be consistent, the two microphones should (and certainly may) be positioned at the same spot.

Hearing Zone versus Area Monitoring

In addition to effects produced by positioning of microphones near or away from the person being monitored, significant discrepancies can result from comparing personal hearing zone to area monitoring, as was discussed previously. The basic configurations of the instruments tend to encourage hearing zone monitoring for the dosimeter and area monitoring for the SLM. However, neither instrument is uniquely restricted to one measurement mode alone. Used comparably, the instruments should give consistent results.

Differential Susceptibility to Shock and Vibration Artifact

All microphones are sensitive to shock and vibration, and it is possible that an unprotected dosimeter microphone might be so subjected. Spurious recordings of impulse peaks might result from inadvertent or intentional activation of the microphone. All dosimetry measurements should collect detailed time-history data. Inspection of these can identify suspect peaks for further investigation. Hand-held SLMs or dosimeters can be used for such investigations to assure avoidance of undesired microphone excitation. (For additional discussion, see *Dosimeter Use, Measurement Artifacts.*)

Response to Impulsive Noise

When impulsive noise is present, there are several reasons why discrepancies may occur. These are:

• Differences in dynamic response: Conventional sound level meters and contemporary dosimeter instrument standards have different dynamic response requirements specified. SLMs are rated by *crest factor* (ANSI S1.4-1983) while dosimeters are rated by *pulse range* (ANSI S1.25-1991). For isolated, randomly occurring impulses, measured average values can be quite different. Furthermore, even precision-grade SLMs do not have adequate transient response capabilities for certain impulses.

- Effect of 4- or 5-dB exchange averaging: The generic instrument diagram of Figure 3.1 shows that SLMs and dosimeters are functionally identical up to the point past the exponential averager. Past that point, data can be selectively processed in different ways. Depending on the application, subsequent averaging can be done with 3-, 4-, or 5-dB ERs. For an SLM and dosimeter that have identical dynamic characteristics up to the selector switch, responses to identical impulses would be identical. Subsequent processing will produce different results for the various ERs.

As previously stated, use of an SLM is not appropriate when impulsive noise is present. In addition to constraints imposed by inadequate transient response by SLMs, obstacles are posed by the limited ability of a human operator to perceive and retain details of sound-level amplitudes and frequencies of occurrence. Now outdated, old reference texts recommended that infrequently occurring high levels of brief duration be ignored.

In the past, erroneous conclusions were drawn regarding accuracy of average levels and noise doses determined with dosimeters versus SLMs for impulsive noise. Even if the problems cited above were circumvented by analyzing noise patterns that did not tax human observations, investigators still reached erroneous conclusions because the effect of exponential averaging, when impulses are processed, was ignored (Rockwell, 1981). This is particularly outstanding when the OSHA-mandated SLOW response is used along with 5-dB exchange-rate processing. (Note OSHA mandates *both* for SLM-derived data, as well as for dosimeters.)

When properly analyzed, results obtained with SLMs or dosimeters having comparable dynamic ranges and responses produce identical results accurate within allowed instrument tolerances. To eliminate visual perception limitations of human operators, experimental verification has been obtained through the use of a video camera and tape recorder to monitor SLM meter readings in detail (Earshen, 1985). Playback of recordings, one frame at a time, enabled accurate meter readings to be made. Analyses of readings obtained with a GenRad 1933 and a B&K 2209 SLM demonstrated that their SLOW dynamic responses were commensurate with the prescribed SLOW response of dosimeters. Furthermore, the noise dose measured by four dosimeters agreed closely with that computed from the video recordings of the SLMs. (Both sets of instruments were exposed to identical transient waveforms.)

Subjective Assessment Via Earphones

The ear/brain system can process information and perform sophisticated discrimination and recognition tasks yet to be equaled by automated instruments or computers. An example is the ability to follow selectively one of several simultaneous conversations in a group gathering. Another example more susceptible to instrument processing is the ability of humans to perceive narrow-band or pure-tone sounds that are substantially lower in amplitude than superimposed interference (10 dB or more). Use of instruments alone may not detect the submerged signal without extensive investigative procedures and use of narrow-band analyzers. In contrast, the detection may be readily accomplished by listening.

It is nevertheless possible to enhance auditory detection and recognition by filtering a sound signal prior to audition. Many general-purpose SLMs have an output jack in the amplifier chain placed after the weighting or octave-band filters. By connecting a high fidelity set of earphones, selectively filtered sound can be monitored. If monitoring is in a noisy environment, a circumaural noise-excluding headset is required. With such instrumentation, a skilled operator having normal hearing can perceive characteristics of noise literally impossible to analyze with an instrument alone. For example, the effect of masking noise can be greatly reduced if the auditioned signal is filtered to discriminate against noise in adjacent bands.

Another example of the utility of listening to headphones is detection of wind- and air-flow noise generated by turbulence at the microphone. This becomes evident by comparing the sound in the earphones with the actual sound directly heard. Yet another is the ability to detect instrument or interference problems if the headphone sound appears distorted, or if interfering noise is heard that is not directly audible in the location where the measurement is made.

Susceptibility to Electromagnetic and Radio Frequency Interference

In industry, measuring instruments are sometimes employed in environments having high levels of electromagnetic fields. Though not highly prevalent, conditions do occur where such fields can interfere with SLM and dosimeter measurements. Performances specified in SLM and dosimeter standards define susceptibility to interference only at power-line frequencies. At frequencies not included, other possible sources of interference are: vehicular ignition systems, arc furnaces, induction heaters, radio frequency heaters and welders, arc welders, communications transmitters, etc. Only recently has a standard pertaining to interference been adopted (ANSI S1.14-1998); thus, information on susceptibility of current instruments is generally not available.

If the suspect interfering source can be turned on selectively, the potential impact on the measurement can be established. The first step is to positively identify the suspect source. The magnitude of the interference should be then evaluated in relation to the magnitude of the acoustic noise being investigated. Consider a test in a quiet office where turning on a hand-held radio or cell phone produces a level of 80 dBA, as measured by a dosimeter or SLM. This may seem large, but considering that levels affecting hearing damage are typically 5 or more decibels higher, the contribution of the interference may not be significant. Normally, the evaluation should be done in the presence of the noise being investigated, where the impact of potential interference will be clearly evident.

When the suspect source is not under the control of the surveyor, a simple check procedure can use an inactivated acoustic calibrator. Such calibrators can provide 10 dB or more of acoustical attenuation when they are covering a microphone. In the majority of cases, electromagnetic interference will produce readings higher than those resulting from acoustical noise alone. When interference is suspected at a particular location, the reading of the instrument should first be noted. The measurement should then be repeated under identical conditions with the unpowered calibrator covering the microphone. A substantial reduction should be observed (at least 10 dB); if not, interference must be suspected and further efforts to locate the source (or electromagnetically shield the instrument) must be made. If a reduction of at least 10 dB is observed, then only a minor contribution from electromagnetic interference exists, and the effect on measurement accuracy will be insignificant.

Caution in applying the foregoing principle must be exercised in cases where high levels of acoustic noise at low frequencies exist. For such low-frequency noise, the attenuation of the calibrator case may not be in excess of 10 dB. To be sure that a particular calibrator has sufficient acoustic isolation, it should be tested (in a low-frequency noise) in an area known to be free of electrical interference. Another means of acoustic isolation would be to cover the microphone with plastic wrapping and then cover it with a ball of modeling clay.

For some types of instruments, substitution of a dummy microphone may be used to check for electrical interference. The dummy microphone has electrical properties identical to an active microphone, but has no response to sounds. In effect, this results in total elimination of an acoustical input. Therefore, the remaining responses observed must be entirely contributed by electromagnetic interference.

Very infrequent occurrences have been reported where indicated sound levels have been *reduced* by the presence of electromagnetic interference. Other occurrences have been reported for microprocessor-based instruments where interference has locked up computation. Attention should be given to detecting such effects. If they occur, measurements with alternate instruments should be explored.

Significant changes in shielding the instruments are only possible at the manufacturing level. There is, however, a simple procedure that can be effective in some instances when hand-held radios are used. At close separation distances, even very small changes of distance (e.g., inches) can have a large effect on interference levels. For the hand-held radios, it is recommended that the dosimeters and microphone be positioned on the opposite side of the body from that of the hand holding the radio.

Resolution of Frequency Analyses

In planning frequency analysis of noise, it is important to heed the prior admonitions that such analyses require skill and understanding of acoustical engineering and instrument applications. If measurements are to be made, one must first establish the extent to which analyses will be detailed (Bies and Hansen, 1996; Randall, 1991). It is generally true that the analyses should be performed at the coarsest resolution in frequency compatible with the end objective. Figure 3.8 provides some instructive examples.

To determine the spectral characteristic of a source with increased details, measurements can be repeated with filters of successively smaller bandwidth. Figure 3.8 shows results for octave, 1/3 octave, 4%, and 2-Hz constant bandwidth analyses. Each successive measurement reveals greater detail about the source. It can, however, be a mistake to go immediately to the highest possible resolution in frequency because unwieldy quantities of data will be rapidly accumulated. Also, high resolution in frequency requires a noise source with a reasonably stable noise spectrum. A good starting point is to use octave or 1/3 octave-band filters. As more detail becomes necessary, higher resolution analyses can be conducted.

Concluding Remarks

Measurement of noise, generation of integrated noise measures, and application of results to protect hearing and minimize annoyance under regulatory control all require the exercise of judgment (Rock, 1986). The design, operation, and application of instruments are inextricably tied to the metrics relating noise

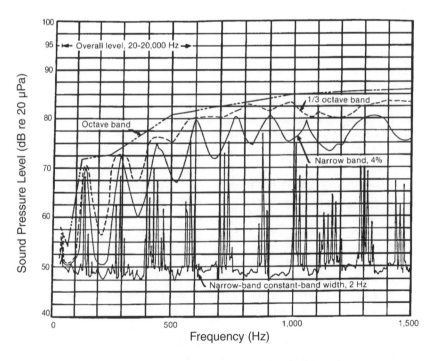

Figure 3.8 — Results of analyzing a noise source with multiple bandwidth analyzers.

exposure to physiological and psychological effects, and to value judgments implicit in regulatory practices. Comprehension of hearing damage processes involving exposure to noise is limited and based on empirical data. Regulatory and professional practices are founded on available scientific knowledge, thus, changes in the future can be reasonably expected as knowledge expands. Practitioners are cautioned to judge proposed changes on the basis of strong supporting evidence regarding the underlying physiological processes. Considering that the proliferation of computer resources and high performance instruments facilitate generation of elegant, impressive reports, extreme care must be exercised to assure that form does not overwhelm substance.

The reader has been besieged in this chapter with widely varying topics having crucial impact on the design and employment of noise-measuring instrumentation. The objective has been to place in perspective the measurement challenge faced by the industrial hygienist. Measurement of noise, assessing its effects on humans, formulating and interpreting regulatory practices, and effecting control and mitigation require extensive professional training and experience. Of these, analysis of spectral composition of noise and design of engineering controls are the most removed from the normal training and background of industrial hygienists. They should be addressed with caution.

It has been shown that sound level meters and their extensions, dosimeters, frequency analyzers, graphic level recorders, and integrating/averaging SLMs, are related in prescribed ways. For regulatory compliance, it is especially important that final noise measures that are generated must be based on commensurate data. To produce such data, when different types of instruments are used, each must conform to performance specifications that are stated or inferred by regulatory agencies.

Given sufficient understanding of objectives and constraints, many noise measurements can be performed with a basic SLM. It should, however, be evident from the text that modern instruments are available that greatly improve the efficiency and versatility of measurements. When the workers' noise environment has large variability in time and over space, instruments capable of real-time processing and tracking the worker become essential.

Computers are now well established, both as components of measurement instrumentation and as means to store, retrieve, and process industrial hygiene information. Though it is possible to use basic SLMs for noise measurements and to store, retrieve, and process information without computers, the industrial hygienist must identify the true objectives for measurements and information handling before starting a program. In selecting instruments and means for utilizing and managing resulting information, the total costs associated with the overall objectives must be examined and used as a basis for selection. Focusing on individual component costs alone can lead to erroneous conclusions. For example, the cost of an SLM, clipboard, paper, and pencil is deceptively low compared to computerized instruments and data-handling systems. When personnel costs associated with manual monitoring and data processing are factored in, conclusions about cost versus effectiveness are drastically changed.

References

ACGIH (1998). "1998 TLVs® and BEIs®, Threshold Limit Values and Biological Exposure Indices," American Conference of Governmental Industrial Hygienists, Cincinnati, OH.

Anon. (1998). "Dynamic Measurement Instrumentation Buyer's Guide," *Sound and Vibration* 32(4), 29–35.

ANSI (1978). "Specification for Personal Noise Dosimeters," S1.25-1978, Acoustical Society of America, New York, NY.

ANSI (1983). "Specification for Sound Level Meters," S1.4-1983 (R1997), Acoustical Society of America, New York, NY.

ANSI (1984a). "Preferred Frequencies, Frequency Levels, and Band Numbers for Acoustical Measurements," S1.6-1984 (R1997), Acoustical Society of America, New York, NY.

ANSI (1984b). "Specification for Acoustical Calibrators," S1.40-1984 (R1997), Acoustical Society of America, New York, NY.

ANSI (1986). "Specification for Octave-Band and Fractional-Octave Band Analog and Digital Filters," S1.11-1986 (R1998), Acoustical Society of America, New York, NY.

ANSI (1991). "Specification for Personal Noise Dosimeters," S1.25-1991 (R1997), Acoustical Society of America, New York, NY.

ANSI (1995). "Method of Measurement of Sound Pressure Levels in Air," S1.13-1995, Acoustical Society of America, New York, NY.

ANSI (1997). "Specification for Integrating-Averaging Sound Level Meters," S1.43-1997, Acoustical Society of America, New York, NY.

ANSI (1998). "Recommendations for Specifying and Testing the Susceptibility of Acoustical Instruments to Radiated Radio-Frequency Electromagnetic Fields, 25 MHz to 1 GHz," S1.14-1998, Acoustical Society of America, New York, NY.

Bies, D. A. and Hansen, C. H. (1996). *Engineering Noise Control*, Chapman & Hall, London, UK.

Bishop, D. E., and Schomer, P. D. (1991). "Community Noise Measurements," in *Handbook of Acoustical Measurements and Noise Control*, edited by C.M. Harris, McGraw Hill Book Co., New York, NY, pp. 50.1–50.24.

Botsford, J. H. (1967). "Simple Method for Identifying Acceptable Noise Exposures," *J. Acoust. Soc. Am. 42*, 810–819.

Bruce, R. D. (1988). "Field Measurements: Equipment and Techniques," in *Noise and Vibration Control*, edited by L.L. Beranek, Institute of Noise Control Engineers, Washington, DC, pp. 74–99.

Bruel, P. V. (1981). "Sound Level Meters, The Atlantic Divide," *Proceedings Inter-Noise '81*, edited by V. M. A. Peutz and A. de Bruijn, Noise Control Foundation, Poughkeepsie, NY, pp. 1099–1104.

Earshen, J. J. (1980). "On Overestimation of Noise Dose in the Presence of Impulsive Noise," *Proceedings Inter-Noise '80*, edited by G. C. Maling, Jr., Noise Control Foundation, Poughkeepsie, NY, pp. 1007–1010.

Earshen, J. J. (1985). "Noise Dosimeter Transient Response Characteristics," *J. Acoust. Soc. Am. 78*, Suppl. 1, S4.

Earshen, J. J. (1997). "Noise Exposure Measurement Options," National Hearing Conservation Association Conference, February 1997, Orlando, FL.

Fletcher, H., and Munson, W. A. (1933). "Loudness, Its Definition, Measurement, and Calculations," *J. Acoust. Soc. Am.* 5, 82–108.

Hassall, J. R. (1991). "Noise Measurement Techniques," in *Handbook of Acoustical Measurements and Noise Control*, edited by C.M. Harris, McGraw Hill Book Co., New York, NY, pp. 9.1–9.22.

IEC (1985). "Integrating-Averaging Sound Level Meters," International Electrotechnical Commission, IEC 804:1985, Geneva, Switzerland.

ISO (1987). "Acoustics—Normal Equal-Loudness Level Contours," International Organization for Standardization, ISO 226:1987(E), Geneva, Switzerland.

ISO (1990). "Acoustics—Determination of Occupational Noise Exposure and Estimation of Noise Induced Hearing Loss," International Organization for Standardization, ISO 1999:1990, Geneva, Switzerland.

James, R. R. (1992). "Employee Sound Exposure Profiling," Hearing Conservation Conference, April 1992, Cincinnati, OH.

Johnson, D. L., Marsh, A. H., and Harris, C. M. (1991). "Acoustical Measurement Instruments," in *Handbook of Acoustical Measurements and Noise Control*, edited by C. M. Harris, McGraw Hill Book Co., New York, NY, pp. 5.1–5.21.

Kamperman, G. W. (1980). "Dosimeter Response to Impulsive Noise," in *Proceedings Inter-Noise '80*, edited by G. C. Maling, Jr., Noise Control Foundation, Poughkeepsie, NY, pp. 1011–1014.

Kuhn, G. F. (1986). "Comparison Between A-Weighted Sound-Pressure Levels in the Field and Those Measured on People or Manikins," *J. Acoust. Soc. Am.* 79, Suppl. 1, 2.

Kuhn, G. F., and Guernsey, R. M. (1983). "Sound Pressure Distribution about the Human Head and Torso," *J. Acoust. Soc. Am.* 73, 95–105.

Kundert, W. R. (1982). "Dosimeters, Impulsive Noise, and the OSHA Hearing Conservation Amendment," *Noise Control Engineering J.* 19(3), 74–79.

Marsh, A. H., and Richings, W. V. (1991). "Measurement of Sound Exposure and Noise Dose," in *Handbook of Acoustical Measurements and Noise Control*, edited by C. M. Harris, McGraw Hill Book Co., New York, NY, pp. 12.1–12.18.

NIOSH (1998). "*Criteria for a Recommended Standard-Occupational Exposure to Noise*," National Institute for Occupational Safety and Health, Pub. 98-126, Cincinnati, OH.

Randall, R.B. (1991). "Acoustical and Vibration Analysis," in *Handbook of Acoustical Measurements and Noise Control*, edited by C.M. Harris, McGraw Hill Book Co., New York, NY, pp. 8.1–8.21.

Rock, J. C. (1986). "Can Professional Judgement be Quantified?" *Am. Ind. Hyg. Assoc. J.* 47(6) (Guest Editorial), A-370.

Rockwell, T. H. (1981). "Real and Imaginary OSHA Noise Violations," *Sound and Vibration* 15(3), 14–16.

Royster, L. H. (1997). "The Frequency of Abnormal Sound Levels When Using Noise Dosimeters to Establish the Employee's Daily TWA and the Resulting Effects on the Measured Noise," AIHA Conference and Exposition, 1997, Dallas, TX.

Schultz, T. J. (1972). *Community Noise Ratings*, Applied Science Publishers, London, UK.

Seiler, J. P. (1982). "Microphone Placement Factors for One-Half Inch Diameter Microphones," M.S. thesis, U. of Pittsburgh Graduate School of Public Health, Pittsburgh, PA.

Seiler, J. P., and Giardino, D. A. (1994). *The Effect of Threshold on Noise Dosimeter Measurements and Interpretation of their Results*, U.S. Dept. of Labor Mine Safety and Health Administration Information Report IR-1224, August 1994.

Suter, A. (1983). "The Noise Dosimeter on Trial," *Noise/News 12*(1), 6–9.

Yeager, D. M., and Marsh, A. H. (1991). "Sound Levels and Their Measurement," in *Handbook of Acoustical Measurements and Noise Control*, edited by C. M. Harris, McGraw Hill Book Co., New York, NY, pp. 11.1–11.19.

The Noise Manual, revised 5th edition, edited by E.H. Berger,
L.H. Royster, J.D. Royster, D.P. Driscoll, and M. Layne
©2003 American Industrial Hygiene Association

Anatomy and Physiology of the Ear: Normal and Damaged Hearing

4

W. Dixon Ward,
Larry H. Royster* and Julia Doswell Royster*

Contents

---

* Editors following the death of W. Dixon Ward

The Normal Ear

The human ear is the special organ that enables man to hear, transmitting the information contained in sound energy to the brain, where it is perceived and interpreted. Because the deleterious effects of excessive noise occur within the ear, some knowledge of its structure and function is desirable. Figure 4.1 shows a cross-section of the ear indicating its three main components: outer, middle, and inner.

The action of the ear involves three stages: (1) modification of the acoustic wave by the outer ear (the pinna and earcanal leading up to the tympanic membrane, or eardrum), (2) conversion of the modified acoustic wave to vibration of the eardrum that is transmitted through the middle ear to the inner ear, and (3) transformation of the resulting mechanical movement of the basilar membrane to nerve impulses in the inner ear (Henderson and Hamernik, 1995).

Outer Ear

The outer ear consists of the pinna (commonly called "the ear," reference Figure 4.2) and the external auditory canal (earcanal), an open channel leading directly to the tympanic membrane (eardrum). Both the pinna and the external auditory canal modify the acoustic wave, so that the spectrum of the sound impinging on the eardrum is not quite the same as the sound that originally reaches the pinna. The pinna is to some extent a "collector" of sound; because of the exact dimensions of the convolutions of the pinna, certain sound frequencies are amplified, others attenuated, so that each individual's pinna puts its distinctive imprint on the acoustic wave progressing into the auditory canal. This information is used in the recognition and localization of sounds.

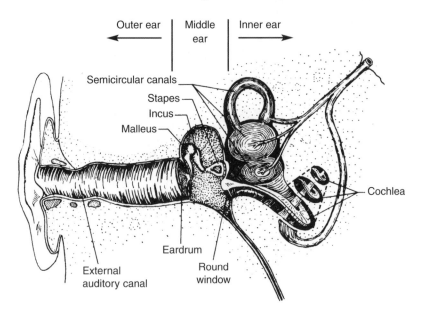

Figure 4.1 — Cross-sectional view of the human ear.

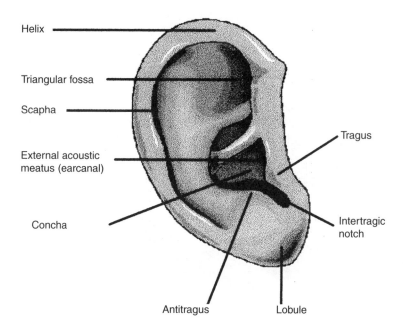

Figure 4.2 — Locations of different parts of the outer ear.

A very important role in modifying the acoustic wave is played by the auditory canal. Because of its shape and dimensions, it greatly amplifies sound in the 3-kHz region. The net effect of the head, pinna, and earcanal is that environmental sounds in the 2 – 4 kHz region are amplified by 10 to 15 decibels. It is not surprising, therefore, that sensitivity to sounds is greatest in this region and that for sound spectra that are uniform across frequency, noises in this frequency range are the most hazardous to hearing.

Middle Ear

The eardrum, or tympanic membrane, in addition to providing protection against invasion of the middle and inner ears by foreign bodies, vibrates in response to the pressure fluctuations in the sound wave, and this vibration is transmitted by the small bones of the middle ear — the malleus, incus, and stapes (hammer, anvil, and stirrup, respectively as shown in Figure 4.1) — to the oval window at the entrance to the inner ear (as shown in Figure 4.3a). The function of the middle ear is to efficiently transform the motion of the eardrum in air to motion of the stapes, which must in turn drive the fluid-filled inner ear. When sound travels from one medium (or structure) to another that exhibits a different acoustic impedance, part of the energy will be transmitted and the remainder will be reflected. Without the transformer action (impedance matching) of the middle ear, only about 1/1000th of the acoustic energy in air would be transmitted to the inner-ear fluids, a loss of 30 decibels.

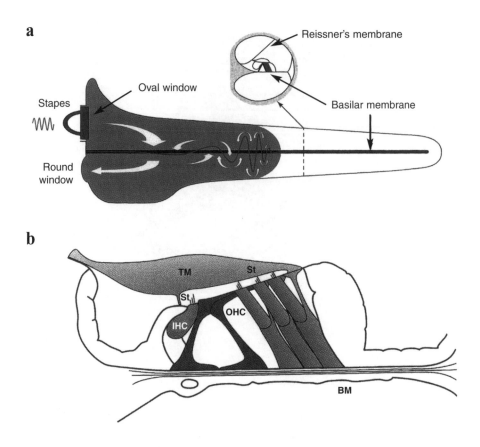

Figure 4.3 — Motions along the basilar membrane (a) and a cross-sectional view of the Organ of Corti (b) depicting the basilar (BM) and tectorial membranes (TM) along with inner (IHC), outer hair cells (OHC), and stereocilia (St). Reprinted from Nobili, R. et al. (1998), with permission.

The middle ear enhances the energy transfer in two main ways. First, the area of the eardrum is about 17 times as large as that of the oval window (where the stapes footplate attaches), so that the effective pressure (force per unit area) is amplified by this amount. Second, the ossicles are so constructed that they produce a lever action that further amplifies the pressure. As a result, most of the energy entering the normal ear via the eardrum is transmitted to motion of the stapes and hence to stimulation of the inner-ear system.

The middle ear also has two muscles that are attached to the malleus and the stapes: the tensor tympani and the stapedius muscles, respectively. These muscles, which act in opposite directions, are activated by the process of vocalization (speaking or singing) and by loud sounds (for the stapedius) or events eliciting a startle reaction (for the tensor tympani). Activation stiffens the whole middle-ear system, thereby reducing the transmission of low-frequency energy (1500 Hz and below). Although this muscular activity shows partial adaptation with time, it provides some protection against sustained high-intensity noise. However, because about a tenth of a second is needed for the muscles to reach full contraction, they are of no importance in protection against sudden impulsive noises such as gunfire.

A third important feature of the middle ear is the eustachian tube, a channel that is connected to the nasal air passages by means of a valve that is supposed to open periodically in synchrony with swallowing. When this does not occur, a difference between the static middle-ear pressure and the outside pressure may develop, so that the eardrum may be displaced inward or outward; in either case, the efficiency of the middle ear is reduced, and less energy will be transmitted to the inner ear.

Inner Ear

The sensory receptors actually responsible for the initiation of neural impulses in the auditory nerve consist of approximately 4000 inner hair cells that are assisted by approximately 12,000 outer hair cells in each cochlea. The cochlea, deeply embedded in the temporal bone, is a tube coiled in the form of a snail with three spiral turns. Two membranes, Reissner's membrane and the basilar membrane, run the length of this spiraling tube, dividing it into three compartments: scala vestibuli, scala media, and scala tympani. The basilar membrane and the complex of sensory and supporting cells that rest upon it comprise the Organ of Corti.

Figure 4.3a shows a traveling wave, input by the stapes, along the basilar membrane. The arrows indicate the direction of fluid flow at the instant the wave is shown. Figure 4.3b presents a cross-sectional view of the Organ of Corti. The inner hair cells (IHC) and outer hair cells (OHC), which rest on the basilar membrane (BM), are the most important components of the Organ of Corti. The tectorial membrane (TM), lies above the fine hairs (stereocilia) that extend from each hair cell. Only the tallest stereocilia (St) from each of the three rows of OHC are firmly anchored at their apex. With respect to the IHC, none of their stereocilia actually contacts the TM.

In the normal cochlea, initiation of neural impulses occurs when the basilar membrane moves up or down; because a shearing force is developed between the tectorial membrane and the Organ of Corti, the stereocilia of the hair cells will be bent, and neural impulses will be developed in one or more of the sensory neurons that are connected mostly to the IHC. The initiation of neural impulses depends primarily on the single row of IHC, although the three rows of OHC serve to facilitate neural activity.

The OHC are motor cells that respond to variation in potential; they change length at rates unequaled by other motile cells. The forces generated by these cells can change the characteristics of the hearing system in terms of the level of hearing sensitivity and frequency selectivity. The ear cannot simply be described as a passive frequency analyzer; rather it is an active nonlinear filter that allows specific characteristics of an acoustic signal to be passed on to the acoustic nerve (Nobili et al., 1998).

Now let us return to the progression of acoustic energy to the inner ear. When the stapes footplate, set in the oval window that opens onto the scala vestibuli, begins to move in response to an incoming wave, the resulting pressure is transmitted into the fluids of the inner ear almost instantaneously. However, since these fluids, unlike air, are essentially incompressible, the footplate can actually move inward only if something else moves outward. In the cochlea this is accomplished by an elastic membrane, the round window, which separates the scala tympani from the middle ear (see Figure 4.3a). When the oval window is pushed inward into scala vestibuli, the round window will bulge outward, and vice versa. This involves the actual movement of fluid from scala vestibuli to scala tympani, so there is a corresponding force tending to deflect the basilar membrane toward scala tympani. This force initiates a wave of movement that begins near the oval window (at the basal end of the cochlea) and travels away from it (toward the apex of the cochlea, or apically). The basilar membrane, loaded by the Organ of Corti, has different physical characteristics along its length, being narrower and stiffer at the base than at the apex. The amplitude of this wave of movement gradually builds up to a maximum and then rapidly dies out as shown in Figure 4.3a. The distance at which the movement reaches its maximum value depends on the speed with which the oval window is moved: If it were pushed in very rapidly, as would be the case with high-frequency acoustic stimulation, the maximum displacement will occur near the base of the cochlea; with slower movement, as in low-frequency stimulation, the maximum will occur farther toward the apex.

Deflection of the basilar membrane and the resultant bending of the hair cells leads to initiation of nerve impulses, as described earlier. There are two main characteristics of the pattern of neural firing aroused by a pure tone. First, the firing of any particular fiber will be synchronized with the frequency of the tone: If the pure tone has a frequency of 100 Hz, then the neural unit will be fired at most 100 times each second, or once every 10 milliseconds. If it fails to be fired by one maximum of the input wave, it may be fired by the next one; in any event, successive firings will be separated by 10 msec, or 20, or 30, or 40, and so forth. Second, the firing will be localized along the basilar membrane, high frequencies giving rise to maximum activity near the base, and low frequencies giving rise to maximum activity near the apex, although this pattern becomes increasingly broad as the intensity of the sound is raised.

Both the place of maximum neural excitation and the frequency (timing) of neural discharge represent information about the acoustic stimulus that somehow is used by the central nervous system in interpreting the overall pattern of neural activity. Place is more important for high frequencies, timing for low frequencies, with both principles apparently being active in the range from 200 to 4000 Hz. However, the process whereby a particular pattern of activity in 31,000 different nerve fibers, transmitted from the cochlea to the cerebral cortex via several intermediate neural centers, gives rise to a specific sensation is still largely unknown.

Pathology of Hearing Damage

Normal hearing, then, requires a normal outer, middle, and inner ear. Differences or malfunction of the outer- and middle-ear systems will vary the amount of sound energy conducted to the inner ear; the term conductive hearing loss is therefore applied to such conditions. Employment-related conductive hearing losses are not common, although they may occur occasionally as the result of an accident (eardrum rupture or ossicular chain disarticulation by a head blow, an explosion, or a rapid pressure change in a decompression chamber, or penetration of the eardrum by a sharp object or fragment of metal).

Because conductive hearing losses can generally be reversed, either by medical or surgical means, they are not as serious as sensory hearing losses associated with irreversible damage to the inner ear (which causes distortion as well as loss of sensitivity), or neural deficits due to damage to higher centers of the auditory system. Noise produces primarily sensory hearing loss. Study of the inner ears of experimental animals exposed to noise has shown that after even moderate exposures, subtle effects may be observed such as twisting and swelling of the hair cells, disarray of the stereocilia, detachment of the tectorial membrane from the stereocilia, and reduction of enzymes and energy sources in the cochlear fluids; these are all conditions that would reduce the sensitivity of the hair cells. The system at this point is in a state of auditory fatigue. In order to initiate neural activity, more acoustic energy must enter the cochlea than before the noise exposure, and the mechanical motion of the cochlear partition must be greater.

As the severity of the noise exposure increases, these changes increase in degree and eventually become irreversible: stereocilia are broken at their base, become fused into giant cilia or disappear; hair cells and supporting cells disintegrate; holes appear in the reticular lamina, thereby allowing intermixture of fluids within the cochlea, which poisons the hair cells; and ultimately even the nerve fibers that innervated the hair cells disappear (Henderson and Hamernik, 1995). Figures 4.4a and 4.4b illustrate the appearance of a normal cochlea and one severely damaged by noise exposure, respectively.

These effects are accentuated in acoustic trauma, the immediate permanent aftermath of a single noise exposure of relatively short duration but of very high intensity, such as an explosion. In this case the picture that emerges is that of a system that has been vibrated so violently that its "elastic limit" has been exceeded. In addition to rupture of the reticular lamina, attachments of the various elements of the Organ of Corti are broken, and hair cells may be torn completely from the basilar membranes on which they normally rest.

Auditory Sensitivity

Irreversible damage to the cochlea as described above cannot, in humans, be measured directly. Loss of hearing has therefore traditionally been assessed by measuring auditory sensitivity, the ability to detect weak pure tones. The auditory absolute threshold is defined as the level of a sound that can just be heard, in quiet surroundings, some specific percentage of the time, generally 50%. In the normal ear, the localization of activity on the basilar membrane described earlier is accentuated in the case of weak tones. It appears that a weak 1000-Hz tone, for example, will actually trigger the

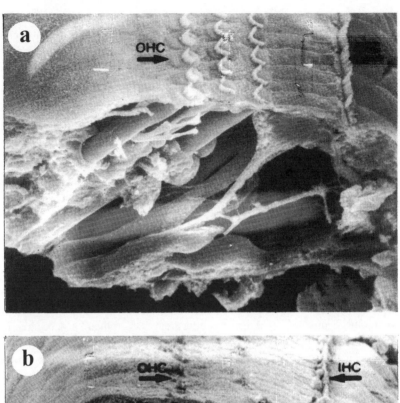

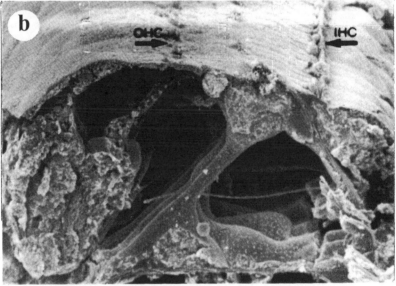

Figure 4.4 — Scanning electromicrographs of (a) the normal cochlea, and (b) a cochlea severely damaged by noise exposure. Note the disarray of the hairs of the inner hair cells (IHC) and the near-complete destruction of outer hair cells (OHC). Photographs courtesy I. Hunter-Duvar.

release of energy stored in the Organ of Corti, but only in a very restricted area, so that only the receptors tuned to 1000 Hz, and not those tuned to 800 or 1200 Hz, for example, will respond vigorously. Because of this specificity of response, the auditory capability of an individual ear is usually assessed by measuring its sensitivity to these weak pure tones, even though there is only a moderate correlation between hair-cell destruction and loss of auditory sensitivity in individual ears.

The auditory threshold depends on all the conditions under which it is tested, including the test procedure used, the instructions to the listener, the listener's tendency to guess when in doubt, the duration of the test tone, the amount of internal and external noise that may mask the test tone, and even on the procedure that is used to determine the intensity of the tone that was just barely heard. Two classical examples of sensitivity to long sustained pure tones, in which two different methods were used to specify the average sound level that was just heard by a group of people with presumably "normal" hearing, are shown in Figure 4.5. For the solid curve, thresholds were determined by having the test persons listen with open ears in a completely quiet room (Berger, 1981; Sivian and White, 1933). The average values shown represent

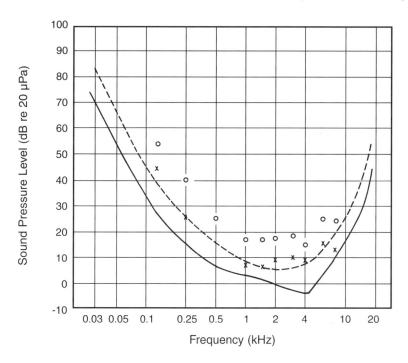

Figure 4.5 — Auditory thresholds associated with "normal" hearing. The solid curve indicates sensitivity to pure tones for young adults listening with open ears (minimum audible field), the dashed curve listening with earphones (minimum audible pressure). Circles represent the SPLs that defined 0 dB HL under the old audiometric standard (ASA, 1951), and the crosses denote 0 dB HL under the present reference values (ANSI, 1969; ISO, 1964).

the just-audible sound levels measured at the position of the center of the listener's head, but with the listener removed from the situation; such a sensitivity curve is called a minimum audible field (MAF). The dashed curve, by contrast, represents an experiment in which the tones were presented to the listeners by means of an earphone (Sivian and White, 1933). The sound level in this case is the output of a microphone at one end of an "artificial ear," a device designed to simulate the human ear, when the earphone, positioned at the other end of this artificial ear is driven with the same electrical signal that produced a just-audible signal. This sensitivity curve is called a minimum audible pressure (MAP) curve. For these persons, sounds down to 20 Hz and up to 20 kHz in frequency can be heard, although only at high intensities at those extremes. The range of hearing is therefore often said to be 20 to 20,000 Hz.

In neither of the curves of Figure 4.5 do the values indicate the actual sound level at the eardrum, that is to say, the sound being transmitted to the cochlea. The MAF curve is influenced by the amplification of the signal provided by the resonance characteristics of the head and outer ear. Putting on an earphone not only eliminates the effect of head and pinna on the sound, but also alters the resonant characteristics of the earcanal. More recent measurements of the sound level actually present at the eardrum indicate that field and earphone listening result in the same values for threshold, except at low frequencies, where MAP measures are elevated because wearing an earphone increases the level of low-frequency body noises, which may produce some masking (Killion, 1978).

Loudness

Because the auditory mechanism is less sensitive to extremely high and low frequencies, the loudness of audible sounds in those frequency regions will in general be less than that of middle-frequency sounds of the same sound pressure level. For example, Figure 4.5 implies that a tone of 40 dB SPL will be heard clearly by a normal individual when its frequency is 1000 Hz, but will be barely audible if its frequency is 100 Hz, and will arouse no sense of loudness at all if it is a 50-Hz tone, as it will be inaudible. Therefore if it is desired to develop an instrument that will indicate the loudness of sounds rather than their physical intensities, it will be necessary to somehow de-emphasize the sound components having these extreme frequencies. In order to accomplish this, equal-loudness contours must be determined: curves that indicate the intensity and frequency of tones judged to be equal in loudness. If one can assume that, at threshold, all tones sound equally loud, both of the curves of Figure 4.5 are equal-loudness contours — one for earphone listening, the other for listening with open ears. Other contours can be determined by asking listeners to adjust a tone of some particular frequency to have the same loudness as a standard tone of a fixed frequency and intensity.

Equal-loudness contours for above-threshold tones are shown in Figure 4.6. These are open-ear field data, because most judgments of loudness of noises in real life are not made while wearing earphones. Each contour defines tones that have been judged equal

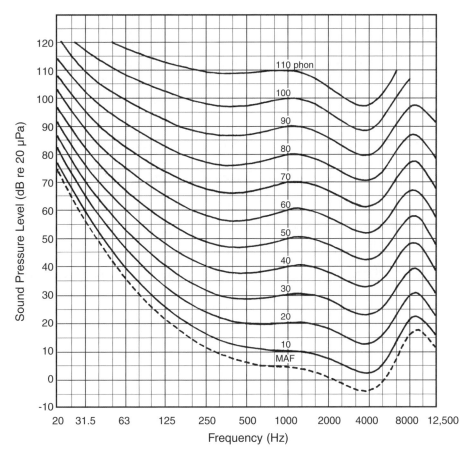

Figure 4.6 — Equal-loudness contours of pure tones for field (open-ear, binaural, frontal incidence) conditions. The numbers indicate the loudness level, in phons, of the tones that fall on each contour. This figure is from ISO 226:1987 and has been reproduced with the permission of the International Organization for Standardization, ISO. This standard can be obtained from any member body or directly from the Central Secretariat, ISO, Case postal 56, 1211 Geneva 20, Switzerland. Copyright remains with ISO.

in loudness to a 1000-Hz tone of a specific SPL; the *loudness level* of a particular contour, in *phons*, is numerically equal to the SPL of the 1000-Hz standard. Thus for instance, the following tones each have a loudness level of 40 phons: 1000 Hz at 40 dB SPL (by definition), 4000 Hz at 32 dB SPL, 8000 Hz at 48 dB SPL, 100 Hz at 51 dB SPL, and 64 Hz at 70 dB SPL. So if the sound-measuring device is supposed to indicate the relative loudness of rather weak sounds it must exhibit a frequency weighting filter that is the inverse of the equal loudness curves. That is, it will reduce the 50-Hz components of the sounds by 24 decibels (64 – 40), the 100-Hz components by 11 dB, and the 8000-Hz components by 8 dB, while the 4000-Hz components should be amplified by 8 dB, relative to 1000 Hz. The A-frequency weighting network on the standard sound-level meter does approximate these adjustments (the 40-phon curve, see Chapter 2), so that when the sound-level meter is set at "A weighting," the reading on the meter, in dBA, will indicate the overall loudness level of the weak sound in question.

If the contours for equal loudness were parallel, A-weighting could serve to indicate the relative loudness of sounds at all levels. Unfortunately, Figure 4.6 shows clearly that the auditory system of humans does not behave in this way. By the time the level of components reaches 100 dB SPL, the contours become flatter in the low frequencies while still retaining the greater efficacy of the 3000 – 4000-Hz components associated with the resonance of the outer ear at these frequencies. So to evaluate the overall loudness of very loud sounds, a different weighting network, the C-weighting, one that provides little attenuation or amplification of different frequencies, is used.

As indicated in Chapter 3, the use of A-weighting to characterize the relative hazard of high-intensity complex noises would appear paradoxical at first glance, since A-weighting was designed to assess weak sounds. However, the hazardousness of a sound is not indicated by its loudness. Empirically, low frequencies are less hazardous, for whatever reason, even at 100 dB SPL, than the higher frequencies, and of the various weightings available on the sound-level meter, A-weighting happens to de-emphasize the low frequencies by approximately the right amount.

It is worth noting that loudness levels indicate only an ordinal relation between sounds. If one sound, X, has a loudness level of 50 phons, while another sound, Y, has a loudness level of 40 phons, then we know only that X will be judged louder than Y, but not how much louder; it is not true, for instance, that the louder is $50/40 = 1.25$ times as loud as the softer. Similarly, the fact that the 50-phon sound has about 10 times as much intensity as the 40-phon sound does not imply that it would be judged to be 10 times as loud. Most people would say that two sounds that differ in loudness level by 9–10 phons have a loudness ratio of about 2:1. That is, a sound that is 10 phons higher in loudness level than another will appear to be twice as loud. The *sone* has been defined as the loudness perceived by a normal listener when presented with a 1000-Hz tone at 40 dB SPL (40 phons), so a 50-phon sound would have a loudness of 2 sones, a 60-phon sound a loudness of 4 sones, an 80-phon sound loudness of 8 sones, and so on.

Masking and the Critical Band

It is only at low intensities that a particular pure tone will have no effect at all on the perception of another tone. The cochlear activity associated with a 1000-Hz tone will interfere with the ability to detect a tone of slightly higher or lower frequency, either masking it completely (i.e., making it inaudible) or, if audible, reducing its loudness (partial masking). At SPLs of 40 – 50 dB or lower, masking has been shown to be relatively constant over a narrow range of frequencies above and below a particular frequency, a range that represents a "sphere of influence" of that sound called the "critical band." The most direct demonstration of the critical band is provided by examining the effect of noises of different bandwidths on the audibility of a tone in the middle of the noise range. If a white noise (having equal energy at all frequencies) is filtered so that the noise to be used for masking consists of only the frequencies between 980 and 1020 Hz, the masked threshold of a 1000-Hz tone will be some value. If the width of the filtered white noise is now doubled by changing the band limits to 960 and 1040 Hz, the masked threshold will increase by 3 dB, indicating that the additional elements (980 – 960 Hz and 1040 – 1020 Hz) contributed as much to the masking as those from 980 to 1020 Hz. Further increases in the bandwidth will continue to raise the threshold until the critical bandwidth is reached; at this point further addition of frequencies will not produce an increase in threshold. At 1000 Hz, this critical band is about 200 Hz wide, so that when the masker increases beyond 900 – 1100 Hz, no increase in masking will occur. Empirically, the critical band has a width of about 100 Hz for frequencies up to 500 Hz, and around a third of an octave for higher frequencies.

At SPLs above 50 dB, the concept of the critical band is of limited value for predicting the masked threshold of a particular tone. Sounds outside the critical band now can influence the threshold, whether the masker is lower in frequency than the maskee (upward spread of masking) or higher (remote masking), though the amount of masking is greater for frequencies nearer the maskee. This is consistent with the fact that the higher the intensity of a sound, the broader the pattern of motion of the cochlear structures.

Additional forms of masking, if defined as the interference of one sound with perception of another, do exist. These include *central masking* (elevation in threshold in one ear by sounds in the other), *forward masking* (the influence of a sound that has already ceased, although the phenomenon is better characterized as "adaptation"), and *backward (in time) masking*, in which a sound interferes with the perception of another that has already occurred.

Hearing Level

Returning to the measurement of auditory threshold, it must be noted that sound levels at the eardrum are difficult to measure, and MAF determinations require a large sound-free room, so threshold measurements have been, and probably will continue to be, expressed in terms of MAP, although in a somewhat indirect manner. Instead of sound levels relative to a reference pressure of 20 μPa and a reference intensity of 10^{-12} W/m^2 (both of which are implied by the term "dB SPL") as in Figure 4.5, an individual's sensitivity is expressed in terms of hearing level (HL): the number of decibels by which, to be heard, a pure tone at a particular frequency must be raised above a reference SPL, specific to the test frequency concerned. Audiometers, the instruments with which thresholds are measured using earphones, are calibrated so that when the HL dial is set to zero, this reference level is generated by the earphone. To illustrate, if the reference sound pressure at a particular frequency were 10 dB SPL, and a worker requires an SPL of 25 dB to hear this tone, then s/he could do so only when the HL dial was set at 15 dB or higher; the worker would be said to have a threshold of 15 dB because s/he can just hear a tone presented at 15 dB HL. For reasons that have been lost in antiquity, the graph relating HL to frequency, the audiogram, is so constructed that 0 dB (hearing ability corresponding to the reference) is at the top of the graph, with HL increasing in the downward direction. Therefore the audiogram is, in essence, an inverted representation of the relative auditory sensitivity of the individual ear as a function of frequency.

0 dB HL: Typical Hearing or Normal Hearing?

Only one major step would appear to remain in order to standardize measurement of auditory sensitivity: selection of suitable values of SPL to represent 0 dB HL at the audiometric frequencies. Unfortunately, this has not proven to be as easy to accomplish as it might appear, because of a lack of agreement, among those who set standards, as to whether this hearing standard should represent "typical" hearing or "normal" hearing. The first attempt to establish 0 dB HL adopted the former objective, using values of the average threshold SPLs of 20- to 29-year-old persons in a random sample of the U.S. population tested in 1934 (Beasley, 1938). These values for frequencies from 125 to 8000 Hz are shown by the open circles in Figure 4.5, and are referred to as "0 dB HL (ASA, 1951)" because they were approved by the American Standards Association in 1951. For several reasons, including the fact that nobody was eliminated from this sample, so that people with damaged hearing were included, these average sound levels were somewhat higher than they otherwise would have been. Consequently, most young adults showed negative values of HL — that is, they had hearing that was better (more sensitive) than 0 dB on the audiometer dial. For example, the group whose MAP thresholds are shown by the dashed curve in Figure 4.5 would have negative average HLs at all frequencies relative to the ASA 1951 reference data indicated by the circles.

The fact that young people with "normal" ears showed negative values of HL, coupled with the fact that in those days "HL" was often referred to as "Hearing Loss," was a source of distress to those who believed that 0 dB HL should mean "normal: in its unsullied state; absolutely unaffected by anything." How, they complained, can anyone have better than normal hearing? So in 1954 British standardizers proposed a new set of reference SPLs that were about 10 dB lower than ASA 1951, and these were adopted, with minor changes, by the International Standards Organization in 1964 and by the American National Standards Institute in 1969 (and reaffirmed in later revisions of the audiometer standard, currently ANSI S3.6 - 1996). These reference SPLs (ISO, 1964 or ANSI, 1969) are shown in Figure 4.5 by the crosses. It must be noted that both the ASA and ISO/ANSI values apply only to one particular earphone+cushion combination, namely, TDH-39 earphones with MX41/AR cushions; for other earphones and/or cushions, the SPLs that produce an equivalent effect may differ by a few decibels.

Unfortunately, the ISO/ANSI norms are modal values of thresholds, gathered under optimum laboratory conditions, of a very highly selected group of young, intelligent listeners. Excluded from consideration were not only persons who, as far as could be determined by questionnaire, might be suffering from hereditary hearing loss, or had been exposed to high-intensity noises or various diseases known to sometimes produce deterioration of hearing, but also those whose sensitivity was, for unknown reasons, worse than the modal values of other listeners — that is, they were rejected "because they obviously are not normal." The result of this bootstrappery is that the ISO audiometric standard is nearly as unrealistically stringent as the ASA is unrealistically lax, so that now the typical young adult with no prior exposure to industrial noise will show positive values of HL (though negative thresholds still occur).

This fact is particularly crucial when audiometric survey data are used to infer whether or not a particular work environment is causing damage to hearing. For example, if a group of 18- to 25-year-old workers in a particular noise environment is found to have an average HL (ISO/ANSI) of 3 or 4 dB, this is completely typical (here, in the sense of "to be expected"), and hazard has not been demonstrated.

The same uncertainty over the hazard associated with a particular work environment is even more accentuated in the case of older workers. Loss of auditory sensitivity can result from many causes other than industrial noise; therefore, the older the group of workers in question, the greater will be the net effect of these other causes, and greater must the average HL be in order that the industrial noise exposure be judged hazardous. Just how great the HLs must be to justify the inference that a hazard exists is a problem that forces a closer examination of the causes of hearing loss.

Presbycusis, Sociacusis, and Nosoacusis

Hearing loss may be categorized not only as being conductive, sensory, or neural, but also in terms of possible cause. One suitable system of classification (Ward, 1977) divides sensory hearing loss into the following:

(1) presbycusis: loss caused by the aging process per se;

(2) noise-induced hearing loss, which must itself be divided into

    (2a) occupational hearing loss caused by work-related noise exposure,

    (2b) sociacusis: losses that can be ascribed to noises of everyday life;

(3) nosoacusis: losses attributable to all other causes, such as hereditary progressive deafness; diseases such as mumps, rubella, Meniere's disease; ototoxic drugs and chemicals; barotrauma; and blows to the head.

It is commonly assumed, for simplicity, that these losses are additive, although there is good evidence that this strict additivity is incorrect, especially when any one cause or their sum is expected to produce a loss of 40 dB or more.

Although these causes of sensory hearing loss can be studied separately in laboratory animals, it is obvious that in humans they are inextricably mixed. Therefore, in order to determine how much damage has been produced by a particular industrial noise environment, the hearing of the workers concerned must be compared with that of a control group of individuals who have never worked in a noisy industry, but who are matched to the workers not only in age but also in histories of exposure to all other influences.

Ideally, for each noise-exposed worker there should be a control subject who is the same age, has the same noisy hobbies, has been exposed to the same diseases and industrial chemicals, does as much hunting as the worker in question, and so on. However, selecting these control subjects is so expensive that it has seldom been attempted. So in practice, the thresholds of the workers are compared either with those of a non–industrial-noise-exposed group of employees of the same age in the same industry or with a set of HL norms that purport to indicate "typical" thresholds for persons of that age who have experienced no workplace noise exposure.

Use of the non–noise-exposed employees of the same industry has the advantage that they will have been tested with the same procedure and under the same conditions as the noise-exposed workers. In addition, since they live in the same area, some of the sociacusic and nosoacusic influences may, on the average, be nearly the same for both groups. However, this equivalence should not be taken for granted. All persons must be administered the same questionnaire, so that it can be shown that it is not mere conjecture that the groups are, on average, comparable in exposure to chain saws, gunfire, loud music, mumps, ototoxic drugs, head blows, and so on. Experience has shown that they will not be comparable if there is a disparity in gender, as an example, between the groups: the average man typically has been exposed to more sociacusic influences than the average woman, so they will inevitably have higher average HLs than women at frequencies above 1000 Hz. Care must therefore be taken to ensure that the control group has the same proportion of males as the noise-exposed workers; to compare the hearing of male steel workers with that of female secretaries would clearly be inappropriate.

The other alternative for making a comparison of HLs, that is to say, using a set of normative curves, is even more likely to lead to invalid results. For maximum validity, it must be the case that all details of the audiometric procedures used to test the noise workers are the same as those used to test the persons on whom the norms were established: type of audiometer, instructions, procedure, order of frequencies tested, criterion for defining threshold, and so forth (see Chapter 11 on audiometry). In addition, sociacusic and nosoacusic influences in the two groups are unlikely to be equivalent. Of the many proposals for "age-corrected HL norms" that have been suggested, most differ from each other in the rules of exclusion of individuals on the basis of their history. Indeed, often these rules were so vague (e.g., "otological abnormality" or "excessive exposure to gunfire") that an equivalent exclusion rule cannot be applied to the data from the industrial workers.

Age Correction Curves

In an attempt to circumvent these problems, at least in part, Spoor (1967) derived average "age correction" curves based on several studies on non–industrial-noise-exposed populations by comparing the median hearing of a particular age group to that of the 20-year-old individuals in the same study. The main assumption here is that in the 20-year age group, all of the populations really should have the same median thresholds, having been exposed on average to the same sociacusic and nosoacusic influences, so that any differences in measured HLs between the 20-year-olds in different surveys may be attributed to differences in audiometric calibration or technique.

The age correction curves for males derived by Spoor from all the data then available are shown in Figure 4.7. Although these smoothed curves are sometimes called "presbycusis corrections," they did involve also some sociacusis and nosoacusis. Subsequent studies of non–industrial-noise-exposed populations have generally substantiated their accuracy; indeed, they have been adopted internationally (ISO, 1984). Unfortunately, however, these curves have often been erroneously indicated to be

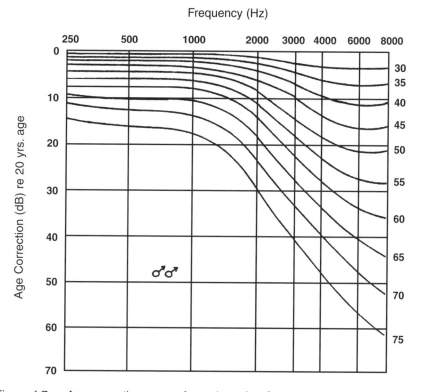

Figure 4.7 — Age-correction curves for males, after Spoor (1967). The ordinate indicates the difference between the Hearing Threshold Level of a person of a given age (parameter) and that of an average 20-year-old, as a function of frequency. These correction values incorporate not only the effects of the aging process per se, but also assume a moderate amount of sociacusis and nosoacusis.

"Hearing Levels" (rather than "age corrections relative to age 20"). The two will be the same only if the median 20-year-old in the non–industrial-noise-exposed population concerned, tested with the audiometric procedures actually involved, has HLs of precisely 0 dB. As indicated earlier, this will be the case only for audiometric tests, under optimum conditions, of a highly screened group of listeners. All actual field studies of non–industrial-noise-exposed 20-year-olds have found median HLs of at least 3 – 4 dB at most test frequencies. It is clear, therefore, that tables or graphs that purport to show realistic HLs to be expected in a group of workers at a particular age if their industrial noise exposures have had no effect must consist of age corrections added to the presumed HLs of the 20-year-olds.

One set of expected hearing level norms are shown in Figures 4.8 and 4.9. These curves are based in large part on a NIOSH study (Lempert and Henderson, 1973) of both non–noise-exposed workers and those whose industrial noise exposure could be determined. Efforts were made to minimize sociacusic and nosoacusic influences by

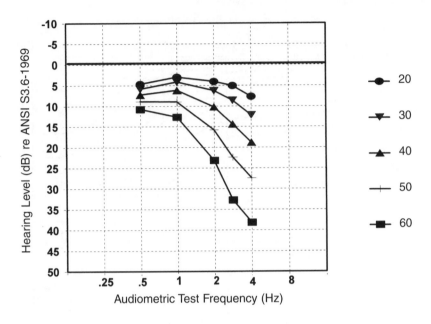

Figure 4.8 — Age-related expected HLs in an industrialized society, for males not exposed to workplace noise (OSHA, 1983). Average of right and left ears.

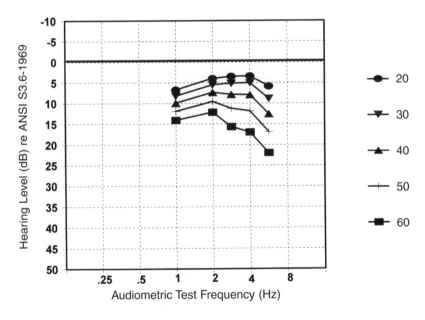

Figure 4.9 — Age-related expected HLs in an industrialized society, for females not exposed to workplace noise (OSHA, 1983). Average of right and left ears.

using fairly specific exclusion principles. The contours of Figures 4.8 and 4.9 are based on data that do not include workers whose questionnaire answers suggested exposure to military or farm noise; excessive gunfire; noisy hobbies such as motorbike riding, machine workshop activity, drag racing, and rock-and-roll music playing; head trauma; use of ototoxic drugs; history of ear surgery or recent middle ear infection; or a conductive hearing loss.

For all practical purposes, therefore, Figures 4.8 and 4.9 represent the expected values of HL for men and women who have not been exposed to industrial noise or significant values of sociacusic or nosoacusic influences. These "age corrections," expressed in tabular form, have been incorporated into federal rules regulating noise exposure as nonmandatory Appendix F of OSHA Rule 29 CFR 1910.95 (see Appendix I).

An additional set of expected norm thresholds for different gender and race groups are included in ANSI S3.44-1996 (see Chapter 17). ANSI S3.44 includes, as Annexes A, B, and C, three different populations that may be selected to use as the age-related component in predicting the total hearing thresholds for a population or for an individual on a statistical basis. Annex A (ANSI-A) is a highly screened population that is assumed to represent only presbycusis effects. The two sets of controls included as Annexes B (ANSI-B) and C (ANSI-C) are assumed to include, in addition to the presbycusis effects, all other potential hazards to hearing except the effects of on-the-job noise exposures. Such a population includes items (1), (2b), and (3) of the system defined by Ward (1977). The reference data presented in ANSI S3.44 as Annex C were

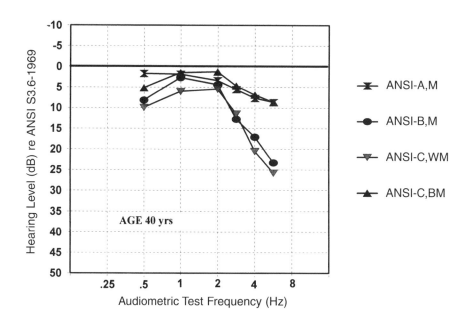

Figure 4.10 — Hearing levels for males at age 40 taken from ANSI S3.44-1996: databases ANSI-A,M; ANSI-B,M; ANSI-C,WM; and ANSI-C,BM.

developed specifically as a comparison group for use in estimating the effects of significant on-the-job noise exposures (Royster et al., 1980; Royster and Thomas, 1979).

Presented in Figure 4.10 are the age-related hearing levels for the four different population choices in ANSI S3.44 for 40-year-old males. Note especially at the higher audiometric test frequencies the large differences in thresholds between the ANSI-A,M (male), and the ANSI-C,BM (black male) and ANSI-C,WM (white male) populations. Also note that the ANSI-C,BM population exhibits thresholds at the higher frequencies that are very close to the ANSI-A,M presbycusis data. Figure 4.11 presents a similar set of data for the female populations included in ANSI S3.44-1996. Differences among HLs for the female populations are smaller but do exist.

To emphasize the effect of gender and race in selecting a set of controls, Figure 4.12 presents the data included in ANSI-C for different gender and race groupings at age 40. For this set of reference data, it is obvious that blacks have significantly better age-effect thresholds than whites, and females exhibit lower thresholds than their male counterparts of the same race, at the higher frequencies.

It cannot be emphasized too strongly that appropriate age-effect reference data should be used in inferring how much hearing loss can be attributed to workplace noise exposure. Unless HLs were actually measured at age 20 (or when workers began employment) the most probable value of the initial HL is given by the age-20 contours selected from the potential reference populations presented in Figures 4.7 through 4.11, and *not* 0 dB at all frequencies. This is essential in assigning "blame" for hearing losses.

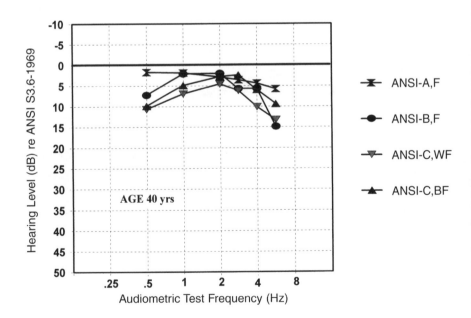

Figure 4.11 — Hearing levels for females at age 40 taken from ANSI S3.44-1996: databases ANSI-A,F; ANSI-B,F; ANSI-C,WF; and ANSI-C,BF.

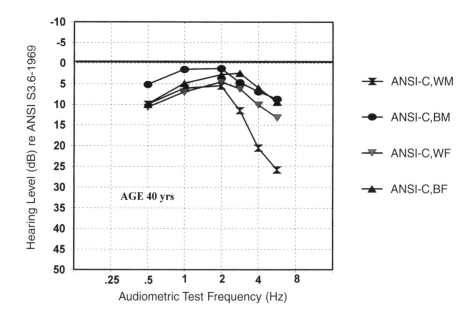

Figure 4.12 — Hearing levels for males and females at age 40 by race taken from ANSI S3.44-1996.

For examples of how the selection of the reference population might impact the process of assigning blame for hearing loss, see Chapter 17.

References

ANSI (1969). "Specifications for Audiometers," ANSI S3.6-1969, Acoustical Society of America, New York, NY.

ANSI (1996a). "Determination of Noise Exposure and Estimation of Noise-Induced Hearing Impairment,"ANSI S3.44-1996, Acoustical Society of America, New York, NY.

ANSI (1996b). "Specification for Audiometers," ANSI S3.6-1996, Acoustical Society of America, New York, NY.

ASA (1951). "American Standard Specification for Audiometers for General Diagnostic Purposes," Z24.5-1951, Acoustical Society of America, New York, NY.

Beasley, W. D. (1938). "National Health Survey (1935–36), "Preliminary Reports, Hearing Study Series, Bulletins 1–7," U.S. Public Health Service, Washington, DC.

Berger, E. H. (1981). "Re-examination of the Low-Frequency (50–1000 Hz) Normal Threshold of Hearing in Free and Diffuse Sound Fields," *J. Acoust. Soc. Am.* 70(6): 1635–1645.

Henderson, D., and Hamernik, R. (1995). "Biologic Bases of Noise-Induced Hearing Loss," in Occupational Hearing Loss, edited by T. C. Morata, and D. Dunn, *Occupational Medicine: State of The Art Reviews 10*(3), 513–534.

ISO (1964). "Standard Reference Zero for the Calibration of Pure Tone Audiometers," R389, International Organization for Standardization, Geneva, Switzerland.

ISO (1984). "Acoustics — Threshold of Hearing by Air Conduction as a Function of Age and Sex for Otologically Normal Persons," ISO 7029, International Organization for Standardization, Geneva, Switzerland.

ISO (1987). "Acoustics — Normal Equal Loudness Contours," ISO 226, International Organization for Standardization, Geneva, Switzerland.

Killion, M. C. (1978). " Revised Estimate of Minimum Audible Pressure: Where Is the Missing 6 dB," *J. Acoust. Soc. Am. 63*(5), 1501–1508.

Lempert, B., and Henderson, T. L. (1973). "Occupational Noise and Hearing 1968–1972," NIOSH Report TR-201-74. NTIS No. PB-232 284, Cincinnati, OH.

Nobili, R., Mammano, F., and Ashmore, J. (1998). "How Well Do We Understand the Cochlea?" *Trends in Neurosciences 21*(4), 159–167.

OSHA (1983). "Occupational Noise Exposure; Hearing Conservation Amendment; Final Rule," Occupational Safety and Health Administration, 29CFR1910.95 *Fed. Regist. 48*(46), 9738–9785.

Royster, L. H., Driscoll, D. P., Thomas, W. G., and Royster, J. D. (1980). "Age Effect Hearing Levels for a Black Nonindustrial Noise Exposed Population (NINEP)," *Am. Indust. Hyg. Assoc. J. 41*(2), 113–119.

Royster, L. H., and Thomas, W. G. (1979). "Age Effect Hearing Levels for a White Nonindustrial Noise Exposed Population (NINEP) and Their Use in Evaluating Industrial Hearing Conservation Programs," *Am. Indust. Hyg. Assoc. J. 40*(6), 504–511.

Sivian, L. J., and White, S. D. (1933). "On Minimum Audible Sound Fields," *J. Acoust. Soc. Am. 4*, 288–321.

Spoor, A. (1967). "Presbycusis Values in Relation to Noise Induced Hearing Loss," *Internat. Audiol. 6*, 48–57.

Ward, W. D. (1977). "Effects of Noise Exposure on Auditory Sensitivity," in *Handbook of Physiology, Vol. 9: Reactions to Environmental Agents*, edited by D.H.K. Lee, American Physiological Society, Bethesda, MD, pp. 1–15.

The Noise Manual, revised 5th edition, edited by E.H. Berger,
L.H. Royster, J.D. Royster, D.P. Driscoll, and M. Layne
©2003 American Industrial Hygiene Association

Auditory and Nonauditory Effects of Noise

5

W. Dixon Ward,
Julia Doswell Royster* and Larry H. Royster*

Contents

---

* Editors following the death of W. Dixon Ward

Introduction

The most undesirable effect of exposure to noise is generally agreed to be permanent hearing loss. The primary goal of any occupational noise control program is usually the prevention of this damage, even though there are other effects of noise, both concomitant with and subsequent to exposure, that are also relevant to occupational health and productivity. However, an aggressive noise control program may produce desirable side effects such as an increase in productivity, a decrease in accident rate, and/or an improvement in worker morale. "Prevention of hearing loss," however, is a rather vague objective, and so it is necessary to examine the concept in somewhat greater depth.

Social Handicap

Basically, what society would like to avoid is production of hearing loss sufficient to cause social handicap. Unfortunately, there is no widespread agreement on how "social handicap" is to be defined and measured. Until the middle of this century, a hearing loss was regarded as handicapping, and therefore compensable as the result of an occupational hazard, only if it was disabling, in the sense that it led to a loss in earning power of the individual, a situation whose existence could be relatively easily and unambiguously determined. Since then, however, there has been a gradual acceptance of the principle that a worker is entitled to compensation for any material impairment suffered as a result of employment; thus, handicap has come to be any condition that interferes with everyday living (see Chapter 18).

A typical example of this viewpoint is the definition of handicap adopted by the American Academy of Otolaryngology: "an impairment sufficient to affect a person's efficiency in the activities of daily living" (AAO, 1979). Since "activities of daily living" include so much, yet vary so widely from one person to another, recent practice has been to attempt to measure auditory handicap in terms of a reduction in the individual's ability to "understand ordinary speech," a simplification that ignores, among other things, the social significance of the perception of warning signals, sounds of nature, and music.

However, even this simplification fails to solve the problem, because there is no accepted definition of "ordinary speech." Speech consists of messages of various degrees of complexity and redundancy, spoken by talkers differing in age, gender, ethnic background, education, and dialect, at a large range of sound levels, in quiet and in the presence of a near-infinite variety of interfering noises. Any test that claims to measure an individual's ability to understand ordinary speech would have to include enough test items to provide a representative range of all these parameters providing that agreement could be reached on what "representative" means. No such direct test has yet been developed.

In view of the lack of a direct measure of social handicap, auditory dysfunction has traditionally been assessed indirectly in terms of threshold sensitivity to pure tones, because a relation does exist between pure-tone thresholds and the ability to detect specific speech sounds in quiet. Most of the information-carrying energy in speech occurs in the range from 500 to 4000 Hz (approximately 50% of the content being below 1800 Hz), and so the ability to detect speech sounds will depend primarily on auditory sensitivity in this region. The perception of vowels depends mainly on the lower frequencies and consonants on the higher frequencies. Hence the avoidance of social handicap is at present equated with preservation of normal sensitivity to certain pure tones.

Before examining the relation between noise exposure and a change in pure-tone sensitivity, it is worthwhile to summarize other manifestations of damage to the auditory system, if only to make it clear why hearing threshold sensitivity continues to be the main indicator of damage.

Tinnitus

A common accompaniment of a loss of ability to hear weak pure tones is, in a sense, its opposite: the hearing of sounds that do not exist in the external environment. Tinnitus, a ringing, buzzing, roaring, or other sound in the ears, often occurs in conjunction with hearing loss. Although intractable tinnitus is often distressing to the individual concerned, measurement of its loudness is difficult. Furthermore, only seldom does noise cause a permanent tinnitus without also causing hearing loss. So although the person with tinnitus should probably be considered more impaired than someone with the same amount of hearing loss but without tinnitus, the question of how much more impairment a particular tinnitus represents has not yet been answered. For guidance on assessing the impact of tinnitus on individuals, see Axelsson and Coles (1996).

Paracusis

Some sounds may be heard, but heard incorrectly. Musical paracusis exists when the pitch of tones near a region of impaired sensitivity due to noise is shifted; that is, a tone is heard, but one having an inappropriate pitch. Unfortunately, direct measurement of paracusis is possible only in highly musical persons, so the phenomenon has received little attention. If the paracusis is greater in one ear of a given individual than in the other, then binaural diplacusis will be found: a particular tone will give rise to different pitches in the two ears, and the magnitude of this difference can be inferred by having the listener adjust the frequency of a tone in one ear to match the pitch of a fixed tone in the other. Again, however, noticeable degrees of paracusis or diplacusis that are attributable to noise exposure occur only in conjunction with a considerable loss of sensitivity, so the importance of paracusis per se in determining social handicap is still unknown.

Speech Misperception

A complete failure to hear certain speech sounds, or phonemes, is not the only effect on speech perception caused by a loss of pure-tone sensitivity: some phonemes may be heard, but heard incorrectly. As indicated earlier, phonemes differ in frequency composition, vowels having energy predominantly in the low-frequency range, consonants in the high-frequency range (above 1500 Hz), with more subtle differences among specific closely related consonants such as *s, f, sh* and *th*, or *p, k* and *t*. A particular configuration of hearing loss (an example is shown in Figure 5.1) may result in only part of the

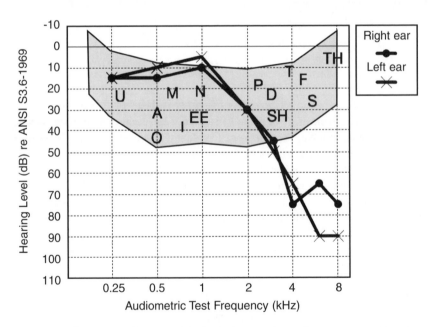

Figure 5.1 — Sample audiogram with shaded range for ordinary conversational speech levels, showing speech sounds for which the spectral peaks are inaudible for the illustrated high-frequency hearing loss (left and right ear threshold values depicted with bold lines). Inaudible peaks are those with hearing level values less than (i.e., to the right of) the thresholds shown.

spectral energy being perceived so that the individual hears — often very plainly — the wrong phoneme. For example, in a study designed to find differences between auditory characteristics of ears with high-frequency losses presumably caused by steady noise and those of ears with similar losses caused by gunfire (none was found), individuals with a high-frequency loss that began at 2000 Hz consistently heard an initial 't' as a 'p'; for example, when given the word 'tick,' with no opportunity to see the lips of the speaker, they almost always responded 'pick' when forced to choose between 'tick' and 'pick' (Ward et al., 1961). [Although visual speechreading cues can increase speech perception accuracy, it is often inconvenient or impossible to watch the speaker.] It is fairly obvious that the individual who mistakenly believes that something has been correctly heard is usually much worse off than someone who knows that a particular message has not been understood. So in recent years, this misperception of speech sounds has received increasingly greater attention. Eventually some standardized consonant-confusion test may be adopted as part of a battery of speech tests designed to assess the evanescent "ability to understand ordinary speech." As yet, however, no such test has met wide acceptance.

Physiological Measures of Damage

Physiological indicators of damage, including destruction of the hair cells of the inner ear, have already been described in Chapter 4. As indicated there, these characteristics cannot be observed directly in the intact human, so it is necessary to infer the presence of damage from occupational noise by less direct methods. The foregoing discussion should have made it clear why the indicator of choice is a permanent change for the worse in sensitivity to pure tones — that is to say, an increase in hearing threshold level (HTL) that constitutes a noise-induced permanent threshold shift (NIPTS). This NIPTS component is over and above changes that can be attributed to the action of the noises of everyday life (sociacusis), to results of infections and otological diseases and of blows to the head (nosoacusis), and to the aging process per se (presbycusis), as discussed in Chapter 4 and illustrated in Figures 4.8 through 4.12.

Otoacoustic Emissions

Otoacoustic emissions (OAEs) may provide a clinically useful early warning of incipient cochlear damage. OAEs are very low-level sounds originated by the outer hair cells (OHC); they can be detected by sensors placed in the external earcanal. Although some individuals exhibit spontaneous OAEs, all persons with normal OHCs exhibit OAEs in response to both click and tonal stimuli (Hall and Mueller, 1997). Because OAEs are produced by the OHCs, which are the ear structures showing the earliest damage by noise exposure, OAE magnitude provides a measure of OHC function. Therefore, it has been proposed that OAEs can be used to detect individuals who are highly susceptible to noise before significant PTS occurs. Human studies do show OAE changes following noise exposures causing TTS (Attias and Bresloff, 1996; Kvaerner et al., 1995), and investigations of OAE applications in HCPs are in progress. However, since OAE testing equipment is not readily available in industry, and since careful audiometry can detect TTS or developing PTS before it becomes significant, the use of OAEs in occupational settings remains only a future possibility. In contrast, OAE testing is becoming common in clinical settings, and the technique may be useful when functional hearing loss is suspected in claims for compensation since it does not depend on subject cooperation. However, OAE results are not yet adequately frequency-specific to suffice as the sole test for NIHL (Attias et al., 1998).

NIPTS

At first glance, the problem of protection of workers against noise-induced hearing loss might appear to be simple: it is necessary only to determine the relation between noise exposure and the resultant NIPTS and then limit exposures to those that produce less than some tolerable amount of NIPTS. Unfortunately, the simplicity is quite illusory, as both "noise exposure" and "NIPTS" are multidimensional, and the question of how much NIPTS is "tolerable" is still undecided.

Noises exist in infinite variety. They may be sinusoids of constant or varying frequency (whistles, sirens), narrow bands of frequencies (the hiss of escaping steam), broad bands of frequencies ("static" on the radio, most occupational noises), impulses (explosive release of gas, as in gunfire), or impacts (a hammer striking a steel plate, as in drop forges), of short or long duration. They may be steady or fluctuating regularly or irregularly in level. In short, noise exposures vary widely in frequency, intensity, and temporal pattern.

Similarly, the PTS produced by a given noise exposure will depend not only on the audiometric frequency evaluated, but also on a host of characteristics of the individual whose NIPTS is being measured. Some of the most important of these individual characteristics that contribute to differences in susceptibility to NIPTS are described below.

Susceptibility

The Sound Conduction Mechanism

Other things being equal, the more acoustic energy that reaches the inner ear, the greater will be the effect (this may be one of the few principles that can be stated unequivocally about noise damage). Therefore the structural characteristics of the external and middle ears, which determine how much power finally is transmitted into the cochlea, must play a role: characteristics such as the size and shape of the pinna and the length of the auditory canal, the area of the eardrum and of the footplate of the stapes, the mass of the ossicles, and the strength of the middle-ear muscles that contract in the presence of high-intensity noise and thereby reduce the transmission of sound. Collectively these factors control not only the energy reaching the cochlea, but also its frequency spectrum; the difference between the diffuse-field spectrum outside the ear and the spectrum at the eardrum is the head-related transfer function (HRTF). In the lower frequencies the HRTF is primarily influenced by diffraction around the torso; at frequencies above 1 kHz, the acoustics of the outer ear are the major contributor. Hellstrom (1996) concluded that the HRTF is a better predictor of noise-induced hearing loss (NIHL) than is earcanal volume or earcanal length per se.

Characteristics of the Inner Ear

Once the sound has reached the cochlea, specific properties of the inner ear, both structural and dynamic, must also play some role in determining susceptibility. For example, the stiffness of the cochlear partition, the thickness of the basilar and tectorial membranes, the blood supply to the cochlea, the rate of oxygen metabolism, and the density of afferent and efferent innervation doubtless have some effect.

Gender

In industries in which men and women work side by side, audiometric surveys invariably show that, on average, women have significantly better hearing than men (e.g., Berger et al., 1978; Welleschik and Korpert, 1980). This could be due to average differences in some of the dimensions of the auditory system listed above, in which case women would have to be considered to have "tougher" ears. On the other hand, it could simply indicate that women generally are exposed to fewer and less severe sociacusic influences, especially gunfire. Or it may bear some relation to absentee rate, which is higher in women than men, so that the former actually get less exposure. It is also possible that women are ordinarily freer to leave a noisy job if the noise bothers them, so that only the least susceptible women are the ones who continue to work and hence get included in the survey. Whatever the underlying cause, the difference between men and women cannot be ignored in assessing the effectiveness of hearing conservation programs.

Skin Color

Comparison of the hearing of white and black workers in the southeast section of the U.S. has indicated that blacks have significantly better average hearing than whites (Royster et al., 1980). It is not yet clear whether this difference is due to a disparity in sociacusic exposure (e.g., whites shoot rifles more often than blacks) or instead implies that dark skin pigmentation, presumably correlated with similar pigmentation in the cochlea, somehow makes the individual less susceptible to damage, or again it could simply be that the HRTF differs for the populations being studied. [However, age-effect hearing levels differ by race in the absence of occupational noise, indicating that the race difference reflects more than simply differential susceptibility to NIPTS. See Chapters 4 and 17 for more information.]

Age and Experience

Although common sense might suggest that the "young, tender" ear of a teenager would be more easily damaged than the ear of a middle-aged worker, it is just as sensible to propose that the "young, resilient" ear of the teenager should be less easily damaged. There is little evidence for either view, although Hetu et al. (1977) have shown that susceptibility to temporary hearing loss is no different in 12-year-olds than in adults. Similarly, no convincing experimental support exists for the notion that the ears of persons who have worked in noise have gradually become "toughened" and therefore more resistant to NIPTS (Welleschik and Raber, 1978). In short, age per se does not seem to be a factor that directly influences the amount of hearing loss produced by a given occupational noise exposure.

Initial HTL

Age does have an indirect influence because the effects of presbycusis, sociacusis, and nosoacusis do increase with age, so that the older the individuals are at the beginning of the exposure being studied, the higher will be their median HTLs, as shown in Chapter 4. To the extent that a change in HTL depends on its initial value, then, age will be an important factor. The evidence in this case indicates that the higher the initial HTL, the smaller will be the TTS from a particular exposure (Ward, 1973), and presumably the same will be true of NIPTS. If it were not — that is, if PTS were independent of the initial HTL — then the growth of PTS with time of exposure would be linear; for example, if 1 year of exposure to a particular noise produced 15 dB of PTS, then 2 years would produce 30 dB, 3 years 45 dB, and so on. Instead, both animal and human studies agree that growth is exponential — rapid at first, then slowing down (Burns and Robinson, 1970; Glorig et al., 1961; Herhold, 1977). So the initial HTL is an important population parameter.

Body Conditioning

As is the case for nearly all human ills, attempts have been made to link susceptibility to hearing loss with every type of bad habit such as smoking, deficiencies or excesses in dietary substances such as Vitamins A, C, or B12, iron, lead, and magnesium, artificial food additives, use of social drugs or stimulants, poor posture, lack of exercise, over-exercise, promiscuity, sexual inactivity, and so on. Such studies, however, are seldom adequately controlled, so no causal relationship between any of these characteristics and susceptibility to hearing loss has been established.

A similar situation exists for a plethora of other conditions and substances that could possibly affect vulnerability to damage from noise. Despite dozens of studies of such factors as blood pressure, blood cholesterol, pregnancy, phase of menstrual cycle, head position, body temperature, emotional state in general, attitude toward the sound, and mental activity, no clear-cut effect on vulnerability has emerged. The same is true for diseases such as diabetes, thyroid malfunction, leprosy, or silicosis.

Interaction of Noise and Other Noxious Agents

There are, however, a few substances that have been shown to be related to vulnerability. For the most part, these are substances that directly affect the auditory threshold themselves, such as ototoxic drugs like kanamycin, gentamicin, cisplatin, and aspirin, or a few industrial chemicals such as toluene, paint solvents, and carbon disulfate. Carbon monoxide is the substance that has been most clearly demonstrated to act synergistically with noise in producing hearing damage — that is, a combination of simultaneous exposure to carbon monoxide and noise will produce a greater PTS than the sum of the effects of the noise alone and the carbon monoxide alone. Anoxia also aggravates NIPTS. Finally, vibration may also increase the amount of damage. For further information, including additional substances that may interact with noise, see Fechter (1995), Franks and Morata (1996), Cary et al. (1997), and Morata (1998). The American Conference of Governmental Industrial Hygienists (ACGIH) is considering whether current exposure limits are adequate to minimize the possibility of potentiating NIHL via combined exposures. ACGIH's documentation (1999) recommends audiometric monitoring for exposures to noise in combination with toluene, lead, manganese, or n-butyl alcohol, and lists other substances under current investigation for ototoxicity.

Ameliorative Agents

A search for substances or conditions that will reduce vulnerability to NIPTS has in general been unsuccessful, except for conditions that reduce the transmission of sound to the cochlea, thus producing a conductive hearing loss. From a practical viewpoint, the question is ordinarily moot; even if some protective substance existed, few persons would have the foresight to use it before a hazardous noise exposure. Consequently, most efforts have been directed toward determination of substances that will hasten the process of recovery from acoustic trauma, presumably by increasing blood flow to the cochlea by means of vasodilation, increasing microcirculation, or raising the oxygen tension.

Of several dozen substances whose action has been studied, the only one that consistently produces a greater or faster recovery from acoustic trauma is oxygen, either pure or in the form of carbogen (95% oxygen and 5% carbon dioxide). However, the ameliorative effect is much smaller when carbogen is given after the exposure than when it is given simultaneously. So at present there seems to be no ameliorative agent that can be used to reduce NIPTS; although a few substances showed promise in a single study, their efficacy was not confirmed by subsequent studies.

Predicting NIPTS Is Complex

It is clear that no simple answer can be expected to questions as vague as "How much PTS is caused by noise exposure?" Questions must be posed that are much more specific, "How much PTS at frequency F is caused, in the average individual of gender G, skin color C, and initial HTL of H dB, by an exposure for time T to a noise of spectrum S at an intensity level L and temporal pattern P?" None of the infinite number of forms of this question has yet been convincingly answered by use of actual occupational audiometric measurements. Indeed, even if attention is confined to the simplest workshift temporal patterns possible — that is to say, a daily 8-hour exposure to a steady noise of a fixed spectrum, 5 days/week, 50 weeks/year — the data are inadequate. In many studies, the audiometric measurements have been characterized by faulty technique (such as the use of screening audiometry), by the presence of masking noise in the testing room, or by temporary threshold shifts that inflated the inferred PTSs. In nearly all cases, the initial HTL of the workers was unknown, so that occupational-noise-induced PTS (ONIPTS) had to be estimated either (1) by comparison of HTLs with those of a control group of non–noise-exposed workers whose age, gender history of exposure to nosoacusic influences, and habitual exposure to sociacusic agents were, on average, like those of the noise-exposed workers, (2) by assuming that the workers would have had hearing that was typical for their age if they had not been exposed to the occupational noise, or even (3) by pretending that the measured HTLs represent ONIPTSs directly, thus neglecting completely the effects of presbycusis, nosoacusis, and sociacusis. However, it is noted that in recent years the use of models that include age-effect options, such as the ANSI S3.44 model (see Chapter 17), have been very successful in predicting the existing thresholds for occupationally noise-exposed populations. Implicit in these models is the assumption that NIPTS stops accumulating when noise exposure ceases; this is consistent with clinical consensus about noise-exposed humans (ACOM, 1989) as well as animal research (Borg et al., 1995). (For discussion of clinical judgments regarding individual cases, see Chapter 11.)

ONIPTS from Steady-State Noise

Despite the foregoing inadequacies of estimates of ONIPTS from steady exposure, a compilation by Passchier-Vermeer (1968) of these estimates displayed a surprising degree of consistency. Figure 5.2 shows the inferred ONIPTS at 4 kHz, the frequency most severely affected by noise, after 10 years of 8 h/day exposure to a steady occupational noise environment whose A-weighted level is given on the abscissa.

The use of A-weighting to reduce different spectra to a single index reflects evidence that low- and very-high-frequency components of a noise are less hazardous than the middle frequencies, in the sense that temporary effects are less severe. Of the weightings available on a standard sound lever meter, A-weighting comes closest to reflecting the actual relative hazard. The reason for the greater hazard of the middle frequencies may be complex, but at least a large part of the story is the fact that the outer earcanal is resonant at about 3 kHz. This fact, coupled with the observation that the frequency most affected by a narrow-band noise may be half an octave or so above the frequency of the noise itself, is responsible for the fact that the 4-kHz region is the first and the most affected for the average individual by the broadband noises that are typical of industry. However, in situations where the noise environment is not broadband, then an analysis of the audiometric data can identify dips at frequencies other than 4 kHz that correlate with the distribution of acoustic energy in the work environment (Royster, Royster, and Cecich, 1984).

Additional support of the original findings based on noise exposures determined using A-weighting is indicated by the two open circles, in Figure 5.2, which represent the results of two subsequent large-scale studies of workers employed in levels of 80 to 90 dBA. Robinson et al. (1973) found that textile workers from an 83-dBA environment showed a mean loss about 5 dB greater than a control group who worked in 70 dBA or below, and a later study involving several industries demonstrated an average loss of 11 dB in workers whose daily A-weighted exposure levels were about 87 dB (Yerg et al., 1978).

Figure 5.3 shows the entire set of functions similar to that of Figure 5.2 for all of the audiometric frequencies normally tested. These functions indicate that, after 10 years of daily 8-h exposures to a steady noise, (1) 80 dBA is innocuous; (2) 85 dBA will result in an average ONIPTS of around 10 dB at the most noise-sensitive audiometric frequencies of 3, 4, and 6 kHz, which is the smallest change in HTL that can be regarded as significant in the individual ear; (3) only at 90 dBA and above do the average ONIPTSs reach values that will, when added to the inevitable changes in HTL caused by presbycusis, sociacusis, and nosoacusis, produce a loss of hearing that can be noticed by the worker in question.

Burns and Robinson (1970) published results of a careful study of 759 workers in various industries, again those with a uniform environment. Figure 5.4 shows what they consider to be the median HTLs at 4 kHz to be expected in workers exposed to various levels for up to 45 years, assuming that they began work with 0 dB HTL. Inspection of these contours confirms the implication of Figure 5.2 that 80 dBA produces a negligible effect; the contours also indicate that ONIPTS, at least at this frequency, grows rapidly in the first few years of exposure, reaching a near-asymptote by

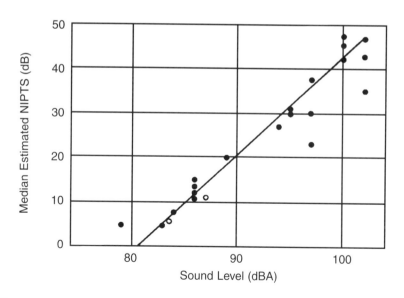

Figure 5.2 — Estimated occupational-noise-induced permanent threshold shift at 4 kHz produced by 10 years or more of exposure to noise at the indicated A-weighted level, 8 hours/day, 250 days/year. Solid points indicate values calculated from the literature by Passchier-Vermeer (1968), and open points represent subsequent studies by Robinson et al. (1973) and Yerg et al. (1978).

the end of 10–15 years, after which HTLs continue to deteriorate no more rapidly in these workers than in the "no noise" group. This is illustrated for the NIPTS component of hearing loss in Figure 5.5, and for total hearing loss in Figure 5.6; progression of NIPTS after 10 years broadens the notch toward lower frequencies, and the addition of age-related hearing change exacerbates this effect while increasing total loss.

The ONIPTSs in Figure 5.4 are somewhat smaller than is indicated in Figure 5.2, but this might be expected in view of the fact that the Burns and Robinson data did exclude some sources of error that were operating in the studies summarized by Passchier-Vermeer, particularly temporary threshold shifts. Therefore the somewhat smaller values of Figure 5.4 are probably more valid. A comparison of all major damage risk data is provided by National Research Council Committee on Hearing, Bioacoustics, and Biomechanics (CHABA, 1993).

These data are consistent with the adoption of 90 dBA as the 8-hour exposure limit in many countries, although 85 dBA is advocated by those who feel that a 15-dB loss at high frequencies is too great to be tolerable, arguing that the figure of 15 dB is only an average that does not take individual differences into account. When the average ONIPTS is 15 dB, some workers will have lost 20 dB, and a few even 30 dB. In order

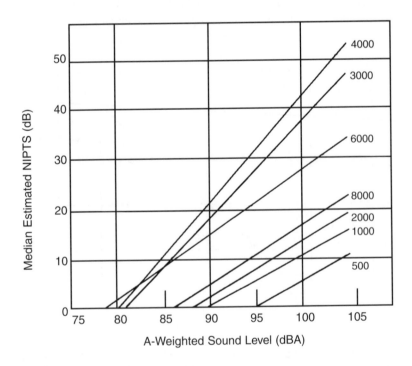

Figure 5.3 — Estimated occupational-noise-induced permanent threshold shifts at various frequencies produced by 10 years or more of exposure to noise at the indicated A-weighted level, 8 hours/day, 250 days/year. After Passchier-Vermeer (1968).

to protect the most susceptible individuals, it is contended, the exposure limit should be set at a value lower than merely what is necessary to protect the average worker from the average amount of ONIPTS.

For the case in question, that is, where the median HTL shows a shift, this line of argument is reasonable. It is, however, sometimes extended to situations where the median NIPTS (for example at 85 dBA across the audiometric test frequencies of 500, 1000, and 2000 Hz) is zero. That is, it is still argued that because even though the average person is not affected by 85 dBA, the most susceptible persons will be. But the only way in which a shift in the HTLs of the most susceptible individuals would not influence the median HTL would be if these high susceptibility persons are exclusively those whose initial HTLs already lie above the median — that is, only if already-damaged ears are the only ones that will be affected. There is, however, no evidence to support this assumption. Indeed, the opposite seems to be true. That is, the public health survey (PHS) data show that the median HTL shifts slightly more with age than the worst 10 percentile (Glorig and Roberts, 1965), which implies that ears with best hearing are the most likely to be affected (Ward, 1976). Basing exposure standards on some hypothetical "10th percentile most susceptible" individuals is therefore unjustified.

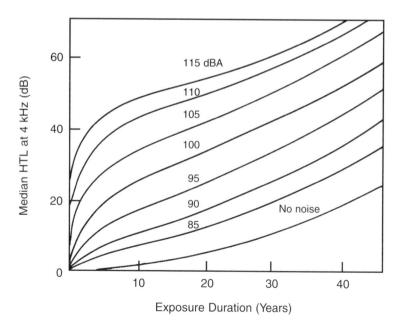

Figure 5.4 — Curves illustrating the expected growth of hearing threshold level at 4 kHz with years of exposure (8 hours/day, 250 days/year) to industrial noise at the indicated A-weighted levels. After Robinson (1970).

135

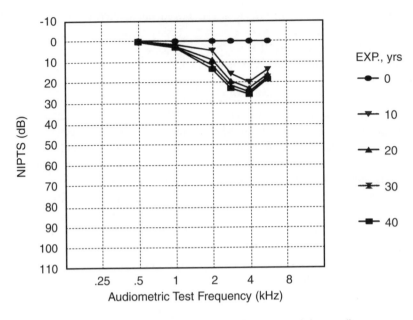

Figure 5.5 — Predicted NIPTS (based on ANSI S3.44-1996) for median ears exposed to an L$_{A8hn}$ of 95 dBA for 0–40 years.

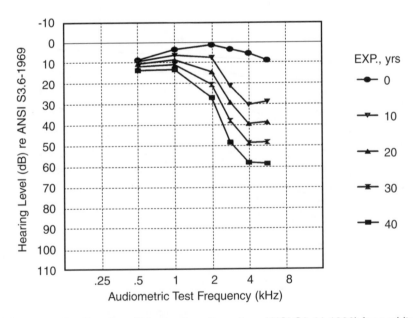

Figure 5.6 — Predicted total hearing loss (based on ANSI S3.44-1996) for a white male population of median susceptibility at ages 20–60 years with 0–40 years of noise exposure at an L$_{A8hn}$ of 95 dBA.

ONIPTS from Nonsteady Noise

Unfortunately, Figures 5.3 to 5.4 represent nearly the total quantitative knowledge about noise exposure and ONIPTS in humans: the loss in hearing that can be attributed to many years of daily 8-hour exposure to a steady noise, in an average worker with originally normal hearing (i.e., HTLs of 15 dB or less) who is also exposed to an average amount of sociacusic and nosoacusic influences during those years. It is also known that acoustic trauma from a short exposure at a very high level — for example, a single firecracker — can cause the same loss, at least at high frequencies, as do 10 years of daily 8-h exposure at 110 dBA. Most of the exposures that actually occur in industry lie somewhat between these simple extremes. Occupational noises often fluctuate regularly or irregularly, and impact and impulse components may exist. Even when the noise itself is steady, the worker's job may require movement among several areas, so that the exposure involves a number of different levels. For assessing the hazard associated with these nonsteady exposures, information such as that shown in Figures 5.3 and 5.4 is of little use, unless some scheme can be devised by which a particular exposure, no matter how complicated, can be expressed as being equivalent in effectiveness to some particular steady noise. That is, the exposure in question, if repeated every day, would eventually lead to some value of ONIPTS. If this ONIPTS at 4 kHz were, say, 15 dB, then from Figure 5.4, the noise exposure could be said to possess an effective level of 90 dBA in terms of its ability to cause hearing damage, because an 8-hour exposure to a steady 90 dBA would produce the same 15-dB ONIPTS.

Two major approaches have been taken in attempts to develop a simple scheme that will permit the determination of this effective level. The equal-energy approach is an example of attempts to equate exposures on the basis of their physical characteristics directly, while the equal-TTS approach is based on an assumed correlation between permanent and temporary effects of noise exposure.

Equal Energy

The equal-energy approach (Robinson, 1970) makes the assumption that damage depends only on the daily amount of A-weighted sound energy that enters the ear of the worker, and that the temporal pattern during the day is irrelevant. Since this energy is the product of the A-weighted sound intensity and its duration, then in order to hold constant the energy of two exposures and hence, presumably, their effect on hearing, it is only necessary to balance a difference in intensity by a corresponding difference in duration (in the opposite direction). Thus if an exposure of 90 dBA for 8 hours (480 minutes) is established as the tolerable daily noise dose, the equal-energy theory postulates that an exposure of 110 dBA (20 dBA higher than, and hence possessing 100 times the intensity of a 90-dBA noise) will be just as hazardous if its duration is 4.8 minutes (480/100). Furthermore, the hazard would be the same whether this exposure came in one 4.8-minute burst or instead consisted of 10 bursts, each of 0.48 minute in duration, spread over the 8-hour workday.

The equal-energy principle leads to the so-called "3-dB rule" in evaluation of the hazard associated with exposures involving levels other than whatever level is adopted as the permitted 8-hour limit. In such a system, the duration must be cut in half when the A-weighted level is increased by 3 dB, because a 3-dB increase means that the intensity has doubled. If 90 dBA for 8 hours constitutes the permitted daily exposure, then the 3-dB rule postulates that 93 dBA can be tolerated for 4 hours, 96 dBA for 2 hours, 99 dBA for 1 hour, and so on, regardless of the distribution of the energy within the workday. All of these exposures are said to possess an equivalent level of 90 dBA. In formal terms, the 8-hour equivalent level L_{A8hn} of an exposure is that level that, if maintained for 8 hours, would transfer the same amount of energy to the ear as the actual exposure.

The equal-energy principle makes calculation of the presumed effective level of the exposure relatively easy, because in this case the effective level is simply the equivalent level. Even if the exposure is hopelessly complicated, involving multiple or continuously varying levels, it is easy to construct devices — dosimeters — that will integrate the instantaneous intensity over time in order to provide a measure of the total energy, expressed either in terms of L_{A8hn} or as a fraction or multiple of the standard daily dose.

Because of this simplicity, the 3-dB rule has been adopted by the International Standards Organization. In ISO 1999 (1990), it is stipulated that exposure be measured only in terms of L_{A8hn}, a step that automatically commits prediction of hazard to use of the equal-energy principle.

Unfortunately, the equal-energy theory has one serious shortcoming: it is true only for single steady uninterrupted exposures, because the daily temporal pattern does play a role in ONIPTS. In the chinchilla it is not the case, for instance, that a 48-min exposure to 110 dBA and a series of 40 1.2-min bursts, also at a level of 110 dBA but separated by 10.8 min of quiet, will produce the same damage (Ward et al., 1983). The short bursts produce much less damage. The short-term recovery that occurs during the quiet periods of the intermittent exposure serves to reduce the permanent effect (Ward, 1991a). By contrast, the second principle that has been used to evaluate nonsteady exposures, the equal-TTS approach, does not make the mistake of ignoring temporal pattern.

Equal TTS

The equal-TTS approach to the problem of determining the equivalence of steady and time-varying noise exposures is based on the hypothesis that daily exposures that produce the same temporary effects will eventually produce the same permanent effects. For example, if the standard daily dose produces a temporary threshold shift of 15 dB, then any daily exposure that also produces a 15-dB TTS should ultimately cause the same ONIPTS as does the standard daily dose. Because TTS, which can be studied under controlled conditions in the laboratory, behaves fairly lawfully, it is a relatively simple matter to determine combinations of level, duration, and temporal pattern that produce the same TTS as the standard daily noise dose. There is, of course, a wide variety of TTS measures that could be used in these determinations.

The TTS index about which the most empirical information existed in 1964, when Working Group 46 of the NAS/NRC Committee on Hearing, Bioacoustics and Biomechanics (CHABA) attempted to develop standard equal-effect contours, was TTS_2, the TTS at a particular audiometric frequency measured 2 minutes after the termination of the noise. So this index was the one they used. However, indices that would appear to have equal face validity include, in addition to TTS at any arbitrarily chosen time following exposure other than 2 minutes, the maximum value of TTS that occurred at any time during the exposure, or the time required for complete recovery, or the integral over time of the instantaneous value of TTS from the beginning of the exposure to the end of the exposure or to the time of complete recovery (Kraak, 1982). Even more complicated possible measures would include those involving integration of each of the foregoing indices over frequency.

The most well-known set of formal damage-risk contours based on TTS, however, are those finally proposed by CHABA (Kryter et al., 1966; Ward, 1966). These are contours that, according to the most accurate data then available, showed the levels and durations of octave-band noises that would just produce a TTS_2 of 10 dB at 1000 Hz or lower frequencies, 15 dB at 2000 Hz, or 20 dB at 3000 Hz or higher frequencies, in the average normal-hearing person. These are values of TTS_2 that are produced by 8-hour continuous exposures to noises of 85 to 95 dBA, depending on the actual spectrum of the noise in question. Equal-effect contours were developed not only for exposures shorter than 8 hours but also for those involving various patterns of intermittence. The end result was a set of ten graphs that, if used properly, would indicate whether or not a particular exposure was more hazardous or less hazardous than the standard daily dose of 8 hours at 90 dBA, provided that TTS_2 were in fact a valid indicator of relative hazard. The procedure for using these contours was so complicated that they were never widely used.

Nevertheless, the implication of these contours was clear: intermittence did reduce the TTS, quite drastically under certain conditions, so that considerably more energy could be tolerated in an intermittent exposure without exceeding the TTS_2 criterion described above. Melnick (1991) has reviewed the TTS approach to damage-risk criteria.

OSHA Regulations

Consequently, when it became necessary to establish exposure standards for industry in the U.S., the committee that attempted to find a simplification such that equivalent exposure could be calculated easily, but without assuming the validity of the equal-energy theory, decided that the best way to take into account the reduction of hazard associated with intermittence was to use a trading relation of 5 dB per halving of exposure time instead of 3 dB (Intersociety Committee, 1967). In effect, this use of a fixed trading relation again ignores the effect of different patterns of intermittence; the only difference is that use of the 5-dB rule trading relation tacitly assumes that any exposures involving levels above 90 dBA will be intermittent, indeed involving only short noise bursts distributed relatively evenly throughout the workday. Thus with a standard daily dose of 90 dBA for 8 h, a 4-h exposure at 95 dBA was defined as equivalent in effect, as was a 2-h exposure at 100 dBA, and so on, up to a 15-min exposure at 115 dBA. Each of these exposures is said to possess a time-weighted average (TWA) of 90 dBA, a metric that is analogous to the L_{A8hn} that results when the exposure is evaluated using the 3-dB trading relation. This formula for assessing the effective level of short or intermittent exposures was adopted as a federal regulation for government contractors under the Walsh-Healey Act of 1969 (U.S. Department of Labor, 1969) and was then extended to all firms engaged in interstate commerce under the Occupational Safety and Health Act (OSHA) of 1970 (U.S. Department of Labor, 1971). It is still in force, and dosimeters that can evaluate exposures in terms of TWA instead of (or in addition to) L_{A8hn} are commercially available.

It is noted that one of the reasons that the OSHA criteria prohibit unprotected exposures above 115 dBA is that the 5-dB trading ratio would not adequately protect individuals from very high-level, short-duration noise exposures. If a 3-dB trading ratio is used, as specified by the American Conference of Governmental Industrial Hygienists (ACGIH, 1999), then unprotected exposures up to 139 dBA are allowed for very short durations, even with an 85-dBA criterion level.

Other Equivalent-Exposure Regulations

The history of development of exposure standards indicates that only simple formulas for assessing equivalence of exposures will be acceptable to those who must establish and enforce these standards. Unfortunately, however, damage to hearing clearly depends on the intensity, duration, spectrum, and temporal pattern of the noise in a very complex manner, and so no simple solution can possibly be correct, although it may be acceptably accurate over a limited range of exposures.

The 3-dB rule of ISO and ACGIH overestimates the hazard of intermittent noises. On the other hand, use of the 5-dB rule of OSHA will underestimate the hazard of a steady noise with a duration less than 8 hours, especially at the highest levels. The contrast between them is greatest at 115 dBA, the highest level permitted for more than 1 second in the OSHA regulation: the 15 minutes permitted by OSHA is nearly 10 times as long as the 1.6 min allowed by the equal-energy principle (assuming a criterion level of 90 dBA for 8 hours). A compromise between these two systems was adopted by the armed forces of the U.S., who for many years employed a 4-dB-per-halving rule (U.S. Air Force, 1973), which is still used by the U.S. Navy, although other branches of the service have changed to the 3-dB exchange rate. Some German investigators have proposed that a 6-dB rule would be appropriate (Kraak, 1982).

Another possibility that might be explored is to use one trading relation over the range of the lower levels and another for the higher levels, or even to adopt a fixed curvilinear relation that, although difficult to use directly, would pose no problem to a dosimeter. Still another solution would be to accept the 3-dB trading relation of equal energy because it is the most protective, but develop a set of correction factors, to be applied to the L_{A8hn}, that depend on the degree of intermittence of the exposure. Which of the foregoing solutions will eventually prove to be the best is obviously still to be determined, and because very few measurements of actual auditory damage from steady versus intermittent exposures in experimental animals have been made, the decision as to which is the most valid lies far in the future.

Real-World Noise Environments

Notwithstanding the above discussion debating the 3-dB vs. 5-dB rules, the real question involves the ultimate impact of the choice on occupationally noise-exposed populations. Therefore a few general guidelines are in order:

1. If the noise exposure environment is essentially steady-state (that is, if observed sound levels vary less than 5 dB), then very little difference will be observed between the daily measured OSHA TWA and the L_{A8hn} equivalent energy value.

2. If the noise environment is one that varies between several levels during the day, with up to 10 dB variation, and the general background level is 85 dBA or higher, then the L_{A8hn} value will be 2–4 dB higher than the corresponding OSHA TWA value.

3. If the noise environment is one where the employees are intermittently exposed to high levels (say greater than 100 dBA) alternating with low background levels below 80 dBA, then the L_{A8hn} value will be significantly greater than the corresponding OSHA TWA value. More importantly, under the existing OSHA criteria the affected employees might not have to be included in the hearing conservation program, whereas they would be (and probably should be) included if a 3-dB exchange rate were utilized.

4. Finally, the use of a 3-dB exchange rate will result in the inclusion of a significantly greater number of the noise-exposed population with very low TWAs (80–85 dBA) that would not have to be included under the OSHA criteria. Based on the information presented early in this chapter with respect to the limited damage to hearing that is expected for these low noise exposures, the required inclusion of this population in the hearing conservation program is not justifiable.

ONIPTS from Pure Tones

Some regulatory schemes prescribe a 5- or 10-dBA reduction in permitted level if the noise has a pronounced pure-tone component. The weight of evidence, however, indicates that a pure tone in the frequency range most important to hearing conservation — that is to say, from 1000 Hz to 4000 Hz — produces no greater damage than does an octave-wide band of noise at the same sound level (Ward, 1962). Although a pure-tone component may increase the annoyingness of a noise, it does not constitute an additional hazard to hearing.

ONIPTS from Very High Sound Levels

Most exposure regulations have some sort of ceiling on the instantaneous peak sound level, a limit that may not be exceeded regardless of duration. This concept of a "critical intensity" — an acoustic intensity that would produce some irreversible damage no matter how short the exposure might be (Ruedi, 1954) — is somewhat consistent with the notion of an "elastic limit" of the structures in the inner ear that if exceeded leads to immediate mechanical injury to tissues (Price, 1992). The existence of such a "breaking point" would be implied by a sudden increase in the rate of growth of damage in a series of experimental exposures at increasing intensity but fixed duration. Such abrupt changes are indeed found, but the implied critical intensity nevertheless still depends on the duration: the shorter the exposure, the higher the indicated critical intensity. For example, in the case of the shortest of acoustic events, single impulses such as those produced by explosions, although the critical level inferred from measurements of TTS following exposure to impulses of about 1 millisecond duration is somewhere around 140 dB SPL (McRobert and Ward, 1973), exposures to the pulses of less than 100 microseconds generated by a cap gun imply that the critical level for them is somewhat higher. Conversely, for reverberant sounds such as the impact noises generated by drop forges, somewhere around 130 dB may represent a critical level.

Just where, if anywhere, this type of limit should be placed is still undecided. Although the present OSHA regulations state: "Exposure to impulsive or impact noise should not exceed 140 dB peak sound pressure level" (U.S. Department of Labor, 1971), this number was little more than a guess when it was first proposed in the CHABA document (Kryter et al., 1966), and no convincing supportive evidence has since appeared. While 140 dB may be a realistic ceiling for impact noises, it is inappropriate for impulses, so exposure limits in which the permitted peak level increases as the duration of the pulses becomes shorter should continue to be used (CHABA, 1968).

It should be noted that the present OSHA exposure limit has a de facto second critical level for continuous noise: 115 dBA. Although 15 min of exposure at 115 dBA are permitted, more than 1 second at any higher level is forbidden, even though a 1-second exposure at 130 dBA would represent a dose considerably less than 1% of the permitted daily dose if the 5-dB-per-halving trading relation were extended to include exposures at that level. Indeed, in the ISO equal-energy system, 3 seconds of exposure would be permitted at 130 dBA, as this would be the energy equivalent of 90 dBA for 8 hours. This 115-dB ceiling is therefore highly artificial, as no human data even suggest that permanent damage can be caused by 3 seconds or less of 130 dBA.

Documented cases of acoustic trauma have resulted from sources such as the earpiece ringers of early cordless telephones (Orchik et al., 1987), rail car retarders, release of high-pressure valves, and impulses such as firecrackers and explosions. Very susceptible ears may sustain acoustic trauma from exposures such as 1 minute at 130 dB, 10 seconds at 135 dB, or 2 seconds at 140 dB (Ward, 1991b). However, the spectrum of the noise is important: military weapons research confirms that very-low-frequency energy is less damaging (Patterson et al., 1993) while energy around 3000 Hz is most hazardous (Price, 1981). Although in occupational settings most acoustic trauma incidents are accidental, hearing conservation personnel should identify likely causes of such accidents and educate workers about these special hazards.

Infrasonics and Ultrasonics

It has been known for some time that very high sound pressure levels in the infrasonic region (below 20 Hz) and in the ultrasonic region (generally frequencies above 20 kHz) can present a hazard to the workforce. With respect to these two types of noise exposure, ACGIH provides some guidance regarding human exposure to acoustic energy in these two frequency ranges (ACGIH, 1999). In the infrasonic range, ACGIH recommends in a notice of intent to establish TLVs® that in the 1/3 octave bands (OBs) from 1 to 80 Hz, the sound pressure levels should not exceed 145 dB. In addition, the overall unweighted sound pressure level should not exceed 150 dB. For noise exposures in the upper range of auditory perception and in the ultrasonic range, ACGIH specifies ceiling value SPLs of 105 dB for 1/3 OBs from 10 – 20 kHz, 110 dB for 25 kHz, and 115 dB up to 100 kHz. If the ultrasonic source is not coupled to the skin, then from 25 to 100 kHz, the allowed ceiling values may be increased by 30 dB. ACGIH also specifies maximum 8-hour TWAs for 1/3 OBs of 10 – 20 kHz, while acknowledging that annoyance may occur at lower levels.

Professionals interested in protecting a noise-exposed population from infrasonic and ultrasonic energy should consult Chapter 3 and Chapter 10. Earmuffs, in particular, provide very little protection in the infrasonic region. Note that because the A-weighting filter specifications do not exclude ultrasonic energy, A-weighted measurements for OSHA purposes can be high in environments with ultrasonic transducers although there is no hazard to hearing.

Nonauditory Effects of Noise

Noise levels and exposures that pose no hazard to hearing may nevertheless produce some undesirable effects — interference with speech communication as discussed earlier, sleep, or task performance — and these lead to annoyance, irritation, or even rage on the part of some individuals. That noise can mask speech and thereby disrupt communication is obvious (see Chapter 14). While disruption of communication can be fairly well predicted by the level and temporal pattern of interfering sounds, this is not true for sleep or task performance. The main conclusion from myriad laboratory and field studies over the past 40 years is that people differ drastically in the resultant degree of annoyance. Some people are disturbed by the barking of a neighbor's dog that is barely audible, while others can tolerate living next to a busy freeway, railroad line, or airport.

The interference with normal sleep and the correlated annoyance caused by noise depends on many factors, including the level and duration of single events, the pattern of intermittence, the spectrum of the noise, the meaning of the noise to the sleeper, and the time of night. As in the case of ONIPTS, however, the Gordian knot has been cut by ignoring intermittence completely and proposing exposure limits mainly based on total A-weighted energy.

Noise levels sufficiently high to be hazardous to hearing if maintained long enough also effect some changes in the cardiovascular and psychophysiological systems: increase in heart rate, blood pressure, catecholamines, adrenalin secretion, vasoconstriction of the extremities, and dilation of the pupil of the eye. There is little evidence that these changes have any enduring aftereffects unless exposures are considerably higher than 85 dBA. Unexpected sounds produce the startle reaction. Severe exposures produce or augment the stress reaction of the body, and perhaps have an effect on the immune system, although the evidence for the latter is only fragmentary. For details of the above effects, see the report of the Health Council of the Netherlands (1996).

A careful analysis of the literature relative to protection of the public from noise exposures that can disrupt sleep, communication, task performance, and productivity has recently been prepared for the World Health Organization (Berglund and Lindvall, 1995). This document reaches the following conclusions. Noise measures based only on energy summation are inadequate for the characterization of most noise environments, particularly when health assessment and prediction are concerned, and durations for L_{eq} measurements depend upon the activities involved. One must measure the maximum values of noise fluctuations, preferably combined with a measure of the number of noise events, and assess whether the noise contains a large proportion of low-frequency components. For dwellings, recommended guideline values inside bedrooms are 30 dB L_{Aeq} for steady-state continuous noise and 45 dB L_{AFmax} for a single noise event. To prevent a majority of the populace from being seriously annoyed during the daytime, the equivalent level from steady continuous noise in outdoor living areas should not exceed 55 dB L_{Aeq}. During the night, outdoor levels should not exceed 45 dB L_{Aeq} so that people may sleep with bedroom windows open. It is recommended that in schools, the level should not exceed 35 dBA during teaching sessions. For hearing-impaired children, a still lower level may be needed. The reverberation time in the classroom should be about 0.6 sec, and preferably lower for hearing-impaired children. In hospitals during nighttime, the recommended values for wardrooms are 30 dB L_{Aeq} together with 40 dB L_{AFmax}.

It must be mentioned that too low an ambient noise level will annoy some people, as sounds that they wish not to hear are not masked. Clearly, there is no universal solution to the problem of noise pollution. Annoyance will always be a problem, no matter how well other effects are controlled.

References

AAO Committee on Hearing and Equilibrium (1979). "Guide for the Evaluation of Hearing Handicap," *J. Am. Med. Assoc. 241*, 2055–2059.

ACGIH (1999). *1999 TLVs® and BEIs®: Threshold Limit Values for Chemical Substances and Physical Agents*, American Conference of Governmental Industrial Hygienists, Cincinnati, OH.

ACOM (1989). "Occupational Noise-Induced Hearing Loss," Am. College of Occup. Med., Noise and Hearing Conservation Committee, *J. Occup. Med. 31*(12), 996.

Attias, J. and Bresloff, I. (1996). "Noise Induced Temporary Otoacoustic Emission Shifts," *J. Basic Clin. Physiol. Pharm. 7*(3), 221–233.

Attias, J., Bresloff, I., Reshef, I., Horowitz, G., and Furman, V. (1998). "Evaluating Noise Induced Hearing Loss with Distortion Product Otoacoustic Emissions," *Brit. J. Audiol. 32*, 39–46.

Axelsson, A., and Coles, R. (1996). "Compensation for Tinnitus in Noise-Induced Hearing Loss," Ch. 33 in *Scientific Basis of Noise-Induced Hearing Loss*, edited by A. Axelsson, H. Borchgrevink, R.P. Hamernik, P.-A. Hellstrom, D. Henderson, and R.J. Salvi, Thieme, NY, 423–429.

Berger, E. H., Royster, L. H., and Thomas, W.G. (1978). "Presumed Noise-Induced Permanent Threshold Shift Resulting from Exposure to an A-weighted L_{eq} of 89 dB," *J. Acoust. Soc. Am. 64*, 192–197.

Berglund, B., and Lindvall, T. (eds.). (1995). *Community Noise* (Archives of the Center for Sensory Research volume 2 issue 1), Stockholm University and Karolinska Institute, Stockholm, Sweden. Available from World Health Organization, Geneva, Switzerland.

Borg, E., Canlon, B., and Engstrom, E. (1995). "Noise-Induced Hearing Loss: Literature Review and Experiments in Rabbits," *Scand. Audiol. Suppl. 40*, 1–147.

Burns, W., and Robinson, D. W. (1970). *Hearing and Noise in Industry*, Her Majesty's Stationery Office, London, England.

Cary, R., Clarke, S., and Delic, J. (1997). "Effects of Combined Exposure to Noise and Toxic Substances — Critical Review of the Literature," *Ann. Occup. Hyg. 41*(4), 455–465.

CHABA (1968). "CHABA Proposed Damage-Risk Criterion for Noise (Gunfire). Report of Working Group 57," National Academy of Sciences — National Research Council Committee on Hearing, Bioacoustics, and Biomechanics, Washington, DC.

CHABA (1993). *Hazardous Exposure to Steady-State and Intermittent Noise*, National Academy Press, Washington, D.C.

Fechter, L. D. (1995). "Combined Effects of Noise and Chemicals," *Occupational Medicine State of the Art Reviews 10*(3), 609–622.

Franks, J. R. , and Morata, T. C. (1996). "Ototoxic Effects of Chemicals Alone or in Concert with Noise: a Review of Human Studies," Ch. 35 in *Scientific Basis of Noise-Induced Hearing Loss*, edited by A. Axelsson, H. Borchgrevink, R. P. Hamernik, P.-A. Hellstrom, D. Henderson and R. J. Salvi, Thieme, NY, 437–446.

Glorig, A., and Roberts, J. (1965). *National Center for Health Statistics: Hearing Levels of Adults by Age and Sex*, Vital and Health Statistics, PHS Publication Number 1000 Series 11, No. 11, Public Health Service, U.S. Government Printing Office, Washington, D.C.

Glorig, A., Ward, W. D., and Nixon, J. (1961). "Damage Risk Criteria and Noise Induced Hearing Loss," *Arch. Otolaryngol. 74*, 413–423.

Hall, J. W. III, and Mueller, H. G. III (1997). "Otoacoustic Emissions (OAE)," Ch. 5 in *Audiologists' Desk Reference Volume 1: Diagnostic Audiology Principles, Procedures, and Practices*, Singular Publishing Group, Inc., San Diego, CA, 235–287.

Health Council of the Netherlands (1996). "Effects of Noise on Health," Ch. 3 of *Noise and Health*, as reprinted in *Noise News International 4*(3), 137–150.

Hellstrom, P.-A. (1996). "Individual Differences in Peripheral Sound Transfer Function: Relationship to NIHL," Ch. 10 in *Scientific Basis of Noise-Induced Hearing Loss*, edited by A. Axelsson, H. Borchgrevink, R. P. Hamernik, P.-A. Hellstrom, D. Henderson and R. J. Salvi, Thieme, NY, 110–116.

Herhold, J. (1977). "PTS beim Meerschweinchen nach Langzeitbelastung mit stationa rem Larm," (PTS in the Guinea Pig after Extended Exposure to Steady Noise) in Larmscjadem Fprscjimg 1977 (Proceedings, Meeting 14 April 1977 of Working Group "Larmschaden" of the KdT). Friedrich-Schiller-Universitat, Jena, Deutsche Demokratische Republik, 29–36.

Hetu, R., Dumont, L., and Legare, D. (1977). "TTS at 4 kHz among School-Age Children Following Continuous Exposure to a Broad Band Noise," *J. Acoust. Soc. Am. 62*, Suppl. 1, S96.

ISO (1990). "Acoustics — Determination of Occupational Noise Exposure and Estimation of Noise-Induced Hearing Impairment," International Organization for Standardization, ISO 1999: 1990, Geneva, Switzerland.

Intersociety Committee (1967). "Guidelines for Noise Exposure Control," *Am. Ind. Hyg. Assoc. J. 28*, 418–424.

Kraak, W. (1982). "Investigations on Criteria for the Risk of Hearing Loss Due to Noise," in *Hearing Research and Theory,* Vol. 1, edited by J. V. Tobias and E. D. Schubert, Academic Press, New York, 187–303.

Kryter, K. D., Ward, W. D., Miller, J. D., and Eldredge, D. H. (1966). "Hazardous Exposure to Intermittent and Steady-state Noise," *J. Acoust. Soc. Am. 39*, 451–464.

Kvaerner, K. J., Engdahl, B., Arnesen, A. R., and Mair, I. W. S. (1995). "Temporary Threshold Shift and Otoacoustic Emissions after Industrial Noise Exposure," *Scand. Audiol. 24*, 137–141.

McRobert, H., and Ward, W. D. (1973). "Damage-Risk Criteria: The Trading Relation Between Intensity and the Number of Non Reverberant Impulses," *J. Acoust. Soc. Am. 53*, 1279–1300.

Melnick, W. (1991). "Human Temporary Threshold Shift (TTS) and Damage Risk," *J. Acoust. Soc. Am. 90*, 147–154.

Morata, T. C. (1998). "Assessing Occupational Hearing Loss: Beyond Noise Exposures," *Scand. Audiol. 27* (Suppl. 48), 111–116.

Orchik, D. J., Schmaier, D. R., Shea, J. J. Jr., and Emmett, J. R. (1987). "Sensorineural Hearing Loss in Cordless Telephone Injury," *Otolaryngol. Head Neck Surg. 96*, 30–33.

Passchier-Vermeer, W. (1968). "Hearing Loss Due to Exposure to Steady-State Broad Band Noise," (IG TNO Report 35) Delft, Netherlands.

Patterson, J. H. Jr., Mozo, B. T., and Johnson, D. T. (1993). "Actual Effectiveness of Hearing Protection in High Level Impulse Noise," in *Noise and Man '93: Noise as a Public Health Problem, Proceedings of the Sixth International Congress*, edited by M. Vallet, Institute Nationale de Recherche sur les Transports et leur Securite, Arcueil Cedex, France.

Price, G. R. (1981). "Implications of a Critical Level in the Ear for Assessment of Noise Hazard at High Intensities," *J. Acoust. Soc. Am. 69*, 171–177.

Price, G. R. (1992). "Importance of Spectrum for Rating Hazard: Theoretical Basis," Ch. 31 in *Noise-Induced Hearing Loss*, edited by A. L. Dancer, D. Henderson, R. J. Salvi, and R. P. Hamernik, St. Louis, Mosby Year Book, 349–360.

Robinson, D. W. (1970). "Relations Between Hearing Loss and Noise Exposure, Analysis of Results of a Retrospective Study," in *Hearing and Noise in Industry*, edited by W. Burns and D. W. Robinson, Her Majesty's Stationery Office, London, England.

Robinson, D. W., Shipton, M. S., and Whittle, L. S. (1973). "Audiometry in Industrial Hearing Conservation – I," NPL Acoustics Report Ac 71, National Physical Laboratory, Teddington, England.

Royster, L. H., Royster, J. D., and Cecich, T. F. (1984). "An Evaluation of the Effectiveness of Three Hearing Protection Devices at an Industrial Facility with a TWA of 107 dB," *J. Acoust. Soc. Am. 76*(2), 485–497.

Royster, L. H., Royster, J. D., and Thomas, W.G. (1980). "Representative Hearing Levels by Race and Sex in North Carolina Industry," *J. Acoust. Soc. Am. 68*, 551–566.

Ruedi, L. (1954). "Actions of Vitamin A on the Human and Animal Ear," *Acta Otolaryngol. 44*, 502–515.

U.S. Air Force (1973). "Hazardous Noise Exposure: Air Force Regulation 161-35," USAF, Washington, D.C.

U.S. Department of Labor (1969). "Occupational Noise Exposure," *Fed. Regist. 34*, 7891–7954.

U.S. Department of Labor (1971). "Occupational Safety and Health Standards," *Fed. Regist. 36* (105), part 11.

Ward, W. D. (1962). "Damage-Risk Criteria for Line Spectra," *J. Acoust. Soc. Am. 34*, 1610–1619.

Ward, W. D. (1966). "The Use of TTS in the Derivation of Damage Risk Criteria for Noise Exposure," *International Audiol. 5*, 309–313.

Ward, W. D. (1973). "Adaptation and Fatigue," in *Modern Developments in Audiology*, edited by J. Jerger, Academic Press, New York, NY, 301–344.

Ward, W. D. (1976). "Susceptibility and the Damaged-Ear Theory," in *Hearing and Davis*, edited by S. K. Hirsh, D. H. Eldredge, I. J. Hirsh, and S. R. Silverman, Washington University Press, St. Louis, MO.

Ward, W. D. (1984). "Noise-Induced Hearing Loss," in *Noise and Society*, edited by D.M. Jones and A.J. Chapman, John Wiley and Sons Ltd., London, 77–109.

Ward, W. D. (1991a). "The Role of Intermittence in PTS," *J. Acoust. Soc. Am. 90*, 164–169.

Ward, W. D. (1991b). "Hearing Loss from Noise and Music," paper presented at the 91st convention of the Audio Engineering Society, New York, NY.

Ward, W. D., Fleer, R. E., and Glorig, A. (1961). "Characteristics of Hearing Loss Produced by Gunfire and by Steady Noise," *J. Aud. Res. 1*, 325–356.

Ward, W. D., Turner, C. W., and Fabry, D. A. (1983). "The Total-Energy and Equal-Energy Principles in the Chinchilla," Contributed paper, 4th Int. Congress on Noise as a Public Health Problem, Torino, Italy.

Welleschik, B., and Korpert, K. (1980). "Ist das Larmschwerhorigkeitsrisiko fur Manner grosser als fur Frauen?' (Is the Risk of Noise-induced Hearing Loss Greater for Men than Women?), *Laryngol. Rhinol. 59*, 681–689.

Welleschik, B., and Raber, A. (1978). "Einfluss von Expositionszeit und Alter auf den larmbedingten Horverlust," (The Influence of Exposure Time and Age on Noise Induced Hearing Loss), *Laryng. Rhinol. 57*, 681–689.

Yerg, R. A., Sataloff, J., Glorig, A., and Menduke, H. (1978). "Inter-industry Noise Study; The Effects Upon Hearing of Steady State Noise Between 82 and 92 dBA," *J. Occup. Med. 20*, 351–358.

The Noise Manual, revised 5th edition, edited by E.H. Berger,
L.H. Royster, J.D. Royster, D.P. Driscoll, and M. Layne
©2003 American Industrial Hygiene Association

6

Program Overview and Administration*

Andrew P. Stewart

Contents

---

* Brief portions of this chapter were previously published in "The Comprehensive Hearing Conservation Program," Chapter 12 of *Hearing Conservation in Industry, Schools, and the Military* (1988), edited by D. M. Lipscomb, Singular Publishing Group, San Diego, CA, pp. 203–230.

Overview of Occupational Hearing Conservation Programs

This chapter presents an overview of the elements of a comprehensive occupational hearing conservation program (HCP), as well as a focused look at the administration of HCPs. Readers are referred to the following chapters in this section for in-depth coverage of individual program components. This chapter will introduce the various elements of an HCP, interrelate these elements where possible, and provide an approach to organizing and administering an effective, successful HCP in an occupational setting.[1]

Program Elements

The basic elements of an occupational HCP include the following:

1. Noise surveys and data analysis.
    This phase involves the measurement of workplace noise levels and monitoring of worker noise exposure levels to identify potential noise-hazardous areas and actual worker overexposures. Various noise regulations (those of, e.g., the Occupational Safety and Health Administration [OSHA, 1983], the Mine Safety and Health Administration [MSHA, 1999], and the Department of Defense [DOD, 1996]) provide relevant permissible exposure limits and action levels that govern subsequent HCP activities based on obtained noise data.

2. Education and motivation.
    This phase involves education of workers and management about the workplace noise hazard and how the HCP prevents noise-induced hearing loss, as well as motivation of affected personnel to cooperate with the activities and requirements of the HCP.

3. Noise control.
    This phase involves engineering and administrative noise controls to reduce or eliminate hazardous workplace noise exposures.

4. Hearing protection devices (HPDs).
    This phase involves the selection and fitting of wearable devices to protect the hearing of overexposed workers, including user training and motivation, and management enforcement of consistent correct use of these devices.

5. Audiometric monitoring.
    This phase involves the performance and review of audiometric evaluations, with associated follow-up activity for any significant hearing change detected.

In addition to these five basic functional phases of an HCP, other associated program elements include recordkeeping for each phase, management supervision and "ownership" of the program, professional supervision, ongoing program evaluation, and documentation of regulatory compliance for each program phase. (For a brief

---

[1] Traditionally, writers and regulatory documents have referred to "industrial" HCPs and "employees" protected by those programs; this chapter, in acknowledging the existence and equal importance of military and municipal government HCPs, will refer to "occupational" HCPs and "workers" protected by them.

summary of the most widely used hearing conservation regulation, see OSHA Publication 3074, "Hearing Conservation," revised edition, 1995.)

No one approach to HCP organization and administration will work in all settings. The relative emphasis of program elements may vary, and some may be entirely eliminated. For example, in cases where successful use of engineering or administrative controls has reduced worker noise exposure levels so that no 8-hour time-weighted average (TWA) level meets or exceeds the action level, there may be no need for the audiometric and education phases of the HCP. In such cases, of course, there might be a need for temporary hearing protector use by workers entering usually unoccupied high-noise areas (in cases of extreme sound levels or unusual worker susceptibility or sensitivity to high sound levels), as well as for continuing maintenance attention to ensure that noise controls are effective, and that sound surveys are conducted to document their effectiveness. Also, individual needs and strengths of work facilities and available personnel resources, both professional and nonprofessional, will often necessitate the modification of the "classical" program model to meet local needs. This is completely appropriate, so long as all required program components are included and a suitable melding takes place that results in an effective program that protects the hearing of all noise-exposed workers.

Program Organization and Administration

In order to function effectively and achieve its goals, an occupational HCP, like any other workplace program, must be carefully organized and administered. A plan must be written that specifies program components, and a team of players must be assembled to carry out the various functions of the program. The organization and direction of this team is crucial to the HCP's success. In this chapter we will review the details of how HCPs can be set up so that they *actually work*, that is, so that they really prevent hearing loss in the workplace and not just comply with governmental regulations. A program that contains the minimal programmatic elements required by a regulation (e.g., the OSHA noise rule for general industry or an MSHA or DOD noise regulation) may be construed as being comprehensive from a legal standpoint, but it may be far from effective. Currently, no specific requirements are included in any of the widely used noise regulations to assess or guarantee effectiveness of program operation. A worksite is in compliance with a regulation if its HCP simply incorporates all of the required program components and activities. An effective program, on the other hand, should incorporate various activities and documentations that may not be included in the regulation but are nonetheless important to the proper functioning of an HCP.

The HCP Plan

There are a number of different approaches to managing a successful HCP. The choice of a particular approach will be based on a facility's individual needs and will be successful to the extent that the methods chosen fit the facility's management philosophy and all personnel within the facility are committed to the goals of the HCP. Whether the

HCP is being inaugurated as part of a comprehensive safety and health program in a new work facility or (a more common situation) is being reorganized after a period of lapsed management attention, the initial step is the same. A plan must be developed that states:

1. the activities to be carried out in the program,
2. delegation of responsibility to ensure that each activity is carried out, and
3. minimal specifications for performing the activities effectively.

This must be a working plan, designed with the assistance of the HCP team members who will be involved in each activity, faithful to current facility policy and governmental regulations, and amenable to restructuring as required by real-life experience and changing policy or regulation. Such a plan will serve as a blueprint for effective HCP functioning, even in the midst of changing management priorities and changing personnel. Table 6.1 lists the activities and concerns that might be addressed in a typical HCP plan.

TABLE 6.1
Activities addressed in a basic HCP plan.

Noise Exposure Monitoring
Determination of exposure monitoring criteria and schedules.
Calibration of sound measuring equipment.
Performance of area sound level and individual worker exposure measurements.
Calculation of individual TWA noise exposure levels.
Communication of TWA exposure values to workers.
Maintenance of proper documentation of worker exposure levels and other sound level measurements.

Education and Motivation
Development or selection of appropriate educational materials.
Performance of annual worker HCP educational session.
Assurance that all HCP workers participate in annual educational program.
Provision of ongoing worker education and motivation throughout the year.
Updating and replacement of educational program materials as needed.
Maintenance of proper documentation of educational program activities.

Engineering and Administrative Noise Controls
Identification of dominant sound sources that contribute most to worker noise exposures.
Assessment of feasibility of engineering and administrative noise controls.
Performance of time and motion studies for administrative noise control decisions.
Implementation of feasible noise controls.
Postimplementation assessment of success of controls implemented.
Ongoing assessment of control effectiveness over time.
Specification of acceptable noise output levels for machinery being considered for purchase.
Maintenance of proper documentation of engineering and administrative noise control efforts and results.

— continued on next page —

TABLE 6.1 — continued
Activities addressed in a basic HCP plan.

Personal Hearing Protection
Assessment of HPD attenuation adequacy for workplace noise exposures.
Assessment of HPD convenience and comfort in the work environment.
Determination of appropriate HPDs for specific work areas.
Maintenance of inventory of selected HPDs.
Fitting of HPDs in workers' ears.
Training and motivation of HPD users.
Checking of fit and condition of HPDs on periodic basis.
Enforcement of effective HPD use.
Auditing of HPD use compliance among workers and departments.
Specification of worker HPD use restrictions.
Training and supervision of HPD fitters.
Assistance to workers in solving HPD problems.
Maintenance of proper documentation of HPD fitting, training, and usage.

Audiometric Testing, Audiogram Review and Referral
Training of new and replacement OHCs.
Certification and recertification of OHCs.
Professional supervision of OHCs.
Selection of appropriate audiometric test equipment.
Performance of audiometer calibration checks.
Selection of suitable audiometric test areas.
Documentation of test areas meeting applicable background sound level criteria.
Scheduling and performance of worker audiograms.
Assurance that all HCP workers participate in annual audiometric program.
Administration of noise exposure, general health, and otologic history questionnaires to workers.
Performance of otoscopic inspections.
Timely professional review of audiograms.
Communication to workers of referral recommendations from audiogram review.
Counseling of workers with exceptional audiometric findings.
Implementation of professional reviewer's audiometric referral criteria.
Arrangement for worker referrals.
Follow-up on recommendations made by referral professionals.
Maintenance of proper documentation of audiometric data and review and referral activities.

The HCP Team

Because hearing conservation is clearly a multidisciplinary field and requires the contribution of many different persons, both professional and nonprofessional, a "team approach" is always needed to achieve an effective HCP. This position has been officially adopted by the Council for Accreditation in Occupational Hearing Conservation (CAOHC) and recommended to its course directors as the philosophy to use in developing occupational hearing conservationist (OHC) training courses (Suter, 1993). The make-up of the team will vary from facility to facility, depending on the size of the facility and the number of workers involved. Some of the team members will very likely be professional health, safety, or engineering personnel, such as clinical or occupational audiologists, occupational nurses, occupational or otological physicians, industrial hygienists, safety specialists, and acoustical engineers. Some of these may be outside consultants, and some may be facility staff personnel. But most of these will necessarily be part-time team members, owing to the specificity of their roles and activities. The more active key roles in the site's HCP will be filled by on-site personnel, such as occupational nurses, certified OHCs, human resource staff members, safety directors, production and supply room supervisors, and employee safety committees, and often by only one or two individuals in small- and medium-sized workplaces. And in every facility, no matter its size or place in the corporate structure, there must be a *key individual* (Royster and Royster, 1990) who has been given the responsibility and authority for ensuring that the HCP in that facility functions successfully. This person is (as the designation implies) the most important member of the HCP team, the one at whose desk the "buck stops." This person is approachable and friendly, knows individual employees by name, and is strongly committed to making the HCP work. S/he frequents the production floor, solicits workers' comments and complaints about the HCP, communicates with other team members, and uses all information obtained to communicate to management the need to make program changes. In many small facilities, this individual may be the *only* HCP "team member," at least in a formal sense. However, his/her dedication to this role, if it is strong enough and if management support and responsiveness are sufficient, will assure the program's success. Lack of a key individual at a facility, even if the so-called "HCP staff" numbers many individuals, including outside professional consultants and/or corporate health and safety staff, will almost certainly result in failure of the facility to achieve an effective HCP.

A critical component of the HCP team approach is that team members (including, and especially, production supervisors and other management representatives) be held accountable for their performance. They must be evaluated periodically on how well they carry out their HCP tasks, and the evaluation must be considered in their annual performance appraisals and ultimately reflected in salary adjustments. Similarly, the effectiveness of off-site professionals and consultants must be periodically evaluated in order to determine whether they are strengthening or weakening the HCP. Accountability is central to the effectiveness of the HCP team concept.

Choosing the HCP Team

Since not all facilities that must implement HCPs for their workers have equal access, whether on-site or through corporate or contractor services, to all of the potential HCP team members listed above, a variety of approaches exist to organize and staff HCPs. Generally, they all can be reduced to three basic approaches: (1) a totally local program, (2) a partly local/partly contracted program, and (3) a totally contracted program (an illusory approach which, unfortunately, is sometimes elected by management; see below). The choice of approach will be based on a number of different factors, such as available in-house personnel, level of training and experience of personnel, costs involved, management support, size of the corporation, number of facility locations, facility operating procedures, and availability of outside consultants and resources. Neither of the first two legitimate approaches is necessarily better than the other; the worksite's particular needs and preferences should be the final determinant of team make-up, although often the influence of corporate pressure is a major consideration. But it must be noted that, while the notion of the HCP "team" is quite flexible and its make-up quite different from location to location, no matter what approach is chosen, it is the level of interest and dedication of local personnel, especially that of the key individual, that renders the HCP effective or ineffective. All of the data and recommendations provided (at great cost or at no cost) by corporate or contract professionals must be put to use, day in and day out, to protect workers' hearing; it is up to the responsible person or persons at the worksite to do that, or it won't get done. Just as "all politics is local," so in the end is all hearing conservation local.

Totally Local Program

This method may be selected by manufacturing or mining facilities or by military installations where well-staffed medical and safety departments make it feasible to provide locally for all of the services and activities listed in Table 6.1. This approach requires that the worksite employ a full- or part-time audiologist, otologist, or occupational physician to provide at least audiogram review and referral services and perhaps other audiometric program-related services. Generally, if an audiologist is employed, s/he will perform other duties such as audiometer calibration checks, training of new OHCs, sound surveys and noise dosimetry measurements, and HPD attenuation and effectiveness checks. Some of these tasks can also be performed by other professional or nonprofessional personnel, including occupational health nurses, industrial hygienists, safety specialists, electronics technicians, and engineers. Of course, a key player would be the OHC, whose primary HCP tasks should include audiometric testing, employee education and motivation, and HPD fitting and training. The totally local approach to HCPs, even more than alternate approaches that involve outside consultants, requires careful coordination by management of HCP tasks and responsibilities, and an ongoing system of program quality control and effectiveness evaluation. The role of the key individual is especially important in this type of HCP, where s/he becomes the primary lobbyist of management to maintain ongoing attention to the HCP.

Partly Local/Partly Contracted Program

This appears to be the more commonly selected approach to staffing the HCP team. Facilities using this approach have one or more in-house personnel involved in the program (often a nurse, human resource staff member, or another nonprofessional staff member as a dedicated or part-time OHC, as well as a safety director and, occasionally, an industrial hygienist), but some of the services are provided by outside consultants, whether corporate staff members or contractors. Services commonly provided by consultants may include one or more of the following: noise exposure measurements, noise engineering assessments and controls, mobile audiometric testing, audiometric testing at a local private clinic or office, audiogram review, and otologic referral and examination of workers. Tasks commonly provided by in-house personnel include worker education and motivation, HPD fitting and training, and (often) on-site audiometric testing. Recordkeeping is usually a jointly shared responsibility. If mobile van or local clinical office audiometric testing services are contracted, often the vendor also provides worker education and HPD fitting and training sessions. Audiogram review and referral and HCP supervision and auditing services will normally be performed by a contracted audiologist or physician.

As stated above, outside consultants may be either private contractors or corporate staff members from the corporation that owns the local facility. The latter situation commonly occurs with larger corporations with extensive professional support within the company and with military service bases. It involves delegating responsibility for certain HCP tasks to the appropriate corporate department, such as medical, safety, industrial hygiene, training, human resources, and engineering. This method works well if suitable professionals are available and if they are all well trained in hearing conservation. For example, a number of companies throughout the U.S. have hired corporate or divisional industrial hygiene staffs to provide hazard assessment and remediation and toxicology services, and these professionals generally provide noise exposure measurement and noise engineering services as well. Similarly, a number of corporations have hired audiologists and otologists for the purpose of providing the audiogram review and referral portion of the HCP within the company. Corporate engineers are sometimes given the responsibility for assessing the feasibility of and for carrying out noise engineering controls. The corporate staff approach to HCPs often involves the use of networked computers (or compatible software in local individual computers) in order to facilitate in-house audiogram review and referral services, coordination of sound surveys and dosimetry measurements, noise engineering assessments and designs, and HPD program assessments and recordkeeping. If this capability is available, on-line or mailed diskette transmission of local facility data to corporate professional staff is often a possibility, resulting in potential savings in professional review and recordkeeping time and costs.

The corporate consultant approach to HCPs, although possible for larger corporations and the military services, is quite expensive in terms of salaries, benefits, and overhead costs, and for that reason has not been widely used in the U.S. The more common approach is for local facilities to contract with private consultants or HCP service firms to provide needed services.

Numerous combinations of responsibility for HCP activities are found in companies and industries. For example, sometimes the facility OHC performs baseline and retest audiograms, and an outside consultant provides audiogram review and mobile testing services for annual audiograms. This approach works well in very large facilities, where it would be extremely time consuming for the OHC to perform hundreds or thousands of annual audiograms, but the testing service could complete them easily in several days or weeks. Often, large corporations use a different approach or combination of approaches in various facilities, depending on the availability of corporate and private contractor consultants. However, in all situations where some HCP services are contracted out, just as in the case of totally local programs, it is essential that there be a *key individual in each facility* who has responsibility for "bird-dogging" the entire HCP. This person must be knowledgeable in all areas of the program and must work to integrate the various functions and activities so that the HCP effectively protects all workers' hearing. Constant attention and follow-up activity are needed, as well as frequent communication with outside consultants to ensure that activities are proceeding in a timely fashion and that the goals of the HCP are being met. The key individual at the facility may be the OHC who performs worker audiograms, or the safety director who performs the annual worker training program, or the industrial hygienist who performs noise exposure measurements — it really doesn't matter who this person is so much as how committed s/he is to the goals of the HCP. For example, if s/he is unwilling or is incapable of being a thorn in management's side, if that is required in order to advance the needs of the HCP, s/he should not function as the facility's key HCP individual.

Totally Contracted Program

A third approach to staffing HCPs is sometimes selected by facilities (usually smaller companies) wishing to avoid the expense and personnel resource problems associated with in-house HCPs. These facilities lack the necessary staff support and training to provide HCP services to workers and, instead, contract with outside consultants to provide all needed activities. These may be provided on-site (in the facility), as with mobile audiometric testing and worker training sessions, or they may take place off-site in the consultant's office or clinic. Generally, only audiometric testing, HPD fitting, worker education sessions, and audiogram review and referral lend themselves easily to off-site programming; and having even these services provided away from the facility entails numerous problems related to insurance, transportation costs, and expenses associated with increased worker time away from the job. Activities associated with noise surveys and engineering noise controls naturally must take place on-site at the facility.

Here it must be stressed that this approach is often really a nonapproach. Facilities that elect it have, knowingly or unknowingly, abrogated their responsibility to provide an effective HCP. It is impossible to "buy an HCP off the shelf" and opt thereby for no local staff involvement. There *will* be local involvement, but it will result in an inferior program because of the management mindset betrayed by the decision to "farm it all out." Someone at the facility must deal with the audiometric results returned by the consultant, including complying with regulatory requirements for employee notification and follow-up activity. Someone must educate and motivate workers about hearing conservation throughout the year. Someone must fit HPDs and train users in their proper use and care. Someone must motivate and enforce correct HPD use. Often, no one at the facility has the training or experience (or even the interest) to carry out any of these tasks, and so they will not be performed. The resulting HCP, no matter the quality of the contracted services, will be ineffective.

Coordinating the HCP Team

Once the HCP team has been selected and duties assigned to each member, how is the program coordinated to ensure that assigned duties are performed and the program is effective in preventing occupational hearing loss? This is where the key individual plays a critical role. No matter what his or her formal title, the key individual ensures the success of the HCP by acting as the de facto coordinator of the program, the person who intimately knows the program's functions and limitations and acts as the conduit of information up and down the chain of communication in order to effect needed changes. This person must have management's full support and authority, so that when HCP problems occur they are dealt with expeditiously and thoroughly. It matters less *who* this person is than *that* s/he exists and accepts the responsibility of coordinating the HCP. Too many HCPs are rudderless and ineffective because of the lack of a key individual.

The program coordinator must ensure that each of the activities specified in the facility's HCP plan is carried out in a timely, effective, and economical manner. This involves selecting personnel to perform each task, arranging for or providing training where it is needed, communicating management's philosophy and expectations to contractors, supervising the ongoing activity of all team members, evaluating the performance of team members, and reporting HCP performance to management. Regular face-to-face meetings of team members are essential, as these provide opportunities to schedule program activities at times acceptable to all concerned (ensuring their successful completion), share experiences and problems, report on program evaluations, propose changes or improvements, and shift staff responsibilities as needed. Such meetings can result in solutions that will actually work, since all concerned team members have shared in their generation. In-house team members (such as the facility manager and department superintendents, occupational nurses, OHCs, human resource staff members, safety directors, engineers, production supervisors, and worker safety committee members) should meet together as often as necessary to ensure smooth operation of the HCP (e.g., monthly, at least until program effectiveness is well established, and then perhaps every other month or quarterly). The entire team (including outside contractors, to the extent feasible) should meet annually for an HCP review and problem-solving session.

A key part of coordinating the HCP is managing the services of outside contractors, whether corporate or private. Since these individuals generally provide specific, defined services (e.g., sound surveys, audiograms, worker-training sessions) their understanding of the facility's total HCP dynamics and needs is usually limited. Often, their contribution to the HCP is offered "in a vacuum," without much effort to coordinate it with other program components. If the contract service is not carefully integrated into the overall HCP, the information provided will, as often as not, remain unused or at least underused. This important follow-up task is the responsibility of the key individual. Information obtained must be shared with other in-house HCP team members in a timely fashion, and implications for program modification must be explored at the next team meeting. For example, sound survey or dosimeter data must be carefully digested, with a view toward implementing reasonable engineering or administrative noise controls or perhaps altering the inclusion of certain departments in the HCP. Similarly, the results of annual audiogram reviews must generate certain follow-up activities, such as audiometric retests and refitting of HPDs, which depend for their accomplishment on the understanding and cooperation of other team members, especially line supervisors. Recommendations resulting from annual HCP audits often involve actions by other team members.

HCP Recordkeeping

Recordkeeping is often a weak link in an HCP. Management and staff usually give more attention to setting up the program, staffing it, arranging for the mechanics of it, buying the necessary equipment and supplies, and so forth, than they do to documenting the activities and results of the program. Governmental noise regulations rarely specify recordkeeping requirements detailed and specific enough to result in effective documentation. An adequate recordkeeping system is one that documents all significant interactions between noise-exposed workers and HCP team members, as well as between workers and the noise environment in which they work. Records are sufficient when it is possible to trace workplace and off-the-job activities that could result in deteriorated hearing, as well as make recommendations for diagnostic or therapeutic follow-up, and monitor worker compliance with these recommendations. In short, recordkeeping is the glue that holds the HCP together. It links each of the component phases of the HCP with every other part by documenting what has been done and, by implication, what still needs to be done. For example, the noise survey and data analysis phase results in records of worker noise exposure, which in turn suggest engineering and administrative control efforts that need to be considered, as well as worker audiometric testing and education/motivation sessions that need to be scheduled. Each of these latter activities results in data that need to be recorded, which may lead in turn to still other activities, such as referral for otologic evaluation, HPD refitting/retraining, and possible recording on injury/illness logs.

Although recordkeeping is sometimes considered to be a separate functional phase of the HCP, in reality it is an integral part of each of the different phases of the HCP. The program supervisor or key individual must ensure that each HCP team member maintains the records associated with his/her duties and that all relevant records are available on a need-to-know basis to other team members, except as limited by confidentiality requirements.

For specific recommendations about HCP records that should be maintained, see the following chapters in this section on particular HCP phases and activities.

Besides facilitating the performance of HCP activities and thereby fostering protection of workers' hearing, another important use of recordkeeping is documenting activities and program results in order to protect the company or facility from liability with regard to regulatory compliance, worker compensation, or third-party legal activities. If, for example, an OSHA inspection takes place, the only way for the facility to demonstrate that all required program activities were performed is through adequate documentation of activities related to the five phases of the HCP. If a worker files a compensation claim alleging occupational hearing loss, the only way for the facility to mount an adequate defense or to be aware of the legitimacy of the claim is, again, through adequate documentation of what was done during the worker's tenure of employment.

However, in recordkeeping, as in all phases of the HCP, it is important to remember that the interactions to be recorded are between people, and not forms (Gasaway, 1985). The prevention of occupational hearing loss is a dynamic, interactive process in which people communicate with people; it is the effectiveness of that communication that determines how successful the HCP is in preventing hearing loss.

Evaluation of HCP Effectiveness

An effective HCP is one that accomplishes the goals established for it. The primary goal of an occupational HCP must be the prevention (or, at least, limitation) of permanent hearing loss associated with exposure to occupational noise (Royster and Royster, 1990) or to other environmental factors in the workplace that may interact with noise, such as ototoxic industrial chemicals (Morata et al., 1993) or vibration (Hamernik et al., 1989; Pekkarinen, 1995). Other goals may be formulated in addition to this primary goal, such as compliance with governmental regulations, reduction of worker stress and absenteeism, reduction of accidents due to workplace noise levels, and reduction of the facility's liability to worker compensation claims for occupational hearing loss (Franks et al., 1996). However, the primary goal remains the prevention of noise-induced permanent threshold shift (NIPTS) caused by workplace exposure.

Historically, two approaches have been used to assess the effectiveness of HCPs. The more basic approach is to define a program as effective if it can be shown to be *complete*, that is, if it contains all of the elements thought to be necessary for an effective HCP, and if all of its required tasks have been completed. For example, did all workers included in the HCP receive their annual audiograms and attend the annual educational program this year? Has a system been set up to evaluate the need for and feasibility of engineering and administrative control of worker noise exposure? Has the key individual received requested feedback from workers about the new earplug they were evaluating? Or, most basically, have all aspects of the relevant regulations, such as the OSHA hearing conservation amendment (OSHA, 1983) been fulfilled? Checklists and audits are usually used to answer these and other "completeness" questions. This approach may provide much useful information. However, even if a program is well-structured, complete, and all of its scheduled activities "up-to-date," its success in preventing hearing loss is not assured.

Various sources of contamination and error, as well as inferior quality of implementation, exist in HCPs, which may confound the efforts of program personnel to carry them out effectively (Melnick, 1984). These may be found in all elements of the HCP and include such factors as inaccurate assessment of worker noise exposures; improper fitting and use, as well as deterioration, of hearing protectors; calibration errors in noise measuring equipment and audiometric instruments; excessive ambient noise levels in audiometric areas; the use of nonstandard or erroneous audiometric techniques and omission of required or recommended audiometric follow-up activities; and insufficient education and motivation of workers and OHCs. These and similar problems will prevent an apparently complete HCP from being effective.

The second approach, which arose out of the inadequacies of the first, is to define a program as effective if it works, that is, if it succeeds in preventing or limiting the occurrence of NIPTS in a worker population. In recent years a number of researchers (e.g., Melnick, 1984; Royster and Royster, 1990) have stressed the need for some type of outcome measurement using empirical, objective data to document whether an HCP is successful in this way. They generally have used the findings of the annual audiometric evaluations as the source of these data, analyzing the variability of group audiometric data to discover trends for departments or other groups of workers (see draft American National Standard S12.13-1991, *Evaluating the Effectiveness of Hearing Conservation Programs*, for recommended procedures). The advantage of this approach is that HCP problems can be detected early, before a large number of individual workers have developed significant changes in hearing.

It is clear that, although programmatic completeness is an essential element of effective HCPs, criteria based on evaluation of worker audiometric results must also be used to establish definitively the effectiveness of individual HCPs. For an in-depth discussion of assessing HCP effectiveness, the reader is referred to Chapter 12.

Containing Costs and Maximizing Effectiveness: The Current Dilemma

As with every other facet of modern life, occupational HCPs must somehow be shown to "pay their way," or, in a time of downsizing and program elimination, they will become unacceptable in the workplace. This will occur despite the fact that HCPs are "required" by federal or state occupational health and safety regulations. That is to say, management (whether industrial, military, or municipal) will find a way to eliminate what they view as a "wasteful" or "unnecessary" or "too-costly" enterprise, not by ending the program through a formal policy but by attenuating its effectiveness through neglect or fiscal constraint. Simply put, the HCP will continue to exist on paper, but its viability will be severely reduced. And the most unfortunate aspect of this scenario is that the HCP costs nearly as much to carry out poorly as to carry out well. The fixed costs are there, but there is no payback for an inferior program; in fact, the opposite is true: losses may increase because of increased prevalence of worker hearing loss. An effective program pays back its costs in the prevention of hearing loss.

There is no question that most HCP activities in the workplace are inherently expensive. Engineering and administrative noise controls, noise monitoring equipment, audiometric testing, program supervision, worker referral for follow-up clinical examination, equipment calibration, HPD provision — all of these cost money and produce no direct, immediate profit to the facility. Unless management can be persuaded that the indirect benefits are tangible and ultimately profitable for the facility, in terms of reduced medical costs, reduced incidence of compensation claims, reduced liability to regulatory citations, enhanced worker satisfaction with consequently reduced absenteeism, and so forth, they will gradually reduce their support of the HCP, particularly if no "catastrophic" outcomes (compensation claims, regulatory citations) have occurred within recent corporate memory.

So the task for the OHC, the HCP supervisor, and the key individual in the facility is to be as fiscally responsible as possible with the delivery of HCP services, without sacrificing necessary activities or oversight in a way that would lead to program ineffectiveness. How can this be done? First, the supervisor should carefully distinguish between required HCP activities, activities needed for effectiveness, and superfluous activities. Many program activities are sold to management by inexperienced, naive professionals as being "required by OSHA" or "essential to a complete HCP" that really are neither required nor essential. Examples include unnecessary audiometric tests, unnecessary medical or audiological referrals, unnecessarily large selections of HPDs, high-cost custom-molded earplugs or other expensive HPDs, and unnecessary or unworkable engineering noise control efforts. It is important that facility management and HCP staff realistically examine the credentials and experience of service providers and (if in doubt) obtain more than one estimate for products and services to be sure that they are buying what they actually need and that what they buy will really enhance the effectiveness of the HCP. Nor should they mistakenly assume that the least-costly bid for services is going to turn out to be the most cost-effective. Inferior or inadequate services, which have to be repeated or which result in increased loss liability, may ultimately prove to be far more costly than the inexpensive up-front fee suggested.

To assist facility HCP team members in selecting high-quality services from contract providers, the facility's HCP plan should specify the quality of contracted services expected and how the key individual is to evaluate whether this quality has been achieved. For example, if a local clinic is providing audiometric tests, the HCP plan should specify the need for its providing reliable data (i.e., test–retest reliability of measured hearing threshold levels should be high in the absence of hearing change), as well as providing documentation of annual checks of audiometer calibration and testing area sound level adequacy, and documentation of training/certification status of all OHCs used by the clinic. If a consulting audiologist is used to review audiograms, the HCP plan should specify this individual's credentials and/or experience, as well as the types of data analysis and referral activities expected, and acceptable turn-around time for audiogram review.

Second, fiscal responsibility is enhanced by insisting on the continuing need for competent professional supervision of the entire HCP. This includes oversight of the audiometric program by a knowledgeable, experienced audiologist or otolaryngologist (nothing will reduce the effectiveness and credibility of the HCP more quickly in both management's and workers' estimation than a poorly run, inaccurate audiometric program) and oversight of the entire HCP by the same professional or by an experienced industrial hygienist, safety professional, or other qualified individual. This is particularly important to remember in a time when many facilities are turning to the use of in-house computer-assisted review of audiometric, worker education, and HPD programs as a means of achieving HCP cost savings. The use of an internal software program to sort and store worker health data cannot obviate the need for informed professional interpretation of test results and guidance with regard to regulatory compliance, referrals, and other follow-up activity.

Third, sometimes enormous cost savings can be realized by careful attention to the use and care of HPDs by workers. In many facilities, workers use even premolded earplugs only once or twice before "losing" or disposing of them, while in reality even foam earplugs can often be re-used several times if properly handled. Providing workers with carrying cases for earplugs can minimize their loss, workers can be trained and motivated to take better care of their HPDs, and thus the annual cost of HPDs can be substantially reduced.

Other examples could be produced, but those cited above are sufficient to demonstrate the need for responsible use of facility resources in conducting the HCP. Such efforts by HCP team members will maintain respect for and adequate fiscal support of the program on the part of management without compromising effectiveness.

Conclusion

Although an effective HCP cannot be guaranteed by the presence of each of the elements discussed in this chapter, any program that does not include all of the elements will doubtlessly be ineffective. The actual degree of effectiveness experienced by a work facility in its HCP depends on: (1) the commitment of management to adequate implementation; (2) the enthusiasm and interest shown by OHCs, whether local facility or service provider employees, in each worker's hearing health; (3) the competence and dedication of each member of the HCP team, particularly that of the key individual; and (4) the motivation of each worker to protect his or her own hearing. Hearing conservation can work, but it must be vigorously and continuously supported by management, affected workers must do their part, and the program must include the technical assistance of experienced, competent professional supervisors to guarantee that each aspect of the HCP is functioning with maximum effectiveness. If this is accomplished on a wide scale in industry, the military, and municipalities, occupational hearing loss with its needless costs in human and economic terms will eventually become a thing of the past.

References

ANSI (1991). "Draft American National Standard for Evaluating the Effectiveness of Hearing Conservation Programs," American National Standards Institute, S12.13-1991, New York, NY.

DOD (1996, April 22). *DOD Hearing Conservation Program*, Department of Defense, Instruction 6055.12.

Franks, J. R., Stephenson, M. R., and Merry, C. J. (eds.). (1996). *Preventing Occupational Hearing Loss — A Practical Guide*, National Institute for Occupational Safety and Health, Cincinnati, OH.

Gasaway, D. C. (1985). *Hearing Conservation: A Practical Manual and Guide*, Prentice-Hall, Englewood Cliffs, NJ.

Hamernik, R. P., Ahroon, W. A., and Davis, R. I. (1989). "Noise and Vibration Interactions: Effects on Hearing," *J. Acoust. Soc. Am.* 86(6), 2129–2137.

Melnick, W. (1984). "Evaluation of Industrial Hearing Conservation Programs: A Review and Analysis," *Amer. Ind. Hyg. Assc. J.* 45, 459–467.

MSHA (1999). "Health Standards for Occupational Noise Exposure; Final Rule," Mine Safety and Health Administration, 30 CFR Part 62, 64 *Fed. Reg.*, 49548–49634, 49636–49637.

Morata, T. C., Dunn, D. E., Kretschmer, L. W., Lemasters, G. K., and Keith, R. W. (1993). "Effects of Occupational Exposure to Organic Solvents and Noise on Hearing," *Scand. J. Work Environ. Health 19*(4), 245–254.

OSHA (1983, March 8). "Occupational Noise Exposure; Hearing Conservation Amendment; final rule," 29CFR 1910.95 *Fed. Regist. 48*(46), Occupational Safety and Health Administration, 9738–9785.

OSHA (1995). *Hearing Conservation* (OSHA 3074), Occupational Safety and Health Administration, revised edition, Washington, DC.

Pekkarinen, J. (1995). "Noise, Impulse Noise, and Other Physical Factors: Combined Effects on Hearing," *Occ. Med.: St. of the Art Rev. 10*(3), 545–559.

Royster, J. D., and Royster, L. W. (1990). *Hearing Conservation Programs: Practical Guidelines for Success*, Lewis Publishers, Inc., Chelsea, MI.

Suter, A. H. (1993). *Hearing Conservation Manual* (3rd ed.), Council for Accreditation in Occupational Hearing Conservation, Milwaukee, WI.

The Noise Manual, revised 5th edition, edited by E.H. Berger,
L.H. Royster, J.D. Royster, D.P. Driscoll, and M. Layne
©2003 American Industrial Hygiene Association

7

Noise Surveys and Data Analysis

Larry H. Royster, Elliott H. Berger, and Julia Doswell Royster

Contents

Introduction

An Overview

The primary goal of this chapter is to present guidelines for the development and execution of the sound survey phase of a company's hearing conservation program (HCP) (L.H. Royster, Royster, and Berger, 1982). Information contained in other chapters of this manual should also be considered, especially Chapter 3 on instrumentation, as well as the general literature (Alpaugh, 1975; ANSI S12.18-1994; ANSI S3.44-1996; ANSI S12.19-1996; ANSI S1.14-1998; Gasaway, 1985; Wells, 1979). In the opinion of the authors of this chapter there does not exist any one set of sound survey guidelines that would be appropriate for all of U.S. industry (Royster and Royster, 1984). Therefore, it is important for the sound surveyor to consider existing local characteristics, needs, and constraints before developing and implementing the sound survey phase of the HCP.

The sound surveyor must also keep in perspective the importance of the sound survey efforts as they relate to the other phases of the company's HCP. To allow one phase of the HCP, such as the sound survey phase, to operate independently of all other phases could result in the sound surveyor's spending excessive funds on possibly unnecessary equipment or other items, when these monies could have been put to better use in presenting educational programs, purchasing more adequate audiometric instrumentation, or obtaining additional training for other HCP personnel.

Political Considerations

The sound surveyor must recognize that the organization's political climate is determined by the sum of many individual motivations. It is important that the sound surveyor be sensitive to this climate, listening to the concerns of all affected parties but not making early commitments or expressing preliminary judgments of environmental conditions or actions that may be necessary.

The sound surveyor should take the time to express an outwardly friendly attitude toward the workers and give them an opportunity to voice opinions or concerns about the noise environment. This approach will minimize both the misuse of survey equipment and the likelihood that resentful workers may find ways to significantly alter the noise environment. It will also provide helpful insights and information. The result will be more accurate estimates of actual worker time-weighted average (TWA) or 8-hour equivalent equal-energy (L_{A8hn}) noise exposures.

The sound surveyor should also consider the interests and needs of other personnel who may be affected by the results of the survey. For example, production managers and front-line supervisors are often concerned about the potential impact that the sound survey findings may have on production efficiency and annual personnel evaluations. Management, supervisors, and employees in work areas without a requirement to wear hearing protection devices (HPDs) will be watching with interest. Each of these groups has been observed to attempt to alter the noise environment, such as by shifting noisy work efforts to second or third shifts when sound surveys are not normally conducted in order to project a more favorable climate.

The preceding discussion indicates the importance of considering the political implications of a sound survey. A moderate effort on the political front during the planning stage can have a very positive impact, and can potentially reduce the level of effort necessary to obtain reasonable results. The sound surveyor should have a strong interest in the content of the company's educational and motivational program (Chapter 8). This phase of the HCP, if properly developed and implemented, can assist the sound surveyor in preparing the workers not only to cooperate in sound surveys, but also to participate in the company's overall HCP (Royster and Royster, 1985a).

Why Conduct a Sound Survey?

Estimate Potential Noise Hazard

The principal reason for conducting a sound survey should be to establish the noise environment's potential for producing a permanent noise-induced hearing loss. Therefore, it is essential that the sound survey result in a database of noise exposures (TWAs or L_{A8hn} values) that accurately describes the population's noise exposure hazard (see *Statistical Factors in Sound Survey Methods*).

OSHA and Other Government Regulations

One of the strongest incentives over the past several years for U.S. industry to conduct sound surveys has been federal regulations, the most important being the Occupational Safety and Health Administration (OSHA) Noise Standard and subsequent Hearing Conservation Amendment (HCA) (OSHA, 1974, 1983, and 1984; see Appendix I). The sound survey database provides information to satisfy the legal requirements of OSHA and other similar governmental regulations.

Input to Company's HCP

Sound survey results will also be utilized in making important decisions with respect to all phases of the company's HCP, especially identification of the employees that will be included in the program. In addition, decisions made by management in selecting the HPDs to be offered to employees are often based on the findings of the survey, especially for work environments where the employee's TWA is 100 dBA or higher (Royster and Royster, 1985b).

The sound survey results assist management in selecting work areas for possible engineering noise control efforts. The information generated by sound surveys is also useful when one attempts to explain to workers why they are, or are not, included in the HCP, or why a particular HPD is not acceptable for a workstation or job classification. The results of the sound survey must be considered in making other administrative decisions, such as selecting a location for the audiometric test facility.

In summary, the sound survey provides the necessary information to solve many of the typical problems that arise on a day-to-day basis in running an effective HCP.

ACGIH and Foreign Regulations

In addition to the OSHA regulations, the American Conference of Governmental Industrial Hygienists (ACGIH, 1998) has published threshold limit values (TLVs®) for noise exposure that differ significantly from the present OSHA criteria. Some industrial hygienists and companies may wish to adhere to these criteria, which are more stringent in some situations. ACGIH's threshold limit values use 85 dBA as the criterion level (representing 100% noise dose for an 8-hour exposure) and a 3-dB exchange rate (equal energy) rather than the OSHA 90-dBA criterion level and 5-dB exchange rate. Therefore, if a company elects to adopt the ACGIH noise exposure criteria, survey instrumentation must be set up to collect the necessary information. The noise criteria of many foreign countries also use the 3-dB exchange rate and a 85-dBA criterion level. Since many U.S. companies have total or partial foreign ownership or operate divisions outside the U.S., professionals in the U.S. are often asked to compare the impact of using foreign noise-exposure criteria vs. the existing OSHA criteria. The sound survey data, if properly collected, can provide the necessary information to address this question. For a discussion of how different criteria might impact the worker's estimated daily noise exposure, see *Different Noise Exposure Criteria and Exchange Rates*.

Workers' Compensation Claims

Workers' compensation for noise-induced hearing loss provides an additional incentive for U.S. industry to conduct sound surveys (Berger, 1985). However, the type of data required for OSHA compliance and for workers' compensation purposes may differ. As an example, when conducting a sound survey to satisfy OSHA requirements, the surveyor is mainly attempting to determine the employees' TWAs, i.e. equivalent daily exposures. As a consequence, when TWAs are less than the 85-dBA OSHA action level, the sound surveyor may fail to maintain detailed data about the sound-level variations. However, in at least one state, employees can file for compensation for noise-induced hearing loss if they are exposed to a sound level of 90 dBA or greater regardless of the duration of exposure. Of course, the fact that a TWA is below 85 dBA does not ensure that it came only from exposure to sound levels below 90 dBA. Therefore, failure to maintain adequate data for environments that do not mandate employee inclusion in the company's HCP could lead management to overlook potential problem areas, and can also result in inadequate documentation for legal purposes.

Safety Considerations

Another reason for conducting sound surveys is to investigate potential safety hazards related to employee communication and detection of warning signals. At all employee workstations, whether or not the daily TWA exceeds the OSHA action level, it is essential that acoustic warning signals be detectable above the background sound level. The data measured during the sound survey can provide part of the information necessary to estimate the adequacy of the company's audio warning and/or communication system (see Chapter 14).

Recently it has become very popular for employees to request to listen to personal radios. Management should decide whether the use of personal radios will create a potential hazard such as hearing damage, speech interference, or masking of warning signals. Again, sound survey data will be required to estimate the potential noise hazard and provide recommendations to management (see Chapter 10, *Recreational Earphones*).

Special Requests

The individual responsible for conducting the company's sound surveys is frequently asked to make additional measurements that may seem a little bit out of the ordinary, but are still part of the job. Examples include requests to survey the production manager's office or conference room, the company cafeteria, a special secretarial office, or even the engine room of the boss' private boat. Other examples include sound surveys of the computer room and similar areas that are not normally included in the company's HCP. Although some of these requests may seem inappropriate, the sound surveyor should keep in mind that checking the sound levels in these areas and explaining the results is in fact an extension of the company's educational program. These types of requests indicate an acceptance, on the part of management and other individuals requesting the service, of the fact that noise has its effects and that hearing conservation is important (see Chapters 5, 13, 14, and 15).

Classifying Noise Exposures

Recommended Classification Scheme

The basic goal of most sound survey efforts is to determine TWAs for selected workers or for a particular workstation or job classification. There may be other goals as well, such as determining the typical L_{OSHA} values for work areas or identifying the dominant noise source(s) for engineering noise control purposes, but for the majority of sound surveys conducted in general U.S. industry, the goal is employee noise exposure determination. In order to provide guidance for the sound surveyor as to the level of effort and accuracy necessary in determining employee TWAs, a practical scheme for grouping the predicted TWAs is desirable.

The classification scheme recommended is presented as Table 7.1 (Royster and Royster, 1985b). Five exposure ranges are used for grouping the estimated TWAs. Schemes that employ a larger number of categories will not ensure better employee protection and at the same time are cumbersome and impractical, often creating unnecessary administrative burdens on the nurse, the audiometric technician, and other company personnel. Additional reasons for limiting the classification scheme to the five ranges shown in Table 7.1 include the inability to estimate real-world TWAs more accurately than the defined 5-dBA ranges, the variability of real-world levels of protection provided by HPDs, and the fact that approximately 97% of all industrial TWAs are less than 100 dBA (Bruce et al., 1976; OSHA, 1981; Royster and Royster, 1994).

TABLE 7.1
Recommended scheme for classifying TWAs for sound survey purposes.

TWA (dBA)	Classification
84 or below	A
85–89	B
90–94	C
95–99	D
100 or above	E

We have observed situations and reviewed reports in which individuals spent far too much time and resources in trying to determine employee TWAs to an accuracy of ±1 dBA. The use of a classification scheme such as the one presented in Table 7.1 should minimize attempts by sound surveyors to predict TWAs to unreasonable levels of accuracy, and will therefore encourage more effective utilization of the limited available resources.

The lowest exposure classification A, as shown in Table 7.1, includes nonhazardous work areas such as computer rooms and similar work environments in which the employees may request hearing conservation related information and HPDs, but where the employees should not be identified as part of the company's HCP for OSHA purposes (although they should be classified as part of the company's wellness program or some other similar non-OSHA related classification category, see Chapter 11). As pointed out in Chapter 10, it may be necessary to limit use of hearing protection for some employees in low noise level environments based on safety considerations (communication problems, warning signal detection, etc.). However, exposure classification A might also include employees who have brief exposures to higher sound levels which might require mandated use of HPDs while the affected workers are in HPD-required areas.

Exposure classification E includes all noise exposures of 100 dBA or greater because the magnitude of real-world protection provided by most HPDs (see Chapter 10) is potentially inadequate at this level. As a consequence, HCP personnel should give special attention to the portion of the workforce classified as E.

Alternative Classification Schemes

The Blanket Approach

An alternative to a multi-tier classification scheme is to assign one TWA range to all employees in a given department or plant, based upon some statistical or administrative criteria (see *Statistical Factors in Sound Survey Methods*). To illustrate this type of blanket approach, assume that in a production area, 70% of the employees are exposed to TWAs of 85–89 dBA, while 20% have exposures less than 85 dBA and the remaining 10% have exposures in the 90–94 dBA range. The blanket approach would require that the total work area be classified as C, with the result that all employees working in this area would be included in the HCP and be required to wear HPDs. On each employee's audiometric record, a TWA of 90–94 dBA would be indicated.

Now, what are the potential problems and benefits from using the blanket approach? The most obvious problem is that a large segment of the workforce would be required to wear HPDs in relatively low levels of noise. For these workers, a significant percentage could experience communication-related problems, especially employees with significant preexisting hearing loss. Workers who are forced to use HPDs in the lower noise level environments, i.e., TWAs less than 90 dBA, are more resistant to using HPDs, with the result that management typically will not strictly enforce HPD use (Royster and Royster, 1985b).

A second problem with the blanket approach is that the noise hazard will be exaggerated for a high percentage of the workforce, leading the affected employees to believe that the hazard is greater than it really is. If the employees who are not wearing their HPDs properly are not being flagged by the HCP's audiometric evaluations because the noise hazard was overstated, then these workers will begin to downplay the importance of the HCP. If in the future they are moved to work areas where the noise hazard is in fact serious, they may fail to participate willingly in the company's HCP as a consequence of their previous experience.

A third problem created by the blanket approach concerns the need to indicate a TWA on each employee's audiometric record. Since the blanket approach may result in an inflated TWA, the company's ability to properly judge the level of protection being provided to the noise-exposed work force (by type of HPD utilized, comparison of hearing threshold level changes, etc.) through analysis of the company's audiometric database (see Chapter 12), may be significantly reduced. Thus, for example, when an employee exhibits a significant threshold shift (or a standard threshold shift [STS] if minimum OSHA requirements are followed), the professional who reviews employee audiometric records may be misled into assuming that the indicated shift is due to on-the-job noise exposures, when in fact the shift may have been due to off-the-job noise exposures or other factors such as advancing age (presbycusis).

Now, what are some of the benefits of the blanket approach, or classifying the work area as C? The benefit most often stated is avoidance of the problem encountered when an employee who works in an area classified as B is identified as having a significant threshold shift and is required to utilize HPDs. The comment is, "The supervisor can't enforce the use of hearing protection by one or two employees when the majority of the employees do not have to use them." Although there is some truth in this statement, it is also true that in many work areas different employees must use different types of safety equipment (such as eyeglasses and safety shoes) depending on the piece of machinery operated or the location of their respective workstations. Why should the use of hearing protection be different? Why penalize, in this instance, 90% of the workers because of the hesitation to enforce the use of hearing protection by 10% of the workers?

A second stated benefit is that the single-number area classification scheme reduces the administrative difficulty of establishing several different classifications for the work area and having to record appropriate values on different employees' audiometric records each year. A third benefit is that the sound survey effort is significantly reduced since the number of samples needed is lowered.

Individual Worker Classification

The opposite of blanket classification of work areas is the establishment of a TWA for each worker. This approach is often justified based on the assumption that an accurate knowledge of each worker's TWA will result in significantly greater protection for the workforce. In addition, sound surveyors have attempted to imply that an "accurate" estimation of each worker's TWA will somehow result in a lower potential compensation cost to the company for on-the-job noise-induced hearing loss. However, data in the literature and experience of the authors do not support these claims. Indeed, the models that exist for predicting hearing loss, level of protection provided by HPDs, and the hazardousness of a particular noise environment have no greater accuracy than the 5-dBA ranges found in Table 7.1.

Another reason for limiting the effort expended in predicting individual TWAs is the significant mobility and changing noise exposure histories of employees who switch between job functions on a yearly, weekly, and even daily basis. Unless a realistic classification scheme is selected, HCP personnel will spend unnecessary effort and funds in trying to define TWAs to a much greater accuracy than is warranted.

General Classification Guidelines

To illustrate our recommendations, consider the following two examples. Assume that on a work floor 80% of the workforce is exposed to an estimated TWA of 85–88 dBA and is scattered throughout the total work area, while 10% have TWAs below 85 dBA and 10% have TWAs estimated in the 91–94 dBA range. The 10% of the work force with TWAs of 91–94 dBA are located at workstations at one end of the production area and their workstations are associated with specific pieces of production machinery. These two groups were defined by sampling a sufficient number of employees (half- or whole-shift samples) to ensure that the database will include at least one TWA in the top 20% of the distribution (95% confidence level; see *Statistical Factors in Sound Survey Methods*).

For this work environment, the following classification procedure was implemented. The 80% of the population with an estimated TWA of 85–88 dBA was classified as B. The specific job functions where the 10% of the employees exhibited TWAs in the 91–95 dBA range were classified as C. These employees in area C were required to wear hearing protection and their area and machines were appropriately posted. In addition, all employees who entered the posted area were required to wear HPDs regardless of the time spent in the area.

Obviously the remaining problem is what to do with the 10% of the employees who exhibited TWAs less than 85 dBA. Since they were working in a potentially harmful noise environment and their workstations were not clearly isolated from other workstations in the same area where higher TWAs had been established, they could be exposed to higher TWAs. Therefore, their job functions were classified as B so that they would be a part of the company's HCP, receive annual audiometric evaluations, and be required to utilize HPDs if they exhibited a significant threshold shift. However, this group should be considered officially enrolled in the company's wellness program, not in the OSHA HCP. That way if one or more develop an OSHA STS (due to the aging process, pathology, recreational noise exposure, etc.) the company does not have to deal with it from the perspective of logging hearing loss on the OSHA Form 200 (see Chapter 11).

As a second example consider two adjoining work areas with the following characteristics. In production area 1, 80% of the employees exhibited TWAs in the 90–94 dBA range and 20% had TWAs in the 87–89 dBA range. In production area 2, 90% of the employees exhibited TWAs in the 86–89 dBA range, while 5% exhibited TWAs in the 90–94 dBA range and 5% exhibited TWAs in the 80–84 dBA range.

Production area 1 was classified as C and all employees working in this area were required to wear HPDs, including the employees working in one end of the room who exhibited TWAs in the 87–89 dBA range. Based on the information received from the foreman in this area as to the general mobility of his work force, it was decided that the potential for exposure to the higher sound levels was significant. Therefore, requiring the use of HPDs by all employees was recommended.

Production area 2 was classified as B for all employee work areas exhibiting TWAs less than 90 dBA, including the 5% of the employees who exhibited TWAs less than 85 dBA. At those workstations where the workers' estimated daily TWAs were 90 dBA or above, the stations were classified as C and posted as HPD-required work areas where the use of hearing protection was enforced for all persons entering the areas.

An additional problem, not necessarily unique to this production facility, involves the frequency of employee movement between jobs within each department and between departments. This type of mobility creates questions regarding the proper classification to be placed on the employee's audiometric record. In general, it is recommended that the employee's highest noise exposure classification during the past year be utilized so that the employee's highest potential exposure category is indicated to the audiogram reviewer.

Different Noise Exposure Criteria and Exchange Rates

The criterion level and exchange rate differ between the present OSHA noise criteria and those of ACGIH. Whereas OSHA defines 100% noise dose as an exposure to 90 dBA for 8 hours, ACGIH defines a 100% noise dose as an exposure to 85 dBA for 8 hours. Whereas OSHA requires the use of a 5-dB exchange rate, ACGIH requires that a 3-dB (equal energy) exchange rate be used. Based soley on the two criterion levels defined, one would assume that the ACGIH criteria would be more protective. However, OSHA requires that an employee be included in a HCP when the measured noise dose is 50%, or a TWA of 85 dBA. Therefore, if the employee's noise environment is fairly steady over the work shift, both criteria would be equally protective, all other aspects being the same.

Consequently, the main difference between OSHA and other existing noise-exposure criteria is the specified exchange rate and how it impacts the employee's measured noise exposure. Recall that the exchange rate defines how one trades off level vs. time in computing the equivalent daily noise exposure level. Referring to Table G-16a in Appendix I, OSHA allows 8 hours at 90 dBA, 4 at 95, 2 at 100, etc. In other words, level and time are traded at a rate that doubles the allowed exposure time for each 5-dB decrease in level. (Also see Chapter 3, Figure 3.5.) On the other hand, the ACGIH criteria use a 3-dB exchange rate which allows 8 hours at 85 dBA, 4 at 88, and 2 at 91, etc. For each doubling of the allowed exposure time, the allowed noise level decreases by 3 dB.

As an example of how the exchange rate impacts TWA and L_{A8hn}, consider a worker who is exposed to a constant level of 100 dBA for 1 hour and assume that during the remaining 7-hour period the noise level is below 80 dBA **(this is not a typical industrial noise exposure)**. Using a 5-dB exchange rate, this would result in a TWA of 85 dBA (100 dBA for 1 hour, 95 for 2, 90 for 4, and 85 for 8). Using the ACGIH required 3-dB exchange rate, we would obtain an L_{A8hn} of 91 dBA (100 dBA for 1 hour, 97 for 2, 94 for 4, and 91 for 8). Therefore the difference in the daily 8-hour noise exposure would be 91 minus 85, or 6 dB.

However, normally one would expect that the employee's noise exposure during the other 7 hours is significant, let's say equal to 90 dBA. Then the predicted daily equivalent levels would be as follows (reference Equations 3.8 and 3.13):

$$\text{TWA} = 16.61 \log[(1/8) \times (1 \times 10^{(100/16.61)} + 7 \times 10^{(90/16.61)})]$$
$$= 92.3 \text{ dBA, and} \tag{7.1}$$
$$L_{A8hn} = 10 \log[(1/8) \times (1 \times 10^{(100/10)} + 7 \times 10^{(90/10)})]$$
$$= 93.3 \text{ dBA.} \tag{7.2}$$

The impact of this background noise, when combined with the higher level for 1 hour, results in a difference in the predicted daily 8-hour levels of only 1 dB. Therefore, one should conclude that only for special noise exposure situations (as described in the previous paragraph) would the resulting measured daily noise exposures differ by 5 dB or more. Examples of such noise exposures where this might occur would be jet engine mechanics who might work for short durations next to running jet engines, production mechanics who work inside and up close to machinery for limited time periods at very high noise levels (125 dBA and higher), maintenance personnel who every so often blow-down high pressure steam lines (140–165 dBA), etc.

Figure 7.1 represents a collection of 129 industrial noise exposure samples for which the predicted L_{OSHA} and $L_{Aeq,T}$ data were available (Petrick et al., 1996, Royster and Royster, 1994). The vertical axis presents the measured $L_{OSHA,T}$ values and the $L_{Aeq,T}$ minus $L_{OSHA,T}$ difference values are shown plotted on the horizontal axis. A linear regression line fitted to the data is shown as the sloping

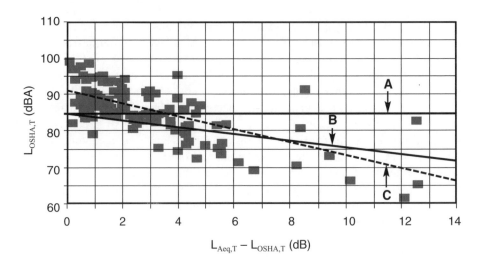

Figure 7.1 — L_{OSHA} level vs. the difference between the $L_{Aeq,T}$ and L_{OSHA} for 129 noise samples from several industrial sites.

dashed line (C). The y-intercept is at 91.2 dB and the slope is -2.0 dB per difference dB. This configuration is typical of other industrial sound survey databases that the authors have investigated. The region between the solid horizontal line (A) at 85 dBA and the sloping solid line (B) represents the region in which a data point would be moved up to or above the 85 dBA line if a 3-dB exchange rate had been used. For the employees represented by these points, use of the 3-dB exchange rate would require them to be included in the HCP. For this database, the average $L_{Aeq,T}$ - L_{OSHA} value is 2.8 dB.

In general U.S. industry the expected typical increase in the measured 8-hour daily equivalent exposure value, if a 3-dB exchange rate were used, is between -1.0 to +3.0 dBA. The minus value (TWA higher than L_{A8hn}) comes about when employees work longer than 8 hours a day and the equivalent daily 8-hour level is calculated. As an example, assume an employee works 16 hours a day and the noise level is steady at 90 dBA. Then the TWA would be 95 dBA and the L_{A8hn} would be 93 dBA. Therefore the 3–5 exchange rate difference would be -2 dB.

Presurvey Considerations

Types of Sound Surveys

For discussion purposes, we have elected to group sound survey efforts into three categories as shown in Figure 7.2. The three categories defined are the basic, detailed, and engineering noise control surveys. The type of survey selected will depend upon several factors.

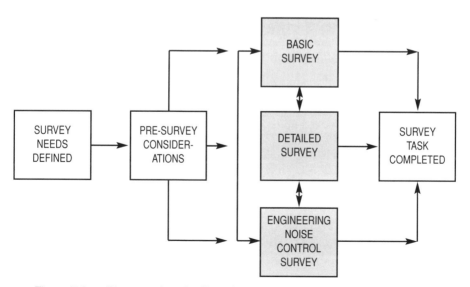

Figure 7.2 — Diagram of work efforts for the sound survey phase of the HCP.

When an environment is surveyed for the first time, usually a basic sound survey will be conducted. This includes a general screening survey of all plant areas to identify environments that do not present any noise hazard to the worker. For job environments where the employees' TWAs are easily defined (noise levels and exposure periods are readily determined), the data collected during the basic survey may be sufficient. The basic sound survey may be thought of as a short-term effort. That is, for a plant of 200–300 employees, the basic sound survey would not require more than 1 day to conduct.

If, during the basic sound survey, worker and/or job classifications are found where high variability in the sound levels requires either the use of a different survey instrument (such as a dosimeter) or more detailed sampling, then a detailed sound survey would be scheduled. Likewise, if a work area is surveyed where several job classifications exist and the measured sound levels and/or employee task or mobility variations indicate that different classifications may be required (Table 7.1), then a detailed sound survey would be called for. Therefore a detailed sound survey, as the name implies, should be thought of as a survey effort that will require several days to complete (collect the data, run the analysis, and write an acceptable report). However, after the initial basic sound survey has been conducted, then follow-up detailed sound surveys of worker positions or areas could be completed over shorter time periods.

If, during the basic sound survey, a noise source is identified that is a main contributor to the employee's TWA or to the area's noise level, or which is creating a communication or annoyance problem, then a noise control survey may be scheduled. The need for a noise control survey is often established by observations made during a sound survey or as a consequence of analyzing the resulting data. A noise control survey of a new piece of production equipment may also be requested by management based on sound data supplied by the equipment's manufacturer. In other words, the noise control survey may originate from more than one source. The main objective of the engineering noise control survey is the collection of data for noise control purposes.

Although we have elected to separate sound survey efforts into three categories for discussion purposes, practically speaking any one industrial sound survey effort will most likely include objectives that reflect more than one survey classification. Also, as Figure 7.2 indicates, there exist many feedback paths among the three classifications. As an example, after the recommendations from an engineering noise control survey have been implemented, it is often necessary to conduct a new survey to establish the effect on the workers' TWAs.

Educational and Informative Efforts

It is important that employees, supervisors, and management be informed of the purpose and approximate dates of all sound surveys. The reasons for notification include the need to point out to personnel the necessity for measured sound levels to be representative of normal production activities, to ensure that the production equipment will not be down for routine maintenance on the planned day of the survey, and to arrange for appropriate personnel to be available to assist the surveyor. It is not at all uncommon for the surveyor to arrive at the site only to find out that one or more of the major noise sources are down for maintenance. Even if the surveyor properly communicates with management, some planned surveys may have to be rescheduled due to such factors as equipment failure, modification of production plans, and personnel-related problems.

Seasoned sound surveyors know that the success of their efforts can depend significantly on the experience and competence of the company personnel (including the noise-exposed employees) who assist in collecting the data. Often an experienced mechanic or machine operator can provide more effective assistance in identifying a noise source than a company engineer, safety director, etc., who is remote from production efforts. Failure to properly inform all affected parties of the planned sound survey and associated needs can create a situation where the personnel required to conduct a successful sound survey are either on vacation or not available due to other scheduled duties.

"Smell the Roses"

An inexperienced sound surveyor will tend to show up on the morning of the scheduled survey, calibrate the equipment, and immediately proceed to start monitoring the noise environment. The result is often a need to resurvey part or all of the plant because of the failure to note some critical fact, such as improper operation of a significant contributor to the noise environment. It is always advisable to leave the instruments in the office area that will serve as the sound survey headquarters and, as the saying goes, "smell the roses" before actually beginning the survey. It is amazing the amount of additional information that can be gained by walking through the work area to be surveyed with pencil, paper, and clipboard and recording general observations and the insights offered by persons working in the area prior to conducting the actual survey. By becoming familiar with the environment, the surveyor will be able to more effectively relate the movement of the dial and/or digital indicator to the contributing noise source(s). The authors have had the opportunity to resurvey a few production areas several times over a period of weeks. It has been interesting the extent of additional pertinent sound survey information that was gained during each additional visit. Such observations raise significant concerns with respect to the accuracy of one-time sound survey findings.

Collect Information on Environment and Equipment

The surveyor should attempt to collect as much information as possible about the work environment and equipment prior to the scheduled sound survey. This includes, for example, all past sound surveys conducted either by plant personnel or outside consultants. Even if less than desirable survey procedures were followed in obtaining these data, they often provide useful information about the noise environment, production equipment modifications, and dominant noise sources.

If practical, the surveyor should discuss the noise survey with an approachable supervisor and/or machine operator. These individuals will often remember information about equipment modifications that seemed to affect the production area noise level. It is essential that the surveyor not be critical of the information given by these nonprofessionals. Although their statements may not be technically correct, they can supply important information which may be useful for noise control or sound survey purposes.

As an example, one of the authors investigated a noise problem that was created by a paper-drive train at a paper production facility. The plant engineering staff had studied the problem for several weeks prior to the planned survey and had failed to establish the cause of the noise problem. Management stated that the noise problem had surfaced approximately 2 years earlier.

Since experience had shown the need to communicate with supervisors and machine operators, an informal meeting with the area supervisor and head machine operator was scheduled on the morning of the planned survey. During the meeting, the machine operator asked the surveyor why he was interested in talking to nonengineers. Upon being told that front line supervisors and machine operators had been very helpful in the past in identifying possible noise sources, the machine operator volunteered the information that the noise level had changed when the engineering staff had made modifications to the hydraulic lines in order to increase the production capacity of the system. Although the modifications did not succeed, the system had not been returned to its original condition.

The company engineering staff was questioned about the earlier equipment modifications and the information provided by the machine operator was confirmed. The lines were eventually restored to their original design and the noise levels returned to an acceptable level. Although for general interest purposes a few sound readings were made during the initial visit to the plant, they were not necessary to achieve a solution of the noise problem.

When conducting a sound survey for noise control purposes, the surveyor should assemble general noise control information concerning the type of equipment being studied. Possible information sources include the equipment manufacturer and pertinent trade journals and publications. A search of the Internet will yield a multitude of potentially useful noise control information as well as source listings provided by some professional journals (Anon., 1998 and 1999). Sometimes it will be found either that no noise control option is available or that past noise control efforts have not been successful, making any quick fix unlikely.

Prepare a Check-Off List and Survey Outline

No matter how experienced the surveyor may be, it is important that a check-off list be developed and utilized. Otherwise it is easy to overlook items that may be needed during the survey. A representative list might include:

- noise dosimeter(s) and/or sound level meters (SLMs) and possible backup unit(s),

- calibrator and calibration forms,
- tape or digital recorder for recording noise samples for later analysis or for educational purposes,
- extra batteries and connector cables,
- clipboard, paper, survey forms, extra pencils, markers and/or pens, camera and film, or a digital camera,
- floor and equipment location identifiers (tape measure, chalk, masking tape, pens, and/or markers),
- copies of floor plans of the facility being surveyed, or sketches if scale plans are unavailable,
- calculator, laptop computer, and reference publications or tables that might be needed,
- personal HPDs, safety glasses, safety shoes, etc.,
- belt or lightweight vest to support each noise dosimeter, and safety pins or Velcro strips to secure the noise dosimeter's microphone cable when needed,
- windscreen for windy work areas or for microphone-damaging environments and a thin plastic cover for use in protecting the instruments from other harmful environments,
- watch, timer, and flashlight,
- one or two foam sheets for use in isolating the equipment from vibrating surfaces or for easing the load on aged knees when conducting "low level" sound surveys,
- a small voice recorder for recording pertinent comments before, during, and after the survey,
- finally, for the really technically oriented sound surveyor, a global positioning system device for locating measurement positions in the field and a laptop direct satellite-internet connection to be used to immediately down load the data files and retrieve useful information from a home-based system.

Even if the sound surveyor is highly experienced, it is recommended that a brief summary of the objectives of the survey be prepared and that a preliminary outline of procedures be developed. When sound surveys are conducted at locations that have not been measured previously, the prepared outline will often have to be altered due to unforeseen circumstances. Regardless, more effective sound surveys will be conducted by following a planned outline than by using the do-it-as-you-go approach.

Instrumentation Considerations

Noise Dosimeters and/or SLMs

In choosing equipment one should ask the question, "Why are we interested in establishing a TWA?" There are several potential responses to this question. First, if the concern is solely for the potential of the noise environment to produce a permanent noise-induced hearing loss, the estimated TWA will be compared to some damage-risk criterion. Virtually all such databases have been established using an SLM (A-weighted, slow response), with the resultant TWAs correlated with the magnitude of hearing damage exhibited by the noise-exposed population. Therefore, when hearing damage risk is being assessed, one might argue that the most appropriate instrumentation and survey procedures to utilize would be those identical to the ones involved in conducting the original damage-risk studies. Second, if the concern is to match the results that have or would be obtained by the OSHA inspector, then the most appropriate equipment would be that used by the inspector (mainly noise dosimeters). Third, if the concern is a multipurpose use of the data (including any present or future compensation or other legal needs), then the equipment of choice would be a noise dosimeter that provides not only the overall information that is commonly desired (noise dose and equivalent TWA) but a detailed histogram of the individual's noise exposure during the work day, sound level exceedance information, percentile distributions, etc.

Several products are presently available that provide within the same instrument both SLM and noise dosimetry capabilities, and also the capacity for later recall of the level-versus-time histories of worker noise exposures. The ability to review the employee's daily level-versus-time history can afford significant insight as to the noise sources that contributed to the measured TWA, and can also enhance other areas of the HCP such as educational presentations (DiBlasi et al., 1983).

Therefore it is recommended that sound survey instrumentation include at least one noise dosimeter and one integrating SLM (or a dual capability instrument) with appropriate calibrators. Other useful instruments that should be considered are octave or 1/3 octave-band filters, recorders, and real-time analyzers (see Chapter 3). If placed in willing and capable hands, multifunction instrumentation can produce additional data not obtainable with a basic noise dosimeter and/or SLM, that can enhance the hearing conservation effort.

Equivalence of Measurements

The limited number of reports in the literature concerning the equivalence of TWAs measured using SLMs and noise dosimeters indicate that TWAs obtained with the two types of instruments differ (Erlandsson et al., 1979; Jones and Howe, 1982; Shackleton and Piney, 1984; Walker, 1979). Typically the data obtained using noise dosimeters are higher by 0.5–2.0 dBA. As an additional point of reference and also as a demonstration of the adequacy of the SLM sampling methodology (random sampling over a complete work shift), 79 North Carolina OSHA inspection sound surveys conducted over a period of several weeks were analyzed. The procedure followed by these inspectors involved measuring the TWAs of three employees at each plant site using SLMs (full-shift sampling, at least 30 random samples) and noise dosimeters. The mean difference between the 237 pairs of TWAs was 0.6 dBA, with the noise dosimeters indicating the higher TWAs.

Several reasons have been given for the observation that noise dosimeters tend to predict higher TWAs, including the ability of the dosimeter to more accurately follow the noise signature over time and predict the employee's time-level history for conditions such as when employees move in close to their equipment, and the effect that the employee's body has on the sound field when the dosimeter's microphone is placed on the shoulder (Giardino and Seiler, 1996; Svensson, 1978).

When the dosimeter's microphone is placed on a wearer's shoulder in a nondirectional pink noise field, the average increase in the measured A-weighted sound level is 0.4 dBA (GenRad, 1976). However, if the employee is working in a directional sound field, such as when the employee is in close to a noise source, the effect of the body on the sound field may be as high as 2 dBA for perpendicular incidence. When conducting a sound survey the authors typically will record the on- vs. off-the-shoulder readings for a couple of noise dosimeters in typical noise areas 10 times each. The observed difference gives a reference as to the potential on-the-body effect expected in the environment being evaluated. The differences observed have always been within the range indicated above, typically an increase of 0.5–1.0 dBA for the on-the-shoulder readings.

It should also be pointed out that sophisticated instrumentation is not always needed. Companies with very hazardous noise environments have established effective HCPs based on sound surveys conducted by company nurses or audiometric technicians who borrowed a type 2 SLM (see Chapter 3) and calibrator for a limited time period each year. The critical factors are the dedication of the personnel to the job at hand, a willingness to take the time to obtain good data, and the HCP policies related to other phases of the program, not the instrumentation.

Expected Accuracy

The sound surveyor often asks, "What is the level of accuracy to be expected from SLMs and noise dosimeters when conducting real-world sound surveys?" Obviously the level of accuracy expected depends not only on the instruments' inherent accuracy but also on the sampling and data analysis procedures employed (Bruel, 1983; Delany et al., 1976; Peterson, 1980).

One indication of the expected error range for field-utilized sound survey instrumentation is given by OSHA's policy of issuing citations for a violation of the noise standard or HCA only when the estimated TWA exceeds the target level by 2 dBA (OSHA, 1984). That is, the stated 85-dBA TWA action level for initiating an HCP becomes 87 dBA. This 2-dBA allowance is assumed by OSHA to account for errors in calibration and equipment accuracy as well as variability in measurement techniques. Of course the obvious follow-up question is why use a + 2-dB correction rather than a - 2-dB correction. If protection of the noise-exposed population were the primary concern, then one could easily argue that the action level should be 83 instead of 87 dBA!

The first author has in the past routinely checked the accuracy of the SLMs, noise dosimeters, and calibrators for one state OSHA program (Royster, 1980). The type 2 SLMs and calibrators, usually in groups of 10, were given pre- and postcalibration evaluations. Initially the instruments were cross calibrated prior to any adjustments. That is, each calibrator was utilized to check the output for all SLMs before it was compared to the laboratory reference standard and adjusted as needed. A final postcalibration check was completed by again recording the SLM outputs using each calibrator. The standard deviation for the recorded precalibration SLM output sound levels across instruments and calibrators was typically 0.3–0.5 dB. After each calibrator's output had been checked, and adjusted as necessary, and each SLM had been readjusted using each unit's calibrator, the standard deviation for the recorded SLM outputs normally decreased to 0.2–0.3 dB. These findings, based upon single-frequency pure-tone calibration at 1000 Hz, suggest a high degree of accuracy.

Contamination of Data

The question of potential contamination of data measured using acoustic instrumentation, especially noise dosimeters, has been raised in the past. The type of fears expressed include employees intentionally hitting the microphone, hollering into the microphone, blowing across the microphone, and other inadvertently induced artifactual inputs.

In order to investigate the potential for such measurement artifacts to significantly impact the measured employee's daily TWA, a series of experiments was carried out (Royster, 1997a). Each of the experiments involved a Larson-Davis 700 type 2 integrating sound level meter and noise dosimeter with the microphone placed in a pink-noise free-field at 85 or 90 dBA. In the first experiment the microphone was thumped with a finger as hard a possible once every 10 minutes over a period of 1 hour (the equivalent of 48 thumps during an 8-hour work day). The resulting measured $L_{OSHA,1h}$ was compared to the level that would have resulted if no thumping of the microphone had occurred. In a second experiment, an individual hollered as hard as practical from a distance of around 2–3 cm from the microphone, again once every 10 minutes for a period of 1 hour. Third, a male individual blew as hard as was practical at the microphone from a distance of 2–3 cm once every 10 minutes for a period of 1 hour.

For each of the above types of potential sound-field contaminations, the impact on the anticipated daily TWA for the two different background noise conditions is shown in Table 7.2. The greatest impact on the daily TWA was for blowing across the microphone, which caused an increase in level of 2.0 dB (for a situation where 48 such events occur over 8 hours at the lower 85-dBA background condition).

TABLE 7.2
Effect of contamination on the daily TWA for different
background noise levels (48 events per 8 hours).

Source	Background Noise Level (dBA)	
	85	*90*
Thumping	0.3	0.2
Hollering	0.8	0.4
Blowing	2.0	1.0

One additional condition was evaluated where two noise dosimeter wearers shifted the microphone, including undoing and re-pinning the safety pin holding the microphone, between the centers of their left and right shoulders. The shifting occurred once every 10 minutes over a period of 1 hour in an environment where the sound level was approximately 85 dBA and broad-band in nature. No variation in the meter's output histogram (15-second intervals) was detectable over the measurement period.

In addition to the above investigations, noise dosimeter records (consisting of detailed 1-minute histogram samples) for 74 surveys from five different industrial plants (crystal production, food processing, tire manufacturing, paper production, and aluminum smelting) were reviewed (a total of 28,542 samples). Across all samples, only seven potentially abnormal histogram samples were noted. Removing these potentially contaminated samples would not have changed the resulting daily measured noise dose.

The bottom line is that the contamination of noise dosimetry data is not a significant issue at least when the OSHA noise-exposure criteria are utilized (Giardino and Seiler, 1996). This conclusion can easily be backed up by simply evaluating the appropriate math expressions or by downloading the histogram outputs (intervals, percentiles, etc.), and manually removing the suspected contaminating samples to observe the impact on the measured TWA during the actual sampling period. However, the use of ACGIH's noise criteria would result in a significant increase (by about 1 to 3 dB) in the levels shown in Table 7.2.

The Basic Sound Survey

Purpose of the Basic Sound Survey

The main objective of the basic sound survey is to identify those areas where the worker TWAs are 85 dBA or higher (meeting OSHA's action level). The findings may suggest the need for a more detailed sound survey effort. However, for those workstations where the noise levels are steady and therefore are easily measured, and where employees are not highly mobile across areas, the findings from the basic sound survey effort may be sufficient to adequately establish appropriate TWAs.

A second objective of the basic sound survey is to identify areas where employee TWAs are below the action level and to determine typical sound level ranges for these workstations. Obviously one may ask why these areas should be surveyed and documented. In addition to the reasons presented earlier (see *Why Conduct a Sound Survey?*), production equipment is often moved from one location to another, or equipment may be replaced by units with different noise characteristics. The existence of sound survey data allows management to estimate the effect of these equipment modifications on existing noise levels. Also, the basic sound survey should identify work areas where the sound levels could create annoyance and possible communication difficulties. Employees who are located in these typically non-HCP areas often consider the noise in the area "loud" and as a result may question management as to the need to be furnished HPDs and/or be included in the company's HCP. The data collected during the basic sound survey will assist company personnel in providing an adequate response to these and other related questions.

At one plant where an employee was injured and was not able to obtain immediate assistance due to a background noise level which was slightly less than 85 dBA, the employee filed a lawsuit contending that the environmental conditions, which were under the control of the employer, were directly responsible for the additional personal injury that was incurred by the delay. The employee also claimed that the employer had not properly documented the environmental noise conditions where the accident occurred and that the employer should have been aware of the potential environmental hazard. This claim of neglect on the part of management would have been avoided if company personnel had properly documented the noise environment for all areas of the company, not just the one production area that exhibited TWAs higher than 85 dBA.

A third objective of the basic sound survey may be the establishment of a sound level database that can be utilized to broadly estimate TWAs for workers and/or job descriptions in order to stratify the workforce for statistical sampling purposes (that is, to divide the workforce into groups of similar noise exposures, as described in the section *Statistical Factors in Sound Survey Methods*).

During the basic noise survey, all areas at the plant site under study should be visited and the existing noise levels documented. Areas that are often overlooked are computer rooms (where the noise exposures can approach a TWA of 85 dBA) and isolated work areas where the sound level is 90 dBA or higher (pump and generator stations, and postal rooms with paper-shredders or mail-handling machinery).

Information to Record
Supporting Information
During the time of the initial visit to the plant site and production areas, data concerning the typical work schedules for the general plant population should be collected. This information, along with the noise level samples obtained during the basic sound survey, will allow the surveyor to establish estimates of the expected TWAs for all fixed workstations or areas. During or following the initial visit to the plant facility, floor plan drawings should be obtained, if available, or general sketches of the work areas should be completed. Photographs or slides of the areas and equipment studied are also very useful in recalling information at a later time. However, most managers are concerned about the taking of pictures, even though photographs have been found to be very helpful in analyzing the data later on and in reporting the findings to management. The authors have had very little resistance to photography if the undeveloped film or digital file is left with management for development and/or review prior to sending the photos to the consultants.

The sound surveyor should always keep in mind that data acquired during any visit to the plant facility not only will be utilized by company hearing conservation personnel but could end up in a court of law. Therefore the recommended guidelines for recording, maintaining, and reporting noise survey data as discussed herein should be followed.

Calibration Data

Most likely the initial data to be recorded will be the results of the pre-survey calibration of the SLM. In calibrating the instrumentation, adjustments should not be made unless the deviation in the indicated sound level from the expected reference level exceeds ± 0.25 dB. If it is necessary to adjust a unit, the magnitude of the adjustment should be recorded on the calibration form or stored as part of the digital data file.

Prior to calibrating acoustic instrumentation, it is important that the equipment have an opportunity to approach the work area's ambient temperature. It should not have been left in the trunk of a car when the outside temperatures are either very low or very high. If the equipment is being transported to the plant facility, it is good practice to take the equipment to the plant the day before the survey or store the equipment in the motel room the night before in order to keep the equipment at a reasonable temperature prior to the survey. Extremes of temperature have been known to significantly affect the stability of sound measurement instrumentation.

When calibrating an SLM, a complete set of instrumentation descriptors and set-up conditions should be recorded. In the case of survey instruments that have a removable microphone, the sound surveyor should check that the microphone has not been replaced by a different unit unless the surveyor is the only individual who has access to the equipment.

During the calibration of an instrument, record the date, time, serial numbers of the meter or dosimeter and its microphone (if a separate unit) and calibrator, and the battery check status. The sound surveyor should always begin with fresh batteries in each unit because the cost of the batteries is trivial compared to the cost of lost or invalid survey data.

At the conclusion of the noise survey, a postsurvey calibration of the instrument should be completed. If the instrumentation's reference level has shifted more than ± 0.5 dB (which is very unusual unless extreme environmental conditions are encountered or an instrumentation failure has occurred) then the data collected during the survey should be questioned. Any unusual equipment responses or environmental conditions encountered such as rain, high or low temperatures, strong electromagnetic fields (ANSI S1.14-1998), etc., should be indicated in the general response section of the calibration data form. Finally, the surveyor should sign and date all data log sheets using a nonerasable pen.

Survey Sound Level Data

During the basic sound survey, the data obtained should be sufficient to estimate TWA ranges for all personnel. If the sound surveyor is fortunate, the measured sound levels will be "relatively constant," defined as regular variations in the SLM's indicator on slow response setting over a range of no more than 8 dBA. For this type of noise environment, the sound surveyor can estimate the average position of the meter by eye (Delany et al., 1976). This procedure should provide an estimate of the L_{OSHA} within approximately 1 dBA of the true value (Christensen, 1974). If the sound level is not regular or is varying over a range greater than 8 dBA, then it is recommended that a sampling methodology be utilized in order to satisfactorily predict worker TWAs.

An alternate method of sampling utilizes an integrating/averaging SLM having the appropriate dynamics and exchange rate. This type of instrument and mode of use simplify the problem of eyeball averaging and interpretation of the dial. However, keep in mind that a satisfactory sampling strategy must be followed.

Several different types of data log forms, or computer files, are available that may be used to record sound level measurements. Such forms provide different options for recording data including the date, SLM type, serial number, time or duration of survey samples, microphone type and serial number if the microphone is not an integral part of the instrument, meter response selection, and environmental conditions such as temperature and wind conditions and if a windscreen for the microphone is being used. It is important that, if special circumstances are observed during the sound survey (such as an expected major noise source turned off or operating or in a manner not typical of normal production), such observations must be indicated on the form in the space allocated.

There should be a place on all forms, or computer print outs of the data, for the surveyor to sign the form. DO NOT at a later time transfer the recorded data to a clean form and throw away the original data. During legal hearings involving sound surveys, there have been instances of the data submitted being brought into question because the original forms could not be produced. It is difficult to explain how a sound survey could have been conducted in a dirty production area and the data end up on a clean log form or, even more unbelievably, one on which the recorded values have been typed! It is normally absolutely essential that the surveyor(s) take notes during a survey and these notes should be kept with the sound survey data files.

Collecting the Data

Area Survey

Since one purpose of the basic sound survey is the identification of locations where more detailed surveys will be necessary, it is recommended that a check of the area L_{OSHA} be conducted initially. Assuming that the work environment sound level is relatively constant (less than 8 dBA variation in the SLM's indicator), then during the basic sound survey a minimum of 10 sound level samples should be recorded, spaced over a typical workday. Measurements should be recorded at a height of approximately 1.5 meters above the work surface. In interpreting the movements of the SLM indicator, the surveyor should observe the instrument's response for approximately 5 seconds for each measurement and estimate the indicated mean sound level value.

It is also recommended that sample C-weighted sound levels be recorded. The difference between C-weighted and A-weighted levels has applications in the selection of HPDs and in certain noise control procedures.

Workstation/Job Description Survey

If the findings of the area survey predict an L_{OSHA} of 85 dBA or higher, then the surveyor may elect to schedule a future detailed sound survey of the area to determine the employees' TWAs. However, even if the findings of the area sound survey do not result in an L_{OSHA} of 85 dBA or higher, it will still be necessary to survey specific workstations and various job descriptions in the area, because area sound levels are typically lower than actual worker TWAs. Exceptions occur in special instances such as when the time spent away from the noisy work area by a worker is significant or when the dominant noise source is located at a sufficient distance from the employee's workstation. Typically as the workers approach their workstations the sound level will increase, unless of course the dominant noise source is coming from a different machine or other sources. As a consequence, a survey of an employee workstation will normally exhibit a higher TWA than the L_{OSHA} for the overall area.

In conducting sound surveys close to dominant noise sources, it is necessary that the surveyor give particular attention to the effects of the type of microphone utilized on the measurements obtained, potential effects of the surveyor's body, and the changing characteristics of the sound field (Chapter 3).

Evaluating the Data for OSHA Purposes

Area Survey

In order to demonstrate the calculation procedures for determining the area's typical L_{OSHA} and predicted noise dose and TWA for employees stationed there, two examples are presented. We begin by presenting the overall general methodology of calculating the area's L_{OSHA}, noise dose, and predicted TWA.

For discussion purposes we will assume that the employee's noise level time exposure profile presented as Figure 7.3 is for an area measurement. To determine the equivalent TWA we could first determine the noise dose using the following equation (the reader is encouraged to refer to Chapters 2 and 3 [especially Equations 3.17 and 3.18], the OSHA noise regulations [Appendix I], and Table 7.12 at the end of this chapter where all the equations referenced herein are summarized):

$$D = [(C_1/T_1) + (C_2/T_2) +... + (C_N/T_N)] \times 100, \text{percent} \qquad (7.3)$$

where D is the noise dose in percent, C_N is the exposure duration for the Nth sound level, and T_N is the corresponding allowed duration to accumulate 100% noise dose at this same sound level. For the OSHA criteria, Table G-16a in Appendix I presents values of T_N for different sound levels. However, if the exposure duration needed is not listed, then it can be calculated using the following equation:

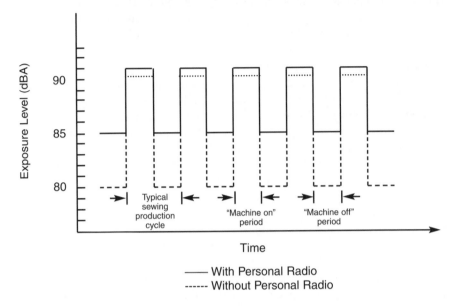

Figure 7.3 — Representation of the noise exposure of a typical employee with and without the use of a personal radio.

$$T_N = (8)/(2^{(L_N - 90)/5}), \text{ hours} \tag{7.4}$$

where L_N is the sound level corresponding to the C_N exposure duration.

Once the noise dose has been determined, the equivalent TWA can be found by evaluating the following equation:

$$TWA = 16.61 \log[D(\%)/100(\%)] + 90, \text{ dBA} \tag{7.5}$$

where D is the noise dose for the whole shift. If the sampled period is for a time period **less than** the actual work shift, then the associated dose must be corrected to a whole-shift equivalent value, that is,

$$D(H^*) = D(\text{measured}) \times H^*/T^*, \text{ percent} \tag{7.6}$$

where T^* is the actual period sampled and H^* is the whole shift duration. (Also see section, *Detailed Sound Survey: Atypical Exposures*.)

However, different approaches to the problem are possible. One would be to directly determine the equivalent L_{OSHA} sound level for the sampled period of noise exposure using the following equation:

$$L_{OSHA,T} = 16.61 \log\left[(1/T) \times \sum_{i=1}^{N} (t_i \times 10^{(L_i/16.61)})\right], dBA \qquad (7.7)$$

where t_i is the duration of exposure to level L_i and T is the total sample duration. Then, knowing the OSHA allowed exposure time for the $L_{OSHA,T}$ level calculated, the employee's noise dose and equivalent TWA can be determined using Equations 7.3 through 7.6.

Unfortunately, the hearing conservationist is often presented sound survey data calculated using different approaches and different formats. Therefore it is important that the surveyor understand how the data were developed and be able to verify their accuracy.

In presenting sound survey data, values should be rounded off to the nearest dB, since measurement accuracy does not justify reporting the values to tenths of a decibel. However, in order that the reader can verify the sample calculations presented herein we have elected to maintain at least one figure beyond the decimal. An exception to this procedure of rounding off occurs during the actual calculation process. It is always necessary to maintain an additional number of digits to prevent accumulating large errors in the final predicted values, and this is especially true when evaluating numbers raised to a power such as in Equation 7.7.

EXAMPLE 7.1, TWA Computation for Multiple Samples

A basic sound survey was carried out in a synthetic fiber production department by sampling the area noise environment 10 times during a typical workday. The measurements recorded were as follows: 106, 107, 105, 106, 107, 107, 105, 104, 106, and 105 dBA. To determine the TWA for employees in this area it is necessary to take into account the employees' duration of noise exposure. For the production area studied, the employees worked 9-hour (540-minute) shifts each day excluding time for lunch and breaks. The lunch and break areas exhibited sound levels less than 80 dBA.

The noise dose for this work area can be calculated using Equation 7.3. Each (C_N/T_N) ratio may be calculated using either hours or minutes provided a consistent set of units is used for each ratio. The problem can be approached by treating each sound level measurement as a separate noise exposure of 540/10 = 54 minutes, yielding 10 C_N/T_N terms to evaluate in Equation 7.3. Alternatively the equivalent sound level over the work period can be estimated as follows. If the range of the measured sound levels (maximum reading minus the minimum reading, or for this problem 107-104 = 3 dBA, is 5 dBA or less, then the OSHA equivalent 9-hour sound level can be approximated by calculating the arithmetic average of all the samples, or 105.8 dBA (Irwin and Graf, 1979; Peterson, 1980).

Therefore the employees working in the area are exposed to an area equivalent L_{OSHA} of 105.8 dBA for 9 hours (540 minutes) each day. According to the information presented in Table G-16a, Appendix I, the employees are allowed a daily exposure to a noise level of 105.8 dBA for 53.7 minutes. The user may interpolate between the values given or use Equation 7.4. The evaluation of Equation 7.3 provides the daily noise dose in percent, or

$$D = (540/53.7) \times 100 = 1006\%.$$

The corresponding equivalent daily TWA can now be determined either by using the information presented in Table A-l, Appendix I, if the noise dose is equal to or less than 999%, or by evaluating Equation 7.5:

$$TWA = 16.61 \log(1006/100) + 90 = 106.7 \text{ dBA}.$$

(Note that the TWA of 106.7 dBA, based on 8 hours of exposure, should not be confused with the 105.8 dBA $L_{OSHA,9h}$ value.)

Comments on Example 7.1:

The OSHA criteria for allowed noise exposures is based on an 8-hour work day. For many industrial job functions, the actual work day will be either more or less than 8 hours and the period sampled will not be equal to the time worked. If scheduled break periods are in areas where the equivalent sound level is less than 80 dBA, then the noise exposures during breaks are not included in the calculated TWA. However, if the break and lunch areas are in workplace areas with sound levels at or above OSHA's threshold of 80 dBA, the noise exposure accumulated during those periods must be included (see section on *Detailed Sound Survey: Atypical Exposures*).

EXAMPLE 7.2, TWAs Computed by Two Methods

The following set of measured sound levels was obtained in a textile spinning department by sampling the middle of the aisles at various times during the first shift: 90, 92, 91, 94, 90, 93, 92, 91, 90, and 92 dBA. Assume that for this work area the employees work an 8-hour day including a 30-minute lunch period and two 15-minute breaks in areas where the sound levels are less than 80 dBA. Therefore the daily duration of noise exposure is taken to be 7 hours or 420 minutes.

Approach A: One approach to determining the TWA for employees in this work area involves first determining the noise dose by treating each measurement as representative of its equivalent portion of the total noise exposure time, or (420 minutes/10 samples) = 42 minutes per sample, or C_N. To determine T_N we can utilize the data presented in Table G-16a, Appendix I, or use Equation 7.4. Then the noise dose is calculated using Equation 7.3:

$$D = (42/480 + 42/364 + 42/418 + ... + 42/364) \times 100$$
$$= 1.095 \times 100 = 109.5\%.$$

Now to convert the noise dose in percent to an equivalent TWA use Equation 7.5, or

TWA = 16.61 log(109.5/100) + 90 = 90.7 dBA.

Approach B: An alternate approach to determining the area's TWA would be to first determine L_{OSHA} for the 7-hour work shift, then use Equation 7.3 to obtain the area's noise dose, and finally Equation 7.5 to determine the TWA.

As pointed out in Example 7.1, if the range of the sound level measurements (maximum value minus the minimum value) is 5 dBA or less (for this problem 94-90=4 dBA), then the OSHA equivalent 7-hour sound level can be approximated by calculating the average of all 10 samples, or 91.5 dBA.

However, if one desires to follow a formal procedure, or if the range in measured sound levels is greater than 5 dBA, then Equation 7.7 may be used with T = 7 hours, or

$$L_{OSHA} = 16.61 \log\{(1/7) \times [(7/10) \times 10^{(90/16.61)} + (7/10) \times 10^{(92/16.61)} + ...$$
$$+ (7/10) \times 10^{(92/16.61)}]\} = 91.6 \text{ dBA}.$$

(Since a 7-hour equivalent L_{OSHA} value is being calculated, each measurement must be given a weight of 7/10 of an hour, or 0.7 hours/sample so that the 10 samples end up representing the total survey time of 7 hours.)

Now that we know the OSHA equivalent 7-hour sound level is 91.6 dBA, we can use Equation 7.3 to determine the daily area noise dose. First from Table G-16a, Appendix I, or Equation 7.4 we determine that for an exposure of 91.6 dBA, the worker is allowed an exposure of 385 minutes. The actual exposure duration is 7 hours or 420 minutes. Therefore using Equation 7.3 we obtain:

D = (420/385) × 100 = 109.2%.

Now using Table A-l, Appendix I, or Equation 7.5 the equivalent TWA is found to be 90.6 dBA, or

TWA = 16.61 log(109.2/100) + 90 = 90.6 dBA.

Comments on Example 7.2:

Two approaches were used to determine the area's TWA. First, each survey sample was treated as an exposure level, and the duration of exposure per sample was set equal to the number of samples divided by the typical employee's daily work duration. This procedure established a daily noise dose which was then converted to an equivalent TWA. The second approach involved first determining the 7-hour equivalent sound level, then calculating the noise dose and corresponding TWA. Either approach is acceptable. However, do not lose sight of the fact that the basic data set provides an indication of the affected employee's noise exposure variability. Such variability should always be reported along with the other findings.

Workstation/Job Description Survey

When measurements are being made in the immediate vicinity of sources, a large range of sound levels will often be observed. For areas in which the sound levels vary by more than 8 dBA, the surveyor will have to estimate the varying levels of exposure and corresponding time durations or else await the completion of the detailed sound survey using a different instrument or a more elaborate sampling plan. Consider the following example.

EXAMPLE 7.3, Effect of Personal Radios on a TWA

Approximately 50% of the employees at a production facility involved in a sewing operation had elected to use personal radios as a way of alleviating job boredom. However, management at this facility had expressed concern over the potentially harmful nature of the additional noise exposure resulting from radio usage (Skrainar, 1985; Skrainar et al., 1985).

Presented in Figure 7.3 is the idealized typical employee noise exposure level (based on sound survey samples) with and without the effect of using a personal radio. Initially the employees' TWA will be calculated without the effect of the personal radio, i.e., for the noise exposure indicated by the dashed line. The employees typically worked a 7-hour day, not including the lunch period and two break periods allowed each day. During the break periods their exposure level was less than 80 dBA.

The idealized "machine-on" and "machine-off" periods as shown in Figure 7.3 are of equal time durations and the corresponding levels of exposure are 80 and 90 dBA respectively. Therefore the daily noise dose is determined using Equations 7.4 and 7.3, or

$$D = [(C_1/T_1) + (C_2/T_2)] \times 100 = [(3.5h/32 \text{ h}) + (3.5h/8h)] \times 100$$
$$= (0.11 + 0.44) \times 100 = 54.7\%.$$

The corresponding daily TWA may then be estimated by evaluating Equation 7.5:

$$TWA = 16.61 \log(D/100) + 90,$$
$$= 16.61 \log(55/100) + 90 = 85.6 \text{ dBA}$$

(without the effect of a personal radio). If the effect of the personal radio is considered, then for the "machine-on" and "machine-off" exposures of 91 and 85 dBA respectively, the following predictions are obtained based on the assumption that the work schedules of the two groups are identical.

$$D = [(3.5h/16h) + (3.5h/7h)] \times 100 = (0.22 + 0.50) \times 100 = 72\%, \text{ and}$$

the TWA is calculated as:

$$TWA = 16.61 \log(72/100) + 90 = 87.6 \text{ dBA}.$$

Therefore the increase in the employees' predicted daily TWA due to the use of personal radios is 87.6 - 85.7 = 1.9 dBA.

Comments on Example 7.3:

We have assumed that the level-time exposure profile presented in Figure 7.3 is typical for both noise-exposed populations. In making this determination it is necessary to sample the noise levels at a sufficient number of employee work-stations. Normally the recording of 10 samples during the **basic sound survey** is sufficient for our purposes (see the following section on *Statistical Factors in Sound Survey Methods*).

The Detailed Sound Survey

Purpose of the Detailed Sound Survey

The data collected during the basic sound survey should identify the work areas, workstations, and/or job classifications at the plant site where more detailed sound survey information is needed. A sketch or drawing of the plant facility should now exist that can be referred to during the detailed sound survey. In addition, some work areas can be classified based on the findings of the basic sound survey. That is, all work areas where the basic sound survey findings clearly predict TWAs less than 85 dBA should now be indicated on the sketch or drawing of the plant facility. It is also recommended for those areas with TWAs less than 85 dBA that the typical and maximum sound levels measured also be indicated on the sketch. In addition, for those areas with TWAs above 85 dBA for which the data obtained during the basic sound survey were sufficient, the appropriate TWA can also be indicated on the drawing or sketch.

The purpose of the detailed sound survey is to complete the classification of the remaining areas. At the conclusion of the detailed sound survey a plant draw-ing or tabulation list should exist that defines the typical TWA ranges for all company employees.

Collecting the Data

Using a Noise Dosimeter

The microphone of the noise dosimeter should be placed roughly in the center of the shoulder on the side that would result in the higher estimated daily noise dose, if known. Otherwise the surveyor should alternate the placement of the microphone between shoulders across the employee population sampled. The microphone holder, if provided by the manufacturer, should be considered. However, our experience is that such devices are sometimes not effective or are cumbersome to use. One such difficulty that arises in practice is that employees often wear very thin and loose T-shirts, or no shirt at all, or other garments that do not readily support some of the dosimeter microphone holders.

For difficult situations, the authors have used safety pins to support the microphone slightly above the surface of the shoulder and ignored the fact that as the worker moved about, the microphone did not always remain in the exact desired orientation. However, the impact on the measured noise dose has been determined to be insignificant. For the smaller and more omni-directional modern-day microphones their orientation is not the critical issue, but that of the position of the microphone on the body. A potential benefit of using a safety pin, or similar device, is that it is very reliable and also if the cable were to get caught in equipment, this type of connection would easily tear away.

An attempt should be made to attach the microphone lead-in cable to the worker's clothes in order to minimize the possibility that the cable may get caught in the production equipment or on other protruding objects. When practical, an even safer solution is to route the cord underneath the shirt or top clothing.

The noise dosimeter case should be firmly attached to the employee's belt in a position that will minimize interference with work and susceptibility to impact. If the unit is placed up front it may be uncomfortable to the employee when s/he leans toward the equipment. The position that seems to work the best is slightly behind either the left or right side. If the employee normally does not wear a belt and the clothing will not adequately support the unit, then a supporting belt may have to be provided. An alternate method for supporting the dosimeter and microphone is to use a vest made of a light breathable fabric. This procedure mitigates problems caused by extra belts which may, for some job functions, hinder the workers in carrying out their job assignments.

A critical factor to be considered when the subject is in a very cold environment is protection of instrumentation batteries from temperatures which will derate them or even render the instrument inoperative. An effective solution is to place the dosimeter inside protective clothing at locations where body heat will keep it at a satisfactory temperature.

Ideally, the data collection should start and stop at the beginning and end of the work shift. This implies that both the employee(s) who will be sampled and the sound surveyor are willing to arrive soon enough before the work shift begins so that the worker can be fitted with the dosimeter and properly instructed in its use, and to stay long enough beyond the end of the work shift so that the dosimeters can be collected and the employees questioned about their use. However, some employees will refuse to report to work early or stay later unless they are paid for the extra time. If this compensation cannot be provided, the dosimeter must be fitted after the work shift begins and collected before it ends, in which case it will be necessary to account for the reduced sampling time relative to the length of the typical employee's workday.

The surveyor should keep in mind that for many job assignments the employees' noise exposures during the first and last 15 minutes to an hour of the work shift, as well as during the morning and afternoon work breaks and the lunch period, can vary significantly, either higher or lower. The noise dosimeter can provide a histogram **(which in the authors' opinion should always be obtained)** of the employee's noise exposure during the break and lunch periods, if sampled. Barring any knowledge to the contrary, it is best to assume that the short nonsampled periods would provide the same dose rate as the sampled period. Still the sound surveyor can take comfort that even if the sound levels vary significantly (± 10 dBA) during the nonsampled short noise exposure periods, the impact upon the predicted TWA would be small. For OSHA enforcement comparison purposes the sound surveyor should attempt to obtain data approaching either a half-shift sample (then double the noise dose) or an entire work period.

Before issuing the unit, the surveyor should carefully instruct the employee in its use and care. Hopefully, as part of the educational phase of the company's HCP the employees will already have been partially informed about the importance of the noise survey and of their responsibilities in ensuring the validity of the results. The instructions to the employee might include how to handle the unit when using the restroom and what to do if the microphone becomes entangled in the equipment, the case is dropped, or the microphone or case is unexpectedly watered down. These types of events are not common when working with informed and educated employee populations.

Employees are hesitant to inform the surveyor of instances that might have affected the dosimeter or the readings obtained. This reluctance is usually not because of hostility, but stems from a fear that if the unit or data were adversely affected while the employee was wearing the unit, then the employee's job status could be jeopardized. The employee is more likely to provide this type of information, at the end of the sampling period when the noise dosimeters are being collected, if the surveyor simply asks a nonintimidating question such as whether any difficulties were experienced while using the unit.

The sound surveyor should generally stay in the vicinity of the employees who are using the noise dosimeters until a reasonable knowledge of the noise level characteristics for the environment has been established. It is recommended that the range of noise dosimeter data obtained be supported by SLM measurements (dBA, dBC, 1/3, or octave-band SPLs to support engineering controls and make decisions about hearing warning signals, etc.). The surveyor will most likely collect these data at the same time that the noise dosimeter samples are being collected.

If the sound surveyor is conducting a resurvey of a production area and obtains data that are significantly different from previous results, the surveyor should be suspicious of the data unless the differences can be accounted for. The noise dosimeter's histogram of the employee's noise exposures (Chapter 3) can go a long way in solving these types of situations. Such records provide valuable information with which to assess the validity of the measured noise exposures. For example, in an area where the maximum sound level is known to be 100 dBA, measurements which show instantaneous rms or short-term average values of 110 dBA certainly demand further investigation.

Recently, the histogram of employees carrying out the same task were observed to exhibit significant differences in the L_{OSHA} as a function of time. One employee's record exhibited periods with levels 15–20 dBA higher than what was believed to be typical. It turned out the management and supervision somehow was not aware that the employee had to go into a soap mixing room where the sound levels were 115–120 dBA for periods of time each shift. In this instance, neither the employee, supervisor, or safety director provided this critical piece of information, but the noise dosimeter histogram did. With industry requiring employees to be more multifunctional and the employees more empowered to make decisions, this type of situation has become more common.

Ordinarily, after the employees in the work area being sampled get over any initial effects of wearing the noise dosimeters and/or seeing other employees using them, they begin to ignore the units (Shackleton and Piney, 1984). Sound survey findings obtained during the adjustment period should be analyzed for atypical values and potential impact before they are included in the permanent sound survey database.

The general guidelines for recording data discussed in the previous section should be followed when obtaining noise dosimeter data. It is important to know the times at which sampling began and ended, and the name, location and/or job classification associated with the survey sample. Any pertinent information that may bear on the predicted TWA should be indicated in the general remarks section of the data log. Examples would include unusual down times for the operator's equipment or other significant noise sources in the work area, the need for the employee to spend an unusual amount of time at a location not normally visited, and any other information that the surveyor feels could affect the data. Of course, all noise dosimeter units should be calibrated before and after each daily use.

Using a Sound Level Meter

Whereas the use of a noise dosimeter to record sound survey data is relatively straightforward, employing an SLM to obtain estimates of the actual TWA for those types of environments that could not be sampled during the basic sound survey is more difficult. This is especially true when the employee's exposure exhibits significant variations in the measured sound levels or when workers are highly mobile.

In sampling the sound field in the vicinity of a worker, the SLM's microphone should ideally be placed at the position in the sound field where the center of the noise-exposed employee's head would normally be, but with the employee removed from the immediate area. However, in many instances this is not a practical approach, since the operator may need to be present for the machine to operate properly. In such cases, the microphone is positioned approximately in the vicinity of the worker's ear that is receiving the highest noise exposure. Ear placement differences are only critical if the sound field changes significantly with the position and orientation of the microphone. These restrictions and potential problems require the operator of the SLM to scan the vicinity of the worker's work location(s) to ensure that the typical noise exposure level is obtained.

A simple way to detect significant variations in the instrument's indicated sound level associated with position is to move the microphone around. When significant variations (greater than ± 2 dBA in the hearing zone of the employee, see Chapter 3) are detected, the surveyor should attempt to place the microphone within 6 inches of the worker's ear. Keep in mind that the hearing zone of the employee is not typically a stationary volume. When dominant noise sources exist, then small movements of the employee's hearing zone can result in significant changes in the L_{OSHA} level. In selecting the orientation of the microphone with respect to the sound field, the surveyor should consider the manufacturer's recommendations.

When using an SLM, the surveyor must not allow the operating condition of pertinent noise sources (either up or down) to affect a preselected sequence for taking samples. Often sound surveyors neglect to record the level at a workstation if the equipment is not running at the scheduled sample time, even though the observed occurrence was normal for the production area being surveyed. Such a procedure will bias the data toward higher TWAs than should have been recorded or would have been indicated if a noise dosimeter had been employed. However, if the equipment is experiencing an unusually high number of failures, then this fact should be noted on the survey form.

If the sound surveyor is using an SLM and is surveying several workers in an area, then the periods between measurements usually will be sufficiently randomized due to environmental constraints placed on the surveyor's movements.

Sampling Procedure Examples

Presented in Table 7.3 are the results of a sound survey by an OSHA inspector of a supervisor at a sawmill using both a noise dosimeter and an SLM. The SLM sampling format shown and data analysis outline presented in this section are similar to the format that has been used in the past by some OSHA inspectors (OSHA, 1981). Initially the employee was fitted with the noise dosimeter and properly instructed in its use and care. Then the SLM sampling was initiated. Since other employees were sampled during the same time period, the time intervals between the SLM measurements naturally varied over the sampling period.

In order to determine the employee's TWA, each measured sound level is initially given a relative weighting. The relative weighting is equivalent to the noise dose (divided by 100%) that would have been predicted if the affected employ-

TABLE 7.3
Exposures for one employee as measured by SLM and by dosimeter.

JOB FUNCTION: Supervisor/Sawmill Operation
WORKSHIFT: 08:00 to 16:30 with one half-hour for lunch and no scheduled break periods

Time (HH:MM)	Sound Level (dBA)	Relative Weighting	Comments/Tasks
08:37	84	0.44	Planer coasting to a stop
09:08	78	0.00	Change over on planer
09:46	110	16.00	Standing next to top head
09:50	106	9.19	Standing next to top head
09:55	88	0.76	Planer idling-sharpening head
09:56	108	12.13	Next to planer, making adj.
09:59	87	0.66	Planer coasting to stop
10:08	64	0.00	Changing heads
10:14	64	0.00	Changing heads
10:20	106	9.19	Standing next to planer, 2'
10:27	87	0.66	Planer coasting to stop
10:32	67	0.00	In maintenance room
10:42	106	9.19	Planer running, making adj.
10:55	70	0.00	In maintenance room
11:22	107	10.56	~2' from planer, running
12:20	105	8.00	~2' from planer, running
12:24	—	—	Lunch
13:01	101	4.59	Feeding planer
13:10	100	4.00	Feeding planer
13:30	88	0.76	Planer down
Summed Rel. Wts. =		86.13	

No. of samples = 19 (not including the lunch sample)
DOSIMETER DATA: Time On: 7:54 Time Off: 13.37 Readout (%) = 320.7

ee had been exposed to the measured sound level for 8 hours. As an example, refer to Table 7.3 for the first sound level measured (08:37–84 dBA). Interpolation of Table A-1, Appendix I, provides a relative weighting for this measurement of 44%/100% or 0.44. The relative weighting can also be calculated using the inverted form of Equation 7.5, or

$$\text{relative weighting} = D(\%)/100\% = 10^{[(L_A - 90)/(16.61)]}, \qquad (7.8)$$

where L_A is the measured A-weighted sound level. For this example,

$$\text{relative weighting} = 10^{[(84-90)/(16.61)]} = 10^{(-0.361)} = 0.44.$$

When Equation 7.8 is being used, the contribution of sound levels below 80 dBA is zero since the L_{OSHA} threshold is 80 dBA.

Once the relative weightings for all samples are determined, the average of all the relative weightings is determined. For the example problem presented,

(86.13/19) = 4.53. This number multiplied by 100% is now the estimated employee's daily noise dose, or 453%.

If desired, the TWA can be determined either from Table A-l, Appendix I, or by using Equation 7.5:

TWA = 16.61 log(453%/100%) + 90 = 100.9 dBA.

The duration of the sampling period when using the **noise dosimeter** was from 07:54 to 13:37, including a 30-minute lunch period, or an actual sampling time of 5.1 hours in noise. Since this supervisor typically worked 8 hours per day (excluding lunch), the TWA for the measured dose of 320.7% is:

TWA = 16.61 log[(320.7/100) × (8/5.1)] + 90
= 101.6 dBA.

Even though for this problem the SLM-measured sound levels indicate very strong variability during the work period sampled (effectively 30 dBA), the predictions of the employee's TWA by the two sampling instruments are very close, within 0.4 dBA. The reason that the SLM agrees so well with the noise dosimeter findings is that a sufficient number of samples at appropriate times were obtained. Also it is important to note that we have assumed that the data collected during the sampled periods are in fact representative of the employee's noise exposure during the period that was not sampled.

We could have employed a different approach in determining the TWA for the SLM measurements presented in Table 7.3. First assume that the time period between each measurement is constant (which is not exactly the case for this actual example). Since the duration of the sampling period was from 08:37 to 13:30 less 30 minutes for lunch, or 263 minutes, and 19 samples were obtained, the sampling time represented by each sample is 263/19 or 13.8 minutes. Therefore in using Equation 7.7, t_i = 13.8 minutes and T = 263 minutes. However, since the duration between each level is assumed constant, it would also be correct to let t_i = 1 and T = 19 units of time.

Now evaluating Equation 7.7 yields:

L_{OSHA} = 16.61 log{(1/19) × [1 × $10^{(84/16.61)}$ + 0(less than 80) +
1 × $10^{(110/16.61)}$ + 1 × $10^{(106/16.61)}$ +... +1 × 10(88/16.61)]}
= 100.9 dBA.

Sound Exposure Profiling

All of the sound surveying procedures discussed previously have had as their main focus the determination of the daily noise dose for an individual or for groups of employees that have a common work assignment (department X, sewing machine operators, etc.). The TWAs thus established are assumed to represent the expected distribution of daily noise exposures for all workers located within a department or belonging to a common job category. The sampling methodology utilized with noise dosimeters or SLMs provides both a variability of the samples for an individual employee and a variability of the predicted TWAs across employees. **It is as important to know the variability of the noise exposure hazard (TWA ranges) as it is the variability in the level of protection provided by hearing protection devices.**

It is typically assumed that the distribution of measured noise exposures are representative of the long-term noise exposure of the affected employees. However, to have a truly long-term indication of the affected employee's noise exposure, the sound survey should be repeated every 2 to 3 years until the stability of the measured noise exposures over time is established.

A different approach to determining the employee's daily noise exposure is sound exposure profiling. Profiling involves applying task-based noise exposure procedures to identify worker tasks. Then for each of these tasks, the corresponding L_{OSHA} (or if an employee works at just one task for 8 hours, the task's equivalent TWA) is established. Once all possible worker tasks and their equivalent L_{OSHA} values have been put into a computer model, each worker's expected daily TWA can be estimated if his/her work cycle can be defined, if all related tasks carried out by the worker are included in the computer model, and if the worker carries out the job function in the same manner as the model assumes.

The pros and cons of sound exposure profiling have been studied and this procedure presently has its supporters and critics (Behar and Plener, 1984; Buringh and Lanting, 1991; Driscoll, 1997a, 1997b; Elmaraghy and Baronet, 1980; Erlandsson, et al., 1979; Hager, 1998; Jackson-Osman et al., 1994; Kugler, 1982; Reggie, 1953; Royster, 1995; Royster, 1997b; Shackleton and Piney, 1984; Simpson and Berninger, 1992; Stephenson, 1995). For further discussion of the details of sound exposure profiling, the reader is referred to the paper by Hager (1998).

If all overexposed employees in a plant are included in the hearing conservation program, adequately audiometrically monitored, and the maximum L_{OSHA} is less than around 93–95 dBA, then it makes little difference what format for determining the employees' noise exposures is utilized. Still, it is the opinion of the authors of this chapter that the gold standard today is the utilization of noise dosimeters placed on individual employees. In addition to providing the basic information normally obtained (noise dose, maximum rms level, exposure percentiles, etc.) it also yields a time history (histogram) of the individual's noise exposure over the day surveyed and a database of variability in exposures across different workers in the same job category.

Sound Level Contours

Presented as Figure 7.4 is a sound level contour for one floor of a turbine building (DiBlasi et al., 1983). Sound level contours may be used to illustrate to workers the degree of the noise hazard in their work areas, as an easy-to-understand educational tool during presentations to management, to identify dominant noise sources in a work area or even to identify hot spots (high noise levels) close to dominant noise sources, and to approximate the employees' noise dose.

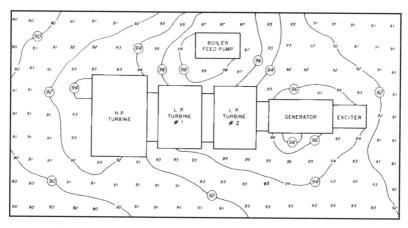

Figure 7.4 — Sound level contours — operating level turbine building (from DiBlasi et al., [1983]).

To construct a sound level contour, the surveyor starts with plant layout drawings or sketches. Then measurement positions are selected on approximately 3-meter grid patterns. If the spacing of the facility's main structural beams is only slightly different from 3 meters, then the existing spacing should be utilized. When the observed sound levels vary significantly, then a smaller grid pattern will be necessary. In the neighborhood of dominant noise sources, it usually will also be necessary to shorten the grid spacing due to the more rapidly changing sound level.

The in-close contour lines drawn in Figure 7.4 are based on 2-dBA changes in the measured sound level. However, if the purpose of creating sound level contours is for employee or management educational purposes, we would recommend that the contours be drawn based on 5-dBA increments in order to be consistent with the classification scheme presented in Table 7.1.

Detailed Sound Surveys: Atypical Exposures

At the present time OSHA inspectors typically only check an employee's noise exposure on the day or days of the inspection. The goal is to establish one or more employees' daily noise exposures in terms of noise dose or the equivalent TWA. Then the measured value is compared to the OSHA noise standard to see if overexposure exists. However, damage risk criteria (predicting how much hearing loss is anticipated for different noise exposures) are based on the assumption that the surveyed employee will be exposed to equivalent 8-hour daily noise exposures, 5 days a week, year after year, over a working lifetime. It should therefore be obvious that an employee who is exposed to a TWA of

90 dBA for only 1 day a week for 5 years will not have the same noise exposure and potential hearing loss that would have occurred if the same employee had been exposed to the same hazard for 5 days a week for the same time period.

The goal of this section is to present a recommended set of procedures for dealing with atypical noise exposures from an inhouse perspective. Hopefully, OSHA will in the future adequately address all of the issues raised in this section.

Daily Noise Exposures

In dealing with measuring or predicting **daily** TWA values, one must use a 5-dB exchange rate if compliance with the OSHA noise standard is the primary goal. If on the other hand compliance with the somewhat more conservative ACGIH noise-exposure criteria is the goal, then a 3-dB exchange rate (equal energy) is required. In this section we will be dealing only with OSHA compliance situations.

Equivalence of Noise Hazard

Consider three different employees who have three different noise exposure periods in three different industrial work environments on the same day: the first employee is exposed to an L_{OSHA} of 100 dBA for 2 hours, the second employee is exposed to an L_{OSHA} of 90 dBA for 8 hours, and the third employee is exposed to an L_{OSHA} of 87.1 dBA for 12 hours. All three employees, if wearing a noise dosimeter, would have accumulated a total noise dose of 100 percent, and by OSHA definition would have been exposed to the same potential noise hazard, a TWA of 90 dBA. For these examples if the employees had been wearing a noise dosimeter for the entire work periods, then the instrument would indicate the correct accumulated noise dose and their equivalent TWAs could either be read directly from Table A-1 in Appendix I, or could be calculated using Equation 7.5 where as noted the reference time period is always 8 hours.

Partial Shift Sampling
Less than 8-hour Shifts

It is frequent that the noise dosimeter sample does not extend across a full-shift period or that the employee works only a half-shift. In such cases it is necessary to adjust the dosimeter indicated noise dose in order to predict the expected full-shift noise dose.

Assume that an employee works a 480-minute workday, is given two 15-minute breaks (one in the morning and one in the afternoon), and is allowed 30 minutes for lunch (see Figure 7.5). Further, assume that the employee's noise exposure is relatively constant during the noise exposure periods and that during the breaks and lunch the noise exposure level is below 80 dBA. In other words, the employee is exposed to the noise environment 210 minutes in the morning and in the afternoon or a total of 420 minutes each day.

Then let's assume that the noise dosimeter is placed on the employee 30 minutes after s/he starts to work and it is removed as the employee goes to lunch

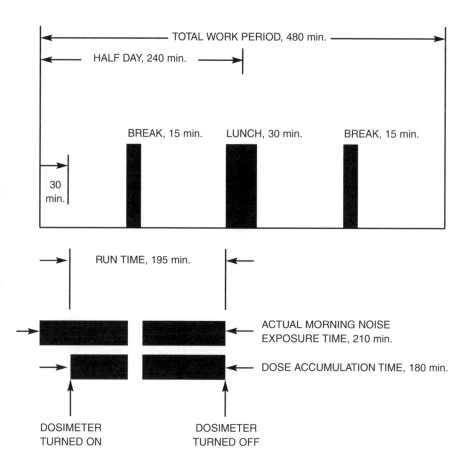

Figure 7.5 — Employee noise exposure cycle and dosimeter utilization period (4- and 8-hour workdays).

(a run time of 195 minutes). Next assume that the dosimeter indicates an accumulated noise dose of 40%. The time period over which the dosimeter was accumulating the noise exposure (noise dose) was 240 (half day) -30 (placement of the dosimeter) -15 (break) -15 (half of lunch) = 180 minutes. Assuming the L_{OSHA} level is constant on the average during the exposure period, the *dose rate* (DR) is 40/180, or 0.222% dose per minute of exposure. Since this employee is actually exposed 420 minutes a day to the noise, then the expected daily noise dose, if the noise dosimeter had been worn by the employee over the total work period, would be 420 minutes of daily exposure × 0.222% per minute of exposure, or 93.2 percent. Using Table A-1, in Appendix I, the equivalent TWA is 89.5 dBA, or using Equation 7.5,

TWA = 16.61 log(93.2/100) + 90 = 89.5 dBA.

After a partial-shift sample the surveyor must take care in interpreting the output of the unit, especially any TWA value it may provide based upon the sample dose. Upon removal of the noise dosimeter (for the limited period sampled), the projected TWA for a dose of 40% would have been 83.4 dBA (Table A-1). That is, an L_{OSHA} value of 83.4 dBA, if constant for 8 hours, would produce the same noise dose as measured.

The indicated TWA at the time of removal of the noise dosimeter could also have been calculated as follows:

$$TWA_{indicated} = 16.61 \log(40/100) + 90 = 83.4 \text{ dBA}$$

Since the DR was 0.222 %/min. over the sampling period, the dosimeter would have indicated an L_{OSHA}, based on a percentage exposure computed as (480 × DR)/100 = 4.8 × DR, of:

$$L_{OSHA} = 16.61 \log(4.8 \times DR) + 90 = 90.5 \text{ dBA} \tag{7.9}$$

Since the actual noise exposure period is 420 minutes, the indicated L_{OSHA} value can be converted to the equivalent TWA as follows:

$$TWA = 16.61 \log[(420/480) \times 10^{(90.5/16.61)}] = 89.5 \text{ dBA}$$

Although, as indicated above, several approaches are available for computing the employee's daily TWA, it is strongly recommended that the surveyor use the first approach outlined above (based on predicting the total daily accumulated noise dose) at least until s/he becomes very comfortable with the equations.

Now, assume that this employee only worked up to the lunch period, in other words a half-day. His DR stays the same, 0.222 %/min. Therefore the employee's predicted total dose accumulation would be 0.222 × 210, or 46.6%. The equivalent TWA can be calculated using Equation 7.5, or

$$TWA = 16.61 \log(46.62/100) + 90 = 84.5 \text{ dBA}$$

This is 5-dB less than the full-day TWA which is what one would expect since the OSHA criteria uses a 5-dB trading relationship.

12-hour Shifts

Consider the noise exposure profile shown in Figure 7.6 which is assumed for one employee's 12-hour shift. The noise dosimeter is placed on the employee after s/he has been on the job 30 minutes and it is removed 60 minutes before the end of the work shift. Therefore it would have indicated 570 minutes of run time that would include 60 minutes on breaks and lunch. Again we assume that the noise exposure is approximately the same during the nonsampled periods as it is during the time periods sampled.

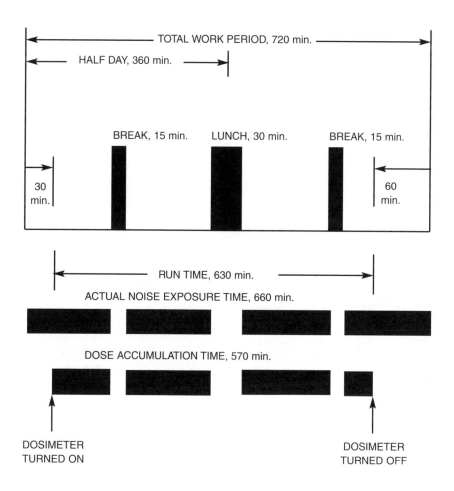

Figure 7.6 — Employee noise exposure cycle and dosimeter utilization period (12-hour workday).

For this problem, assume that the noise dosimeter indicates a noise dose of 95% when removed from the employee. Upon removal, the noise dosimeter should have indicated a TWA of 89.6 dBA because it assumes that the sample dose is the full-day dose. Note that the noise dosimeter does not know that additional noise exposure must be accounted for. Now the DR is 95%/570 minutes or 0.167%/minute. If the noise dosimeter had been sampling during the entire shift, then it would have indicated a noise dose of .167 x 660 or 110.2 percent. Using Table A-1 or Equation 7.5, determine the employee's predicted TWA, or

$$TWA = 16.61 \log(110.2/100) + 90 = 90.7 \text{ dBA}$$

Using the established DR, the instrument-indicated L_{OSHA} value can be determined using Equation 7.9, or

$$L_{OSHA} = 16.61 \log(4.8 \times DR) + 90 = 88.4 \text{ dBA}$$

The TWA can also be predicted by assuming that this level of L_{OSHA} existed over the complete noise exposure cycle and then normalized to 8 hours as required by OSHA, or

$$TWA = 16.61 \log[(660/480) \times 10^{(88.4/16.61)}] = 90.7 \text{ dBA, or}$$

the same value that was obtained using the established DR.

Weekly, Monthly, or Yearly Equivalent Values
Less than 40-hour Weekly Noise Exposures

Assume that an employee is exposed to an L_{OSHA} of 90 dBA for 8 hours, or an equivalent TWA of 90 dBA, which is an equivalent noise dose of 100 percent, only once each week. Now most likely if the OSHA inspector surveys this employee on the day of overexposure to noise, some form of citation would be possible. However, clearly this employee is not presented with a significant potential hearing hazard if during the remaining 4 days s/he is exposed to noise levels that are below OSHA's 80 dBA threshold. A weekly average exposure would better indicate the actual hazard to hearing.

In finding the equivalent noise exposure level averaged over days, weeks, or years, one should use a 3-dB exchange rate (energy average) rather than using OSHA's required 5-dB exchange rate that is used in determining the daily equivalent TWA. The scientific community is in general agreement that when averaging noise exposures over days, weeks, etc., an equal energy exchange rate should be used. However, when it comes to averaging varying sound levels over a single day's workshift, significant disagreement exists. One of the reasons for the selection of a 5-dB exchange rate by U.S. OSHA was to allow for some correction for breaks (intermittency) in the typical daily noise exposure cycle. For example, it was assumed that a typical worker would be expected to take at least five breaks (break periods, lunch, bathroom trips, etc.) during the workday (see Chapter 5).

For the example defined the equivalent weekly exposure level would be calculated as follows:

$$L_{Awkn} = 10 \log[(1/5) \times 10^{(90/10)}] = 83.0 \text{ dBA,} \qquad \text{(7.10)}$$

where the OSHA predicted (1-day) noise exposure of 90 dBA averaged on an equal energy basis is equivalent to a weekly daily TWA of 83.0 dBA.

More than 40-hour Weekly Noise Exposures

At times some employees may work significant overtime where their weekly noise exposures exceed the typical 40-hour noise exposure cycle. Let's assume that an employee works a 7-day work week and that the OSHA determined daily TWAs (using the 5-dB exchange rate) are as follows: 90, 95, 85, 80, 90, 95, and 90 dBA. Since our damage risk criteria is based on 40-hour weekly noise exposures, the predicted equivalent weekly noise exposure must also be based on 40 hours.

Therefore to calculate this employee's (or population's) equivalent daily noise exposure normalized to a 5-day week by averaging the energy addition of the total exposure, with division by five to compute the exposure per day on a 5-day basis, proceed as follows:

$$L_{Awkn} = 10 \log \{(1/5) \times [10^{(90/10)} + 10^{(95/10)} + 10^{(85/10)} + 10^{(80/10)} + 10^{(90/10)} + 10^{(95/10)} + 10^{(90/10)}]\} = 92.9 \text{ dBA} \qquad (7.11)$$

One can quickly check the math by noting that the employee had 2 days with a noise exposures of 95 dBA. Since 2 days at 95 is equivalent to, on a 3-dB equal energy basis, 92 dB over 4 days, the 92.9-dBA answer is obviously in the ballpark.

Yearly Averaged Values

Assume that an employee worked in department A for 20 weeks where the TWA was 95 dBA. Then s/he moved to a new department for 10 weeks where the measured TWA was 90 dBA. Then s/he was promoted and worked out of an office for the remainder of the year where the background noise level was 65 dBA, or below OSHA's 80-dBA threshold for hearing conservation purposes. For this equivalent noise exposure calculation the standard yearly noise exposure cycle is assumed to be 48 weeks.

The yearly average noise exposure is calculated as follows:

$$L_{Ayrn} = 10 \log[(1/48) \times (20 \times 10^{(95/10)} + 10 \times 10^{(90/10)})] = 91.8 \text{ dBA}$$

Again we can check our answer by noting that 95 dBA for 20 weeks is equivalent to 92 dBA for 40 weeks. Including the remaining 18 weeks of non-noise exposure results in a drop of the yearly averaged value by only 0.2 dB.

For additional examples of averaging yearly TWA values over some assumed number of years of noise exposure, the reader is referred to Chapter 17.

A note of caution: When finding a weekly or yearly equivalent value, some limit should be placed on the maximum daily TWA that can be included in the averaging process. Realizing that we are averaging OSHA TWAs (daily exposures) **which are based on a 5-dB exchange rate**, and that scientific data are very scarce on this topic, the authors prefer to err on the safe side and suggest that averaging not take place for employees who exhibit daily TWAs greater than 99 dBA, or for employees who experience maximum rms sound levels greater than 120 dBA for any time duration. In such instances, these employees' noise exposures should be evaluated using the present OSHA single-day criteria.

The Engineering Noise Control Survey

Purpose of the Engineering Noise Control Survey

Since the focus of this handbook is on practical guidelines for implementing effective HCPs, this section is not intended to provide a complete education on how to conduct general engineering noise control surveys (see Chapter 9 for additional recommendations). Rather its purpose is to provide fundamental guidelines for the surveyor and other hearing conservation personnel in identifying major noise sources for which engineering noise control measures may be implemented.

Experience has demonstrated that effective noise control sound surveys can be conducted and resulting noise control projects implemented by individuals with only limited, if any, formal training in noise control techniques. Examples include company nurses and audiometric technicians who, armed only with a basic SLM and calibrator and the benefits of a 2-hour noise control lecture presented during an audiometric technician training course, have carried out effective noise control programs. They simply followed recommended procedures for identifying the dominant noise sources, then brought together company personnel with the best available background experience and technical capability to solve the noise problem, and finally, they stayed with the project until it was completed.

It is important to realize that for most U.S. production facilities with noise problems, personnel with formal training in noise control technology are not available and most likely never will be, simply because of financial constraints. Therefore it is critical that hearing conservation personnel be educated in the basic sound survey techniques for noise control purposes by all means possible.

Determining the Dominant Noise Sources

During the basic and detailed sound surveys, it is likely that several possible dominant noise sources will have been identified. In some instances the dominant noise source will be obvious. An example would be a work area where the typical sound level is less than 90 dBA except when a grinding operation is in progress, at which time the level rises to 95 dBA. A second example would be a workstation where the measured sound level is 90 dBA with the equipment in operation but where the level at the workstation and surrounding area is typically less than 85 dBA with the equipment inoperative.

Area Sources

We will discuss two stages in the identification of noise sources. The first is to determine the *dominant noise source*, or sources in one production area.

Examples of dominant noise sources include individual pieces of production equipment, HVAC (heating, ventilating and air conditioning) systems, and conveyor systems. The second stage of the survey concerns the identification of the *component noise sources* within each dominant source. As an example, the initial survey in a production area might identify two or more packaging machines and a vibrator bowl as the dominant noise sources. Then a follow-up study of each of these noise sources would attempt to identify the component sources for each piece of equipment. After the dominant and the component noise sources have been identified and their respective contributions to the employee's TWAs determined, a priority list for noise control efforts can be established.

Once the noise survey has established typical exposure levels, the surveyor needs to set up a team who will work together to identify the dominant noise sources. Ideally this team should include an experienced machine operator, a mechanic and other personnel who the surveyor feels could contribute to the successful completion of the planned study. Personnel from the "professional" pool, such as management, engineers, or designers, are not desirable members unless they can contribute to the solution or are needed for political reasons.

The ideal procedure consists of running the individual machines in the production area in an incremental sequence while making measurements at specified locations. The surveyor may also elect to run each individual piece of equipment separately and to add together the effects of each successive piece of equipment.

It is desirable to conduct the noise control surveys at a time when all production equipment and other noise sources can be turned off, so that the surveyor has complete control of the background sound level. However, this situation often exists only during the second or third shift, the weekend, or once a year when the plant is closed down for annual maintenance. As a result, the surveyor will find it necessary to be flexible in selecting the survey periods. Fortunately, with today's modern acoustic survey equipment, a sample analysis of the configuration under study (A-frequency weighted value, 1/3 octave-band SPLs, etc.) can be obtained in 3 to 5 seconds (depending on the equipment's duty cycle) and immediately stored in the instrument's data bank.

To demonstrate the procedures for implementing the two different types of noise control surveys, an actual noise source identification effort at a dairy packaging facility is discussed. For the dairy production area surveyed, the background sound level was measured as less than 70 dBA as indicated in Table 7.4. If the background sound level is not at least 10 dBA below the level being evaluated, the measured level should be corrected for the effects of background noise. First, each piece of production equipment was operated independently, resulting in the equipment sound levels indicated by the center column in Table 7.4. As a check on the sound measurements made for individual machine operations, all four pieces of equipment were put into operation successively and the sound level measurements repeated as each additional machine was added.

The sound level data measured for each of the four pieces of equipment can now be combined to check the sound level measured for all units operating simultaneously by using the following equation:

TABLE 7.4
Sound level measurements made to define the dominant machine noise sources at a dairy packaging plant workstation.

Noise Source	Sound Level (dBA)	
	Individual Piece of Equipment	Addition of New Pieces of Equipment
Background	<70	<70
Machine No. 1	80.4	80.8/(No. 1)
Machine No. 2	83.0	85.0/(1+2)
Machine No. 3	83.4	87.3/(1+2+3)
Machine No. 4	83.0	88.6/(1+2+3+4)
All Units Operating	88.6	

$$L_{sum} = 10 \log \sum_{i=1}^{N} (10^{(L_i/10)}), \tag{7.12}$$

where L_{sum} is the level resulting from the contributions of all levels, $L_i=1$ to N (see Chapter 2, Equation 2.11).

For the data presented in Table 7.4, column two, the combined calculated level for all four machines running simultaneously, neglecting the contribution from the background sound level, is

$$L_{sum} = 10 \log(10^{(80.4/10)} + 10^{(83/10)} + 10^{(83.4/10)} + 10^{(83/10)})$$
$$= 88.6 \text{ dBA}$$

The calculated overall level for the sources measured individually agrees with the data obtained by running all units at the same time.

For some production lines, it is not practical to operate individual pieces of equipment because the product of one machine is needed by a following machine in order for the latter to operate properly. Under these circumstances it is necessary to repeat the sound survey measurements as new pieces of production equipment are turned on. This situation produces the type of measurements shown in the third column of Table 7.4.

In order to determine the contribution by each unit, each successive sound level is subtracted from the previous level. This is because the sound level measured for Machines No. 1 and No. 2 operating simultaneously (column three, [1 + 2]) includes the contributions from both machines. As a consequence, the effect of Machine No. 1 plus background must be subtracted from the effect of both machines operating simultaneously in order to obtain the sound level attributed to Machine No. 2 operating independently.

Subtraction of levels can be achieved using the following equation:

$$L_{sub.} = 10 \log[10^{(L_1/10)} - 10^{(L_2/10)}], \tag{7.13}$$

where L_1 is the larger of the two levels being subtracted. Also see Chapter 2, Equation 2.13.

For the data presented in column three of Table 7.4, if 80.8 dBA is subtracted from 85.0 dBA, using Equation 7.13, we obtain:

$$L_{sub.} = 10 \log(10^{(85.0/10)} - 10^{(80.8/10)}), \text{ or}$$

$$L(\text{Machine No. 2}) = 83.0 \text{ dBA}$$

A word of caution: theoretically, there are certain types of acoustic signals (nonrandom with respect to phase) for which Equations 7.12 and 7.13 are not valid (Irwin and Graf, 1979). However, during 30 years of conducting noise surveys the authors have encountered only two instances where the above equations did not produce reasonable predictions. In both of these cases, the dominant noise sources were HVAC fans that exhibited strong approximately equal pure-tone, or single-frequency, components.

Contributors to Each Dominant Noise Source

Assuming that the dominant noise sources have been identified, the surveyor may attempt to identify the contributing noise sources for each dominant source. For this type of sound survey effort, it is essential that a qualified machine operator and mechanic be available. It is relatively simple to operate many production units for sound measurement options, but it is usually an order of magnitude more difficult to operate the various components of the equipment independently.

The procedure for studying the individual machine components is very similar to the procedures described above. However, it is the norm when studying contributing noise sources within a piece of production equipment that the measurements be made as components of the unit are sequentially disengaged or as the machine is reengaged. In general it is recommended that both approaches be utilized and the findings compared, if practical.

A word of caution is again warranted at this point. When the sound levels created by the contributing noise sources of a production unit are studied by either disengaging or reengaging the machine's components, greater differences between the predicted (for individual machine components) and measured overall machine sound levels are to be expected due to the fact that the mechanical loads within the piece of equipment are changing. However, the effect of changing loads on the actual contribution made by individual machine components is normally minor in comparison to the overall effect of these individual sources.

As an example, consider the sound survey findings for a dairy packaging machine as presented in Table 7.5. For this packaging machine, it was relatively easy to turn on the major machine components, since the test procedure followed was similar to the normal start-up procedure. The resulting measured sound levels are presented in the second column. Each level presented in the second column includes the effect of all previous machine components.

To predict the contribution to the measured sound levels from each of the components of the machine, the procedure described in the previous section may be used: that is, by using Equation 7.13 to subtract the contribution from the background + air sources from the measured level for the background + air + vacuum pump. The result is 65.9 dBA, as shown in the third column of Table 7.5. Continuing in this manner would result in an estimation of the contribution of each component of the machine to the overall sound level (83 dBA). A final check on the calculations can be obtained by using Equation 7.12 to combine the calculated contributions from each component as given in the third column of Table 7.5.

The data presented in the third column of Table 7.5 clearly identify the jaw mechanism of this packaging machine as the principal noise source. Now we are able to ask the basic question: If the principal source can be adequately controlled, what would be the net effect on the measured level at the measurement position in question? This can be estimated by adding the contributions of the remaining noise sources (background, air, vacuum pump, burners, idle speed, and defoamer) to the predicted controlled level for the jaw mechanism. Using Equation 7.12 we find a predicted level of 76 dBA (excluding the jaws). If the level contributed by the jaw mechanism were reduced by 10 dBA, the predicted sound level then would be 72 + 76 = 77.4 dBA, or a reduction of 5.6 dBA.

Noise source identification procedures have been used extensively in industrial environments by various company personnel with a reasonable degree of success. Obviously there are pieces of production equipment for which the best noise control consultants have not been able to achieve a cost-effective solution. Therefore one would not normally expect less trained personnel using the procedures outlined above to succeed in such situations. However, experience has

TABLE 7.5

Defining the contributing noise sources for a dairy packaging machine.

Equipment Condition	Sound Level (dBA)	
	Measured Level	Calculated Individual Component Levels
Background + Air	62.7	62.7
+ Vacuum Pump	67.6	65.9
+ Burners	73.6	72.3
+ Idle Speed	74.0	63.4
+ Defoamer (pump)	76.0	71.7
+ Jaws (clamping mechanism)	83.0	82.0
Combined Levels	—	83.0

also clearly demonstrated that solutions to problems do not necessarily come only from the formally educated population.

Information to Record

The basic guidelines for recording sound survey data described in previous sections of this chapter also apply to sound surveys aimed toward noise control applications. However, additional information such as room volume, surface characteristics, relative location of source(s) and receivers, and other pertinent information as discussed in Chapter 9 should be recorded on the noise control sound survey data sheet. It should be noted that information recorded during noise control surveys may also be useful supporting data when the time comes for management to demonstrate to OSHA that a sincere noise control effort is part of the company's HCP.

Special Noise Control Surveys

There are several situations in which the surveyor will find it necessary or desirable to obtain additional information concerning the frequency range over which the acoustic energy radiated by a noise source or sources is located. One use for such information is to more effectively select materials to be used in the construction of an enclosure to control the noise radiated by a source. A second purpose would be to investigate the sound at a workstation in order to obtain a better estimate of the attenuation needed by HPDs (Chapter 10).

The initial instrument of choice to obtain this type of additional information is the octave-band analyzer (Chapter 3). Instruments of greater frequency resolution, such as the 1/3 and 1/10 octave-band filters, are also available and can be very useful especially when dealing with room noise problems, speech or signal detection issues and community noise problems (see Chapters 13, 14, and 15).

In order to demonstrate the use of an octave-band analyzer, a sample problem is presented. Shown in Table 7.6 are the findings of an octave-band analysis at one employee's workstation. While collecting these data, the surveyor noticed that the noise created by a major source seemed to contribute most of its acoustic energy over a narrow range of frequencies. In addition, it was observed that the machine would cease operation for short time periods. Therefore the surveyor decided to conduct a second octave-band analysis but with the SLM's FAST response option selected in order to be able to quickly measure the sound levels when the noise source was not operating. The estimated octave-band SPLs with and without the noise source in operation are presented as steps (1) and (2) in Table 7.6. A comparison of the two sets of measurements verified the surveyor's subjective evaluation in that the only change observed was in the 250-Hz octave band.

Once the contribution from the suspected noise source has been estimated, then the potential effect of a noise control application can be calculated. Begin by assuming that the noise control application would reduce the employee's exposure level at the octave-band center frequency of 250 Hz from 100 dB to 90 dB. Next apply the A-weighting corrections (Chapter 3) as indicated by step

TABLE 7.6
Weighted and octave-band sound pressure levels, and the
predicted effects of spectrum modification.

Step No.		Octave-Band Sound Pressure Levels (dB)							
		Octave-Band Center Frequency (kHz)							
		.063	.125	.250	.5	1.0	2.0	4.0	8.0
(1)	spectrum/								
	100.4 dBC, 93.3 dBA	75	70	100	85	80	85	80	70
(2)	modified spectrum	75	70	90	85	80	85	80	70
(3)	A-weighting corr.	-26	-16	-9	-3	0	+1	+1	-1
(4)	(2) + (3)	49	54	81	82	80	86	81	69
(5)	$L_A = 10 \log (10^{(49/10)} + 10^{(74/10)} + \dots 10^{(69/10)}$ (Equation 7.12)								
	= 89.6 dBA								
(6)	Reduction in L_A = 93.3 - 89.6 = 3.7 dBA								

(3) to calculate the A-weighted octave-band SPLs shown as step (4). Then the A-weighted octave-band SPLs are summed using Equation 7.12, as indicated by step (5), to predict a controlled exposure level of 89.6 dBA.

Finally the predicted reduction in the A-weighted sound levels as a result of the assumed noise control modification is obtained. This is achieved by subtracting the initially measured weighted sound levels from the predicted sound levels after adequate noise controls are implemented, step (6), or 3.7 dBA.

The procedures described for identifying dominant and component noise sources in the previous sections only provided an A-weighted sound level for each source investigated. However, it is sometimes advantageous to know the range of frequencies over which each dominant or component noise source contributes acoustic energy. Presented in Figure 7.7 are the octave-band SPLs for different component noise sources of a packager. The frequency resolution provided by the octave-band filter yields additional insight into the component contributors of this machine.

At the two lower octave-band frequencies investigated (63 and 125 Hz), the major component is the background + air + burners + vacuum pump combination. At the octave-band center frequencies of 250, 500, and 1000 Hz the defoamer dominates, and at the remaining three higher frequencies the idle mechanisms (motors, drive train, conveyors) dominate. Since the A-weighted filter places the most emphasis on the frequency range covered by the octave-band center frequencies of 500 to 4000 Hz, the defoamer and idle noise sources should be given top priority for noise control. In order to provide a visual picture reflecting the effect of A-weighting on the measured octave-band SPLs and thus implying which component should be reduced, the A-weighted octave-band SPLs are often plotted.

The results of the octave-band analysis reveal that overall the packager radiates acoustic energy predominantly at and below 1000 Hz. This fact implies that constraints exist on the options that would be most effective for noise control. As an example, if the measured reverberant field (normally several of these units are

216

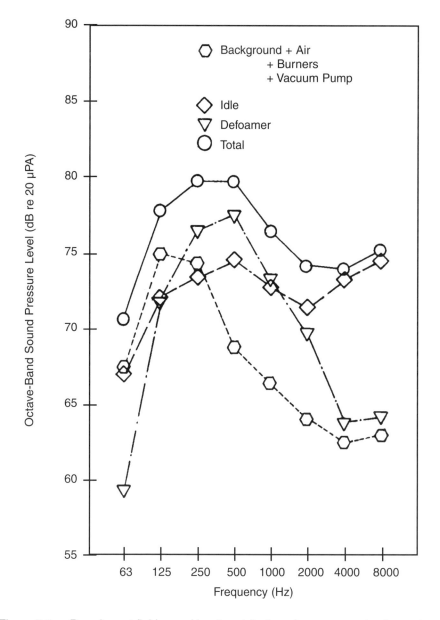

Figure 7.7 — Reverberant-field sound level contributions for components of a packaging machine.

located in an acoustically hard room) is to be controlled by the application of acoustical absorbing materials, then the fact that the octave-band analysis reveals that this machine radiates significant acoustic energy at the lower frequencies dictates that the materials used must be capable of significant absorption at low frequencies. Of course, the C-weighted minus A-weighted sound pressure level from the basic or detailed sound survey data would also indicate, though with less accuracy, that most of the acoustic energy is located in the lower frequencies.

Statistical Factors in Sound Survey Methods

Drawing Conclusions from Sample Data

Sound surveyors hope to make inferences about (draw conclusions from) the noise exposures of employee populations by measuring exposures for small samples of workers. Well-planned sampling procedures and statistical analysis of sound survey data can help describe the noise environment and employees' noise exposures with more confidence than by looking at raw sound level measurements. This section will briefly summarize the application of selected statistical techniques to answer questions such as whether any employees' TWAs ever exceed the 85-dBA OSHA action level or the 90-dBA OSHA Permissible Exposure Limit (PEL), or what the range of the 95% confidence interval around the mean TWA for workers in a particular job classification might be. Readers should consult additional references to determine whether the techniques presented are appropriate for their specific situations and to obtain more detailed guidelines, especially for sampling procedures (DiBlasi et al., 1983; Leidel et al., 1977; Natrella, 1963; Snedecor and Cochran, 1967; USAEHA, 1994; Wells, 1979).

Noise exposure measurements vary due to random factors which are accounted for statistically as well as due to systematic factors which introduce undesired additional variability or bias into measurement values (Leidel et al., 1977). Random factors can include hour-to-hour fluctuations in the noise levels produced by processes or machinery, worker mobility and/or job task changes, and unavoidable measurement error.

Nonrandom systematic sources of variation may include calibration errors, technical errors in measurement procedures, and systematic changes in exposure level due to noise source factors (regular production changes) or employee factors (removal of noise control mufflers, taking extra-long breaks, etc.), as well as measurement system factors (differences between individual SLMs or between SLMs and noise dosimeters). Statistics cannot distinguish between random and nonrandom sources of variation, so measurement mistakes must be avoided and systematic source or employee factors must be identified and handled through sampling techniques (Mellott, 1978).

Because of the variability of the noise environment, each worker has a distribution of exposures over time rather than a fixed exposure level. Similarly, a group of workers performing the same job has a different distribution of exposures than that for any one group member. The entire population of workers across job classifications also has a distribution of noise exposures that is probably even wider. We want to select the distribution which is appropriate for answering the question we have in mind, then describe that distribution based on measurements for a sample of its members. For example, we infer the range of the population mean from the measured sample mean. Because the sample is only a fraction of its parent population and due to measurement uncertainty, inferences are not exact, but rather are probability statements made at a specified level of confidence.

Sampling Considerations

One requirement for valid statistical inferences is random sampling, which means that at each stage in sample selection each available member of the sampling group has an equal chance of being selected for measurement. However, the sound surveyor must be familiar enough with the production processes and employees' job tasks to choose appropriately homogeneous sampling groups. For example, in one production department including workers with job classifications Y and Z, if job-Y workers stay on one end of the room operating the loudest machinery while job-Z employees are on the opposite end doing a quiet task, then the surveyor should take each job category as a separate sampling group rather than combining the entire department into a single sampling group.

By choosing homogeneous sampling groups and then selecting subjects randomly within groups (stratified random sampling), the surveyor can reduce the variability of noise exposure measurements within the sample, and thereby make more accurate estimates of the population characteristics. DiBlasi et al. (1983) discuss the selection of stratified groups based on a noise contour map of the plant developed from SLM noise measurements (see their section 4 and our section, *Sound Level Contours*).

Besides group member selection, other critical sampling factors include the timing and duration of measurements and the number of observations sampled. If noise levels are relatively steady, then simple random selection of sampling times is adequate. See section 4.4.3. of DiBlasi et al. (1983) for detailed instructions on how to randomly select sampling times. However, if periodic variations are suspected (such as morning vs. afternoon, first shift vs. second shift, or summer vs. winter), then stratified random sampling should be used to draw proportionate numbers of sampling times from each period. DiBlasi et al. suggest spreading the measurements out over a duration of 3–6 months in order to tap random noise level fluctuations over time.

The duration of each measurement sample and the number of observations comprise a cost–benefit trade-off. Greater numbers of observations allow the confidence interval around the mean exposure to be defined more narrowly, but at a greater cost of time and effort. However, if on each day of sampling an instrument is used to measure two or more observations, then the sample size may be increased at little extra cost if it can be assumed that the noise environment does not vary systematically through the workshift. For example, one dosimeter can be used to obtain 4-hour doses for two separate employees (which can then be converted to expected 8-hour doses), rather than an 8-hour dose for a single worker. Sampling a larger number of employees has the added advantage of reducing the influence of any "extreme outlier" observations which might be included in the sample, such as a worker who always leans closer to the machinery than average, or an employee who leaves the workstation more than average. Extreme cases cannot be neglected, but in very small samples it may not be clear which values are more typical. A word of caution: for the individual who leans in closer to the equipment, his or her hair cells will receive more damage if s/he exhibits the same susceptibility to noise as the remaining employees. Therefore, caution must always be a concern when eliminating or ignoring an outlier data point.

In contrast, if typical noise doses over time for small groups of individuals are of interest, then the same workers can be sampled repetitively. Behar and Plener (1984) sampled the same employees each day for an entire work week, then integrated the daily TWAs using the OSHA 5-dB exchange rate to find an equivalent weekly exposure. Although OSHA currently does not address the issue of time-weighted average levels for periods longer than a single day, this idea has some merit. However, it would be more protective to use the equal-energy 3-dB exchange rate to combine TWAs across days, as was shown previously in this chapter.

Unrelated to OSHA purposes, the sound surveyor may need to establish long-term equivalent exposures in order to evaluate the adequacy of worker protection. An equivalent exposure level is needed if one wishes to predict potential hearing damage by using a model of noise-induced hearing loss such as the ANSI S3.44-1996 standard (see Chapter 17).

The most common question sound surveys can answer is whether selected workers' exposures exceed criterion values such as the TWA = 85 dBA action level, the TWA = 90 dBA PEL, or a company policy criterion such as a restriction in choice of HPDs at or above TWA = 95 dBA. To determine whether certain job classifications of workers should be placed in an HCP, the surveyor should sample the persons most likely to have the highest noise exposures. If only a few employees are at high risk, then the TWAs can be measured for each one (a nonrandom survey of the complete subpopulation rather than a sample). However, if the individuals most at risk cannot be singled out intuitively, then the surveyor should randomly sample the entire job classification group. Leidel et al. (1977) have provided a guide for the sample size required to attain the desired probability (90% or 95%) of including within the sample at least one of the employees with the highest exposures (either the top 10% of exposures or the top 20% of exposures) based on groups of differing size. The required sample sizes for the 95% confidence level are presented as Tables 7.7a and 7.7b. As group size increases the required sampling proportion decreases. These tables can also be used to determine the minimum sample size for other stratified groups which clearly will be included in the HCP, though of course larger samples always provide better inferences.

The OSHA Hearing Conservation Amendment (HCA) is based on exposures for individuals, requiring that every person whose TWA equals or exceeds the

TABLE 7.7a
Sample size needed to ensure at the 95% confidence level that
the sample will include one or more observations for employees with
exposures in the top 10% of the distribution
(from Leidel et al., 1977).

Size of Group (N)	12	13-14	15-16	17-18	19-21	22-24	25-27	28-31	32-35	36-41	42-50	∞
Required no. of measured employees (n)	12	12	13	14	15	16	17	18	19	20	21	29

TABLE 7.7b
Sample size needed to ensure at the 95% confidence level that
the sample will include one or more observations for employees with
exposures in the top 20% of the distribution
(from Leidel et al., 1977).

Size of Group (N)	7-8	9-11	12-14	15-18	19-26	27-43	44-50	51-∞
Required no. of measured employees (n)	6	7	8	9	10	11	12	14

action level of 85 dBA (a dose of 50%) on any single day be placed in an HCP. Therefore, use of a statistical sampling procedure in place of individual exposure monitoring does not alleviate the employer's responsibility to identify all those workers with doses of 50% or more. However, if the employer's classification of employees into the HCP based on statistical sampling and analysis is at least as protective as the hearing conservation amendment requires, then individual monitoring can be replaced by appropriate sampling of the population.

Use of Confidence Intervals

Because the sample mean is based on only a portion of the entire population, and because noise exposures vary, the mean TWA for a sample cannot be applied indiscriminately. If other individuals had been sampled, or if the same persons had been sampled at other times, then a different sample mean would likely have been obtained. **Setting a confidence interval around the sample mean gives the surveyor a range of values within which the true population mean is expected to lie. Usually the 95% confidence level is used, meaning that of all possible samples, only 5% would give intervals that would fail to include the true population mean.** However, the choice of a confidence level is not dictated by scientific principles, and the surveyor may apply a more or less strict confidence level in certain cases.

Depending on the purpose, the surveyor may need to use a two-sided confidence interval or a one-sided confidence interval. Two-sided confidence intervals extend both below and above the sample mean to indicate the range of mean values expected for different samples. One-sided confidence intervals, which extend only in one direction either below or above the sample mean, are used when only the larger or smaller expected mean values are of interest. One-sided confidence intervals are therefore useful in evaluating criterion compliance questions.

The use of confidence intervals to make criterion compliance decisions is illustrated in Figure 7.8. If the upper confidence limit (UCL), or limit of the upper one-sided confidence interval, on the mean TWA for a single worker is below 85 dBA, then the sampled exposures could be considered in compliance, and the employee might not need to be included in an HCP (case A). If the UCL on the mean TWA is above the 85-dBA criterion, then the employee should be included in the HCP, whether the particular sample's mean TWA is above the cut-off (case B) or below the cut-off (case C). However, Leidel et al. (1977) sug-

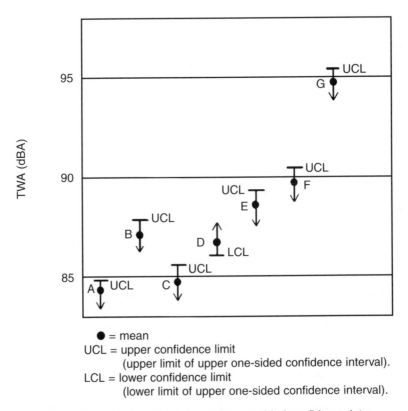

● = mean
UCL = upper confidence limit
      (upper limit of upper one-sided confidence interval).
LCL = lower confidence limit
      (lower limit of upper one-sided confidence interval).

Figure 7.8 — Comparison of boundaries for one-sided confidence intervals to criterion levels relevant for making compliance decisions (adapted from Leidel et al., 1977).

gest that an OSHA compliance officer probably should not issue a citation for noncompliance unless the lower confidence limit (LCL), or limit of the one-sided lower confidence interval, is above the cut-off (Case D). Cases E and F illustrate similar uses of the UCL in relation to the 90-dBA cut-off for mandatory HPD utilization: HPDs would not be mandatory for case E, but would be for case F. Case G illustrates a situation in which the UCL is above a TWA of 95 dBA, so the company might adopt a policy of requiring only HPDs with high real-world attenuation in this case.

Use of Tolerance Limits

In addition to determining confidence intervals for the mean TWA, it is helpful to calculate an upper tolerance limit of the TWA distribution below which a selected percentage of observations would fall (see Natrella [1963], section 2.5). Typically the tolerance limit is calculated to include 75%, 90%, 95%, or 99% of the observations. If the TWA distribution being analyzed is exposures over time for an individual, then the 90% tolerance limit indicates the level below which the person's TWA would be expected to fall 90% of the time. If the distribution under consideration is for a homogeneous group of employees in a situation where every worker performs every potential task on an equal basis, then the 90% tolerance limit indicates the level below which there is a 90% probability that any individual's daily exposure would fall. However, if different individuals in a group were more likely to perform one task than another, then the tolerance limit would not be interpretable with respect to individuals, but only for the group of worker exposures.

Tolerance limits may be useful in making criterion compliance decisions for groups of workers. Although OSHA does not address the issue of infrequent exposure, the company might want to adopt a policy that employees would be placed in the HCP only if their probability of receiving a dose of 50% or more exceeded some selected level, such as 5% or 10%. Such a strategy is in use at one large U.S. Army installation (Henry, 1992).

Example Statistical Applications

The following paragraphs outline steps in analyzing sound survey data for homogeneous sampling groups of employee exposures by completing sample calculations for an example situation.

A basic sound survey in one department of a paper manufacturing plant yielded L_{OSHA} values of approximately 85–95 dBA, so a detailed sound survey was scheduled. Thirty employees worked in the area on each shift, and there were no known production differences between shifts. Using Table 7.7a to determine the minimum sample size needed to ensure at the 95% confidence level that at least one employee with exposure in the top 10% would be included, the surveyor found that a minimum of 18 of the 30 employees should be sampled. In order to check for unsuspected exposure differences between shifts, 18 workers on first shift and 18 on second shift were sampled, for a total N of 36.

Basic Descriptive Techniques
Step 1: Tabulating the Data
The TWAs obtained are tabulated below in ascending order for each shift with associated frequencies. As shown below, the ranges of TWA values for each shift are similar, and in each case most observations fall between 87 dBA and 90 dBA. A *t*-test of the difference between the two means showed no significant difference (see Snedecor and Cochran [1967], Chapter 4), so the data for the two shifts were combined.

First Shift		Second Shift	
TWA (dBA)	Frequency	TWA (dBA)	Frequency
82	1	83	1
85	2	85	1
86	1	87	2
87	3	88	2
88	2	89	5
89	1	90	3
90	3	91	1
91	1	93	2
92	2	94	1
93	1		
95	1		

mean = 88.7 dBA
standard deviation = 3.3

mean = 89.1 dBA
standard deviation = 2.7

Step 2: Plotting the Data
Drawing a simple histogram or frequency polygon of the exposure values is a basic step in data evaluation. A graph allows the surveyor to see if the distribution is bimodal (having two peaks), which might mean that dissimilar job classifications or time periods had mistakenly been combined in a single sampling group. A histogram for our sample data, shown as Figure 7.9, displays an approximation of a normal distribution's bell-shaped curve.

Step 3: Calculating the Sample Mean
The mean of the sample's TWA values (the simple arithmetic average value) is an indicator of the center of the data. For a normal distribution, as shown in Figure 7.10, the mean is the exact center of the distribution.

$$\text{mean} = \sum_{i=1}^{N} x_i / N \qquad (7.14)$$
$$= (82+83+...+94+95)/36$$
$$= 88.92 \text{ dBA}$$

Step 4: Calculating the Sample Standard Deviation
The standard deviation is an indicator of the spread of the data around the mean. For normal distributions with a wide range and great variability, the stan-

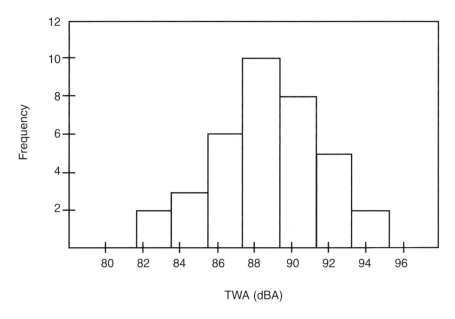

Figure 7.9 — Frequency histogram for the sample TWA data.

dard deviation is larger than for more tightly clumped normal distributions. However, for all normal distributions the standard deviation defines the proportion of observations under various parts of the bell-shaped curve that is assumed to represent the true noise exposure distribution. As shown in Figure 7.10, 68.2% of observations fall within the range from one standard deviation below the mean to one standard deviation above the mean, and 95.4% fall within plus or minus two standard deviations of the mean.

To calculate the standard deviation (SD) of the sample TWA values:

$$SD = \sqrt{\left[\sum_{i=1}^{N} x_i^2 - (\sum_{i=1}^{N} x_i)^2/N\right] / (N-1)} \qquad (7.15)$$

$$= 2.96 \text{ dBA}$$

Step 5: Checking for Normality of the Distribution

The surveyor can check the normality of the sample's distribution by applying the Chi-square goodness-of-fit statistic. If the normality assumption is met, then parametric statistical techniques can be used to analyze the sample; if not, then a log transformation may be applied to the data in an attempt to achieve normality before proceeding (Snedecor and Cochran,1967 [section 3.12]; DiBlasi et

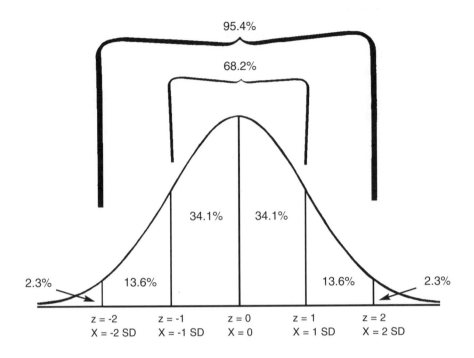

Figure 7.10 — Unit normal distribution showing the percentages of observations within one and two standard deviations of the mean, and corresponding standard scores (Z scores).

al., 1983 [Appendix B]). The Chi-square statistic is based on the difference in expected class frequencies between a normal bell-shaped distribution and a flat distribution. The range of values is divided into classes, each of which must have an expected frequency of at least 5 observations (i.e., the total number of observations divided by the number of classes must be 5 or greater when class boundaries are chosen to have equal probabilities). The number of observations within each class is counted, and the statistic is calculated using the observed and expected frequency counts per class. If the calculated Chi-square does not exceed a critical value then the distribution may be considered normal at the 95% confidence level.

Class boundaries for the Chi-square statistic are computed from standard scores, or Z scores, found in Table 7.8. Z scores represent the distance from the score to the mean in terms of the standard deviation (score mean / SD). The unit normal distribution has a mean of 0 and SD of 1; this special distribution is used to define the proportion of observations with values of Z or less, as shown in Table 7.8.

Calculating the Chi-square goodness-of-fit statistic: If we divide 36 observations into 7 classes we would expect 5.14 observations per class.

TABLE 7.8
Cumulative unit normal distribution showing the proportion P of the population with standard Z scores ≤ the indicated Z values. The value of P for a negative Z score (-Z) equals 1.0 minus the value of P for the corresponding positive Z. For example, for a probability of 0.143, look up 1–0.143 = 0.857 in table to find Z = 1.07, and hence for 0.143, Z = –1.07
(from Natrella [1963], Table A-1).

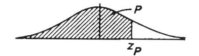

Z_p	.00	.01	.02	.03	.04	.05	.06	.07	.08	.09
.0	.5000	.5040	.5080	.5120	.5160	.5199	.5239	.5279	.5319	.5359
.1	.5398	.5438	.5478	.5517	.5557	.5596	.5636	.5675	.5714	.5753
.2	.5793	.5832	.5871	.5910	.5948	.5987	.6026	.6064	.6103	.6141
.3	.6179	.6217	.6255	.6293	.6331	.6368	.6404	.6443	.6480	.6517
.4	.6554	.6591	.6628	.6664	.6700	.6736	.6772	.6808	.6844	.6879
.5	.6915	.6950	.6985	.7019	.7054	.7088	.7123	.7157	.7190	.7224
.6	.7257	.7291	.7324	.7357	.7389	.7422	.7454	.7486	.7517	.7549
.7	.7580	.7611	.7642	.7673	.7704	.7734	.7764	.7794	.7823	.7852
.8	.7881	.7910	.7939	.7967	.7995	.8023	.8051	.8078	.8106	.8133
.9	.8159	.8186	.8212	.8238	.8264	.8289	.8315	.8340	.8365	.8389
1.0	.8413	.8438	.8461	.8485	.8508	.8531	.8554	.8577	.8599	.8621
1.1	.8643	.8665	.8686	.8708.	8729	.8749	.8770	.8790	.8810	.8830
1.2	.8849	.8869	.8888	.8907	.8925	.8944	.8962	.8980	.8997	.9015
1.3	.9032	.9049	.9066	.9082	.9099	.9115	.9131	.9147	.9162	.9177
1.4	.9192	.9207	.9222	.9236	.9251	.9265	.9279	.9292	.9306	.9319
1.5	.9332	.9345	.9357	.9370	.9382	.9394	.9406	.9418	.9429	.9441
1.6	.9452	.9463	.9474	.9484	.9495	.9505	.9515	.9535	.9535	.9545
1.7	.9554	.9564	.9573	.9582	.9591	.9599	.9608	.9616	.9625	.9633
1.8	.9641	.9649	.9656	.9664	.9671	.9678	.9686	.9693	.9699	.9706
1.9	.9713	.9719	.9726	.9732	.9738	.9744	.9750	.9756	.9761	.9767
2.0	.9772	.9778	.9783	.9788	.9793	.9798	.9803	.9808	.9812	.9817
2.1	.9821	.9826	.9830	.9834	.9838	.9842	.9846	.9850	.9854	.9857
2.2	.9861	.9864	.9868	.9871	.9875	.9878	.9881	.9884	.9887	.9890
2.3	.9893	.9896	.9898	.9901	.9904	.9906	.9909	.9911	.9913	.9916
2.4	.9918	.9920	.9922	.9925	.9927	.9929	.9931	.9932	.9934	.9936
2.5	.9938	.9940	.9941	.9943	.9945	.9946	.9948	.9949	.9951	.9952
2.6	.9953	.9955	.9956	.9957	.9959	.9960	.9961	.9962	.9963	.9964
2.7	.9965	.9966	.9967	.9968	.9969	.9970	.9971	.9972	.9973	.9974
2.8	.9974	.9975	.9976	.9977	.9977	.9978	.9979	.9979	.9980	.9981
2.9	.9981	.9982	.9982	.9983	.9984	.9984	.9985	.9985	.9986	.9986
3.0	.9987	.9987	.9987	.9988	.9988	.9989	.9989	.9989	.9990	.9990
3.1	.9990	.9991	.9991	.9991	.9992	.9992	.9992	.9992	.9993	.9993
3.2	.9993	.9993	.9994	.9994	.9994	.9994	.9994	.9995	.9995	.9995
3.3	.9995	.9995	.9995	.9996	.9996	.9996	.9996	.9996	.9996	.9997
3.4	.9997	.9997	.9997	.9997	.9997	.9997	.9997	.9997	.9997	.9998

Class boundaries (6 boundaries yield 7 classes):

Boundary 1: Probability expected = 1/7 = 0.143
From Table 7.8 the Z score for p = 0.143 is -1.07
Class boundary = mean + (-1.07 × SD)
= 88.92 + (-1.07 × 2.96) = 85.74

Boundary 2: Probability expected = 2/7 = 0.286
From Table 7.8 the corresponding Z score = -0.57
Class boundary = 88.92 + (-0.57 × 2.96) = 87.24

Boundary 3: Probability expected = 3/7 = 0.429
From Table 7.8 the corresponding Z score = -0.18
Class boundary = 88.92 + (-0.18 × 2.96) = 88.38

Boundary 4: Probability expected = 4/7 = 0.571
From Table 7.8 the corresponding Z score = 0.18
Class boundary = 88.92 + (0.18 × 2.96) = 89.45

Boundary 5: Probability expected = 5/7 = 0.714
From Table 7.8 the corresponding Z score = 0.56
Class boundary = 88.92 + (0.56 × 2.96) = 90.58

Boundary 6: Probability expected = 6/7 = 0.857
From Table 7.8 the corresponding Z score = 1.07
Class boundary = 88.92 + (1.07 × 2.96) = 92.09

Observed frequency counts for each class are:
Class 1 (less than 85.75): 5 observations
Class 2 (85.76 to 87.23): 6 observations
Class 3 (87.24 to 88.39): 4 observations
Class 4 (88.40 to 89.45): 6 observations
Class 5 (89.46 to 90.58): 6 observations
Class 6 (90.59 to 92.09): 4 observations
Class 7 (92.1 and above): 5 observations

CAUTION: If data are rounded to whole units (grouped data), it is best to modify the class boundary to conform to the real limits of the measurement units, which are the midway mark between integers (Steel and Torrie, 1980). Otherwise empty classes may occur if some class widths are too narrow to include any integer values. After modifying the boundaries and combining too-narrow classes with adjacent classes, determine the Z scores of the new boundaries using Table 7.8, calculate the expected number of observations for the new classes from the Z scores (same procedure as Step 10), and proceed with the Chi-square computation.

$$\text{Chi-square} = \sum_{i=1}^{N} (\text{observed} - \text{expected})_i^2/\text{expected}$$

$$= [(5-5.14)^2 + (6-5.14)^2 + ... + (5-5.14)^2]/5.14$$
$$= 4.857/5.14$$
$$= 0.94 \text{ with 4 degrees of freedom (df)}$$
$$(\text{df} = \text{number of classes minus three})$$

Determine the critical value of Chi-square from Table 7.9. For 7 classes and 4 df the value is 9.49. Compare the calculated value to the critical value. The critical value is greater than the calculated value, so the distribution may be considered normal at the 95% confidence level.

Making Inferences
Step 6: Setting a Confidence Interval Around the Mean
The 95% confidence interval around the mean is the range of values constructed from the sample in such a way that it has a 95% chance of including the population mean. Since the value of sample means depends on the composition of each sample, the confidence interval around the mean gives the surveyor a way of estimating the true population mean. The alpha probability of error is split into an alpha/2 probability that the true mean will be lower than the confidence interval's lower bound (2.5% error in the lower tail) and an equal probability that the true mean will be higher than the confidence interval's upper bound (2.5% error in the upper tail).

TABLE 7.9
Critical values for the Chi-square goodness-of-fit test for normality of a distribution, using the 95% confidence level. If the calculated Chi-square value for the sample does not exceed the critical value, then the sample may be considered a normal distribution (adapted from Snedecor and Cochran [1967], Table A 5).

Number of Classes	Degrees of Freedom	Critical Value of Chi-Square
4	1	3.84
5	2	5.99
6	3	7.81
7	4	9.49
8	5	11.07
9	6	12.59
10	7	14.07
11	8	15.51
12	9	16.92
13	10	18.31
14	11	19.68
15	12	21.03
20	17	31.41

Calculating a Two-Sided Confidence Interval for the Mean:

Choose desired confidence level and determine the value of the term 1 - (alpha/2).

> Desired level of confidence is 95%
> alpha = 1 - .95 = .05
> 1 - (alpha/2) = 1 - .025 = .975

Determine t for the 1 - (alpha/2) column and the correct df from Table 7.10.

> Find the t value for the .975 column in Table 7.10.
> The df is (N - 1), or 35 for our example.
> The $t_{.975}$ with 35 df = 2.032 (interpolated)

Determine upper bound of confidence interval:

$$UCL = mean + t \, (SD / \sqrt{N}) \qquad \qquad \textbf{(7.16)}$$
$$= 88.92 + 2.032(2.96/6)$$
$$= 89.92 \text{ dBA}$$

Determine lower bound of confidence interval:

$$LCL = mean - t(SD / \sqrt{N}) \qquad \qquad \textbf{(7.17)}$$
$$= 88.92 - 2.032(2.96/6)$$
$$= 87.92 \text{ dBA}$$

Therefore we are 95% confident that the true mean of the TWA values for the population sampled falls within the range from 87.9 to 89.9 dBA (rounded).

Step 7: One-Sided Upper Confidence Interval for the Mean
If the surveyor is concerned only about how high a value the mean might take, then a one-sided upper confidence interval is appropriate. Rather than splitting the 1-alpha probability of the mean's falling outside the confidence interval between an upper tail and a lower tail, the entire 5% probability of error is placed in the single upper tail. This provides a more conservative test which may be appropriate for compliance-related decisions (as discussed in reference to Figure 7.10).

Calculating One-Sided Confidence Interval Above the Mean:

Choose desired confidence level and determine the value of the term (1 - alpha).

> We will use the 95% confidence level, that is an alpha of .05.
> Therefore,
> 1 - alpha = 1 - .05 = .95

Determine t for the 1 - alpha column and correct degrees of freedom from Table 7.10.

> Find the t for the .95 column and 35 df in Table 7.10.
> Interpolated between 30 df and 40 df, t = 1.691

TABLE 7.10
Percentiles of the *t* distribution
(from Natrella [1963], Table A-4).

df	$t_{.90}$	$t_{.95}$	$t_{.975}$	$t_{.99}$	$t_{.995}$
1	3.078	6.314	12.706	31.821	63.675
2	1.886	2.920	4.303	6.965	9.925
3	1.638	2.353	3.182	4.541	5.841
4	1.533	2.132	2.776	3.747	4.604
5	1.476	2.015	2.571	3.365	4.032
6	1.440	1.943	2.447	3.143	3.707
7	1.415	1.895	2.365	2.998	3.499
8	1.397	1.860	2.306	2.896	3.355
9	1.383	1.833	2.262	2.821	3.250
10	1.372	1.812	2.228	2.764	3.169
11	1.363	1.796	2.201	2.718	3.106
12	1.356	1.782	2.179	2.681	3.055
13	1.350	1.771	2.160	2.650	3.012
14	1.345	1.761	2.145	2.624	2.977
15	1.341	1.753	2.131	2.602	2.947
16	1.337	1.746	2.120	2.583	2.921
17	1.333	1.740	2.110	2.567	2.898
18	1.330	1.734	2.101	2.552	2.878
19	1.328	1.729	2.093	2.539	2.861
20	1.325	1.725	2.086	2.528	2.845
21	1.323	1.721	2.080	2.518	2.831
22	1.321	1.717	2.074	2.508	2.819
23	1.319	1.714	2.069	2.500	2.807
24	1.318	1.711	2.064	2.492	2.797
25	1.316	1.708	2.060	2.485	2.787
26	1.315	1.706	2.056	2.479	2.779
27	1.314	1.703	2.052	2.473	2.771
28	1.313	1.701	2.048	2.467	2.763
29	1.311	1.699	2.045	2.462	2.756
30	1.310	1.697	2.042	2.457	2.750
40	1.303	1.684	2.021	2.423	2.704
60	1.296	1.671	2.000	2.390	2.660
120	1.289	1.658	1.980	2.358	2.617
∞	1.282	1.645	1.960	2.326	2.576

Entries originally from Table III of *Statistical Tables* by R.A. Fisher and F. Yates, 1938, Oliver and Boyd, Ltd., London.

231

Determine the upper bound of the one-sided confidence interval above the mean.

$$UCL = mean + t(SD / \sqrt{N}) \qquad (7.16)$$
$$= 88.92 + 1.691(2.96/6)$$
$$= 89.75 \text{ dBA}$$

Therefore we are 95% confident that the true population mean is less than a TWA of 89.8 dBA (rounded).

Step 8: One-Sided Tolerance Limit of TWA Distribution

Table 7.11 presents the K factors for the 95% confidence level; these K factors correspond to Z scores, but with adjustments for the uncertainty associated with sample size, as shown by the left-hand N column. By using a tolerance limit the surveyor may determine at the selected level of confidence the TWA level below which a selected proportion P of the distribution falls, meaning that only 1 - P % of employee exposures would be greater than the tolerance limit.

Calculating an Upper Tolerance Limit for a Distribution:

Choose the desired confidence level.

We will use 95% confidence level.

Choose the percentage of the population distribution which you wish to be included within the tolerance limit.

We will choose 90%, or the .90 proportion.

Find the appropriate K value in Table 7.11 for the selected proportion and the appropriate df.

The K factor for the .90 proportion and N=35 (close enough to our true N of 36) is 1.732.

$$\text{Upper tolerance limit} = mean + K(SD) \qquad (7.18)$$
$$= 88.92 + 1.732(2.96)$$
$$= 94.05 \text{ dBA}$$

Therefore we can say that there is a 95% probability that 90% of worker TWAs will fall below 94 dBA (rounded), while the remaining 10% would be at this level or higher.

Step 9: Proportion of Observations Above a Certain Value

Often the surveyor is more interested in knowing what percentage of TWAs fall at or above a predetermined criterion (such as 85 dBA or 90 dBA) than in setting a tolerance limit for a certain proportion of the distribution. By converting criterion values such as 85 dBA or 90 dBA into standard scores or Z scores (expressed in units of the sample standard deviation), the surveyor can predict

TABLE 7.11

K factors for one-sided tolerance limits for normal distributions, using the 95% confidence level, where N is the number of observations in the sample and P is the selected proportion of the population (from Natrella [1963], Table A-7).

N / P	0.75	0.90	0.95	0.99	0.999
3	3.804	6.158	7.655	10.552	13.857
4	2.619	4.163	5.145	7.042	9.215
5	2.149	3.407	4.202	5.741	7.501
6	1.895	3.006	3.707	5.062	6.612
7	1.732	2.755	3.399	4.641	6.061
8	1.617	2.582	3.188	4.353	5.686
9	1.532	2.454	3.031	4.143	5.414
10	1.465	2.355	2.911	3.981	5.203
11	1.411	2.275	2.815	3.852	5.036
12	1.366	2.210	2.736	3.747	4.900
13	1.329	2.155	2.670	3.659	4.787
14	1.296	2.108	2.614	3.585	4.690
15	1.268	2.068	2.566	3.520	4.607
16	1.242	2.032	2.523	3.463	4.534
17	1.220	2.001	2.486	3.415	4.471
18	1.200	1.974	2.453	3.370	4.415
19	1.183	1.949	2.423	3.331	4.364
20	1.167	1.926	2.396	3.295	4.319
21	1.152	1.905	2.371	3.262	4.276
22	1.138	1.887	2.350	3.233	4.238
23	1.126	1.869	2.329	3.206	4.204
24	1.114	1.853	2.309	3.181	4.171
25	1.103	1.838	2.292	3.158	4.143
30	1.059	1.778	2.220	3.064	4.022
35	1.025	1.732	2.166	2.994	3.934
40	0.999	1.697	2.126	2.941	3.866
45	0.978	1.669	2.092	2.897	3.811
50	0.961	1.646	2.065	2.863	3.766

what percentage of workers' TWAs would fall above the selected criterion. This procedure is discussed by DiBlasi et al. (1983) (see their section 6). Unlike the tolerance limit procedure, the use of Z scores includes no allowance for sample variation, **so there is no level of confidence associated with these predictions.** The mean and standard deviation of the sample are simply taken as the true values for the population, when actually they are only estimates.

Table 7.8 (from Natrella's [1963] Table A-1) gives the Z scores at or below which differing proportions of a distribution are expected to fall based on the normal distribution concept with no adjustments made to correct for the uncertainty of variance estimation due to the sampling process and sample size.

Predicting the Proportion of TWAs at 90 dBA or Higher:

Since Table 7.8 gives Z values at or below which a proportion falls, the desired criterion of 90 dBA must be changed to a value just barely smaller: 89.9 dBA. The table will tell us the proportion of the distribution up through this value, and the remainder falls above it (that is, at 90 dBA or above).

Convert the criterion score to a Z score.

$$Z = \text{(criterion - mean)} / \text{SD} \qquad (7.19)$$
$$= (89.9 - 88.92)/2.96$$
$$= .33$$

Locate the obtained Z score in Table 7.8.
Find the value of Z_p in the left column. Then go across the .3 row to the .03 column to find the P value of Z=.33

Read the associated proportion P from the table.
The table value for Z=.33 is P=.6293.

Therefore, 62.9% of employee TWAs are estimated to fall below 90 dBA, while 37.1% are estimated to fall at 90 dBA or above.

Step 10: Predicting the Proportion of TWAs in a Range
The surveyor may want to know what percentage of employees' TWAs are in the range 85–89.9 dBA. This answer is obtained using the same method as Step 9. The proportion in the range 85–89.9 dBA is simply the proportion which falls at 89.9 dBA or below (from step 8) minus the proportion which falls at 84.9 dBA or below.

Predicting the Proportion of TWAs from 84.9 dBA to 89.9 dBA:

Convert the criterion value 84.9 dBA to a Z score.

$$Z = \text{(criterion - mean)}/\text{SD}$$
$$= (84.9 - 88.92)/2.96$$
$$= -1.36$$

Locate the obtained Z in Table 7.8. For negative Z scores the proportion P equals one minus the P value for a positive Z of the same size. Therefore we look up the P for Z=+ 1.36.

Read the proportion P for the desired Z.
The P for Z=+1.36 is .9131.
The P for Z=-1.36 is (1.00 - .9131) or .0869.

The proportion P in the desired range 84.9 dBA through 89.9 dBA is P for 89.9 (from Step 9) minus P for 84.9.

$$P = .6293 - .0869$$
$$= .5424$$

Therefore 54.2% of the employee TWAs are estimated to fall in the range from 85 dBA through 89.9 dBA.

Using Statistics

The preceding section has provided basic statistical methods for consideration by the sound surveyor in analyzing the database and making predictions based on its content. However, it is essential that the surveyor seek out and study the suggested references in order to apply these techniques properly. The availability of personal computer software for statistical applications (SAS Institute Inc., 1999; Statistical Graphics Corporation, 1999) greatly increases the sound surveyor's ability to use statistics easily and quickly in decision making, provided of course that the software user understands the procedures being applied and can check their validity.

Above all, a practical understanding of the environment from which the data originates is absolutely essential if meaningful predictions are to result. Do not ignore the obvious simply because some governmental regulation, agency, consultant or statistical procedure says otherwise.

Report Preparation

Goals

There are at least three important facets of a successful sound survey report. First, the objective(s) established prior to the survey must be clearly stated in the report and the text should demonstrate the attainment of those objectives. Second, the report should be complete. A sufficient condition for completeness is the requirement that if at a later date an individual different from the author(s) of the report desired to reproduce the report's contents, such could be accomplished provided that the characteristics of the noise environment had not changed. Third, the sound surveyor's report should exhibit an acceptable level of technical writing skill (Brusaw et al., 1976).

Potential Audience

When preparing the report it is important to keep in mind not only the survey objectives but also the potential audience. It is an unfortunate waste of resources to spend days conducting a sound survey only to fail to communicate the findings effectively or to antagonize the reader of the report. As a consequence, management, the key individual or company nurse may fail to properly implement the recommendations of the report.

For example, if the sound surveyor forgets that the primary reason for writing the report is for OSHA compliance purposes, essential information to demonstrate compliance may not be included or adequately covered. However, if the report is to be used as evidence in workers' compensation proceedings, the sound surveyor must keep in mind that different data may be necessary depending upon the state in which the claim has been filed. It is sad to say, but the most likely reason that a sound survey report will come under study at a future date is that some legal process occurs for which the data in the reports is critical to its outcome.

Consider the Political Constraints

When preparing a report careful consideration must be given to local management attitudes and potential political constraints. It is unfortunate, but true, that at some plant sites the report will have to be written so as to account for political realities. For example, it may be advisable to purposely omit from the report any mention of breakdowns of the enforcement of HPD utilization in some production areas, or it may be wise to praise a production manager who has not strongly supported the company's HCP. Failure to do so could irritate the reader, who in turn might limit any potential for further sound survey efforts, noise control modifications, or other projects at the plant site. Still, ultimately the surveyor may be held accountable for his/her decisions regarding the inclusion of such information.

In other words, playing the political game effectively is extremely important in achieving significant progress in the hearing conservation area. The sound survey report through its indicated findings and resulting recommendations is in fact a very important document. The effective sound surveyor will make maximum utilization of this fact.

Report Format

A general outline for preparing a sound survey report is as follows (Brusaw et al., 1976):

    (a) Title Page
    (b) Summary/Acknowledgments
    (c) Background /Introduction and Purpose
    (d) Data Collection/Measurement Methods/Instrumentation
    (e) Analysis of the Data
    (f) Results/Conclusions/Recommendations
    (g) References
    (h) Appendix

This list provides guidance for topics that could be included in a sound survey report. However, the two most important sections of the report will be the summary/acknowledgments and results/conclusions/recommendations sections. The types of individuals who normally implement the findings often will take the time necessary to read only these two sections. Therefore if these sections are not suitably short (limited to two to three pages) the report may be effectively ignored or suppressed by top management. Likewise, if the sound surveyor makes too many recommendations, management may be overwhelmed and disregard the report. Therefore in writing these two sections remember to be brief, striving for the most important actions desired and being sure to keep the political implications in mind.

Do not forget to acknowledge all individuals who provided special assistance during the sound survey. This takes very little effort and can reap substantial benefits.

The background and/or introduction sections should include information such as why the survey was conducted, the time period over which the survey took place, and any other pertinent information that would help the serious report reader to better understand the contents.

A detailed discussion of the data collection and measurement procedures should be included. The instrumentation used and calibration procedures followed should be adequately explained. It is important that the reader of the report be able to verify the analysis of the data that is provided. If the analysis involves detailed calculations and many tables, then only sample calculations and tables should be included in the main body of the report. The remaining information should be put in a section of the appendix where it can be found and studied if desired, but where it does not detract from the readability of the report.

If the surveyor selects materials and equations extensively from other reports, then proper references should be included. Likewise, if the author elects to use a sound survey procedure different from what is commonly understood as acceptable practice, then the source of the justification for a modified procedure should be stated. Reports that include statements such as "The existing noise levels will cost the company $2,000,000 over the next 10 years," without adequate supporting evidence, will fail to achieve credibility during the review process.

If the sound survey was conducted in order to investigate a special problem, such as a request from a manager concerning the noise levels in a computer room, then it is acceptable to limit the reporting effort to a letter-report format or some other type of short summary document. This type of report would typically include a one- or two-paragraph description of the type of sound survey conducted and its purpose, a summary of the findings and any recommendations that would be appropriate.

If the sound survey report is a result of a regularly scheduled repeat survey, then it is not necessary to include information such as sound level measurement, calculation, data recording procedures, etc., that have not changed since the previous survey. The surveyor should simply refer to an earlier report that is on file which described the procedures followed. The periodic update report should include the findings from this most recent survey and point out clearly any significant changes from prior results that have been found in accordance with the general classification scheme presented earlier in Table 7.1.

Recordkeeping

It is important for the surveyor to establish reasonable estimates of employee TWAs, at least within the framework of the classification system defined herein. However, it is equally important that adequate records of the actual measurements upon which the TWA estimations are based, including the results of calibration checks, be maintained.

Recall that the sound survey findings are not only for the purpose of satisfying governmental regulations, such as those of OSHA, but that they also may end up being involved in other legal proceedings such as workers' compensation hearings and negligence suits against the company. With respect to workers' compensation claims for noise-induced hearing loss, the records should be kept for the duration of employment for an employee plus the potential length of time after employment during which the employee could file a claim.

As a consequence, it is recommended that all sound survey records be kept indefinitely. Unfortunately, some companies that had conducted annual sound surveys for several years elected to discard all earlier survey findings or to keep only those obtained every 5 or 10 years. The inability to establish the employee's TWA over lengthy employment periods, during which significant noise-induced hearing loss is possible, can be used against the company during compensation hearings. All data should be kept on file until such time as it is determined that no potential future use of the data exists.

Guidelines for Updating Survey Findings

The decision as to the need for updating sound survey findings will normally depend upon several factors including: (1) the severity of the potential noise hazard, (2) possible significant changes in the predicted TWAs for specific job classifications, (3) changes in TWAs for workers or work locations due to production changes, (4) observed abnormal numbers of significant threshold shifts for the noise-exposed population, (5) the finding of a potentially ineffective HCP based on audiometric database analysis (ADBA) procedures (see Chapter 12), (6) legal requirements specified by federal or state OSHA programs, workers' compensation boards or governing agencies, and union contracts, or (7) the sound survey guidelines as specified by the company's HCP policy manual.

Our general recommendations for conducting and updating sound survey findings are as follows: For all employees, workstations, and job classifications for which the initial sound survey yielded a TWA classification of B or C (85 to 94 dBA, Table 7.1), a follow-up survey should be conducted 2 years later and if the distributions of TWAs in the database are similar, then sound surveys can be conducted every 5 years.

Where the predicted TWA classification is D or higher (95 dBA), it is recommended during the first 3 years of sound surveying that annual sound surveys be conducted. Again if the TWA distributions of the initial three surveys are similar, then the surveys can be conducted every 5 years. The TWA distributions are considered to be similar if less than 25% of the TWAs have changed one letter and if no classification grade has changed more than one letter.

There are several important reasons for conducting more frequent annual surveys when the TWAs are in the D or higher classification. First, the potential of the noise environment to produce significant noise-induced hearing loss over a relatively short time span, 6 months to a year, increases significantly for noise-susceptible employees whose predicted TWAs approach or exceed 95 dBA (ANSI S3.44-1996). Second, the guidelines for conducting sound surveys presented herein do not recommend extensive daily sampling procedures in order to correct possible low estimated TWAs at the time of the initial sound survey. Third, the level of protection by HPDs expected in real-world environments is only around 10 dB.

For plant locations, job descriptions, and worker locations that were initially classified as A, it is recommended that a basic sound survey be conducted every 5 years unless of course there is reason to suspect that the characteristics of the noise environment have changed significantly, for example because of the addition of new equipment.

The End

The recommendations and guidelines presented herein are a framework for the development and implementation of an effective sound survey program. Real-world constraints necessarily play an important part in determining the final form of the company's sound survey efforts. As long as the participants in the HCP are being adequately protected from on-the-job noise exposures, then the sound survey efforts, along with all other hearing conservation efforts, must be judged adequate. Finally, it is often true that the impact of the sound surveyor's efforts on the level of protection provided by the company's HCP will depend not only upon the technical skills exhibited, but also upon the level of concern expressed by this individual toward the noise-exposed population.

Acknowledgments

The authors wish to acknowledge general industry for the sound survey experience gained over the years which provided the background knowledge needed for this chapter. The authors are also grateful for the help of all those persons who contributed to this chapter by criticizing draft versions.

Special thanks is due to Dr. Peter Bloomfield, Professor in the Department of Statistics, North Carolina State University, for his review of the statistical section of the chapter.

TABLE 7.12
List of equations used in Chapter 7 and where appropriate,
the ACGIH equivalent equation.

$TWA = 16.61 \log[(1/8) \times (1 \times 10^{(100/16.61)} + 7 \times 10^{(90/16.61)}] = 92.29$ dBA	**(7.1)**
$L_{A8hn} = 10 \log[(1/8) \times (1 \times 10^{(100/10)} + 7 \times 10^{(90/10)})] = 93.3$ dBA	**(7.2)**
$D = [(C_1/T_1) + (C_2/T_2) + ... + (C_N/T_N)] \times 100$, percent	**(7.3)**
$T_N = (8)/(2^{(L_N - 90)/5})$, hours	**(7.4)**
(**) $T_N = (8)/(2^{(L_N - 85)/3})$, hours	
$TWA = 16.61 \log[D(\%)/100(\%)] + 90$, dBA	**(7.5)**
(**) $L_{A8hn} = 10 \log[D(\%)/100(\%)] + 85$, dBA	
$D (H^*) = D(measured) \times H^*/T^*$, percent	**(7.6)**
$L_{OSHA,T} = 16.61 \log[(1/T) \times \sum_{i=1}^{N} (t_i \times 10^{(L_i/16.61)})]$, dBA	**(7.7)**
(**) $L_{Aeq,T} = 10 \log[(1/T) \times \sum_{i=1}^{N} (t_i \times 10^{(L_i/10)})]$, dBA	
relative weighting $= 10^{[(L_A-90)/(16.61)]}$	**(7.8)**
(**) relative weighting $= 10^{[(L_A-85)/(10)]}$	
$L_{OSHA} = 16.61 \log(4.8 \times DR) + 90$, dBA	**(7.9)**
(**) $L_{eq} = 10 \log(4.8 \times DR) + 85$, dBA	
$L_{Awkn} = 10 \log [(1/5) \times 10^{(90/10)}] = 83.0$ dBA	**(7.10)**
$L_{Awkn} = 10 \log \{(1/5) \times [10^{(90/10)} + 10^{(95/10)} + 10^{(85/10)} + 10^{(80/10)}$	
$+ 10^{(90/10)} + 10^{(95/10)} + 10^{(90/10)}]\} = 92.9$ dBA	**(7.11)**
$L_{sum} = 10 \log \sum_{i=1}^{N} (10^{(L_i/10)})$,	**(7.12)**
$L_{sub.} = 10 \log (10^{(L_1/10)} - 10^{(L_2/10)})$.	**(7.13)**
mean $= \sum_{i=1}^{N} x_i / N$	**(7.14)**
$SD = \sqrt{[\sum_{i=1}^{N} x_i^2 - (\sum_{i=1}^{N} x_i)^2/N)] / (N-1)}$	**(7.15)**
Chi-square $= \sum_{i=1}^{N}$ (observed - expected)$_i^2$/expected	
UCL $=$ mean $+ t$ (SD $/ \sqrt{N}$)	**(7.16)**
LCL $=$ mean $- t$ (SD $/ \sqrt{N}$)	**(7.17)**
Upper tolerance limit $=$ mean $+ K(SD)$	**(7.18)**
$Z = $ (criterion - mean) $/$ SD	**(7.19)**

(**) ACGIH criteria equivalent equation

References

ACGIH (1998). *1998 TLVs® and BEIs®: Threshold Limit Values for Chemical Substances and Physical Agents*, Cincinnati, OH [http://www.acgih.org].

Alpaugh, E. L. (1975). "Sound Survey Techniques," in *Industrial Noise and Hearing Conservation*, edited by J. B. Olishifski and E. R. Harford, National Safety Council, Chicago, IL.

Anon. (1998). "Materials Buyer's Guide," Sound and Vibration, Bay Village, OH, 32(7), 36–42 [SV@mindspring.com].

Anon. (1999). "Instrumentation Buyer's Guide," Sound and Vibration, Bay Village, OH, 33(2), 26–33 [SV@mindspring.com].

ANSI (1994). "Procedures for Outdoor Measurement of Sound Pressure Level," ANSI S12.18-1994, Acoustical Society of America, New York, NY.

ANSI (1995). "Measurement of Sound Pressure Levels in Air," ANSI S1.13-1995, Acoustical Society of America, New York, NY.

ANSI (1996). "Determination of Occupational Noise Exposure and Estimation of Noise-Induced Hearing Impairment," ANSI S3.44-1996, Acoustical Society of America, New York, NY, [http://asa.aip.org].

ANSI (1996). "Measurement of Occupational Noise Exposure," ANSI S12.19-1996, Acoustical Society of America, New York, NY.

ANSI (1998). "Recommendations for Specifying and Testing the Susceptibility of Acoustical Instruments to Radiated Radio-Frequency Electromagnetic Fields, 25 MHz to 1 GHz," American National Standards Institute S1.14-1998, Acoustical Society of America, New York, NY.

Behar, A., and Plener, R. (1984). "Noise Exposure—Sampling Strategy and Risk Assessment," *Am. Ind. Hyg. Assoc. J. 45*(2),105–109.

Berger, E. H.(1985). "EARLog#15-Workers' Compensation for Occupational Hearing Loss," *Sound and Vibration 19*(2), 16–18.

Bruce, R. D., Jensen, P., Jokel, C. R., Bolt, R. H., and Kane, J. A. (1976). "Workplace Noise Exposure Control—What Are the Costs and Benefits?" *Sound and Vibration 10*(9), 12–18.

Bruel, P. V. (1983). "Sound Level Meters—The Atlantic Divide," *Noise Control Engineering J. 20*(2), 64–75.

Brusaw, C. T., Alred, G. J., and Oliu, W. E. (1976). *Handbook of Technical Writing*, St. Martin's Press, NY.

Buringh, E., and Lanting, R. (1991). "Exposure Variability in the Workplace: Its Implications for the Assessment of Compliance," *Am. Ind. Hyg. Assoc. J. 52*, 6–13.

Christensen, L. S. (1974). "A Comparison of ISO and OSHA Noise Dose Measurements," *Technical Review 4*,14–22, Bruel & Kjaer Instruments, Denmark.

Delany, M. E., Whittle, L. S., Collins, K. M., and Fancey, K. S. (1976). "Calibration Procedures for Sound Level Meters to be Used for Measurements of Industrial Noise," *NPL Acoustics Report Ac. 75*, Teddington, Middlesex, UK.

DiBlasi, F. T., Suuronen, D. E., Horst, I. J., and Bradley, W. E. (1983). "Statistics Audio Dosimeter Guide for Use in Electric Power Plants," Empire State Electric Energy Research Corporation, New York, NY.

Driscoll, D. P. (1997a). "Sound Survey Options, Forum #1," in *Proceedings of 22nd Annual Hearing Conservation Conference* (held in Orlando, February 1997), National Hearing Conservation Association, Denver, CO.

Driscoll, D. P. (1997b). "Accuracy and Practical Implications of Noise Exposure Assessment," *American Industrial Hygiene Conference and Exhibition*, Roundtable 233, Dallas, TX.

Elmaraghy, R., and Baronet, C. N. (1980). "An Effective Noise Diagnosis Scheme for Industrial Plants," *Sound and Vibration 14*(9), 14–18.

Erlandsson, B., Hakansson, H., Ivarsson, A., and Nilsson, P. (1979). "Comparison Between Stationary and Personal Noise Dose Measuring Systems," *Acta Otolaryngol. Suppl. 360*, 105–109.

Gasaway, D. C. (1985). *Hearing Conservation: A Practical Manual and Guide*, Prentice-Hall, Englewood Cliffs, NJ.

GenRad (1976). *Instruction Manual: Type 1954 Personal Noise Dosimeter*, GenRad, Concord, MA.

Giardino, D. A., and Seiler, J. P. (1996). "Uncertainties Associated with Noise Dosimeters in Mining," *J. Acoust. Soc. Am. 100*(3),1571–1576.

Hager, L. D. (1998). "Sound Exposure Profiling: A Noise Monitoring Alternative," *Am. Ind. Hyg. Assoc. J. 59*, 414–418.

Henry, S. D. (1992). "Characterizing TWA Noise Exposures Using Statistical Analysis and Normality," in *Proceedings: 1992 Hearing Conservation Conference*, Univ. of Kentucky Office of Engineering Services, Lexington, KY, 65–68.

Irwin, J. D., and Graf, E.R. (1979). *Industrial Noise and Vibration Control*, Prentice-Hall, Inc., Englewood Cliffs, NJ.

Jackson-Osman, P. A., Wynne, M. K., and Kasten, R. N. (1994). "Prediction of 8-Hour Noise Dose from Brief Duration Samples," *J. Am. Acad. Audiol. 5*, 402–411.

Jones, C. O., and Howe, R. M. (1982). "Investigations of Personal Noise Dosimeters for Use in Coalmines," *Ann. Occup. Hyg. 25*(3), 261-277.

Kugler, B. A. (1982). "Development and Validation of Shipboard Noise Exposure Data Acquisition Procedures," NTIS AD-A115 272, Washington, DC.

Leidel, N. A., Busch, K. A., and Lynch, J. R. (1977). "Occupational Exposure Sampling Strategy Manual," HEW (NIOSH) Publication No. 77-173, U.S. Department of Health, Education, and Welfare, Cincinnati, OH.

Mellott, F. D. (1978). "Noise Exposure Sampling: Use with Caution," *Inter-Noise 78*, Noise Control Foundation, Poughkeepsie, NY, 953–956.

Natrella, M. G. (1963). *Experimental Statistics*. National Bureau of Standards Handbook 91, U.S. Department of Commerce, Washington, DC.

OSHA (1974). "Occupational Noise Exposure; Proposed Requirements and Procedures," *Fed. Reg. 39*, 37773–37778, U.S. DOL, Occupational Safety and Health Administration, Washington, DC.

OSHA (1981). "Occupational Noise Exposures; Hearing Conservation Amendment," *Fed. Reg. 46*, 4078–4179, U.S. DOL, Occupational Safety and Health Administration, Washington, DC.

OSHA (1983). "Occupational Noise Exposure: Hearing Conservation Amendment; Final Rule," *Fed. Reg. 48*, 9738–9785. U.S. DOL, Occupational Safety and Health Administration, Washington, DC.

OSHA (1984). "Chapter VI—Noise Survey Data," OSHA Industrial Hygiene Technical Manual, OSHA Instruction CPL 2-2.20A, Office of Health Compliance Assistance, U.S. DOL, Occupational Safety and Health Administration, Washington, DC.

Peterson, A. P. G. (1980). *Handbook of Noise Measurement Ninth Edition*, GenRad, Inc., Concord, MA.

Petrick, M. E., Royster, L. H., Royster, J. D., and Reist, P. (1996). "Comparison of Daily Noise Exposures in One Workplace Based on Noise Criteria Recommended by ACGIH and by OSHA," *Am. Ind. Hyg. Assoc. J. 57*, 924–928.

Reggie, A. S. (1953). " Employee Movement Analysis," *Occup. Health & Safety 52*(3), 31–36.

Royster, L. H. (1980). "Calibration of OSHA-NC Noise Survey Instrumentation," in *Noise Survey Procedures—Phase II* edited by L.H. Royster. Proceedings of a special session at the fall 1980 meeting of the N.C. Regional Chapter of the Acoustical Society of America, D.H. Hill Library, NCSU, Raleigh, NC.

Royster, L. H. (1995). "Comparisons of Sound Survey Methodology Options, Their Respective Levels of Appropriateness, and Guidelines for Choosing," AIHA Conference and Exposition, June, Dallas, TX.

Royster, L. H. (1997a). "The Frequency of Abnormal Sound Levels When Using Noise Dosimeters to Establish the Employee's Daily TWA and the Resulting Effects on the Measured Noise Exposures," AIHA Conference and Exposition, May, Dallas, TX.

Royster, L. H. (1997b). "Sound Survey Options—Your Point of View," Forum #1, Sound Survey Options, *22nd Annual NHCA Conference Proceedings*, Orlando, FL.

Royster, L. H., and Royster, J. D. (1984). "Hearing Protection Utilization: Survey Results Across the USA," *J. Acous. Soc. Am., Suppl. 1*, 74, S5.

Royster, L. H., and Royster, J. D. (1985a). "An Overview of Effective Hearing Conservation Programs," *Sound and Vibration 19*(2), 20–23.

Royster, L. H., and Royster, J. D. (1985b). "Hearing Protection Devices," Chapter 6 in *Hearing Conservation in Industry*, edited by A. S. Feldman and C. T. Grimes, Williams and Wilkins, New York, NY.

Royster, L. H., and Royster, J. D. (1994). "The Impact of Using a 3-dB vs 5-dB Exchange Rate on the Predicted Employee's 8-h Equivalent A-weighted Sound Pressure Level in Three Different Types of Industrial Work Environments," *J. Acoust. Soc. Am. 96*, 3273.

Royster, L. H., Royster, J. D., and Berger, E. H. (1982). "Guidelines for Developing an Effective Hearing Conservation Program," *Sound and Vibration 16*(5), 22–25.

SAS Institute Inc. (1999). *The SAS System for Windows*, SAS Institute Inc., Cary, NC.

Shackleton, S., and Piney, M. D. (1984). "A Comparison of Two Methods of Measuring Personal Noise Exposure," *Ann. Occup. Hyg. 28*(4), 373–390.

Shadley, J., Gately, W., Kamperman, G. W., and Michael, P. L. (1974). "Guidelines for a Training Course in Noise Survey Techniques," National Academy of Sciences—National Research Council Committee on Hearing, Bioacoustics, and Biomechanics, Report of Working Group 70, Office of Naval Research Contract No. N00014-67-A0244-0021.

Simpson, T. H., and Berninger, S. (1992). "Comparison of Short- and Long-Term Sampling Strategies for Fractional Assessment of Noise Exposure," in *Proceedings: 1992 Hearing Conservation Conference*, Univ. of Kentucky Office of Engineering Services, Lexington, KY, 69–73.

Skrainar, S. F. (1985). "The Effects on Hearing of Using a Personal Radio in an Environment Where the Daily Time-Weighted Average Sound Level is 87 dBA," M.S. Thesis, N.C. State University, Raleigh, NC.

Skrainar, S. F., Royster, L. H., Berger, E. H., and Pearson, R. G. (1985). "Do Personal Radio Headsets Provide Hearing Protection?" *Sound and Vibration* 9(5), 16–19.

Snedecor, G. W., and Cochran, W. G. (1967). *Statistical Methods*, sixth edition, The Iowa State University Press, Ames, lA.

Statistical Graphics Corporation (1999). STATGRAPHICS, Statistical Graphics System, STSC, Inc., Englewood Cliffs, NJ [www.sgcorp.com].

Steel, R. G. D., and Torrie, J. H. (1980). *Principles and Procedures of Statistics—A Biometrical Approach* (2nd Edition), McGraw-Hill, New York, NY.

Stephenson, M. R. (1995). "Noise Exposure Characterization via Task Based Analysis," Hearing Conservation Conference III/XX Proceedings, National Hearing Conservation Association, 63-71[www.hearingconservation.org].

Svensson, J. (1978). "Dosimeter Response to Impulsive Noise—Measurement Errors and Their Consequences," in *Proceedings of Inter-Noise 78*, edited by W.W. Lang, Noise Control Foundation, Poughkeepsie, NY, 225–228.

USAEHA (1994). "Noise Dosimetry and Risk Assessment," USAEHA TG No. 181, U.S. Army Environ. Hyg. Agency, Aberdeen Proving Ground, MD, 21001–5422.

Walker, D. G. Jr. (1979). "Noise Control Efforts in the Dairy Packaging Industry," M.S. Thesis, N.C. State University, Raleigh, NC.

Wells, R. (1979). "Noise Measurements: Methods," in *Handbook of Noise Control*, edited by C. M. Harris, McGraw Hill Book Company, New York, NY.

The Noise Manual, revised 5th edition, edited by E.H. Berger,
L.H. Royster, J.D. Royster, D.P. Driscoll, and M. Layne
©2003 American Industrial Hygiene Association

8 Education and Motivation

Larry H. Royster and Julia Doswell Royster

Contents

Introduction

Making It Worthwhile

Of the five phases of an effective hearing conservation program (HCP) (L. H. Royster et al., 1982) the education and motivation phase is least likely to receive the attention it deserves. Education is often viewed as an introductory task to be completed during orientation for new workers, or as an annual nuisance in which canned presentations are repeated to disinterested employees in order to achieve regulatory compliance. Experience has proven that poorly planned and carelessly implemented educational activities are of little value. In contrast, if educational content is relevant and presented with sincerity, the resulting impact on participation in an HCP can be significant. Ongoing motivational efforts establish and nurture the safety culture needed for effective HCPs.

In this chapter we have drawn on our own experiences and those of other hearing conservationists, plus observations from interviews with HCP personnel at industrial and military sites throughout the U.S. (J. D. Royster and Royster, 1990), to provide a broad approach to education and motivation in HCPs. Due to the variety of occupational noise environments, hearing conservationists should seek information from numerous knowledgeable sources and critically assess whether it applies to their own sites. The reader is strongly encouraged to tailor the overall approach of this chapter by including materials specific to the local site when developing educational programs.

The Key Individual

Effective HCPs share common characteristics (L. H. Royster et al., 1982), including strict enforcement of proper hearing protection device (HPD) utilization (Hager et al., 1982), the presence of a key individual who is personally responsible for all aspects of the HCP (Spindler et al., 1979), active communications among all levels of personnel involved in the HCP, and the availability of potentially effective HPDs for the existing work environments. The key individual who sincerely believes in the value of hearing conservation is the critical factor in HCP success, and education and motivation are the main tools the key individual uses to develop and maintain the program. If an existing HCP lacks cohesion and support, an interested person can initiate change by educating top management to understand the benefits to be gained by improving the HCP. By showing management that a more effective HCP will give the company a return on its investment, the interested person will probably be given the authority to become the key individual — the coordinator for all five phases of the HCP.

What's in It for Me?

Each HCP phase requires cooperative participation by other personnel, and the key individual must use education and motivation to elicit this involvement by stimulating participants' self-interest. If an HCP is administered solely on the basis of adhering to company policy, then supervisors and employees may not participate in efforts to save their hearing. Most of us are suspicious when some-body in authority tells us that something is for our own good, and hearing con-servation is no exception to this human tendency. Managers, supervisors, and employees need educational and motivational information that will show them that their active involvement in the HCP on a daily basis is worthwhile because of the resulting benefit to themselves.

The involvement of the sincere key individual in education, motivation, and each other phase of the HCP will go a long way toward changing employees' attitudes and behaviors regarding hearing conservation. It is especially important for the in-house key individual to be involved in each phase if outside consultants are used for audio-metric testing or other HCP services. Employees will not believe in the sincerity of the company if HCP phases are farmed out to consultants without the active supervision of the key individual. Whether the key individual runs the entire HCP in a small com-pany or supervises in-house or external personnel to administer the HCP for a large firm, the unifying drive provided by the key individual can make the program succeed.

Educational Concepts

An Approach to Education
Program Guidelines

An ongoing educational process will include regular activities to present and review HCP information throughout the year. Although the formats used will vary from group meetings to one-on-one con-versations and from commercial films to plant newsletter articles, certain principles apply to each technique (Stapleton and Royster, 1981a):

- Keep it simple.
  - Keep it short.
    - Keep it meaningful.
      - Keep it motivating.

Never forget that industrial employees typi-cally are not eagerly awaiting new knowledge about hearing conservation. Taking a break from the job may be the only reason workers perceive for attending a safety meeting. Therefore, the key individual must design the content of educational activities to capture employees' attention and interest.

The most successful educational efforts are simple in content and presentation. Terminology should be easy to understand, and details that are not directly appli-cable to the employee's daily life should be omitted. The anatomy and physiol-ogy of the cochlea are commonly overemphasized, and discussion of unseen internal areas of the ear only makes the risk of noise-induced hearing loss seem more remote. Similarly, the details of noise monitoring and OSHA time-weighted averages (TWAs) may be unnecessarily confusing (unless employees are seeking knowledge about their noise hazard).

Educational efforts should be short in length. For group presentations an ideal time is 15–25 minutes, with 30 minutes as an upper limit. If an otherwise acceptable film is too long, employees will be dozing in the darkness before the end. If a newsletter article exceeds a few paragraphs, potential readers may stop at the headline. We have encountered companies that group annual education about all Occupational Safety, and Health Administration (OSHA) required topics into marathon sessions lasting an entire day — a good way to guarantee wandering attention!

In order to keep the message meaningful the key individual must pare the content down to the core of ideas that are relevant to employees' daily functioning. Stress only the facts employees need to know: the risk of noise-induced hearing loss, the phases of the HCP, how the individual can prevent hearing loss, and the employer's HCP policies.

Beyond summarizing the facts about the HCP, the educational message must focus on motivating the employee to participate fully in the program. Although some employers actually offer rewards for HCP participation or penalties for nonparticipation, the ultimate motivator is avoidance of progressive hearing loss. The educational program must make the real-life problems associated with hearing loss so clear to employees that they are willing to work to avoid these difficulties, which include impaired spoken communication ability, inability to hear safety warnings, social isolation from friends and family, and reduced enjoyment of leisure activities.

Timing Is Important

In order to maintain the good will of the foremen, supervisors, and managers whose support is needed for an effective HCP, always remember to (Stapleton and Royster, 1985):

- Minimize the effect on production.
- Maintain flexible program schedules.

Production managers need advance notice of educational activities so they can schedule substitute workers if necessary. The key individual should work with department supervisors to arrange audiograms and HPD fitting sessions at times that minimize the impact on production. In scheduling group meetings the educator should be flexible enough to meet production needs by holding programs whenever needed — for example, after second shift.

Another critical aspect of timing involves the regularity of educational activities:

- Keep education going continuously.

Employees tend to forget the importance of the HCP if audiometric monitoring and educational sessions are held during only 1 month out of the year. Spread out the range of activities throughout the entire year to remind workers about hearing conservation over and over in different ways (Karmy and Martin, 1982).

Who Needs Training?

The personnel who need education and motivation are: first, top management and consultants; next, HCP team members including audiometric technicians and personnel who issue and reissue HPDs; then foremen and front-line supervisors; and finally the employees who wear HPDs and others included in the HCP (Else, 1982; Stapleton and Royster, 1981a). By involving each level of the hierarchy in sequence when an HCP is initiated or revised, the company's program can be organized and implemented down through the administrative channels. Feedback at each stage helps to keep the HCP realistic in its goals and clear in its enforcement policies. Although a variety of educational activities are recommended in the following section, the basic method for organizing or reorganizing an HCP involves group meetings to allow communication among the personnel involved.

Management

Even though top management may delegate responsibility for the details of an HCP, it is essential that managers have basic knowledge about the requirements for an effective HCP and the policies needed to handle administrative problems that might develop. In our experience managers typically ask "what if . . .?" questions concerning hypothetical situations that involve potential company liability (L. H. Royster and Royster, 1981).

Education for management should stress the effects of noise on employee health and productivity, the firm's legal obligations for the HCP, the cost–benefit analysis for an effective HCP, and why strict guidelines for HCP procedures will minimize problem incidence. Above all, managers must understand that their active support (rather than nominal policy approval) is needed to establish an effective HCP. For example, HPD enforcement will be taken seriously by supervisors, foremen, and employees only if managers wear HPDs when they visit production areas, however briefly (Else, 1981). To keep management involved, annual HCP update meetings should be held to show the progress of the program and discuss any needs for improvement or policy change.

Consultants

Any consultants who assist in any aspects of the HCP should be educated concerning company health and safety rules to avoid potential conflicts with company policy. For example, before establishing a referral relationship with physicians or audiologists, the key individual should meet with them to outline company policies and describe the operation of the HCP, the physical work environment, the degree of noise hazard, and HPD utilization requirements. By educating the consultants, the key individual will avoid problems such as receiving a doctor's note that a worker should discontinue wearing HPDs due to discomfort, since the doctor would understand from the company's training that the employee must wear HPDs to keep his or her job.

HCP Team Members

The HCP team members are the people who implement the five phases of the program: sound surveys, education and motivation, engineering controls, hearing protection, and audiometric monitoring. These personnel need specific training in how to carry out their duties (unless professional education has already prepared them).

Training in sound level meter and noise dosimetry techniques is available at professional association meetings. Training for audiometric technicians is offered through a 20-hour curriculum specified by the Council for Accreditation in Occupational Hearing Conservation (CAOHC), with certification available to those who complete it successfully. CAOHC courses familiarize trainees with all phases of the HCP and include some hands-on practice in fitting HPDs.

Specific training in HPD fitting is less widely available, although it may be offered in conjunction with meetings of the American Industrial Hygiene Association (AIHA), the National Hearing Conservation Association (NHCA), or the National Safety Council (NSC). Videotapes of HPD fitting techniques are available from some HPD manufacturers, and some written guidelines are available (Berger and Royster, 1988; NHCA, 1996a, b; J. D. Royster and Royster, 1994). The audiologist who reviews employees' audiograms or another consultant who assists with the HCP may be able to provide hands-on training in HPD fitting. It is desirable for the staff members who perform audiometric testing to check HPD fit, even if they are not the initial fitters. In large plants HPD reissuing may need to be performed by numerous persons in locations throughout the plant. HPD utilization and audiometric monitoring are the most critical phases of the HCP, and these contacts with employees provide the best opportunities to motivate them and answer their questions and concerns.

HCP team members should meet together to review the company's HCP policies, to develop uniform and mutually agreeable HCP procedures, and to receive extra educational background not provided in whatever formal training courses they have attended. Some sites offer no education for HPD issuers and reissuers unless they also perform audiometry; however, neglecting training for these personnel almost guarantees that HPD utilization will be ineffective. The key individual should stress to the staff involved with audiograms and HPDs just how important their roles are to the success of the HCP so they will appreciate their responsibility and take it seriously (J. D. Royster and Royster, 1984). Annual refresher sessions are needed to ensure continued uniformity in the audiometric and hearing protection phases.

Foremen or Front-Line Supervisors

Because foremen and/or supervisors receive questions from workers about hearing conservation, they need more detailed information than their employees do in order to handle concerns and complaints. If they lack answers for workers' questions, they will probably adopt an "anything goes" attitude to eliminate further embarrassing questions. Educational sessions for supervisors also provide an opportunity for them to discuss HCP policies and clarify their roles in implementing the program. **The duty that falls most heavily on foremen is daily enforcement of HPD utilization.** They especially need to understand why poorly fitted, improperly worn, or inconsistently worn HPDs are ineffective. They should be explicitly taught how to recognize when employees are wearing HPDs improperly and how to correct the problem. Management should take account of supervisors' HCP efforts in their performance evaluations in order to emphasize the supervisors' responsibilities in the HCP.

Employees

Finally, educational programs should be held for all employees included in the HCP. Face-to-face education is cost-effective because it reduces the occurrence of potential administrative problems by familiarizing workers with HCP phases and policies. It also helps develop the beliefs and attitudes that employees need in order to take an active part in protecting their own hearing by wearing HPDs properly. Though other types of educational activities can be used effectively to remind workers of hearing conservation needs, annual group meetings are essential to provide question-and-answer periods and interaction with managers who can demonstrate the company's commitment to the HCP.

Special Groups

Separate programs for target groups may help to accomplish specific goals. For example, if a new HCP is being established or an older HCP modified, a special program for union officials at the affected plant may smooth the transition. Educational sessions may be needed for engineering staff members to begin a new emphasis on noise controls. Similarly, vendors of equipment could be trained as a group when the company establishes requirements for noise levels on new equipment purchases.

Formats for Education

Depending on the size of the company and the nature of the production process, educational efforts will take different forms. A combination of educational activities scattered throughout the year will be most effective (Harford, 1978). Several options are described in this section.

One-on-One Encounters

The educational opportunities that are most effective in reaching employees to change attitudes and behaviors occur during personal contacts when workers receive their annual audiograms and are fitted or refitted with HPDs (Esler, 1978; J. D. Royster and Royster, 1990). Immediately after taking the audiogram the employee is usually interested in his or her own health status and is especially receptive to constructive feedback. While the audiogram may be reviewed later by a professional, delayed written feedback after the review will have much less impact than a few immediate words from the technician. The verbal feedback should describe the amount and direction of hearing change and relate the results to HPD fit and utilization on and off the job. Praise for the conscientious HPD wearer will reinforce good habits, and concerned warnings for employees with hearing threshold shifts will stimulate improvement. The employee is also most likely to ask questions in this individual setting.

Today's hearing conservation computer software enables the technician to show the employee graphs of his or her own hearing as soon as the audiogram is completed. This capability is the best motivational tool available and should definitely be used! Three results are useful to compare: baseline hearing thresholds, current thresholds, and expected thresholds for the individual's age, gender, and race.

Uninformed management often forbids feedback to employees about their hearing because they fear that employees will be more likely to file for workers' compensation or participate in other legal activities if they realize their hearing is poor. However, when management attempts to hide information from the employees, they will perceive that the HCP is a farce. Potential compensation costs can best be reduced by establishing an effective HCP, and employee feedback is one of the best ways to develop and maintain an effective HCP.

Note that some employers instruct the audiometric technician to deliver the employee's entire annual educational experience at the time of the yearly audiogram. It is a challenge to ensure consistency in covering all the needed information with each employee, and it is usually more efficient to cover topics that apply to all employees in a group setting. The time of the audiogram appointment should be reserved for topics specific to the individual employee's audiometric trends, HPD fit and use, etc.

Small Employee Groups

After one-on-one encounters, the next most effective educational format involves meetings with small groups of 10 or fewer workers. Many industries encourage group meetings of the foreman and his working group to handle personnel difficulties, production problems, and health and safety issues. Such small meetings of a production group are ideal for HCP education because the members share a common work experience and normally feel free to ask questions.

Companies that contract with mobile audiometric testing services may use the time period between audiometric evaluations to present information to the small group of employees who are waiting. This is acceptable as a part of the educational effort, but the employees would still need another opportunity later to ask questions in a less restricted time frame.

Regular Safety Meetings

Many industries hold periodic safety meetings to inform workers about safety and health issues, reinforce company policies, and obtain employee feedback about procedures. If properly structured, these meetings can be excellent educational experiences. The group size should not exceed around 30 people, and the meeting should be run as a two-way discussion, not a one-way lecture. Only if the educator has charismatic communicative abilities can member interaction be achieved in a large group. Safety meetings are an especially good way to foster employee involvement by distributing samples of new HPDs for trial use, then discussing workers' reactions to the product at the next meeting.

Self-Education

Various programs on hearing conservation are available that can be set up for employees to view at will during their breaks, while waiting for their audiometric tests, or during other free periods. Such products based on slide-cassette technology have been sold for many years. The newest products are offered via compact disk, read only memory (CD-ROM) for computer-based instruction in which the employee may test his or her knowledge by answering questions on-screen. Self-education is a useful way to present basic information, but interactive educational opportunities are always necessary to project management's involvement in the HCP and to answer specific workers' questions. Another danger of relying on self-scheduled audiovisuals is that management may start relying on them completely. They may even expect employees to watch them during their own unpaid time. Education necessary to protect the safety and health of the worker should always occur during company time, even if the work schedule must be extended to accommodate the training.

Handout Materials

The least effective employee educational technique is the distribution of pamphlets or similar handout materials. It is impractical to create an easily understood short booklet that would cover all the needed topics and would be so interesting that all employees would take the time to read it. If given to the general worker population without explanation, pamphlets are usually discarded unread. Therefore, we strongly discourage written materials as the primary educational method. However, booklets can supplement live educational presentations, especially if the presenter refers to parts of the material. Written materials may also provide a review for new hires. Personnel who need more detailed knowledge, such as foremen or HPD reissuers, may appreciate materials they can keep for reference.

Bulletin Boards and Company Publications

One way to reinforce HCP educational efforts is to run hearing conservation stories in the plant newsletter or to post HCP-related items on bulletin boards. These posters and articles serve as reminders, not as primary educational efforts. The material should convey a positive message about how HPD utilization and other HCP efforts relate to the worker's greater enjoyment of home life, social occasions, and hobby activities. As an example, post the results of quarterly department-specific survey findings of HPD utilization and/or annual audiometric results, or post the documented hearing losses and their effects on life activities of individuals (without identifying references) due to long-term-noise exposure or as a consequence of immediate acoustic trauma (electrical box explosion, high pressure line exhaust or explosion, etc.).

Using Educational Aids Effectively

Carefully chosen visual/auditory aids are useful in stimulating employee interest during group educational presentations. Straight lecture formats are undesirable because the educator will invariably have trouble keeping the employees' attention in a classroom atmosphere. The best approach is to use aids to illustrate basic concepts and rely on live presentations for company-specific information, discussions, and questions. Fortunately, today there is an abundance of materials available for consideration as referenced in this chapter and Appendix II.

Slide Shows or Slide-Cassette Programs

Slide formats allow the educator more freedom than movies or videotapes because their content can be modified more easily and the presentation can be interrupted more readily. Individual slides in a purchased program can be replaced with shots of the company's own work environments, audiometric facilities, sound survey procedures, and personnel. The educator can easily pause to insert extra information or emphasize specific topics.

Commercial slide-cassettes can be found to suit the needs of most companies, but custom programs can also be developed using local talent for production purposes (Stapleton and Royster, 1981b). The key individual should resist political pressure to feature powerful managers in a custom program since their degree of popularity may detract from the message being presented. Management can best express interest in the HCP by personally attending educational sessions to say a few unstructured words of support.

Videotape Programs

Video equipment offers the same custom options as slide-cassettes with the added feature of movement on the screen. Moving pictures do project situations

255

more realistically than still slides, but greater technical expertise is needed to produce a videotape that would compare favorably in quality with commercial films to which audiences are accustomed (Stapleton and Royster, 1981a). However, if a firm is willing to invest the time and has the resource personnel, producing an educational video is an excellent choice.

Movies

There are numerous commercial films available about hearing conservation that are suitable for use in educational programs. Before purchasing a film, the key individual should review several choices to select one that emphasizes the topics and point of view most compatible with the company's needs. For example, from a study of HPD use across the U.S., it was determined that only about 12 percent of the workforce who were wearing HPDs chose to wear earmuffs instead of earplugs or semi-inserts (L. H. Royster and Royster, 1984a). However, some films show nearly all employees wearing earmuffs because they are obvious to the camera. Older movies tended to overemphasize the anatomy of the hearing mechanism. To ensure that the general tone of the film will be appealing to employees, it is wise to ask representative workers to help in the review and selection process for all educational materials.

Computer-Based Programs

With the ever-increasing numbers of faster and cheaper computers with very large memory storage capabilities, computer-based health and safety materials have exploded onto the commercial market. In addition, searching the Internet provides access to many topics related to hearing conservation. Some HPD providers and professional associations have posted on their home pages not only product information but also excellent educational materials. Today, it is quite easy for an interested employee with Internet access to become better educated in the specifics of various HCP phases than many professionals are. However, this fact should not be viewed with concern. In fact, management should encourage this computer-based self-education process by employees.

Indeed, management should make available worker noise exposure data, OSHA TWAs by departments, and other related data so employees can discover for themselves the magnitude of the noise hazard. Of course management may hesitate to make HCP-related data available to employees. However, they must recognize that today many national societies and agencies are making available noise exposure data for different occupational and nonoccupational environments as well as other related types of information. Therefore, if management does not publish factual employee-related noise exposure data then the employees might obtain a false impression via the Internet or other sources, of the magnitude of the local noise-exposure hazard.

The computer-based information highway is here to stay. During HCP-related educational activities it cannot be neglected.

Educational Variety

No matter how terrific an educational aid may be, repeated use will dull its impact. It is fine to use the same material year after year in orientations for new employees, but the annual programs for current workers should be varied to minimize boredom and inattention. If the company cannot afford to maintain several educational aid choices to alternate among plant locations, local industries are wise to loan and borrow materials from each other to provide variety.

Warning Labels

In the last few years a lot of attention has been given to the concept of warning labels. The legal process has driven manufacturers of all types of equipment and products to develop lengthy warning labels, not primarily for the purpose of educating the public, but to reduce the manufacturer's liability by covering all known hazards that could lead a user to file a civil suit for failure to warn of a potential hazard. Such suits have been filed by employees against the suppliers of noisy hand tools or equipment used in their workplaces. Therefore, the question regarding the educational or motivational benefit of warning labels or signs in the occupational environment needs to be addressed by management.

Human factors research indicates that written warnings are not effective with respect to changing protective behavior against long-term hazards. Our experience in HCPs is that labeling of tools or equipment that produce potentially hazardous noise levels is of no significant value because they take no account of the duration of use, the background sound environment, the user's hearing ability, other hazards, etc. In fact, labels indicating that the user should wear HPDs have created safety hazards related to audibility (reference Chapter 14). Labels affixed to equipment or signs posted in work areas may be helpful if they are (1) based upon site-specific sound survey data, (2) integrated into the overall HCP of the site by educating employees about their meaning, and (3) backed up by real HCP policies. When used within the context of the HCP, such signs remind employees that HPDs are required during the operation of particular equipment, while performing designated tasks, or when entering specified areas. Without meaningful enforcement of HPD use, such reminders are futile.

Who Makes the Best Educator?

From our experience, the best person to facilitate educational sessions is the one whom employees recognize as being sincerely interested in their well-being, familiar with the local plant environment, and available to help with any problems they may encounter concerning the HCP. A strong background of knowledge about all aspects of hearing conservation is also desirable, but not at the expense of the former characteristics.

In-House Personnel

Obviously, the key individual fits the description above. However, other local personnel are also good choices as educators to help the key individual with sessions and/or hold programs at times when the key individual is not available. The effective educators we have observed include nurses, audiometric technicians, industrial hygienists, safety directors, personnel directors, and other staff associated with the HCP. Although in-house personnel may not be polished speakers who are comfortable in front of groups, their actual involvement in implementing the HCP makes up for any lack of speaking skills if they are committed to protecting the employees' hearing. Their everyday involvement in audiometric evaluations, HPD issuing and reissuing, HPD utilization checks, noise monitoring, etc., gives them a credibility that external consultants normally do not exhibit.

External Educators

Many companies have found it necessary and/or beneficial to use outside consulting services to fulfill some of the requirements of an HCP. In considering potential external educators for the workforce, company personnel must be careful to select someone whose knowledge and opinions are consistent with the company's own particular situations. An educator's credibility is quickly destroyed if s/he makes statements that are contrary to the worker's own experiences. For example, a lecturer once stressed the need to use white earplugs, which could be easily checked for cleanliness and cautioned against using plugs with attached cords, since cords could be a safety hazard if they became caught in machinery. These opinions had some merit, but not for the specific audience being addressed: dairy employees who had to wear brightly colored plugs on cords to facilitate finding HPDs that might fall into the milk supply! As this anecdote indicates, company personnel may need to educate the educator about the particular requirements of their production processes.

One sensible way to evaluate potential outside educational sources is to contact two or three industries that have used their services. If similar industries were satisfied with the service they received, this feedback can help make the choice. Several types of sources should be considered in choosing an outside educator, as described below.

Hearing conservation consulting firms offer the services of professionals such as audiologists, industrial hygienists, and acoustical engineers. These firms may not wish to be involved in the education phase of a company's HCP unless they are involved in other phases of the program as well.

State and federal agencies such as OSHA, Mine Safety and Health Administration (MSHA), Departments of Health, Industrial Commissions, etc., often have personnel who do hearing conservation educational programs free of charge with the goal of helping companies develop their own in-house capabilities. Our past experience with these services has generally been favorable. Often management hesitates to utilize government agencies' services because of fear that the educator will tip the enforcement branch of the agency to visit if any problems are observed. Although this potential exists, we are not aware of major problems in this area. If other industries in the area have used the agency's services successfully, we believe there is no reason to avoid these free resources.

Trade associations are increasingly involved in making educational presentations and even developing custom educational aids appropriate to the needs of their particular client industries.

Industrial insurance carriers often have extensive health and safety expertise to share with client industries. Assess carefully the advice of carriers. Although some firms have truly outstanding records of developing educational materials, some agents have attempted to prevent the flow of information to employees out of fear of increasing the number of hearing-loss-related compensation claims filed.

Manufacturers and distributors of HCP products often have effective reference materials available and may have knowledgeable personnel who could serve as educators. However, the company must evaluate the depth of expertise and the prejudices of these sources before accepting their services. A few manufacturers periodically offer detailed free seminars on hearing conservation topics that can be extremely beneficial for HCP personnel such as the key individual, audiometric technicians, HPD issuers and reissuers, and those involved in noise monitoring. Some companies offer extensive educational materials via the Internet.

Suggested Content for Educational Programs

In planning program content for managers, the primary HCP implementation personnel, and employees in the HCP, the key individual must take account of the perspective of each group. The educational and motivational impact can be maximized by modifying the presentation to appeal to their specific interests and needs. In selecting material for each type of presentation, identify the behaviors desired for the target group and choose materials that will provide the information and motivation to stimulate those particular behaviors (Feeney and Nyberg, 1976).

Programs for Management

The goal of education for top management is to develop their committed support of the HCP. Too often HCPs are established as token efforts to satisfy regulatory requirements; such programs are typically ineffective because there is no driving power behind the motions. To improve a poor HCP or maintain a good one, education must start at the top each year. Even if management support has already been achieved, the key individual must renew it by combating potentially adverse influences that also affect management. Legal and medical advisors, trade or business associations, and internal company officials may downplay the importance of hearing conservation or suggest changes that would hurt the program (L. H. Royster and Royster, 1981). Once a year the key individual should present a 1- to 2-hour program for managers to update them on the status of the HCP and show areas that deserve praise or need improvement. Because the profit incentive drives management, educational programs for management must show how an effective HCP benefits the company without costing significantly more than an ineffective program. Suggested topics (L. H. Royster and Royster, 1981) for the education of management include:

1. *Effects of noise on hearing and productivity*, stressing the reduction in administrative problems when the HCP policies are clearly established and consistently followed, and the increase in production efficiency due to easier communication between employees with normal hearing.
2. *Requirements for an effective HCP*, focusing on characteristics of good programs and any areas in which the company's HCP currently falls short — especially regarding the leadership role of management in rewarding the primary HCP personnel for proper program implementation and the modeling role of management in setting an example of HPD utilization for employees.
3. *Compliance with regulations*, including the elements of the noise regulations established by OSHA and other regulatory agencies and the proper procedures and documentation to be followed for future use in potential workers' compensation cases.
4. *Reduction of fears* by answering "What if . . .?" questions and formulating solutions to problems previously identified as concerns of management by interviews or questionnaires prior to the educational session.
5. *Estimated HCP costs*, or presenting actual costs if available.
6. *Estimated compensation costs* with and without an effective HCP as determined by analyzing the company's audiometric database.
7. *Expected and achieved benefits of the HCP*, including a progress update on various departments.

Due to the 1- to 2-hour time limit, each topic should be presented concisely with a written executive summary available for managers' later reference. If questions arise that are not resolved to the educator's satisfaction, these problem topics should receive prompt follow-up and special attention in the next year's meeting.

Programs for Primary HCP Personnel

Included in this group are the nurses, safety/health personnel, audiometric technicians, supervisors, and HPD fitters/reissuers (if different) who actually carry out the phases of the HCP on a daily basis. The target behavior for this group is competent and consistent performance of their assigned functions in making the HCP succeed. Although some small group meetings might be necessary to cover particular aspects such as audiometric testing methods, it is important for all these personnel to meet face-to-face and communicate about their cooperation in achieving an effective HCP. Each person needs to understand the importance of his/her own responsibilities in the overall success of the HCP and to appreciate the purpose of other individuals' responsibilities. We have observed too many cases, especially in larger firms, where the HCP failed because of a lack of communication and cooperation among resource personnel responsible for separate parts of a program. They lacked the unifying supervision of a key individual.

Primary HCP personnel need some reward for good performance beyond the simple satisfaction of knowing they are protecting the employees from excessive noise exposures (who may not always appreciate their efforts). Therefore, it is desirable to have a management representative attend the meeting to express the company's commitment to the HCP and to outline how the performance evaluations of the primary HCP personnel will include their HCP responsibilities.

Suggested content for the educational program for primary HCP personnel is similar to that for employees (next subhead section), but with greater detail of coverage so that these personnel will feel confident to answer workers' questions. They should be provided with a written compilation of common questions and associated responses, as well as written materials for review and reference (Berger, 1982a, 1982b, 1983).

Programs for Employees

The content for a general educational program for employees, such as the orientation for new hires or the company's annual group meeting, should include the areas described below, with emphasis distributed according to the needs of the particular industry and the conditions under which training is done. Within each topic, focus on the points that are directly related to the target behaviors desired from employees: wearing HPDs properly on the job and for leisure activities, and participating in the company's audiometric evaluations.

Effects of Noise and Initial Motivation

Coverage of the physical damage to the cochlea should be minimized in order to stress the resulting social and psychological handicaps that accompany hearing loss. We all value the activities that good hearing allows us to enjoy, so the educational program should emphasize our dependence on the sense of hearing in everyday life. Concentrate on the effects of hearing loss on interpersonal communication, enjoyment of formal group situations (such as meetings and religious services), and participation in recreational activities. Cochlear hair cells themselves are abstract and unappreciated, but they are worth saving because of their value in terms of these everyday activities. Although the payback for protecting one's hearing is normally long term rather than immediate, there are also some current benefits of wearing HPDs, so the educational program should mention the reduction of fatigue, headaches, and tension as side benefits (see also Chapter 1). The noise hazard areas of the production facility should be identified, as well as off-the-job hearing hazards such as gunfire, chain saws, snowmobiles, etc.

Hearing Protection Devices

In most industries, HPDs are the primary method of protecting the worker from noise; therefore, the educational program should spend at least one third to one half of its allotted time covering HPD utilization. Topics must include the purpose of HPDs, how they work, the necessity for a good fit to achieve protection, the ineffectiveness of substitute materials such as cotton or user-modified HPDs with ventilation holes, etc. The types and styles of HPDs available within the company for different ranges of employee exposure levels should be reviewed, emphasizing the company's willingness to work with the employee to find a good personal choice. It is important that the typical reasons given for not wearing HPDs be addressed during the group educational session (Forrest, 1982; J. D. Royster and Royster, 1984).

The educator should keep in mind that the oldest employees are usually the most resistant to wearing HPDs, with the youngest workers also showing more resistance than middle-aged employees (Chung et al., 1982; Foster, 1983; L. H. Royster and Royster, 1985).

Even though fitting and use instructions should be covered on an individual basis at audiometric test time, review information is needed concerning the correct methods for HPD utilization and care. If the group is small enough, HPD fit for individuals should be checked by the educators and fellow employees.

Finally, the company's procedures for obtaining replacement HPDs should be outlined, stressing the need to return to the fitter if a different style or size is desired for better comfort or greater convenience. Examples of deteriorated or worn HPDs can be shown to demonstrate the need for regular replacements. In developing the specific content of the educational program concerning HPDs, the educator should review Chapter 10 of this manual.

Audiometric Evaluations

Employees' attitudes toward their annual audiometric tests vary, and the educational program should attempt to counteract workers' fears, prepare them for the test (Morrill, 1984), and present the audiometric evaluations as a health and safety benefit. Employees may fear audiometric testing and even refuse to be tested if they believe that their jobs, pay rate, or chance for advancement will be jeopardized if they show a hearing loss. The educator must address these concerns by showing how the audiogram results are evaluated and how they are used to trigger follow-up action to provide greater protection from noise for employees with threshold shifts, but not to terminate or demote workers.

Audiometric testing also serves as a health benefit to employees by detecting nonoccupational hearing problems for which medical attention is needed. Although a detailed explanation of audiometric test frequencies and threshold determination methods is not necessary, workers do need an understanding of the normal hearing changes with age and gender, how noise-induced hearing loss adds to age-effect changes, and how increasing degrees of loss and progression of loss to lower frequencies mean increased difficulty in communication and activity enjoyment. It is useful to display posters showing the audiogram format and examples of age-effect hearing loss and noise-induced damage so that employees can refer to them as needed after their annual audiograms.

The Company's HCP Policies

The five phases of the firm's HCP should be summarized, with an assurance that the company is meeting or exceeding its legal responsibilities to the workers through the scope of the program. A representative of management should be present to voice the company's commitment to an effective HCP and to outline the respective responsibilities of the company and the employees. Although positive consequences for wearing HPDs are needed (see motivation section), we feel that the absolute enforcement of proper HPD utilization is also necessary for an effective HCP. Therefore, the negative consequences of failure to wear HPDs must be presented, with management's assurance that the enforcement policies will be followed.

Questions and Answers

Employees should be encouraged to ask questions freely during the educational program, with a special portion set aside only for this purpose. It is strongly recommended that prior to conducting an educational session, the educator should study the most commonly asked questions and develop appropriate responses (Berger, 1982a; Berger, 1982b; Berger, 1983; Mellard et al., 1978; L. H. Royster and Royster, 1981). For questions that arise after the educational session is over, employees should be told to consult their supervisors for simple concerns, with supervisors referring larger problems to the key individual.

Final Motivation

The educational program should close on a positive note. In summary statements, the educator should re-emphasize the company's interest in protecting the workers' hearing and solicit their full participation and cooperation in the HCP so that they will avoid the handicapping results of noise-induced hearing loss. If the educator does not have a true belief in the value of the HCP, employees will quickly sense this insincerity and will discount the educational and motivational message. Therefore, it is important for the educator to retain enthusiasm and self-motivation in order to be effective in motivating employees. Varying the structure of the educational sessions and using visual aids will make the job of presenting the programs less tiring for the educator.

Motivational Concepts

A variety of motivational concepts should, of course, be used within the formal educational programs, but motivation and educational reinforcement are also an ongoing process throughout the HCP's yearly cycle of audiometric testing, HPD reissuing, and informal contacts between the primary HCP personnel and the employees. In fact, all the policies and procedures followed in implementing the HCP should be established with motivational principles in mind. Most performance discrepancies (such as failure to wear HPDs properly or failure to effectively use enclosures installed to reduce equipment noise) are not due to a lack of skill or knowledge, but rather to the existence of barriers preventing that behavior or to a lack of positive consequences for the behavior (Feeney and Nyberg, 1976; Mager and Pipe, 1970).

In an excellent short reference book, Mager and Pipe (1970) outline steps for analyzing performance problems to identify potential solutions. Assuming employees have the skills needed to perform their jobs, solutions involve removing any punishments or obstacles for the desired behavior and arranging rewarding consequences. In hearing conservation the ultimate goal of preventing hearing loss is a rather abstract and distant avoidance of a negative situation, not a positive event. The immediate steps toward the goal, such as wearing HPDs, may be annoying and are seldom obviously rewarding; indeed failure to wear HPDs may be rewarding in itself due to greater comfort. The HCP should be structured to make wearing HPDs as easy and comfortable as possible and to provide frequent positive reinforcement for proper utilization, plus punishment for failure when all else fails (Cluff, 1980; Hager et al., 1982; L. H. Royster and Holder, 1982).

Recent research has investigated whether the frequency and adequacy of hearing protection use can be explained by theories such as health promotion (Pender, 1987) or protection motivation (Prentice-Dunn and Rogers, 1986). Some people are motivated to wear HPDs in order to avoid the immediate annoyance and stress of noise exposure (Melamed et al., 1994). However, the best general predictor appears to be the individual's perceived capability to use HPDs in spite of any difficulties (termed "self-efficacy"), which is inversely related to perceived barriers to HPD use such as discomfort and inconvenience (Lusk et al,, 1995; Melamed et al., 1996). Such research can guide the choice of topics to cover in real-world educational and motivational efforts. Other investigators confirm the conventional wisdom that management leadership and the establishment of a "safety culture" are critical for effective hearing conservation as well as all other health and safety programs (Franks et al., 1996; Leinster et al., 1994). Both views provide useful input for enhancing HCP effectiveness through education and motivation.

The following sections present a variety of motivational ideas that can be employed during HCP activities.

Motivation for Wearers of Hearing Protection
Annual Audiometric Findings

The contact between the audiometric technician and the employee at the time of the annual audiogram provides the single best opportunity to influence that individual's behavior regarding the HCP. As mentioned earlier, the worker's interest is aroused regarding the audiogram results due to concern for his or her general health, worries about hearing loss as a sign of aging, anxiety that a hearing impairment would affect job security or promotion opportunities, and fear that the presence of a potentially stigmatizing hearing loss (which the employee may already be aware of) would become known to fellow workers. For these and other reasons the employee will be receptive to constructive comments about his/her hearing status after the audiogram (Hager et al., 1982; L. H. Royster et al., 1982).

In some HCPs, two individuals work as a team, with one person doing the testing while another person obtains auditory history updates before the audiogram and gives feedback about the results. Although it is not necessary to give the worker a copy of the audiogram at this time, it is extremely beneficial for a visual comparison of past and current threshold levels, plus the expected age-related thresholds, to be shown to the employee. Graphic display of these thresholds on a computer screen is the best method of feedback; the visual display makes the amount of hearing improvement or decline much more concrete to the employee than numerical threshold values.

The content of the feedback information provided about hearing status will depend on company policy, the experience and knowledge of the technician, and the degree of change in light of the employee's medical and auditory history. The audiologist or physician who reviews the audiograms can provide specific guidance on immediate feedback (see Chapter 11). If the current thresholds indicate stable or improved hearing, the employee should be commended and urged to continue to wear HPDs during noise exposures. If the worker makes comments about off-the-job situations in which s/he wears HPDs, the technician must be especially sure to praise him/her and to make a note about the comments for future use in educational/motivational activities for other employees. For example, one worker described how she wore HPDs when she accompanied her husband for his weekend target shooting, even though she herself did not shoot. Anecdotes such as this are the best source of effective program materials, and they are free for the asking if the HCP personnel will simply take time to observe and record them.

If the audiogram results show significantly declining thresholds, the technician should caution the employee about the changes, inquire about situations involving unprotected noise exposure on or off the job, check the adequacy of HPD fit and the employee's HPD placement skills (see next topic), and ask whether there are any problems with the current HPDs that discourage their consistent use. If HPD difficulties are reported, different options should be tried to find the best protector for the individual. The employee may also express general concerns, such as there being no use in wearing HPDs since s/he already has a substantial hearing loss. The technician must emphasize the need to save the remaining hearing, expressing personal concern and the company's concern for the individual. Any fears about termination or demotion must be counteracted if the worker is cooperating in the HCP efforts. However, if an employee refuses to wear HPDs, the technician must emphasize that company policy requires termination or transfer to a non-noise job (possibly at lower pay) for safety/health infractions involving the HCP.

Individual Evaluation of HPD Adequacy

If an employee shows significant threshold shifts or expresses concern that his/her HPDs do not seem to give good protection, the audiometer can be used to give an approximate indication of the effectiveness of earplugs as worn. As long as the earplug style does not protrude far enough from the earcanal to contact the audiometric earphone speaker, just measure the individual's hearing thresholds at 500–1000 Hz with and without the earplugs. The difference in measured thresholds gives an estimate of the amount of protection provided in the work environment (see Chapter 10 for additional details). Special earplug test devices are commercially available. If the indicated attenuation is insufficient, it will be necessary to reinstruct the worker regarding proper insertion methods and/or consider whether a different size earplug, or an alternative style of HPD is needed.

The process of determining HPD adequacy and showing the difference in attenuation between a poorly fitted or incompletely inserted plug and a proper choice can be very motivating to employees because it demonstrates that properly used HPDs can really make a difference. This demonstration may convince the dubious or inconsistent wearer that HPDs are worth the bother. On the other hand, some conscientious workers may worry that their HPDs are inadequate because of the misconception that they should not be able to detect speech or any other sounds through well-fitted HPDs, even though the educational program has covered this topic. In this case the effectiveness check will reassure them that their protectors are doing their job and have reduced the noise to nonhazardous levels, even though it is still audible. The results of HPD effectiveness checks for individuals can be described in posters for bulletin boards, vu-graphs for group educational meetings, or newsletter articles to spread their impact to other workers in the HCP.

Examples of Permanent and Temporary Hearing Loss

Most occupational audiometric databases contain examples of permanent and temporary noise-induced hearing loss from on-the-job as well as off-the-job sources. These audiograms, plus personal comments from the affected individuals, are excellent motivational material because they relate to real-life experiences of fellow employees. Of course, the identity of the affected persons must be kept anonymous by removing all identifying information, unless the permission of employees has been obtained. Some individuals are willing to identify themselves as victims of hearing loss and give brief motivational talks to groups of coworkers to encourage them to protect their own hearing. This type of personal testimony is the best endorsement the HCP can receive. For example, we visited a newspaper printing facility in which the press room foreman, a 30-year veteran who lost his hearing before the introduction of the HCP, personally educated new hires about the handicaps of hearing loss and spoke each year to the press crew as a group to remind them how his life was affected by his impairment, motivating them to wear their HPDs conscientiously.

When creating a poster or newsletter article from an individual's example of hearing loss, the writer should include the age and gender of the affected person, the probable noise source that caused the damage, and the effects of the loss on everyday activities. Such effects can be noted from comments individuals make to the audiometric technician, as shown by these examples:

- *"I get angry at my daughters when they don't speak clearly."*

- *"I have a hard time understanding the words when the church choir sings or listening to the lyrics of a song on the radio in my car."*

- *"Sometimes I feel like people avoid talking with me because they have to repeat what they say so much."*

Males frequently mention difficulty understanding female speakers, especially their daughters or grandchildren, because of the higher frequency and lower acoustic power levels of female voices. Discussion of these types of personal comments brings the effects of noise close to home, showing employees why their hearing matters to them enough to protect it.

Hearing aids should be discussed to counteract the impression that if a hearing impairment occurs, then an aid will make things good as new. Although modern hearing aids have improved tremendously, and many individuals with high-frequency loss can benefit more from amplification now than in the past, an aid does not improve poor hearing to the degree that eyeglasses correct poor vision. Therefore, no one can afford to be careless about HPDs. The best way to communicate the limitations of hearing aids is to present the experiences and opinions of several employees who wear them.

It is also useful to present examples of temporary threshold shifts (TTSs) from occupational noise or hobby exposures such as guns or chain saws to demonstrate that single noise exposures really do affect the ears temporarily, with permanent damage gradually accumulating. Examples of TTS are often observed during the annual audiometric test sessions, and these provide effective motivational materials, especially if the results of retesting are available to show improved thresholds after TTS recovery. Special TTS studies can be done to measure thresholds at the beginning and end of a work shift to convince workers that inadequate HPD utilization (poor fit and/or placement) will

result in temporary loss (L. H. Royster, Royster, and Cecich, 1984; Zohar, 1980; Zohar et al., 1980).

In promoting HPD utilization it is effective to tell workers about nonoccupational and even non–noise-related uses of HPDs, especially those with a twist of humor. One employee noted that he wore his earplugs more off the job than at work; asked why, he responded that they were the best thing he knew to shut out the noises produced by his kids and neighbors. The spouses of habitual snorers also find HPDs indispensable. We have heard that closed-cell foam earplugs are good fishing tools if placed below a sinker on the line, so that the foam's buoyancy will gradually wiggle the bait up through the water. Knowing extra uses of HPDs may make employees take a fresh look at them, as well as adding interest to the educational program.

Annual ADBA Findings for Departments or Groups

It is difficult to motivate some employees on an individual basis because they do not like to be singled out and are more comfortable receiving information as part of a group. In addition, group feedback provides an extra incentive for the group to compare their status in HCP activities with other departments. One way to achieve group feedback is to post the findings of annual audiometric database analysis (ADBA) procedures (see Chapter 12) in simple bar graph formats. However, care must be exercised to present a range of acceptable results that all departments can achieve, rather than simply comparing raw numbers for different groups. Population differences in characteristics such as age, gender, race, mean hearing level, etc., could affect the results regardless of the degree of protection being achieved.

Influence by Management and Peers

There is no more effective way to destroy the employee's interest in the HCP than for top managers, middle managers, primary HCP personnel, or front-line supervisors to enter noise hazard areas without wearing HPDs (Else, 1981; Hager et al., 1982; Lofgreen et al., 1982). The reason these individuals usually give for their failure to observe HCP rules is that they will be in the area only a short time. Although they may not be overexposed based on the OSHA criteria, they need to show support for the HCP by wearing HPDs for any noise exposure. The appearance of managers in the formal educational programs to advocate HCP policies will reinforce their support for the program, but their everyday participation will mean much more to employees. In fact, management should ask workers to require any individual who enters a noise hazard area without protection to leave and obtain HPDs before returning (Sadler and Montgomery, 1982). We are familiar with successful HCPs in which employees feel comfortable in asking their supervisors and visiting guests to put on HPDs or leave the area.

In addition to the formal company hierarchy, there is also an informal pecking order within work groups that can be tapped to the advantage of the HCP (Carroll, 1982; Cluff, 1980). If the lead man or woman within a group (the one who commands the respect of coworkers) can be motivated to set a good example by wearing HPDs properly, then others will follow. Such employees are especially appropriate to be asked to try out new types of HPDs that are under consideration for purchase; they can collect feedback from their work groups about the practicality and comfort of new styles, as described in the next subsection.

Employee Input into HCP Decisions

When employees help to select their own protective equipment they are more willing to wear it in spite of inconveniences (Else, 1982). For example, one safety director obtained special disposable respirators because employees had complained about the trouble involved in washing their current style. After a trial period with the new respirators, the employees decided that they preferred the old ones after all and were content to wear them because they had made the decision. Similarly, if an individual who complains of discomfort with HPDs is given individual attention and allowed to try each permissible option the company has to offer, then the worker will feel more satisfied even if a degree of discomfort remains (L. H. Royster and Royster, 1985).

Wise key individuals know that they can pick up good information about the practicality of various HPDs for their own work environments by spending a few minutes chatting informally with employees and foremen in the coffee break area. Employees are often able to offer potential solutions for problems they experience in HPD utilization.

On a more formal basis, it is good policy to introduce any changes in the HPD selection or other aspects of the HCP in safety meetings in order to ask workers for their feedback and suggestions about implementation of the change (Komaki et al., 1980). For example, if operators are consulted before the construction of noise-controlling machinery enclosures, the enclosures can usually be designed to permit efficient machine operation as well as noise reduction (J. D. Royster and Royster, 1990). Some industrial managers frown upon this type of employee involvement, but it is a good way to foster worker support for HCP policies and procedures.

Examples of Worn and Abused HPDs

One of the more difficult daily challenges for the hearing conservationist is to motivate employees to replace HPDs that are worn out or deteriorated, and not to abuse their protectors by making creative modifications (Gasaway, 1984; Ohlin, 1981; L. H. Royster and Holder, 1982; L. H. Royster and Royster, 1985). Attitudes regarding HPD replacement vary dramatically from plant to plant; in some sites workers throw away reusable HPDs on a daily basis because they are unconcerned about company costs, while in other environments employees hang onto HPDs designed for short-term use for weeks or months because they are cost conscious. Individual employees may become attached to their HPDs, especially custom-molded types or polyvinyl chloride (PVC)-based earplugs that have shrunk to a very comfortable size, and keep them for years (examples we know of include 7-year-old custom molds and 8-year-old V-51Rs). Overextended use of most HPDs normally reduces their effectiveness due to material deterioration.

269

In addition to inevitable wear, intentional HPD abuse can be a significant prob-lem, especially when HPD utilization is enforced without any display of sincere company interest in the HCP.

Obviously HPD wearers need training and motivation regarding HPD replacement and proper HPD care. The audiometric technician should always inspect the worker's HPDs at audiogram time to detect wear or abuse (Else, 1982; L. H. Royster and Royster, 1985). Deteriorated or modified HPDs should be saved so that they can be used as examples in educational programs. If pos-sible, it is useful to do protection level checks of the old HPDs versus new ones to document the poorer effectiveness of the old HPDs, then make motivational posters from the results. Often abuse occurs when employees are attempting to reduce discomfort or relieve difficulties in communicating with coworkers or detecting machinery noises (Acton, 1977); in these cases selection of a more appropriate HPD for the individuals can increase their willingness to wear HPDs.

In addition to one-to-one counseling with identified offenders, departmental uti-lization compliance records should be compiled by making unannounced walk-through surveys, then posting the results. Replacement HPDs should be issued on the spot to any employee found wearing deteriorated HPDs (Guild, 1966). Periodically the HCP personnel should set up a replacement station at the plant entrance at shift-change time to distribute new HPDs to those who need them. HCP procedures should also be designed to ensure that it is easy and convenient for employees to obtain new HPDs throughout the year by making them available at numerous locations within the plant.

Reward Programs for HPD Utilization

Positive reinforcement behavioral techniques have been reported as successful in increasing HPD acceptance. Zohar (1980) summarized two studies in which token economies were established in industrial settings to reward HPD wearers. In each case utilization rose to 90% after the reward program began and remained at that level after token distribution was discontinued, indicating that the technique had fostered a new group norm. Once employees are induced to try HPDs through token economies, the benefits of regular wear may begin to moti-vate HPD utilization (Lofgreen et al., 1982). In contrast, others conclude that poorly conceived reward programs may backfire if employees feel they are being manipulated (Franks et al., 1996).

Extra Hearing Screening

As a special health benefit to promote the HCP, it is easy to utilize the audio-metric technicians to screen the hearing of non–noise-exposed plant personnel and especially the children and spouses of employees. Screenings can be held on weekends or during annual employee picnics or other parties. Usually out of 100 children screened, 4–5 will be identified with temporary or permanent hearing loss. Referral of these children to appropriate medical, audiological, and educa-tional resources for needed treatment can result in a degree of appreciation for the company's HCP, which can be achieved in no other way. Of course, it is recommended that such screening programs be implemented under the guidance of a local audiologist, otolaryngologist, or company physician.

Motivation for Primary HCP Personnel

The audiometric technicians, safety/health personnel, and supervisors who implement the day-to-day procedures of the HCP carry the main burden of the program. The motivation of these personnel is critical to the HCP's success. If they do not perform their tasks consistently and with sincerity, then the program's credibility is destroyed.

Reinforcement by Upper Management

Management must be convinced to make HCP implementation an integral part of the job descriptions upon which the primary HCP personnel are evaluated during annual performance appraisals. On a more frequent basis, managers should inquire about the status of the HCP and any implementation difficulties being encountered, and should praise the primary HCP personnel for their efforts.

Annual ADBA Results

The findings from annual ADBA procedures can provide a concrete reward to HCP personnel. If the HCP is effective, mean hearing levels may improve for several years due to the learning effect (reference Chapter 12). If results differ by department, the personnel will see where they have succeeded and where they need to concentrate additional effort.

Group Communication

Primary HCP personnel should be assembled for progress updates often enough to ensure that there is good communication and cooperation among the individuals who carry out various phases of the program (L. H. Royster et al., 1982). If the key individual facilitates a process of expressing appreciation for tasks well done and offering helpful suggestions for areas that need improvement, the personnel will see why their performance matters and will develop a team spirit of cooperation.

Motivation for Management

Managers do not need to know many details about the operation of the HCP, but they do need summary information about the status of the program.

Annual ADBA Results

Presentation of summary graphs illustrating ADBA findings for various divisions or departments will help the key individual obtain continued or increased resources and support for the HCP. Demonstration that the results are actually poorer for the section in which the supervisor is less committed to the HCP will stimulate the responsible manager to re-emphasize the importance of HCP duties to that subordinate. Suppose, for example, that HPDs with greater attenuation are needed in a certain area, but they are more expensive. Differential ADBA results for that area can generate approval of the extra budget allocation to offer the needed protectors. If the ADBA results look good, then management can be encouraged to praise the primary personnel responsible, and the key individual can thank the general managers and production manager for their part in supporting the successful program.

Company's Liability for Workers' Compensation Claims

The one topic that always gets managers' attention is the company's liability for workers' compensation claims. Although occupational hearing loss has accounted for only a fraction of 1 percent of the compensation industry currently pays (Berger, 1985), the expense would be greater if more eligible workers filed claims. Establishment of an effective HCP will result in declining potential compensation costs for several years due to the learning effect (L. H. Royster and Royster, 1984b) and to reduced long-term costs as hearing loss in younger workers is prevented. The immediate learning-related decline in potential costs can be several orders of magnitude greater than the annual cost of administering the HCP, demonstrating to management the concrete benefit of the program.

A Safer and More Acceptable Working Environment

Increased safety awareness and job satisfaction often mean increased profit to the company, and an effective HCP can help accomplish these ends (Lofgreen, 1982; Staples, 1981; and see Chapter 1). The available data that attempt to relate injury rates with the implementation of an effective HCP suggest that workers who are adequately protected from noise will exhibit significantly fewer injuries than workers who are not properly protected or who have not been sufficiently motivated to wear the available HPDs (Cohen, 1976; Komaki et al., 1980; Schmidt et al., 1982). This relationship may well be an expression that employees who have been motivated to wear HPDs are also more likely to be safety-conscious in other ways. Regardless of the underlying reason for this observation, the relationship between effective HCPs and reduced injury rates can be used to generate management support for the HCP. Injury data are typically compiled in most companies, so it is usually not a major effort for the key individual to correlate ADBA findings with department or plant injury rates.

HPD Utilization and Effectiveness Studies

Key individuals often ask how to convince management to make decisions related to HPD utilization, such as to allow the use of a more comfortable, lightweight earmuff that has a lower Noise Reduction Rating (NRR) than the heavyweight clunker approved by the company's untrained professional or an outside consultant, or to remove from inventory an HPD which seldom gives an acceptable fit, or to strengthen lax HPD enforcement. If the audiometric data are computerized, ADBA procedures can provide the data needed to accomplish these purposes.

To motivate management to improve HPD enforcement, first try to determine the reason for laxity. If upper managers simply do not believe that enforcement is poor, HPD utilization surveys may provide the needed evidence that a problem exists. If management allows laxity because of fear of confronting employees about HPD utilization, then ADBA results would probably be needed to demonstrate to management the real costs they are incurring through a lack of enforcement.

We have observed that managers typically place unwarranted emphasis on small differences in NRRs between protectors when deciding which types will be allowed, when actually the real-world protection offered by different types varies less than the NRRs would suggest (reference Chapter 10). Regardless of the NRR, it is the employee's willingness to wear the HPD properly and consistently during noise exposures that determines the actual protection received. Therefore, management may actually achieve greater real-world protection by offering a lower NRR device that is more comfortable. Temporary threshold shift studies of groups of employees wearing the two HPDs can provide the needed data to demonstrate the lower NRR HPD actually provides better or equal real-world protection levels (L. H. Royster et al., 1984).

Audiometric Findings for Managers

One indirect motivational tool for influencing management is to include them in the HCP by administering annual audiograms and providing HPDs for their use off-the-job as well as when they visit production areas. Typically a significant percentage of managers will exhibit mild to severe high-frequency hearing loss, often from military noise exposure or other nonoccupational exposures. Counseling from the key individual about how wearing HPDs could have prevented such a loss will help managers see the benefit of the program. Unfortunately, some managers may be reluctant to take part in audiometric testing because of embarrassment about their hearing loss. However, if top level management participates in the program, their example will encourage others to follow.

Regulations

There is no doubt that the dominant force behind the implementation of HCPs in U.S. industry has been federal and state safety and health regulations. Workers' compensation regulations also influence management, but not nearly to the same degree as OSHA or other regulations. Should regulations be used to motivate management to act or employees to participate more fully in the company's HCP? The answer is both no and yes.

No, federal regulations should not be used as the main source of motivation to get employees to wear HPDs or to force management to establish an HCP or modify HCP policies. This approach would probably result in failure on both fronts. The employee will not use HPDs because government agencies think s/he should. Management will develop an HCP on paper to comply with the regulations but will probably not be willing to allocate the necessary financial or administrative commitment necessary to establish an effective HCP.

Yes, both employees and management need to be familiar with relevant federal and state regulations that relate to workplace safety and health, including the minimum requirements for compliance. Knowledge of each party about the responsibilities placed on workers and on management will provide a degree of motivation for both. However, the main emphasis should be on the background information that resulted in the creation of the regulation: the effects of noise and what is necessary for adequate protection of the employee.

Although management needs to understand details of the regulation to evaluate the HCP for compliance with it, the educator must make certain not to convey the idea that the primary reason for HCP implementation is meeting regulatory requirements. When this happens, the educator has failed to motivate management properly.

Putting It All Together

At the beginning of this chapter we referred to the multiplicity of different types of work environments across the U.S. where workers and management are involved in industrial HCPs with the common goal of protecting the hearing of the noise-exposed populations. Due to this diversity we do not feel that it is possible to promote a fixed educational program format for workers or management that would adequately satisfy the needs of general industry. Similarly we do not feel that any single motivational strategy exists that would be appropriate for all situations.

Because of this variety of needs in industrial HCPs with respect to education and motivation we have attempted to cover these two subjects broadly, leaving the formulation of the particular educational and motivational activities for any one plant facility up to the responsible key individual.

Our goal in writing this chapter is to provide the educator with general guidelines and ideas for developing ongoing programs to educate employees and managers and motivate them to be interested in and involved in the HCP. However, in the final analysis the level of success achieved by you, the educator, will be determined by factors beyond our range of influence — mainly by your commitment to hearing conservation and the health and safety of the work force for which you are responsible.

References

Acton, W. I. (1977). "Problems Associated with the Use of Hearing Protection," *Ann. Occup. Hyg. 20*, 387–395.

Berger, E. H. (1982a). "EARLog #8 — Responses to Questions and Complaints Regarding Hearing and Hearing Protection (Part I)," *Occup. Health Saf. 51*(1), 28–29.

Berger, E. H. (1982b). "EARLog #9 — Responses to Questions and Complaints Regarding Hearing and Hearing Protection (Part II)," *J. Occup. Med. 24*(9), 646–647.

Berger, E. H. (1983). "EARLog #10 — Responses to Questions and Complaints Regarding Hearing and Hearing Protection (Part III)," *Prof. Saf. 28*(3), 14–15.

Berger, E. H. (1985). "EARLog #15 — Compensation for a Noise Induced Hearing Loss," E-A-R, Indianapolis, IN.

Berger, E. H., and Royster, J. D. (1988). "An Earful of Sound Advice About Hearing Conservation," E-A-R, Indianapolis, IN.

Carroll, B. J. (1982). "An Effective Safety Program Without Top Management Support," *Prof. Saf. 27*(7), 20–24.

Chung, D. Y., Gannon, R. P., Roberts, M. E., and Mason, K. (1982). "Hearing Conservation Based on Hearing Protectors: A Provincial Project," in *Personal Hearing Protection in Industry*, edited by P.W. Alberti, Raven Press, New York, NY, 559–568.

Cluff, G. L. (1980). "Limitations of Ear Protection for Hearing Conservation Programs," *Sound and Vibration 4*(9), 19–20.

Cohen, A. (1976). "The Influence of a Company Hearing Conservation Program on Extra-Auditory Problems in Workers," *J. Saf. Res. 8*(4), 148–162.

Else, D. (1981). "Hearing Protection: Who Needs Training?" *Occup. Health 33*(9), 451–453.

Else, D. (1982). "Hearing Protection Programme Establishment," in *Personal Hearing Protection In Industry*, edited by P. W. Alberti, Raven Press, New York, NY.

Esler, A. (1978). "Attitude Change in an Industrial Hearing Conservation Program: Comparative Effects of Directives, Educational Presentations and Individual Explanations as Persuasive Communications," *Occup. Health Nursing 26*(12), 15–20.

Feeney, R. H., and Nyberg, J. P. (1976). "Selling the Worker on Hearing Conservation," *Occup. Health Saf. 435*(4), 56–59.

Forrest, M. R. (1982). "Protecting Hearing in a Military Environment," *Scand. Audiol.*, Suppl. 16, Proceedings of the Oslo International Symposium on Effects of Noise on Hearing, edited by H.M. Borchgrevink, Norway, 7–12.

Foster, A. (1983). "Hearing Protection and the Role of Health Education," *Occup. Health 35*(4), 155–158.

Franks, J. R., Stephenson, M. R., and Merry, C. J. (1996). "Preventing Occupational Hearing Loss — A Practical Guide," DHHS (NIOSH) Publication No. 96–110, National Institute for Occupational Safety and Health, Cincinnati, OH.

Gasaway, D. C. (1984). "'Sabotage' Can Wreck Hearing Conservation Programs," *Natl. Saf. News 129*(5), 56 – 63.

Guild, E. (1966). "Personal Protection," in *Industrial Noise Manual, Second Edition*, American Industrial Hygiene Association, Fairfax, VA, 84–109.

Hager, W. L., Hoyle, E. R., and Hermann, E. R. (1982). "Efficacy of Enforcement in an Industrial Hearing Conservation Program," *Am. Ind. Hyg. Assoc. J. 43*(6), 455–465.

Harford, E. R. (1978). "Industrial Audiology," in *Noise and Audiology*, edited by D. M. Lipscomb, University Park Press, Baltimore, MD, 299–327.

Karmy, S. J., and Martin, A. M. (1982). "Employee Attitudes towards Hearing Protection as Affected by Serial Audiometry," in *Personal Hearing Protection in Industry*, edited by P. W. Alberti, Raven Press, New York, NY, 491–509.

Komaki, J., Heinzmann, A. T., and Lawson, L. (1980). "Effect of Training and Feedback: Component Analysis of a Behavioral Safety Program," *J. Appl. Psycol. 65*, 261–270.

Leinster, P., Baum, J., Tong, D., and Whitehead, C. (1994). "Management and Motivational Factors in the Control of Noise Induced Hearing Loss (NIHL)," *Annals Occup. Hyg. 38,* 649–662.

Lofgreen, H. (1982). "The Human and Economic Benefits of Hearing Protection in the Plant," *Canadian Occup. Saf. 20*(6), 2–3, 9.

Lofgreen, H., Holm, M., and Tengling, R. (1982). "How to Motivate People in the Use of Their Hearing Protectors," in *Personal Hearing Protection in Industry,* edited by P. W. Alberti, Raven Press, New York, NY, 485–490.

Lusk, S. L., Ronis, D. L., and Kerr, M. J. (1995). "Predictors of Hearing Protection Use Among Workers: Implications for Training Programs," *Human Factors 37*(3), 635–640.

Mager, R. F., and Pipe, P. (1970). *Analyzing Performance Problems or 'You Really Oughta Wanna',* Fearon Publishers, Belmont, CA.

Melamed, S., Rabinowitz, S., Feiner, M., Weisberg, E., and Ribak, J. (1996). "Usefulness of the Protection Motivation Theory in Explaining Hearing Protection Device Use Among Male Industrial Workers," *Health Psychology 15,* 209–215.

Melamed, S., Rabinowitz, S., and Green, M. S. (1994). "Noise Exposure, Noise Annoyance, Use of Hearing Protection Devices, and Distress Among Blue-Collar Workers," *Scand. J. Work Environ. Health 20,* 294–300.

Mellard, T. J., Doyle, l. J., and Miller, M. H. (1978). "Employee Education—the Key to Effective Hearing Conservation," *Sound and Vibration 12*(1), 24–29.

Morrill, J. C. (1984). "Instructions and Techniques for Reliable Audiometric Testing," *Occup. Health Saf. 53*(8), 64–68.

NHCA (1996a). "A Practical Guide to Selecting Hearing Protection," National Hearing Conservation Association, Denver, CO.

NHCA (1996b). "A Practical Guide to Fitting Hearing Protection," National Hearing Conservation Association, Denver, CO.

Ohlin, D. (1981). "User Training and Problems," in *Proceedings of Noise-Con 81,* edited by L. H. Royster, N. D. Stewart, and F. D. Hart, Noise Control Foundation, New York, NY, 131–136.

Pender, N. J. (1987). *Health Promotion in Nursing Practice* (2nd edition), Appleton and Lange, Norwalk, CT.

Prentice-Dunn, S., and Rogers, R.W. (1986). "Protection Motivation Theory and Preventive Health: Beyond the Health Belief Model," *Health Educ. Research 1,* 153–161.

Royster, J. D., and Royster, L. H. (1984). "Hearing Protection Practices, Problems and Solutions," *Asha 26*(10), 77.

Royster, J. D., and Royster, L. H. (1990). *Hearing Conservation Programs: Practical Guidelines for Success,* Lewis Pub., Chelsea, MI.

Royster, J. D., and Royster, L. H. (1994). "Practical Tips for Fitting Hearing Protection," *Hearing Instruments 45*(2), 17–18.

Royster, L. H., and Holder, S. R. (1982). "Personal Hearing Protection: Problems Associated with the Hearing Protection Phase of the Hearing Conservation Program," in *Personal Hearing Protection in Industry,* edited by P. W. Alberti, Raven Press, New York, NY, 447–470.

Royster, L. H., and Royster, J. D. (1981). "Educational Programs for Management and the New OSHA-US Noise Regulations (Educational Requirements)," in *Proceedings of a Special Session on Hearing Conservation Programs — The Educational Phase,* edited by L. H. Royster, D. H. Hill Library, North Carolina State Univ., Raleigh, NC.

Royster, L. H., and Royster, J. D. (1984a). ''Hearing Protection Utilization Survey Results Across the USA.'' *J. Acoust. Soc. Am. 26*, Suppl. 1, S 43.

Royster, L. H., and Royster, J. D. (1984b). "Making the Most Out of the Audiometric Data Base," *Sound and Vibration 18*(5), 18–24.

Royster, L. H., and Royster, J. D. (1985). "Hearing Protection Devices," in *Hearing Conservation in Industry*, edited by A. S. Feldman and C. T. Grimes, Williams and Wilkins, Baltimore, MD.

Royster, L. H., Royster, J. D., and Berger, E. H. (1982). "Guidelines for Developing an Effective Hearing Conservation Program," *Sound and Vibration 16*(1), 22–25.

Royster, L. H., Royster, J. D., and Cecich, T. F. (1984). "An Evaluation of the Effectiveness of Three Hearing Protection Devices at an Industrial Facility with a TWA of 107 dB." *J. Acoust. Soc. Am. 76*(2), 485–497.

Sadler, O. W., and Montgomery, G. M. (1982). "The Application of Positive Practice Overcorrection to the Use of Hearing Protection," *Am. Ind. Hyg. Assoc. J. 43*(6), 451–454.

Schmidt, J. W., Royster, L. H., and Pearson, R. G. (1982). "Impact of an Industrial Hearing Conservation Program on Occupational Injuries," *Sound and Vibration 16*(1), 16–20.

Spindler, D. E., Olson, R. D., and Fishbeck, W. A. (1979). "An Effective Hearing Conservation Program," *Am. Ind. Hyg. Assoc. J. 4*(7), 604–608.

Staples, N. (1981). "Hearing Conservation — Is Management Short Changing 'Those at Risk'?" *Noise and Vib. Control Worldwide 12*(6), 236–238.

Stapleton, L., and Royster, L. H. (1981a). ''Educational Programs for Hearing Conservation,'' in *Noise-Con 81 Proceedings*, edited by L. H. Royster, N. D. Stewart, and F. D. Hart, Noise Control Foundation, New York, NY, 153–156.

Stapleton, L., and Royster, L. H. (1981b). "Educational Programs for Hearing Conservation," *Proceedings of a Special Session on Hearing Conservation Programs — The Educational Phase*, edited by L. H. Royster, D. H. Hill Library, North Carolina State University, Raleigh, NC.

Stapleton, L., and Royster, L. H. (1985). ''The Education Phase of the Hearing Conservation Program," *Sound and Vibration 19*(2), 29–31.

Zohar, D. (1980). "Promoting the Use of Personal Protective Equipment by Behavior Modification Techniques," *J. Saf. Res. 12*(2), 78–85.

Zohar, D., Cohen, A., and Azar, N. (1980). "Promoting Increased Use of Ear Protectors in Noise Through Information Feedback," *Human Factors 22*(1), 69–79.

Supplemental Reading

Barker, D. M., Driscoll, D. P., and Florin, J. D. (1982). "Programming Hearing Conservation for a Large Corporation," *National Safety News*, November, 54–57.

Berger, E. H. (1981a). "EARLog #6 — Extra-Auditory Benefits of a Hearing Conservation Program," *Occup. Health Saf. 50*(4), 28–29.

Berger, E. H. (1981b). "EARLog #7 — Motivating Employees to Wear Hearing Protection Devices," *Sound and Vibration 15*(6), 10–11.

Carroll, C., Crolley, N., and Holder, S. R. (1980). "A Panel Discussion of Observed Problems Associated with the Wearing of Hearing Protection Devices by Employees in Industrial Environments," in *Proceedings of a Special Session on the Evaluation and Utilization of Hearing Protection Devices (HPDs) in Industry*, edited by L. H. Royster, D. H. Hill Library, North Carolina State Univ., Raleigh, NC.

Gasaway, D. C. (1984). "Motivating Employees to Comply with Hearing Conservation Policy," *Occup. Health Saf 53*(6), 62–67.

Harris, L. A. (1980). "Combating Hearing Loss Through Worker Motivation," *Occup. Health and Saf. 49*(3), 38–40.

Lusk, S. L., Ronis, D. L., Kerr, M. J., and Atwood, J. R. (1994). "Test of the Health Promotion Model as a Causal Model of Workers' Use of Hearing Protection," *Nursing Research 43*, 151–157.

Lutz, G. A., Decatur, R. A., and Thompson, R. L. (1973). "Psychological Factors Related to the Voluntary Use of Hearing Protection in Hazardous Noise Environments," U.S. Army Med. Res. Laboratory Report No. 1,006, (NTIS-AD-777520), Fort Knox, KY.

Maas, R. (1970). "The Challenge of Hearing Protection," *Indus. Med. 39*(3), 29–33.

Maas, R. (1971). "Compliance with OSHA on Hearing Conservation," *J. Environ. Control and Saf. Manage.* (December), 11–13.

Maas, R. B. (1972). "Industrial Noise and Hearing Conservation," in *Handbook of Clinical Audiology*, edited by J. Katz, Williams and Wilkins Co., Baltimore, MD, 772–818.

Pelausa, E. O., Abel, S. M., Simard, J., and Dempsey, I. (1995). "Prevention of Noise-Induced Hearing Loss in the Canadian Military," *J. Otolaryng. 24*, 271–280.

Wilkins, P. A., and Acton, W. l. (1982). ''Noise and Accidents A Review," *Ann. Occup. Hyg. 25*(3), 249–260.

Woodford, C. M., and Lass, N. J. (1994). "Hearing Conservation in Hunter Education Programs," *Am. J. Audiol. 3*(2), 8–10.

Wyman, C. W. (1969). "Industrial Hearing Conservation, Administration and Human Relations Aspects," *Natl. Saf. News 99*(5), 65–70.

Zohar, D. (1980). "Safety Climate in Industrial Organizations: Theoretical and Applied Implications," *J. Appl. Psychol. 65*(1), 96–102.

The Noise Manual, revised 5th edition, edited by E.H. Berger,
L.H. Royster, J.D. Royster, D.P. Driscoll, and M. Layne
©2003 American Industrial Hygiene Association

9

Noise Control Engineering*

Dennis P. Driscoll and Larry H. Royster

Contents

* Portions of this chapter are adapted with permission from: International Labour Organization,
©1998: *Encyclopedia of Occupational Health and Safety*, 4th Edition, Volume II, edited by J. M.
Stellman, PhD, ILO Publications, Geneva, Switzerland, Chapter 47, pages 47.8–47.10, authored
by D. P. Driscoll.

Introduction

Make no mistake about it: acoustics is an exact science. On the other hand, noise control is truly an art form. *Merriam-Webster* defines "art form" as the "conscious use of skill and creative imagination" (Anon., 1991). The word "skill" stresses the technical knowledge, such as the science of acoustics, while the "creative imagination" refers to the ingenuity on the part of the individual(s) devising the control measures.

There are many options available to control noise and consequently reduce employee noise exposures. Toward an effective noise control program, it is important to know the following:

- How sound is generated.
- How to identify the source(s) of noise.

- What options are available for treating the source, path, and/or receiver.
- How to determine the benefits and costs of noise control.
- What appropriate noise control products and resources are available for selection and procurement.
- What other methods are available to reduce worker noise exposure.

The information listed above is critical for controlling noise during the procurement and design phases for new equipment and facilities, as well as handling existing noise problems. A comprehensive and cost-effective noise control program can be designed and implemented by combining knowledge of acoustics with an understanding of the manufacturing equipment and process, including all production and maintenance constraints. Without an informed approach, the likelihood of success and the effective use of resources will be tenuous at best.

Many noise control challenges lend themselves to straightforward solutions. With an understanding of the principles of noise control and proper use of acoustical materials, both occupational health professionals and plant engineers can make significant progress in reducing equipment noise levels and employee noise exposures. However, there are situations where the acoustical environment is too complex, or the professional overseeing the noise control program simply does not have sufficient time, so outsourcing the project often occurs. As a result, a noise control engineer may be retained to conduct the detailed survey, identify the sources, design the engineering controls, and develop a plan of action for a client company. In this situation, the program administrator needs to have a good working knowledge of acoustics and noise control to effectively manage and/or assist the consultant, plus direct the implementation of various recommendations. To assist the reader in developing an understanding of acoustics, as applied in real-world industrial environments, or to enhance one's existing experience, this chapter focuses on the practical aspects of noise control engineering.

Benefits and Costs of Noise Control

As with other phases of a hearing conservation program (HCP), noise control should play an integral part in achieving the goals of the HCP. Moreover, the noise control phase can potentially influence the effectiveness of the HCP to a degree that other phases cannot. That is, if the application of a noise control modification to equipment reduces the employee's time-weighted average (TWA) below 85 dBA and the level is maintained, then the potential for significant on-the-job noise-induced hearing loss is, for all practical purposes, eliminated, and the other phases of the HCP become unnecessary. However, as long as a potential noise hazard exists, then other means of protecting employees included in the HCP can potentially fail, and often do.

It is also valuable to consider that when noise control measures reduce employee noise exposures below the action level of 85 dBA (or better yet 83 dBA as a measure of safety), the other HCP elements including audiometry, training, hearing protection, recordkeeping, and all the associated ongoing administrative costs can be saved, which in many instances may fully offset the initial costs of noise controls.

Noise Control Program Goals

Primary Short-Term Goal

A question that is often asked prior to implementation of a noise control solution is, "What magnitude of noise reduction in the employee's TWA is possible, and is it worth doing?" That is, if an employee's TWA can be reduced by 3 dBA using noise control, should it be achieved? The answer depends on the number of employees affected, the cost and benefits of the application, the extent of the noise hazard, and pertinent regulatory requirements. In a following section guidelines are presented for use in comparing the relative merits of alternative noise control projects. However, here we are asking the broader question as to the magnitude of noise reduction that is possible before a noise control application should be considered.

One relevant factor is the level of real-world protection expected from hearing protection devices (HPDs). In Chapter 10, data are presented that suggest that the typical level of protection from HPDs in real-world environments is approximately 10 dBA. Because it is more difficult to protect employees exposed above 95 dBA, the primary short-term goal of the noise control program should be to reduce all higher employee TWAs to 95 dBA or less. In meeting this goal, management should identify all dominant noise sources (reference Chapter 7) and establish a relative priority rating for each source (see section *Setting Priorities*). If the employee's TWA significantly exceeds 95 dBA, the potential for noise control solutions should be addressed.

Noise Control Feasibility

Obviously there will be some dominant noise sources for which a practical and economic solution is not within the limitations of company resources. For these situations it will be necessary to rely on HPDs (with wearer choices limited to appropriate devices) or a combination of HPDs with administrative exposure control options, plus a stricter audiometric monitoring program than is required for employee TWAs less than 95 dBA (reference Chapters 11, 12, and 17).

Primary Long-Term Goal

The primary long-term goal of the noise control program should be to identify and treat dominant noise sources that contribute to employees' TWA noise exposures that are equal to or greater than 85 dBA. Management should establish a priority rating for each of these noise sources and implement an ongoing noise control program to reduce workers' TWAs below 85 dBA. In the long term, reducing TWAs to nonhazardous levels will eliminate the need for other hearing conservation efforts (saving their costs) and eliminate potential workers' compensation liability for occupational hearing loss.

It must be recognized that the resources for any one company are limited and therefore should be efficiently utilized in solving all types of possible health and safety problems based on their priority. In some fiscal years the available resources might have to be spent on solving a potentially highly toxic chemical employee exposure problem instead of solving a potentially less dangerous noise problem. However, it is management's responsibility to justify the internal prioritization process and demonstrate that the employees' hearing is being adequately protected (reference Chapter 12 on audiometric database analysis procedures). In addition, it is helpful in the long term to include at least some noise control efforts in every budget cycle, so that over time TWAs will be lowered.

Reduced NIPTS

Presented as Figure 17.1 (see Chapter 17) is the expected noise-induced permanent threshold shift (NIPTS) for the typical (median susceptibility) industrial population that is exposed to a 95-dBA TWA for durations of up to 40 years. These data were calculated using the American National Standards Institute (ANSI) S3.44-1996 model for predicting a population's expected NIPTS, which is very similar to the version in the international standard ISO 1999 (ISO, 1990). Note that if the same data are calculated for the most susceptible 5% of the population, then the predicted magnitude of NIPTS is typically increased by at least 10 dB or greater for exposure durations of at least 10 years.

Obviously, based on the data in Figure 17.1, if the magnitude of the employee's daily TWA is lowered through engineering noise control, then the potential for the employee to obtain on-the-job noise-induced hearing loss is reduced, or even eliminated. Prevention of hearing loss benefits both employee and employer.

One comment made by some managers when they are presented data as shown in Figure 17.1 is that because of high employee turnover rates in their industry, long-term hearing loss is unimportant. That is, their employees will change jobs before they develop a hearing loss for which the company could be held responsible or a communication impairment that might impact the company's production capability. But where do the new employees come from to replace the poorly protected workers who make up the high-turnover population? They tend to come from similar types of industries with similar noise environments. If management of these preceding companies also failed to protect their employees from noise, then the newly hired employee will in fact exhibit hearing loss trends that follow along similar NIPTS curves, as shown in Figure 17.1, as did the employee who left. Therefore, it is to management's (company and employees) benefit to consider both the long-term as well as the short-term implications of not properly protecting the noise-exposed workforce.

Other managers express a concern only for possible impacts of their decisions on the company's profit margins in the near future, during the period of time that they expect to be in their present administrative position. If the impact from their decisions will not occur until after they move on, then why care? Fortunately, many U.S. industries are realizing the pitfalls of this short-sighted management attitude. However, the data presented in Figure 17.1 also indicate that, at the higher noise exposures, significant noise-induced hearing loss can occur in 6 months or less, demonstrating the importance of an effective HCP.

Reduced Compensation Cost

It has been shown that the percentage of a population that exhibits a potentially compensable noise-induced hearing loss, as indicated by the audiometric database, can be significantly reduced in the short term (in some cases by 50%) by adequately protecting the workers from harmful on-the-job noise exposures (L.H. Royster and Royster, 1984). If the noise-exposed workers are completely removed from the noise hazard through engineering controls, so that the need for an HCP is eliminated, then the incentive to file for compensation by the workforce will also be reduced as the population ages. On the other hand if the workers are exposed to high noise levels throughout their work careers, being educated regularly through HCPs about the effects of noise, and exposed to the experiences of their fellow employees filing for compensation, they are more likely to expect appropriate compensation upon retiring or leaving the company.

An additional potential future benefit is that dollars spent on noise control solutions today will result in a lower operating cost to the company in the future. This is due to eliminating the effects of noise in future years that would have occurred had not the noise controls been implemented. The reduction in cost is due to expected inflation and the anticipated significant increase in the future of the magnitude of compensation awards for on-the-job noise-induced hearing loss. There exist today many U.S. companies that are facing considerable potential liability costs due to their failure to adequately assess the long-term cost in the safety and health areas.

Improved Communications

Anyone who has attempted to communicate in an environment where the TWA is 85 dBA or higher is aware of the additional vocal and attention effort necessary, even with normal hearing. The additional effort required to communicate and the potential for misunderstanding communications would be expected to result in an increase in the number of production errors made and accidents recorded, which can have a negative impact on the employer–employee relationship, as well as on productivity (also see Chapter 1). The authors are aware of several instances where employee complaints regarding their inability to communicate in the noisy workplace forced management to modify the work environment in order to reduce the existing noise levels. In some of these instances the noise levels were significantly below the present regulatory requirements.

284

Reduced Absenteeism and Injuries

Research shows that many factors affect absenteeism, including dust, heat, fumes, noise, personnel problems, etc. However, the studies that have attempted to relate noise exposure either directly or indirectly to employee absenteeism have produced data that suggest that the cost to industry due to noise-related absenteeism is significant (Ashford et al., 1976; Cohen, 1976): approximately 13 times the projected cost due to employees' filing for compensation for noise-induced hearing loss (also see Chapter 1).

In one study where the population's TWA of 92–96 dBA was reduced through the use of HPDs, the noise-exposed population's injury rate was reduced 50% (Schmidt et al., 1982). The study findings also implied that if the same level of reduction had been obtained through effective engineering noise controls an even greater effect would have been expected. This assumption was based on the observation that the magnitude of the reduction in the injury rate for different populations strongly correlated with the level of protection being provided by the HPDs. That is, the group that was realizing the most protection from their HPDs exhibited the lowest injury rate.

If a population's injury rate is reduced, then the level of absenteeism would also be expected to fall. Workers who are injured (but remain on the job) function less efficiently than is normal, resulting in lower productivity on their part and a net cost to the company. Therefore a major motivator for management to control the employees' noise exposure through engineering controls is the reduction of the company's injury and absenteeism rates.

Cost Must Be Considered

The available studies in the literature indicate the cost of noise control to achieve a TWA of 90 or 85 dBA is strongly dependent on the type of industry studied (Bruce, 1979; Bruce et al., 1976; Gibson and Norton, 1981), though variation within industries also occurs. Most likely these difficulties in estimating costs are the reason the literature provides limited guidance on this subject. If reducing the noise of production equipment is expected to increase productivity, then management may well implement the noise control program. If on the other hand the noise control solution (such as placing the machine in an acoustic enclosure) decreases productivity by 5%, then management may decide the cost outweighs the benefits.

Should noise control always result in a short-term dollar savings before a noise control program can be justified? Obviously not! However, since the available dollars for safety and health programs within any one company are limited, it is essential that adequate consideration be given to the cost–benefit aspects of all noise control projects.

Noise-Reduction Cost-Control Planning Options

In considering the cost related to implementing a noise control program, management should examine various options in the following order:

1. Controlling noise at the equipment or facility design stage.
2. Purchasing new production equipment.
3. Retrofitting and/or modifying the existing production equipment and environment.

It is assumed that the cost for achieving an acceptable TWA for a typical production environment increases significantly as the selected option goes from 1 to 3.

Unfortunately the available literature does not provide adequate data to state with confidence the increase in cost from having to modify the existing equipment and/or environment as compared to the cost of achieving the same level of control at the design stage. However, noise control consultants have often quoted a ratio of 10/1. The actual cost ratio in some situations could be significantly higher. This is due to the tendency for some retrofit engineering noise controls to have a negative impact on productivity. If retrofit noise controls decrease productivity, then over time the cost of these controls can be much higher than noise control at the design stage.

At one plant where a new production process was brought on-line for testing purposes, the noise created by the ventilation system was not only above 85 dBA throughout the production area, but exhibited a very annoying pure-tone component in the 500-Hz octave band. The facility designers had neglected to specify in-line acoustic attenuators for the ventilation system, although they should have known it would create a noise problem. The unionized employees assigned to this new production area strongly protested to management. Consequently, over a period of 3 weeks the duct work had to be rerouted, specially modified attenuators fabricated, duct wrapping materials purchased, the products flown by special airline services to the plant site, and the system modified by work crews working around the clock. In the end, what should have been a $15,000 noise control effort up front during the design stage turned into one that resulted in a net retrofit cost of $150,000 to $200,000.

Setting Priorities

In determining the relative priority of implementing the noise control measures, one should consider the employee exposures, the occupancy of the space, potential hearing-damage risk, type of environment, potential for success, and effect on productivity. Obviously, the desired result is to obtain the maximum employee noise-exposure reduction for each dollar invested. The noise control rating scheme presented below may be useful toward development of a prioritized list of all noise control options.

The noise control priority factor (NCPF) is defined as:

$$\text{NCPF} = \frac{\text{NE} \times \text{LD} \times \text{EC} \times \text{SF} \times \text{PF}}{\text{CK}} \qquad (9.1)$$

Where:

NE: Number of employees affected by source(s).
LD: Potential for noise to produce significant damage.
EC: Environmental characteristics factor.
SF: Problem solution potential success factor.
PF: Productivity factor.
CK: Estimated cost of controls (per thousand dollars).

To determine the NCPF for each noise control project the following steps are presented:

Step 1 — Determine NE, the number of affected employees:
Include the number of employees affected by one or more of the identical dominant sources which are being examined.

Step 2 — Look up LD, the potential for hearing-damage risk:
Select the LD number using Table 9.1 below, knowing the employee's long-term TWA. If the employee's TWA is between two values, use the higher LD factor. Note: NIPTS is the abbreviation for noise-induced permanent threshold shift.

TABLE 9.1
Potential NIPTS damage factor.

TWA (dBA)	LD Factor
80	0.1
85	1
90	4
95	10
100	40
105	100
110	400
115	1,000
120	4,000
125	10,000
130+	100,000

Step 3 — Select EC, the environmental characteristic:
Use Table 9.2 to select the appropriate combination of EC factors. Note: The final EC factor will be the product of each factor that applies.

TABLE 9.2
Environmental factors.

EC	Type of Environment
1	Broadband or high-frequency noise
2	Low-frequency emphasis
2	Pure tone exists
2	Impulsive or impact type noise

287

Step 4 — Determine SF, the potential for success:
> For this factor you will first need to determine your expectations for success, i.e., low expectation, 50–50 chance, or high expectations. Then look up the appropriate rate factor in Table 9.3.

TABLE 9.3
Potential for success factor.

SF	Potential for Success
0.1	Low expectations
1	50-50 chance
10	High expectations

Step 5 — Determine PF, the productivity factor:
> Table 9.4 lists the PF for various affects. Look up the appropriate value and use this number in the NCPF formula.

TABLE 9.4
Productivity factor.

PF	Impact on Productivity
0.1	Decrease production
1	No effect
10	Improve production

Step 6 — Obtain CK, the estimated cost for controls per thousand U.S. dollars ($):
> Either through the noise control product representative, a consultant, or internal means, determine the estimated cost for the noise control option(s), per thousand dollars (i.e., the CK for a $12,000 noise control is $12,000/$1000 = 12). Management should not forget that the cost of some noise control options will result in the need to maintain the treatment on a regular basis in order to sustain the level of noise control achieved. Therefore the cost of maintenance should be included in the cost factor. At least one source has estimated the cost of maintenance at 5% per year (Ashford et al., 1976). Therefore a reasonable total cost of controls would be the actual cost plus the cost of maintaining the noise control solution for 10 years. Once this CK value is determined, plug this number into the NCPF formula.

When multiple dominant noise sources exist, Equation 9.1 is used to determine the NCPF value for each source's noise control measure. To prioritize these items for noise control they need to be rank ordered from the highest NCPF value through the lowest. From this list a long-term noise control plan and budget may be developed, based on the available funds.

EXAMPLE 9.1, Determining Noise Control Priority Using the NCPF

A manufacturing plant has two dominant noise sources, described as follows:

Area 1 is the packaging department with air exhaust from a case packer as the dominant noise source. The source is primarily low-frequency noise and the worker's TWA is 90 dBA. There are a total of 20 employees exposed to this source. Control of this air noise source can be accomplished through installation of an exhaust silencer. This control will have no impact on production, and the probability of success is high. A silencer manufacturer provided a cost estimate for the silencer of approximately $1900.

Area 2 is a woodworking shop and the dominant noise source is a wood chip collection fan that exhibits a strong pure tone. The most reasonable control option is to construct a plenum chamber and attach it to the exhaust duct, thus allowing noise to be exhausted into an unoccupied area outside the shop. In this shop are 5 workers whose TWAs all equal 90 dBA. Implementation of this control will have no effect on production and the likelihood for success is rated at a 50–50 chance. Facilities maintenance estimates the cost for this plenum will be approximately $500.

First the NCPF is determined for Area 1, the packaging department. Based on the information provided, the following input factors were determined for use in Equation 9.1:

NE = 20   (Given information.)
LD = 4   (From Table 9.1 for a TWA of 90 dBA.)
EC = 2   (From Table 9.2 for primarily low-frequency noise.)
SF = 10   (From Table 9.3 for high expectations or potential for success.)
PF = 1   (From Table 9.4 for no effect on production.)
CK = 2.85 ($1900/$1000)[(1.0)+(0.05)(10)] = 2.85. (Note: the [(1.0)+(0.05)(10)] expression includes the initial unit cost plus a 5% additional cost for maintaining the control over a 10-year period.)

Therefore, the NCPF = [(20)(4)(2)(10)(1)/2.85] = 561 (Area 1)

For the second area, the NCPF is calculated from the following input data:

NE = 5   (Given information.)
LD = 4   (From Table 9.1 for a TWA of 90 dBA.)
EC = 2   (From Table 9.2 for pure tone.)
SF = 1   (From Table 9.3 for a 50–50 chance for success.)

PF = 1   (From Table 9.4 for no effect on production.)
CK = 0.75 ($500/$1000)[(1.0)+(0.05)(10)] = 0.75.

Therefore, the NCPF = [(5)(4)(2)(1)(1)/0.75] = 53 (Area 2)

Because Area 1 has the higher NCPF, the case packer exhaust silencer is the first priority for implementation, assuming all other relevant issues are equal.

Techniques for Conducting an Engineering Noise Control Survey

The protocol for conducting a noise control survey will depend on the goals and objectives for the survey. If the goal is to modify or treat the source of noise, then significantly more detailed information will be required as compared to the amount of data needed to simply treat the sound transmission path. Controlling noise at the source requires identification of the origin or source of noise, and definition of its acoustical properties (i.e., frequency spectrum, sound level versus time, etc.). This provides the necessary information for design of the controls. Treatment of the sound transmission path does not always require clear identification of the root cause of noise, but instead relies heavily on the frequency spectrum and room characteristics to provide the information needed to select the acoustical materials.

The initial objective of all noise control projects should be directed toward treating the source. However, if additional noise reduction is required over and above what treatment of the source can provide, or should modification of the source be cost-prohibitive, then the next step would be to implement a path treatment that will effectively prevent excessive noise from reaching the receiver. The survey procedures described below are developed to address the primary objective of treating the source. The same information collected will also apply for controlling noise along the sound path and at the receiver's position. However, detailed 1/3 octave-band or narrow-band data needed for source identification are not necessary for path and/or receiver treatment. Measurement of the full octave-band spectrum is usually sufficient to allow selection of the appropriate noise control products for this type of noise control treatment.

How Sound Is Generated

To control noise at the source and get to the root cause of the problem, it is useful to have an understanding of how noise is generated. Noise is created for the most part by mechanical impacts, high-velocity air, high-velocity fluid flow, vibrating surface areas of a machine, and quite often by vibrations of the product being manufactured. As a result, in industries such as metal fabrication, glass manufacturing, food processing, mining, etc., it is the interaction between the product and machine that creates the noise.

Mechanical impact noise is common in equipment utilizing air valves or solenoids, punch press devices, riveting operations, application of impact and percussive pneumatic hand tools on metal structures, etc. For example, air valves are often used to move a mechanical part, such as a push-rod or ram used to insert product into a carton, or cartons into a case pack. Each time the air valve is employed the push-rod extends and retracts, which in turn causes a structural impact at both ends of the its stroke. The more the driving force or harder the impact, the more noise is generated. Similarly, when valves are driven by compressed air, many times these devices exhaust high-pressure air directly to the atmosphere causing a shearing action between the relatively still air and the high-velocity exhaust. The end result is a very high noise level, typically on the order of 95–120 dBA with significant high-frequency energy, usually in the 1-kHz band and higher depending on the air pressure and configuration of the discharge port.

Another source of noise occurs when a liquid or gaseous medium is transported by pipelines. In piping systems excessive fluid-flow velocities can create noise within the medium that radiates directly through the pipe wall, or creates vibrational energy within the pipe wall. It is common for the vibrational energy within a pipe wall to be carried to a point or surface that radiates the structure-borne vibration as sound. Similarly, process equipment can generate vibratory energy, which is transferred to a surface that may be an efficient radiator of sound. An example would be an electric motor directly mounted to a metal casing. Although the casing is not the origin of the acoustical energy, it becomes a sounding board, typically radiating noise that exhibits a resonant tone based on the vibrational characteristics of the panel or surface area.

Identifying the Origin or Source of Noise

One of the most challenging aspects of noise control is identification of the actual source. In an industrial setting there are usually multiple machines operating simultaneously, which can make it difficult to identify the dominant contributor(s) to an employee's noise exposure. This is especially true when a standard sound level meter (SLM) is used to evaluate the acoustical environment. The SLM typically provides a measure of the sound pressure level (SPL) and/or the overall A-weighted sound level at a specific location, and the measured level is the result of energy created by all operating noise sources. Therefore, it becomes incumbent upon the surveyor to employ a systematic approach that will help separate the individual noise sources and their relative contributions to the noise environment. The following techniques may be used to help identify the origins or sources of noise:

- Measure the overall A-weighted sound level and frequency spectrum, and graph the data.
- Measure the sound level, in dBA, versus time.

- Compare frequency data from similar equipment, production lines, etc.
- Isolate components with temporary controls, or by turning on and off individual items whenever possible (see Chapter 7).

One of the most effective methods for locating the source is to measure the frequency spectrum. Once the data are obtained, it is useful to graph the results to see the frequency characteristics of the source. For most noise abatement problems, the measurements can be accomplished with either full (1/1) or 1/3 octave-band filters used with an SLM. The advantage of a 1/3 octave-band measurement is that it provides more detailed information about the noise emanating from a piece of equipment.

Figure 9.1 exhibits a comparison between 1/1 and 1/3 octave-band measurements conducted near a nine-piston pump. As depicted in this figure, the 1/3 octave-band data more clearly identify the pumping frequency and a few of its harmonics. If one only used octave-band data, as depicted by the 9 solid rectangular tick marks in Figure 9.1, then it becomes more difficult to diagnose what is occurring within the pump. In this example the fundamental pumping frequency is at 200 Hz, which is clearly depicted with the 27 points in the 1/3 octave-band spectrum but not in the octave-band format. It is common for pump noise to contain many discrete tones with the maximum noise occurring several harmonics above the fundamental frequency. In this figure the maximum SPL occurs at 1000 Hz, which is evident in both sets of data. However, unless the cause is known, it is difficult to determine the necessary noise control for treating the source. The 1/3 octave-band measurements in this instance provide the requisite information.

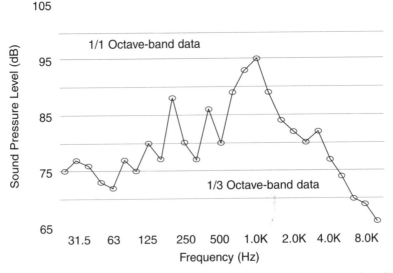

Figure 9.1 — Comparison between 1/1 and 1/3 octave-band data for a nine-piston pump.

Figure 9.2 exhibits a comparison between the 1/3 octave-band spectrum measured 3 feet from the crossover pipe of a liquid chiller compressor and the level measured approximately 25 feet away. This latter position represents the general area where employees typically walk through the room. For the most part the compressor room is not routinely occupied by workers. The only exception exists when maintenance workers are repairing or overhauling other equipment in the room. Besides the compressor, there are several other large machines operating in this area. To assist with the identification of the primary noise sources, several frequency spectra were measured near each of the equipment items. When each spectrum was compared to the data at the position in the walkway, only the crossover pipe of the compressor unit exhibited a similar spectral shape. Consequently, it may be concluded that the crossover pipe is the primary noise source controlling the levels measured at the employee walkway. As depicted in Figure 9.2, it is often possible to graphically compare frequency data measured near individual sources to the data recorded at employee workstations, or other areas of interest, to identify the dominant noise sources and determine their relative contribution to the employee's daily noise exposure.

When the sound level fluctuates, with cyclic equipment and/or production, it is useful to measure the overall A-weighted sound level versus time. With this procedure it is important to observe and document which events are occurring at specific times. Figure 9.3 exhibits the sound level measured at the operator's workstation over one full machine cycle. Since we are interested in viewing the sound level as it is rapidly changing with time, the "fast" response setting of the SLM is utilized (see Chapter 3 for additional information on instrumentation settings).

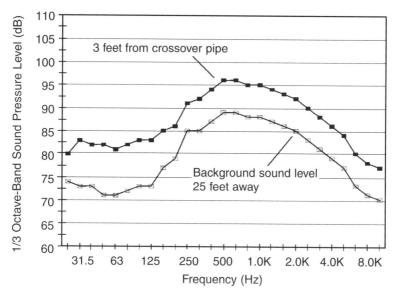

Figure 9.2 — Comparison of crossover-pipe SPL versus the background SPLs 25 feet from source.

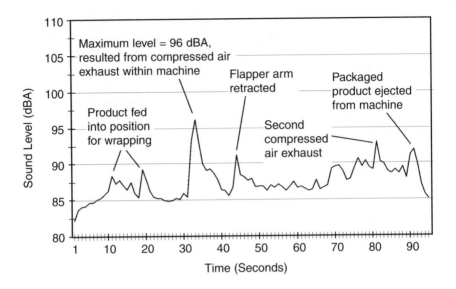

Figure 9.3 — Workstation for packaging operator measured over one machine cycle.

The process depicted in Figure 9.3 represents a product-wrapping machine, which has a cycle time of approximately 95 seconds. The maximum noise level of 96 dBA occurs during the release of compressed air, 33 seconds into the machine cycle. The other important events are also labeled in the figure, which permits the identification of the sources and relative contribution of each activity during the full wrapping cycle.

In industrial settings where there are multiple process lines with the same equipment, it is useful to compare the frequency spectra for like equipment and processes to identify differences in sound level that may be easily remedied with effective maintenance or other adjustments. Figure 9.4 depicts this comparison for two similar process lines, both of which manufacture the same product and operate at the same speed. Part of the process involves the use of a pneumatically actuated device that punches a 1/2-inch hole in the product as a final phase in its production. Inspection of this figure reveals that Line #1 has an overall sound level 5 dBA higher than Line #2. In addition, the spectrum depicted for Line #1 contains what appears to be a fundamental frequency and some of its harmonics that do not appear in the spectrum for Process Line #2. Consequently, it is necessary to investigate the cause of these differences. Often significant differences will be an indication of the need for maintenance, as was the situation for the final punch mechanism of Process Line #1. However, once maintenance corrects this problem, additional noise control measures are still required (for both process lines), if the objective is to reduce the overall noise level significantly below 100 dBA.

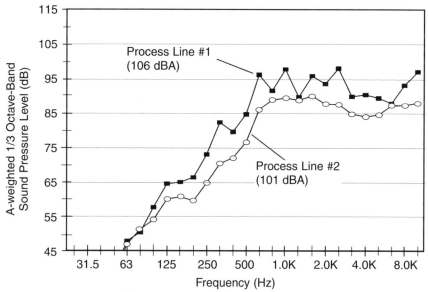

Figure 9.4 — Final punch operation for identical process lines operating at the same speed.

As mentioned above, an SLM typically provides a sound level that is composed of acoustical energy from one or more noise sources. Under optimum measurement conditions, it would be best to measure each equipment item with all other equipment turned off. Although this situation is ideal, it is not always practical. To work around this condition, it is often effective to use temporary control measures that will provide some short-term noise reduction to allow measurement of another source. Some materials that can provide a temporary reduction include plywood or cardboard enclosures, acoustical blankets, silencers, and barriers. Often permanent application of these materials will create long-term problems such as heat buildup, interference with the operator's access or product flow, or a costly pressure drop associated with improperly selected silencers. However, for assisting with the isolation of individual components, these materials can be effective as a short-term control.

Another method for isolating a particular machine or component is to turn on and off different equipment, or sections of a production line. To effectively conduct this type of diagnostic analysis the process must be capable of functioning with the item turned off. All valid data may then be rank ordered by magnitude of each sound level, as one method to help prioritize sources for engineering noise control. (Also, see section in Chapter 7, *Determining the Dominant Noise Sources.*) For this procedure to be legitimate it is critical that the manufacturing process not be affected in any manner. If the process is affected, then it is entirely possible the measurement will not be representative of the noise level under normal conditions.

Noise Control Options

After the cause or source of noise is identified, and its radiation to employee work areas is understood, the next step is to decide what the available noise control options are. The standard model used for control of almost any health hazard is to examine the various options as they apply to the source, path, and receiver, as depicted in Figure 9.5. In some situations control of one of these elements will be sufficient. However, under other circumstances it may require the treatment of more than one element to attain acceptable results.

The decision makers responsible for the design and implementation of engineering controls must take into account significant physical, economic, production, and worker operational constraints. For example, consider the problem of excessive noise generated by the cooling fan on an electric motor. Space limitations may dictate that changing out the fan blade with a more aerodynamic type is the only available option at one plant; however, at another location with an identical motor it may be more practical to simply install an air-intake silencer to treat the sound path. The first motor in this example had a physical constraint, which was limited space to fit an intake silencer over the cooling fan. The second motor did not have any space limitations, and using a commercially available silencer was deemed to be the best noise control option.

Reducing noise at the source should be the first step in the noise control program. Source modification addresses the root cause of a noise problem, whereas

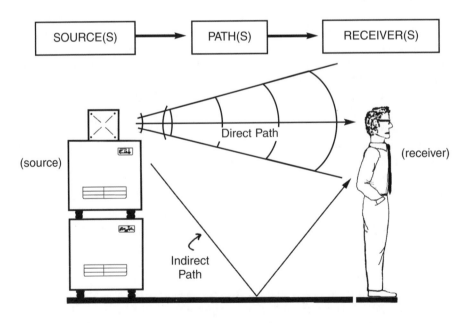

Figure 9.5 — Noise control model.

control of the sound transmission path with partial barriers, enclosures, and surface treatments only addresses the symptoms of noise. Reducing noise at the source is the best long-term noise control measure, since the cause is eliminated, and over time it can often be the least expensive solution. In those situations where there are multiple sources within a machine, and the objective is to treat the source, it will be necessary to address all noise-generating mechanisms on a component-by-component basis.

Obtaining Noise Control Products and Materials

One problem often encountered by health and safety professionals, as well as plant engineers, is how to obtain information on noise and vibration control products. Table 9.5 presents a comprehensive listing of many product manufacturers categorized by material and system type. It is worth noting that this table is not a complete listing of all manufacturers, and is not an endorsement of any company. To facilitate communication and review of their product lines the Internet website address is included for each manufacturer. Quite often it is easiest to contact one or more of the manufacturers, describe the noise problem, all your operational and production constraints, and the control measure you are contemplating; then a representative can identify which of their products are appropriate for your particular situation. In addition, the product representative can provide a price quote for consideration.

Source Treatment

There are at least four means by which the source of noise can be treated: (1) modification, (2) retrofit, (3) substitution, and (4) relocation. Each of these noise control options are described below.

Modification

Modification of the source can only be accomplished after a clear determination of the cause has been achieved. Recall from an earlier discussion that noise results from mechanical impacts, high-velocity fluid flow, vibrating surface areas of a machine, etc., and at times by the product being manufactured.

Modify Mechanical Impacts

When mechanical impacts are the cause of employee overexposure to noise, the control options to investigate are reducing the driving force, reducing the distance between impacting parts, dynamically balancing rotating equipment, and maintaining equipment in good working order.

For example, the operation of punch presses often creates significant mechanical impact noise. Research indicates that the maximum noise level typically occurs during the punching operation, and is dependent upon the stock material thickness and shear strength, speed of impact, and distance or clearance between the punch and die. Source modification can be achieved during tool alignment. Locating the workpiece as near to the end of the punch's stroke as possible will

TABLE 9.5
Noise and vibration control product manufacturer guide.[1]

This guide lists manufacturers noise and vibration control products grouped into the following categories:

- Sound Absorptive Materials
- Sound Absorptive Systems
- Sound Barrier Materials
- Sound Barrier Systems
- Composite Materials

- Composite Systems
- Vibration Damping Materials
- Vibration Isolation Systems
- Silencers

Each category contains a series of numbered divisions for specific product identification. To locate the noise and vibration control product of interest, first refer to the main categories listed above then go to the detailed listings to locate the manufacturers that offer that particular product. The numbers after each company listing indicate the products they offer. Finally, each manufacturer and website address are provided.

Sound Absorptive Materials	*Sound Absorptive Systems*	*Sound Barrier Materials*
1. Felts	1. Ceiling Systems	1. Pipe Lagging
2. Foams	2. Masking Noise Generators	2. Plain and Mass-Loaded
3. Glass Fiber	3. Panels	Plastics
4. Mineral Fiber	4. Unit Absorbers	3. Sealants and Sealing
5. Perforated Sheet Metal	5. Wall Treatments	Tapes
6. Spray-On Coatings		4. Sheet Glass, Metal, and
7. Wall Treatments		Plastic

Sound Barrier Systems	*Composite Materials*	*Composite Systems*
1. Curtains	1. Barrier/Fiber Composites	1. Curtains
2. Doors	2. Barrier/Foam Composites	2. Enclosures/Quiet Rooms
3. Operable Partitions	3. Masonry Units	3. Open-Plan Partitions
4. Panels		4. Panels
5. Seals		5. Quilted Composites
6. Transportation Noise		6. Roof Decks
Barriers		
7. Walls		
8. Windows		

Vibration Damping Materials	*Vibration Isolation Systems*	*Silencers*
1. Active Dampers	1. Active Isolators	1. Active Attenuators
2. Adhesives	2. Bases	2. Ducts
3. Constrained-Layer	3. Cable Isolators	3. Duct Silencers
Composites	4. Elastomeric	4. Electric Motor Silencers
4. Coatings	5. Floating Floors	5. Fan Silencers
5. Sheets	6. Machinery Mounts	6. Filter Silencers
6. Tapes	7. Pipe Connectors	7. General Industrial
	8. Pneumatic	Silencers
	9. Seismic	8. High-Pressure Exhaust
	10. Steel Springs	Silencers
	11. Vibration Dampers	9. Intake and Exhaust
		Silencers
		10. Pulsation Dampers
		11. Splitter/Louvre Silencers

— *continued on next page* —

[1] Table adapted with permission from: "Buyer's Guide to Products for Noise and Vibration Control," *Sound & Vibration*, 1999, 33(7).

TABLE 9.5 — continued

Product Manufacturer and Website Address	Sound Absorptive Materials	Sound Absorptive Systems	Sound Barrier Materials	Sound Barrier Systems	Composite Materials	Composite Systems	Vibration Damping Materials	Vibration Isolation Systems	Silencers
Acoustic Systems www.acousticsystems.com		1-3		1-4, 6-8		2, 4			
Allied Witan Company (no website available)	5							8	6, 8-10
Blachford, Inc. www.blachford.com	2, 3		2	6	2		6		
Burgess-Manning, Inc. www.nitram.com									6-10
E-A-R Specialty Composites www.earsc.com	2	3	1,2,4	1,6	1,2	1,5	3,6	4,6, 11	
Eckel Industries, Acoustic Div. www.@eckelacoustic.com	2,3	1, 3-5		2,4, 6-8	2	2,4	6		9
Empire Acoustical Systems www.empireacoustical.com	4,5	1-3, 5		1-4, 6-8		2-4			2-5, 7-9,11
George Koch Sons, Inc. www.kochllc.com	4,5	3	4	2,4, 6-8		2-4			
Illbruck, Inc. www.illbruck.com	2,7	1,3	1	1,4, 7	2	1,2, 4,5			
Industrial Acoustics Company www.industrialacoustics.com	5,7	1, 3-5		2-4, 6-8		2,4			1,3,5 7-9,11
Jamison Door Company www.jamison-door.com				2,8					
Kinetics Noise Control, Inc. www.kineticsnoise.com	2,3, 5,7	1-3, 5	1,2	1,2, 4,6-8	1,2	1,2, 4,5	4,6	2,4-7, 9-11	2-5, 7-9,11
Lord Corporation www.lordcorp.com							1	1-2, 4-6,11	
McGill AirPressure Corp. www.mcgillairpressure.com	2, 5-7	3-5	1-2	1-8	1-2	1-6	5		3,5,7, 9,11

— continued on next page —

TABLE 9.5 — continued
Noise and vibration control product manufacturer guide.

Product Manufacturer and Website Address	Sound Absorptive Materials	Sound Absorptive Systems	Sound Barrier Materials	Sound Barrier Systems	Composite Materials	Composite Systems	Vibration Damping Materials	Vibration Isolation Systems	Silencers
Mason Industries, Inc. www.mason-ind.com								2, 4-11	
3M www.3m.com							2,3, 5,6		8
Maxim® Silencer Division www.beairdindustries.com									5-9
Overly Manufacturing www.overly.com			2						
Silvent, Inc. www.silvent.com									8
Soundcoat Inc. www.soundcoat.com	1,2, 6,7	3,5	1,2	6	2	2	2-4, 6,7	4	
Stock Drive Products www.sdp-si.com							6	2-4, 6,11	
The Proudfoot Company www.noisemaster.com	2,3, 7	2-5	1,2	1,4, 6	1-3	1,2, 4,5			
United Process, Inc. www.soundseal.com	1-4, 6,7	1-5	1,2, 4	1,2, 4,7	1,2	1-5	3,4, 6		
Universal Silencer www.universal-silencer.com									1,3, 5-9

reduce the velocity of impact. Empirical data show this realignment can provide as much as a 10-dBA reduction in noise depending upon the initial setting (Petrie, 1975). Petrie has also shown that the effects of clearance are a function of the material properties and the type of fracture taking place. His data reveal that changing the angle of the punch can provide a significant noise reduction. Even a 2° to 4° angle can yield a noise reduction as much as 8 dBA. This effect occurs because the magnitude of force required to complete the punch cycle is reduced and spread out over a longer period of time, as illustrated in Figure 9.6. Granted the change in time is on the order of milliseconds; however, lengthening the event and reducing the force results in a significant decrease in noise.

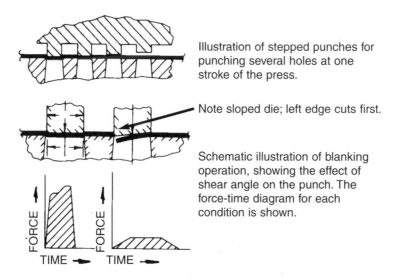

Illustration of stepped punches for punching several holes at one stroke of the press.

Note sloped die; left edge cuts first.

Schematic illustration of blanking operation, showing the effect of shear angle on the punch. The force-time diagram for each condition is shown.

Figure 9.6 — Illustration of how to reduce the force required by increasing the duration of the work cycle.

Another consideration in reducing the driving force is that noise levels typically increase as the speed of the machine increases. For example, Figure 9.7 exhibits the 1/3 octave-band spectrum recorded at the employee's typical location near a piece of rotating machinery operated at three different speeds. As depicted in the figure, the overall sound level is approximately 83 dBA at a speed of 40 revolutions per minute (rpm). This level increases to roughly 84 dBA at 70 rpm, then significantly increases to 91 dBA when the speed is raised to 88 rpm. The data for each of these measures also clearly show the significant increase in acoustical energy that occurs between 200 Hz and 4000 Hz as the machine is operated at 88 rpm. The production schedule at the facility calls for this machine to operate at 88 rpm 4 days per week over an entire 8-hour shift, which results in a TWA of 90 dBA for the equipment operator. The operator is then assigned miscellaneous nonproduction duties on the fifth day. However, if this equipment is operated at 70 rpm, the same volume of product could be manufactured over 5 days, and the resultant TWA for the operator would be approximately 83 dBA. Therefore, by reducing the operational speed, and spreading out the work over 5 days, the operator's TWA can be reduced from 90 dBA to 83 dBA, which is below the action level for inclusion in the HCP. This reduction in TWA may be a benefit to the employees who operate this equipment, and to the company in the form of potentially fewer machine failures, reduced maintenance costs, and reduced risk and liability resulting from lower noise exposures. However, beside these benefits and potential cost savings, it is important also to consider the increased cost, if any, to manufacture a product before implementing noise controls (see *Benefits and Costs of Noise Control*).

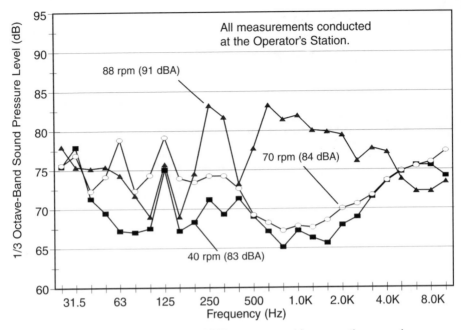

Figure 9.7 — Comparison of SPL versus machine operating speed.

It is important that manufacturing equipment be operated at the lowest speed setting necessary to satisfy production demands and routine or normal maintenance requirements. Quite often an employee will increase the speed without understanding the associated side effects such as an increased noise level, extra maintenance costs, etc. Therefore, the employee education and training component of the HCP needs to include a discussion about the noise control program, specific information such as the required machine settings to limit noise exposure, and a candid explanation as to why employee cooperation is needed to maintain the success of noise controls over time.

Optimize High-Velocity Fluid Flow
Excessive noise due to high-velocity fluid flow is a common problem with valves in process pipes. Quite often pipelines transporting a gas, vapor, or liquid can radiate high levels of noise due to acoustical energy within the medium being transported, and/or vibratory energy transferred to the pipe wall, especially when the pipe ring frequency or other resonant modes are excited. When the energy is in the medium, the noise actually radiates through the pipe wall. It is also common for acoustical energy to propagate along the entire length of pipe when turbulent flow exists. When vibratory energy is transmitted to the pipe wall, the pipe itself will often be an efficient radiator of noise, plus the energy may be transferred to other surfaces such as floors and panels (via clamps, hangers, etc.) that are capable of radiating airborne noise.

Modifying the source of noise within process pipelines is best achieved in the design stage. However, design changes to existing systems may be implemented during a process shutdown, such as times when periodic maintenance is scheduled. Figure 9.8 illustrates a modification to a piping system designed to minimize turbulence, and its associated loud shrieking noise.

To help ensure noise levels will be less than 85 dBA due to high velocity of the medium being transported through a pipe, the following rule of thumb may be useful: design the system such that the flow velocity in feet per second (fps) does not exceed 100 times the square root of the specific volume (in cubic feet per pound) for gases and vapors. For a liquid medium, flow velocities equal to or less than 30 fps are recommended. When viscous or laminar flow exists within the liquid, it is not anticipated that any excessive noise will result. Laminar flow exists whenever the Reynolds Number is less than 2000 (Van Houten, 1991). Conversely, when the Reynolds Number is greater than 4000, turbulent flow may be expected, which can cause excessive noise within the liquid and vibrations in the pipe wall. It is important to note that either laminar or turbulent flow can exist when the Reynolds Number falls between the limits described above. Finally, to prevent cavitation noise, the system needs to be designed such that the pressure

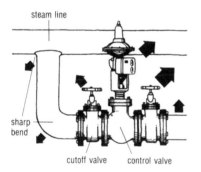

Example

A branch of a steam line has three valves which produce a loud shrieking sound. The branch has two sharp bends which also produce a lot of noise.

Control measure

A new branch is created with softer bends. Tubing pieces are placed between the valves, so that turbulence will be reduced or eliminated before the stream reaches the next valve.

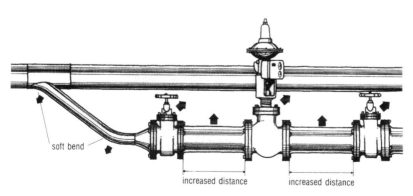

Figure 9.8 — Modifying a piping system to minimize turbulent flow and noise.

downstream of a valve is greater than the vapor pressure of the liquid. This type of design prevents bubbles from being formed in the liquid medium, which often collapse, causing cavitation and resulting noise.

Optimize High-Velocity Air Flow

High-velocity air from compressed air and pneumatic devices such as air valves, solenoids, or air cylinders is one of the most common noise sources within manufacturing equipment. These air systems are used to actuate or move components within a machine. For example, these devices allow packaging machinery sections to extend and retract as they move product, unload cartons from a magazine, etc. In addition, debris is often blown off work benches with air guns, and compressed air is used to eject product from a conveyor line.

As the high-velocity compressed air is mixed with the relatively still air in the atmosphere, excessive turbulence is created, which in turn generates a high noise level. The most dramatic noise level occurs whenever compressed air encounters a sharp object, such as an edge of the machine's casing or the product itself. It is important to note that the intensity of sound is proportional to the 8th power of the velocity of air flow (Lord et al., 1980). Consequently, the first step in controlling compressed air noise is to reduce the air velocity to as low a value as is practical. This adjustment can result in noise reductions on the order of 5–20 dBA. For example, consider the data exhibited in Figure 9.9 that were measured at a machine which forms the tapered necks on aluminum cans. This equipment utilized a compressed air line to move cans into position. The initial setting of the

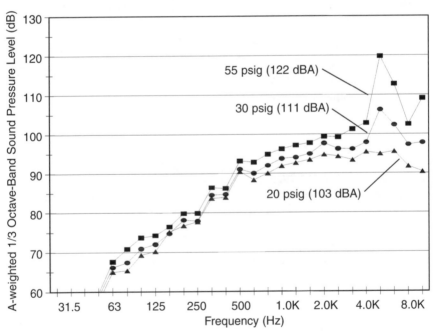

Figure 9.9 — Necker can infeed with air pressure set to 55, 30, and 20 psig.

air regulator was 55 pounds per square-inch gauge (psig), which was desired by the equipment operator, and resulted in an overall noise level of approximately 122 dBA, 1 foot from the source. By throttling down the air pressure to 30 psig, the sound level dropped to 111 dBA at the same measurement location. Finally, the air pressure was set to 20 psig, which was recommended by the equipment manufacturer, and the noise level decreased to 103 dBA.

Simply by re-setting the air pressure to the manufacturer's specification of 20 psig the initial sound level was reduced by 19 dBA and the equipment still functioned properly. This control measure did not cost anything; in fact, there was an energy savings in the company's cost for producing compressed air. Granted 103 dBA is still excessively high, but it is a dramatic improvement over the original condition. As previously mentioned, it is vital to educate and train all machine operators and maintenance personnel on the need to run their equipment at the optimum settings needed for production. It is common for an operator to unnecessarily increase the air pressure in an attempt to deliver more power, when in fact this higher pressure setting likely does not increase production, but almost always generates a significantly higher noise level.

Modify Surface Areas

Vibrating surface areas, such as a machine's casing, can be a source of noise. The key toward controlling this source is to identify the cause of vibration and either treat the source or interrupt the transmission path. A thorough noise control survey will assist with identification of the source of vibration. However, there will be times when vibration measurements are needed to confirm the origin. Some options for treating surface-radiated noise include stiffening sections (for large low-frequency sources such as the casing of large fans), making openings or perforations in the panels to reduce radiation efficiency (as illustrated in Figure 9.10), isolating the panels from the vibrating source, and applying vibration-damping materials. These latter two items are discussed in the following section on retrofit options.

Retrofit

Retrofitting a noise source involves installation of one or more commercially available noise-control products or systems. For example, the noise and vibration control product manufacturer guide presented earlier in Table 9.5 includes a comprehensive list of noise and vibration control product manufacturers, divided by product and system type. Most of these products are employed as a form of source treatment, to be positioned either on or relatively close to the actual source of noise.

Vibration Damping

In instances where production parts impact thin panels, such as the sides and bottom of drop bins, vibratory feeders (used to orient parts), walls of enclosures, circular lightweight particle transfer tubes, etc., the proper application of a damping treatment to these surfaces can be beneficial in controlling their vibrations.

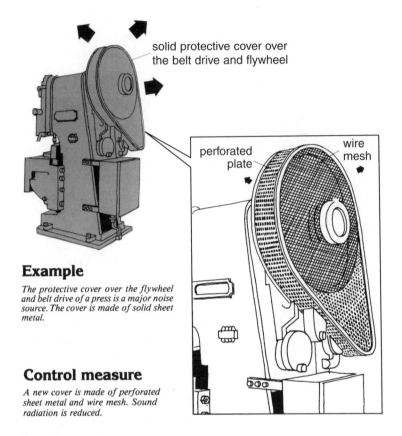

solid protective cover over
the belt drive and flywheel

perforated
plate

wire
mesh

Example

The protective cover over the flywheel and belt drive of a press is a major noise source. The cover is made of solid sheet metal.

Control measure

A new cover is made of perforated sheet metal and wire mesh. Sound radiation is reduced.

Figure 9.10 — Modifying surface area to reduce noise radiation efficiency.

Two basic types of damping techniques are shown in Figure 9.11. They are commonly referred to as free- and constrained-layer treatments. A free-layer (extensional) approach involves placing some form of energy-dissipating material on one or both sides of a vibrating surface. The constrained-layer procedure is similar to the free-layer control measure except an additional panel is placed on top of the treatment. This provides shearing of the damping material and normally improves the energy-dissipating effectiveness of the application. The following comments and guidelines given in this section are only for the free-layer type of treatment.

The type of material used can be either a preformed material, such as a PVC damping sheet with a stick-on or magnetic-holding backing material (E-A-R, 1988) or it can be painted or sprayed on. In instances where very thin or very lightly damped surfaces need to be controlled, duct tape, or a similar material, can be used as a temporary or possibly permanent treatment in controlling the unwanted vibrations.

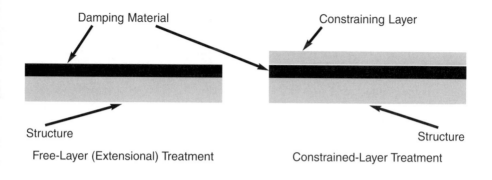

Damping Material   Constraining Layer

Structure   Structure

Free-Layer (Extensional) Treatment   Constrained-Layer Treatment

Figure 9.11 — Two different viscoelastic damping treatment configurations.

It is important to note that regardless of which side of a panel or structure damping materials are applied to, the noise reduction capabilities are essentially the same. For example, in the mining industry it is common to reduce the size of rock by grinding it in stages through a series of cone crushers. The product is typically conveyed and dropped into metal hoppers that, when left undamped, virtually ring like a bell. To damp the hopper panels, vibration-damping material is applied to the *exterior* surface areas of the casing. Logic may lead one to think that the damping material needs to be applied to the *interior* surface, since that is where the rock impacts the hopper casing. But as you might expect, the interior application will undoubtedly wear off in a relatively short period of time. Keep in mind the application of vibration-damping material reduces the ability of metal panels and structures to vibrate and subsequently radiate airborne noise. So in practice applying the material externally or internally serves a similar purpose.

It should be pointed out that damping material characteristics in general are frequency- and temperature-dependent. In addition damping materials exhibit two properties that play a significant role in determining the amount of noise reduction achieved when the material is applied to a surface: the material's loss factor, η, and dynamic modulus, E (E-A-R, 1988; Rao, 1995; Ungar, 1988; Ungar and Sturz, 1991). The material's loss factor is a measure of how efficiently a material dissipates the energy stored in it and the material's dynamic modulus is a measure of the material's stiffness that comes into play when trying to predict the damping material's effectiveness when applied to a surface.

Two simplifying assumptions are made herein: (1) the effectiveness of damping materials as applied in real-world situations occurs over the frequency range that impacts the measured A-weighted noise reduction, and (2) the temperatures of the surfaces being treated are in the range of 10–20 degrees Celsius. Therefore, this section presents a simplified view of the application of damping materials in real-world situations. It is intended to educate the reader about the fundamental concepts, and how to avoid basic errors in applying damping treatments. To learn the ins and outs of damping the reader should obtain and study the references listed and/or other similar sources.

For a damping treatment to be successful, certain conditions must be satisfied:

1. The panel being treated must be capable of creating noise if noise control is the primary reason for treating the panel.
2. The structure must be vibrating at one of its natural frequencies or normal modes of vibration.
3. When the panel (or system) vibrates and it has been treated, the energy stored and hence dissipated in the material applied to the panel must be a significant percentage of the total energy stored in the panel.

Panel Vibration Creates Significant Noise: Many structures can be observed vibrating at low frequencies. However, unless they are capable of coupling to the surrounding medium (air) they will not produce significant noise. Treating non–noise-radiating surfaces should not be expected to have any effect on the employee's noise exposure.

It is somewhat difficult to give definitive guidelines about when a surface will produce noise that impacts measurements made using the A-weighting filter for the type of surfaces encountered in real-world environments. However, a good rule of thumb is that $(2\pi/\lambda)r$ should be equal to or greater than around 0.5 for a panel in a large rigid structure (no openings) and 1.5 for a panel in a structure that exhibits significant openings (Olson, 1957), or;

$$(2\pi/\lambda)\, r \geq 0.5 \text{ to } 1.5, \tag{9.2}$$

where λ is the wavelength in meters associated with the frequency of the wave under consideration, $2\pi/\lambda$ is the wave number, and r is the effective radius of the plate in meters. To find the effective radius of our vibrating piston-equivalent of the panel, convert the area of the panel to a piston with a radius, r, that gives the same area (that is, the area of the plate equals πr^2 in square meters). If the panel is very narrow and long, then consult Olson (1957) for further guidance.

The wavelength of sound is given by Equation 2.2 (see Chapter 2). In air, the relationship between the frequency of the wave, f, and its speed in air, c (approximately 344 m/s) is as follows:

$$c = f\lambda \tag{9.3}$$

Since it can usually be assumed that the speed of sound is constant, the product $f\lambda$ is also constant. Therefore at a frequency of 100 Hz, the wavelength is approximately 3.4 m (11 ft), at 1000 Hz, 0.34 m (1.1 ft), and at 10,000 Hz, .034 m (0.11 ft).

EXAMPLE 9.2, Determining if Panel Vibration Warrants Damping Treatment

Assume that we are assessing the need to control side-panel vibrations for an enclosure with significant gaps around the panels. The effective radius of the side panels is 0.34 m (1.1 ft). Plugging Equation 9.3 into Equation 9.2 we get

$$(2\pi/\lambda)r = (2\pi f/c)r \geq 1.5, \tag{9.4}$$

Solving for f,

$$f \geq 1.5\, c/2\pi r = (1.5 \times 334)/(2 \times 3.14 \times 0.34) = 235\ Hz \tag{9.5}$$

Therefore if the panel vibrates at a frequency much lower than 235 Hz it should not be a significant radiator of acoustic energy, and it is unlikely that damping treatment would be effective.

Panel Is Vibrating at a Natural Frequency: When a panel (or any system that exhibits distributed mass, stiffness, and damping) vibrates, its motion for a stimulus of a given level and frequency of excitation depends upon the relationship of the exciting frequency to its natural frequency. The amplification ratio for steady-state vibrations (a motor or machine running at a constant speed) presents the ratio of the magnitude of a system's response, X, to the static deflection (X_o) that would occur if the exciting unbalance force were statically applied to the supporting springs (see Figure 9.12). The amplification response is plotted in Figure 9.12 versus the frequency ratio, (ω/ω_o), where ω is

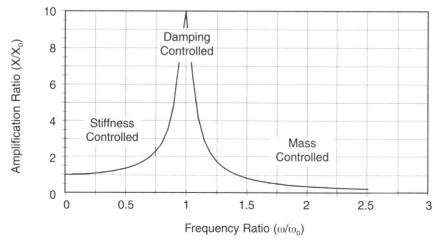

Figure 9.12 — Amplification ratio for a single-degree-of-freedom system.

309

the angular frequency, in radians per second, of the unbalanced exciting force and ω_o is the system's natural frequency.

For this simple single-degree-of-freedom system, the system is stiffness-controlled when it is excited significantly below its natural resonant frequency, i.e., the magnitude of the system's displacement is controlled predominantly by the system's stiffness. If it is being excited at or around its resonant frequency, the system is damping-controlled. Finally, if it is being excited significantly above its resonant frequency, it is mass-controlled. In order for a damping treatment to be effective, the panel must be vibrating at one of its resonant frequencies, where the magnitude of vibration is almost totally controlled by the damping in the system, and the panel must exhibit very low levels of damping, due either to internal damping, and/or damping provided by other supporting structures. Figure 9.13 illustrates a damping application used to reduce noise radiated by a saw blade.

It is important to note that after treating vibrating panels with damping sheets or spray-on materials, it is typical to observe some increase in the transmission

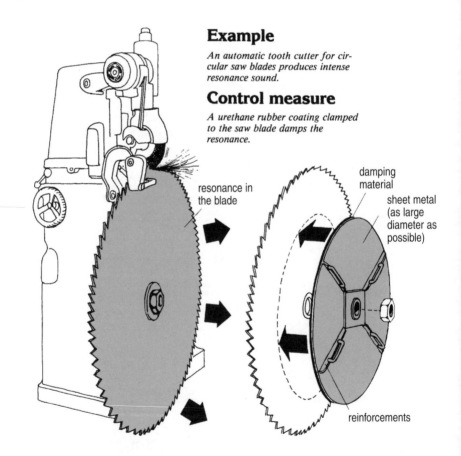

Example

An automatic tooth cutter for circular saw blades produces intense resonance sound.

Control measure

A urethane rubber coating clamped to the saw blade damps the resonance.

resonance in the blade

damping material

sheet metal (as large diameter as possible)

reinforcements

Figure 9.13 — Vibration-damping application used to reduce saw blade noise.

loss of the panel. In many observed instances the panel was not vibrating in the damping-controlled region and the increase in transmission loss was simply due to the fact that the mass of the panel was increased. That is, if you double the mass of a panel that is being driven in its mass-controlled region, an additional 6 dB of transmission loss is realized. This type of application of damping materials is costly and should be avoided. The application of other much less expensive materials would be just as effective in controlling the noise transmitted by the panel.

Energy Stored in the Damping Material Must be Significant: An estimation, and only an estimation, of the *noise reduction* (NR) expected from applying some form of damping material to a system may be obtained by evaluating the following Equation:

$$NR = 10 \log (\eta_A/\eta_B), \quad dB \tag{9.6}$$

where η_A is the system's loss factor after treatment and η_B is the system's effective loss factor before treatment. Equation 9.6 indicates that if it is possible to obtain a η_A/η_B ratio of 10, then 10 dB of noise reduction would be expected, or if a ratio of 100 was obtained, then 20 dB of noise reduction is predicted. However, the important point is that since η_A is normally not greater than around 0.1, η_B must be small in order to achieve significant NR.

The loss factor, η_A in Equation 9.6, cannot be directly equated to the published η values for different materials as usually provided by suppliers. For an actual structure,

$$\eta_A = (\eta, \text{ energy stored in the damping material})/$$
$$(\text{total energy stored in system}) \tag{9.7}$$

Consider the two different applications of damping materials to the structure shown in Figure 9.14. When a pinned-pinned (clamped-clamped, etc.) structure vibrates, most of its stored energy for a given motion is in the middle two-thirds

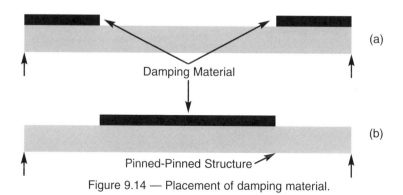

Figure 9.14 — Placement of damping material.

of the structure. Therefore, if one can treat only 50% of the surface area, due to cost or other constraints, the most effective treatment would be as shown in Figure 9.14b. Treating the whole structure would be inefficient since the material placed on the outer 50% of the structure would experience very little flexure and would not contribute significantly in reducing radiated noise.

An approximation of the effective loss factor for a vibrating system that consists of a single panel to be treated in its entirety with a damping material is as follows (Ungar, 1988):

$$\eta_A = \eta / \{1 + [1/(e_2 h_2 (3 + 6h_2 + 4h_2^2))]\}, \tag{9.8}$$

where $e_2 = E_2/E_1$ is the ratio of the dynamic modulus of the viscoelastic layer to the stiffness (Young's) modulus of the material being treated (stiffness ratio), and $h_2 = H_2/H_1$ is the ratio of the thickness of the viscoelastic layer to that of the material being treated (thickness ratio). A plot of Equation 9.8 is presented as Figure 9.15. The vertical axis is the loss factor ratio (η_A/η) and the horizontal axis is the thickness ratio (H_2/H_1). As an example, if E_2/E_1 equals 0.01, then for a thickness ratio of 1, the loss factor ratio read from the vertical left axis is approximately 0.11.

Presented in Table 9.6 are approximate damping properties for some common materials used to build machine structures and for a couple of commercially available damping materials. The given real-world values attempt to include the effects of end supports on panel vibrations to highlight application limitations in

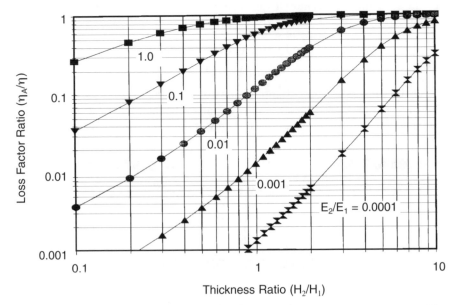

Figure 9.15 — Estimated loss factor ratio for a free-layer viscoelastic treatment, with the stiffness ratio (E_2/E_1) as a parameter.

TABLE 9.6
**Some common building materials and commercially available
damping materials, and their estimated loss factor and moduli at
1000 Hz at room temperature.**

Material	Loss Factor, η		Modulus, E (N/m^2)
	Specified	Real-World	
Aluminum	0.0001	0.001 – 0.005	7×10^{10}
Glass	0.001	0.01	6.5×10^{10}
Plexiglas or Lucite	0.005	0.01	
Steel	0.0001	0.001 – 0.01	20×10^{10}
Plywood	0.01	0.03	
E•A•R 2003*	0.8 – 0.9	0.7 – 0.6	4×10^{9}
E•A•R Isodamp*	0.6 – 0.7	0.5 – 0.6	8×10^{9}

* For additional information on damping products see Table 9.5, and for other material characteristics see Appendix III.

applying commercial damping materials to real-world structures. They also serve as a reminder to the user that the real-world effective damping constants of metal and wood structures can be significantly greater then the material's specified internal damping values.

EXAMPLE 9.3, Estimating NR Expected From Applying Damping Material

Assume that a vibrator bowl that orients plastic caps for half-gallon milk containers has been determined to be a significant noise source. Observation of the vibrator during fill-up and beginning of passage of the caps through the exit chute indicates that the primary noise source is the impact of the caps with the sides of the exit chute. It has been established by measuring and plotting the A-weighted levels created by the vibrator bowl that the exit noise is approximately 10 dB above the remaining combined vibrator bowl noise sources. Therefore it has been decided that damping the exit chute should have a significant impact on the employee's noise exposure.

The exit chute is constructed of 1/8-in steel. The treatment selected is an application of the C-2003 damping material (see Table 9.6). This material has a thickness of 3/10 in. Therefore the thickness ratio is 2.4 and the ratio of moduli, reference Table 9.6, is 0.02. From Figure 9.15 the loss factor ratio is estimated to be 0.6. Therefore, assuming that η is 0.65 (from manufacturer's specifications), η_A is found to be $(0.6)(\eta)$ or approximately equal to 0.39. Using Equation 9.6, an approximation to the expected noise reduction can be calculated (assuming that the loss factor of the steel including the effects of its mounting is 0.01) as follows:

$$NR = 10 \log (0.39/0.01) = 15.9 \text{ dB}$$

The actual real-world NR achieved was approximately 10 dBA.

Concluding Comments: In applying damping materials, it is common to predict large noise reductions for the particular noise source being controlled. However, practicing professionals rarely find a noise source that lends itself to the application of damping materials that results in more than 1–3 dBA of reduction in the affected employee's daily TWA. The reason is that rarely is a resonating structure the dominant noise source. The authors of this chapter have encountered only a few such situations. One example was the operator at a transfer station where hard plastic pellets were being conveyed from a holding bin to waiting trucks through thin aluminum pipes. In this instance a reduction in the operator's daily TWA greater than 10 dBA was achieved.

Keep in mind the three basic requirements presented earlier for the successful application of damping materials: the vibrating surface must be capable of creating noise in the first place; the vibrating surface must be vibrating at one or more resonant frequencies; and the energy stored and hence dissipated in the applied damping material in relation to the total energy stored in the vibrating system must be significant.

Vibration Isolation

It is common for industrial equipment to incorporate rotary or oscillatory mechanisms that create unacceptable vibrations in the supporting structures. These motions can result in individual vibration or noise problems or they can adversely affect the production process (reduced product quality, early component failures, limitations on equipment's ability to measure critical metrics, etc.). The result is often a need to solve the problem by reducing the transmitted vibrations to acceptable levels. This is referred to as vibration isolation. This section will address only the most elementary aspects of vibration isolation. The readers are referred to the following additional references: Kelly (1996); Rao (1995); Stock Drive Products (1984); Thomson (1981); Ungar and Sturz (1991).

Assume that a piece of equipment as shown in Figure 9.16 is to be placed on a production floor next to a vibration- and noise-sensitive measuring instrument. In addition assume that the production floor is located over a basement where recordkeeping employees carry out routine office activities. It has been determined that both of these activities are negatively impacted by the noise and vibrations created by this piece of equipment.

For discussion purposes the equipment can be represented as a rigid body on four vibration isolators (springs), one at each of its corners as shown in Figure 9.17. This rigid body system exhibits 6 degrees of freedom, that is, it can in theory vibrate in six different modes: three translational vibrational modes along the x, y, and z axis, or $X_x(t)$, $X_y(t)$, and $X_z(t)$, and three rotational vibrational modes around the x, y, and z axis, or $\theta_x(t)$, $\theta_y(t)$, and $\theta_z(t)$ motions. To simplify the problem, we will limit discussion to the motions along the vertical axis, $X_y(t)$. However, the reader should be prepared to also deal with torsional vibrations around the longitudinal axis (x), or $\theta_z(t)$ as well as other modes of vibration. Motions in the vertical direction are easy to visualize. However, in real-world environments vibrations resulting from torsional motions are just as common.

Figure 9.16 — High-speed punch press mounted on pneumatic rubber mounts. (Courtesy of *Sound and Vibration.*)

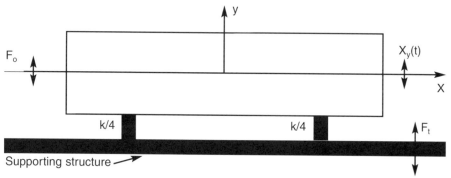

Figure 9.17 — Vibration isolation model of a rigid body on four spring isolators. $X_y(t)$ is the magnitude of the model's vertical vibration, F_o is the system's unbalanced force, F_t is the transmitted force to the supporting structure and k is the total stiffness of the selected vibration isolators.

Vertical Motions: The governing equation of motion for vertical vibrations (y axis only) is as follows:

$$m\ddot{X}(t) + C\dot{X}(t) + kX(t) = F(t), \text{ N} \qquad (9.9)$$

where m is the mass of our system (kg), C is the appropriate viscous damping coefficient (N-s/m), k is the system's total spring constant (N/m), F(t) is the system's steady-state exciting force (N), and ($\ddot{}$) and ($\dot{}$) denote respectively the second and first derivatives with regard to time. For this system, the transmissibility or transmission ratio of the isolator (T_r) is defined as the ratio of the magnitude of the force transmitted to the base to that of the exciting force, or

$$T_r = F_t/F_O = \left[\frac{1 + \left(2\zeta \dfrac{\omega}{\omega_n}\right)^2}{\left\{1 - \left(\dfrac{\omega}{\omega_n}\right)^2\right\}^2 + \left(2\zeta \dfrac{\omega}{\omega_n}\right)^2} \right]^{1/2} \qquad (9.10)$$

where F_t is the magnitude of the force transmitted through the vibration isolators to the support base or floor, F_O is the magnitude of the exciting force, ζ is the ratio of the system's viscous damping constant (C) to its critical damping value, C_c (Rao, 1995), ω is the speed at which the system is running (rad/s), and ω_n is the system's natural frequency (rad/s).

Assuming that we ignore the damping associated with our springs or vibration isolators (the ideal condition in that the presence of damping in the isolators increases the force transmitted to the base), then Equation 9.10 reduces to:

$$T_r = F_t/F_O = 1 / [1 - (\omega/\omega_n)^2] \qquad (9.11)$$

However, since T_r is ≤ 1 only when ω/ω_n is zero (static condition) or $\geq \sqrt{2}$ (where vibration isolation occurs), Equation 9.11 is typically written as:

$$T_r = F_t/F_O = 1 / [(\omega/\omega_n)^2 - 1] \qquad (9.12)$$

Mechanical vibration isolation is defined as,
$$R = (1 - T_r), \qquad (9.13a)$$

or in percent as:

$$R = 100\% \times (1 - T_r), \% \qquad (9.13b)$$

where R is the system's isolation in percent. Since the range of T_r values that are of interest is typically very large, it is sometimes helpful to utilize the logarithmic scale in evaluating Equations 9.10, 9.11, or 9.12. Therefore we define a transmissibility level as follows:

$$T_r = 20 \log (F_t / F_O), \quad dB \qquad (9.14)$$

A plot of Equation 9.14 using the F_t/F_O ratio defined by Equation 9.10 (including isolator damping) is shown in Figure 9.18. Note that 0 dB corresponds to the condition when the force transmitted to the base, F_t, equals the unbalanced force, F_O. This condition occurs for the static case, $\omega = 0$, and when the ratio of the mechanism's running speed to the isolated system's natural frequency is equal to 1.41 ($\omega/\omega_n = \sqrt{2}$). Also note that to achieve vibration isolation, that is the force transmitted to the base is less than the applied force, the mechanism's exciting frequency (running speed) must be greater than $\sqrt{2}\omega_n$. For a running speed assumed to be three times the system's natural frequency ($\omega/\omega_n = 3$ as indicated by the arrow in Figure 9.18) T_r is -9.3 dB. Also note in this figure that if the isolators exhibit damping (typically the case), such as ζ equal to 0.5, the transmissibility is increased in the vibration isolation region. That is, less vibration isolation is achieved (see Equation 9.10). This negative impact of damping in the isolators at frequencies above $\omega/\omega_n = \sqrt{2}$ is the primary reason that metal, or other types of undamped springs, are often the preferred option when very high levels of vibration isolation are required.

Isolation is achieved (transmissibility is reduced) by blocking unbalanced mechanical forces by making the system's resonant frequency much lower than

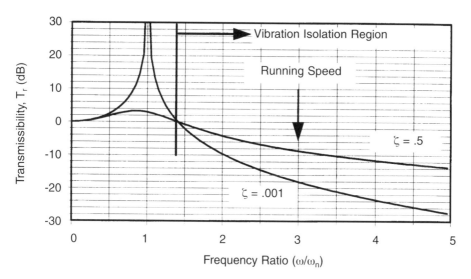

Figure 9.18 — Transmissibility for a single-degree-of-freedom system with and without damping.

the frequency associated with the unbalanced force that is being controlled. That is, the system is operated in the mass-controlled region, so that its mass is actually being used to block the unbalanced forces or moments.

System's Natural Frequency: For the rigid-body equivalent system shown in Figure 9.17, the fundamental natural frequency for undamped vibrations along the y axis is defined by Equation 9.15:

$$\omega_n = \sqrt{\frac{k}{m}}, \quad \text{rad/s} \tag{9.15}$$

where k and *m* were previously defined.

The static deflection of vibration isolators is as follows:

$$k = W/\delta, \quad \text{N/m} \tag{9.16}$$

where k was previously defined, W is the total weight of the mechanism, and δ is the deflection of each isolator when it is loaded statically. Recall that the mass and weight of the mechanism are related as:

$$m = W/g, \quad \text{kg} \tag{9.17}$$

where g is 981 cm/s^2 in the metric system and 386 in/s^2 for the English system of units if deflection in cm or inches is required.

If Equations 9.15, 9.16, and 9.17 are combined, a useful equation that relates the system's natural frequency to the static deflection achieved is obtained as follows:

$$\omega_n = \sqrt{\frac{g}{\delta}}, \quad \text{rad/s} \tag{9.18}$$

What is normally known about a machine is its lowest or most problematic rotational speed [revolutions/minute (rpm), cycles/second (Hz) etc.]. The system's rotational speed in Hz, f, is related to ω (rad/s) as follows:

$$\omega = f\,2\pi, \quad \text{rad/s} \tag{9.19}$$

Using Equations 9.15 and 9.19,

$$f_n = (1/2\pi) \times \sqrt{\frac{k}{m}}, \quad \text{Hz} \tag{9.20}$$

Or using Equations 9.16 and 9.19,

$$f_n = (1/2\pi) \times \sqrt{\frac{g}{\delta}}, \quad Hz \qquad (9.21)$$

Equation 9.21 is plotted in Figure 9.19. As an example, if a mechanism is placed on vibration isolators and a static deflection of 0.254 cm (0.1 in) is observed across all supporting isolators, then the isolated system will have a natural frequency of approximately 10 Hz.

By combining Equations 9.12, 9.13a, 9.19, and 9.21 an additional useful equation can be developed that relates the static deflection of the isolators, the required isolation, and the speed of the system in rpm as follows:

$$\delta = (91.2) \, g \, (1/N^2)(2-R)/(1-R), \qquad (9.22)$$

where N is the system's rotational speed in rpm.

A plot of Equation 9.22 using English units is shown as Figure 9.20. As an example, assume that we have a compressor whose motor is running at 3600 rpm and is the dominant source of vibration for the system. Locating 3600 rpm on the vertical axis (the solid line) and moving across to the intersection of the 99% isolation curve, it is found that the static deflection of the vibration isolators to give us the required isolation is approximately 0.33 in. To achieve 90% isolation, a deflection of approximately 0.033 in. is required.

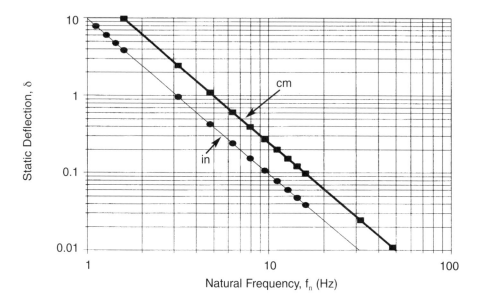

Figure 9.19 — Relationship between system's natural frequency and static deflection. Vertical axis in cm or inches depending upon which line is selected.

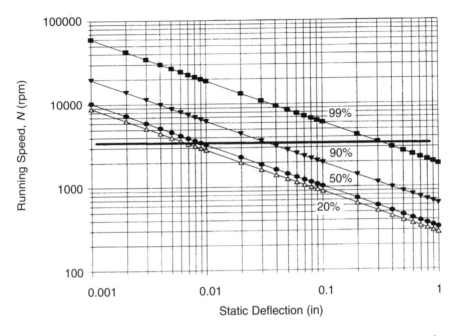

Figure 9.20 — Running speed versus static deflection for four different degrees of isolation. Horizontal line applies to example dicussed in text.

Vibration Isolation Steps: The basic simplified steps in isolating a mechanism are as follows:

1. Determine the frequency (running speed) of the machine, mechanism, etc., to be isolated.
2. Determine the equivalent weight(s) (mechanism specifications) that each isolator must support.
3. Select the percent of isolation, R, desired and then calculate the equivalent transmissibility, T_r using Equation 9.13b.
4. Knowing T_r, calculate the required ω/ω_n value using Equation 9.12.
5. Knowing ω/ω_n from (4), calculate the system's required static deflection from Equation 9.18 (note it must be the same for each isolator) or use Figure 9.19 or its equivalent.
6. Then knowing the system's weight and required isolator static deflection (system's natural frequency), select an isolator using a supplier's information charts such as shown in Figure 9.21.

Cautions and Guidelines:

1. Recall that the same required static deflection must be attained on all isolators (equipment may exhibit uneven loads at different isolator attachment points).
2. The most common mistake made in selecting vibration isolators is the failure to take into consideration the stiffness of the supporting structure

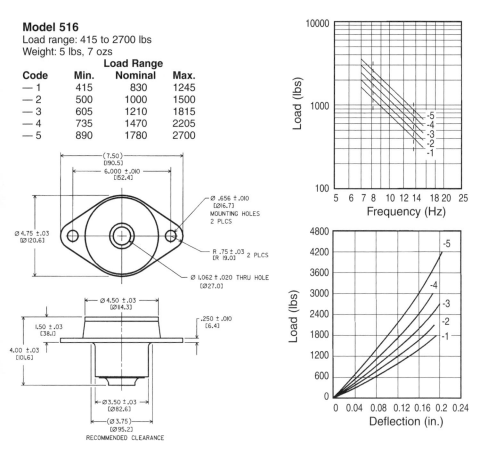

Model 516
Load range: 415 to 2700 lbs
Weight: 5 lbs, 7 ozs

Code	Min.	Load Range Nominal	Max.
— 1	415	830	1245
— 2	500	1000	1500
— 3	605	1210	1815
— 4	735	1470	2205
— 5	890	1780	2700

Figure 9.21 — Supplier's data sheet for selection of vibration isolator (Courtesy of Barry Controls).

(shown in Figure 9.17). A soft supporting structure can, and often has, resulted in underisolation, which can *actually increase* the vibration amplitude of the supporting structure. This problem can typically be solved by either overisolating the system or putting the mechanism on an inertia block (a large mass) and isolating the combined system. As a rule of thumb the supporting structure should exhibit a stiffness that is at least 10 times the stiffness of each isolator.

3. Consider the stability of the system (for large isolator deflections). It should be obvious that if one places a motor on springs that exhibit a large static deflection, then the system may simply fall over or turn over when turned on. One way of solving this problem is by placing the motor on a large inertia base and isolating the combined system.

4. Do not forget that damping in the isolators increases transmissibility and reduces the percent of isolation achieved.

5. Note that an isolated system has more flexibility and will vibrate at larger amplitudes than did the unisolated system. This fact is important when alignment between isolated and unisolated systems is critical. One solution to this problem is to locate all system components on the same inertia block and isolate the complete system.

6. Do not forget to isolate all connecting points, if important, to the mechanism being isolated. It may be of little value to isolate a pump while at the same time leaving solid pipe connections between the pump and supporting structures. Not only will these connections transmit vibrations to the supporting structure, but the increased mobility of the mechanism will put additional stresses on the rigid connections. Figure 9.22 illustrates an application using flexible pipe connectors and spring isolators on a compressor system.

7. If several vibration sources are located within the piece of equipment that is causing a vibration problem, the chances are good that if you isolate the dominant source with the lowest excitation frequency, then the remaining sources will also be adequately controlled.

Example

Cooling systems may be serious sources of noise as a result of intense pressure shocks in the liquid from compressors.

Control measure

Compressors may be vibration isolated with steel springs. In addition, flexible connections should be used for all inlet and discharge pipes.

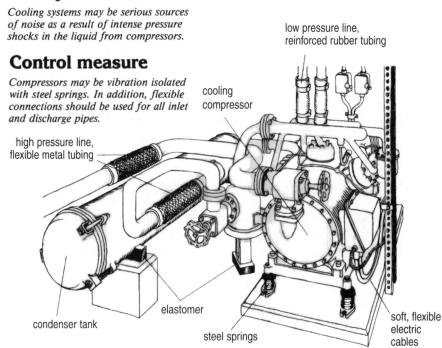

Figure 9.22 — Application of flexible connectors to reduce compressor inlet and discharge noise.

8. Consider the potential effects that the environment (cleaning fluids, oils, ozone, etc.) might have on the performance and endurance of the selected isolators.
9. Although it has been assumed that the system referenced herein exhibits only rigid-body resonant frequencies, and that the isolators and their supports exhibit only stiffness, both systems can exhibit natural frequencies. When this is the case, then the simplified approach presented in this section most likely will not result in an acceptable solution.
10. Seldom has the process of isolating a piece of equipment resulted in a significant reduction in the worker's noise exposure. Vibration-generated problems are more commonly associated with noise-annoyance issues. Therefore, do not expect that isolating all production equipment will significantly reduce employee noise exposures.

Dealing with the Isolated Mechanism's Natural Frequency: As indicated in Figure 9.18, the largest degree of isolation is obtained when isolators with very little damping, such as steel springs, are utilized. The downside of using undamped springs is for systems that pass through resonance, such as occurs when a motor comes up to speed, or slows down. The most critical situation often occurs when the system's power is removed, and due to the inertia of a system, its speed slowly diminishes and as a consequence it spends extended time around the critical resonant region. Therefore it is necessary to provide some means of control over the system's amplitude of vibration as it passes through resonance.

Methods of controlling the system include designing in a speed-braking system, using vibration isolators that have excursion limits, or selecting isolators that exhibit sufficient damping to limit the motions of the system. Shown in Table 9.7 are several different types of isolator materials and their respective damping values ($\zeta = C/C_c$). Note that although a highly damped isolator can effectively control the amplitude of vibration when the system goes through resonance, it also significantly reduces the degree of isolation achieved.

TABLE 9.7
Damping factors, ζ, for commonly used isolator resilient media.

Material	Damping Factor (approximate) $\zeta = C/C_c$
Steel Spring	0.005
Elastomers:	—
Natural Rubber	0.05
Neoprene	0.05
Barry Hi Damp	0.15
Barry LT	0.11
Friction Damped Springs	0.33
Metal Mesh	0.12
Air Damping	0.17
Felt and Cork	0.06

EXAMPLE 9.4, Selection of Proper Vibration Isolation

Assume that a compressor is creating a vibration-generated noise problem for the accounting employees downstairs. The compressor runs at 1800 rpm, or at a frequency of 30 Hz or $30 \times 2\pi = 188.5$ rad/s. The mass of the compressor is 2000 kg. Therefore the force applied to the springs (i.e., the load) is 2000 kg $\times$ 9.81 m/s^2 = 19,620 N $\times$ 0.2248 lb$_f$/N = 4411 lb$_f$. Referring to Figure 9.17, since we have 4 isolators under our mechanism, the load per isolator is 4905 N or 1103 lb$_f$. Although an unbalanced rotating element around the x axis as indicated in Figure 9.17 would exert a force along both the y and z axis, it is assumed that the only force component that needs controlling is along the y axis.

For this problem it has been determined that 90% isolation of the compressor's unbalanced forces should be sufficient. Referring to Equation 9.13b, to achieve 90% isolation requires a T$_r$ of 0.1. Solving Equation 9.12 for the value of ω/ω_n that results in a T$_r$ of 0.1, yields 3.31, or the required system's natural frequency must be 1/3.31 times the running speed of the compressor, or 30/3.31 = 9.1 Hz = 56.9 rad/s.

Knowing the required resonant frequency, utilize Equation 9.15 or 9.21 (or Figure 9.19) to determine the required static deflection of the vibration isolators to be installed under the equipment (all four isolators must deflect the same amount). The required deflection is approximately 0.3 cm or 0.11 in. Note that it has not been necessary to actually calculate the stiffness of the isolators in order to initially solve the problem. However, the stiffness could be determined using Equation 9.16. Keep in mind that k represents the total stiffness of the four supporting isolators.

In selecting an isolator, refer to suppliers' data sheets such as shown as Figure 9.21. For this system, utilize the bottom right graph that provides the expected static deflection for different versions of the Model 516 isolator under different loads. For a weight per isolator of approximately 1100 lb$_f$, the Model 516-2 unit will provide at least the static deflection required (more is okay). Alternatively, if the top right graph is utilized for a weight per isolator of 1100 lb$_f$, a resonant frequency of approximately 9–10 Hz is indicated.

The manufacturer has also provided in Figure 9.21 a suggested maximum and minimum loading range. In selecting isolators, care should be exercised not to exceed the operating ranges of the isolators specified.

Recommendations: Industrial hygienists and other professionals interested in solving vibration isolation problems are encouraged to obtain copies of some of the previously cited references in order to achieve a broader understanding of vibration isolation. Some suppliers of vibration isolation equipment have useful websites where existing problems can be defined and solutions recommended (see Table 9.5). Of course when critical pieces of equipment are being treated (such as electric turbine generators, major ventilation systems located within the bounds of office buildings, etc.), the opinions of trained professionals should be obtained. The authors have found that the workbook by Kelly (1996) and the handbook produced by Stock Drive Products (1984) (both contain CDs) are very useful information sources.

Silencers

Silencers are devices inserted in the path of a flowing medium to reduce the downstream sound level. For many industrial situations, the "flowing medium" is air. These devices are necessary because gas flow through a system is performing useful work and cannot be significantly altered. Silencers are known by many other names depending on the application. Internal combustion engines typically use *mufflers*, high-pressure gas systems often use *attenuators*, and heating-ventilation-air-conditioning (HVAC) systems use *sound traps, plenum chambers, baffles*, etc. They all accomplish the same goal, so we will use the term "silencer" to refer to all these devices.

Silencers can be classified as either *dissipative* or *reactive*. Dissipative silencers use porous sound-absorbing materials surrounding the primary air flow passage. Often the air moves straight through the silencer with little or no change in direction, so the pressure drop for a dissipative silencer is generally acceptable for most industrial applications. The noise reduction provided by a dissipative silencer covers a wide range of frequencies, primarily in the middle to high frequencies. The principal method of sound attenuation is by absorption.

Special applications may require facings on the absorbent material due to the environmental conditions inside the pipe or duct. If the air flow is high-velocity, the fibrous packing in the silencer may be drawn into the pipe and slowly disintegrate over a period of time. Mylar® or Tedlar® thin-film facings can be installed over the sound absorbent materials inside the silencer, if constructed properly, with only a slight decrease in attenuation at the high frequencies. Clean-room environments may require protective facing over the absorbent material to maintain high air quality. Often the facing also resists absorption of oil and chemical contaminants that may be in the air stream. It is important to identify possible interactions of the facing material(s) and contaminants in the air stream. High-temperature applications such as pressurized steam pipelines may require special absorptive materials or surface films to withstand harsh environments.

Figure 9.23 exhibits a dissipative silencer for a positive displacement rotary blower. There are two dissipative elements in this design: an inner cylinder of absorptive material and a layer inside the outer shell. The absorptive material consists of a 2-inch layer of fiberglass covered with perforated metal to help hold the fiberglass in place. The perforation also prevents fragments or glass fibers from separating and being drawn into the air stream. When dirt and moisture exist in the medium the perforation will help protect the underlying absorption material. However, caution must be exercised to ensure that the perforated holes do not plug up over time. This would degrade the effectiveness of the absorption material. It is recommended that, as part of the facility's routine maintenance program, all silencers used in contaminated air streams be periodically inspected and cleaned or replaced when necessary. HVAC silencers often

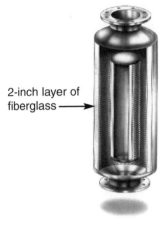

2-inch layer of fiberglass ⟶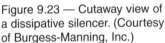

Figure 9.23 — Cutaway view of a dissipative silencer. (Courtesy of Burgess-Manning, Inc.)

Figure 9.24 — Cutaway view of a rectangular duct silencer for HVAC applications. (Courtesy of McGill AirPressure Corporation.)

use multiple elements or baffles mounted parallel to each other to increase the absorptive surface in the air stream. The configuration shown in Figure 9.24 works well for the rectangular ducts found in many ventilation systems.

Reactive silencers use sound reflections and large impedance changes (area variations) rather than absorptive materials to reduce noise in a pipe or duct. When sound waves expand from an inner perforated tube into the large outer cavity, the waves reflect off the sides of the silencer shell and interfere with incoming sound waves. The result is noise reduction in a narrow frequency range depending on the size of the outer shell. Reactive silencers work most effectively at low frequencies, but by combining a number of different expansion chambers inside the same outer shell, the silencer can attenuate noise over a wide range.

The construction of reactive mufflers includes several components. The primary sound attenuation is accomplished by using multiple expansion chambers. Some designs use side-branch resonators or tuned resonators to attenuate sound over a very narrow range of low frequencies. The inlet pipe of the silencer is typically perforated with many small holes to allow the sound waves to expand from the inlet tube into the expansion chamber volumes. Finally, bends are used to provide additional attenuation and to allow a long flow path through the silencer while maintaining a more compact overall length.

Some reactive silencers result in a greater pressure drop in the system than dissipative silencers, but they are more effective at reducing low-frequency sounds. Reactive silencers can be used in high-temperature and/or corrosive environments, such as combustion engine exhaust systems or steam pipelines, where sound absorptive materials may quickly deteriorate. The reactive silencer shown in Figure 9.25 has many chambers of different volumes fed by tubes of differing lengths and diameters to provide *insertion loss* (IL) over a wide range of frequencies. However, silencers as illustrated in Figure 9.25 create a larger pressure drop than silencers that contain a straight-through air path.

Many reactive mufflers can extend their high-frequency range by adding dissipative elements to the existing reactive chambers. Most often, a layer of fiberglass or absorptive material is added inside the outer shell of the silencer. Combination silencers are very effective in providing noise reduction at both low and high frequencies in one compact package. They may not be practical for high temperature or contaminated air streams, but they work well in other applications.

When selecting a silencer for a particular application, there are three performance criteria that must be satisfied simultaneously: (1) the acoustical performance of the silencer, (2) the aerodynamic considerations to minimize flow disturbances and satisfy back-pressure constraints, and (3) physical limitations of size and weight. Octave-band SPLs and corresponding sound levels in dBA for the pipe or duct are required to select the correct type of silencer and to determine the required IL. A silencer with high IL will generally have a high pressure drop as well, so select a silencer with only the amount of noise reduction needed. Determination of the silencer pressure drop depends on the temperature and pressure of the gas in the system and the flow rate through the silencer.

Physical constraints on the size, shape, and weight of the silencer must be determined before selection is made. Silencer manufacturers supply many different configurations to fit various applications. Before looking at the acoustical performance of a silencer, choose one that can maintain sufficient flow, to allow the medium to easily pass through the system. To determine silencer size, find the velocity range from the list of equipment presented in Table 9.8.

Figure 9.25 — Cutaway view of a reactive silencer. (Courtesy of Burgess-Manning, Inc.)

TABLE 9.8
Using equipment type to select silencer size.
(Courtesy of Burgess-Manning, Inc.)

Recommended Silencer Operating Velocities

General
Size to match connection size if within the recommended velocities and specified system pressure drop.

Engine Intake Silencers

4 cycle, 4 or more cylinders	4000 to 6000 fpm
2 cycle, 2 or more cylinders	4000 to 6000 fpm
4 cycle, 2 or 3 cylinders	2000 to 3000 fpm
2 cycle, 1 cylinder	2000 to 3000 fpm
4 cycle, 1 cylinder	1000 to 1500 fpm

Rotary Positive Blower Silencers
5500 to 6000 fpm

Centrifugal Blower Silencers
3000 to 7500 fpm
Max. ΔP — 2 to 4" H_2O

Reciprocating Comp. Intake Silencers
Single Acting — 2000 to 3000 fpm
Double Acting — 4000 to 5000 fpm

Engine Exhaust Silencers

Low Speed <350 rpm	4000 to 7000 fpm
Med. Speed 350 to 1200 rpm	6000 to 9000 fpm
High Speed >1200 rpm	8000 to 10,000 fpm
Turbo Charged All rpm's	5000 to 10,000 fpm
Maximum Allowable	12,000 fpm

Fan Silencers
4000 to 7500 fpm
Max. ΔP — 2 to 4" H_2O

Vent & Blowdown Silencers

Intermittent	15,000 to 20,000 fpm
Continuous	10,000 to 15,000 fpm
Maximum Allowable	25,000 fpm

Vacuum Pump Separators
4000 to 6000 fpm

Steam Ejector Silencers
Match connection or larger
4000 to 8000 fpm
Max. ΔP — 5 to 10" H_2O

EXAMPLE 9.5, Silencer Selection

It is recommended that a fan silencer provide 4000–7500 feet per minute (fpm) flow velocity to maintain a specified pressure drop. Now, find the total gas flow in *actual cubic feet per minute* (ACFM) from the manufacturer's specifications for the equipment in question. If gas flow is given at *standard cubic feet per minute* (SCFM), it can be corrected to actual operating conditions of temperature and pressure using the appropriate formulas provided in Table 9.9. Using Figure 9.26, find the gas flow on the horizontal axis of the graph. For example, assume a flow velocity of 5000 cfm. Move vertically on the 5000 cfm line to reach the velocity range of interest. In this case, the range is 4000–7500 fpm found in the middle region of the graph. The recommended silencer size is determined by finding the nearest slanting line crossing the vertical cfm line and falling within the velocity range. If no lines fall in the velocity range, pick the line nearest the *lower* limit of the range. In this case, the silencer size is 12 inches in diameter. It is acceptable to move up or down one pipe size to match existing

TABLE 9.9
Formulas for flow conversions and pressure drop calculations used for silencer selection. (Courtesy of Burgess-Manning, Inc.)

Flow Conversions

Air & Gas

$$ACFM = SCFM \times \frac{T(°R)}{530} \times \frac{14.7}{P(Abs)} \times \frac{28.93}{M.W.}$$

$$ACFM_2 = ACFM_1 \times \frac{T_2}{T_1} \times \frac{P_1}{P_2}$$

$$ACFM = \frac{\#/hr}{60} \times \frac{10.72}{M.W.} \times \frac{T(°R)}{P(Abs)}$$

Steam

$$ACFM = \frac{\#/hr}{60} \times \text{Spec. Vol. @ Flow Conditions}$$

$$T(°R) = °F + 460$$
$$P(Abs) = PSIG + 14.7$$

Pressure Drop Calculations

$$\Delta P \text{ "}H_2O = C\left(\frac{V}{4005}\right)^2 \times \frac{M.W.}{28.93} \times \frac{P_o}{14.7} \times \frac{530}{T_o}$$

Where
C = Silencer Pressure Drop Coefficient (Exit Loss Not Included)
V = Silencer Flow Velocity (Ft/Min) (ACFM ÷ Nozzle Size in Sq Ft)
P_o = Operating PSIA (PSIG + 14.7)
T_o = Operating Temp. °R (°F + 460)
M.W. = Molecular Weight

Conversion Factors
"H_2O × .0361 = PSI
"H_2O × .07355 = "Hg
"Hg × .4908 = PSI

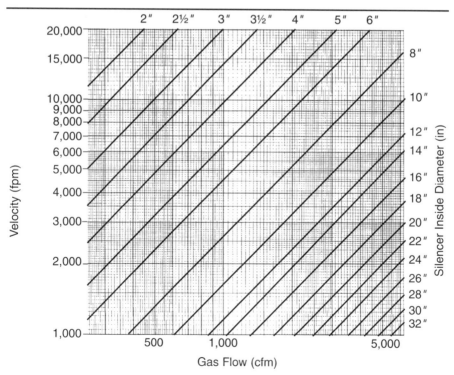

Figure 9.26 — Velocity chart for silencer signing and selection.

pipes or ducts if the slanted line falls below the maximum flow velocity for that type of equipment. However, if matching existing pipe or duct diameters, then caution must be exercised to ensure the resultant pressure drop does not exceed the equipment manufacturer's specifications.

Pneumatic and Compressed Air Systems

Excessive noise caused by high-velocity air is one of the most serious and common noise problems throughout manufacturing plants. In addition, this type of noise is probably one of the most elementary and simplest sources to control. In the *Source Modification* section, it was mentioned that the first step toward controlling compressed air noise is to reduce the air velocity as much as is practical. However, should this step not provide sufficient noise reduction, or should it be impractical, then the second step is to retrofit the air source with a commercially available silencer.

Selection of the appropriate device is critical for this control measure to succeed over time. If the source is a solenoid valve, air cylinder, air motor, etc., that simply exhausts compressed air to the atmosphere, then a simple diffuser-type silencer will suffice. Figure 9.27 exhibits several types of these silencers. The device in Figure 9.27a is a porous or sintered metal (typically brass) that distributes the air flow over a large area at a reduced velocity. The disadvantage to this device is that it can cause unacceptable back pressure. Therefore, it is important when selecting a diffuser silencer that the pressure-loss constraints for the particular application are satisfied. Figure 9.27b depicts another type of diffuser silencer, which is made of a porous polymer material. Both these silencers can be routinely washed out and reused, which is especially important should the compressed air include lubricants or simply be contaminated with dirt or other debris. Finally, Figure 9.27c shows a muffler designed for small pumps (e.g., diaphragm pumps) and other air-operated machinery. This particular muffler is designed to minimize back pressure by use of a porous composite material. All these silencers can provide up to 15–30 dB of IL.

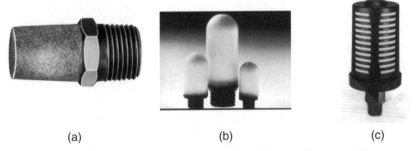

(a)   (b)   (c)

Figure 9.27 — Examples of pneumatic diffuser silencers. Figure 9.27a – sintered brass silencer (Courtesy of Allied Witan Company). Figure 9.27b – porous polymer silencer for high thrust applications (Courtesy of McGill AirPressure Corporation). Figure 9.27c – pneumatic muffler used for many pumps, motors, and other air-operated equipment (Courtesy of 3M Company).

For compressed air systems that do perform a service, such as ejecting parts, blowing off debris, etc., there are a number of devices available for retrofit at the point of discharge. Figure 9.28 shows a muffler specifically designed for compressed air used to eject parts from a machine or conveyor line. Many machines contain open-ended air lines that use compressed air to assist production in various ways. It is common for machine operators to take an existing air line and crimp or pinch the opening down to increase the blowing force, as illustrated in Figure 9.29. As a retrofit to open-ended air lines, the silencer shown in Figure 9.30 may be installed. This nozzle silencer divides the single-stream compressed air into smaller streams, which helps reduce the noise. Also, the nozzle creates a focused conical blowing pattern dependent upon the exact service it needs to perform. Figure 9.31 depicts an air gun application with a dual-flow nozzle, which provides a significant noise reduction compared to a single stream of compressed air.

For process operations that require a broad stream of compressed air, such as air knives used to blow off moisture or solid debris, a series of flat nozzle silencers should be utilized. Figure 9.32a exhibits a flat nozzle, and Figure 9.32b shows an air knife application using several flat nozzles connected to a manifold

Air Ejector Muffler   Ejector Muffler Insert

Figure 9.28 — Air ejector muffler and ejector muffler insert. (Courtesy of Allied Witan, Co.)

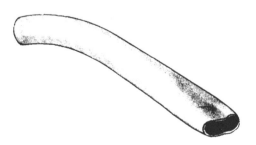

Figure 9.29 — Open port compressed air line crimped or pinched to increase air velocity. (Courtesy of Silvent, Inc.)

331

Figure 9.30 — Pneumatic silencer for open-ended compressed air lines. (Courtesy of Silvent, Inc.)

pipe. In Figure 9.32b, the noise level was reduced 12 dBA after installation of these flat nozzles.

Another typical application for compressed air is blow-off guns or air guns. These tools come in a variety of sizes and shapes, and can generate noise levels of 90–115 dBA, depending on the velocity of air and the surface area it contacts. The major problem with air guns is that, like other pneumatic or compressed air systems used to drive and motivate machinery, equipment operators will often increase the air pressure in an attempt to create more blow-off power. Recall from an earlier discussion that the intensity of noise is proportional to the 8th power of the air velocity. As a result, a higher pressure setting will significantly increase the noise level. In addition, when a compressed air silencer is installed on machines, many operators will remove or defeat this device to maintain the perception of more power they have grown accustomed to, which is based on their subjective assessment of the sound level. To prevent unnecessary or unauthorized air adjustments by the process or equipment operators, air-pressure regulators should be set and locked to ensure that they cannot be modified without supervision's consent, and operators should be educated and trained that the power is adequate.

Figure 9.33 shows a variety of devices manufactured for retrofitting blow-off lines. These nozzles provide significant noise reduction and decreased energy consumption, yet supply sufficient air pressure to satisfy production demands.

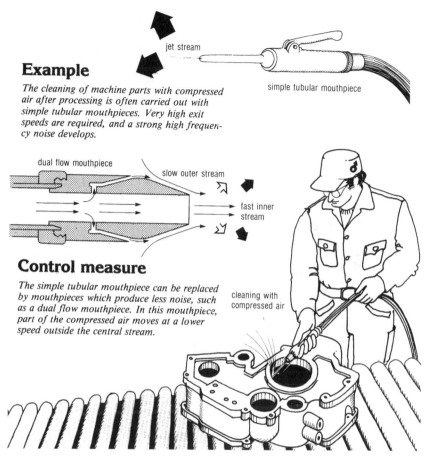

Example

The cleaning of machine parts with compressed air after processing is often carried out with simple tubular mouthpieces. Very high exit speeds are required, and a strong high frequency noise develops.

Control measure

The simple tubular mouthpiece can be replaced by mouthpieces which produce less noise, such as a dual flow mouthpiece. In this mouthpiece, part of the compressed air moves at a lower speed outside the central stream.

Figure 9.31 — Dual-flow air nozzle for control of compressed air noise.

Typically, these air blow-off nozzles will generate noise levels on the order of 75–84 dBA, when used in accordance with the manufacturer's requirements.

Silencers for pneumatic or compressed air systems normally require routine inspection, maintenance, and/or replacement, since these silencers will plug up with debris, be removed by operators, or occasionally become damaged over time. However, with a proper commitment to follow the procedures described above, there are no technical reasons for the existence of excessive high-velocity air noise in manufacturing facilities.

Noise Cancellation

The technology of noise cancellation, also known as active noise control, has been around since the early 1930s. However, it was not until technological advances in adaptive signal processing in the early 1980s that noise cancellation

Figure 9.32a — Flat nozzle silencer for fixed installations. (Courtesy of Silvent, Inc.)

Figure 9.32b — Air knife silencer arrangement utilizing 20 flat nozzles. (Courtesy of Silvent, Inc.)

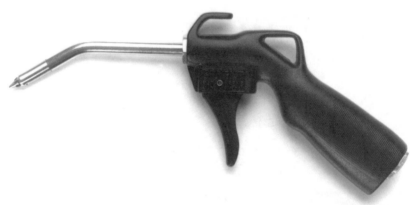

Blow-off gun for high blowing force and low air consumption.

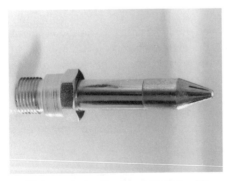

Slot nozzle for fixed installation with space limitations.

Stainless steel nozzle for extremely high blowing force.

Figure 9.33 — Low-noise blow-off devices. (Courtesy of Silvent, Inc.)

developed beyond the early concepts. Figure 9.34 illustrates a basic system used today to cancel noise in a ventilation duct or pipe. The system is configured such that an input microphone (reference sensor) detects the incoming sound wave (disturbance) and relays its signal to the controller. The controller assesses the signal and determines what anti-noise signal is required to cancel the sound wave. Next, a loudspeaker located downstream of the input microphone receives a signal from the controller and radiates a sound wave 180° out-of-phase. The net result is a canceling effect. Since there may be some residual noise left over, a second microphone (error microphone) is often located further downstream. The second microphone samples the modified signal and sends it back to the controller, which adjusts the loudspeaker output accordingly to ensure optimal results. A significant feature of a noise cancellation system over quasi-static reactive attenuators is that it can compensate for changing parameters such as fan velocity and changes in the blade passage frequency. Because the system functions with real-time data, the controller actually hunts down dominant pure tones and generates the appropriate canceling sound wave. This feature potentially could be superior in performance to use of a passive reactive silencer that has a fixed frequency range for optimal insertion loss or attenuation.

This sounds great, so why isn't noise cancellation the panacea for all noise problems? The answer lies in the fact that noise cancellation has a number of technical, physical, and cost constraints. For noise cancellation to work effectively it requires that the maximum cross-sectional dimension of the pipeline or duct be $\lambda/4$ or less, in order to force the sound wave to propagate in the system as a two-dimensional plane wave. For example, if the wavelength of the tone propagating down a duct is 2.2 feet (500 Hz), then the maximum diameter of the duct must be less than 0.55 feet. Noise cancellation usually works best for narrow-band signals, and it works primarily for signals below 500 Hz. For example, fans, blowers, and propellers that generate blade passage frequencies (pure tones) and harmonics at relatively low frequencies are potential candidates for noise cancellation. For high-frequency noise and for all noise sources existing in three-dimensional (3-D) space, the basic science simply does not exist for noise cancellation to be practical. Further explanation regarding the technological and physical limitations is beyond the scope of this chapter. For additional information the interested reader is referred to Gordon and Vining (1992).

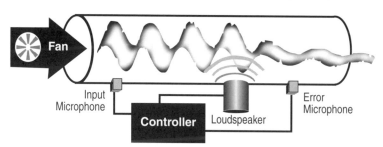

Figure 9.34 — Active cancellation of noise in a duct. (Courtesy of Digisonix, a Division of Nelson Industries, Inc.)

Because noise cancellation can significantly attenuate low-frequency pure tones, by far the most practical industrial application is reduction of pure tones which radiate to a nearby community from exhaust stacks and ducts. For example, a paper-manufacturing plant employed eight large vacuum pumps that discharged 100,000 cfm of vapor-laden air through a 54-inch exhaust stack located on the roof of the facility as shown in Figure 9.35. The exhaust noise contained three distinct pure tones below 100 Hz, which were reduced approximately 20 dB by the installed noise cancellation system, as exhibited in Figure 9.36 (note that six of the eight loudspeakers attached to the stack are visible in Figure 9.35). As a result, the "rumble" noise disturbing the residents within a 1–2 mile radius of the plant was successfully eliminated.

Noise cancellation systems have several advantages over passive silencers: negligible pressure loss, excellent low-frequency attenuation, insensitivity to moisture- and particulate-laden air, and the elimination of pure tones. The major disadvantages are no noise reduction capabilities above approximately 1000 Hz, and periodic maintenance for the components. Because noise cancellation systems excel in areas where passive silencers struggle to be effective, and vice versa, it is becoming more common to see an active noise control system used in conjunction with a standard passive silencer. This combination provides the user with significant attenuation over a relatively wide range of frequencies.

2 of 8 loudspeakers

Figure 9.35 — Noise cancellation system for a vacuum pump application using eight noise-cancelling loudspeakers. (Courtesy of Digisonix, a Division of Nelson Industries, Inc.)

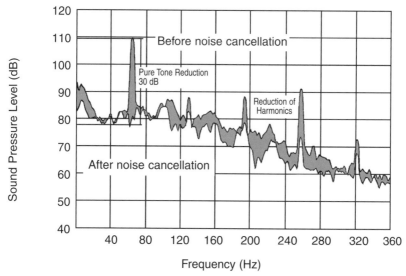

Figure 9.36 — Comparison of frequency spectra before and after activation of noise cancellation system. (Courtesy of Digisonix, a division of Nelson Industries, Inc.)

Substitute for the Noise Source

Assuming the cause or source of noise is identified, another noise control option is to explore quieter alternatives for achieving production needs. For this approach the answers to the following questions should be investigated with the goal of reducing noise:
1.   Is there alternative equipment available?
2.   Are there different materials available?

It is becoming more common for machine manufacturers to have more expensive versions of equipment that are designed to operate at a reduced noise level. For example, gears, bearings, fans, and motors come in a variety of sizes, shapes, and operating speeds, and a wide range in cost. It is common for the less expensive units to have the highest acoustical output or noise level. The quieter units typically cost more due to the increased precision and tighter tolerance limits required to produce them. Therefore, the manufacturer of the noise source, as well as its competitor companies, should be contacted for information regarding the availability of comparable but quieter equipment.

If similar equipment is not available, then use of a different design or device may be an option. For example, belt drives generally produce less noise than gears, belt conveyors are quieter than roller conveyors, and mechanical part ejectors typically create less noise than compressed air sources. Figure 9.37 exhibits a comparison between the use of compressed air to eject parts from a turret versus a mechanical pick-up device. As shown in the figure, the mechanical device is 10 dBA quieter than the air source. Inspection of Figure 9.37 reveals that the

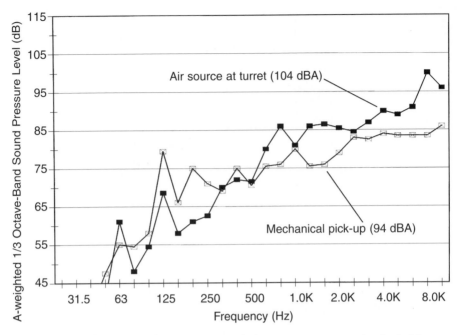

Figure 9.37 — Sound level from compressed air source versus mechanical pick-up at turret.

mechanical pick-up device contains higher sound levels in the low frequencies; however, this effect is more than offset by the significant noise reduction in the higher frequencies.

As for using different materials, the key is to use products that contain high internal-damping properties. For example, Figure 9.38 shows a comparison between the sound levels radiated from a stiff nylon material versus cast steel. The device is an anti-backup dog used to prevent conveyor lines from inadvertently slipping backwards. By substituting this highly damped material the overall sound level was reduced by 5 dBA.

In addition to substitution of equipment and/or materials, selecting quieter items should also be included during the initial design and purchase stage of future process equipment. Additional information regarding purchase specifications for noise control is presented later in the section entitled *Controlling Noise During Design and Procurement Stages*.

Path Treatment

Before considering various path treatments, it is important to know how sound propagates in a confined space or room compared to how sound travels outdoors (see Chapter 2, Physics of Sound and Vibration, for additional discussion). Figure 9.39 illustrates the decrease in sound pressure level (SPL) as distance from the sound source increases (plotted as log *r*). When discussing sound

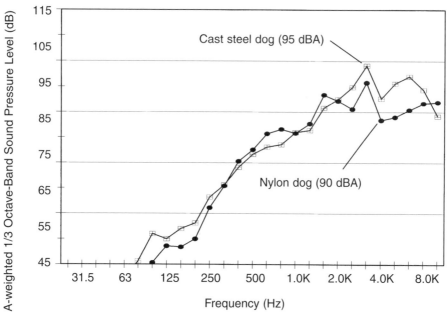

Figure 9.38 — Comparison of the sound level between steel versus nylon anti-back-up dog.

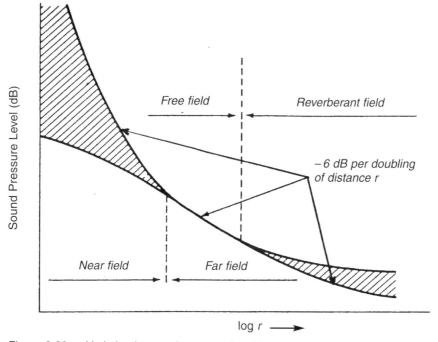

Figure 9.39 — Variation in sound pressure level in an enclosed space with increasing distance from source.

propagation, the sound field is divided into two regions: (1) the *free field*, where sound travels in a straight line from the noise source to the receiver with no interruptions, and (2) the *reverberant field*, where reflections from the surfaces of a room or enclosure add significantly to the sound level produced directly by the noise source. The boundary between free field and reverberant field changes according to the reflectivity of the room surfaces, so the transition between the two sound fields does not occur at a fixed distance from a noise source. Measurements taken in real-world reverberant fields will typically exhibit less decrease in SPL as a function of distance than is illustrated in Figure 9.39.

When SPL measurements are taken very close to a noise source, the sound waves radiating from the contributing sources have not combined into a uniform sound field as described above. The irregular shapes of the radiating sources/surfaces produce variations in SPL until the measurement position has been moved away by a distance of approximately one or two times the longest dimension of the noise source. Figure 9.39 shows the wide variation in SPL at distances very near the noise source (hatched area on left). This measurement region is known as the *near field*. At greater distances from the source, the measurements become uniform and predictable as the individual sound waves combine into a regular pattern of propagation, as if coming from a single source. This measurement region is known as the *far field*. Measurements taken outdoors with no reflecting surfaces will display, in theory, a 6-dB decrease in SPL for every doubling of the distance from the noise source in the far-field measurement region (free-field condition). However, it is important to note that for outdoor measurements, ground conditions, wind, elevation variations, etc., can have a significant impact on the actual increase/decrease which is observed (see Chapter 15, Community Noise, for a more detailed discussion on outdoor sound measurement).

Measurements taken inside a room or enclosure for a dominant noise source will exhibit the 6-dB decrease in SPL for each doubling of distance from the source that is characteristic of the far-field measurement region, but the plateau in sound level characterized by sound reflections in the reverberant field will also have to be considered (see Figure 9.39). The result is a transition zone between the near-field measurements taken very close to the source and the plateau in sound level characterized by sound reflections in the reverberant field. When the sound field is predominately caused by the source, this region is called the *direct field*.

Because measurements taken in the unrestricted far field fall off at 6 dB per doubling of distance from the source, an SPL measurement taken at one position can be used to approximate the SPL at a second location within the same far-field region. These predictions are also valid in the transition zone of the far field discussed above. However, measurements taken in the near field typically show too much variation in SPL at the first location to make predictions of the SPL at other distances from the source.

Recall that Equation 2.21 defined the relationship between sound power level (PWL) and SPL for enclosed spaces, as reproduced below:

$$L_p = L_W + 10 \log \left[\frac{Q}{4\pi r^2} + \frac{4}{R} \right] + k, \quad \text{dB}, \quad \text{re } 20 \text{ }\mu\text{Pa} \tag{9.23}$$

Where,

L_p = sound pressure level referenced to re 20 μPa
L_W = sound power level referenced to 10^{-12} W
Q = directivity factor (dimensionless)
r = distance from source, ft or m
R = room constant, ft^2 or m^2
k = constant factor = +10.5 for English units, and 0 for metric units

It is important to keep in mind that PWL is the acoustical energy produced by the "cause" of a disturbance, and for a very high percentage of noise sources is constant regardless of the environment in which the source is located. The SPL is the measured "effect" of that sound energy at an observation point away from the source. The first term inside the brackets ($Q/4\pi r^2$) describes the effect of directivity and placement of the noise source, and the second term inside the brackets (4/R) describes the effect of the reverberant sound field. Equation 9.23 characterizes the sound field at any given point throughout an ideal room regardless of the relative magnitude of the direct and reverberant sound fields. Figure 2.5 (presented in Chapter 2) depicts a generalized graph of Equation 9.23. This figure is useful for determining the SPL in a large room or enclosure relative to the PWL as a function of the resulting Q, distance r from the center of a source, and the room constant R.

Equation 9.23 also allows the user to separate the contributions of sound energy from the direct and reverberant fields when determining the overall SPL inside a room. For any noise source, the direct field predominates close to the source and its reverberant field predominates at some distance away. Even non-dominant sources in a production area will exhibit a direct field close to the source, which is often the space occupied by workers.

The effect of adding walls and a ceiling around a noise source causes reflections that create a reverberant sound field. The reverberant sound field depends primarily on the size of the room, and the reflectivity of the floor, walls, ceiling, and any other surfaces. When sound waves strike a surface (like a wall), a portion of the energy is absorbed, another portion passes through the wall, and the remainder is reflected back into the room. The part that passes through the wall is discussed later on in this chapter.

Sound Absorption

The *absorption coefficient*, α, is used to describe the ability of a surface area to absorb sound energy. By definition, the absorption coefficient is the ratio of acoustical energy absorbed by a surface to the acoustical energy incident on the surface. The coefficient is a dimensionless quantity that in theory varies from a value of 0 to 1, with 0 corresponding to all the incoming sound being reflected (no absorption) and 1 when all the sound striking the surface is absorbed (no reflection). Despite the fact the an absorption coefficient of 1 is the maximum theoretical value, it is possible to obtain values slightly above 1 (i.e., 1.05), which is often a function of the mounting technique used during the test procedure (NIOSH, 1980). In general, thick, soft, porous, and fuzzy materials are good sound absorbers, while rigid, hard, smooth surfaces are poor sound absorbers.

It is critical to note that materials or surfaces do not absorb sound equally at all frequencies. Because high-frequency sound waves have relatively short wavelengths and low frequencies relatively long wavelengths, most materials absorb sound more effectively at high frequencies than they do at low frequencies. This variation by frequency must be accounted for when calculating the sound absorption for a given type of material in a room. Figure 9.40 shows the variation in absorption coefficient of 6 lb/ft³ fiberglass as a function of frequency and material thickness, i.e., low-frequency sound absorption increases with increasing thickness. Most common building materials and all acoustical materials have been tested at a wide range of frequencies to allow the determination of their ability to absorb sound energy in a variety of situations (NIOSH, 1980).

Textbooks and acoustical journals often provide tables showing the absorption coefficients for building materials such as plywood, drywall, or glass, etc. (Lord et al., 1980; NIOSH, 1980). Representative absorption coefficients for some

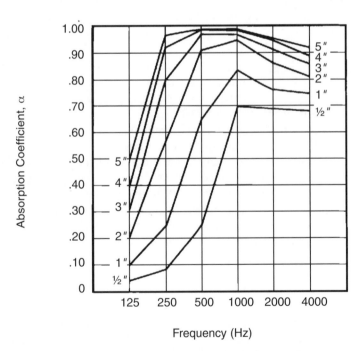

Figure 9.40 — Variation in the absorption coefficient of 6 lb/ft³ fiberglass as a function of frequency, with material thickness as a parameter.

TABLE 9.10
Typical sound absorption coefficients for common building materials
(AIMA, 1974; NIOSH, 1980).

Material		Sound Absorption Coefficient Octave-Band Center Frequencies (Hz)						
		125	250	500	1000	2000	4000	NRC
Ashes dumped, loose (2.5	11" thick	.90	.90	.75	.80			
lb water per cu ft)	3" thick	.25	.55	.65	.80	.80		0.70
Brick, unpainted		.02	.02	.03	.04	.05	.05	0.04
Brick, unglazed		.03	.03	.03	.04	.05	.07	0.04
Brick, unglazed, painted		.01	.01	.02	.02	.02	.03	0.02
Carpet, heavy, on concrete		.02	.06	.14	.37	.60	.65	0.29
Same, on 40 oz hairfelt or foam rubber		.08	.24	.57	.69	.71	.73	0.55
Same, with impermeable latex backing on 40 oz hairfelt or foam rubber		.08	.27	.39	.34	.48	.63	0.37
Concrete block, coarse		.36	.44	.31	.29	.39	.25	0.36
Concrete block, painted		.10	.05	.06	.07	.09	.08	0.07
Fabrics:								
Light velour, 10 oz per sq yd, hung straight, in contact with wall		.03	.04	.11	.17	.24	.35	0.14
Medium velour, 14 oz per sq yd, draped to half area		.07	.31	.49	.75	.70	.60	0.56
Heavy velour, 18 oz per sq yd, draped to half area		.14	.35	.55	.72	.70	.65	0.58
Floors:								
Concrete or terrazzo		.01	.01	.02	.02	.02	.02	0.02
Linoleum, asphalt, rubber or cork tile on concrete		.02	.03	.03	.03	.03	.02	0.03
Wood		.15	.11	.10	.07	.06	.07	0.09
Wood parquet in asphalt on concrete		.04	.04	.07	.06	.06	.07	0.06
Foams:								
1", 2 lb/cu ft polyester		.23	.54	.60	.98	.93	.99	0.76
2", 2 lb/cu ft polyester		.17	.38	.94	.96	.99	.91	0.82
Glass:								
Large panes of heavy plate glass		.18	.06	.04	.03	.02	.02	0.04
Ordinary window glass		.35	.25	.18	.12	.07	.04	0.16
Glass fiber:								
1", 3 lb/cu ft		.23	.50	.73	.88	.91	.97	0.76
1", 6 lb/cu ft		.26	.49	.63	.95	.87	.82	0.74
Gypsum Board, ½", nailed to 2 × 4's 16" o.c.		.29	.10	.05	.04	.07	.09	0.07
Marble or glazed tile		.01	.01	.01	.01	.02	.02	0.01
Openings:								
Stage, depending on furnishings				.25 to .75				
Deep balcony, upholstered seats				.50 to 1.00				
Grills, ventilating				.15 to .50				
Plaster, gypsum or lime, smooth finish on tile or brick		.01	.02	.02	.03	.04	.05	0.03
Plaster, gypsum or lime, rough finish on lath		.14	.10	.06	.05	.04	.03	0.06
Same, with smooth finish		.14	.10	.06	.04	.04	.03	0.06
Plywood paneling, ⅜" thick		.28	.22	.17	.09	.10	.11	0.15
Sprayed-on acoustical material, 1" cellulose applied to metal lath, 2.5 lb/cu ft		.47	.90	1.10	1.03	1.05	1.03	1.02
Water surface, as in a swimming pool		.008	.008	.013	.015	.020	.025	0.01

common building materials can be found in Table 9.10. In addition, manufacturers of acoustical materials can supply data sheets that show the theoretical absorption coefficients at each octave-band center frequency (see Table 9.5 for list of manufacturers).

Typically, a room will have different materials on the walls than on the floor or ceiling. Each material has different frequency-dependent absorption coefficients, so it is helpful to calculate the *average absorption coefficient*, $\overline{\alpha}$, of the room as a whole. The average absorption coefficient at each frequency is given by Equation 9.24 and is found by multiplying the absorption coefficient of each surface times the area of that surface to get its absorption in Sabins. The absorption values for each surface are added up to find the total absorption for all the surfaces in the room. Finally, the total absorption of the room is divided by the total surface area to get the average absorption coefficient at each frequency. Mathematically, $\overline{\alpha}$ is calculated as follows (also see Equation 2.17):

$$\overline{\alpha} = \frac{\sum_{i=1}^{n}(S_1\alpha_1 + S_2\alpha_2 + ... + S_n\alpha_n)}{\sum_{i=1}^{n}(S_1 + S_2 + ... + S_n)} \qquad (9.24)$$

where,

S_n = area of the nth surface in the room, ft^2 or m^2
α_n = absorption coefficient of the nth surface in the room

This formula is simply an area-weighted average of the absorption coefficient per octave-band frequency for each room surface. For Equation 9.24 to be a good approximation of the actual value, the absorption must be distributed somewhat evenly over the surfaces.

EXAMPLE 9.6, Determining Average Absorption Coefficient

Consider the following example: determine the average absorption coefficient at 1000 Hz for a room 40 ft by 70 ft with a 12-ft ceiling. The effective average absorption coefficients at 1000 Hz for the surfaces are as follows:

Floor: $\alpha = 0.1$
Ceiling: $\alpha = 0.7$
Walls: $\alpha = 0.2$

Step 1: Determine area of each room surface
Floor area, $S_{floor} = (40 \text{ ft})(70 \text{ ft}) = 2800 \text{ ft}^2$
Ceiling area, $S_{ceiling} = (40 \text{ ft})(70 \text{ ft}) = 2800 \text{ ft}^2$
Wall area, $S_{walls} = (2)(40 \text{ ft})(12 \text{ ft}) + (2)(70 \text{ ft})(12 \text{ ft}) = 2640 \text{ ft}^2$

Step 2: Plug appropriate values into Equation 9.24 for average absorption coefficient at 1000 Hz:

$$\overline{\alpha}_{1kHz} = \frac{(S_{floor}\alpha_{floor} + S_{ceiling}\alpha_{ceiling} + S_{walls}\alpha_{walls})}{(S_{floor} + S_{ceiling} + S_{walls})}$$

$$\overline{\alpha}_{1kHz} = \frac{[(2800)(0.1) + (2800)(0.7) + (2640)(0.2)]}{(2800 + 2800 + 2640)} = \frac{2768}{8240}$$

$$= 0.34 \text{ (dimensionless)}$$

Because sound absorption is frequency-dependent, this calculation must be repeated for each additional frequency of concern, as identified during the noise control survey.

The reverberant field component, indicated in Equation 9.23, is 10 log (4/R), where R is the *room constant*. The expression for R, previously given in Equation 2.16, is:

$$R = \frac{S\overline{\alpha}}{1 - \overline{\alpha}}, \quad \text{ft}^2 \text{ or m}^2 \qquad (9.25)$$

The term R takes into account the total absorption at each frequency provided by the surfaces inside the room of concern. The average absorption coefficient is used to calculate the total *absorption* (a) of the room surfaces in the numerator of the equation ($a = S\overline{\alpha}$, in Sabins, ft^2 or m^2). Equation 9.25 demonstrates that increasing the average absorption coefficient causes a corresponding increase in the room constant. The result is a room with a lower reverberant sound field.

It is the experience of the authors that theoretical calculation of room constant, using only the surface areas for a space (floor, walls, and ceiling) will underestimate the real-world or effective room constant in industrial settings. This difference is due to the fact that additional sound absorption is provided by the process equipment and other objects in the room. To account for these objects it is recommended that the calculated room constant be significantly increased. As a rule of thumb, for typical industrial environments, R should be increased by at least 25%.

Room constant is expressed in units of area, such as ft^2 or m^2. These units apply because the total absorption of the room is the product of the average absorption coefficient, which has no units, and the total surface area of the room, which has units of area. Therefore, room absorption is described as "x square feet" of absorption. This can be confusing because it is not the same number as the total surface area of the room, and it is important to keep these two quantities separate. To avoid confusion, absorption units are often given the name Sabins or metric Sabins to replace square feet or square meters, respectively.

EXAMPLE 9.7, Estimating Room Constant

Using Equation 9.25, determine R, the room constant, at 1000 Hz for a room 40 ft by 70 ft with a 12-ft ceiling height, and the average absorption coefficient for the room at 1000 Hz is 0.34 (from previous example). The room is a typical industrial environment with several machines located throughout the area.

$$R_{1kHz} = \frac{S\bar{\alpha}_{1kHz}}{1 - \bar{\alpha}_{1kHz}} = \frac{(2800 + 2800 + 2640)(0.34)}{(1 - 0.34)} = \frac{2802}{0.66} = 4245 \text{ ft}^2$$

$R_{1kHz} = 4245 \text{ ft}^2$ is the calculated room constant for the floor, walls, and ceiling surfaces. If desired, this value may simply be used as the final result. Alternatively, to account for sound absorption by the equipment at 1000 Hz, it is recommended the value for R_{1kHz} be multiplied by 1.25 (25% increase), which provides a more realistic assessment of the effective room constant at this frequency of concern. The effective room constant then becomes $(4245 \text{ ft}^2)(1.25) = 5306 \text{ ft}^2$.

The practical importance of the relationship between direct and reverberant sound fields becomes evident when deciding on how to reduce the sound level in a room. Suppose that the direct-field term at a given location in the room is much larger (≥ 10 dB) than the reverberant-field term, i.e., $10 \log (Q/4\pi r^2) \gg 10 \log (4/R)$ in Equation 9.23. Assuming $10 \log (4/R)$ can be neglected, Equation 9.23 reduces to:

$$L_p = L_W + 10 \log \left[\frac{Q}{4\pi r^2} \right] + k, \quad \text{dB}, \quad \text{re } 20 \text{ }\mu\text{Pa} \qquad \textbf{(9.26)}$$

If the reverberant term dominates, i.e., $10 \log (4/R) \gg 10 \log (Q/4\pi r^2)$, then treating the room surfaces with additional absorption may be effective. In this case Equation 9.23 becomes:

$$L_p = L_W + 10 \log \left[\frac{4}{R} \right] + k \quad \text{dB}, \quad \text{re } 20 \text{ }\mu\text{Pa} \qquad \textbf{(9.27)}$$

To predict the change in SPL in the dominant reverberant field as a result of changing the absorption on one or more surfaces, the room constant calculations are repeated using new values of the absorption coefficients corresponding to the new materials placed in the room. Then, the change in SPL is calculated using the Equation 9.28.

$$\Delta L_p = L_2 - L_1 = 10 \log \left(\frac{R_1}{R_2} \right), \quad \text{dB} \qquad \textbf{(9.28)}$$

Where,

$\Delta L_p =$ change in SPL
$L_1 =$ original SPL
$L_2 =$ predicted SPL
$R_1 =$ original room constant
$R_2 =$ predicted room constant

It is important to keep in mind that changes to the absorptive material will only significantly affect the SPL at distances away from the noise source. For rooms that are relatively hard ($\overline{\alpha} < 0.4$) with a dominant reverberant field, adding sound absorption can benefit those workers exposed at a distance from a number of sources, provided these workers are not within the near field of their own production equipment. However, if the operator of a machine is in its near field, changes in room absorption may provide minimal or no reduction in the SPL at that position.

Now we have all the information needed to determine the overall SPL, or A-weighted sound level inside an enclosed space. Consider the following example:

EXAMPLE 9.8, Estimating A-weighted Sound Level When Relocating Equipment

A fairly small machine is to be relocated within a manufacturing plant. The A-weighted sound level measured 3 feet from the machine located on the floor near the center of the original room (Room 1) is approximately 90 dBA. Besides this small machine, Room 1 is empty. There is a need to estimate what the new sound level at 3 feet will be for this equipment in the room (Room 2) where it is scheduled to be relocated. Note: the machine will again be positioned on the floor in the center of the room. The dimensions of Room 1 are 20'×20'×8' with a concrete floor, lay-in acoustical ceiling panels, and 3/8th-inch thick plywood walls. Room 2 is currently void of any mechanical equipment. This room has dimensions of 30'×30'×10' and also has a concrete floor, but the ceiling is ½-inch gypsum board and walls are unpainted brick.

To start, obtain the sound absorption coefficients for the surface areas in Room 1. Lines 1–3 in Table 9.11 completed below present these data for the frequency range of interest, which is 125–4000 Hz. Equation 9.24 is applied to estimate the average absorption coefficient per octave-band center frequency (see line 4). The surface areas for Room 1 are 400 ft² for both the ceiling and floor, and 640 ft² for the walls. Line 5 is the room constant per octave band, which is determined using Equation 9.25. Since the equipment is placed on the floor, the directivity factor, Q, is assumed to be equal to 2. The SPLs measured approximately 3 feet from the machine in Room 1 are given in Line 6. To estimate the PWL per octave band, Equation 9.23 is utilized, and the resultant levels are shown in line 7. Recall from an earlier discussion, that the PWLs for the machine will remain the same in Room 2, assuming no equipment adjustments or modifications are made.

Lines 8–10 give the sound absorption coefficients for the various surfaces in Room 2, which are all obtained from Table 9.10. The average absorption coefficients per octave band are calculated and shown in line 11. For Room 2 the walls have a surface area of 1200 ft^2, and the ceiling and floor areas are 900 ft^2 each. The room constants are then estimated and presented in line 12. Using the estimated PWLs from line 7 and Equation 9.23, line 13 contains the estimated SPLs per octave band for Room 2. The A-weighting correction factors given in line 14 are obtained from Table 3.1, and then applied to the data in line 13 to produce the A-weighted sound levels per octave-band shown in line 15. Finally, Equation 2.12 is utilized to estimate the overall sound level in Room 2, which is approximately 93 dBA. Therefore, it may be assumed there will be approximately a 3-dBA increase in the sound level 3 feet from this machine if it is relocated to Room 2.

TABLE 9.11
Completed for 125–4000 Hz.
Computations described in Example 9.8.

Line No.	Description	Octave-Band Center Frequency (Hz)					
		125	250	500	1000	2000	4000
Room 1:							
1	α for concrete floor (see Table 9.10)	0.01	0.01	0.02	0.02	0.02	0.02
2	α for ceiling panels (provided by supplier)	0.08	0.25	0.74	0.95	0.97	0.99
3	α for 3/8" thick plywood panels (Table 9.10)	0.28	0.22	0.17	0.09	0.10	0.11
4	ᾱ for Room 1 (see Equation 9.24)	0.15	0.17	0.29	0.31	0.32	0.33
5	Room constant, R (ft^2) (see Equation 9.25)	254	295	588	647	678	709
6	Octave-band SPL (dB) at 3 feet	89	89	86	84	83	81
7	Octave-band PWL (dB) (Equation 9.23)	93	94	92	90	89	87
Room 2:							
8	α for concrete floor (see Table 9.10)	0.01	0.01	0.02	0.02	0.02	0.02
9	α for ½" gypsum board ceiling (see Table 9.10)	0.29	0.10	0.05	0.04	0.07	0.09
10	α for brick walls, unpainted (see Table 9.10)	0.02	0.02	0.03	0.04	0.05	0.05
11	ᾱ for Room 2 (see Equation 9.24)	0.10	0.04	0.03	0.03	0.05	0.05
12	Room constant, R (ft^2) (see Equation 9.25)	333	125	93	93	158	158
13	Octave-band SPL (dB) (Equation 9.23)	88	92	90	88	86	84
14	A-weighting correction (see Table 3.1)	-16	-9	-3	0	+1	+1
15	A-weighted octave-band sound level (dBA)	72	83	87	88	87	85

Note: since both the original and new rooms were void of additional equipment, it is concluded the calculated room constants are representative of the effective or real-world values, so no additional increase in R is necessary.

Noise Reduction Coefficient

To compare the sound-absorbing capability of similar materials, an average of the absorption coefficients of each material at 250, 500, 1000, and 2000 Hz (rounded to the nearest 0.05) is often used. This metric is termed the *noise reduction coefficient* (NRC), and is calculated as follows:

$$NRC = \frac{\alpha_{250} + \alpha_{500} + \alpha_{1000} + \alpha_{2000}}{4} \tag{9.29}$$

When using the NRC, caution must be exercised as this is an average value. Should the noise source contain significant acoustical energy in the very low (< 250 Hz) or very high frequencies (> 2000 Hz), it is best to utilize the sound absorption coefficients at each octave-band frequency of interest for different materials. However, if the noise source is broad-band and contains its maximum SPLs from 250–2000 Hz, use of the NRC is a practical and reasonable approach.

EXAMPLE 9.9, Comparing the Use of Sound Absorption Coefficients and NRC

A control valve is scheduled to be totally enclosed in a relatively small box. To minimize the buildup of sound inside the enclosure, two sound-absorbing materials are being considered. These materials are 1-inch thick polyester foam (2 lb/ft³), and 1-inch thick glass fiber (3 lb/ft³). From Table 9.10 these materials have the following sound absorption coefficients:

Material	Sound Absorption Coefficient Octave-Band Center Frequency (Hz)					
	125	250	500	1000	2000	4000
1-inch thick polyester foam (2 lb/ft³):	0.23	0.54	0.60	0.98	0.93	0.99
1-inch thick glass fiber (3 lb/ft³):	0.23	0.50	0.73	0.88	0.91	0.97

For the polyester foam the NRC is (Equation 9.29):

$$NRC = \left(\frac{0.54 + 0.60 + 0.98 + 0.93}{4}\right) = 0.76$$

The NRC for the glass fiber is:

$$NRC = \left(\frac{0.50 + 0.73 + 0.88 + 0.91}{4}\right) = 0.76$$

The frequency spectrum of the control valve is as follows:

	Octave-Band Center Frequency (Hz)							
	63	125	250	500	1000	2000	4000	8000
Octave-band SPL (dB)	79	82	88	94	85	84	82	80

Both materials have an NRC equal to 0.76, but which one is best suited for lining the enclosure? Inspection of the control valve's frequency data reveals that the primary frequencies of concern range from 250–2000 Hz, with the highest SPL occurring at 500 Hz. Although both materials can potentially provide a high degree of sound absorption, it is recommended that 1-inch-thick glass fiber be used since it has a greater sound absorption coefficient at the dominant frequency of 500 Hz.

Transmission Loss, Noise Reduction, and Insertion Loss

Consider the options in controlling noise that passes from one space (room or enclosure) through the structure to an adjoining space. As mentioned previously, materials that are useful in absorbing sound energy are soft, porous, and have a relatively low weight per unit area. On the other hand materials that are effective in attenuating sound energy are typically more dense. Dense and heavy materials are poor transmitters of acoustic energy, and therefore resist passing sound waves.

The *transmission coefficient*, τ, is defined as the ratio of the sound energy transmitted through a wall unit area to the sound energy incident on the wall. Just like the absorption coefficient, the transmission coefficient is dependent on frequency. Generally, most materials transmit low-frequency sound more efficiently than high-frequency sound, so the transmission coefficient is larger for low-frequency sound.

A more useful measure of a panel's ability to attenuate sound is given by the *transmission loss* (TL) of a partition. The equation for TL is as follows:

$$TL = 10 \log \left(\frac{1}{\tau} \right), \quad dB \tag{9.30}$$

TL values generally range from 0 dB (no attenuation of sound through the panel) to over 60 dB for an excellent sound-attenuating structure. Transmission loss measurements have been made for many types of common building materials. Manufacturers of acoustical materials have product data sheets containing TL data, while textbooks and construction handbooks often have TL data for the most common building materials. Representative TL values are presented in Tables 9.12a and 9.12b. As shown in these tables, more massive or dense partitions generally have higher TL values.

A single-number rating system has been introduced that rates a panel's TL across frequency, and allows for an easy comparison among different types of partitions or structures. The *sound transmission class* (STC) can help architects and builders who do not normally work with acoustical measurements evaluate conditions such as speech privacy of a wall or door panel. The STC uses TL data in 1/3 octave-bands from 125–4000 Hz as defined in ASTM E413-87 (1999) to arrive at the single-value number. A detailed discussion on how to calculate the STC is beyond the scope of this chapter; however, it is important to understand how to use the number because it is often utilized as a specification for acoustical

TABLE 9.12a
Sound transmission loss for various materials.

Material	Weight lb/ft²	kg/m²	125	250	500	1000	2000	4000	8000	STC
Lead										
1/32 in thick	2	9.8	22	24	29	33	40	43	49	35
1/64 in thick	1	4.9	19	20	24	27	33	39	43	29
Plywood										
3/4 in thick	2	9.8	24	22	27	28	25	27	35	26
1/4 in thick	0.7	3.4	17	15	20	24	28	27	25	25
Lead vinyl	0.5	2.4	11	12	15	20	26	32	37	22
Lead vinyl	1.0	4.9	15	17	21	28	33	37	43	26
Steel										
18-gauge	2.0	9.8	15	19	31	32	35	48	53	31
16-gauge	2.5	12.2	21	30	34	37	40	47	52	39
Sheet metal (viscoelastic laminate-core)	2	9.8	15	25	28	32	39	42	47	33
Plexiglas										
1/4 in thick	1.5	4.9	16	17	22	28	33	35	35	27
1/2 in thick	2.9	20.5	21	23	26	32	32	37	37	31
1 in thick	5.8	28.3	25	28	32	32	34	46	46	34
Glass										
1/8 in thick	1.5	7.3	11	17	23	25	26	27	28	25
1/4 in thick	3	14.6	17	23	25	27	28	29	30	29
Double glass										
1/4 x 1/2 x 1/4 in			23	24	24	27	28	30	36	29
1/4 x 6 x 1/4 in			25	28	31	37	40	43	47	36
5/8-in gypsum										
On 2 x 2-in stud			23	28	33	43	50	49	50	38
On staggered stud			26	35	42	52	57	55	57	47
Concrete, 4 in thick	48	234	29	35	37	43	44	50	55	42
Concrete block, 6 in	36	176	33	34	35	38	46	52	55	40
Panels of 16-gauge steel, 4-in absorption, 20-gauge steel			25	35	43	48	52	55	56	46
Cardboard (double wall/corrugated sheet with 1″ acoustical foam attached)										
combination	0.28	1.37	3	4	6	10	15	20	25	11

TABLE 9.12b
Sound transmission loss of general building construction and structures
(NIOSH, 1973; NIOSH, 1980).

Material of Structure	Weight lb/ft²	Transmission Loss (dB) Octave-Band Center Frequencies (Hz)					
		125	250	500	1000	2000	4000
DOORS							
1-¾", wood	4.8	23	28	27	25	30	32
1-¾", wood with 16 ga steel facing	6.8	25	27	32	35	28	30
1-¾", flush wood door	8.1	25	33	34	35	42	46
2-½", acoustical door with 12 ga. steel facing	21.9	43	49	48	55	57	44
10", acoustical door with damped metal absorbing insulation, hardware and gasketing as core	36	47	53	61	64	64	65
GLASS LAMINATES							
¼" plain surface structure, core of cast acrylic plastic	1.45	16	17	22	28	33	35
¼" glass with Saflex polyvinyl Butyral interlayer		29	34	35	33	41	
Two layers of ¼" glass with 0.045" Butacite core	6.2	25	20	27	33	29	36
Two layers of ¼" plate glass with ½" air space	6.2	25	20	27	33	29	36
WALL CONSTRUCTIONS							
3-¾" wall with ½" gyp board on both sides of 2-½" steel studs (no insulation)		17	28	37	45	43	42
4-⅜" wall with 2 layers of ½" gyp board on one side and 1 layer of ½" gyp board on the other side of 2-½" steel studs (no insulation)		28	31	46	51	52	47
5-⅜" wall with ⅝" gyp board on resilient channel on one side and ⅝" gyp board on the other side of wood studs with 3-½" fiberglass	6.4	29	40	51	55	48	58
10-¾" wall with 2 layers of ⅝" gyp board on both sides of a double row of wood studs with 3-½" insulation		37	47	55	63	67	67
3-¼" wall with perforated 18 ga. galvanized steel C-liner with glass fibers sealed in polyethylene bags and channel wall 20 ga. galvanized steel	4.7	18	20	29	38	40	45

performance of a partition or wall. Table 9.12a, which contains TL data for various materials, also presents the STC rating for these materials in the far right-hand column. Note how close (+/- 2 dB) the STC value is to the TL at 1 kHz for most materials.

The TL and/or STC rating of partitions using common construction techniques may not provide sufficient sound attenuation for many applications. A higher TL and STC for a partition can be achieved using a combination of materials in a "sandwich" arrangement. Most of the commercially available enclosure panels or partition walls employ a combination of materials to increase the TL. Figure 9.41 depicts a composite panel. Composite panels often use at least two lightweight surfaces separated by a layer of sound-absorbing material or an air space. The "sandwich" panel has a higher TL than each individual surface because sound energy losses occur each time an impedance mismatch is encountered by the penetrating sound wave. Composite structures attenuate sound energy very well without the weight of a very dense or massive material. This explains why composite panels are very effective in controlling sound through doors where weight is a major factor in their construction.

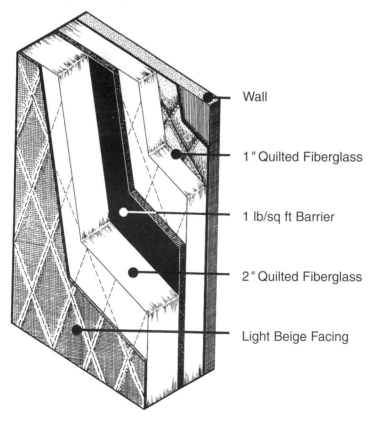

Figure 9.41 — Illustration of a composite transmission loss panel. (Courtesy of McGill AirPressure Corporation)

The TL of the composite panel is not the sum of the value of the TL for the two outside surfaces because the two surfaces are never completely decoupled. However, when the separation nears λ/4, the resultant TL approaches the sum of each panel's TL (Lord et al., 1980). Sound energy can pass through the composite material by means of flanking paths where the material is mounted to the structure of the partition. This degrades the TL of the composite panel. The manufacturer data sheets or specifications need to be consulted to obtain the composite TL for their particular product line. However, it is important to note that most manufacturer's data are for idealistic environments, and the rated TLs are rarely achieved in real-world environments. It is the experience of the authors that 5–10 dB should be subtracted from the manufacturer's TLs to approximate the real-world TLs.

There will often be times when it is necessary to install a door and/or windows in an enclosure wall. To achieve the same TL of the original panel, it is necessary to select door and window systems having at least the same TL as the panel. However, there will be times when the TL values will be less than the original wall. Figure 9.42 may be used to estimate the new TL of the combined structure.

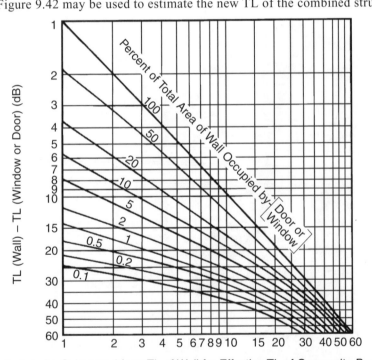

Decibels to be Subtracted from TL of Wall for Effective TL of Composite Barrier

Figure 9.42 — Composite transmission loss of walls with windows and/or doors. Note: start with the difference in TL between the wall and windows or doors on the vertical axis. Traverse horizontally until intersecting the appropriate percent to total area of wall occupied by windows or doors curve. Drop down vertically to obtain the decibels to be subtracted from the wall's TL on the horizontal axis. The resultant value is the effective TL of the composite wall.

EXAMPLE 9.10, Determining the Composite Transmission Loss of an Enclosure Wall with a Window

A window is inserted into an enclosure wall. The TLs of the wall and window are 33 dB and 18 dB, respectively. The window occupies 10% of the total surface area of the wall. What is the composite TL?

Step 1. Determine the difference between the TLs of the wall and window.

TL (wall) - TL (window) = 33 - 18 = 15 dB

Step 2. Use Figure 9.42 and locate 15 dB along the vertical axis.

Step 3. Draw a horizontal line until it intersects with the 10% total area occupied curve.

Step 4. Extend a vertical line down until it meets the horizontal axis, which is at a point slightly above 6 dB.

Step 5. Subtract the approximate 6 dB resultant from the TL of the wall to obtain the effective TL of the composite wall.

TL (composite wall) = TL (wall) - 6 dB = 33 - 6 = 27 dB

Machine enclosures normally exhibit significant gaps around pipe penetrations, small cracks due to wear and tear, and openings for operator access and product flow. As a result, the actual TL for the panel on the side of the enclosure with the opening will be degraded. To account for various openings in an enclosure panel, Figure 9.43 can be used to estimate the actual TL. For example, in Figure 9.43 if a wall section has a rated TL of 38 dB, and an opening on the order of 0.1 percent of the surface area of the wall exists, the actual TL of the wall will be 28 dB. There is a 10-dB loss in potential TL due to this relatively small opening. Therefore, to maintain the acoustical integrity of an enclosure, it is critical to seal all cracks, gaps, and penetrations. When it is necessary to have openings to permit product flow, it is desirable to install an acoustically lined tunnel or chute over items such as the conveyors. This will decrease the reduction in enclosure TL due to these openings.

Another method to determine the sound attenuation between rooms is by measuring the *noise reduction* (NR) provided by the common partition separating the rooms. NR is found by measuring the difference between the SPL in the source room and the receiving room (see Equation 2.14). These measurements are typically conducted away from the partition of interest, approximately 3 feet from both sides. The difference in SPL between room 1 and room 2 is the NR provided by the partition or wall. The NR achieved by a partition or a sound enclosure can be higher or lower than the TL of the material used in the construction of the enclosure. Although both are quoted in decibels, NR is a measurement in the sound field of the room, while the TL is an indication of the wall's inherent ability to block sound. Consequently, TL is independent of room acoustics, whereas sound-field measurements take into account any sound flanking paths and each room's acoustical properties that can influence the NR of the installed barrier.

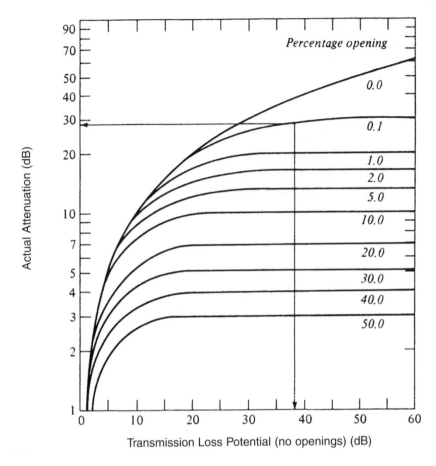

Figure 9.43 — Effects of openings on the potential transmission loss of panels. (From Lord et al. [1980], p. 287.)

In some noise control applications we are interested in determining the IL that results from placing an enclosure around a source. IL refers to the difference in sound levels measured *at the same location* before and after a noise control treatment has been installed (see Equation 2.15). For example, a baseline measurement is taken 10 feet from a machine before any noise control treatment is added. After a sound enclosure has been installed, the measurement is repeated at the same location. If the noise levels dropped from 110 dBA to 85 dBA, then the enclosure has an IL of 25 dBA.

Acoustical Enclosures

One of the most common sound path treatments is the utilization of an acoustical enclosure, as depicted in Figure 9.44. As contrary as it sounds, an enclosure can be the most effective noise control, yet the least desirable option. A well-

Figure 9.44 — Custom machine enclosure. (Courtesy of McGill AirPressure Corporation)

built and maintained total enclosure (no openings) can provide an IL of 20–40 dBA. On the other hand, enclosures often (1) restrict visual and physical access by production and maintenance personnel, (2) cause internal heat buildup which can cause fires, (3) result in long-term internal surface contamination of the enclosure panels from oil mist or other airborne particulate, and (4) incur damage due to normal wear and tear. It is common for employees to reject and defeat enclosures which constrain production or are impractical. Finally, it is also common for enclosures to be dissembled and reassembled by maintenance personnel. They are rarely put back together as tightly as their original condition, which degrades the original noise reduction of the system. However, assuming you can live with and/or successfully address all these constraints, enclosures can be an effective noise control path treatment.

The development of enclosures is well advanced in today's marketplace. Both off-the-shelf and custom-made enclosures are available from several manufacturers, which are listed in Table 9.5 (see *Composite Systems* in the table). Toward procurement of the appropriate system it is necessary for the buyer to provide information as to the overall noise level and the octave-band frequency data, the dimensions of the equipment, the noise reduction goal, the need for product flow and employee access, and any other operating constraints to assist the supplier with selection of the appropriate materials. The vendor will then be able to use this information to recommend a stock item or fabricate a custom enclosure.

In many situations it may be more economical to design and build an enclosure instead of purchasing a commercial system. In designing enclosures, several factors must be taken into consideration if the enclosure is to prove satisfactory from both an acoustical and a production point of view. Specific guidelines for enclosure design are as follows:

1. *Enclosure Dimensions:* There are no critical guidelines for the size or dimensions of an enclosure. The best rule to follow is bigger is better. It is critical that sufficient clearance be provided between the noise source

and enclosure panels to permit the equipment to perform all intended movement without contacting the enclosure and to allow for efficient ventilation, lighting, maintenance, etc.

2. *Enclosure Panels:* The IL provided by an enclosure is dependent upon the materials used in the construction of the panels and how tightly the enclosure is sealed. To provide assistance with the selection of the appropriate materials for the enclosure panels, the following rules of thumb are offered (Miller, 1979):

    A. For an enclosure with no internal absorption:
$$TL_{reqd} = IL_{desired} + 20 \qquad dBA,$$

    B. For an enclosure containing approximately 50% coverage of the internal surface area with sound absorption:
$$TL_{reqd} = IL_{desired} + 15 \qquad dBA,$$

    C. For an enclosure with 100% coverage of the internal surface area with sound absorption material:
$$TL_{reqd} = IL_{desired} + 10 \qquad dBA,$$

    In these expressions TL_{reqd} is the transmission loss required of the enclosure, and $IL_{desired}$ is the insertion loss desired to meet the abatement goal.

3. *Seals:* For maximum TL all enclosure wall joints must be tight-fitting. Openings around pipe penetrations, electrical wiring, etc., should be sealed with a flexible and nonhardening mastic such as silicon caulk. One of the most important people around during the construction or installation of an enclosure is the individual with a caulking gun.

4. *Internal Absorption:* To absorb and dissipate acoustical energy the internal surface area of the enclosure should be lined with acoustical absorbing materials. The manufacturer's published absorption data provide the basis for matching the material thickness and the absorption coefficients at each frequency to the source frequencies with the highest SPLs. The product vendor or manufacturer can also assist with selection of the most effective material.

5. *Protection of Absorption Material:* To prevent the absorptive material from getting contaminated, a splash barrier should be applied over the absorptive lining. This should be of a very light material, such as one-mil plastic film. The absorptive layer can be retained if necessary with expanded metal, perforated sheet metal, hardware cloth, or wire mesh. However, the retaining material should have at least 25% open area.

6. *Enclosure Isolation:* It is important that the enclosure structure be separated or isolated from the equipment to ensure that mechanical vibrations are not transmitted to the enclosure panels, which can re-radiate noise. When parts of the machine do come in contact with the enclosure, it is important to include vibration isolation at the point of contact to minimize any potential transmission path. If the floor vibrates due to motion or movement of the machine, then vibration isolation should be used under the machine.

7. *Product Flow:* As with most production equipment, there will be a need to move product into and out of the enclosure. The use of acoustically lined channels or tunnels can permit product flow and provide acoustical absorption. To minimize the leakage of noise, it is recommended that the length of all passageways be three times longer than the inside width of the largest dimension of the tunnel or channel opening.

8. *Worker Access:* Doors and windows may be installed to provide physical and visual access to the equipment. It is recommended that ideas regarding the location and size of all doors and windows be solicited from machine operators, which not only makes the design more practical, but also improves the likelihood employees will accept the enclosure system. It is important that all windows have nearly the same transmission loss properties as the enclosure walls. If the window TL is less than the wall TL, then the procedure described in Figure 9.42 should be used to estimate the composite TL. The resultant TL should be re-evaluated, as described in item 2 above to ensure the $IL_{desired}$ is still satisfied. If not, then an alternative window system should be used to meet the desired IL. All access doors must tightly seal around all edges. To prevent operation of the equipment with the doors open, it is recommended that an interlocking system be included that permits operation only when the doors are fully closed. To facilitate access some industries support enclosures on hydraulic lifts that can quickly move the enclosure out of the way.

9. *Ventilation of Enclosure:* In many enclosure applications, there will be excessive heat buildup. To pass cooling air through the enclosure, a quiet blower with sufficient air movement capability should be installed on the outlet or discharge duct. Finally, the intake and discharge ducts should be lined with absorptive material.

10. *Fire Prevention:* Keep in mind that although most sound-absorbing materials are listed as fire resistant, fires can and have occurred when a spark and/or excessive internal heat ignites dust, oil mist, etc., that accumulates on the surfaces of the material. In addition, potentially harmful gas may be released by the burning material, depending on its chemical composition. Therefore, if fire is a potential problem for your equipment and enclosure application, it is recommended that an internal fire prevention or suppression system be installed, as is often dictated by local fire codes.

Acoustical Partial Barriers

An alternative sound transmission path treatment is to use a partial acoustical barrier to block or shield the receiver from the direct sound path. A partial barrier is a solid partition or wall inserted between the noise source and receiver. For a partial barrier to be effective it is critical for the receiver to be located in the direct field and not in the reverberant field. By blocking the direct line-of-sight to the source, the sound waves mainly reach the receiver by reflection off various surfaces in the room, and diffraction at the edges of the barrier. Thus, a "shadow zone" is created on the receiver's side of the barrier, and the overall noise level is reduced at the receiver's location. Figure 9.45 depicts a simplified end view of a partial barrier.

The effectiveness of a partial barrier is a function of its relative location to the noise source and receiver, its effective dimensions, and the frequency spectrum of the noise source. To maximize the potential insertion loss, the barrier should be located as close as practical to either or both the source and receiver. As a guide, the width of the barrier on either side of the noise source should be at least twice its height (Lord et al., 1980). Finally, the partial barrier should not contain any openings or gaps, which can significantly reduce its effectiveness. If necessary to include a window for visual access to the equipment, then it is important to use Figure 9.42 and the TLs for the wall and window to determine the barrier's effective composite TL.

The chart shown in Figure 9.46 can be used to estimate the insertion loss of a partial barrier. To use this chart, first determine the effective height (H), which is the distance from the top of the barrier to the line-of-sight between the source and receiver, as depicted in Figure 9.45. From the chart, the insertion loss at each frequency is determined using the angle (θ) in Figure 9.45.

EXAMPLE 9.11, Estimating Barrier IL

Determine the attenuation of a partial barrier having an effective height, H, of 4 feet and an angle (θ) of 30°. First, calculate the wavelength for the frequencies of interest using Equation 2.2, and record the value in Line 2 of Table 9.13. For this example, c=1130 ft/sec is used. Then, calculate H/λ and record the value in Line 4. Finally, determine the estimated noise reduction at each frequency from Figure 9.46 and record the result in Line 6.

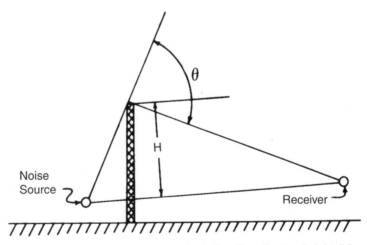

Figure 9.45 — Noise barrier geometry including the effective height (H).

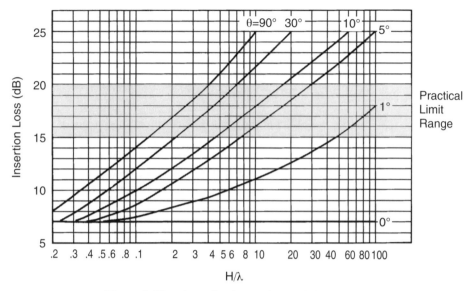

Figure 9.46 — Insertion Loss of a partial barrier.

TABLE 9.13
Insertion Loss through the use of a partial barrier.

Line		Octave-Band Center Frequency (Hz)							
		63	125	250	500	1000	2000	4000	8000
1	Speed of Sound– ft/sec (c)	1130	1130	1130	1130	1130	1130	1130	1130
2	Wave Length– feet (λ)	18	9	4.5	2.2	1.1	0.56	0.28	0.14
3	Barrier Effective Height–ft (H)	4	4	4	4	4	4	4	4
4	H/λ	0.22	0.44	0.89	1.8	3.6	7.0	14	29
5	Angle of Deflection– degrees (θ)	30	30	30	30	30	30	30	30
6	Insertion Loss (dB) (From Figure 9.46)	7	9	11	14	17	20	23	25

It is critical to note the attenuation values in Table 9.13 represent estimated values under a free-field condition. For all practical purposes, diffraction effects will limit the maximum IL provided by a partial barrier to approximately 15–20 dB. In highly reverberant environments, the reverberant sound field determines the maximum IL achievable. For example, the sound level in a totally reverberant field will be the same with or without the barrier in place. As the reverberant field becomes less predominant, the benefit of the partial barrier is increased. Equation 9.31 may be used to estimate the distance from the sound source where the transition from the direct to reverberant sound field in a room takes place (Miller and Montone, 1978).

$$r = 0.14 \ \sqrt{\overline{\alpha}S} \ , \text{ft or m} \qquad\qquad (9.31)$$

where,

$r =$ receiver's distance from the source, ft or m
$\overline{\alpha} =$ average sound absorption coefficient
$S =$ total surface area in the room, ft^2 or m^2

For example, with an $\overline{\alpha}$ of 0.5 at the critical or dominant frequency of concern and an S of 12,200 ft^2, the transition zone exists approximately 11 ft from the source. Therefore, to achieve an IL benefit at the frequency of concern (up to the practical limit of 15–20 dB) the receiver must be no further than 11 ft from source with the barrier located between the source and receiver. If more than one dominant frequency exists, then the distance from the source to the transition zone for each frequency of concern needs to be approximated. The shortest distance estimated should be used as the practical limit from source to receiver for which the partial barrier will provide a measure of IL.

Alternatively, the transition zone between regions of the direct and reverberant sound fields may be estimated using Figure 2.5 as a function of the room constant (see Chapter 2). The reader must exercise caution when estimating the IL for a barrier in a semi-reverberant sound field, as the actual IL values will likely be less than those predicted using Figure 9.46.

Acoustical Pipe Lagging

Airborne noise generated by piping systems is typically the result of one or more of the following factors:

- Noise in the liquid or gaseous medium being transported, which radiates excessive acoustical energy through the pipe wall.
- Vibratory energy transferred to the pipe wall, which may be an efficient radiator of noise, and/or transfer of the vibratory energy to other surface areas capable of radiating airborne noise.

Both forms of noise can result when vibratory energy is transmitted into the medium and/or the pipe wall by (1) the operation of rotating equipment, such as compressors, pumps, etc., (2) control valves, (3) excess velocity or turbulent flow within the medium being transported, and (4) the movement of solid particles (i.e., resin pellets), all of which transfer vibratory energy into the medium and/or the pipe wall.

Acoustical pipe lagging or insulation is the most effective noise control option whenever the pipe wall itself radiates a dominant portion of the acoustical energy being transmitted, and an in-line silencer is deemed to be infeasible. Pipe lagging consists of wrapping the exterior surface area of a pipe with a sound absorption material (i.e., high-density fiberglass), then covering the absorptive material with a jacketing material that has high sound transmission loss properties. Figure 9.47 depicts two standard configurations for pipes. In effect a tight-fitting cylindrical enclosure is formed around the pipe line. Figure 9.48 exhibits

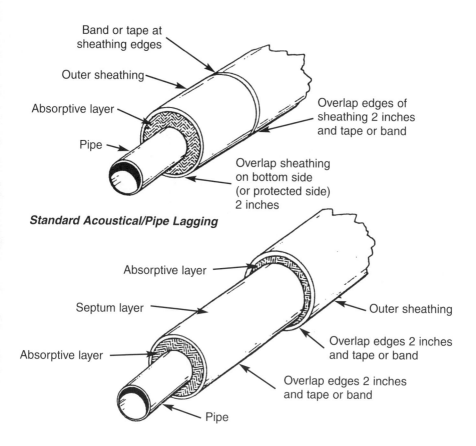

Band or tape at sheathing edges

Outer sheathing

Absorptive layer

Pipe

Overlap edges of sheathing 2 inches and tape or band

Overlap sheathing on bottom side (or protected side) 2 inches

Standard Acoustical/Pipe Lagging

Absorptive layer

Septum layer

Absorptive layer

Outer sheathing

Overlap edges 2 inches and tape or band

Overlap edges 2 inches and tape or band

Pipe

Pipe Lagging Using Sandwiched Septum Layer for High Noise Level Reductions

Figure 9.47 — Standard configurations for acoustical pipe lagging.

Figure 9.48 — Commercially available foam and high-density vinyl lagging system, and a lagged pipe section.

363

a commercially available foam and high-density vinyl lagging system, and a lagged pipe section.

For short pipe runs that incur major discontinuities, applying the lagging treatment to the localized sections will usually be sufficient. On the other hand, when a majority of the acoustical energy exists within the medium being transferred (and not the pipe wall), the primary noise control option is to install an in-line silencer. However, it is worth noting that if an in-line silencer is infeasible, then it will be necessary to lag the entire length of pipe run to effectively control the noise from this source. One technique to determine which condition exists is to conduct sound level measurements along an extended length of the pipeline. If the sound level drops significantly across a valve, branch, or any other discontinuity, then it is reasonable to conclude the dominant energy is contained in the pipe structure. However, should the sound level remain relatively constant, it is clear most of the sound energy is radiating from the medium being transferred.

Example

Turbulent fluid flow within pipes produces sound which can be radiated from the pipes and even transmitted to the building structure.

Control measure

In addition to reducing the turbulence in the pipe, the pipe can be covered with sound-absorbing material. The vibrations can be isolated from the wall or ceiling with flexible connecting mechanisms.

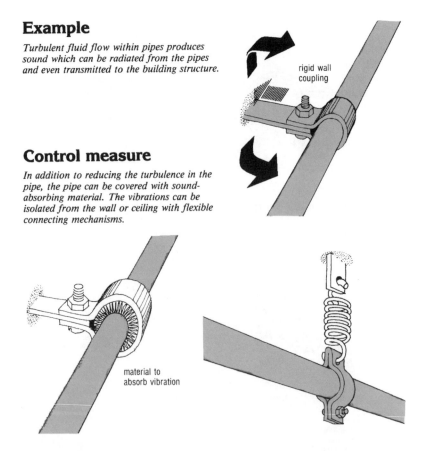

Figure 9.49 — Flexible pipe supports for equipment with pipe connections.

TABLE 9.14

Recommended pipe lagging treatments to achieve a required noise reduction.

Noise Reduction Required (dBA)	Character of Frequency Spectrum	Minimum Recommended Lagging Treatment*
5	High Frequency: ≥ 2000 Hz	2-in. thick, 2–3 lb/ft^3 absorptive material.
5	Medium to Low Frequency: < 2000 Hz	4-in. thick, 2–3 lb/ft^3 absorptive material.
10	High Frequency: ≥ 2000 Hz	Two 2-in. layers, 2–3 lb/ft^3 single foil-faced absorption with facing inside.
10	Medium to Low Frequency: < 2000 Hz	Same as preceding except add 0.5 lb/ft^2 sheet outer covering.**
20	All Frequencies	Two 2-in. layers of 2–3 lb/ft^3 absorptive material with 0.5–1.5 lb/ft^2 septum between each layer.
>20	All Frequencies	Same as preceding except add 0.5–1.0 lb/ft^2 sheet under outer shell.

* Note: all absorptive materials should be covered with 26-gauge steel or 17-gauge aluminum or equivalent.
** Lead or lead-loaded vinyl recommended.

Besides lagging the pipe when significant structure-borne vibration is a dominant source, care needs to be exercised to ensure that the vibratory energy is not transferred to pipe supports, clamps, etc., as illustrated in Figure 9.49, that can then transfer the vibratory energy to a point or surface area that can efficiently radiate airborne sound or noise. Vibration isolators at each pipe hanger or support may likely be required.

If the predicted octave-band SPLs for the piping system design are known, then these data should be used to select the optimum absorption material. Alternatively, with only the predicted noise level (in dBA), the desired noise reduction may be determined using the information in Table 9.14, which provides recommended pipe lagging configurations to achieve a required level of noise reduction (API, 1973). Specific guidelines for acoustical lagging are as follows:

1. *Configuration:*

   A. Provide a layer of resilient absorptive material between pipe surface and outer shell of treatment. A layer of absorptive material must be included; otherwise, there will not be anything to absorb and dissipate the radiated sound energy.

   B. Avoid any mechanical coupling between pipe surface and outer shell of treatment; otherwise a transfer of vibratory energy may occur, which in turn enables the outer shell to become a source of sound radiation.

   C. Seal all edges and joints airtight. Even extremely small openings (on the order of 0.1% of the total surface area) can significantly degrade the noise reduction achieved. Recall Figure 9.43 for small openings in enclosures — it also applies to the installation of acoustical lagging.

   D. Use special materials for high temperature applications. Select a material that will not burn, melt, or decay from exposure to heat and/or moisture.

   E. To help absorb low-frequency energy a relatively thick layer of absorption material is required (on the order of 4–6 inches). Recall that low frequencies have relatively long wavelengths when compared to higher frequencies, which makes it difficult to attenuate noise at these lower frequencies.

   F. Some means to avoid accumulation of condensation should be included for cold piping. Usually drain taps are sufficient to remove condensation.

2. *Fabrication:*

   A. For thin-shelled pipes (wall thickness of 0.25 inch or less) apply a layer of damping material or compound (available from vibration isolation manufacturers listed in Table 9.5) directly on the pipe's outer surface. This damping treatment is not recommended for thicker pipe walls (> 0.25 inch), as the wall thickness is usually sufficient to minimize any potential resonant tones.

   B. Wrap the pipe with absorptive material as recommended in Table 9.14.

   C. Encapsulate the acoustical absorption material with lightweight metal sheathing.

   D. Overlap outside shell edges, compress the acoustical material slightly, and bond the joints.

   E. Fill in irregular cross-sections with loose acoustical material to provide a uniform outer surface.

Receiver Treatment

The final option for reducing a worker's noise exposure is to acoustically treat the space or area where the employee works, or simply relocate his or her primary workstation into a relatively quiet area. Receiver treatment is most practical for job activities such as product inspection or quality control, and equipment monitoring stations where employee movement is confined to a relatively small area and significant time is spent at one location. In this situation, an acoustical booth or shelter may be installed to isolate the employee(s) and provide relief from excessive noise levels. To maximize the benefit of isolation booths the background sound level in the booth needs to be less than 80 dBA. This is because sound levels below 80 dBA are less than the threshold level for accumulating noise dose and TWA noise exposures, and as a result the estimated noise exposures can often be significantly reduced (see Chapter 7). Should speech intelligibility or communication (by telephone or face-to-face), or mental concentration be a concern for workers inside a control booth, then a better design goal is to limit the sound levels to ≤ 60 dBA (see Table 13.3 and Chapter 14).

EXAMPLE 9.12, Estimating Reduction in Noise Exposure (or TWA) Due to Installation of an Acoustical Control Room or Booth

Table 9.15a presents a noise exposure estimate for a process operator at a packaging facility. This table may be set up in any spreadsheet software, using the formulas, permissible exposure levels, reference durations, criterion and threshold levels as mandated by the Occupational Safety and Health Administration's (OSHA's) Occupational Noise Standard (see Appendix I), the Mine Safety and Health Administration (see Table 16.3, in Chapter 16, Standards and Regulations), or any alternative criteria desired. The sound level data and exposure time estimates are determined by the user, then entered into the appropriate columns. Since the example facility is under OSHA jurisdiction and this agency requires an 80-dB threshold level for hearing conservation purposes and a 90-dB threshold for engineering controls and mandatory use of hearing protection, Table 9.15a contains separate columns for reporting the percent dose per source under both criteria. The job activity presented in the table shows the process operator is exposed to 92 dBA for 120 minutes at the packaging station, 98 dBA for 45 minutes at the glue gun station, 91 dBA for 155 minutes where the product feeds into the machine, 87 dBA for 80 minutes at the carton magazine, and 80 minutes per shift cleaning up and on breaks with an average sound level of 65 dBA. As reported toward the bottom of the table, the TWAs calculate to 91 dBA and 90 dBA for the 80-dB and 90-dB threshold levels, respectively.

TABLE 9.15a
Estimated total daily noise dose and TWA from exposure to different sound levels for different durations (OSHA 5-dB exchange rate).

Job Title: Process Operator
Department: Packaging
Shift Length: 8 hours

Source Number	Job Activity/ Location	Sound Level (dBA)	Reference Duration (Minutes)	Exposure Time (Minutes)	HCP % Dose Per Source	Eng. Controls % Dose Per Source
1	Packaging Station	93	317	120	37.89	37.89
2	Glue Gun Station	98	158	45	28.42	28.42
3	Product Infeed	91	418	155	37.09	37.09
4	Carton Magazine	87	728	80	11.00	0.00
5	Clean-up/Breaks	65	NA	80	0.00	0.00

Hearing Conservation Program – ref. 29 CFR 1910.95 (c) Total Accumulated Dose: 114% Calculated TWA: 91 dBA

Engineering Noise Controls – ref. 29 CFR 1910.95 (b)(1) Total Accumulated Dose: 103% Calculated TWA: 90 dBA

TABLE 9.15b
Estimated total daily noise dose and TWA from exposure to different sound levels for different durations (OSHA 5-dB exchange rate).

Job Title: Process Operator
Department: Packaging
Shift Length: 8 hours

Source Number	Job Activity/ Location	Sound Level (dBA)	Reference Duration (Minutes)	Exposure Time (Minutes)	HCP % Dose Per Source	Eng. Controls % Dose Per Source
1	Packaging Station	93	317	120	37.89	37.89
2	Glue Gun Station	98	158	45	28.42	28.42
3	Product Infeed - inside booth	75	NA	140	0.00	0.00
4	Product Infeed - outside booth	91	418	15	3.59	3.59
5	Carton Magazine	87	728	80	11.00	0.00
6	Clean-up/Breaks	65	NA	80	0.00	0.00

Hearing Conservation Program – ref. 29 CFR 1910.95 (c)

Total Accumulated Dose: 81%
Calculated TWA: 88 dBA

Engineering Noise Controls – ref. 29 CFR 1910.95 (b)(1)

Total Accumulated Dose: 70%
Calculated TWA: 87 dBA

Analysis of the process operator's job activities revealed that it would be possible to install an acoustical control booth near the product infeed location. No other station or job activity was a candidate for an isolation booth. Once installed, the operator was able to spend approximately 180 minutes per day inside the booth, and was only required to spend 15 minutes per day immediately outside attending to various production tasks. Table 9.15b presents the new noise exposure estimate for this routine. As a result, the TWAs were reduced to approximately 88 dBA and 87 dBA using the OSHA mandated low- and high-threshold levels, respectively. So the need for mandatory engineering controls were eliminated through this receiver treatment.

Using a spreadsheet routine as presented in Tables 9.15a and 9.15b is recommended as a means to gauge the amount of noise reduction obtainable under different scenarios. To construct such a shelter, the previously described guidelines for enclosure design should be consulted. Alternatively, Table 9.5 provides a comprehensive list of manufacturers of commercially available quiet rooms.

Administrative Noise Controls

This noise control option should include any managerial decision that impacts worker noise exposures in a positive manner. The most effective administrative noise control is to prevent the noise source from entering the plant. This can be accomplished through implementation of an effective "buy-quiet" program. For this program it is necessary to have a well-written and clear purchase specification for noise, which limits the acoustical output of equipment purchased for installation. In addition, there are also ways of designing the facility and the process to account for noise. The buy-quiet program must also include mechanisms to validate equipment noise levels and procedures to force equipment manufacturers to replace or correct equipment failing to satisfy the noise criteria.

Should a high-noise environment already exist within a plant, and there are either no feasible engineering noise controls available or it is important to quickly reduce worker noise exposures while controls are being implemented, significant noise reduction can be achieved by administratively controlling the duration of selected worker job activities. For example, rotating two or more employees through a job activity with high-noise levels actually distributes the daily exposure among the participants, thus lowering the overall exposure that would have been received by a single worker. However, rotating employees in this manner will at least double the number of employees exposed to the source(s) of concern, and this procedure should only be implemented if the resultant noise exposures for the affected workers are still at safe or acceptable levels. Both use of a buy-quiet program and worker rotation scenarios are described below.

Controlling Noise During Design and Procurement Stages

The application of written specifications to define the requirements for equipment, its installation, and acceptance is standard practice in today's environment. Within the United States, ANSI has published a standard entitled: "Guidelines for the Specification of Noise of New Machinery" (ANSI S12.16-1992). This standard is a useful guide for writing an internal company noise specification. In addition, this standard provides direction for obtaining sound level data from equipment manufacturers. Once obtained from the manufacturer, the data may then be used by plant designers while planning equipment layouts. Because of the various types of distinctive equipment and tools for which this standard has been prepared, there is no single survey protocol appropriate for the measurement of sound level data by manufacturers. As a result, this standard contains reference information on the appropriate sound measurement procedure for testing a variety of stationary equipment types. These survey procedures were prepared by the appropriate trade or professional organization in the United States responsible for a particular type or class of equipment.

A sample specification for equipment purchases is presented in Figure 9.50. This is a comprehensive guide that specifies maximum acceptable noise levels, and should be sent to equipment manufacturers, to assist them in submitting to the

I. GENERAL

1.1 This specification is a means of establishing the limiting value of noise generated by equipment to be installed in an industrial plant. It also provides a uniform method of conducting and recording noise tests to be made on such machinery.

1.2 This specification describes limits and methods of measuring sound emission in the purchase of equipment. Tests are to be made by the vendor and witnessed by the purchaser unless otherwise specified. Confirming or additional measurements by the purchaser shall be permissible.

2. INSTRUMENTS

2.1 A type 1 or 2 sound level meter, as specified in ANSI S1.4-1983 (R1994), when used alone, measures overall noise levels only.

2.2 An octave band analyzer, as specified in ANSI S1.1 1-1986 (R1998), is the preferred Instrument for measuring broad-band noise by this specification and is used in conjunction with a sound level meter.

2.3 Instruments shall be calibrated as recommended by the instrument manufacturer. Overall calibration of the instruments, including the microphone and internal calibration of the meters, shall be made before and after the test of each piece of equipment.

3. NOISE TESTS

3.1 Ordinarily, the test will be made at the factory or in a test room provided by the vendor at his expense. The test room should provide conditions free of extraneous sounds that would Interfere with the noise tests.

3.2 Ambient sound levels within the test room should be 10 dB or more below the sound level that prevails when the tested equipment Is in operation. If the ambient levels are not at least 10 dB below the equipment and sound levels, corrections to the data are required as presented in 5.1.

3.3 Unless otherwise specified, equipment tested should be at full load and at anticipated load(s). Loading devices may be provided or specified by the purchaser.

3.4 The placement of the microphone during the test should be such as to protect it from air currents, vibration, electric or magnetic fields, and other disturbing influences that might affect the readings obtained. Positioning of the microphones at ear level and a horizontal distance of 3 feet from the nearest major surface is usually satisfactory. The entire area surrounding the equipment should be explored to ensure that the maximum noise levels are measured.

3.5 Measurements shall be made at a minimum of 6 points approximately 60 degrees apart in the plane specified in paragraph 3.4 and at any operator position(s). Start at the position of maximum noise level. No measurement position shall be located further than 10 feet from adjacent positions. Additional readings will be specified when a directivity pattern is to be established. When multiple, similar units are to be purchased, tests on more than one unit might be requested.

3.6 If desired, noise measuring methods specified in national, international, or trade association standards can be utilized. See ASA Standards Catalog (ASA, 2000) for a comprehensive listing.

Figure 9.50 — Equipment noise specification.

4. RECORDS

4.1 Records of tests for each piece of equipment shall include the information and readings called for in this specification.

4.2 Test results are to be reported to the purchaser for analysis and acceptance before equipment is shipped, unless otherwise specified.

5. SOUND LEVEL SPECIFICATIONS

5.1 The location and orientation of the microphone for measurements of total (ambient plus machinery) and ambient noise levels shall be identical. If either the machine or ambient noise levels fluctuate appreciably, maximum levels shall be recorded. Sound levels shall be measured over a long enough time to account for all fluctuations. If the difference between total and ambient readings is less than 5 dB, the ambient level is unsatisfactory for measuring the noise produced by the machine. If the difference is 10 dB or more, the higher readings are essentially the noise levels generated by the machine. For differences of 5 to 10 dB, the machinery noise levels shall be determined by applying the correction values indicated below.

Correction Values Allowed for High Ambient Sound Levels

Difference between Total and Ambient Noise Levels (dB)	Correction to be Subtracted from Total Sound Level (dB)
3	3
4-5	2
6-9	1
10	0

5.2 For the purpose of this specification, it is assumed that narrow-band or pure tone noise (or both) exists when the noise level of an octave band is at least 6 dB above the noise level in adjacent octave bands. If narrow-band noise as specified here does exist, the noise is more objectionable than broad-band noise upon which criteria are based. Therefore, it should be accounted for by assuming a 5-dB lower criterion for that band.

5.3 Purchase orders should specify maximum acceptable octave band sound power levels re 10^{-12} watt or octave band or A-weighted sound pressure levels re 20 μPa. If sound pressure levels are specified, environmental conditions must also be specified, as discussed in this chapter. Maximum sound pressure levels for both slow and peak (impulse) meter responses should be specified.

6. SPECIAL REQUIREMENTS

6.1 Equipment or locations which create special noise problems not covered by these specifications (such as neighborhood noise) require special consideration. Supplementary specifications and additional descriptions will be necessary.

Figure 9.50 (continued) — Equipment noise specification.

EQUIPMENT NOISE SPECIFICATION

Type of Equipment _____

Manufacturer _____ Vendor _____

Vendor's Order No. _____ Serial No. _____ Shop _____

Purchaser's Project No. _____ Machine No. _____ Order _____

Equipment Specifications: Model No. _____ Serial No. _____

                                 Size _____ Capacity _____

                                 Speed _____ Horsepower _____

Machine Load—% Capacity _____

Test Room—Dimensions   Length _____ Width _____ Height _____

            Material   Floor _____ Wall _____ Ceiling _____

Noise Description   Continuous   Intermittent-Impact

Does Narrow Band Noise Exist   Yes _____ No _____

Octave Band Analyzer   Make _____ Model _____ Serial No. _____

Sound Level Meter   Make _____ Model _____ Serial No. _____

Microphone   Make _____ Type _____ Serial No. _____

Calibration of Instrument   Before _____ After _____

Location of Microphone _____

OCTAVE-BAND RECORDING FORM

Test No.	Time Hrs	Position	Conditions	Sound Pressure Level (dB re 20 μPa)										
					Overall Level	Octave-Band Center Frequency (Hz)								
				dBA		31.5	63	125	250	500	1k	2k	4k	8k

Comments: _____

_____

_____

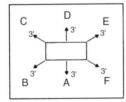

Indicate on the sketch the position of the equipment as placed in the room and orient the machine by identifying features. Note the sound level readings at appropriate locations such as A-F.

Figure 9.50 (continued) — Equipment noise specification.

buyer all SPL or PWL data in a consistent format. Assuming the PWL data are provided, the user may approximate the SPL that will exist within his/her facility using Equation 9.23. Therefore, it is necessary to know the absorptive and reverberant properties for the space where the new equipment will be installed. Should the manufacturer or vendor supply only the SPL data, then it will first be necessary for the user to approximate the PWL, or use other estimating procedures. This initial step requires the supplier to define the acoustical environment in which the data were measured. Once the user knows the reverberant and absorptive properties of the vendor's facility, the PWL may be estimated. With the PWL data, the user can estimate the SPL at his/her plant again using Equation 9.23.

The most proactive approach to control noise in the facility design and equipment procurement stage exists in Europe. In 1985, the 12 member states of the European Community (EC) adopted "New Approach" Directives designed to address a broad class of equipment or machinery, rather than individual standards for each type of equipment. By the end of 1994 there were three "New Approach" Directives issued that contained requirements on noise. These Directives are: (1) Directive 89/392/EEC (EEC, 1989a), with two amendments 91/368/EEC (EEC, 1991) and 93/44/EEC (EEC, 1993a), (2) Directive 89/106/EEC (EEC, 1989b) and (3) Directive 89/686/EEC (EEC, 1989c), with one amendment 93/95/EEC (EEC, 1993b). The first item listed above (89/392/EEC) is commonly called the Machinery Directive. This Directive compels machine manufacturers to include equipment noise control as an essential part of machine safety. As a result, there has been a major emphasis on the design of low-noise equipment since the late 1980s by manufacturers interested in marketing within the EC.

For companies outside the EC attempting to implement a voluntary buy-quiet program, the degree of success achieved is largely dependent upon the timing and commitment of the entire management. The first step in the program is to establish acceptable noise criteria for construction of a new plant, expansion of an existing facility, and purchase of new equipment. For the program to be effective, the specified noise limits must be viewed by both the purchaser and vendor as an absolute requirement. When a product does not meet other equipment design parameters, such as size, flow rate, pressure, allowable temperature rise, etc., it will be deemed unacceptable by company management. Similarly, the decibel limit (noise criteria) must be included with the list of required design parameters; otherwise, the effectiveness of the buy-quiet program will be tenuous at best.

The earlier in the design process that consideration is given to the noise-related aspects of a project or equipment purchase, the greater the probability of success. In many situations the factory designer or equipment buyer will have a choice of equipment types. Knowledge of the noise characteristics of the various equipment alternatives will allow the buyer to specify the quieter ones.

Besides selection of the equipment, early consideration of noise in the equipment layout design is essential. The layout designer must exercise caution and take into account the additive effect of multiple noise sources within a room (see

Chapter 2 for information on decibel addition). Relocating equipment on paper during the design phase of a project is much easier than physically moving the equipment later, especially once the equipment is in operation. A simple rule to follow is to keep machines, processes, and work areas of approximately equal noise level together, and separate particularly noisy and particularly quiet areas by buffer zones having intermediate noise levels.

Validation of noise criteria requires a cooperative effort between company personnel from departments such as engineering, purchasing, industrial hygiene, environmental, safety, and legal. For example, industrial hygiene, safety, and/or environmental personnel may determine the desired noise levels for equipment, as well as conduct sound surveys to qualify equipment. Next, company engineers may write the purchase specification, as well as select quiet types of equipment. The purchasing agent will most likely administer the contract, and rely upon company lawyers for assistance with enforcement. Involvement from all these parties should begin with the inception of the project and continue through funding requests, planning, design, bidding installation, and commissioning.

Even the most thorough and concise specification document is of little value unless the onus of compliance is placed on the supplier or manufacturer. Clear contract language must be used to define the means of determining compliance. Company procedures designed to enact guarantees should be consulted and followed. It may be desirable to include penalty clauses for noncompliance. Foremost in enforcement is the purchaser's commitment to seeing that the requirements are met. Compromise on the noise criteria in exchange for cost, delivery date, performance, or other factors should be the exception and not the rule.

Worker Rotation

Figure 9.51 exhibits a minute-by-minute time-history noise exposure profile for an employee rotating through several workstations. The facility is a glass bottle recycling center. Before any administrative controls were implemented, each worker performed one or two specific tasks throughout the workday. The employee at the waterknife station spent the full shift at this location and had a TWA of 103 dBA, which is an exceptionally high noise exposure. There were two other operators, who split time between the bottle puller, saw assist, and depalletizer (DPL) assist stations. Both workers' TWAs were approximately 91 dBA. Therefore, a job-rotation schedule was devised and implemented to limit the time spent at the waterknife. Five separate tasks, lasting 30 minutes each, were set as the rotation schedule. These assignments were (1) cleanup/break, (2) bottle puller, (3) saw assist, (4) waterknife, and (5) DPL assist. This cycle is repeated three times per workday, with an additional 30 minutes of clean-up added to the end of the shift (see Figure 9.51). As a result, the waterknife operator's TWA is reduced from 103 dBA to 93 dBA, and the two 91 dBA exposures are raised to 93 dBA. Granted the resultant exposure for the waterknife operator is still above 90 dBA; however, this lower exposure presents a more manageable risk as part of the HCP. As for the elevated TWAs for the two

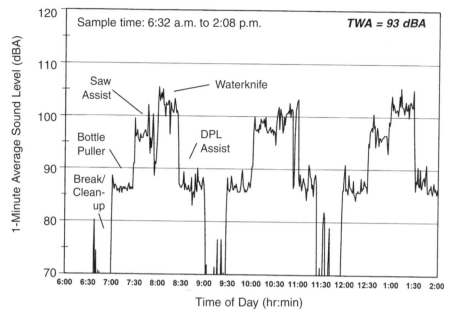

Figure 9.51 — Bottlewash job rotation.

other operators, management decided this was acceptable on a temporary basis until engineering controls could be implemented.

Professional judgment is required when designing a job-rotation schedule. It is critical that no additional workers be added to the HCP, or the affected workers presently exposed to noise do not have their TWAs raised to levels where hearing protection is rendered potentially ineffective. Toward this latter goal, if the resultant TWAs are kept to 95 dBA or less, practically any hearing protector will be sufficient provided the device fits well and is worn consistently and correctly (see Chapter 10, *Recommendations*).

The difficulty with the implementation of administrative controls is that they often require significant training of the operators to ensure the work schedules are followed. In addition, this activity at times may require additional employees, who will also need to be trained on the operation of the equipment. However, the most important factor toward the evaluation of this form of control is the amount of noise reduction that can be effectively achieved as a result of these administrative measures. In other words, do the benefits justify the costs, or would it be better to invest the funds in engineering controls, or other components of the HCP? Therefore, all constraints will need to be examined to determine whether or not this control measure would be feasible or desirable.

Concluding Remarks

Noise control engineering is fundamental to any effort to prevent noise-induced hearing loss among employees. Ultimately, the goal of noise control is to eliminate workplace noise hazards and the need for an HCP. Realistically, however, until employee hearing-damage risk is wholly eradicated, and all engineering control measures are maintained over time, the protection of workers is best achieved through implementation of a comprehensive and effective HCP. The best approach to noise control is to prevent the noise source from entering the facility. For existing problems, all available options for treating the noise source should be exhausted before consideration is given to treatment of the sound transmission path.

This chapter has provided a discussion of the practical aspects of noise control engineering including a priority rating scheme, techniques for conducting a noise control survey, the specifics on commercially available acoustical products, and a buy-quiet plan to provide industrial hygienists, safety professionals, plant engineers, and hearing conservationists with the tools to implement an effective noise control program. The authors trust the material in this chapter will facilitate the reader's successful noise control efforts.

References

AIMA (1974). "Bulletin of the Acoustical and Insulating Materials Associates," Park Ridge, IL.

Anon. (1991). *Webster's Ninth New Collegiate Dictionary*, Merriam-Webster Inc., Springfield, MA.

ANSI (1976). "American National Standard for Specifying Damping Properties of Materials," ANSI S2.9-1976 (R1997), Acoustical Society of America, New York, NY (asastds@aip.org).

ANSI (1992). "Guidelines for the Specification of Noise of New Machinery," ANSI S12.16-1992, Acoustical Society of America, New York, NY.

ANSI (1996). "Determination of Noise Exposure and Estimation of Noise-Induced Hearing Impairment," ANSI S3.44-1996, Acoustical Society of America, New York, NY.

API (1973). "Guidelines on Noise," Medical Research Report EA 7301, American Petroleum Institute, Washington, DC.

ASA (2000). "ASA National Standards Catalog 19-2000," Acoustical Society of America, New York, NY.

Ashford, N. A., Hattis, D., Zolt, E. M., Katz, J. I., and Heaton, G. R. (1976). "Economic/Social Impact of Occupational Noise Exposure Regulations (Testimony Presented at the OSHA Hearings on Economic Impact of Occupational Noise Exposure)," Center for Policy Alternatives, MIT, Cambridge, MA.

ASTM (1999). "Standard Classification for Rating Sound Insulation," E413-87 (1999), American Society for Testing and Materials, West Conshohocken, PA (www.astm.org).

Bruce, R. D. (1979). "The Cost to Industry of Noise Control to Protect Employees' Hearing," *Noise Control Engineering J. 12*(2), 60–73.

Bruce, R. D., Jensen, P., Jokel, C. R., Bolt, R. H., and Kane, J. A. (1976). "Workplace Noise Exposure Control — What Are the Costs and Benefits?" *J. Sound and Vibration 10*(9), 12–18.

Cohen, A. (1976). "The Influence of a Company Hearing Conservation Program on Extra-Auditory Problems in Workers," *J. Saf. Res. 8*(4) 148–162.

E-A-R (1988). "Reduced Frequency Nomograms 'Describe' Material in Application," Tech Review TR 501, EAR Speciality Composites, Indianapolis, IN (www.earsc.com).

EEC (1989a). Council Directive 89/392/EEC of 14 June 1989 on the approximation of the laws of the Member States relating to machinery, OJ No L 183, 29.6.1989, p. 9.

EEC (1989b). Council Directive 89/106/EEC of 21 December 1989 on the approximation of laws, regulations and administrative provisions of the Member States relating to construction products, OJ No L 40, 11.2.89, p. 12.

EEC (1989c). Council Directive 89/686/EEC of 21 December 1989 on the approximation of laws of Member States relating to personal protective equipment, OJ No L 399, 30.12.89, p. 18.

EEC (1991). Council Directive 91/368/EEC of 20 June 1991 amending Directive 89/392/EEC on approximation of the laws of Member States relating to machinery, OJ No L 198, 22.7.1991, p. 16.

EEC (1993a). Council Directive 93/44/EEC of 14 June 1993 amending Directive 89/392/EEC on approximation of the laws of Member States relating to machinery, OJ No L 175, 19.7.93, p. 12.

EEC (1993b). Council Directive 93/95/EEC of 20 October 1993 amending Directive 89/686/EEC on the approximation of laws of Member States relating to personal protective equipment, OJ No L 276, 9.11.93, p. 11.

Gibson, D. C., and Norton, M. P. (1981). "The Economics of Industrial Noise Control in Australia," *Noise Control Engineering J. 16*(3), 126–135.

Gordon, R. T., and Vining, W. D. (1992). "Active Noise Control: A Review of the Field," *Am. Ind. Hyg. Assoc. J. 53*(11), 721–725.

ISO (1990). "Acoustics — Determination of Noise Exposure and Estimation of Noise-Induced Hearing Impairment," International Organization for Standardization, ISO 1999:1990, Geneva, Switzerland.

Kelly, S. G. (1996). *Theory and Problems of Mechanical Vibrations*, Schaum's Outline Series, McGraw-Hill, New York, NY.

Lilley, D. T. (1983). "Understanding Damping Techniques for Noise and Vibration Control," *Plant Engineering 37*(9), 38–40.

Lord, H. W., Gatley, W. S., and Eversen, H. A. (1980). *Noise Control for Engineers*, McGraw-Hill, New York, NY.

Miller, L. N. (1979). "Machinery," in *Handbook of Noise Control*, Second Edition, edited by C. M. Harris, McGraw-Hill, New York, NY, p. 26-4–26-13

Miller, R. K., and Montone, W. V. (1978). *Handbook of Acoustical Enclosures and Barriers*, The Fairmont Press, Atlanta, GA.

NIOSH (1973). *The Industrial Environment — Its Evaluation and Control*, National Institute for Occupational Safety and Health, NIOSH No. 74-117, Cincinnati, OH.

NIOSH (1980). *Compendium of Materials for Noise Control*, Second Edition, National Institute for Occupational Safety and Health, NIOSH No. 80-116, Cincinnati, OH.

Olson, H. F. (1957). *Acoustical Engineering*, Van Nostrand, New York, NY.

Petrie, A. M. (1975). "Press Noise Reduction," in *Proceedings of Internoise 75*, edited by K. Kido, Institute of Noise Control Engineering, Poughkeepsie, NY, 311–314.

Rao, S. S. (1995). *Mechanical Vibrations*, 3rd Edition, Addison-Wesley, New York, NY.

Royster, L. H., and Royster, J. D. (1984). "Making the Most Out of the Audiometric Data Base," *J. Sound and Vibration 18*(5), 18–24.

Schmidt, J. W., Royster, L. H., and Pearson, K. B. (1982). "Impact of an Industrial Hearing Conservation Program on Occupational Injuries," *J. Sound and Vibration 16*(1), 16–20.

Stock Drive Products (1984). *Vibration and Shock Mount Handbook*, Product Catalog 814, New Hyde Park, NY [www.sdp-si.com].

Thomson, W. T. (1981). *Theory of Vibration with Applications*, 2nd ed., Prentice-Hall, Englewood Cliffs, NJ.

Ungar, E. E. (1988). "Damping of Panels," in *Noise and Vibration Control*, edited by L. L. Beranek, McGraw-Hill, New York, NY, 434–475.

Ungar, E. E., and Sturz, D. H. (1991). "Vibration Control Techniques," in *Handbook of Acoustical Measurements and Noise Control, Third Edition*, edited by C. M. Harris, McGraw-Hill, New York, NY, p. 28.1–28.20.

Van Houten, J. J. (1991). "Control of Plumbing Noise in Buildings," in *Handbook of Acoustical Measurements and Noise Control, Third Edition*, edited by C. M. Harris, McGraw-Hill, New York, NY, p. 45.1–45.16.

The Noise Manual, revised 5th edition, edited by E.H. Berger,
L.H. Royster, J.D. Royster, D.P. Driscoll, and M. Layne
©2003 American Industrial Hygiene Association

10 Hearing Protection Devices

Elliott H. Berger

Contents

Introduction

A hearing protection device (HPD) is a personal safety product that is worn to reduce the harmful auditory and/or annoying effects of sound. Hearing protectors are often a method of last resort, when other means such as engineering controls or removal of the person from the noisy environment are not practical or economical.

To a large extent, research and development in hearing protection began during and following World War II in response to the tremendous hearing loss caused by military operations. Hearing conservation programs and use of hearing protection in the military and industry followed in the early 1950s, proliferated in the early 1970s and burgeoned in the 1980s, urged on by federal regulations mandating use of hearing protection in occupational settings (OSHA, 1983). As the number and variety of available products have increased substantially in the past 40 years, their quality has also improved. The crucial variable, however, has remained the same — the wearer, and how s/he fits and uses the device. Thus, much of this chapter focuses upon practical aspects of hearing protector selection and utilization, and the related details of real-world performance. The reader is advised that maximum benefit will be gained by studying and implementing the first five sections of the chapter, through *Enforcement of Utilization*, before proceeding to master the more technical details that are subsequently presented.

Selecting the Right Hearing Protector

In many respects, selecting the right hearing protector is more art than science. Although much of this chapter deals with the science of hearing protection — measuring attenuation, predicting performance, determining effects on speech, and other details, the computations are only estimates, and application to the individual is imprecise. What is often more important is the ability to work with people, the art of education and motivation, supervision and enforcement, attention to detail and ergonomics, and hands-on observation. For greater than 90% of noisy industries the time-weighted average (TWA) exposure levels are ≤ 95 dBA; therefore, all that is needed is 10 dB of actual delivered real-world attenuation, a value that most hearing protectors, when properly fitted, can easily provide.[1] Attenuation then takes on secondary or even tertiary importance, and the hearing conservationist must consider comfort, human engineering, communication needs, climatic and other working conditions, cost, durability, and even styling. A lower attenuation product worn properly and consistently, whenever one is in noise, can often provide more effective protection than higher-rated products that are misused or worn sporadically.

---

[1] For additional comments on assigning HPDs based on their attenuation, see *Real-World Attenuation*.

Prior to 1979, HPD attenuation values were commonly available from manufacturers, but only in the form of octave-band (OB) attenuation data. They were widely ignored because of the difficulty, with the instrumentation of that era, of acquiring and using OB workplace noise measurements together with OB hearing protector data. Although methods of utilizing those data were described in the literature, the OB method of computation (also called the "long" method) was required to perform such calculations, and in the pre-personal-computer era this was a time-consuming process.

The advent of the Noise Reduction Rating (EPA, 1979) or NRR as it is commonly called, and the accuracy and simplicity that it seemed to provide, substantially changed the picture. Attention focused on HPD attenuation values. As a result, manufacturers of hearing protectors highlighted the NRR and a battle of numbers ensued. The NRR became even more entrenched when the Occupational Safety and Health Administration (OSHA) included it as the preferred method for assessing HPD adequacy for compliance with the Hearing Conservation Amendment (OSHA, 1983). In many instances additional key parameters of performance were overlooked in favor of choosing the HPD with the highest NRR. This contributed to wearer dissatisfaction and consequent misuse or even nonuse, resulting in inadequate protection.

A further concern that arose regarding HPD attenuation was how to answer a natural and seemingly straightforward question: How much noise reduction can hearing protectors provide? Regrettably, this has not been easy (see *Real-World Attenuation*), and the application of optimum-performance laboratory data has not provided a useful solution. Consequently the NRR, which is computed from such data, is generally acknowledged to provide virtually no effective guidance in the selection and specification of hearing protection (Casali and Park, 1991). In fact, it can be said with little exaggeration that perhaps the *only* value in the NRR is that it indicates the product with which it is associated has been designed and tested for noise reduction.

Workers will be best served when hearing conservationists' primary emphasis is on learning the basics of properly selecting, fitting, and issuing hearing protectors (Royster, 1995). Comfort, as well as human engineering, should be evaluated personally by prospective buyers and their employees. Program administrators will gain a much improved appreciation of the devices under consideration if they wear them 8 hours per day for a trial period, in order to conduct their own subjective evaluation.

Although OSHA requires an employer to provide a "variety of suitable" HPDs, which has been interpreted to mean at least one type of plug and one type of muff, a preferable approach that provides more choice and "buy-in" from the employee is to offer a minimum of four devices, including at least two earplugs and a muff. Involving employees in the selection process will increase the likelihood of maintaining their participation in the entire program. If after a couple of weeks of daily use, an employee is still experiencing difficulties or discomfort, the protector should be resized and/or refitted, or another hearing protector issued. However, the available selection should not be too large, because the problems of developing uniform high caliber training for a large variety of products can easily become unwieldy. Furthermore, a limited group of devices makes it easier to conduct user trials within the plant, simplifies inventory and spare parts control, and may allow negotiation of higher-volume, lower-cost contracts with suppliers.

Types of HPDs, Features, and Usage Characteristics

Hearing protection devices may be broadly categorized into *earplugs*, which are placed into or against the entrance of the earcanal to form a seal and block sound (insert or semi-insert), *earmuffs*, which fit over and around the ears (circumaural) to provide an acoustic seal against the head, and *helmets*, which normally encase the entire head. Utilization of HPDs in the U.S. shows that about 85% of employees wearing hearing protection on a regular basis for protection from occupational noise select earplugs over earmuffs (Royster and Royster, 1984). It may be presumed that this is due to personal preference and comfort. Anecdotal evidence suggests that in Europe and Australia the balance is more heavily weighted toward earmuffs. For a summary of products sold in the U.S. circa 1993 see Franks et al. (1994), and for a simplified overview of the pros and cons of earplugs vs. earmuffs see Table 10.1.

TABLE 10.1
Simplified overview of comparative features of earplugs and earmuffs.

For detailed guidance in selection see text, especially for semi-insert earplugs which are not specifically covered in this table and are ideal for intermittent-use applications.

Issue	*Earplugs*	*Earmuffs*
Comfort and personal preference	Generally preferable for long use periods, but some are hesitant to put anything in earcanal; acclimatization period may be required.	Generally preferable for multiple applications/removals per day, but some can't tolerate pressure on head; more bulky.
Protection	Highly dependent upon user training, skill, and motivation; well-fitted plugs, especially foam plugs, can provide high levels of protection; see Table 10.2 and Figs. 10.18 and 10.19 for representative performance.	Protection generally more reliable than earplugs with less dependence on user training, skill, and motivation; see Table 10.2 and Figs. 10.18 and 10.19 for representative performance.
Sizing	Some plugs are one-size-fits-most, but proper fitting still must be checked; others come in two or more sizes and generally require assistance in proper sizing.	Generally sold as one-sized devices, but individual fit must be checked to make sure band accommodates head and cups accommodate pinnas.
Ease of fitting	Skill and care in fitting generally required for good protection; foam earplugs more forgiving than other types.	Careful fitting not as critical as with earplugs.

— Continued on next page —

TABLE 10.1 — continued
Simplified overview of comparative features of earplugs and earmuffs.

For detailed guidance in selection see text, especially for semi-insert earplugs
which are not specifically covered in this table and are ideal for intermittent-use applications.

Issue	Earplugs	Earmuffs
Compatibility	Long hair, glasses, earrings, and safety gear don't interfere with seal; doesn't affect hair style.	Seal against head can be broken by long hair, eyeglass temples, caps, other safety gear, etc.; can disturb hair styles.
Use in tight spaces	Ideal for tight spaces.	Can interfere with movement in tight spaces.
Monitoring use	Can be difficult for supervisor to assure devices in use; assessment of fit requires in-depth discussion and personal contact.	Use of device can be checked visually at a distance; assessment of fit requires in-depth discussion and personal contact.
Use in hot environments	Preferable to earmuffs but some experience sweat buildup in earcanals.	Generally uncomfortable with high likelihood of sweat buildup under cups.
Use in cold environments	Can be worn under caps or ear mufflers and easily inserted before entering the cold; can't be easily inserted with cold or gloved hands.	Cups can provide warmth in winter but cushions can harden if stored in the cold; can be adjusted while wearing gloves.
Storage, portability, and loss	Easy to store and carry, or wear around neck on cord when not in use; easy to lose or misplace.	Bulky to store and carry, but can be worn on belt clip; not as easily lost or misplaced.
Sabotage	Subject to cutting, puncturing, or whittling to improve comfort at expense of protection.	Subject to band stretching for relief of pressure or drilling cups for ventilation, both at the expense of protection.
Ear infection or cerumen buildup	Cannot be used in presence of earcanal infection, impacted wax, or other earcanal medical conditions.	May be usable in presence of minor external ear infection (with medical supervision), can be used regardless of impacted wax; not suitable for use with pinna or circumaural skin conditions.

Earplugs

Earplugs are made from materials such as slow-recovery closed-cell foam, vinyl, silicone, elastomer formulations, spun fiberglass, and cotton/wax combinations. They may be grouped into the categories of foam, premolded, formable, custom molded, and semi-insert as shown in Figure 10.1.[2] Earplugs in general, with the exception of most formable earplugs, are available either with or without limp connecting cords, or on flexible spring-loaded head or neck bands. Both

---

[2] The earplug categories used herein differ slightly from prior systems used by this author and others. The categories have evolved to better match the mix of products in the marketplace today.

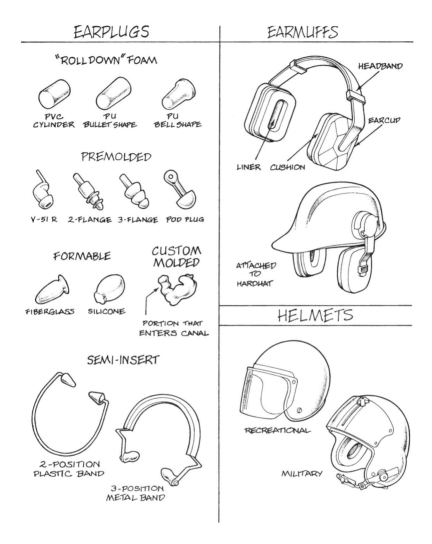

Figure 10.1 — Personal hearing protection devices.

cords and bands help prevent loss, reduce contamination, and simplify storage by permitting placement around the neck while not in use. Cords may also be attached to the webbing of hard hats. Additionally, earplugs specifically targeted toward the food-processing industry may include imbedded or attached metal parts to provide detectability in production operations.

A number of premolded earplugs and some foam earplugs come in more than one size to properly fit virtually the entire population. Records of the size issued to each employee should be maintained at the fitting station and consulted when plugs are reissued to replace those that are lost or worn out. Experience suggests that for premolded earplugs, approximately 2% – 10% of the population will require different sizes for the left and right ears. As a general rule, the more sizes in which a particular premolded plug is manufactured, the greater will be the likelihood that different ones are required for each ear. The use of unmatched sizes for the two ears can pose a problem for those devices that are color coded to indicate size since some employees may be hesitant to wear two different colored plugs. With foam earplugs the likelihood of requiring different sizes for the two ears is minimal.

Because of the dexterity required to maneuver small objects such as earplugs into tight convoluted cavities like earcanals, variations in finger size and shape can also assume importance. Many workers have wide, blunt fingers with short nails, or have missing digits. Hence, the hands rather than the ears sometimes determine the choice of protector (Guild, 1966).

Foam Earplugs

Foam earplugs, also called "roll-down" foam earplugs to emphasize the fact that such products must be rolled and compressed prior to insertion, are made from either slow-recovery polyvinyl chloride (PVC) or polyurethane (PU) material. Neither the PVC or PU foams are of the open-cell "acoustical" variety that is applied to architectural and machinery surfaces to absorb sound. Rather, they are much higher density closed-cell materials (0.1 to 0.2 g/cm^3) that are intended as sound barriers. PVC plugs are commonly punched from sheets of foam and have simple contours with parallel sides and cylindrical or hexagonal footprints, about 14 mm in diameter and 18 – 23 mm in length. PU plugs are typically molded, and thus can have shapes like bullets or bells, and include ridges or asymmetrical features. However, the shapes serve more of a cosmetic than a functional purpose. Despite the fact that PVC and PU earplugs provide similar amounts of attenuation, they feel and behave differently in the hand and in the earcanal. PVC materials are less susceptible to absorption of moisture and thus provide a more stable recovery from roll down, over a wider range of humidity. Although PU plugs feel softer in the hand, this does not translate to better comfort in the ear and in general the balance of sufficient stiffness for proper insertion and adequate softness for improved comfort is better optimized in the PVC-based plugs.

Since their introduction in 1972, foam earplugs have become the most widely used type of hearing protection because of the high levels of comfort and attenuation they provide. They can be difficult for some wearers to insert fully and properly because of the roll down that is required, but because of their conformability, even a partially fitted foam earplug generally affords a good acoustic seal and will deliver attenuation equal to or greater than other types of HPDs (see *Real-World Attenuation*). The decision between the two materials is often best based on personal preference. Although foam earplugs have been frequently labeled as disposable, they can in fact be reused. Typical rates range from one pair/day to one pair every 2 or 3 days, but use for a week or more has been reported without experiencing any problems.

For many years foam earplugs were only available in a single size that fitted most, but not all earcanals. This simplified dispensing, recordkeeping, and inventory problems, but when using such products additional attention was required for wearers with extra-small and extra-large earcanals to make sure that the plugs were not too tight or too loose, respectively. Today some brands come in more than one size to cater to user comfort. Because of the conformability of well-designed foam earplugs, often they can still be used as a single-size product if only the larger size is selected. But, it is not advisable to use the smaller sizes of those products as single-sized foam earplugs, because attenuation can then suffer for those with the largest earcanals.

Since these plugs usually require manipulation by the user prior to insertion, during which time the hands should be relatively clean, they may not be the best choice for environments in which HPDs have to be removed or reinserted during a work shift by employees whose hands are contaminated with caustic or irritating substances or sharp or abrasive matter (see *Initial Ear Examination and HPD Hygiene*).

Premolded Earplugs

Premolded earplugs are formed from flexible materials, including foam, into conical, bulbous, or other shapes, often with flanges or sealing rings, which are typically affixed to or enshroud a flexible stem for handling and insertion. They usually are available in a range of sizes to fit most ears and are pushed into place in the earcanal, whereupon an acoustic (pneumatic) seal is made against the canal walls.

Premolded plugs are available with from zero to five flanges. Generally, the greater the number of flanges, the fewer sizes are required to fit the population, although a recently introduced one-sized pod plug (see Figure 10.1), consisting of a hemispherical foam dome on a flexible stem (i.e., in effect a single-flange design) provides an interesting hybrid between the comfort of a roll-down foam plug and the ease of insertion of a premolded earplug. Nonfoam flangeless premolded plugs normally are more difficult to size correctly and are more prone to work loose during use. Swept-back flanges that tangentially contact the earcanal walls generally provide improved comfort over those flanges that perpendicularly intersect the earcanal, contacting the flesh primarily at their perimeter edges.

One of the first widely used premolded earplugs, the V-51R, was developed during World War II and is still marketed today. It is a single-flanged PVC insert available in five sizes, and sold under various brand names. When the intent is to use this device for an entire noise-exposed population, all five sizes must be stocked, in which case about 95% of adult males can be adequately fitted (Blackstock and von Gierke, 1956). The distribution of sizes required for a given population may vary since earcanal dimensions are a function of the racial and sexual characteristics of the population, with black females having the smallest earcanals and white males the largest. Black males and white females fall in between. For example, in a predominantly black female population nearly 40% might be expected to use an extra-small V-51R, whereas this number would be nearer to 5% for a white male group (Royster and Holder, 1982).

PVC-based premolded earplugs such as the V-51R are prone to shrinkage, cracking, and hardening from exposure to body oils, perspiration, and cerumen. Safe use periods for plugs of this type have been estimated to be 3 months, but even more frequent replacement may be required (Royster and Royster, 1984). Silicone-based premolded earplugs are the longest lived, but regardless, premolded earplugs should be examined periodically and replaced when hardened, deformed, or if their size has changed substantially.

Unlike foam earplugs premolded earplugs do not require a roll down. This is advantageous for those who lack the dexterity to accomplish such a procedure, and it also confers potential hygienic benefits when the roll downs must be accomplished with hands that are befouled with gritty or potentially irritating substances. However, because of the difficulty of obtaining and maintaining a good acoustic seal with premolded earplugs across a wide range of users, real-world performance has been shown to be limited for most products in this category (see *Real-World Attenuation*).

Formable Earplugs

Formable earplugs may be manufactured from malleable materials such as cotton/wax combinations or silicone putty (exposed, or encased in a bladder), and spun fiberglass (colloquially called mineral wool, eardown, or Swedish wool in reference to the country of origin of the product). Life expectancies vary from single-use products such as some of the fiberglass down products, to multiple-use products such as the malleable wax and silicone, and also include relatively permanent items such as the encased putties.

Although cotton alone is a very poor hearing protector due to its low density and high porosity, when it is combined with wax, noise protection can be obtained. The wax tends to make the devices somewhat messy to use, especially in warm conditions. These plugs are not common in industrial hearing conservation programs, but they have attained popularity in the consumer market since their introduction in the late 1920s. Since these materials lack elasticity, they do not expand in place like the foam earplugs and thus are simply packed into the entrance of the earcanal. Consequently, they may lose their seal as a result of jaw and neck motion and require frequent reseating (Ohlin, 1975).

Fiberglass down was first available in the late 1950s as a strip of batting that required tearing and folding into a cone for insertion. More recently, it has evolved into versions that are encased in polyethylene sheaths, and include a lightweight foam core and end cap. These modifications have improved the likelihood of properly fitting these types of plugs and have also reduced the possibility of small sections or pieces of the fiberglass breaking off from the main body of the plug and remaining in the earcanal. However, these plugs are among those most prone to work loose with time and thus users should be alerted to the potential need for frequent reseating.

Malleable silicone earplugs have become popular since the 1980s, but primarily in the consumer market. They are similar to the cotton/wax products, but more malleable, easier to use, and the inherent stickiness of the silicone causes them to adhere better to the earcanal entrance and concha. They are a popular product to use to prevent water from entering the earcanal during swimming and bathing, but have had little application in occupational hearing conservation.

Custom Molded Earplugs

Custom molded earplugs are most often manufactured from silicone putties although some are available in vinyl or acrylics. Earcanal impressions are made using a viscous material with a consistency varying from that of thick syrup to soft putty. The material is mixed with a curing agent and is then injected into the earcanal using a large syringe, or in a less preferable technique, formed into a cone and manually packed into the canal. The material then cures, normally for about 15 minutes. When cured, the impression may itself become the HPD. In some cases it may be dipped into a coating before use. This improves its durability and also increases the tightness of the fit; extra coatings can increase attenuation, but at the expense of comfort. With other products, impressions are returned to the supplier who then uses them to make subsequent negative and positive molds in order to create the final custom earplug.

Most earmolds fill a portion of the earcanal as well as the concha and pinna. The canal portion makes an acoustical seal to block noise whereas the concha/pinna portion principally functions to hold the HPD in position. Incorrect insertions are easily detected and the molds have little chance of working loose with time. Some molds are manufactured with only a canal impression present on the supposition that leaving a greater area of the pinna uncovered makes them cooler and more comfortable. Unfortunately, this may also impair the retention of these molds in the canal, allowing them to more readily lose their seal during use.

Considerable skill and time are required to take individual impressions for each employee. Contrary to popular belief, the fact that custom earmolds are user-specific and intended to fit only the canal for which they are manufactured does not ensure that they will provide better protection than other well-fitted earplugs. In fact, often the opposite is the case (see *Attenuation Characteristics of HPDs*). This observation may be partially explained by reference to Figure 10.2, which depicts five different earmolds, all of which are impressions of one subject's right earcanal, and all of which were manufactured by "experienced" fitters. Note the significant differences in the canal portion of the impressions, the part of the earplug that is responsible for sealing the ear and blocking sound.

A worthwhile aspect of individually molded earplugs is that they are manufactured personally, for only one employee, so that they are "customized." This customization can be effectively utilized as an incentive for motivating employees to wear their HPDs. Another positive aspect of custom molds (those with both a canal and concha impression) is that due to the way in which they fit into the ear, they are less subject to misinsertion under field conditions.

In spite of the longevity claims made by some manufacturers for their "permanent" earmolds, these devices, like other earplugs, are susceptible to shrinkage, hardening, and cracking with time, and must be periodically reexamined to ensure they are still soft, flexible, and wearable. Changes in body weight of 9 kg (20 lbs.) can affect the fit of a custom mold sufficiently to require a remake

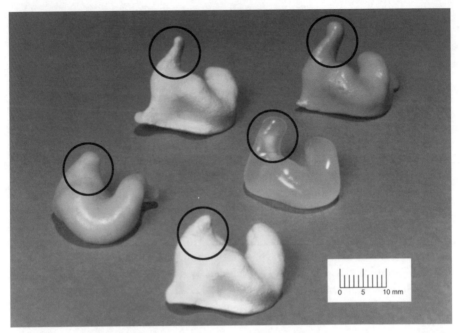

Figure 10.2 — Examples of five custom earmolds for the same earcanal. Circles indicate the portions that enter and seal the earcanal.

(Briskey, 1984). Additionally, since they may be easily lost or misplaced, as can all earplug-type HPDs, their use may create administrative problems since time must be allowed to remanufacture them on an as-needed basis. Employees should be issued alternative backup devices to use in the interim.

Custom earmolds can be very comfortable, but experience has shown that as the molds more fully and tightly fill the canal and therefore more effectively attenuate sound, comfort deteriorates. In practice, there is a limit to how snugly a custom mold can be fitted. Selection of a particular product should be based primarily upon the user's judgment of comfort and the fitter's evaluation of the ease of manufacture of the earmolds. This latter consideration is important since the final performance of the molds is closely linked to the fitter's skill in making the initial impression.

Semi-Insert Earplugs

Another type of earplug is the semi-insert (also called semi-aural devices, canal caps, concha-seated, or banded earplugs). It consists of soft pods or tips that are held in place by a lightweight spring-loaded band. The pods are positioned in the conchae covering the entrances to the earcanals, or fitted to varying depths within the earcanals. Semi-inserts that cap the canal require the force of the band to retain their position and acoustic seal. Semi-inserts that enter the canal behave like earplugs; they seal the ear to block noise without the application of band force.

Semi-inserts are easily removed and replaced, and conveniently hung around the neck when not in use. Their fit is not compromised by safety glasses or hard hats. These devices are usually available with dual position (under the chin and behind the head) or "universal" (dual position plus over the head) bands made from either metal or plastic. The tips can be made from vinyl, silicone, PU foam, or composites such as foam encased in a silicone bladder, and may have a bullet, mushroom, conical, or flanged shape.

Semi-insert devices are principally intended for intermittent use where they must be removed and replaced repeatedly. Examples include: ground crews servicing commercial aircraft, periodic equipment inspection by personnel normally located in sound-treated booths, and supervisor walk-throughs. During longer use periods, the force of the caps pressing against the canal entrance may be uncomfortable, but for those who prefer this type of device for extended use, the better ones can offer adequate protection.

Since semi-insert devices generally cap the canal at or near its entrance, they tend to create the most noticeable occlusion effect and consequently distort the wearers' perception of their own speech more than other types of HPDs (see *The Occlusion Effect*). This may be objectionable to some users. Certain semi-insert devices, particularly those that do not enter the canal, are prone to losing their seal periodically, especially if the band is bumped and caused to skew on the head.

Earmuffs

Earmuffs normally consist of rigid molded plastic earcups that seal around the ear using foam- or fluid-filled cushions (see Figure 10.1), or cushions composed of a fluid-filled bladder combined with a layer of foam. They are held in place with metal or plastic spring-loaded headbands, or by short spring-loaded arms attached to a hard hat. The cups are lined with acoustical material, typically an open-cell foam, to absorb high frequency (>2 kHz) energy within the cup. The headbands may function in a single position only, or may be of the more versatile universal style suitable for use over the head, behind the head, or under the chin. Most manufacturers offer plastic or fabric over-the-head crown straps for use when headbands are worn behind the head or under the chin. This is necessary to maintain proper protection, and to provide a snug, secure, and comfortable fit.

Earmuffs are relatively easy to issue since they are one-sized devices designed to fit *nearly* all adult users. Regardless, earmuffs must be evaluated for fit when initially issued, since not every user can be fitted by all models. Head or ear sizes may fall outside the range that the muff band or cup openings can accommodate. Furthermore, unusual anatomical features such as prominent cheek bones (zygomatic arches) or severe depressions below the pinna and behind the jaw (posterior to the temporomandibular joint) are particularly hard to fit. Asymmetrical earmuffs (i.e., possessing top/bottom or left/right designations) rarely provide useful performance gains over symmetrical designs, and asymmetry increases the likelihood of misuse.

When the use of protective headgear is required, hard hats with attached ear-muffs provide a convenient alternative to the use of headband-attached earmuffs worn behind the head or under the chin. However, hard-hat attached muffs can be more difficult to properly orient and fit since the attachment arms usually do not provide as adaptable an adjustment as do the headband-attached versions, nor can they fit as wide a range of head sizes. For such devices, not only must the attachment arms be properly extended and located, but the helmet's webbing must be adjusted to properly locate the hat on the head. It is also important that users be advised against storing the cushions in the resting position against the hat itself, as the constant pressure on the cushions can lead to permanent defor-mation. Rather, the attachment arms should be snapped out to the "at rest" posi-tion, if available, for storage when not in use.

A number of design parameters affect the attenuation of earmuffs, including cup volume and mass, headband force, area of opening in the cushion, and the materials from which the device is constructed. The effect of volume on earmuff attenuation is shown in Figure 10.3. From 125 Hz – 1 kHz, the larger-volume muffs provide more attenuation than the smaller ones, since in this region the

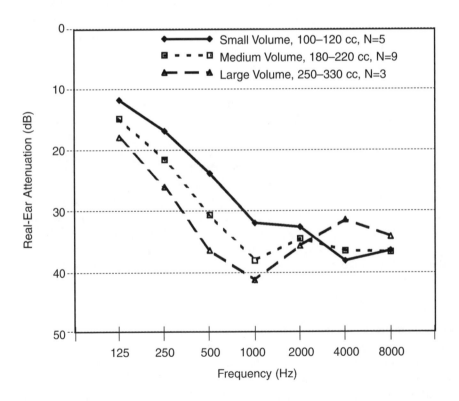

Figure 10.3 — Effect of cup volume on the attenuation provided by earmuffs with foam-filled cushions.

extra volume and mass of their cups are the controlling physical parameters. Above 2 kHz however, larger volume earmuffs tend to be inferior, since their increased shell surface area makes them more susceptible to developing vibrational modes within the cup walls and therefore less capable of blocking higher frequency acoustical energy.

Sufficient force must be exerted so the cushions fully conform and provide a seal against the head. Manufacturers must compromise between high force, which yields good attenuation and poor comfort, and low force, which produces the opposite results. Band force, which can deteriorate with use and age, can be roughly checked by comparing the resting positions of suspect earmuff cups to those of new samples to make sure that the cup-to-cup separation is approximately the same (see Figure 10.4). Alternatively, a relative measurement of the force of two samples can be made by placing a small box or books underneath an electronic balance until the total height equals the average head width of 145 mm, and then taping the assembly together and positioning it on its side. The earmuffs are placed around the assembly with one cup on the book and the other on the scale, being careful to position the new and worn muffs in identical positions for the measurement process. A difference in force between the suspect and reference muff that exceeds 20% suggests the earmuff should be replaced.

Due to the transformation of sound pressure from the outside to the inside of the earmuff cup, the smaller the cup opening (other parameters being held constant), the greater the attenuation. Additionally, smaller diameter cup openings allow cushions to make circumaural contact nearer to the base of the pinna, which in turn tends to increase attenuation since facial contours and jaw and

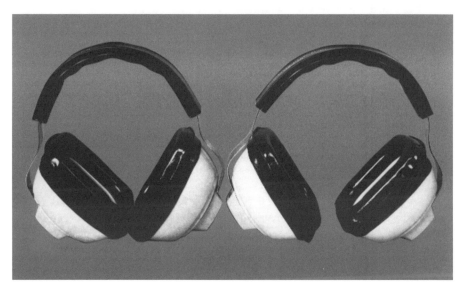

Figure 10.4 — Comparison of undistorted (i.e., as-delivered) and distorted (i.e., stretched) headbands.

neck motion are minimized in those regions. However, when attenuation is increased by decreasing the cup opening, it is often at the expense of comfort and ease of use.

As the cup opening is decreased in size, the difficulty of fitting it over and around the pinna increases. For example, the length and width of male pinnae at the 75th and 95th percentiles are 67 × 38 mm and 74 × 41 mm respectively, whereas some of the higher attenuating earmuffs have openings as small as 58 × 39 mm. For such devices, wearers with even average size ears must be careful to "tuck" them into the cups. Even for cups with more typical openings (65 × 41 mm), employees must still be cautioned to place their pinnas fully within the earmuff cup, since some have been observed resting the cushion on portions of the external ear, which not only reduces comfort but also creates a significant acoustical leak (Riko and Alberti, 1982).

The decision whether to use foam or liquid-filled cushions is somewhat academic in today's market, since both offer similar performance, the fluid-filled cushions providing slightly better protection at the low frequencies, and the foam slightly more at the high frequencies. Tests indicate that for at least one brand of earmuff, both types contour equally well around eyeglass temples (Berger and Kieper, 1997). Liquid-cushion earmuffs weigh 5% – 40% more than their foam counterparts, generally sell at a 10% – 20% premium, and the cushions may be punctured or split, allowing their contents to drain out.

The cover or bladder on both foam and liquid cushions is made of a plastic material that may harden and deform with time due to contact with body oils, perspiration, cosmetics, and environmental contaminants. For this reason, cushions should be examined at least twice yearly and replaced if necessary. Absorbent cushion covers are available to enhance comfort. Paper covers have been shown not to degrade acoustic performance, but fabric covers may reduce attenuation since they impair the seal of the cushion to the side of the head.

Earmuffs are good for intermittent exposures due to the ease with which they can be donned and doffed, and they may be suitable when earplugs are contraindicated (see *Initial Ear Examination and HPD Hygiene*). For long-term wearing it is often reported that earmuffs feel tight, hot, bulky, and heavy, although in cold environments their warming effect is appreciated. It is easy for supervisors to monitor that earmuffs are in use, but their attenuation can be easily compromised by a number of factors (see *Tips for Fitting Hearing Protectors*), so it is not advisable to assume that use is synonymous with "effective protection."

When selecting earmuffs for particular applications, the relatively small differences in attenuation between most popular brands and types suggest that except for extreme noise exposures, where the highest protection possible must be afforded, selection should be based upon other factors. Often smaller, less expensive earmuffs may provide better comfort and therefore will be more readily accepted. In practice, more attenuation can often be gained by assuring that properly maintained earmuffs are worn correctly and consistently than can be assured by buying heavier and perhaps more expensive "high performance" models.

Helmets

Helmets enclose a substantial portion of the head, and are usually designed primarily for impact protection (see Figure 10.1). They are not commonly worn in occupational settings, but more likely will be found in the military for pilot and tank personnel and for sports such as motorcycling. When helmets contain circumaural earcups or a dense liner to seal around the ears, as do the military helmets, they can also provide beneficial amounts of hearing protection (see *Attenuation Characteristics of HPDs*) and in the high frequencies can even provide attenuation beyond the bone-conduction limits normally experienced with conventional hearing protectors (see *Bone and Tissue Conduction*).

Mechanisms and Limits of HPD Attenuation

In the unoccluded ear, the dominant sound path for external sounds is through the earcanal to the eardrum. However, in the occluded ear, four distinct sound paths can be identified as illustrated in Figure 10.5: (1) air leaks, (2) hearing protector vibration, (3) structural transmission, and (4) bone and tissue conduction.

Air Leaks

For maximum protection, HPDs must essentially make an airtight seal with the walls of the earcanal or the circumaural regions surrounding the pinna. Air leakage can reduce attenuation by 5 to 15 dB over a broad range of frequencies depending upon the size of the leak. Typically the loss is most noticeable at low frequencies. This is the key pathway that hearing conservationists and users must attend to as it can dramatically affect the performance of the device. The existence of air leaks due to misfitting or poor condition of HPDs is often what differentiates optimum-laboratory from real-world attenuation.

Hearing Protector Vibration

Due to the flexibility of the earcanal flesh, earplugs can vibrate in a piston-like manner, thus limiting their low-frequency attenuation. Earmuffs, too, vibrate as a mass/spring system, the stiffness of the spring depending upon the dynamic characteristics of the earmuff cushion and the circumaural flesh, as well as the volume of the air entrapped inside the earcup. These actions limit low-frequency attenuation. Representative maximum attenuation values at 125 Hz for earmuffs, premolded earplugs, and foam earplugs, are about 20 dB, 30 dB, and 40 dB, respectively.

Structural Transmission

The exterior surfaces of an HPD will vibrate in response to the forces applied by impinging sound waves. This vibration is transmitted by the protector material to its inner surface where the resultant motion radiates sound, of diminished intensity, into the enclosed volume between the HPD and the wearer's eardrum.

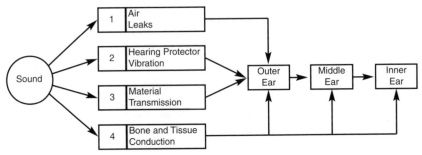

Figure 10.5a — Block diagram of the four sound pathways to the occluded ear.

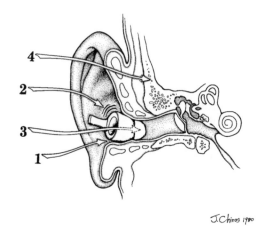

Figure 10.5b — Sound pathways for an earplug.

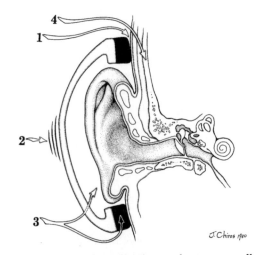

Figure 10.5c — Sound pathways for an earmuff.

The amount of sound reduction is dependent upon the mass, stiffness, and internal damping of the hearing protector material.

Structural transmission through the cup and cushion components of earmuffs is significant, normally providing a limitation to attenuation for frequencies above 1 kHz. Earplug transmission is limited in a similar manner for imperforate materials, however, a special case exists for certain fibrous materials, such as cotton, which are easily permeated by air. In this latter instance attenuation will be very low since sound will penetrate the device as though there were many tiny air leaks.

Bone and Tissue Conduction

Even if HPDs were perfectly effective in blocking the preceding three sound paths, sound energy would still reach the inner ear (see Figure 10.5a) via bone and tissue conduction (BC) through the skull. Energy transmitted in this manner is said to flank or bypass the protector, imposing a limit on the real-ear attenuation that any HPD can provide. However, the level of the sound reaching the ear by such means is approximately 40 – 50 dB below the level of air-conducted sound through the open earcanal, as illustrated in Figure 10.6 wherein dual hearing protector performance is also presented (see *Dual Protection*). The data are

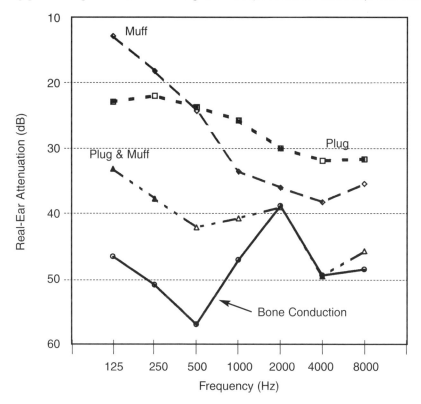

Figure 10.6 — Bone-conduction limits to HPD attenuation and an example of the attenuation provided by an earplug, an earmuff, and the two devices worn together.

from Berger (1983a); they are representative of other estimates that have been published (Schroeter and Poesselt, 1986).

The relative contribution of the energy transmitted via the BC pathways is independent of the noise level in which the hearing protector is worn, thus even though the BC estimates are typically measured with low-level test signals, the results correspond to effects found at higher levels. Unless an HPD's attenuation approaches the BC limits, which is rarely the case in field applications due to air leaks, sound transmitted via the BC paths is normally not of primary concern.

Since the regions of the head around the external ear are only a small portion of the total skull area exposed to sound, covering them with an earmuff is of little significance compared to the other factors controlling hearing protector attenuation. Therefore, the relative performance of plugs compared to muffs is not, in practice, determined by the BC paths, but by factors inherent in the design of HPDs and their interface to the head. Increasing the BC limits to attenuation shown in Figure 10.6 requires completely enclosing the head in a rigid helmet with a face mask, like the flight helmets previously described. Therefore use of earmuffs in conjunction with a hard hat, which covers only part of the head and has many gaps through which the sound energy can penetrate, provides no more protection than the muffs alone.

Fitting and Issuing

The heart of the hearing protection phase of a hearing conservation program (HCP) is fitting and issuing HPDs. This is often where programs fall short. In a nationwide survey conducted in the early 1980s it was found that 58% of those persons issuing HPDs in industrial HCPs *never* evaluated the fit of the HPD for those receiving the devices, and even if they did, 42% of those who issue HPDs (supervisors, tool crib operators, and stock clerks) never received training concerning hearing protection fitting (Royster and Royster, 1984). For hearing conservation practice to improve, this is a key area where programs must be upgraded.

Individual and Group Training

When any type of hearing protector is initially dispensed, the process is best accomplished one-on-one or in small groups with a student/instructor ratio of no more than about 5:1. This is important since the compatibility and fit of protectors *must* be individually checked on each employee. Also to be considered is that the smaller the group, the less likely it is that the trainee(s) will become self-conscious during the fitting process. Allow about 7 – 10 min. for each employee.

HPD training in larger groups is also useful when it occurs in addition to, but not in place of, individual or small-group work. Working with larger classes is a suitable way to provide a review and reminder during the annual educational sessions that are a required part of every hearing conservation program, but is not sufficient in and of itself (Royster and Royster, 1985).

Hearing protectors are typically dispensed in quiet environments away from the noisy workplace. This is primarily a matter of convenience and logistics that obviously makes it easier for the fitter to communicate with the person being fitted. The disadvantage of this approach is that in low (unobjectionable) noise levels the wearer cannot appreciate the beneficial aspects of the noise reduction provided by the HPDs. It is like trying to evaluate sunglasses by wearing them at night or in a dimly lit store.

When noise is used during the fitting process the wearer can listen to it while adjusting the HPDs for the lowest perceived noise level. Recordings of broadband noise or representative industrial sounds can be presented using a portable cassette player. If a noise source is unavailable, the fitter should follow up with employees in their work environment to recheck the fit and suitability of the devices that were dispensed.

Initial Ear Examination and HPD Hygiene

Prior to issuing HPDs the fitter should visually examine the external ear to identify any anatomical factors that might interfere with the use of the protector in question, or medical conditions that might be aggravated. If such medical conditions are present, HPDs should not be worn until professional consultation and/or corrective treatment can be obtained, or the suspected condition has been shown not to constitute a problem. Areas of concern include extreme tenderness, redness or inflammation (either in or around the ears), sores, discharge, congenital or surgical ear malformations, and additionally in the case of earplugs, canal obstructions and/or impacted or excessive cerumen. The effect of the latter condition, however, is difficult to judge since few data are available on the influence of earplugs on the formation, buildup, and possible impaction of cerumen.

As with all clothing and equipment that comes in repeated and intimate contact with the body and the work environment, the cleanliness of HPDs must be considered. HPDs should be cleaned regularly in accordance with manufacturers' instructions, and extra care is warranted for environments in which employees handle potentially irritating substances. Normally, warm water and soap are recommended as cleansing agents. Solvents and disinfectants should generally be avoided.

Earplugs should be washed in their entirety and allowed to dry thoroughly before storage in their carrying containers. Ideally, employees should regularly clean their earplugs, but in actual practice if this is done once or twice a week, that is all that can be expected. Earmuff cushions should be periodically wiped or washed clean. Their foam liners can also be removed for washing but must be replaced before reuse since they do affect attenuation. Earplugs and earmuff cushions should be discarded when they cannot be adequately cleaned or when they no longer approximate their original appearance or resiliency.

Stressing hygiene beyond practical limits, however, can compromise the credibility of the HPD issuer/fitter or can lead to disuse of HPDs. It is often difficult enough to get employees to replace or repair worn-out HPDs, let alone clean them routinely. And in spite of this, information from authorities in the field of audiology and hearing conservation (Gasaway, 1985; Ohlin, 1997), as well as the available epidemiological data (Berger, 1985; Royster and Royster, 1984) suggests that the likelihood of HPDs increasing the prevalence of outer ear infections is minimal.

If an ear irritation or infection is reported, the exact extent and etiology of the problem should be investigated firsthand by medically trained personnel to determine whether the causative agent is an HPD or another predisposing factor. Such factors include excessive cleaning of the ear, recreational water sports, habitual scratching and digging at the ears with fingernails or other objects, environmental contaminants, and systemic conditions such as stress, anemia, vitamin deficiencies, endocrine disorders, and various forms of dermatitis (Caruso and Meyerhoff, 1980). When HPDs are implicated, common causes have been earplugs or even earmuffs that are contaminated with caustic or irritating substances, or embedded with sharp or abrasive matter. In one reported case of earplug contamination (Royster and Royster, 1985), more careful hygiene practices, combined with the use of corded plugs to allow removal without touching the protector, eliminated the problem.

If occurrences of external ear problems develop, it is important to determine whether they are limited to a particular department, operation, or workshift, to one or more brands or types of HPDs, to a change in the HPDs being utilized, to a particular time of year, or if they are perhaps due to some other policies or procedures that may have been modified within the work environment. This will allow a reasoned approach and help to avoid an overreaction which could compromise the HCP, without necessarily resolving the problem at hand.

Tips for Fitting Hearing Protectors

In large part, fitting hearing protectors is a common-sense process that does not require instrumentation or years of training. It does, however, necessitate a hands-on approach and the willingness to work closely with individuals to elicit their feedback, and at the same time to provide ideas, techniques, and direct physical assistance. For all HPDs, even earmuffs and "one-size-fits-all" earplugs, it does demand that the fitter must individually examine each person to be sure that a proper seal can be obtained with the device in question. Additionally, it takes time to get used to hearing protectors, both how they feel and how they sound. Perceptions of comfort change with time, thus a break-in period is advisable for new wearers, especially in the case of earplugs. It may take a week or so to fully adapt to the feeling of wearing hearing protectors and to begin to recognize and appreciate their auditory as well as their nonauditory benefits.

Issuing and training must also include instructions on the need to reseat devices periodically throughout the workshift if they become loose or break their seal, recording the type and size of product fitted to each ear of the employee, instructions on how and where to obtain replacement HPDs or parts, the hazards of user alterations to their HPDs (see *Sabotage*), and documentation of training. When employees need replacement HPDs because they have lost them or worn them out, it is important that if sized devices were initially issued, the records be checked and the same size reissued. If a wearer finds a device uncomfortable an alternative HPD should be provided. When changes are desired in the brand or type of HPD that is being used, the fitter should be consulted and assistance provided.

When fitting hearing protectors, in addition to the obvious need for samples of all types and sizes of HPDs in use at the facility, as well as cleansers and/or disinfectants, and towels or cloths, it will be convenient to have the following available: (1) an otoscope, penlight, or earlight (penlight with a plexiglas tip to direct the beam) to visualize the entrance of the earcanal and circumaural regions, and inside the earcanal to the depth the earplug will be inserted, (2) tweezers with blunt tips to remove earplugs that are inserted too deeply for easy removal, and (3) an earcanal sizing tool to assist in issuing sized earplugs.[3]

Occlusion Effect

Occluding and sealing the ear with an earmuff or earplug increases the efficiency with which bone-conducted sound is transmitted at the frequencies below 2 kHz. Called the occlusion effect, it causes wearers of HPDs to experience a change in their perceived voice quality and other body-generated sounds/vibrations such as breathing, chewing, walking, etc. (Berger and Kerivan, 1983). Of all the tips for fitting HPDs, listening for the occlusion effect is the most widely applicable, since it is suitable for use with nearly all types of hearing protectors. To experience the occlusion effect, plug your ears with your fingers as you read this sentence aloud and note the change in the sound of your voice — its added fullness or resonant bassiness. Other adjectives that have been used to describe the changes in voice quality are deeper, hollow, and muffled.

The occlusion effect can be used as a fit test for either plugs or muffs by asking the wearer to repeat aloud "boom beat" while listening for the change in voice quality that indicates an acoustical seal and the presence of the effect. With earplugs, an alternative approach is to speak with only one ear correctly fitted. The voice should be more strongly heard or felt in the occluded ear (Ohlin, 1975). If this does not occur, the plug should be reseated or resized. When the second ear has been fitted correctly, the effect should be the same in both ears, causing the voice to be heard as though it were emanating from the center of the head.

Some listeners are unable to hear differences in the occlusion effect between their two ears, but most can hear a change in the overall sound of their voice when both ears are sealed. An alternative means of generating a "test signal," and one which some find easier to detect, is to hum (the *hum test*). It is a good way to create sounds of varying pitch and constant level that can be used when listening for the occlusion effect while adjusting the HPD.

---

[3] One commercially available product is the Eargage™ earcanal sizing tool from E•A•R Hearing Protection Products, Indianapolis, IN.

A caveat with respect to the occlusion effect is that although it is a fine way to test the fit of HPDs, its presence is often cited as an objectionable characteristic of wearing hearing protection. As is shown in Figure 10.7, the effect is greatest when the earcanal is covered at its entrance, such as with a semi-insert device. The effect diminishes as earplugs are inserted more deeply or with the use of earmuffs, with the occlusion effect decreasing as earcup volume increases.

Earplugs

When initially dispensing earplugs the fitter should insert at least one plug into the employee's ear so that s/he can experience the feel of a properly seated device. This is especially important because of the reluctance most novice users have of placing anything deeply into their earcanals. With one earplug properly inserted, the person then has an example to match. Ask the wearer to insert the other plug until both ears feel the same and sound equally occluded, and then check both earplugs for proper fit. Once the two plugs have been properly inserted ask the person to remove them both and then insert them one more time for review and additional practice.

For all types of earplugs, with the possible exception of custom earmolds, insertion is easier and more effective for most wearers if the outer ear (pinna) is pulled outward and upward as illustrated in Figure 10.8. Plugs should be inserted into the right ear using the right hand and into the left ear with the left hand. The pinna is grasped with the opposite hand by reaching behind or over the head (see Figure 10.9). This allows the hand fitting the plug to have the best line of approach for proper insertion.

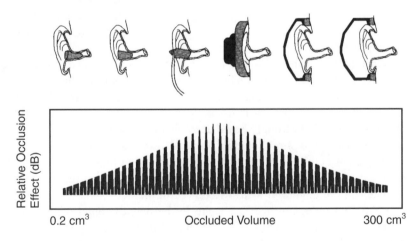

Figure 10.7 — The occlusion effect and its relationship to the fit and type of hearing protector. The effect is minimized with deeply inserted plugs (left side of graph), increases in magnitude as the plugs are withdrawn, peaks when the canal is capped by a semi-insert or the pinna is covered by a supra-aural device such as an audiometer earphone (center two drawings), diminishes as the ear is surrounded by an earmuff, and continues to decline in magnitude as the volume of the earmuff increases (right side of graph).

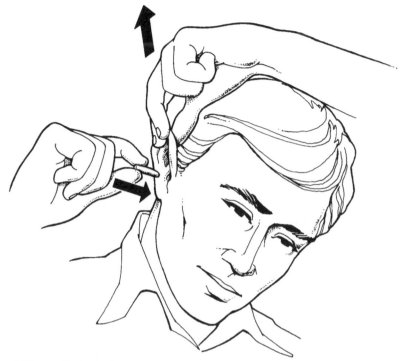

Figure 10.8 — Method of pulling the pinna outward and upward while inserting an earplug.

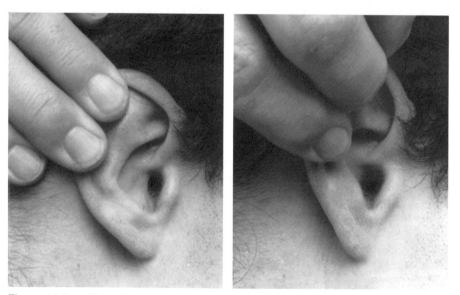

Figure 10.9 — The effect on the lumen (opening) of the canal due to pressing the pinna against the skull (left) vs. pulling it outward and upward (right).

The fitter should determine the best direction in which to pull the pinna to access and enlarge the canal as much as possible. A small focused light source such as a penlight or an earlight is helpful in this regard. Merely pressing the pinna back along the side of the skull is usually not effective. Demonstrate the correct technique by guiding the user's hand to help pull his or her pinna in the proper manner. All wearers should initially use the pinna-pull technique as they learn how to best fit their earplugs, although with time and experience some may find it no longer necessary.

Employees may also require assistance in finding the best direction in which to "aim" the plugs into their canals. Although this will usually be forward and slightly upward, it can vary substantially for different individuals, in some instances even being directed towards the back of the skull. There is little likelihood of hurting the eardrum during insertion since the sensitivity of the earcanal to pressure or pain increases significantly as the eardrum is approached. The discomfort experienced due to touching these deeper portions of the canal will alert the user to stop pushing the plug before a problem can occur.

Once fitted, the noise reduction of earplugs can be tested subjectively by pressing firmly cupped hands (use the palms of the hands, wrists forward, fingers towards the rear) over the ears while listening to a steady noise (the *loudness test*). With properly fitted plugs, the noise levels should seem nearly the same whether or not the ears are covered. Conversely, if the user breaks the seal of one of the two earplugs, a marked increase in the perceived noise level should be noted. If not, it is likely that both plugs were poorly fitted before the seal breakage was attempted.

When dispensing earplugs, fitters will soon learn that people are very conscious of the cleanliness of their earcanals. If cerumen (earwax) adheres to or coats trial earplugs, wearers may be embarrassed. Assure them that earplugs penetrate more deeply into their earcanals than they can or should normally reach when cleaning their ears. Furthermore, a certain amount of cerumen is necessary to provide a protective barrier and emollient for the ear, and it can in fact furnish lubrication to ease and improve the fitting of earplugs (Berger, 1985).

Finally, users must be alerted that earplugs may work loose with time and require reseating. Studies have demonstrated this effect for certain premolded and fiberglass earplugs, but at the same time have shown that foam and custom molded earplug wearers did not experience this problem (Abel and Rokas, 1986; Berger, 1981).

Foam Earplugs

Foam earplugs are compressed for insertion by rolling them into a very thin crease-free cylinder. The cylinder should be as small in diameter as possible, that is, as tightly compressed as can be achieved. Crease-free rolling is accomplished by squeezing lightly as one begins rolling, and then applying progressively greater pressure as the plug becomes more tightly compressed. Be sure to roll (not twist) the plug into a cylinder rather than other shapes such as a cone or a ball (Anon., 1988). Sometimes users flatten a foam plug into a disc and insert it sideways into the entrance of the earcanal, or simply push and squash a plug, tip first into their ear, without ever rolling it down. Such misuse is easy to detect and clearly indicates inadequate instruction, and lack of effort and/or success in properly motivating the employees as well.

After insertion, it may be necessary to hold foam earplugs in place with a fingertip for a few moments until they begin to expand and block the noise. This is not intended to keep them from backing out of the earcanal, since properly inserted foam earplugs do not in fact exhibit such a tendency, but rather is to ensure that the plugs do not slip or dislodge prior to re-expanding enough to hold in place.

Unlike other types of earplugs, foam earplugs should not be readjusted while in the ear. If the initial fit is unacceptable, they should be removed, re-rolled, and reinserted. Furthermore, a large occlusion effect does not usually signify a best fit for foam earplugs since the effect is maximized when they barely enter or cap the canal (see Figure 10.7). In fact, the deeper the insertion (which for foam earplugs is usually associated with improved comfort), the better will be the fit and the attenuation, and the less noticeable and annoying will be the occlusion effect.

The simplest, but least accurate method to assess the fit of a foam earplug, is to visually (for the fitter) or with the fingertips (for the wearer), check the position of the end of the plug relative to the tragus and concha (see Figures 10.10 and 10.11). If the outer end of the plug is flush with or slightly inside the tragus, this generally indicates that at least half of the plug is in the canal and the fit is proper. If most of the plug projects beyond the tragus and out of the concha, the insertion is probably too shallow. Since both tragus-to-earcanal dimensions and earplug length can vary significantly, this check is not a foolproof indicator.

Another test that either the wearer or the fitter can perform is to remove an earplug after it has expanded in the ear for about a minute. If it was well fitted,

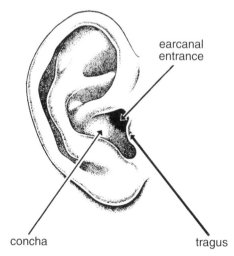

earcanal entrance

concha

tragus

Figure 10.10 — Key features of the external ear.

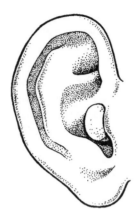 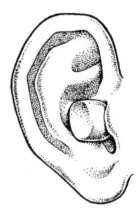

Figure 10.11 — Demonstration of a proper (left) and improper (right) fit of a foam earplug.

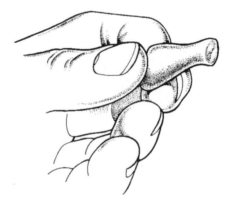

Figure 10.12 — How to "read" a foam earplug after removal from an earcanal in which it has been allowed to expand for about one minute. Primarily useful for medium and smaller earcanals. Example shows a good insertion in which about ½ of the plug is compressed into shape of earcanal, with no evidence of creases, wrinkles, or folds.

it should appear free of creases and wrinkles, and the still partially compressed portion of the plug will indicate that at least one-half of its length had extended beyond the entrance of the earcanal and formed a seal within the canal itself (see Figure 10.12). This test is useful for medium and smaller earcanals since their smaller dimensions cause them to make a more substantial impression on the earplugs. In the larger canals, the plug will show less of an impression upon removal, but because of the larger lumen, insertion is normally easier and less in need of verification.

Premolded Earplugs

When initially inserting premolded earplugs the fitter should be easily able to detect gross errors in sizing. Ear gauges are available from some manufacturers to aid in the sizing process. Plugs that are much too small will tend to slide into the canal without any resistance, their depth of insertion being limited only by the fitter's finger and not the plug itself. Overly large plugs either will not enter the canal at all, or will not penetrate far enough to allow contact of their largest (outermost) flanges with the concha (see Figures 10.13 and 10.14). With certain premolded multiple-flanged earplugs, however, it is unnecessary for the outer-most flange(s) to seal the ear to obtain a proper fit for those with small to extra-small earcanals (see Figure 10.15). Lubrication, such as provided by clean tap water, can often ease the insertion of a premolded earplug, especially until the wearer learns how to most effectively fit the product.

Correctly sizing and fitting premolded inserts will usually require that a balance be established between attenuation and comfort. With care and skill this can be accomplished. A plug that is well seated and appears to make contact with the interior wall of the canal without appreciably stretching the tissues is a good size to begin wearing. When a canal falls between two sizes the larger size plug is not necessarily the best one to choose. Even though it may provide more attenuation, if the plug is not worn or not used correctly due to discomfort, the resultant effective protection may be less than would have been achieved had the smaller more comfortable size been selected.

A properly inserted premolded earplug will generally create a plugged or blocked-up feeling due to the requisite airtight seal. If a seal is present, resistance should be felt when an attempt is made to withdraw the plug from the canal, much like pulling a rubber stopper from a glass bottle (the *tug test*). The seal can be further tested by, without breaking the seal, gently pumping the plug back and forth in the earcanal. When a seal is present, the pumping motion will cause pressure changes in the ear that the wearer should be able to detect (the *pump test*). Because of the pneumatic seal created by properly inserted premolded earplugs, suction is created if they are rapidly removed. This can be uncomfortable, painful, and/or potentially harmful to the ear. Teach wearers to remove plugs slowly, or even to use a slight twisting or rocking motion to gradually break the seals as the plugs are withdrawn. Teach coworkers never to jerk someone else's earplug from his or her ears, whether in seriousness or jest.

Formable Earplugs

Sheathed fiberglass earplugs are inserted by placing them into the earcanal with a slight rocking and twisting motion. A pinna pull (see Figure 10.8) will usually be very helpful in this regard. Since the plugs do not create a pneumatic seal, the tug and pump tests will not be applicable. However, the loudness test can be used.

Malleable plugs, such as the silicone style, are best inserted by kneading them into a spherical shape and them packing them into the entrance of the earcanal while pulling the pinna first in one direction then in another. The inclination to

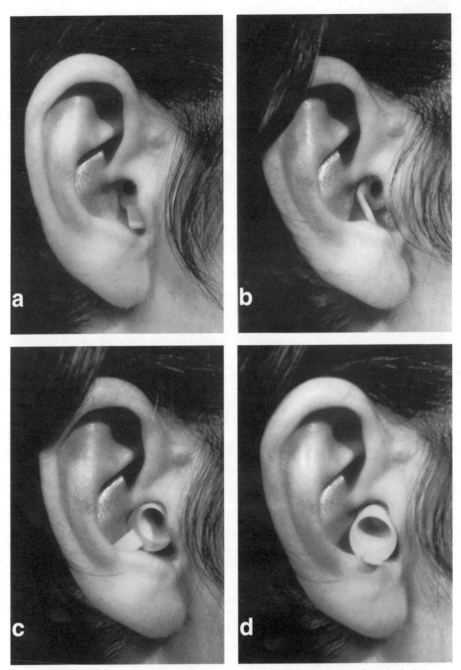

Figure 10.13 — Examples of different insertions of a V-51R earplug for which the preferred orientation for the tab is toward the rear, or occasionally downward, but never forward or upward: (a) plug is undersized or too deeply inserted, (b) properly inserted, (c) properly inserted, and (d) not inserted deeply enough and/or plug is oversized.

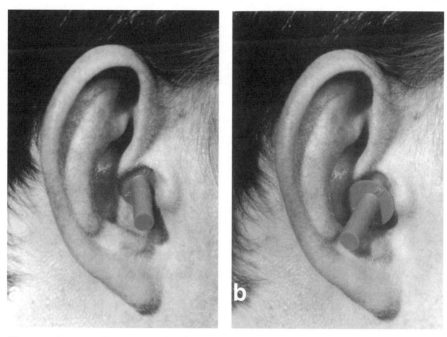

Figure 10.14 — Examples of different insertions of a three-sized three-flanged earplug: (a) properly inserted plug with outermost flange sealing entrance to canal, (b) not inserted deeply enough and/or oversized.

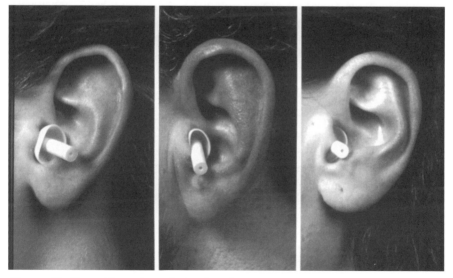

Figure 10.15 — Examples of the fit of a one-sized three-flanged earplug in extra-small (left, third flange just barely enters canal), medium (center, third flange seals canal), and extra-large earcanals (right, back edge of third flange is flush with canal entrance).

knead the mass into a conical shape so that it penetrates the canal will be of no value since the material lacks elasticity and unlike foam will not expand within the earcanal to form a seal. Also, manufacturers of these products warn against tearing off pieces of the presized slug of material since this could result in too small an earplug that could be forced too deeply into the canal and not easily removed.

Custom Molded Earplugs

Taking custom ear impressions is a more intricate and delicate process than the simple fitting of a foam or premolded earplug. Occupational hearing conservationists, technicians, or others who make such impressions should receive special one-on-one training from an audiologist or other professional specifically skilled in such procedures.

One of the most important steps in making a custom impression is that a cotton or foam block, also called an eardam, be placed deeply in the canal prior to injecting the material. This both protects the eardrum and ensures that a better-fitting impression will be taken (Kieper et al., 1991). The better fit results from the dam forcing the viscous material to press outwards against the canal walls to fully fill the earcanal for a tight fit and effective seal. In the absence of a dam the viscous material is permitted to simply flow further into the canal without ever being forced into contact with the canal walls (see Figure 10.2, upper left impression).

Recent research also suggests the benefits of using a moderate- or higher-viscosity impression material in order to provide proper "inflation" of the canal to obtain a suitably tight impression, and there is also evidence that for certain individuals taking an ear impression with the jaw open and teeth gripped on a bite block provides a better impression and more effective acoustical seal (Pirzanski, 1997). Lubrication is often helpful and sometimes necessary to learn how to properly insert a well-made custom molded earplug (see *Premolded Earplugs*, above).

Semi-Insert Earplugs

Semi-inserts generally are simply pushed into place at the entrance to the earcanal although the particulars vary for each product. For example, one type requires rolling and stretching of the pods prior to insertion, another suggests that pulling the pinna may be helpful, and others have specific orientations (left/right, top/bottom) that must be attended to if proper protection is to be attained. The occlusion effect, which is particularly noticeable for this type of product (see Figure 10.7), and the loudness tests are the best methods to check the fit of these devices.

A common complaint with semi-inserts is that when the band is rubbed or knocked an objectionable sound is heard in the ears. This is much more noticeable in less noisy environments. The best solution is careful orientation of the band for a position of least contact or rubbing. With those devices that permit it, this can often mean use in the over-the-head position; however, the most common wearing position for these types of devices is under the chin.

Earmuffs

Contrary to popular belief, earmuffs are not one-size-fits-all devices. The headband may not extend or collapse enough to fit all head sizes, and cup openings may not properly accommodate the largest ears. The contours in the circumaural areas of the wearer's head may be so irregular that the cushions cannot properly seal against them. Although earmuffs are generally easier to fit than earplugs, like earplugs, they must be individually dispensed and checked for fit to acquaint wearers with their features and make sure the HPDs are compatible with their anatomy.

Place the muff on the wearer's head and be sure the cups fully enclose, and are centered about the pinnas, without resting on them. Adjust the headband so that it sits comfortably on the head; the cushions exhibit approximately equal compression at all points around the cup, and the cushions feel to the wearer as though they exert evenly distributed pressure around the ears. Instruct users about the importance of achieving the best possible seal between the earmuff cushions and the side of the head. Caps and other head-worn gear must not interfere with this seal, and long hair should be pulled back and out from beneath the cushions. Some earmuffs have a preferred orientation, either left/right or top/bottom, and this must be pointed out to the employee.

Earmuff performance will be degraded by anything that compromises the cushion-to-circumaural-flesh seal. This includes other pieces of personal protective equipment such as eyewear, masks, face shields, and helmets. Eyeglass temples should fit close to the side of the head and be as thin as practical in order to reduce their impact on the cushions' seal around the ear. The loss in attenuation that temples create, with cushions in good condition, is normally 3 to 7 dB. The effect varies widely among earmuffs and the fit and style of the eyeglasses, with the losses in attenuation being reduced by thinner, closer-fitting temple pieces. Representative data are plotted in Figure 10.16; others have been reported in the literature (Nixon and Knoblach, 1974; Royster et al., 1996).

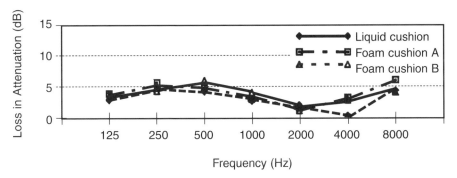

Figure 10.16 — Effects of correctly sized and fitted eyeglasses on the attenuation of three different earmuffs.

Wearing eyeglasses in combination with earmuffs may be uncomfortable for some wearers since earmuff cushions press the eyeglass temples against the skull. The pressure can be relieved by fitting foam pads over the temple pieces, but the increase in comfort may be at the expense of attenuation as has been demonstrated for one commercially available pad product (Berger, 1982). Also, pads do nothing to reduce acoustic leaks caused by over-length temples that break the cushion-to-skull seal behind the ear. However, temple pads should still be considered for use, since the improved comfort they can provide may be crucial in motivating certain employees to wear their HPDs.

Earmuff protection can be roughly checked by asking wearers to listen with earmuffs on while in the noisy environment in which they work. They should be able to detect a considerable difference in the apparent noise level as they lift and break the seal of one earmuff cup and then the other (similar to the loudness test for earplugs). If not, the earmuffs were either grossly misfitted, are in very poor condition, or the noise in which the people work is predominated by the lower frequency sounds for which earmuffs generally give less protection. Most listeners will not be able to detect small to modest degrees of misfit with this test since earmuffs will usually provide enough noise reduction, even when moderately misfitted, to be clearly distinguishable from the no-attenuation (i.e., the lifted-cup) condition.

Sabotage

An important aspect of issuing HPDs is to alert users that if devices are uncomfortable or problematic, they must return to the hearing conservation contact person for refitting, resizing, replacement, or alternative devices. Supervisors should also be alerted to this issue. Wearers must not modify HPDs on their own, since user modifications or willful abuse will generally degrade attenuation (Gasaway, 1984; Royster and Royster, 1985). Many workers don't realize the effects their alterations will have; others realize the effect, but don't understand how this may cause them to lose hearing due to noise.

Examples of employee modification include cutting the length or diameter of foam earplugs to fool the supervisor into thinking the plugs are properly inserted, removing flanges from or punching holes through the body of premolded earplugs or puncturing entrapped air pockets so that the plugs deflate upon insertion to relieve pressure and increase comfort, whittling or eliminating the canal portion of custom earmolds for a looser fit, springing the head or neck bands on semi-inserts or earmuffs, with a consequent reduction in applied pressure and attenuation, and in the case of earmuffs, drilling holes in the cups to promote ventilation and assist the drainage of perspiration. In some instances this latter practice has been so extreme as to include the personalizing of earmuff cups by drilling initials in the cup wall using many closely spaced holes (Riko and Alberti, 1982). In some workplaces employees have been observed with their own creative HPDs such as cotton balls, cigarette butts, cartridge casings, pencil erasers, plastic wrap, and other nonstandard devices. Obviously these and other similar behaviors must be strenuously discouraged.

Enforcement of Utilization

In addition to the proper selection, fitting, and issuing of HPDs, and education and motivation of employees to use their devices (see Chapter 8, Education and Motivation), another fundamental ingredient of the hearing protection phase of an HCP is enforcement of HPD utilization. In fact, some have argued that the rule at the foundation of every successful HCP is simply that everyone who enters a designated noise area must wear HPDs, no matter how brief the visit (Royster and Holder, 1982; Royster and Royster, 1985). To be effective the policy must be absolute. The enforcement must be in deed and in fact; a paper policy that is not enforced does more harm than good. Managers and supervisors need to actively enforce the policy and, similarly, employees should feel free to admonish their bosses and top executives who are found in designated areas without hearing protection.

Enforcement and support by all levels of management sets the tone for the entire program by demonstrating that hearing conservation is important to their company and to their jobs. Enforcement must be firm, equitable, and consistent. A four-step disciplinary process for failure to wear HPDs might consist of: (1) verbal warning, (2) written warning, (3) brief suspension without pay, and (4) termination. Although the latter steps are necessarily a form of discipline, the verbal warning can and should be positive, backed by genuine concern and a willingness to work with the employee to resolve utilization problems. Front-line supervisors should also be held responsible for the performance of their employees and must set a good example by regularly wearing their own HPDs. In union operations the policy should be included in the contract.

Measuring Attenuation

More than a dozen different methods of measuring HPD attenuation have been described in the literature, but only a few have been found practical and reliable, and are commonly utilized (Berger, 1986). They are described below. Whether implemented in the laboratory or the field, these methods yield an estimate of the noise reduction provided by the device(s) in question at a given point in time. Their shortcoming is that they cannot attest to how well the devices work day in and day out when worn for extended periods. Do the HPDs work loose? Are they often removed by wearers to give their ears a break or to hear "better?" Do they become uncomfortable and have to be readjusted?

Alternative methods of assessment based on evaluating the hearing levels of populations of wearers of HPDs provide a bottom-line estimate of effectiveness, i.e., are the HPDs protecting hearing in this population? However, these methods are less useful for describing the exact amounts of protection being provided, since values of protection can only be inferred from such results by also incorporating use of theoretical models relating noise exposure to hearing loss.

Examples of such alternative approaches are audiometric database analysis techniques (ANSI, 1991) as described in Chapter 12, and examination of absolute hearing levels of industrial populations (Bertrand and Zeidan, 1993; Gillis and Harrison, 1993). Regardless of the accuracy of any predictive scheme that matches HPDs to employee noise exposures, if the hearing levels of individuals or groups of employees are deteriorating faster than expected due to aging then additional measures must be instituted. Conversely, if the hearing levels are not changing any more rapidly than those of an appropriately selected nonindustrial noise-exposed reference population, this indicates a job well done.

Laboratory Procedures

Laboratory testing allows the experimenter to exert the most control over the relevant parameters of which s/he is aware, and provides the best assurance of generating repeatable test data. However, even the most accurate of the laboratory methods will only provide data representative of field (real-world) protection to the extent that: (a) the HPD is fit and worn by the subjects in the same manner as by wearers in practice, and (b) the test population is representative of the actual users. Classically, labeled attenuation values and manufacturers' test data have been based upon laboratory testing.

Real-Ear Attenuation at Threshold

The most common, and also one of the most accurate HPD attenuation tests is the measurement of real-ear attenuation at threshold, abbreviated REAT (ANSI S3.19-1974; ANSI S12.6-1997; ISO 4869-1:1990). REAT data are normally viewed as the gold standard to which all other types of measurements of attenuation are compared. Virtually all available manufacturers' reported data have been derived via this method, and it is this procedure that is required by the Environmental Protection Agency to obtain data for computation of the NRR (EPA, 1979).

REAT measures are based upon determination of the difference between the minimum level of sound that a subject can hear without wearing an HPD (open threshold) and the level needed when the HPD is worn (occluded threshold). The difference between these two thresholds, the threshold shift, is a measure of the REAT afforded by the device. Since the test is conducted at a relatively low sound pressure level (SPL), it cannot accurately characterize the performance of HPDs that claim to offer attenuation that increases with sound level. However, for linear HPDs (those not containing valves, orifices, diaphragms, or active electronic circuitry) the attenuation measured by an REAT evaluation will accurately represent the performance of the device regardless of sound level (Berger and Kerivan, 1983; Humes, 1983; Martin, 1979). Application of REAT data to high-level impulsive sources is somewhat questionable (ANSI, 1997; also see *Impulse and Weapon Noise*).

The only procedural artifact that is known to distort REAT results is masking of the occluded thresholds by physiological noise. This spuriously elevates the low-frequency attenuation (below 500 Hz) by a few decibels, with the error increasing as frequency decreases (Berger, 1986; Schroeter and Poesselt, 1986).

Since promulgation of ANSI S3.19 in 1974, U.S. REAT standards have specified that measurements are to be conducted in a diffuse sound field using 1/3 octave bands of noise and a minimum of 10 subjects whose open and occluded thresholds are measured three times each. This results in multiple data points at each frequency, from which a mean attenuation as well as a standard deviation (SD) are computed, the latter parameter providing an indication of the variability in attenuation across subjects and replications. The mean attenuation represents protection that approximately 50% of the test subjects meet or exceed. When it is appropriate to estimate the protection that a greater percentage of the subjects attain, adjustments to the mean may be computed by subtracting one or more standard deviations (see *Using Attenuation Data to Estimate Protection*).

Historically, hearing protector standards committees have been primarily concerned with the precision of HPD attenuation measurements, i.e., the repeatability of the data, both within and between laboratories. Although this is certainly an important issue, experience has shown that undue emphasis on precision and the attendant efforts to involve the experimenter to "control" the testing, with a consequent lack of regard for accuracy or "realism," may lead to data of questionable utility. By definition, "useful" data should predict with a reasonable accuracy the protection that can typically, or even optimally be expected to be achieved in practice (Berger, 1992).

Regardless of the attention of standards writing groups to issues of precision, the variability of test data across multiple measurements within a given facility ("repeatability"), or across measurements made in different facilities ("reproducibility") has been a problem. In one of the earliest interlaboratory REAT studies, the range in NRRs for three different earplugs and one earmuff as tested in eight laboratories was 11 dB or greater (Berger et al., 1982). The international standard on HPD attenuation measurements specifies repeatability ranges of about 3 dB and reproducibility across laboratories of about 5 dB for earmuffs and 7 dB for earplugs (ISO, 1990). The most recent interlaboratory study on this topic (Royster et al., 1996), which led to the development of the new HPD attenuation standard that includes a procedure with improved precision (ANSI, 1997; and see discussion of Method B, below) still found a range of attenuation values for three different earplugs and one earmuff, across four laboratories, of about 5 – 8 dB in the octave-band data. In terms of the NRR, the range was 3 – 8 dB for the earplugs and 2 dB for an earmuff. *With this in mind, it is clearly absurd for purchasers of hearing protectors to place great emphasis on small differences in the NRRs of competing products.*

The latest version of the U.S. standard provides a substantial departure from prior U.S. standards for measuring HPD attenuation in that it recognizes the importance of providing useful data (ANSI, 1997). This is accomplished by defining two testing procedures, Method A (experimenter-supervised fit) and Method B (subject fit). Method A corresponds most closely to tests using prior U.S. standards and is useful in the design of hearing protectors, to provide a theoretical understanding of their performance limitations, and for routine testing for quality assurance purposes. By contrast, Method B is intended to provide "an approximation of the upper limits to the attenuation that can be expected for groups of occupational users" (ANSI S12.6-1997, section 1.2). The standard goes on to acknowledge that properly trained and motivated individuals can potentially attain larger amounts of protection. However, subject-fit values provide a closer correspondence to real-world performance for groups of users than do the experimenter-supervised fit data. An example, from Berger et al. (1998) is shown in Figure 10.17 for a foam earplug.

Since the 1997 standard is recent and since the EPA hearing protector labeling regulation (EPA, 1979) still calls for testing to the 1974 standard, which utilizes a fitting procedure even more artificially controlled than Method A, namely a strict experimenter-fit interpretation of ANSI S3.19, it will likely be some time before users are routinely able to access 1997 Method-B data (see *Attenuation Characteristics of HPDs*).

Microphone in Real Ear

The microphone in real ear (MIRE) method utilizes physical (acoustical) measurements to determine either the difference between the sound pressure levels in the human earcanal, with and without the HPD in place, termed *insertion loss*, or the difference between the sound pressure levels outside and underneath

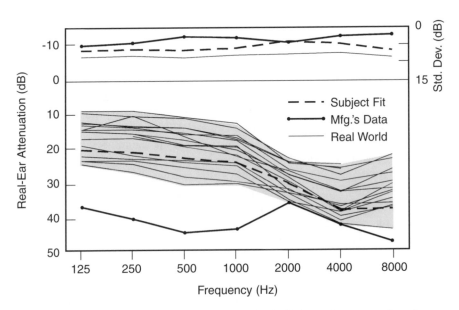

Figure 10.17 — Mean subject-fit data (Method B) for a foam earplug, from an interlaboratory study (Royster et al., 1996) and from manufacturers' labeled values, compared to 16 real-world studies. Individual real-world studies shown by thin lines w/out symbols (lower part of chart); thin line w/out symbols (upper part of chart) is average real-world SDs and shading shows range of real-world data.

the HPD, termed *noise reduction*. Of necessity, insertion loss must be measured sequentially for the protected and unprotected conditions, but noise reduction data can be acquired with two simultaneous measurements and is therefore ideally adapted for field use. For insertion loss, the levels are measured via a miniature or probe microphone at the entrance to, or sometimes in the earcanal itself (ANSI, 1995), and the subject who wears the HPD is required to sit still and behave as an inanimate acoustic test fixture. With the noise-reduction approach, the exterior microphone is generally mounted on the earmuff itself.

The insertion-loss form of MIRE measurements can be viewed as an objective type of REAT since in both cases a difference in levels is measured at one point in the auditory system. MIRE data do not suffer from the physiological-noise contamination discussed above with respect to REAT, and thus tend to show lower (more accurate) values of attenuation in the low frequencies. However, MIRE suffers some of its own artifactual errors in the higher frequencies (Berger, 1986).

MIRE can be used to measure attenuation for all types of hearing protectors including those fitted with electronics; difficulties can arise with earplugs because of the problems of inserting both a microphone and an earplug simultaneously into the earcanal (Smoorenburg, 1996). The MIRE method is much less time consuming than REAT testing and provides a fine alternative, especially when measurements are required at high sound levels such as needed for investigation of the response of HPDs to weapons fire and explosions. Although MIRE values are objective and thus obviate inconsistencies in the subjects' threshold responses, they do include their own measurement error due to positioning of the microphones; MIRE-derived SDs are less than or equal to those found with REAT procedures (Berger and Kerivan, 1983; Berger et al., 1996, Casali et al., 1995).

Acoustical Test Fixture

The acoustical-test-fixture (ATF) method is directly analogous to the MIRE method, except that an inanimate fixture is utilized in place of a test subject (ANSI, 1995). The ATF incorporates a microphone at the approximate eardrum location, and depending upon its sophistication may match the impedance of the eardrum (via a suitable acoustic coupler), and also include simulations of the circumaural and earcanal tissues.

ATF measurements provide a quick assessment for product development, quality control, and acceptance testing of earmuffs and helmets, and have also been successfully used to explore the level-dependent response of passive and active types of HPDs (Dancer et al., 1996). Although substantial gains have been made in ATF modeling and design (Kunov et al., 1986; Schroeter, 1986) measurements of earplugs are still problematical, and for all devices the data are generally not suitable for direct prediction of real-ear attenuation values. In all cases, modeling of the BC pathways is incorporated as a postmeasurement mathematical correction as the difficulties of physically incorporating the necessary mechanical elements to mimic the response of the human skull have not been overcome.

Field Procedures

Generally, the best of the laboratory procedures, modified to reduce cost and improve portability and ruggedness, are those that are utilized under field conditions. The purpose is to derive estimates of the attenuation provided by HPDs in actual use. The REAT procedures that have been most successfully applied are sound-field REAT in a manner similar to existing standards (ANSI S12.6-1997; ISO, 1990), and MIRE with microphones simultaneously mounted on the inside and outside of earmuffs (Stewart and Burgi, 1979). When REAT is implemented in the workplace, the sound field is typically achieved in a smaller acoustical chamber than used in the lab, or established using earphones mounted inside large circumaural cups instead of loudspeakers in a special acoustical chamber (Michael et al., 1976).

When REAT is implemented in the field for research purposes (see *Real-World Attenuation*), workers are usually selected without warning from their place of work, and accompanied to a test site with care taken to ensure they do not readjust their HPDs prior to testing. For the MIRE procedure, the instrumented earmuffs are utilized to simultaneously measure the protected and unprotected noise doses received by the exposed workers during typical work shifts. Since earplugs are the type of HPD most likely to be misused in practice, and since the REAT procedure with circumaural cups (i.e., an earplug tester) is the best suited to measuring that type of product, REAT with circumaural cups has garnered the most attention of the field methods and at least one device to accomplish this is now commercially available (Michael and Letowski, 1997). MIRE procedures with either miniature microphones or probe-based systems have also been used for earplug measurements but it is difficult to avoid spurious results since the microphone or probe will substantially interact with the earplug to alter its insertability and fit.

Today, the most readily implemented field procedure for checking earplug attenuation is use of standard supra-aural audiometric earphones to give an audiogram with and without the earplug in place. Conceptually this is identical to the circumaural earcup procedure described above. However, the earphone and cushion assembly directly rest on the pinna, which causes the procedure to be prone to errors, potentially the greatest being effects on the fit of the earplug either via direct contact between the earphone and earplug, or distortion of the pinna and concha by the earphone-cushion assembly. Nevertheless, the data from one pilot study suggest that a criterion of $\geq$ 15 dB measured attenuation at 500 Hz can provide reasonable assurance (about 94% of the 32 cases that were tested) of achieving a sound-field attenuation of at least 10 dB at 125 Hz and 15 dB or more at and above 500 Hz (Berger, 1984).

Field techniques such as the circumaural-earcup or even the audiometer-earphone procedure can be valuable in providing feedback and training to employees in better fitting their HPDs, furnishing a tool for follow-up on HPD fitting when an OSHA standard threshold shift (STS) is detected, helping fitters gain experience in training HPD users, facilitating department audits of HPD effectiveness, and in developing quantitative information to document effectiveness in HPD training and usage for potential future litigation and workers' compensation.

Attenuation Characteristics of HPDs

The attenuation provided by a hearing protector is strongly controlled by the way in which the wearer and/or test subject fits the device, especially for insert-type devices (Casali and Lam, 1986). This leads to a wide variation in reported attenuation, even in the case of purportedly "optimum-fit" data measured in the laboratory (Berger et al., 1982). Thus, comparison of published attenuation values or single-number ratings such as NRRs and rank ordering of HPDs can be substantially influenced as a function of the data that are being compared. Precise rank ordering is meaningful only for data from a single laboratory. Sources of interlaboratory variability are (1) differences in interpretation and implementation of the standardized measurement methodology, (2) uncertainty of obtaining the proper fit to avoid acoustic leaks, (3) differences in subject selection and training, and (4) differences in data reduction techniques. The repeatability of attenuation measurements in the same laboratory is also subject to variation, but to a lesser extent than between laboratories.

Specification of the attenuation provided by an HPD is fraught with difficulty because of the importance that the user and the method of fit have upon the outcome. With this in mind, data will be provided both for well-fitted hearing protectors under ideal laboratory conditions, as well as for typical field performance. The values reported will be based upon REAT measurements unless otherwise specified. In selecting devices it is important to realize that *differences of less than 3 dB in single-number ratings like the NRR have no practical importance*, and even 4- to 5-dB changes are of questionable significance unless closely controlled data are being compared (see *Real-Ear Attenuation at Threshold*).

Optimum Attenuation

The values reported in this section and in Table 10.2 are representative of those obtained in laboratory-based REAT measurements, wherein the HPDs are either fitted by, or under the close supervision of, the experimenter. Highly motivated and trained individual users can potentially achieve the upper end of the attenuation values listed for each device in Table 10.2, but average values for groups of users will fall near or below the lower end of the ranges and will be better approximated by the data in the section on *Real-World Attenuation*.

There is a wide degree of overlap in attenuation among the various types of earplugs, but in general, slow-recovery foam earplugs provide among the highest overall protection of any single device. Their attenuation can range from 30 to 45 dB at and above 2000 Hz, and depending upon the depth of insertion, from about 20 to 40 dB below that frequency. For many users even the higher values of attenuation can be achieved while still retaining an excellent degree of comfort. Attenuation of custom-molded earplugs can vary widely due to impression-taking and manufacturing procedures, as well as differences in impression materials. On average, the attenuation of these devices tends to be similar to premolded and formable earplugs.

TABLE 10.2

Representative minimum and maximum mean attenuation values of well-fitted hearing protectors under laboratory conditions, in dB. Data are intended to account for brand and testing variability; however, not all manufacturers' reported data or values referenced in the literature will necessarily fall within the ranges cited. All data are from E·A·RCAL Laboratory, except for the shot-blasting helmets (Price and Whitaker, 1986) and fingers (Holland, 1967).

Type of Hearing Protector	Octave-Band Center Frequency (Hz)						
	125	250	500	1000	2000	4000	8000
Foam earplugs*	20 - 40	20 - 40	25 - 45	25 - 45	30 - 40	40 - 45	35 - 45
Premolded earplugs	20 - 30	20 - 30	20 - 30	20 - 35	25 - 35	30 - 45	30 - 45
Formable (fiberglass)	20 - 30	20 - 30	20 - 30	25 - 30	25 - 30	35 - 40	35 - 40
Formable (wax or silicone)	20 - 25	20 - 25	20 - 25	25 - 30	30 - 35	40 - 45	40 - 45
Custom-molded earplugs	15 - 35	15 - 35	15 - 35	20 - 35	30 - 40	35 - 45	30 - 45
Semi-insert earplugs	15 - 30	15 - 30	10 - 30	15 - 30	25 - 35	25 - 45	30 - 45
Earmuffs**	5 - 20	10 - 25	15 - 40	25 - 45	30 - 40	30 - 40	25 - 40
Military helmets	0 - 15	5 - 15	15 - 25	15 - 30	25 - 40	30 - 50	20 - 50
Dual protection (earplugs + earmuffs)	20 - 40	25 - 45	25 - 50	30 - 50	35 - 45	40 - 50	40 - 50
Active noise reduction***	15 - 25	15 - 30	20 - 45	25 - 40	30 - 40	30 - 40	25 - 40
Cotton balls	0 - 5	0 - 10	5 - 10	5 - 10	10 - 15	10 - 20	10 - 20
Motorcycle helmets	0 - 5	0 - 5	0 - 10	0 - 15	5 - 20	10 - 30	15 - 35
Air-fed shotblasting helmets	0 - 5	0 - 5	0 - 5	0 - 15	15 - 25	15 - 30	15 - 25
Finger tips in earcanals	25 - 30	25 - 30	25 - 30	25 - 30	25 - 30	30 - 35	30 - 35

*Attenuation varies with depth of insertion
**With or without communications components
***Closed-cup systems; identical to conventional muffs above 1 kHz

The attenuation of earmuffs usually increases about 8 – 9 dB/octave from 125 Hz to 1000 Hz, and at 2000 Hz, approaches the limit imposed by bone conduction of approximately 40 dB (cf. Figure 10.6). Above that frequency, attenuation averages around 35 dB. For a discussion of the effects of eyeglasses on earmuff attenuation see *Fitting and Issuing, Earmuffs*. In general, earplugs, especially the roll-down foam variety, provide better attenuation than earmuffs below 500 Hz, and equivalent or greater protection above 2000 Hz. At 1000 Hz, larger-volume earmuffs will usually provide attenuation that exceeds that of earplugs.

Helmets, unless they seal well around the head and/or contain internal circum-aural cups, provide very little attenuation, especially at the frequencies below 2000 Hz. As shown in Table 10.2, typical motorcycle helmets provide less than 10 dB of protection below 1000 Hz and only from 10 – 35 dB at the higher frequencies, and air-fed shotblasting helmets perform similarly. Some military helmets (for example, flight helmets), which possess an effective acoustical design, yield attenuation that is comparable to conventional earmuffs, but still about 5 – 10 dB less at the frequencies below 2000 Hz. From 4 – 8 kHz, military helmets can actually surpass earmuff attenuation by about 5 dB if they fully enclose the head, since this decreases bone-conduction transmission at high frequencies.

The data in Table 10.2 suggest that in a pinch, for brief exposures, fingers can provide high levels of protection equivalent to that of most hearing protectors.

Real-World Attenuation

In the latter part of the 1970s researchers began to investigate the amount of protection that HPD users were actually achieving in the workplace, typically called *field or real-world attenuation* (see *Field Procedures*). Field attenuation can be measured on-site by taking instrumentation to the workplace, or off-site, by having employees report with their own HPDs to an independent test facility. By 1997 there were at least 23 available studies providing measurements of real-world attenuation (Berger et al., 1996; Berger et al., 1998). Those studies span greater than 90 industries in seven countries, with a total of approximately 3000 subjects. For additional details and a list of the available field studies see Berger et al. (1996). The data from 22 of the studies, which for purposes of simplification are expressed in terms of the NRR, are summarized using dark bars in Figure 10.18. Similar findings are also apparent in the underlying OB data from which the NRRs are computed as presented in Figure 10.19.

In Figure 10.18 the devices are grouped into two general categories, earplugs on the left, and earmuffs on the right. Devices and/or device types were selected to assure a minimum sample size of greater than 30 subjects (summed across

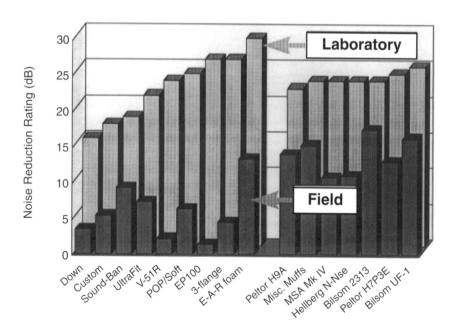

Figure 10.18 — Comparison of NRRs published in North America (labeled values based upon laboratory tests), to real-world "field" attenuation results derived from 22 separate studies.

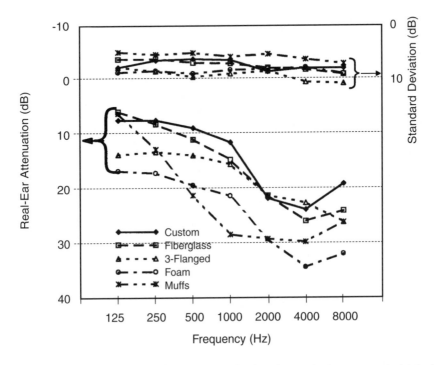

Figure 10.19 — Summary real-world data for hearing protectors separated into five categories (after Berger et al., 1996).

studies) for each data bar shown. For some categories the sample size was very large, as in the case of the E-A-R® foam earplug, for which the data represent 633 subjects from 15 studies. The 90 industries that were studied in order to generate the results shown in Figure 10.18 probably represent today's better HCPs. This presumption is based upon the increased likelihood of finding higher-quality HCPs among companies interested in and choosing to participate in the complicated and time-consuming research of the type required for real-world evaluations (Berger et al., 1996).

For purposes of comparison to the field data, Figure 10.18 also provides the associated labeled NRRs, shown by the lighter-colored bars. The labeled values are based upon manufacturers' published North American laboratory results which, as required by the EPA, are measured according to a strict experimenter-fit interpretation of ANSI S3.19-1974 (see *Real-Ear Attenuation at Threshold*). The procedure calls for determining "optimum performance values *which may not usually be obtained under field conditions*." ANSI S3.19-type data are the only standardized values that regulators and manufacturers currently have available (circa 1998) for labeling and informational purposes.

The most obvious feature of Figure 10.18 is the very poor correspondence between the magnitude of lab and field NRRs. Measured as a percentage of the laboratory-rated attenuation, the field NRRs for earplugs other than foam yield only about 25% of the labeled values, foam provides about 40%, and earmuffs about 60%. However, *not only do the absolute values disagree, but so too do the relative rankings.* Although the labeled values are arranged in ascending order from left to right within each category in Figure 10.18, the same does not hold true for the field data. Furthermore, the labeled values suggest that earplug attenuation is typically equivalent to or greater than that of earmuffs, whereas the field data indicate otherwise. With the exception of the foam earplug, only earmuffs can generally be expected to provide 10 dB or more of real-world protection for 84% of the exposed population.

The labeled NRRs in Figure 10.18 were computed, as per the EPA (1979), by subtracting a 2-SD correction from the mean attenuation values in order to estimate the minimum noise reduction theoretically achieved by *98% of the laboratory subjects (NRR_{98}).* The field data were computed in the same manner except that only a 1-SD correction was included, thus estimating the minimum attenuation achieved by *84% of the actual wearers (NRR_{84}).* A 1-SD correction simplifies examination of real-world data, since the 2-SD correction used in the labeled NRRs (i.e., NRR_{98}s) would cause many field NRRs to become negative numbers. Further justification for using a 1-SD correction with real-world NRRs is based upon practical, psychophysical, and statistical considerations (Berger et al., 1996). Also, see *Using Attenuation Data to Estimate Protection.*

To fully appreciate the meaning of the very small real-world NRRs reported in Figure 10.18, one must recall that the field NRRs in Figure 10.18 do not represent average attenuation values, but rather represent the minimum attenuation achieved by at least 84% of the users. Since the NRR_{84} includes a subtractive 1-SD correction, which usually amounts to 8 dB or more for earplugs, the attenuation achieved by half of the wearers (NRR_{50}), which is computed with a 0-SD correction, is about 8-dB larger. For example, a real-world NRR_{84} of 4 dB is typically equivalent to an NRR_{50} of 12 dB (see Figure 10.20). And, in the example of Figure 10.20, if one asks what is the protection achieved by the top few percent of the wearers, approximately another 2 SDs (16 dB) must be added, yielding an NRR of 28 dB.

Wearing Time

The real-world attenuation reported in the prior section applies under the presumption the device is actually being worn. The effective protection of a device is substantially reduced when it is removed for even short periods of time during the workday, as illustrated via use of the NRR and a 5-dB exchange rate, in Figure 10.21. Similar effects would apply to any other attenuation metric that was utilized, with greater losses in attenuation for computations using a 3-dB instead of a 5-dB exchange rate. An HPD with a nominal NRR of 25 dB that is removed for 15 minutes during an 8-hour workday provides an effective NRR of only 20 dB — a loss of 5 dB. Greater reductions in effectiveness are observed for higher NRRs. Little effect is observed for NRRs of less than 10 dB. The les-

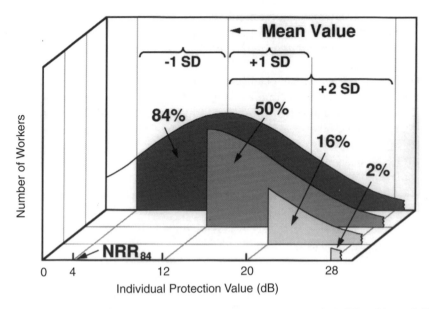

Figure 10.20 — Distribution of individual protection values (NRR with a 0-SD correction) for a group of real-world users with an NRR$_{84}$ of 4 dB, i.e. 84% achieve at least 4 dB of protection. The SD is 8 dB and the average NRR is 12 dB (NRR$_{50}$). Note: 16% and 2% of the users, respectively, exceed NRRs of 20 and 28 dB.

son is: Teach employees that they must don hearing protection prior to entering the noise, and that removal while in the noise can seriously degrade its usefulness. Because of this effect, a device with lower attenuation (and perhaps greater comfort), if worn consistently and correctly, can provide greater protection than a less-regularly applied higher-attenuation HPD.

Dual Protection

Dual protection, such as earplugs worn in combination with earmuffs, helmets, or communications headsets, typically provides greater protection than either device alone (Berger, 1983b; Damongeot et al., 1989). However, the attenuation of the combination is not equal to the sum of the individual attenuation values, as illustrated in Figure 10.6. No empirical or theoretically derived equations are available that can predict the attenuation of an earplug and earmuff combination with sufficient accuracy to be useful. Dual protection is advisable when 8-hour time-weighted-average exposures exceed 105 dBA, but is often problematical for exposures below that level due to interference with communication and the difficulty of motivating and enforcing use of dual HPDs in lower sound levels. The use of dual HPDs is especially recommended when high-intensity noise is dominated by energy at or below 500 Hz since it is in this frequency range that the attenuation of single HPDs will be the least and the potential gains from dual protection are the greatest.

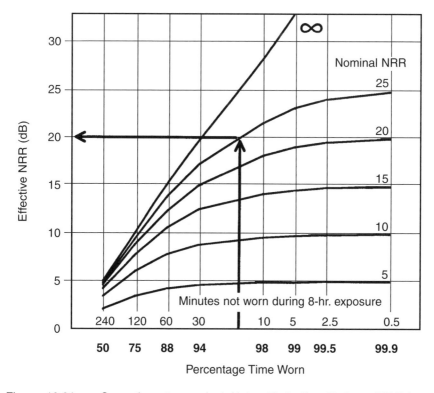

Figure 10.21 — Corrections to nominal Noise Reduction Ratings (NRRs) as a function of the time during a workday that an HPD is not worn, based on a 5-dB trading relationship.

At individual frequencies the incremental gain in performance for dual hearing protection varies from approximately 0 to 15 dB over the better single device, but at 2000 Hz the gain is limited to only a few decibels. Attenuation changes very little when different earmuffs are used with the same earplug, but for a given earmuff the choice of earplug is critical for attenuation at frequencies below 2000 Hz. At and above 2000 Hz, all dual-protection combinations provide attenuation essentially equal to the limitations imposed by the bone-conduction pathways, approximately 40 to 50 dB, depending upon frequency. As a rule of thumb, the OSHA procedure of computing the dual protection by adding 5 dB to the NRR of the more protective of the two devices is a reasonable approximation.

Infrasound/Ultrasound

REAT data are normally limited to the frequency range of 125 Hz to 8 kHz, although some laboratories test to frequencies as low as 63 Hz, especially in Europe. However, a few published reports are available from which to draw conclusions. Well-fitted insert earplugs provide attenuation at infrasonic frequencies (i.e., below about 20 Hz) that is about equal to that in the 125 Hz 1/3 octave band. However, at those same frequencies, earmuffs provide very little protection and may even amplify sound (Nixon et al., 1967; Paakkonen and Tikkanen, 1991). Conventional earplugs and earmuffs generally provide adequate protection at ultrasonic frequencies (i.e., above about 20,000 Hz), with attenuation exceeding 30 dB for frequencies from about 10,000 to 30,000 Hz (Behar and Crabtree, 1997; Berger, 1983a).

Impulse and Weapon Noise

Estimation of the attenuation provided by HPDs in the presence of impulsive noise is one of the most problematic issues in hearing protection. Not only are the damage-risk criteria (DRC) for impulse noise being reexamined, but application of those criteria to hearing-protected ears is questionable since the time history of a waveform under an HPD is so different from the free-field conditions under which the DRC were derived (see Chapter 5). Under the HPD, especially under earmuffs, the waveform lacks the sharp rise time of the incident pulse, and generally also exhibits substantial low-frequency components not present in the original (Dancer et al., 1996).

Although a number of studies (Ylikoski et al., 1995) have used MIRE techniques to measure either the noise reduction (NR) (i.e., the differences in SPLs between sound-field and earcanal mounted mics) or the insertion loss (IL) (i.e., the difference in SPLs between the protected and unprotected conditions as measured in the concha or earcanal), the meaningfulness of the resultant measurements is in question in that such measurements significantly underestimate the actual protection provided in terms of reduction of temporary threshold shift (TTS) (Johnson and Patterson, 1992). MIRE measurements indicate attenuation of peak SPLs for typical earmuffs of about 30 dB for pistol fire, about 18 dB for rifle fire, and as little as 5 dB for cannons. Foam earplugs were found to provide similar attenuation as earmuffs against pistols, but substantially more noise reduction than earmuffs against cannons and bazookas, providing attenuation greater than 15 dB. Dual HPDs were found to afford extra protection especially for low-frequency impulses with attenuation values of 20 to 25 dB (Ylikoski et al., 1995). These values also correspond with predictions based upon use of conventional REAT data applied to the OB spectrum of the impulses (Berger, 1990).

In a recent large-scale multilaboratory European study that examined earmuff performance in industrial impulse noise arising from pneumatic nailers, hammers, punch presses, and plastic explosives, it was concluded that for both active and passive earmuffs, the impulse noise attenuation tends to be larger than or equal to the attenuation found for steady broadband noise at lower levels (Smoorenburg, 1996). In a separate study, Lloyd (1996) drew similar conclusions. The only exception that Smoorenburg observed to this finding was in the case of explosions that were of sufficient intensity to actually cause blast-induced movements of the earcups that would break the acoustic seal of the earmuff against the side of the head. Johnson and Patterson (1992) have documented similar cup motion in the presence of 190-dB explosive impulses intended to mimic high-level military exposures.

A point of confusion has existed since 1979 as a result of mandated wording on the EPA hearing protector label that states,

> "Although hearing protectors can be recommended for protection against the harmful effects of impulsive noise, the NRR is based on the attenuation of continuous noise and may not be an accurate indicator of the protection attainable against impulsive noise such as gunfire."

Many misread this to suggest that HPDs cannot attenuate gunfire. To the contrary, HPDs can and do attenuate gunfire. The purpose of this caution, when drafted, was to suggest that certain HPDs, especially amplitude-sensitive devices (see *Amplitude-Sensitive HPDs*) might work better in gunfire than the REAT-derived NRR would suggest, and indeed, current research supports such a conclusion.

It would seem that the best estimates of HPD performance in impulse and weapon noise should be based on studies of TTS in the presence of actual impulses. Johnson and Patterson (1992) and Dancer et al. (1996) have provided such data and they suggest that a variety of conventional hearing protectors, even with intentional acoustical leaks introduced, can provide sufficient protection for even high-intensity weapon noise. A cautionary note was voiced by Johnson for protection from very high-level impulses (above 180 dB) using earplugs. He observed that the earplug type that was most likely to ensure an acoustical seal in the earcanal was the safest choice, namely a foam earplug (Johnson, 1997). With respect to analytical predictions, application of REAT values to the peak of the impulse with the requirement that it be reduced below 140 dB peak SPL is overly conservative. Instead, Dancer et al. have suggested use of attenuation values based on IL measurements that are applied to exposures assessed in terms of $L_{Aeq,8}$ values as the best conservative approximation at this time.

Using Attenuation Data to Estimate Protection

Depending upon the purpose of the predictions, either laboratory or real-world attenuation data may be used to estimate the protected noise exposures that wearers receive. But first, a word of caution. The accuracy with which the laboratory or real-world data can predict the protection for an individual or group of users is critically dependent upon the correspondence between the test and actual usage conditions — garbage in, garbage out. The accuracy of predictions can be very poor as is clearly shown when optimum-laboratory data are used to estimate protection in a typical hearing conservation program (cf. Figure 10.18). Therefore, excessive emphasis on protection estimates is unwarranted. Furthermore, since the noise-exposure measurements themselves also have a limited precision, some authors recommend categorizing exposure estimates into ranges as wide as 5 dB (see Table 7.1 this text, and Royster and Royster, 1985). At time-weighted average levels below about 95 dBA, it is more important to focus on the facilitation and enforcement of proper HPD usage than on the precise determination of HPD attenuation.

427

Octave-Band Method

Potentially the most accurate computational procedure for using HPD attenuation values to estimate protected exposures is the octave-band (OB) method, also called NIOSH Method #1 (Kroes et al., 1975) or the Long Method, as illustrated in Table 10.3. The inherent accuracy of this method can only be realized if the data that are used for the computations, namely the HPD attenuation values and the OB noise measurements, are reliable (see *Real-World Attenuation* and Chapter 7). This method is used as the gold standard by which other computational methods are judged.

At each frequency, the HPD's mean attenuation value minus an SD correction (2 SDs are used in Table 10.3), is subtracted from the measured A-weighted OB SPLs. The protected levels are then logarithmically summed to determine the A-weighted sound level effective when the protector is worn. This computation requires that the user perform the appropriate calculations for each individual noise spectrum, i.e., the amount of protection afforded cannot be calculated independently of the noise in which the HPD is worn.

TABLE 10.3
Octave-band method of calculating HPD noise reduction.

| | Octave-Band Center Frequency (Hz) | | | | | | | dBA* |
	125	250	500	1000	2000	4000	8000	
1. Measured sound pressure levels	85	87	90	90	85	82	80	
2. A-weighting correction	-16.1	-8.6	-3.2	0.0	+1.2	+1.0	-1.1	
3. A-weighted sound levels (step 1 + step 2)	68.9	78.4	86.8	90.0	86.2	83.0	78.9	93.5
4. Attenuation, typical premolded earplug	27.4	26.6	27.5	27.0	32.0	44.0†	42.2††	
5. Standard deviation x 2	7.8	8.4	9.4	6.8	8.8	7.0†	10.4††	
6. Estimated protected A-weighted sound levels (step 3 - step 4 + step 5)	49.3	60.2	68.7	69.8	63.0	46.0	47.1	73.0

The estimated protection for 98% of the users in the noise environment of this example, assuming they wear the device in the same manner as did the test subjects and assuming they are accurately represented by the test subjects, is **93.5 - 73.0 = 20.5 dBA**
*Logarithmic sum of the seven octave-band levels in the row.
† Arithmetic average of 3150- and 4000-Hz data.
††Arithmetic average of 6300- and 8000-Hz data.

A critical conceptual error that is often made is to presume that the SD correction adjusts laboratory data to estimate real-world values. The actual purpose of the SD is to adjust the mean test data to reflect the attenuation achieved by 84 percent (for a 1-SD correction) or 98 percent (for a 2-SD correction) of the test subjects. The applicability to actual users is determined by the extent to which the laboratory test properly modeled field performance.

Noise Reduction Rating (NRR)

A single-number descriptor is convenient and often sufficiently accurate to estimate protected exposures. The descriptor that has been standardized in the U.S. since 1979, and hence precalculated by manufacturers and provided on their packaging, is the NRR (EPA, 1979). It is an attenuation index that represents the overall average noise reduction, in decibels, that an HPD will provide in an environment with a known C-weighted sound level. The NRR is identical (within 0.5 dB) to NIOSH Method #2, from which it was adapted (Kroes et al., 1975).

The NRR is calculated in a manner analogous to the OB approach, except that a pink noise spectrum (equal energy in each OB) is used instead of the actual noise spectrum (line 1 in Table 10.3), the estimated protected A-weighted levels are subtracted from the C-weighted pink noise and not the A-weighted environmental noise, and an additional spectral safety factor of 3 dB is subtracted. The spectral factor accounts for errors arising from use of pink noise instead of the actual noise spectrum to which the wearer is exposed. As in the OB method, the computation incorporates a 2-SD adjustment for percentage of population protected (i.e., theoretically a 98% protection factor, sometimes explicitly denoted as NRR_{98}). The NRR computed for the example in Table 10.3 is 20.7 dB. For the full computational details of the NRR see EPA (1979) or Berger (1988).

The NRR is used to estimate wearer noise exposures by subtracting it from the C-weighted sound levels as shown in Equation 10.1.

$$\text{estimated exposure (dBA)} = \text{workplace noise level (dBC)} - \text{NRR} \qquad \textbf{(10.1)}$$

The practice of subtracting the NRR from the C-weighted sound level may seem illogical; however, it is justified on theoretical and empirical grounds (Berger, 1988). Considerable accuracy is lost when the NRR is subtracted from A-weighted sound levels, in which case an additional safety factor of 7 dB must be included, as shown in Equation 10.2. This method has been referred to as NIOSH Method #3 (Kroes et al., 1975).

$$\text{estimated exposure (dBA)} = \text{workplace noise level (dBA)} - (\text{NRR} - 7 \text{ dB}) \qquad \textbf{(10.2)}$$

The 7-dB safety factor in Equation 10.2 is a "worst-case" correction that in most situations will overestimate actual C-A differences. As an alternative, one can correct the A-weighted TWA (obtained from dosimeters which are usually only capable of A-weighted measurements) by using a sound level meter to develop a C-A value for typical processes, areas, or job descriptions. This C-A value is added to the A-weighted TWA to calculate an estimated C-weighted TWA or C-weighted workplace noise level, from which the NRR can be subtracted using the procedure of Equation 10.1. To the extent that an accurate C-A value can be estimated, this method will provide enhanced accuracy over use of Equation 10.2 for those situations in which C-weighted exposures are unavailable.

Comparison of OB and NRR Estimates

Due to simplifications inherent in the NRR, errors may arise in using it to estimate protection, but the values are usually sufficiently close to predictions made using the OB method when one considers the inaccuracies in the basic OB data from which either method of computation must proceed. For the example shown in Table 10.3, with environmental C- and A-weighted sound levels of 95.2 and 93.5 dB respectively, the OB-computed protected exposure level is 73.0 dBA, as compared to the NRR-computed values of

$$95.2 \text{ dBC} - 20.7 = 74.5 \text{ dBA, using Equation 10.1, or} \qquad \textbf{(10.3)}$$

$$93.5 \text{ dBA} - (20.7 - 7) = 93.5 - 13.7 = 79.8 \text{ dBA,} \qquad \textbf{(10.4)}$$
using Equation 10.2

Note that the NRR prediction using Equation 10.3 is only 1.5 dB more protective than the OB-computed level, and that there is a substantially larger discrepancy observed using Equation 10.4. This is due to the very conservative 7-dB adjustment required when only A-weighted sound levels are available. NRR predictions will not always agree so closely with the OB method, especially for steeply sloping noise spectra, but the potential errors in use of the NRR and presumed increase in accuracy of the OB method must be balanced against the large discrepancies between laboratory and real-world data. The primary source of error in applying the NRR to estimate user noise exposures is not in its computation from the basic laboratory data, but in the fact that laboratory data are not suitably representative of the values attained by actual users (see *Real-World Attenuation*).

If one can assure the similarity of labeled and real-world attenuation for the particular user or group of users in question, and if one has the OB data available for the noise environment, then the OB method is preferred. In most cases, however, it is unlikely that either or both of these "ifs" will be satisfied. Most users complain of the need to measure C-weighted sound levels for use with the NRR, let alone considering the additional effort involved in conducting an octave-band analysis.

The primary utility of examining the OB attenuation values for an HPD is to be able to make a gross match between the device and the noise spectrum. For example, both the laboratory and real-world data show that if significant low frequency energy is present (125 – 250 Hz), then an earmuff is a poor choice and a foam or perhaps a premolded earplug would be better. Conversely, if significant energy is present around 1 kHz, then an earmuff is preferred. The desire to perform this type of spectrum matching must be tempered by the realization that assigning particular HPDs to particular jobs in a plant is often impractical. It is difficult enough to ensure that HPDs are worn, and worn correctly, without also having to keep devices from being shifted among work areas with different spectral characteristics and having to ensure that certain devices are worn only by employees exposed to particular spectra.

Noise Reduction Rating (Subject Fit) [NRR(SF)]

In 1995 a new rating method was proposed in the U.S. by a multi-organizational Task Force on Hearing Protector Effectiveness convened under the auspices of the National Hearing Conservation Association (NHCA) (Berger and Royster, 1996; Royster, 1995). The rating, designated the Noise Reduction Rating (Subject Fit) [NRR(SF)] was intended to function like the NRR except that it is based upon data derived according to Method B of ANSI S12.6-1997 (see *Real-Ear Attenuation at Threshold*), hence the subject-fit designation. The goal was to produce an easily applied single-number rating that would indeed provide a useful indication of the protection that could be obtained in actual HCPs.

Besides the fact that the NRR(SF) uses different input data (i.e., Method B) than the NRR, it is computationally similar, with the following exceptions:

a) The NRR is computed with a subtractive 2-SD correction, whereas the NRR(SF) is computed with a subtractive 1-SD correction. For subject-fit data of the type provided by Method B, 1 SD amounts to about 3 – 12 dB. Use of a 1-SD correction (smaller, and more appropriate for use with real-world type data), instead of the 2-SD (larger) correction in existing NRRs, offsets to some extent the change from best-fit (NRR) to subject-fit [NRR(SF)] data.

b) Although the actual computations involved in the NRR(SF) and the NRR are nearly the same, the NRR(SF) is based on the Single Number Rating (SNR) procedure in ISO 4869-2 (ISO, 1994), whereas the NRR is based on prior NIOSH work and the EPA labeling regulation (EPA, 1979). Even if the same set of octave-band attenuation values is used as the input data for both single-number calculations, and the same number of SDs are subtracted (for example a 1-SD correction in both instances), small differences between the methods will cause the computed NRR(SF) to exceed the computed NRR by 3.5 dB.

c) The NRR(SF) is intended to be subtracted from A-weighted values (see Equation 10.5).

estimated exposure (dBA) = workplace noise level (dBA) – NRR(SF)   **(10.5)**

By comparison, both the NRR and the SNR are designed to be subtracted from C-weighted values. Although use of single-number ratings with A-weighted values gives rise to a loss in accuracy (see prior two sections of this chapter), the Task Force decided that the increased likelihood of correct application by more users, was the governing concern. To permit use with A-weighted decibels, with no loss in safety, the NRR(SF) includes a built-in 5-dB adjustment, i.e. NRR(SF)s are 5 dB less than SNRs computed from the same set of data. The 5-dB correction is less stringent than the 7-dB value used by NIOSH and OSHA for adjusting the NRR for use with A-weighted decibels [see *Noise Reduction Rating (NRR)*]. The 5-dB adjustment can (and should) be eliminated when the NRR(SF) is ("correctly") subtracted from C-weighted values. In that case, add 5 dB to the NRR(SF) before subtracting it from a dBC exposure measurement, as in Equation 10.6.

estimated exposure (dBA) = workplace noise level (dBC) − [NRR(SF) +5 dB] **(10.6)**

In summary, differences between the new NRR(SF) and the EPA's NRR will vary by product, depending upon the relationship of the EPA experimenter-fit test data to the new Method-B subject-fit data. The divergence will also be affected by the change from a 2-SD to a 1-SD correction, and the constant offset of 3.5 - 5.0 = -1.5 dB as discussed in paragraphs b) and c) above. This means that if both the NRR(SF) and the EPA NRR were computed using Method-B data and with a 1-SD correction, the relationship between the two would simply be NRR(SF) = NRR - 1.5 dB. However, when they are both computed as intended, the new NRR(SF) will be less than the EPA NRR (computed using ANSI S3.19 data and a 2-SD correction), by amounts varying from about 2 – 20 dB, with the differences being less for earmuffs than for earplugs.

As of the date of publication of this book (mid-year 2000) the NRR is still predominant and the NRR(SF) is rarely utilized. Since the EPA lacks the staff at this time to reopen the rule-making process for the hearing protector labeling regulation, it is unlikely the NRR(SF) will become law in the near future. However, users can insist on obtaining such values from hearing protection vendors and through their purchasing power create change on their own.

Other Rating Schemes

Other rating systems in use today include the Single Number Rating (SNR) and the High-Medium-Low (HML) methods, which are popular in Europe; the Sound Level Conversion (SLC$_{80}$), which is popular in Australia; and class systems, which are used in certain Canadian provinces and in New Zealand.

The SNR, which is computed similarly to the NRR, is described in ISO 4869-2 (1994). Like the NRR, it is intended to be subtracted from a C-weighted sound level, but unlike the NRR, the SNR may be computed with a range of SD subtraction factors. Normally 1-SD is preferred instead of the 2-SD embodied in the NRR. When the SNR and NRR are computed using the same SD factor, the SNR will exceed the NRR by a constant value of 3.5 dB due to small inherent computational differences. The SLC$_{80}$ (SAA, 1988) is the other principal single number rating in use today and, although computed in a slightly different manner from the NRR and SNR, it too is subtracted from C-weighted sound levels to predict A-weighted exposures. The subscripted value of 80 on the SLC is explicit indication that it is intended to predict a value of attenuation achieved by 80% of the users.

The HML is a three-number method that rates HPDs with high-, medium-, and low-frequency attenuation indices (ISO, 1994). Since the NRR is a single-number method and the OB procedure can be viewed as a seven-number method (i.e., it uses seven OBs), the three-number HML can be seen to fall between the single-number and the OB procedures, both in terms of complexity of use and accuracy of predictions. Indeed, it has been shown to provide slightly more predictive accuracy than the SNR, and hence a better approximation to the OB computations (Lundin, 1992). However, there is a question whether the limited accuracy of existing laboratory REAT data, from which an NRR or HML would be computed, warrants the additional complexities and potential errors that can arise in application of the HML (Thomas and Casali, 1995; Waugh, 1984).

The HML is utilized by first measuring the C- and A-weighted sound levels to compute the C-A value of the noise spectrum. Depending upon the magnitude of the C-A value, one of two equations is used to compute the Predicted Noise Level Reduction (PNR) which can then be subtracted from an A-weighted noise level to predict the protected A-weighted exposure level. Various nomograms have also been devised to simplify application of the HML.

Class systems assign a class or a grade to a device based upon its laboratory-measured attenuation values. The advantage of classes is that they simplify hearing protector selection, potentially de-emphasize the importance of attenuation, and eliminate the possibility that the user will be tempted to make unwarranted "precision" 1-dB estimates of protection. It is also possible as in the Canadian system, to build in real-world derating factors that are transparent to the user. However, whether the classes are ABC as used in Canada (CSA, 1994) or grades 1 through 5 as in New Zealand and Australia (Backshall and Bellhouse, 1994; SA/SNZ, 1998) specifiers are tempted to always use the "best" grade. Another shortcoming is that differences of 0.1 dB in the attenuation in a single OB can cause a hearing protector to be moved from one grade to another. Although class systems continue in use and their proponents are often persuasive, they have failed to achieve the international popularity of the number rating systems.

Derating HPD Attenuation and OSHA's 50% Factor

The NRR, or the laboratory data from which it is computed, must be reduced (derated) to provide attenuation values more realistic for groups of users in the workplace. The most commonly cited derating factor in use today is an across-the-board value of 50%, although some authors have suggested that a percentage derating should be less for the easier-to-fit HPDs such as earmuffs than for the harder-to-fit devices such as earplugs (Park and Casali, 1991), and NIOSH has also proposed interim type-specific deratings to be used until Method-B data are more widely available (NIOSH, 1998). A potentially more accurate approach is to base estimates on the type of real-world attenuation values cited earlier in the chapter (Figures 10.18 and 10.19), and in the future, to avoid the need for any deratings whatsoever, ANSI S12.6-1997 Method-B data should be selected as the basis for field estimates.

Although no single derating scheme will be accurate for all HPDs, hearing conservationists must choose at least one approach if they wish to use currently available manufacturer's data to develop even rough estimates of workplace performance. Whatever derating is selected, it must also be applied to the OB calculations of attenuation, i.e., if a percentage derating is used for NRRs, that same percentage should also be taken from the OB attenuation values before they are entered in row 4 of Table 10.3. *Under no circumstances should labeled NRRs be used as is for making predictions for groups of wearers.*

OSHA (1983) specifies use of manufacturers' labeled data, derived from laboratory measurements, to assess adequacy of HPDs for the noise exposures in which they are worn. Subsequent to publication of the Hearing Conservation Amendment, OSHA (1987a,b) recognized that laboratory data must be derated. They now recommend reducing published NRRs by 50 percent (for example, an NRR of 24 dB would be reduced to 12 dB), but only for the purpose of evaluating the relative efficacy of HPDs and engineering noise controls (noise reduction through equipment design and modification). The derated NRRs are not applicable for determining compliance with the hearing protection requirements of the Hearing Conservation Amendment. The situation is confusing to both hearing conservationists and OSHA compliance officers alike. See Berger (1993) for details.

The 50% derating cited above has no relationship to the well-known 7-dB correction specified in Appendix B of the Amendment (OSHA, 1983) and discussed above. The 50% derating adjusts labeled values to better reflect real-world performance, whereas the 7-dB correction accounts for use of the NRR with A- instead of C-weighted sound levels. When using both the derating and the correction together, 7 dB is subtracted from the NRR prior to derating by 50%.

Effects of HPDs on Auditory Perception

An important consideration that arises when one recommends the use of HPDs is what effects, if any, they will have on the wearers' ability to verbally communicate, listen to operating machinery, and respond to warning sounds. Many hearing protector manufacturers claim in their literature that their devices will block out harmful high-frequency noise and let speech through, yet employees often complain that they can't talk to their fellow workers or hear their machinery operate when they are using HPDs. Both of these statements contain elements of truth as will become apparent in the following discussion.

Understanding Speech

The level to which a particular sound will be attenuated by an HPD is dependent only upon its incident sound level and its frequency. HPDs cannot differen-

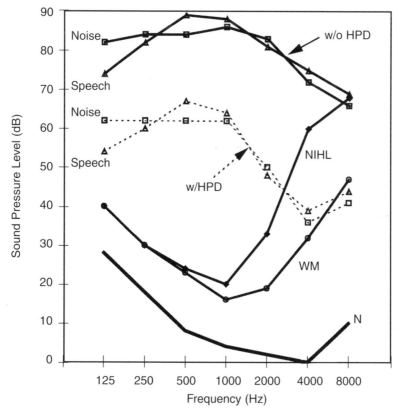

Figure 10.22 — The reduction of speech and noise levels by a representative earplug, and the relationship of those attenuated spectra to hearing thresholds for a normal listener, age 20 [N], a white male, age 45, with somewhat worse than average hearing for his age [WM], and that same white male with noise-induced hearing loss [NIHL]. Where the attenuated noise and speech spectra (upper 4 curves) fall beneath the threshold curve(s), those portions of the signal will no longer be audible.

tiate desired signals such as speech from useless information such as noise; both will be attenuated equally (see Figure 10.22). At any one frequency the resultant signal-to-noise ratio will be unaffected by use of the HPD. But hearing protector attenuation does vary with frequency, generally increasing as frequency increases, so that the frequency balance of the attenuated signal-plus-noise spectrum will differ from the unattenuated condition, generally causing sounds to appear muffled.

Typically HPDs improve the ability of normal hearing listeners to discriminate speech in high-noise-level environments. This occurs because HPDs reduce the overall level of the signal-plus-noise, thus permitting the cochlea to respond without distortion, a characteristic that can only be assured at sound levels well below

90 dBA (Lawrence and Yantis, 1956). The effect is similar to wearing sunglasses on a very bright day. Since the total illumination of the scene is reduced, the eye is allowed to function more effectively and in a more relaxed manner. Metaphorically speaking, HPDs reduce the "acoustical glare" of high-level sounds.

Speech understanding is greatly affected by such factors as a person's hearing sensitivity, signal-to-noise (S/N) ratio, absolute signal levels, visual cues (lip and hand motion), and the context of the message set. Speech discrimination is measured by verbally presenting to subjects one of a number of prepared standardized word lists and determining the percentage of correct responses they achieve. HPD effects can be evaluated by measuring speech discrimination, under constant conditions, with and without the HPDs in place. The results of such tests on normal hearing subjects indicate that HPDs have little or no effect on the speech discrimination of normal hearing listeners in background noise approximately 80 dBA or greater, but will decrease speech discrimination as the noise is reduced below that level (Howell and Martin, 1975; Kryter, 1946; Suter, 1992). For noise levels greater than about 85 dBA, HPDs can improve speech discrimination, depending upon the S/N ratio. See Chapter 14 for specific predictive techniques. As a general rule, it is best if the computed attenuated noise exposure is below the "safe" level of 85 dBA, but not below 70 dBA, thus providing a 15-dB window of optimal performance (ISO, 1996b).

When the listener is hearing impaired, the situation is considerably more complex, and the answer is not as well defined. However, it is clear that HPDs will decrease speech discrimination for hearing-impaired listeners in low to moderate noise situations, with the effects being reduced as both the noise levels and signal-to-noise ratios are increased (Abel et al., 1982; Chung and Gannon, 1979; Rink, 1979; Suter, 1992). The difficulty for hearing-impaired listeners arises from the fact that HPDs may reduce the level of the speech signals below the person's threshold of audibility, especially for important higher-frequency consonant sounds as shown in Figure 10.22. No published studies have been able to unequivocally define the level of hearing loss that would be required before hearing protection will degrade instead of improve speech discrimination in noisy environments; however, a rough estimate based upon the work of Lindeman (1976) would be a hearing threshold level of greater than about 35 dB when averaged across the frequencies of 2, 3, and 4 kHz.

The preceding generalizations can be modified in practice by additional important factors. For example, in real-world environments communications may be either limited in scope and/or accompanied by visual cues, allowing missed words to be "filled in" and intelligibility maintained. Rink (1979) illustrated this fact for subjects listening in a 90-dBA background noise. Both his normal and hearing-impaired listeners were able to maintain speech discrimination scores regardless of the use of hearing protection, as long as visual cues were presented along with the auditory stimuli. Also, Acton (1970) has demonstrated that employees become accustomed to listening in noise, thus performing better with respect to speech discrimination than do laboratory subjects with equivalent hearing levels. Conversely, Howell and Martin (1975) and Hormann et al. (1984) have shown that when the person speaking wears HPDs, speech quality is degraded, adversely affecting communications.

Howell and Martin's observation is at least partially explained by examining the effects of an HPD upon a talker's perception of his or her own voice. The HPD significantly attenuates airborne energy, but has little effect on the BC portion, except in the lower frequencies where perceived voice levels are actually amplified as a result of the occlusion effect (see *Occlusion Effect*). This alters the frequency balance of the information monitored by the talker and causes the loudness of his or her own voice to remain the same or increase slightly, while existing environmental noise levels are significantly reduced. It appears as though the talker's own voice is louder compared to the noise than actually is the case, and speech levels tend to be lowered accordingly, typically by 2 – 4 dB (Howell and Martin, 1975; Kryter, 1946). As a result, employees must be taught that while wearing HPDs in noisy environments, good communications can be ensured only by talking at what seems to be a louder than necessary level.[4]

Responding to Warning and Indicator Sounds

The effects of HPDs on the ability of normal and hearing-impaired users to detect warning sounds are very similar to the findings with regard to speech intelligibility (Suter, 1992). However, when employees are engrossed in alternative tasks and not specifically attending to sounds that may warn them of danger or indicate a machine malfunction, then not only is the question of detection and discrimination of importance, but one must also inquire whether or not the sound will get their attention. Normal inattention may result in elevation of effective thresholds for particular test stimuli by 6 – 9 dB, or even more for certain individuals (Wilkins and Martin, 1978).

One field study was conducted that assessed the effectiveness of intentional (warning horn) and incidental (clinking of metal components spilling from their container) warning sounds under actual factory conditions, using subjects who were employees at the test site (Wilkins, 1980). It was found that there was no significant effect of wearing hearing protection on the ability of either normal or hearing-impaired subjects to detect the horn or the clinking sound when they were specifically awaiting its occurrence. However, when the subjects were distracted by performing their normal job duties, the incidental warning sound (the clink) was less well perceived as a warning by those with substantial hearing loss, and HPDs adversely affected its perception independent of the hearing sensitivity of the wearers. By contrast, even when distracted, there were no significant variations in response rates for the intentional warning sound (the horn) due

---

[4] In quiet, however, the tendency is to speak loudly when wearing HPDs. In such conditions speech effort is determined by the speakers' perception of their own absolute voice level instead of the perceived speech-to-ambient noise ratio. Although the BC component is still amplified by the occlusion effect, the attenuation of the airborne speech is generally of greater significance, and thus the speakers' overall assessment of their speech levels is that they are either acceptable or too low.

to either the hearing ability of the listeners or their ear condition (protected or unprotected).

Since warning sounds may be adjusted in pitch and loudness to achieve optimum perceptibility, results similar to those reported above for the intentional sound should be achievable in most conditions. Warning sounds will be most effective when their primary acoustic output is located below 2 kHz, since hearing-impaired employees will exhibit their largest losses above that frequency and many HPDs will deliver less attenuation below that frequency. See Chapter 14 for computational methods for the design of warning signals. Additional evidence regarding any possible hazards associated with the use of hearing protection while working in noisy environments is provided by two field studies (Cohen, 1976; Schmidt et al., 1982). Those authors demonstrated that implementation of a hearing conservation program that required utilization of HPDs, reduced rather than increased the number of industrial mishaps.

Localization and Distance Perception

Another effect that HPDs can have is to confuse one's ability to locate the direction of origin of sounds (Noble and Russell, 1972; Suter, 1992). Earmuffs tend to interfere with localization accuracy to a greater extent than do inserts that leave the outer ear exposed, and it appears that subjects cannot learn to compensate for their adverse effects (Russell, 1977). However, one study that measured the minimum detectable azimuthal changes of a frontally located sound source (typically <10°) found the higher-attenuation earplug they studied to be more disruptive than a lower-attenuation earplug or than earmuffs (Lin, 1981). There is also some suggestion, but no direct evidence, that HPDs may impair the ability of wearers to judge the distance to a sound source (Wilkins and Martin, 1978).

General Remarks

The preceding data indicate that HPDs can be effectively utilized for the preservation of hearing in high-noise-level environments with minimal negative effects on auditory communications and demonstrated advantages in certain conditions.

For hearing-impaired persons, the utilization of HPDs in lower noise levels should be carefully considered. When substantial hearing impairments are present, especially in the case of hearing-aid users, decisions regarding employment in noisy occupations and/or the use of hearing protection are not clear cut (Berger, 1987; also see *Hearing-Aid Earmolds*). Even with individual counseling, comprehensive audiological workups, and expert consultation, ideal solutions are elusive. It sometimes becomes a decision between preserving an employee's remaining hearing and creating additional communication disadvantages while at work. Application of specialized HPDs (see next section) may be useful. Often warning and indicator sounds can be augmented or replaced by visual or tactile signals to assist the hearing impaired.

Intermittent noise poses a significant problem since HPDs will cause degradation of communications during the intervening quiet intervals. Consequently, earmuffs or semi-aural devices are the preferred type of HPDs for such conditions, since they are easily removed and replaced as the intermittence of the noise may warrant. Unfortunately, passive amplitude-sensitive earplugs cannot normally be recommended, since when the noise is on, it will not generally be of sufficient level to activate their level-sensitive characteristics. Thus the HPD may provide inadequate protection, especially at the lower and middle frequencies. Additionally, although amplitude-sensitive HPDs can improve speech discrimination during the quiet conditions, when noise is present they may actually provide worse speech discrimination than do standard linear protectors.

Specialized HPDs and Other Devices Which Block Sound at the Ear

Special hearing protector designs have been developed to improve auditory perception and speech communication for the noise-exposed wearer. These devices may be categorized into passive HPDs (i.e., without electronics), active HPDs (i.e., incorporating electronics), communication headsets, and other electronic devices that may affect the sound perceived at the ear, such as hearing aids and recreational earphones.

Passive Hearing Protectors
Flat/Moderate Attenuation HPDs
Because the attenuation of conventional HPDs increases with frequency, the wearer's perception of pitch is imbalanced. Not only are sounds reduced in level, and often to a greater extent than is actually required, but they are colored in a spectral sense as well. This problem contributes to protector nonuse where, for instance, a machine operator's auditory feedback from a cutting tool is distorted or a musician's pitch perception is compromised. To counter these problems, flat or uniform attenuation HPDs are designed to impose attenuation that is nearly constant (within a range of about 10 dB) from about 100 to 8000 Hz (see Figure 10.23). Successful techniques generally rely on including an acoustical leak (a specially designed orifice) into the earplug to allow a carefully controlled amount of sound directly into the earcanal. With proper selection of channel diameter and length, as well as other associated acoustical parameters, a flat attenuation profile that avoids overprotection can be obtained (Casali and Berger, 1996).

Properly fit uniform attenuation HPDs provide adequate protection and enhanced hearing perception in low to moderate noise exposures of about 90 dBA or less. Workers with high-frequency hearing losses and professional musicians may find them particularly beneficial. However, for noises having substantial high-frequency energy, uniform attenuation earplugs generally offer less protection than conventional earplugs, since the flat attenuation characteristic is obtained by reducing the high-frequency noise reduction to match that in the lower-frequency range.

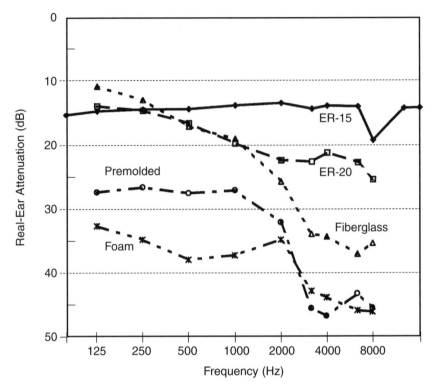

Figure 10.23 — Comparison of the real-ear attenuation of two flat- and moderate-attenuation earplugs [custom-molded ER-15 and premolded ER-20 (also called UltraTech®12 earplug)] to conventional earplugs. The premolded and foam plugs illustrate the typical attenuation plateau from 125–2000 Hz that is representative of conventional earplugs that provide an acoustical seal.

Amplitude-Sensitive HPDs

Because a conventional HPD provides a constant amount of attenuation that is independent of incident sound level, hearing ability is compromised during the quiet periods of intermittent sound exposures. Amplitude-sensitive, also called level-dependent HPDs, address this problem by providing reduced attenuation at low sound levels, with increasing protection at high levels.

Passive HPDs utilize a nonlinear component, such as a valve, diaphragm, or sharp-edged orifice in an earplug, or opening into a duct within an earmuff cup, to effect the change in attenuation. The latter technique takes advantage of the fact that low-intensity sound waves predominantly exhibit laminar airflow and pass relatively unimpeded through the aperture, whereas high-intensity waves involve more turbulent flow and are attenuated due to increasing acoustic resistance (Allen and Berger, 1990).

A critical performance parameter is the transition sound level, normally 110 to 120 dB, above which insertion loss increases at a rate of about 1 dB for each 2- to 4-dB increase in sound level. Because this transition level is so high, passive amplitude-sensitive HPDs are best suited for outdoor impulsive blasts and gunfire exposures, for which recent military studies suggest they may be ideally suited (Dancer at al., 1996). At lower, but still hazardous sound levels, most passive devices exhibit vented earplug behavior, affording very weak protection at frequencies less than 1000 Hz. One exception is an orifice-type earmuff that provides approximately 25 dB attenuation from 400 to 8000 Hz (Allen and Berger, 1990).

Unfortunately, passive amplitude-sensitive devices have been among the most misadvertised products sold today, with claims of dramatic level dependency beginning at levels well below 100 dB. Although not a shred of evidence exists to substantiate such extravagant claims, vented earplugs and passive level-dependent HPDs do provide worthwhile performance improvements for certain applications.

Active Hearing Protectors
Amplitude-Sensitive Sound Transmission HPDs

Active sound transmission hearing protectors consist of modified conventional earmuffs or earplugs that incorporate microphones and amplifiers to transmit external sounds to earphones mounted inside the HPDs. Typically, the amplifier limits sounds to a predetermined (in some cases user-adjustable) earphone level, often at about 85 dBA, until the ambient noise is so intense that direct transmission through the earcup becomes the controlling factor. In comparison to both conventional and to passive amplitude-sensitive earmuffs, sound-transmission earmuffs are more expensive (upwards of $100) but offer a viable alternative for use in intermittent noises, especially those with impulses (e.g., gunfire) or short-duration on-segments. However, in continuous high-level noise, the distortion products of some systems may cause annoyance and compromise hearing ability (Casali and Berger, 1996).

Auditory perception and noise level under sound transmission earmuffs depends on electronic design factors such as: cutoff sound level and sharpness of attenuation transition at this level, system response delay, frequency response and bandwidth, distortion and residual electronic noise, signal/noise ratio at sound levels below the cutoff, and sensitivity to wind effects. Microphone design may be *diotic*, wherein a single microphone in one earcup feeds both earphones, or *dichotic*, in which each earcup has an independent microphone. The latter approach provides better localization performance for situations in which wearers must be able to localize the placement or direction of warning and indicator sounds in their environment.

Active Noise Reduction (ANR)

Recently integrated with conventional passive earmuffs, active noise reduction (ANR) electronics rely on the principle of destructive interference to cancel noise. A microphone senses the sound inside the earcup, which is then passed through a phase compensation feedback filter to a processing circuit and amplifier (Figure 10.24). The resultant "anti-noise" signal is presented through an earphone at equal amplitude to, but 180 degrees out-of-phase with, the original noise, causing energy cancellation.

ANR is most effective against repetitive or continuous noises that are relatively invariant in spectrum and level, which allow the system to stabilize and fine-tune the phase and amplitude parameters needed for cancellation. Furthermore, ANR is limited to the reduction of low-frequency noises below about 1000 Hz, with maximum attenuation of 20 to 25 dB occurring below 300 Hz (McKinley et al., 1996), thus ANR systems are beneficial only in environments with substantial low-frequency energy. Like conventional earmuffs, circumaural ANR devices are susceptible to leakage under the ear cushion, such as caused by eyeglass temples, which can result in a reduction in active attenuation (Casali and Berger, 1996).

The low-frequency effectiveness of ANR is fortuitous in that it compensates the typically high-frequency biased protection of a passive earmuff. If the noise environment consists of only low-frequency energy, open-back ANR headphones (e.g., lightweight supra-aural earphones which rest on the pinnae) may offer communications and perhaps comfort advantages over circumaural ANR earmuffs. However, open-back ANR headphones have the disadvantage of offering no passive protection if the ANR circuit malfunctions.

Attenuation tests for ANR hearing protectors are as yet not standardized. A combination procedure may be required, i.e., real-ear tests (e.g., as per ANSI S12.6-1997) to obtain the muff's passive attenuation and miniature microphone IL tests on a human head (e.g., as per ANSI S12.42-1995) to obtain the active attenuation. The synergistic benefits of both passive and active attenuation in ANR devices come at a relatively high price of about $250 to $1000 per unit.

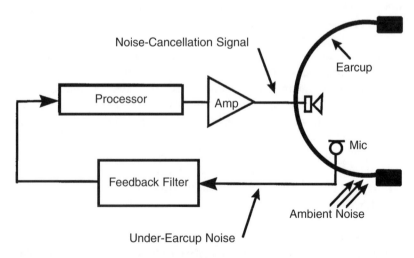

Figure 10.24 — Block diagram of a typical ANR system built into a circumaural earcup.

HPDs with Communication Features

In order to transmit signals to the ear, earphones may be built into both supra-aural and circumaural devices, or can be connected to or molded into earplugs. Usually only circumaural devices or well-sealed earplugs will provide sufficient attenuation to be suitable for use in noisy environments. Devices are available with either wireless (FM or infrared) or wired systems, designed for one- and two-way communications and/or music transmission. Microphones may be of the noise-canceling variety, mounted on the muff cups and positioned in front of the lips, or can pick up the talker's speech via transducers sensitive to bone conduction that are pressed against the neck, or even located in the earcanal, sealed away from the environmental noise by the body of the earplug. The better devices provide specialized electronic circuits to limit the delivered SPLs so that the earphones themselves cannot present signals that could be hazardous to the wearer. Prices for communication headsets range from $100 to $1000 per unit.

In extremely high noise levels, when circumaural communication headsets may not provide sufficient attenuation of ambient noise to permit clear communications, speech intelligibility can be improved by wearing earplugs under earmuff cups. The earplug will reduce both the environmental noise and the desired signal equally, but as long as the headset has sufficient distortionless gain so that its output can be increased in order to overcome the insertion loss of the earplug, the signal-to-noise ratio in the listener's earcanal can be significantly improved.

Another means to potentially improve communications in noise is to integrate ANR into a communications headset. Referring to Fig. 10.24, the desired speech signal is injected via additional circuitry to the processor, where it and the feedback signal are combined and amplified for delivery to the earphone. Since the low-frequency components of the (desired) speech signal are also partly canceled by the ANR, a speech preamplifier is used for compensation. For certain noise situations, ANR headsets result in an improvement in speech intelligibility over the use of passive devices alone (Casali and Berger, 1996).

Hearing-Aid Earmolds

When employees who wear hearing aids work in noise they may request to wear their aids, turned off, in lieu of standard industrial HPDs. This may be due to comfort (since they are accustomed to their custom hearing-aid earmold), or convenience (since their hearing aids are available for use when needed), or reduced attenuation (which may help them hear better under certain conditions), or because they may wish to occasionally use their aids in the noise. The latter is uncommon since it is generally observed that present-day hearing aids are of little value in noisy environments (Gasaway, 1985).

The question is: Can an earmold that is part of a hearing-aid system provide adequate hearing protection? If so, the wearer could quickly and easily turn on and use the aid when needed, and yet turn off the aid and continue wearing it to obtain noise reduction as required.

Berger (1987) examined this question and reported that for the typical vented earmold, and even unvented earmolds depending upon how they were fabricated, attenuation is insufficient for all but the most marginal occupational exposures. However for a tightly fitted unvented earmold or when foam earplugs are used as hearing-aid earmolds, protection equivalent to standard commercially available earplugs is achievable. He suggested that, if possible, it is best to validate the level of protection by asking the audiologist who fitted the hearing aid to estimate its attenuation using sound-field audiometry, i.e., measuring the difference between the individual's unaided, unoccluded thresholds and the occluded thresholds with the aid turned off.

Regardless of the amount of attenuation that is provided by the hearing-aid earmold, the aid itself, which usually supplies from 20 to 50 dB of maximum gain, can potentially cause additional noise-induced hearing loss when used in the presence of sustained high-level noise. Although no definitive answers are available, a prudent recommendation is that employees should never operate their aids without the addition of an earmuff when the sound levels exceed 80 dBA. In such cases hearing-aid gain will most likely need reduction to avoid feedback underneath the earmuff. Whenever hearing aids are worn in noise, careful employee orientation is necessary, and more frequent audiometric monitoring (twice annually) is advised until the stability of the individual's hearing threshold levels can be verified.

Whatever decision is made concerning the suitability of the earmold for use as a hearing protector, the hearing-impaired individual should be protected. Exceptions may include an individual with a hearing loss so severe that the noise is inaudible, or persons with a conductive loss that exceeds in magnitude the attenuation that a hearing protector can provide.[5]

Recreational Earphones

Often, employees request to use stereo earphones for protection against noise while they enjoy the music. The inappropriateness of relying on these recreational devices for hearing protection is illustrated in Figure 10.25 in which the attenuation of a circumaural radio headset and of a more popular set of lightweight foam supra-aural stereo earphones are compared to an industrial earmuff. The foam earphones offer almost no protection. Even the circumaural device provides no more than approximately 20 dB of attenuation at high frequencies, and actually amplifies sounds at others. This protection is inferior to that of a well-designed, properly fitted HPD.

Recreational earphones alone can generate equivalent sound-field noise levels up to approximately 100 dBA. Since they offer so little attenuation, the concern is that employees might turn up the music to mask (i.e., "drown out") the factory noise, with a consequent reduction in their ability to communicate or hear

---

[5] OSHA published regulations do not permit such exceptions, but according to J. Barry (Technical Support IH in OSHA Region #3, 1986) special cases can be discussed with area directors.

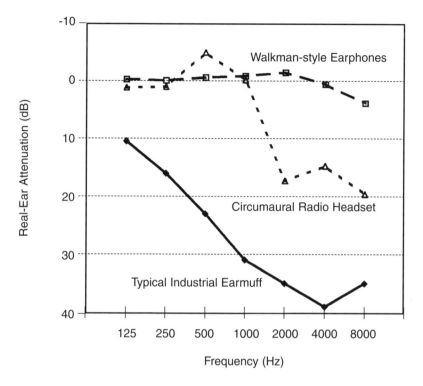

Figure 10.25 — Attenuation of two styles of recreational earphones versus a typical industrial earmuff.

warning sounds and a significant increase in their effective noise exposures. Thus the use of recreational earphones should be prohibited when sound exposures equal or exceed 90 dBA. Between 85 and 90 dBA, their use is problematical. The music they provide may alleviate boredom and increase productivity (Fox, 1971), but they offer little or no protection and can actually increase noise exposures as cited above.

In one industrial study (Royster et al., 1984) with a sound level varying between 80 and 90 dBA (with approximately equal time periods of 5 s at each level), and a TWA of 87 dBA, employees on the average increased their equivalent exposure by only 2 dB as a result of using recreational earphones. However, about 20% of the workers were observed to play their radios at levels of 90 dBA or higher. The authors recommended that employees continue to be allowed use of such devices, with the stipulation that a significant educational effort be directed at the proper use of personal radios, that employees exhibiting permanent threshold shifts of 20 dB or more be prohibited from further use, and that the overall audiometric data base be annually analyzed for relative changes between the hearing of users and nonusers of personal radios.

HPD Standards, Regulations, and Related Documents

United States

Two current U.S. standards pertain specifically to hearing protectors, ANSI S12.6-1997 which specifies REAT evaluations of HPDs, and ANSI S12.42-1995 which specifies MIRE and ATF measurements of HPDs (see *Laboratory Procedures*). A third standard, ANSI S3.19-1974, twice supplanted by newer REAT documents and currently withdrawn, is still specified in the U.S. EPA Hearing Protector Labeling Regulation and thus has substantial impact in the marketplace (see next section). Users often mistakenly presume ANSI REAT standards contain criteria for judging the acceptability of HPDs. In fact, these standards only describe methods for determining hearing protector attenuation. The methods in no way confer approval or attribute a particular degree of quality to the devices being tested.

A Department of Commerce accreditation process also exists and is managed by the National Voluntary Laboratory Accreditation Program (NVLAP). Laboratories can be accredited by the program (as four U.S. facilities currently are) to test according to ANSI S3.19 and ANSI S12.6, and this indicates that they adhere to extensive and specific procedures. It does not however ensure the results on any particular test, nor does it certify the devices that are tested. At this time there are no federal or state agencies or U.S. standards-writing organizations that approve or accredit particular HPDs for sale or use, although there is an existing EPA regulation requiring the labeling of all HPD packaging.

OSHA specifies hearing protector use in the Hearing Conservation Amendment (OSHA, 1983), requiring that HPDs reduce an employee's TWA to 90 dBA or less, and in the case of employees demonstrating standard threshold shifts, to 85 dBA or less. No particular types or brands of HPDs are specifically recommended or proscribed. The method for assessing the amount of reduction to be expected is contained in Appendix B of the amendment (see *Derating HPD Attenuation and OSHA's 50% Factor* and Appendix I). A related government agency, the Mine Safety and Health Administration (MSHA), issues a "Hearing Protector R & D Factor List." The only criterion for having a device placed on the list is that it be tested according to an ANSI REAT standard, and that the data be available in sales literature or on specification sheets.

EPA Labeling Regulation

The Noise Control Act of 1972 empowered the EPA to label all noise-producing and noise-reducing devices. The first and only standard that was promulgated for noise-reducing devices was the hearing protector labeling regulation (EPA, 1979). As with the preceding regulations that have been discussed, it did not specify criteria by which HPDs were to be deemed acceptable or unacceptable. It did however specify that the attenuation of any device or material capable of being worn on the head or in the earcanal, that is sold wholly or in part on the basis of its ability to reduce the level of sound entering the ear, was to be evaluated according to ANSI S3.19-1974. An NRR was to be computed and placed on a label whose size and configuration was specified.

In the early 1980s budget cuts at EPA led to elimination of the noise enforcement division. In recognition of their inability to enforce the regulation, EPA then revoked the product verification testing and the attendant reporting and recordkeeping requirements. However, the remainder of the regulation remains law and in the 20 years since its promulgation it has clearly been recognized in the professional community as providing misleading guidance to potential purchasers of products (see sections on *Selecting the Right Hearing Protector*, and on *Laboratory Procedures*). In 1995 a multi-organizational Task Force on Hearing Protector Effectiveness provided specific recommendations to improve the labeling regulation, including testing according to Method B, Subject Fit of ANSI S12.6-1997 and computation of a revised NRR which has been designated the NRR[SF] (Berger and Royster, 1996; and see *Noise Reduction Rating (Subject Fit) [NRR(SF)]* in this chapter). Although virtually all professional organizations involved in hearing conservation (including AIHA) endorse the recommendations, EPA has not acted, primarily due to lack of funds and staff to review the proposal.

International Standards and Regulations

The International Organization for Standardization (ISO) maintains two working groups ISO/TC43/SC1/WG17 and ISO/TC94/SC12 that are responsible for standards on hearing protection.

WG17 developed the 4869 series, Part 1 on REAT measurements (ISO, 1990), Part 2 on number rating systems (ISO, 1994), and Part 3 on use of an ATF (ISO, 1989). Part 1 corresponds to Method A of S12.6-1997 but differs in details of subject selection and HPD fitting. A new Part 7 that would correspond with subject fit (Method B) is currently being drafted for consideration. There is no current U.S. counterpart to Part 2, although the NRR and the ISO's SNR are roughly equivalent (see *Other Rating Schemes*). Part 3 is similar in some respects to the ATF portion of ANSI S12.42 in that both specify use of ATFs. In addition, WG17 is working on standards for the measurement of level-dependent hearing protectors and the assessment of the performance of HPDs in impulsive environments.

SC12, which is dormant at this time, has produced three draft documents, two pertaining to approval of earplugs and earmuffs (ISO, 1996a; ISO, 1996c), and a third one on selection and use of HPDs (ISO, 1996b). These documents closely correspond to European regulations on the approval of hearing protectors for sale in the European Community. The U.S. has no counterpart to this HPD approval process since the feeling has generally been that for devices of this nature, approvals are best left with the purchasing organizations so as not to burden inexpensive and readily replaceable products with potentially unnecessary testing.

Recommendations

This chapter provides a comprehensive review and analysis of the selection, issuing, measurement, and performance of HPDs, with the most practical materials present earliest in the chapter. It cannot be overemphasized that the greatest impact that the hearing conservationist can have on long-term hearing loss prevention is in the areas of fitting and issuing, training and motivation, and enforcement. In fact, when TWAs are less than or equal to 95 dBA, virtually any hearing protector will suffice as long as it fits well and is worn correctly and consistently (Royster and Royster, 1985). It is only when daily exposures exceed 95 dBA that greater attention should be directed toward real-world attenuation and field estimates of performance, and when they approach and exceed 100 dBA that attenuation criteria for selection become a substantial concern. In such instances ANSI S12.6-1997 Method-B-type data should be utilized for predictions; however, in general, data would suggest that foam earplugs, or earmuffs, or for exposures above about 105 dBA, a combination of the two, should be required.

Hearing protectors are not a panacea and cannot be dispensed indiscriminately, but they can and do work when utilized within the context of a well-defined and properly implemented hearing conservation program (Bruhl and Ivarsson, 1994; Dobie, 1995). Effective implementation refers primarily to the human interaction and not the technical side of the equation. In fact, in one large-scale study that demonstrated hearing conservation effectiveness with a database of well over 100,000 audiograms, no relationship was observed between the changes in observed hearing levels between the groups wearing Class A (most protective) and Class B (less protective) devices (Gillis and Harrison, 1993). This was evident across all industries studied. Such findings support a widely quoted, but often unappreciated axiom — the best hearing protector is the one that is worn and worn correctly. And that protector will be the one that is matched to the noise, the environment, and especially to the person in need of protection.

References

Abel, S. M., Alberti, P. W., Haythornthwaite, C., and Riko, K. (1982). "Speech Intelligibility in Noise: Effects of Fluency and Hearing Protector Type," *J. Acoust. Soc. Am. 71*(3), 708–715.

Abel, S. M., and Rokas, D. (1986). "The Effect of Wearing Time on Hearing Protector Attenuation," *J. Otolaryngol. 15*(5), 293–297.

Acton, W. I. (1970). "Speech Intelligibility in a Background Noise and Noise-Induced Hearing Loss," *Ergonomics 13*(5), 546–554.

Allen, C. H., and Berger, E. H. (1990). "Development of a Unique Passive Hearing Protector with Level-Dependent and Flat Attenuation Characteristics," *Noise Control Eng. J. 34*(3), 97–105.

Anon. (1988). "How to Make Sure You Get the Best Fit From Your E·A·R Plugs," E·A·R/Hearing Protection Products Doc. No. 33002, Indianapolis, IN.

ANSI (1974). "Method for the Measurement of Real-Ear Protection of Hearing Protectors and Physical Attenuation of Earmuffs," S3.19-1974 (ASA STD 1-1975), Acoustical Society of America, New York, NY.

ANSI (1991). "Draft American National Standard, Evaluating the Effectiveness of Hearing Conservation Programs," Draft S12.13-1991, Acoustical Society of America, New York, NY.

ANSI (1995). "Microphone-in-Real-Ear and Acoustic Test Fixture Methods for the Measurement of Insertion Loss of Circumaural Hearing Protection Devices," S12.42-1995, Acoustical Society of America, New York, NY.

ANSI (1997). "Methods for Measuring the Real-Ear Attenuation of Hearing Protectors," S12.6-1997, Acoustical Society of America, New York, NY.

Backshall, D. and Bellhouse, G. (1994). "Hearing Protector Selection Methods," Report to AV/3 Committee of Stds. Australia.

Behar, A., and Crabtree, R. B. (1997). "Measurement of Hearing Protector Attenuation at Ultrasonic Frequencies," in *Proceedings of Noise-Con 97*, edited by C. B. Burroughs, Noise Control Foundation, Poughkeepsie, NY, 97–102.

Berger, E. H. (1981). "Details of Real World Hearing Protector Performance as Measured in the Laboratory," in *Proceedings of Noise-Con '81*, edited by L. H. Royster, F. D. Hart, and N. D. Stewart, Noise Control Foundation, Poughkeepsie, NY, 147–152.

Berger, E. H. (1982). "Various Factors Affecting Laboratory Measured Earmuff Performance: Liquid vs. Foam Cushions, Eyeglasses, Foam Temple Piece, and Fitting Noise," E·A·R/Aearo Technical Rept. 82–26/HP, Indianapolis, IN.

Berger, E. H. (1983a). "Attenuation of Hearing Protectors at the Frequency Extremes," 11th Int. Congr. on Acoustics, Paris, France, Vol. 3, 289–292.

Berger, E. H. (1983b). "Laboratory Attenuation of Earmuffs and Earplugs Both Singly and in Combination," *Am. Ind. Hyg. Assoc. J. 44*(5), 321–329.

Berger, E. H. (1984). "Assessment of the Performance of Hearing Protectors for Hearing Conservation Purposes," *Noise & Vib. Control Worldwide 15*(3), 75–81.

Berger, E. H. (1985). "EARLog #17 — Ear Infection and the Use of Hearing Protection," *J. Occup. Med. 27*(9), 620–623.

Berger, E. H. (1986). "Review and Tutorial — Methods of Measuring the Attenuation of Hearing Protection Devices," *J. Acoust. Soc. Am. 79*(6), 1655–1687.

Berger, E. H. (1987). "EARLog #18 — Can Hearing Aids Provide Hearing Protection?" *Am. Ind. Hyg. Assoc. J. 48*(1), A20–A21.

Berger, E. H. (1988). "Hearing Protectors — Specifications, Fitting, Use, and Performance," in *Hearing Conservation in Industry, Schools, and the Military*, edited by D. M. Lipscomb, College-Hill Press, Boston, MA, 145–191.

Berger, E. H. (1990). "Hearing Protection — The State of the Art (circa 1990) and Research Priorities for the Coming Decade," in *Program and Abstracts of the NIH Consensus Development Conference on Noise and Hearing Loss*, National Inst. of Health, Bethesda, MD, 91–96.

Berger, E. H. (1992). "Development of a Laboratory Procedure for Estimation of the Field Performance of Hearing Protectors," in *Proceedings, Hearing Conservation Conference*, Off. Eng. Serv., Univ. Kentucky, Lexington, KY, 41–45.

Berger, E. H. (1993). "EARLog #20 — The Naked Truth About NRRs," E·A·R/Hearing Protection Products, Indianapolis, IN.

Berger, E. H., Franks, J. R., Behar, A., Casali, J. G., Dixon-Ernst, C., Kieper, R. W., Merry, C. J., Mozo, B. T., Nixon, C. W., Ohlin, D., Royster, J. D., Royster, L. H. (1998). "Development of a New Standard Laboratory Protocol for Estimating the Field Attenuation of Hearing Protection Devices, Part III: The Validity of Using Subject-Fit Data," *J. Acoust. Soc. Am. 103*(2), 665–672.

Berger, E. H., Franks, J. R., and Lindgren, F. (1996). "International Review of Field Studies of Hearing Protector Attenuation," in *Scientific Basis of Noise-Induced Hearing Loss*, edited by A. Axlesson, H. Borchgrevink, R. P. Hamernik, P. Hellstrom, D. Henderson, and R. J. Salvi, Thieme Medical Pub., Inc., New York, NY, 361–377.

Berger, E. H., and Kerivan, J. E. (1983). "Influence of Physiological Noise and the Occlusion Effect on the Measurement of Real-Ear Attenuation at Threshold," *J. Acoust. Soc. Am. 74*(1), 81–94.

Berger, E. H., Kerivan, J. E., and Mintz, F. (1982). "Inter-Laboratory Variability in the Measurement of Hearing Protector Attenuation," *Sound and Vibration 16*(1), 14–19.

Berger, E. H., and Kieper, R. W. (1997). "Comparison of the Real-Ear Attenuation of Liquid vs. Foam Earmuff Cushions Tested in the E·A·RCAL Chamber," E·A·R/Aearo Technical Rept. 97–38/HP, Indianapolis, IN.

Berger, E. H., and Royster, L. H. (1996). "In Search of Meaningful Measures of Hearing Protector Effectiveness," *Spectrum Suppl. 1*, 13, p. 29.

Bertrand, R. A., and Zeidan, J. (1993). "Retrospective Field Evaluation of HPD Based on Evolution of Hearing," in *Noise & Man '93 — Proceedings of the 6th Int. Congr., Noise as a Public Health Problem, Vol. 2*, edited by M. Vallet, Institut National de Recherche sur les Transports et Leur Securite, Arcueil Cedex, France, 21–24.

Blackstock, D. T., and von Gierke, H. E. (1956). "Development of an Extra Small and Extra Large Size for the V-51R Earplug," Wright Air Development Center, Wright-Patterson AFB, OH.

Briskey, R. J. (1984). "NAEL: Fitting Facts, Part VII: Hearing Aid Fitting: Outer Ear and Canal," *Hearing Instr. 35*(9), 30–32.

Bruhl, P., and Ivarsson, A. (1994). "Noise-Exposed Male Sheet-Metal Workers Using Hearing Protectors," *Scand. Audiol. 23*(2), 123–128.

Caruso, V. G., and Meyerhoff, W. L. (1980). "Trauma and Infections of the External Ear," in *Otolaryngology, Volume II*, edited by M. M. Paparella and D. A. Shumrick, W.B. Saunders Co., Phila., PA, 1345–1353.

Casali, J. G., and Berger, E. H. (1996). "Technology Advancements in Hearing Protection Circa 1995: Active Noise Reduction, Frequency/Amplitude-Sensitivity, and Uniform Attenuation," *Am. Ind. Hyg. Assoc. J. 57*(2), 175–185.

Casali, J. G., and Lam, S. T. (1986). "Effects of User Instructions on Earmuff/Earcap Sound Attenuation," *Sound and Vibration 20*(5), 22–28.

Casali, J. G., Mauney, D. W., and Burks, J. A. (1995). "Physical vs. Psychophysical Measurement of Hearing Protector Attenuation-a.k.a. MIRE vs. REAT," *Sound and Vibration 29*(7), 20–27.

Casali, J. G., and Park, M. Y. (1991). "Laboratory versus Field Attenuation of Selected Hearing Protectors," *Sound and Vibration 25*(10), 28–38.

Chung, D. Y., and Gannon, R. P. (1979). "The Effect of Ear Protectors on Word Discrimination in Subjects with Normal Hearing and Subjects with Noise-Induced Hearing Loss," *J. Am. Aud. Soc. 5*(1), 11–16.

Cohen, A. (1976). "The Influence of a Company Hearing Conservation Program on Extra-Auditory Problems in Workers," *J. Saf. Res. 8*(4), 146–162.

CSA (1994). "Hearing Protectors," Canadian Standards Assoc., Z94.2–94, Rexdale, Ontario.

Damongeot, A., Lataye, R., and Kusy, A. (1989). "An Empirical Formula for Predicting the Attenuation Given by Double Hearing Protection (Earplugs and Earmuffs)," *Appl. Acoust.* *28*(3), 169–175.

Dancer, A. L., Franke, R., Parmentier, G., and Buck, K. (1996). "Hearing Protector Performance and NIHL in Extreme Environments: Actual Performance of Hearing Protectors in Impulse Noise/Nonlinear Behavior," in *Scientific Basis of Noise-Induced Hearing Loss*, edited by A. Axlesson, H. Borchgrevink, R. P. Hamernik, P. Hellstrom, D. Henderson, and R. J. Salvi, Thieme Medical Pub., Inc., New York, NY, 321–338.

Dobie, R. A. (1995). "Prevention of Noise-Induced Hearing Loss," *Arch. Otolaryngol. Head Neck Surg. 121*, 385–391.

EPA (1979). "Noise Labeling Requirements for Hearing Protectors," Environmental Protection Agency, *Fed. Regist. 44*(190), 40CFR Part 211, 56130–56147.

Fox, J. G. (1971). "Background Music and Industrial Efficiency — A Review," *Appl. Ergon.* *2*(2), 70–73.

Franks, J. R., Themann, C. L., and Sherris, C. (1994). "The NIOSH Compendium of Hearing Protection Devices," U.S. Dept. of HHS, Pub. No. 95–105, Cincinnati, OH.

Gasaway, D. C. (1984). "'Sabotage' Can Wreck Hearing Conservation Programs," *Natl. Saf. News 129*(5), 56–63.

Gasaway, D. C. (1985). *Hearing Conservation — A Practical Manual and Guide*, Prentice-Hall, Inc., Englewood Cliffs, NJ.

Gillis, H. R., and Harrison, C. (1993). "Hearing Protection — What Is the Best?" *Spectrum Suppl. 1*, 10, 20–21.

Guild, E. (1966). "Personal Protection," in *Industrial Noise Manual, Second Edition*, Am. Ind. Hyg. Assoc., Akron, OH, 84–109.

Holland, H. H. Jr. (1967). "Attenuation Provided by Fingers, Palms, Tragi, and V-51R Ear Plugs," Letter to the Editor, *J. Acoust. Soc. Am. 41*(6), 1545.

Hormann, H., Lazarus-Mainka, G., Schubeius, M., and Lazarus, H. (1984). "The Effect of Noise and the Wearing of Ear Protectors on Verbal Communication," *Noise Control Eng. J. 23*(2), 69–77.

Howell, K., and Martin, A. M. (1975). "An Investigation of the Effects of Hearing Protectors on Vocal Communication in Noise," *J. Sound Vib. 41*(2), 181–196.

Humes, L. E. (1983). "A Psychophysical Evaluation of the Dependence of Hearing Protector Attenuation on Noise Level," *J. Acoust. Soc. Am. 73*(1), 297–311.

ISO (1989). "Acoustics — Hearing Protectors — Part 3: Simplified Method for the Measurement of Insertion Loss of Ear-Muff Type Protectors for Quality Inspection Purposes," International Organization for Standardization., ISO/TR 4869–3:1989, Geneva, Switzerland.

ISO (1990). "Acoustics — Hearing Protectors — Part 1: Subjective Method for the Measurement of Sound Attenuation," International Organization for Standardization, ISO 4869–1:1990(E), Geneva, Switzerland.

ISO (1994). "Acoustics — Hearing Protectors — Part 2: Estimation of Effective A-Weighted Sound Pressure Levels When Hearing Protectors are Worn," International Organization for Standardization, ISO 4869–2:1994(E), Geneva, Switzerland.

ISO (1996a). "Hearing Protectors — Safety Requirements and Testing — Ear-Muffs," International Organization for Standardization, ISO/DIS 10449, Geneva, Switzerland.

ISO (1996b). "Hearing Protectors — Recommendations for Selection, Use, Care and Maintenance — Guidance Document," International Organization for Standardization, ISO/DIS 10452, Geneva, Switzerland.

ISO (1996c). "Hearing Protectors — Safety Requirements and Testing — Ear-Plugs," International Organization for Standardization, ISO/DIS 10453, Geneva, Switzerland.

Johnson, D. L. (1997). "Blast Overpressure Studies," U.S. Army Medical Res. and Materiel Command, DAMD17-93-C-3101, Albuquerque, NM.

Johnson, D. L., and Patterson, J. Jr. (1992). "Rating of Hearing Protector Performance for Impulse Noise," in *Proceedings, Hearing Conservation Conference*, Off. Eng. Serv., Univ. Kentucky, Lexington, KY, 103–106.

Kieper, R. W., Berger, E. H., and Lindgren, F. (1991). "An Objective Assessment of the Effect of Ear Impression Techniques on the Real-Ear Attenuation of Earmolds," *Spectrum Suppl. 1*, 8, p. 24.

Kroes, P., Fleming, R., and Lempert, B. (1975). "List of Personal Hearing Protectors and Attenuation Data," National Institute for Occupational Safety and Health, U.S. Dept. of HEW, Rept. No. 76–120, Cincinnati, OH.

Kryter, K. D. (1946). "Effects of Ear Protective Devices on the Intelligibility of Speech in Noise," *J. Acoust. Soc. Am. 18*(2), 413–417.

Kunov, H., Abel, S. M., and Giguere, C. (1986). "An Acoustic Test Fixture for Use with Hearing Protective Devices," Mt. Sinai Hospital and Univ. of Toronto, final rept. on Contract No. 8SE84-00241, Dept. of National Defence (DCIEM), Toronto, Canada.

Lawrence, M., and Yantis, P. A. (1956). "Onset and Growth of Aural Harmonics in the Overloaded Ear," *J. Acoust. Soc. Am. 28*, 852–858.

Lin, L. (1981). "Auditory Localization Under Conditions of High Ambient Noise Levels With and Without the Use of Hearing Protectors," Masters Thesis at North Carolina State Univ., Raleigh, NC.

Lindeman, H. E. (1976). "Speech Intelligibility and the Use of Hearing Protectors," *Audiology 15*, 348–356.

Lloyd, J. A. (1996). "Assessing Hearing Protectors for Use in Impulsive Noise," in *Proceedings of Inter-Noise 96*, edited by F. A. Hill and R. Lawrence, Inst. of Acoustics, St. Albans, UK, 971–974.

Lundin, R. (1992). "Properties of Hearing Protector Rating Methods," in *Proceedings, Hearing Conservation Conference*, Off. Eng. Serv., Univ. Kentucky, Lexington, KY, 55–60.

Martin, A. M. (1979). "Dependence of Acoustic Attenuation of Hearing Protectors on Incident Sound Level," *Br. J. Ind. Med. 36*, 1–14.

McKinley, R. L., Steuver, J. W., and Nixon, C. W. (1996). "Estimated Reductions in Noise-Induced Hearing Loss by Application of ANR Headsets," in *Scientific Basis of Noise Induced Hearing Loss*, edited by A. Axlesson, H. Borchgrevink, R. P. Hamernik, P. Hellstrom, D. Henderson, and R. J. Salvi, Thieme Medical Pub., Inc., New York, NY, 347–360.

Michael, K., and Letowski, T. R. (1997). "A Field-Monitoring System for Insert Type Hearing Protectors," *Spectrum Suppl. 1*, 14, p. 29.

Michael, P. L., Kerlin, R. L., Bienvenue, G. R., Prout, J. H., and Shampan, J. I. (1976). "A Real-Ear Field Method for the Measurement of the Noise Attenuation of Insert-Type Hearing Protectors," National Institute for Occupational Safety and Health, U.S. Dept. of HEW, Rept. No. 76–181, Cincinnati, OH.

NIOSH (1998). "Criteria for a Recommended Standard — Occupational Noise Exposure, Revised Criteria," Natl. Inst. for Occup. Safety and Health, DHHS (NIOSH) Pub. No. 98-126, Cincinnati. OH.

Nixon, C. W., Hille, H. K., and Kettler, L. K. (1967). "Attenuation Characteristics of Earmuffs at Low Audio and Infrasonic Frequencies," Report AMRL-TR-67-27, Wright-Patterson, AFB, OH.

Nixon, C. W., and Knoblach, W. C. (1974). "Hearing Protection of Earmuffs Worn Over Eyeglasses," Aerospace Medical Research Laboratory, Report No. AMRL-TR-74-61, Wright-Patterson AFB, OH.

Noble, W. G., and Russell, G. (1972). "Theoretical and Practical Implications of the Effects of Hearing Protection Devices on Localization Ability," *Acta Otolaryngol. 74*, 29–36.

Ohlin, D. (1975). "Personal Hearing Protective Devices Fitting, Care, and Use," U.S. Army Environmental Hygiene Agency, Report No. AD-A021 408, Aberdeen Proving Ground, MD.

Ohlin, D. (1997). "Personal Communication."

OSHA (1983). "Occupational Noise Exposure; Hearing Conservation Amendment; Final Rule," Occupational Safety and Health Administration, 29CFR1910.95 *Fed. Regist. 48*(46), 9738–9785.

OSHA (1987a). "OSHA Instruction CPL 2-2.20A, change 2, March 1," in *Industrial Hygiene Technical Manual,* U.S. Government Printing Office, Washington, DC, pp. VI-13–VI-20.

OSHA (1987b). "OSHA Instruction CPL 2.45A, change 12, September 21," in *Field Operations Manual*, U.S. Government Printing Office, Washington, DC, pp. IV-32–IV-34.

Paakkonen, R., and Tikkanen, J. (1991). "Attenuation of Low-Frequency Noise by Hearing Protectors," *Ann. Occup. Hyg. 35*(2), 189–199.

Park, M. Y., and Casali, J. G. (1991). "A Controlled Investigation of In-Field Attenuation Performance of Selected Insert, Earmuff, and Canal Cap Hearing Protectors," *Hum. Factors 33*(6), 693–714.

Pirzanski, C. Z. (1997). "Critical Factors in Taking an Anatomically Accurate Impression," *Hearing J. 50*(10), 41–48.

Price, I. R., and Whitaker, G. (1986). "Noise Exposure During Shotblasting and the Acoustic Properties of Air-Fed Helmets," in *Proceedings of Inter-Noise '86*, edited by R. Lotz, Noise Control Foundation, Poughkeepsie, NY, 559–564.

Riko, K., and Alberti, P. W. (1982). "How Ear Protectors Fail: A Practical Guide," in *Personal Hearing Protection in Industry*, edited by P. W. Alberti, Raven Press, New York, NY, 323–338.

Rink, T. L. (1979). "Hearing Protection and Speech Discrimination in Hearing-Impaired Persons," *Sound and Vibration 13*(1), 22–25.

Royster, J. D., Berger, E. H., Merry, C. J., Nixon, C. W., Franks, J. R., Behar, A., Casali, J. G., Dixon-Ernst, C., Kieper, R. W., Mozo, B. T., Ohlin, D., and Royster, L. H. (1996). "Development of a New Standard Laboratory Protocol for Estimating the Field Attenuation of Hearing Protection Devices. Part I. Research of Working Group 11, Accredited Standards Committee S12, Noise," *J. Acoust. Soc. Am. 99*(3), 1506–1526.

Royster, L. H. (1995). "In Search of a Meaningful Measure of Hearing Protector Effectiveness," *Spectrum 12*(2), p. 1 and 6–13.

Royster, L. H., and Holder, S. R. (1982). "Personal Hearing Protection: Problems Associated with the Hearing Protection Phase of a Hearing Conservation Program," in *Personal Hearing Protection in Industry*, edited by P. W. Alberti, Raven Press, New York, NY, 447–470.

Royster, L. H., and Royster, J. D. (1984). "Hearing Protection Utilization: Survey Results Across the USA," *J. Acoust. Soc. Am. Suppl. 1*, 76, S43.

Royster, L. H., and Royster, J. D. (1985). "Hearing Protection Devices," in *Hearing Conservation in Industry*, edited by A. S. Feldman and C. T. Grimes, Williams and Wilkins, Baltimore, MD, 103–150.

Royster, L. H., Royster, J. D., Berger, E. H., and Skrainar, S. F. (1984). "Personal Audio Headsets and Table Radios in Industrial Noise Environments," *ASHA 26*(10), 77.

Russell, G. (1977). "Limits to Behavioral Compensation for Auditory Localization in Earmuff Listening Conditions," *J. Acoust. Soc. Am. 61*(1), 219–220.

SAA (1988). "Acoustics — Hearing Protectors," AS 1270–1988, Standards Association of Australia, N.S.W., Australia.

SA/SNZ (1998). "Occupational Noise Management Part 3: Hearing Protector Program," AS/NZS 1269.3:1998, Standards Australia and Standards New Zealand, Homebush, Australia.

Schmidt, J. W., Royster, L. H., and Pearson, R. G. (1982). "Impact of an Industrial Hearing Conservation Program on Occupational Injuries," *Sound and Vibration 16*(5), 16–20.

Schroeter, J. (1986). "The Use of Acoustical Test Fixtures for the Measurement of Hearing Protector Attenuation. Part I: Review of Previous Work and the Design of an Improved Test Fixture," *J. Acoust. Soc. Am. 79*(4), 1065–1081.

Schroeter, J., and Poesselt, C. (1986). "The Use of Acoustical Test Fixtures for the Measurement of Hearing Protector Attenuation. Part II: Modeling the External Ear, Simulating Bone Conduction, and Comparing Test Fixture and Real-Ear Data," *J. Acoust. Soc. Am. 80*(2), 505–527.

Smoorenburg, G. F. (1996). "Assessment of Hearing Protector Performance in Impulsive Noise. Final Report," TNO Human Factors Research Inst., Rept. TM-96-CO42, Soesterberg, Netherlands.

Stewart, K. C., and Burgi, E. J. (1979). "Noise Attenuating Properties of Earmuffs Worn by Miners, Volume 1: Comparison of Earmuff Attenuation as Measured by Psychophysical and Physical Methods," Final Report Volume 1 on Contract No. J0188018, Univ. of Pittsburgh, Pittsburgh, PA.

Suter, A. H. (1992). "Communication and Job Performance in Noise: A Review," Am. Speech-Language-Hearing Assoc., *ASHA Monograph No. 28*, Rockville, MD.

Thomas, W. C., and Casali, J. G. (1995). "Instructional Requirements for Using the HML and NRR Methods for Estimating Protected Exposure Levels under Hearing Protectors," Auditory Systems Laboratory, Virginia Tech, Rept. No. 3/1/95-1-HP; ISE Rept. 9502, Blacksburg, VA.

Waugh, R. (1984). "Simplified Hearing Protector Ratings — An International Comparison," *J. Sound Vib. 93*(2), 289–305.

Wilkins, P. A. (1980). "A Field Study to Assess the Effects of Wearing Hearing Protectors on the Perception of Warning Sounds in an Industrial Environment," Inst. of Sound and Vibration Research Contract Report No. 80/18, Southampton, England.

Wilkins, P. A., and Martin, A. M. (1978). "The Effect of Hearing Protectors on the Perception of Warning and Indicator Sounds — A General Review," Inst. of Sound and Vibration Research Tech. Report No. 98, Southampton, England.

Ylikoski, M. E., Pekkarinen, J. O., Starck, J. P., Paakkonen, R. J., and Ylikoski, J. S. (1995). "Physical Characteristics of Gunfire Impulse Noise and its Attenuation by Hearing Protectors," *Scand. Audiol. 24*(1), 3–11.

The Noise Manual, revised 5th edition, edited by E.H. Berger,
L.H. Royster, J.D. Royster, D.P. Driscoll, and M. Layne
©2003 American Industrial Hygiene Association

11

Audiometric Monitoring Phase of the HCP

Julia Doswell Royster

Contents

Purposes of the Audiometric Phase

Audiometric evaluations within a hearing conservation program (HCP) are often thought of as serving merely a *monitoring* function: after hearing threshold levels (HTLs) change from baseline values by an amount defined in the relevant regulations, then HCP personnel react to reduce the potential for further progression of threshold shifts. Requirements of the Hearing Conservation Amendment (HCA) promulgated by U.S. Department of Labor's Occupational Safety and Health Administration (OSHA, 1983a), as well as similar regulations in the U.S. Department of Defense (DoD, 1996) and some other countries, require baseline audiograms (hearing evaluations) to define the employee's hearing ability around the time of entry into the HCP, followed by annual audiograms to detect specified amounts of hearing change which may be related to inadequate protection from noise exposure. Viewed in this limited monitoring perspective, audiograms are an expensive exercise in documenting hearing loss and reacting to it after the fact.

However, in a more proactive view the audiometric monitoring phase offers the employer's best chance to make the entire HCP meaningful and effective for the individual, thereby preventing noise-induced hearing loss (NIHL). This goal can be attained only if the face-to-face contact at the time of the audiogram, as well as subsequent written feedback and follow-up actions, are used for the purposes of *educating and motivating the employee* (Royster, 1985). Individuals who receive useful feedback about their hearing status are much more likely to assume responsibility for preserving their hearing.

The audiometric phase of the HCP serves additional secondary purposes. On a group basis, audiometric results provide data for *assessing overall program effectiveness* (see Chapter 12). Annual audiometry also provides a bonus *health screening benefit* for employees, since non-noise-related ear problems may be identified. Monitoring audiograms also become *legal records* which are relevant to proceedings such as workers' compensation claims and third-party-liability legal actions (see Chapter 18), as well as *revealing the employer's potential liability* in terms of compensation for occupational hearing loss.

Note that this chapter deals primarily with audiometric evaluations conducted within the context of an HCP for employees with noise exposures at OSHA's action level or higher. Many employers offer audiometric evaluations (perhaps less often than annually) as a wellness benefit to employees without occupational noise exposures, or with low exposures below OSHA's action level. Because these employees would not be included in the HCP on the basis of sound surveys (see Chapter 7), OSHA requirements for review and follow-up (as discussed in this chapter) would not apply to these wellness audiograms. However, the chapter's recommendations for ensuring high-quality audiograms would still be relevant.

Benefits of the Audiometric Phase Depend on Its Quality

Too often audiometric monitoring is a half-hearted effort performed only for regulatory compliance: employees are tested swiftly and impersonally by technicians who are uninterested in the outcome, and the workers themselves are told little about the results or their implications. Employees are quick to notice the attitude of disinterest, as well as noisy test booth conditions, careless instructions by the examiner, and other indications of poor test quality. In such a context employees correctly deduce that the employer is giving only lip service to the HCP, and they respond by participating only half-heartedly themselves. Regulatory compliance becomes the primary benefit of the audiometric phase, at a very real cost of the loss of employee confidence in the HCP and unnecessary NIHL.

In contrast, if audiometry is performed carefully by a sincere examiner who shares the results via immediate feedback to the worker (both by verbal comments and by visual illustration of hearing changes using computer displays), who investigates the potential connection between apparent hearing changes and the fit of hearing protection devices (HPDs) and the consistency of HPD use, and who emphasizes preservation of hearing for enjoyment of one's life, then the employee perceives the personal value of his/her own hearing and the benefit of both the audiometric phase and the overall HCP (Gasaway, 1996). In this case the employer gains extra benefits in terms of employee relations and employee participation in HCP efforts. Good-quality audiometric data also provide a sound basis for evaluating HCP effectiveness and assessing potential liability for hearing loss claims, whereas invalid or poor-quality data serve as evidence of an inadequate HCP and increase the employer's risk for hearing loss claims.

In summary, the benefits of the audiometric phase depend on obtaining valid and reliable threshold measurements, and on using them to motivate individual employees and HCP personnel to take hearing conservation efforts seriously.

Options for Implementing the Audiometric Monitoring Phase

Audiograms may be performed through in-house testing programs or through contracting for testing either at a local clinic or by a mobile van service (see Chapter 6). The audiometric phase must be supervised by an audiologist or by a physician for quality control and to meet requirements of OSHA's HCA. Audiograms may be performed under professional supervision by qualified technicians, who may also screen the results under supervision. However, nonroutine results require review by the professional supervisor and potential referral for clinical testing and/or medical examination. The choices made by the employer in selecting among options for obtaining and reviewing audiograms are crucial for HCP effectiveness, particularly the qualifications of the persons administering audiograms and the expertise of the professional who reviews results.

Importance of the Baseline Audiogram

For audiometric monitoring to be fully effective in detecting hearing changes in noise-exposed personnel, baseline audiograms must reflect employees' true hearing status prior to any occupational noise damage on the current job. Ideally, baseline audiograms should be administered before the worker is first assigned to a noisy work area. (This may also be required to meet workers' compensation requirements for identifying pre-existing hearing loss, as discussed in Chapter 18.) OSHA's HCA requires that baselines be administered within 6 months of first noise exposure, but this time window is extended to a year if a mobile testing service is used (with mandatory HPD use after the first 6 months). As discussed in Chapter 17, the rate of growth of noise-induced permanent threshold shift (NIPTS) is most rapid during the earliest months and years of noise exposure. Therefore, if the baseline is delayed for 6 months or more, an inadequately protected employee may already have incurred significant NIPTS before the first audiogram — NIPTS which can never be detected. HCP personnel are urged to perform baselines quickly to provide a true measure of pre-exposure hearing status. If careful HPD fitting and training in proper HPD use are provided together with the baseline audiogram *before* the first noisy job assignment, the employee will learn about protecting his/her hearing from the outset.

Some employers include audiograms as part of an initial employment physical performed off-site. As discussed later, differences in audiometric equipment and methods may cause hearing thresholds measured in such a site to differ systematically (either better or worse) from thresholds in later HCP audiograms, and therefore to be undesirable for use as baselines. In addition, months or years may elapse between hire and first assignment to a noisy job, allowing the possibility that hearing will change in the interim. Rather than relying on an older pre-employment test as baseline, it is best to obtain a baseline audiogram performed just prior to assignment to noise, using the same equipment and methods as later HCP audiograms.

OSHA's HCA required baselines to be established prior to March 1, 1984 for HCPs being newly established at that time. Many employees had worked for many years (often without hearing protection) prior to HCP initiation. Some employers, who had voluntarily implemented audiometric monitoring in earlier years, elected to establish new baselines prior to the 1984 deadline, especially in cases when old audiograms did not meet all of OSHA's specifications. However, OSHA preferred that the earliest valid audiograms be retained as baselines in order to reflect employees' hearing changes over time.

Scheduling Affects Detection of Shifts

NIHL usually develops when repeated temporary threshold shifts (TTSs) gradually result in NIPTS, as described in Chapter 5. Detection of TTS enhances the preventive potential of audiometric monitoring. If employees wear HPDs inconsistently or incorrectly and TTS is found on annual audiograms, this discovery allows the examiner to react to the shifts by refitting HPDs and counseling the employee. In contrast, if audiograms are scheduled only prior to the beginning of the workshift, then TTS cannot be found and shifts will be permanent before they are detected. Some employers prefer to schedule tests prior to the workshift because this reduces the number of retests which need to be performed. However, this reasoning is flawed since, by failing to identify and react to TTS, the employer is increasing the likelihood that NIPTS will develop. The best policy is to schedule annual audiograms throughout the workshift, with no special precautions to wear HPDs more carefully on the day of testing. (Baseline audiograms, in contrast, need to be preceded by 14 hours noise-free, as discussed later.)

In-house vs. Contracted Testing

The number of employees to be included in the HCP and their geographic locations are important considerations in deciding which approach is most practical: equipping and staffing an in-house test facility (either at a fixed site or in a mobile vehicle) versus contracting for services with a mobile testing service or with a local clinic. Advantages and disadvantages of various approaches are summarized as Table 11.1.

Cost is one factor, but cost should not be evaluated without specifying a minimum quality of service, since the employer generally gets what is paid for. The lowest bidder for contract testing probably will not provide the quality of time-of-test employee interaction needed to motivate workers to prevent hearing loss, and may not even provide a technically acceptable test environment or adequately trained personnel. If a contractor is used, the contractor's performance will reflect either well or poorly on the employer and will therefore affect the overall HCP. HCP representatives can interview prospective service providers about important technical aspects of audiometry in mobile testing. However, the interpersonal rapport of the OHC with test subjects and the presence or absence of professional demeanor can only be judged through first-hand observation of a contractor actually providing services, and by interviewing employees who have received audiograms provided by this service. It is very wise to visit a site where the contractor's personnel are doing testing to evaluate their services first-hand (without an arranged appointment if possible), rather than relying on brochures or sales pitches as a basis for decision making. Local clinics vary from expert professionals in private practice, who will personally see each employee, to 24-hour emergency services where the technicians administering audiograms are uncertified and the physician has little knowledge about hearing loss in general, especially NIPTS. Likewise, mobile testing services vary from audiologists personally performing testing and explaining results to single employees, to uncertified technicians overseeing microprocessor testing of 8–10 persons simultaneously.

The employer still bears responsibilities even when audiograms are performed by a contractor. There must still be an on-site *key individual* for the HCP who helps plan the test schedule, ensures that employees receive feedback regarding

TABLE 11.1

Advantages and disadvantages of in-house and external audiometric testing, based on the most common staffing qualifications (in-house testing by OHCs, clinic testing by audiologists, mobile testing by OHCs). Different staff qualifications, such as mobile testing by audiologists who would be able to give immediate professional feedback, would change the comparisons.

	Advantages	Disadvantages
In-house testing	• most flexible time scheduling • OHC can demonstrate employer's commitment to HCP (if sincere) • close control of procedures possible • stable audiometric equipment offers potential for low audiometric variability • immediate feedback can relate to the subject's particular job and noise exposure	• requires investment in audiometer(s) and test booth(s) • calibration and repair expenses • certification for OHC, including 5-year renewals and replacement if staff members leave • time demand on OHC • a professional supervisor must be identified either in-house or externally
Local audiology and/or otolaryngology clinic	• no investment for equipment or OHC • no in-house time demand for testing • potentially credible examiner (if an audiologist familiar with HCPs) • both immediate professional feedback and professional review may be included • same clinic serves as referral source	• transportation time loss • less flexible scheduling • HCP is dependent upon the degree of interest and expertise in HCP issues shown by the local professional and staff • examiner/reviewer has limited knowledge of workplace noise environment and job demands
Local occupational health clinic	• no investment for equipment or OHC • no in-house time demand for testing • immediate physician review may be included (but the expertise varies) • other health testing is available	• transportation time loss • less flexible scheduling • HCP is dependent upon the degree of interest and expertise in HCP issues shown by the clinic staff • OHC and reviewer may be unfamiliar with workplace noise environment and job demands
Mobile testing	• no investment for equipment or OHC • no in-house time demand for testing • delayed professional review and computer data entry are usually included • other health testing may be available • most annual testing can be done in one van visit (excluding absentees, retests)	• potential for high audiometric variability if multiple audiometers are not equivalent • least flexible scheduling • to avoid delays, will probably need alternative test source for new hires, absentees, and retests • OHC and reviewer may be unfamiliar with workplace noise environment and job demands

461

their results, provides needed follow-up for employees with hearing changes, carries out the recommendations of the professional audiogram reviewer, and informs supervisors, managers, and HCP team members about audiometric trends (see Chapter 6).

Qualifications of Personnel

The expertise and interest of the personnel involved in the audiometric phase are critical to its success, since these individuals interact with both noise-exposed employees and management.

Professional Supervisor and Referral Sources

Either an audiologist or a physician must supervise the audiometric phase of the HCP. If the supervisor is a physician, then an otolaryngologist, otologist, or occupational physician is usually more knowledgeable about noise-induced permanent threshold shift (NIPTS) than a physician in an unrelated specialty. Otolaryngologists are physicians specializing in diagnosis and treatment of ear, nose, and throat disorders; otologists further restrict their practice to the ear. Occupational physicians specialize in prevention and treatment of occupational health problems of all types. The most highly qualified physicians are board-certified in their specialties, and specialty certification in hearing conservation for audiologists is under development by the American Speech-Language-Hearing Association (ASHA). Audiologists are advanced-degree professionals who are certified by ASHA (and licensed in most states) to conduct diagnostic audiological evaluations, dispense hearing aids, and provide aural rehabilitation services.

Regardless of credentials, some audiologists and physicians have more training and expertise than others concerning noise exposure measurements, the effects of noise on hearing, and hearing conservation practices, including the selection of appropriate HPDs. The wise employer will select professionals who can demonstrate specific expertise in hearing conservation, as they are most likely to provide correct and practical consulting services either as supervisors for the audiometric phase or as referral sources. Although OSHA allows any type of physician to serve as the audiometric phase supervisor, the employer is less likely to receive satisfactory service from a general practitioner or a physician with an unrelated specialty. Likewise, audiologists with specialized knowledge of occupational hearing conservation will provide superior advice than those with no familiarity with this specialty. Note that the erroneous opinions which may be obtained from professionals without practical experience in hearing conservation may have deleterious effects on the HCP; therefore, it is important to seek a competent professional supervisor. If the choice of professionals is restricted in a remote geographical area, the employer should urge the local practitioners to obtain extra training through independent study and by joining and/or attending workshops sponsored by relevant associations, such as the National Hearing Conservation Association (NHCA) and the American College of Occupational and Environmental Medicine (ACOEM, formerly ACOM).

Hearing Examiners

HCP audiograms, which involve only basic air-conduction threshold measurements, may be performed by technicians under the supervision of an audiologist or physician. OSHA specifies that hearing examiners conducting manual audiometry either must demonstrate competence to the professional supervisor or must be certified by the Council for Accreditation in Occupational Hearing Conservation (CAOHC) via successful completion of a 2.5-day course, earning the designation of occupational hearing conservationist (OHC). CAOHC is a multidisciplinary organization (governed by a board comprising representatives of professional associations of industrial hygienists, occupational health nurses, audiologists, otolaryngologists, occupational physicians, safety professionals, and noise control engineers) which implements training standards for OHCs and grants certification to the course directors who supervise OHC training. OHC certification must be renewed every 5 years via an 8-hour refresher course. CAOHC certification is preferred over OSHA's allowed informal "demonstration of competence" because it guarantees that the hearing examiner has been taught background information about hearing conservation and has passed a written test plus a practicum exam in administering audiograms. In this chapter the hearing examiner will be referred to as the OHC to emphasize the importance of the competence implied by earning and renewing certification.

Unfortunately, OSHA allows microprocessor audiometers to be operated by persons lacking such certification. The employer should question potential service providers regarding the qualifications of hearing examiners and should contract only with services staffed by certified OHCs. An uncertified person operating a microprocessor audiometer may be deficient in knowledge regarding audiometric factors (earphone placement, the importance of proper instructions, and necessary recordkeeping) and hearing conservation concepts (fit-checking of hearing protectors, relevance of noisy off-job activities, etc.). Moreover, some employees are unable to respond validly to microprocessor testing and require manual audiometry, which untrained examiners cannot provide.

Rapport With Employees

To succeed in motivating personnel receiving HCP audiograms, both professionals and OHCs need to demonstrate sincere interest in each noise-exposed individual. They should plan to spend enough time to explain the audiogram results to the employee and counsel him or her about hearing trends in relation to noise exposure and HPD use. The ability to project sincerity and enthusiasm is a quality the employer should seek when choosing in-house personnel to become OHCs or to serve as the key individual overseeing the HCP (Royster and Royster, 1990; also see Chapter 6). These critical personnel can make or break the program; audiometric responsibilities should never be entrusted to personnel who lack interest in the HCP. If a contract service provides testing, the contract personnel need to meet this same standard. Maintaining sincere interest may be a challenge for OHCs employed by testing services, since they perform so many audiograms and may seldom see the same employee twice. However, the employer can and should lay out specific expectations for the amount and quality of employee interaction which is needed in order to achieve the full motivational benefit from the audiometric phase.

Technical Factors for Valid and Reliable Audiograms

Several technical requirements are involved in obtaining good audiometric data. The goal is to minimize measurement variability so that differences in measured thresholds between tests reflect real hearing change rather than measurement error. NHCA has produced a pamphlet summarizing technical aspects of mobile hearing testing (NHCA, 1996), and CAOHC publications also deal with this topic (Suter, 1993).

Audiometric Testing Room

The testing space is usually a commercially fabricated sound-attenuating booth designed to reduce extraneous sounds which could distract the subject or cause masking (the elevation of thresholds due to background noise, as described in Chapter 4). Occasionally an in-house program may possess a room (such as an unused vault) which is adequately quiet for audiometry; however, if the audiometer is inside the test room, then the examiner must be very careful to avoid visual and acoustic distractions. Figure 11.1 shows a commercial test booth with the audiometer located outside.

Ambient Noise

The sound pressure levels (SPLs) inside the testing space are what count. Even a maximum-attenuation booth will not provide an acceptable environment if placed in an area where sound levels are excessive.

Permissible Background SPLs

The standard ANSI S3.1-1999 specifies ambient SPLs which will allow threshold measurements down to a hearing level of 0 dB using different earphones and bone-conduction vibrators. The standard specifies SPLs by octave band (OB) or 1/3 OB; overall A-frequency-weighted or C-frequency-weighted sound level measurements are not sufficient. ANSI-allowed OB SPLs for testing with supra-aural earphones at 500–8000 Hz are given in Table 11.2. Specifications are included for OBs centered below 500 Hz because upward spread of masking can result in threshold elevation in OBs higher than the masking sound itself. In a test environment complying with these specifications the individual being tested should not be distracted by external sounds. Because the 1991 version of ANSI S3.1 is referenced in recommendations discussed below, the bottom row of Table 11.2 shows the values from this earlier standard as well.

Observe that OSHA allows much higher ambient SPLs, as shown by the middle row in Table 11.2. OSHA does not specify SPLs for OBs below 500 Hz,

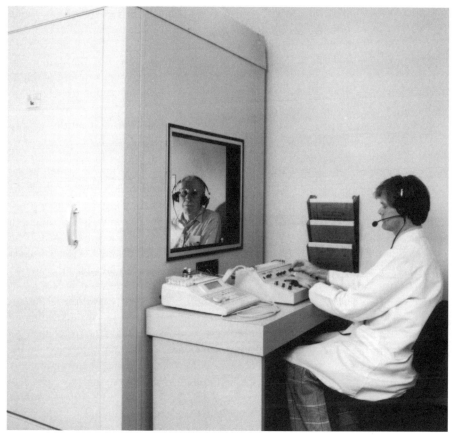

Figure 11.1—A single-subject audiometric booth with audiometer located outside. (Photo courtesy of IAC, Inc.)

TABLE 11.2

Permissible ambient noise levels in audiometric test rooms according to ANSI versus OSHA for threshold measurement at 500–8000 Hz using supra-aural earphones.

	Maximum Permissible Octave-Band SPL (dB)						
	Octave-Band Center Frequency (Hz)						
	125	*250*	*500*	*1000*	*2000*	*4000*	*8000*
ANSI S3.1-1999	49	35	21	26	34	37	37
OSHA's HCA	-	-	40	40	47	57	62
ANSI S3.1-1991*	47.5	33.5	19.5	26.5	28	34.5	43.5

* now obsolete, but referred to in NHCA (1996) recommendations

ignoring the potential for upward spread of masking. A test environment which just barely meets OSHA specifications will result in measurement of spuriously poor hearing levels due to masking across most test frequencies (Franks et al., 1992). In such cases the individuals being tested will typically complain about distracting noises, though they may be unaware if a continuous sound (such as an air conditioner) is creating steady masking. If employees frequently complain about noisy test booths, this is evidence of poor-quality audiometric monitoring.

Surveys of audiometric test booths in permanent in-house locations within industry (Frank and Williams, 1994), as well as samples of booths in mobile testing (Lankford et al., 1999), have documented that most booths in actual use meet the ANSI specifications for OBs centered at 1 kHz and up, but many fail at the 500-Hz OB and below. The NHCA (1996) recommended that OSHA adopt the ANSI S3.1-1991 specifications with the exception of allowing 5 dB more noise at the 500-Hz OB. This recommendation reflects the practical difficulty of meeting the ANSI standard at 500 Hz for mobile testing facilities. Lankford et al. (1999) found that most mobile testing booths which failed the ANSI S3.1-1991 specification at 500 Hz were able to meet the NHCA recommendation. Another rationale for relaxing the 500-Hz OB SPL is that noise-induced hearing loss (NIHL) seldom affects low-frequency thresholds. However, elevation of thresholds at 500 Hz via masking could falsely increase estimations of potential employer liability for worker's compensation or other legal claims based on hearing formulae which include 500 Hz (see Chapter 18) and might increase the incidence of recommended medical referrals for low-frequency hearing loss.

For accurate hearing threshold measurements, HCP personnel are urged to achieve an audiometric test environment which meets ANSI S3.1-1999 or subsequent revisions. SPL measurements to verify compliance with ambient noise specifications should be made in the noisiest situations when testing would be allowed, including operation of all light fixtures, ventilating fans, and other electrical items in the area. For booths with intermittent noise from external sources (such as forklifts passing by on the opposite side of a wall, or tractor-trailer truck traffic outside mobile vans), a vigilant OHC can and should temporarily pause testing during the offending sound. Commercially manufactured monitors are available which continuously sample OB SPLs within the test room to alert the technician or to electronically pause microprocessor-controlled testing if allowed SPLs are exceeded (see Figure 11.6). Again, intelligent selection of the site for the test booth can reduce such transient interference.

Noise-Reducing Earphones and Earphone Enclosures

Figure 11.2 illustrates typical supra-aural audiometric earphones, ER3-A insert earphones, and sample circumaural earphone enclosures. Insert earphones, which are in widespread use in clinical audiometric settings, reduce external noise via a sound-attenuating foam earplug fitted with a central tube which delivers the test signal directly into the earcanal. Note that ANSI S3.1-1999 allows higher ambient OB SPLs when insert earphones are used. However, OSHA has not yet made provisions to allow their use in occupational HCPs

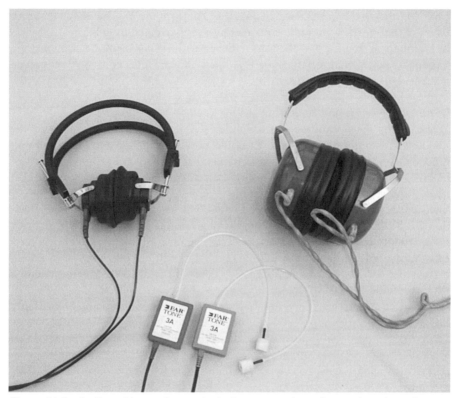

Figure 11.2—Audiometric earphones: typical supra-aural earphones, insert earphones, and noise-reducing enclosures for supra-aural earphones (not recommended).

without some complicated documentation requirements and the possibility of a "de minimis" violation (OSHA, 1993). If OSHA policies change in the future, insert earphones would make it easier for occupational HCPs to achieve compliance with the ambient noise specifications of ANSI S3.1-1999.

Several brands of noise-reducing enclosures for supra-aural earphones are available which are intended to reduce external noise reaching the test subject's ears. The audiometric earphone cushions are mounted within large earmuffs, and the earphones and enclosures can pivot separately. Such devices are not recommended for several reasons (Frank and Williams, 1993; Frank et al., 1997; Franks et al., 1989). They make consistent earphone placement difficult to judge because the examiner cannot verify whether the earphone cushion within the enclosure is coupling directly to the pinna. They do not provide enough attenuation to test 0-dB thresholds in OSHA-allowed OB SPLs. At least some models decrease the repeatability of threshold measurements. Finally, the earphones cannot be calibrated while mounted within these enclosures. Unlike the case with insert earphones, ANSI S3.1-1999 does not allow higher background SPLs

467

based on the use of earphone enclosures. They should not be relied upon in HCPs except possibly to reduce distracting noises created by other test subjects in multiperson booths, and then only if the OHCs using them can demonstrate through experimental verification that they can achieve consistent earphone placement under the enclosures.

Test Booth Selection and Location

The quieter the environment in which the audiometric test room is located, the less noise reduction the booth will have to provide to achieve acceptable background OB SPLs. When it is possible to plan for the location or relocation of an on-site booth, choose an area away from air conditioning compressors, forklift truck routes, slamming doors, ringing telephones, and other noise sources. Measure the OB SPLs in the intended booth location and subtract the specification OB SPLs at each frequency from the measured levels to obtain the required noise reduction. Because manufacturers' noise-reduction data (which are reported as measured in a laboratory setting) probably will not be achieved in a field installation (Franks and Lankford, 1984; Lankford et al., 1981), select a booth which provides 5–10 dB more noise reduction than required. Double-walled booths offer more noise reduction than single-walled units, at a correspondingly higher price. Sample noise reductions for generic types of booths are shown as Table 11.3. Additional guidance is available in published articles (Hirschorn et al., 1982; Singer, 1984), although the allowed ambient OB SPLs from the current ANSI standard must be substituted for the data from now-obsolete standards referenced in these older guides.

The particulars of booth installation are critical; therefore, experienced installers are needed. Structure-borne noise and/or vibration is a less common problem than airborne sound, but it may compromise interior booth SPLs if the attachment of the booth to the supporting building allows flanking paths (Rebke,

TABLE 11.3
Representative lab-measured noise reduction data for single-wall and double-wall commercially manufactured audiometric testing booths (field-measured noise reduction will generally be less).

	Noise Reduction (dB)						
	Octave-Band Center Frequency (Hz)						
	125	*250*	*500*	*1000*	*2000*	*4000*	*8000*
single-wall mini portable booth	15	28	38	46	52	54	>54
single-wall stationary booth	19	30	39	50	58	60	>60
double-wall stationary booth	35	45	55	65	70	75	>75

1992). After installation, verification of achieved OB SPLs within the finished booth is always required; this should be part of the contract agreement.

For mobile audiometric testing vehicles, the booth's noise reduction is fixed, so the temporary parking location of the vehicle must be chosen to ensure that interior OB SPLs in the testing spaces are acceptable. The employer contracting for mobile testing services should assist the contractor in selecting a quiet parking site and should require documentation of the ambient noise inside test booths during actual testing (not merely before testing begins) for each day of operation.

Subjects being tested on mobile vehicles are often distracted by transient vibrations such as those created when persons enter or exit the door of the test vehicle. The employer contracting for such services should evaluate how adequately such distractions are controlled by a prospective vendor. Since conditions may change from year to year, HCP personnel should seek employees' comments regarding the test environment during each annual testing cycle.

Multiperson Test Booths

Some test facilities (most often in mobile testing vehicles) accommodate from 2 to 10 subjects who receive audiograms simultaneously. An example is illustrated in Figure 11.3. In this situation the noises made by other subjects (coughing, moving in their seats, etc.) may be distracting and may create unacceptable

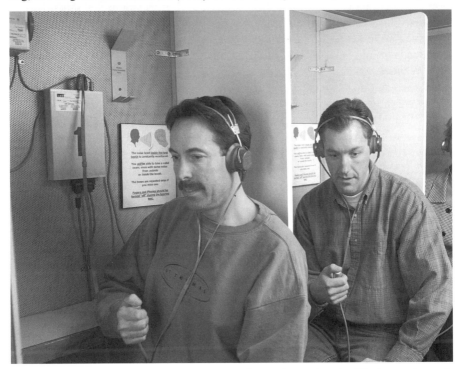

Figure 11.3—Interior of a multiple-subject audiometric test booth. (Photo courtesy of Maico Diagnostics, Eden Prairie, Minn.)

SPLs inside the booth in spite of adequate noise reduction of external sounds. Subjects must be carefully instructed to avoid making noises, and the testing cycles of the audiometers for individual subjects should be designed to keep all subjects busy responding until the last subject has finished the entire audiogram. Test-related sounds, such as clicking by poorly designed subject response handswitches, should be avoided. Subjects with widely discrepant hearing thresholds should not be tested simultaneously because test signals at high presentation levels can leak from under the earphones of a hearing-impaired subject, thereby distracting a nearby subject with better hearing; to avoid this problem, do not simultaneously test (within the same booth) subjects whose thresholds differ by 30 dB or more. The best strategy for eliminating all these problems is to use single-subject test booths rather than multiple-subject test stations within the same booth. Even with single-subject booths in close proximity, however, testing cycles need to be controlled to prevent turnover traffic and booth-door closing noise from distracting other subjects.

Subject Comfort and Attention

The testing space should be large enough that hefty or tall subjects do not feel overly cramped inside; this usually rules out the smallest commercially available sound-attenuating booths. Adequate ventilation and lighting, and a viewing window also help prevent the subject from feeling claustrophobic, although there is a tradeoff between airflow and excess noise from ventilation. The subject should be seated at 90 degrees to the booth's window (not facing the examiner) to prevent inadvertent visual cueing. Installation of a talk-over microphone circuit not only reduces subjects' feeling of isolation but also enhances testing efficiency since the examiner can clarify instructions without opening the booth door and removing the earphones.

Occasionally a subject will experience claustrophobia inside the audiometric test booth. It may be possible to relieve such concerns by demonstrating that the subject can push open the test booth door. However, if the subject remains too anxious to perform a reliable audiogram, the door can be propped slightly open during the test. This deviation from normal testing procedures and conditions should be noted on the audiogram.

Audiometers

For routine audiometry, HCPs typically utilize limited-function audiometers capable of pure-tone air conduction testing only (in contrast to clinical audiometers which include facilities for masking the nontest ear, measuring bone-conduction thresholds, and presenting speech stimuli and special auditory tests). The audiometer allows pure-tone signals at frequencies from 0.5 to 6 kHz (preferably also 8 kHz) to be presented separately to each ear of the subject via earphones at hearing levels covering at least the range of 0 to 90 dB, or preferably -10 to 100 dB. The subject's threshold at each test frequency in each ear is defined as the lowest hearing level at which s/he responds in 50% or more of trials presented following an accepted protocol such as the Hughson-Westlake ascending-descending method (ANSI S3.21-1978 [R 1997]; Suter, 1993). The minimum number of trials needed to establish threshold is three ascending series following an initial descending familiarization procedure, as illustrated in Table 11.4. A fourth ascending series is necessary if the subject responds twice at a higher hearing level and once at a lower hearing level.

Photographs illustrating different types of audiometers are shown as Figure 11.4. *Manual* audiometers require the examiner to operate the frequency dial to select the audiometer's stimulus tone (500 Hz, 1000 Hz, etc.), as well as the presentation level dial (with levels in increments of 5 dB) and the signal presentation switch (stimulus tone on or off) to administer the test. In addition to performing these tasks consistently, the examiner also must identify and document the hearing levels which qualify as thresholds. These tasks were automated during the 1970s when *self-recording* audiometers came into widespread use.

Self-recording audiometers (also called Bekesy audiometers) employ a threshold tracking procedure in which the subject's responses to test signals (via oper-

TABLE 11.4
Sequence of signal presentations involved in measuring audiometric threshold at one test frequency, with a fourth ascending series required because the subject responded once at a lower hearing level (10 dB) and twice at a higher level (15 dB) during the first three ascending presentation series. Each column shows all the signal presentations prior to and including the subject's response, which initiates a new ascending series. The threshold is measured as 10 dB in this example.

Audiometer Hearing Level Control Setting (dB)					
-5	-5	-5	-5	-5	-5
0	0	0	0	⑦ 0 ＼	0
5	5	5	⑤ 5 ＼	⑧ 5 ＼	⑪ 5 ＼
10	10	③ 10 ＼	⑥ [10]	⑨ 10 ＼	⑫ [10]
15	15	④ [15]	15	⑩ [15]	15
20	② [20]	20	20	20	20
25	25	25	25	25	25
① [30]	30	30	30	30	30
35	35	35	35	35	35
40	40	40	40	40	40
45	45	45	45	45	45
50	50	50	50	50	50
55	55	55	55	55	55
60	60	60	60	60	60
65	65	65	65	65	65
70	70	70	70	70	70
Familiarization		Ascending series for threshold determination			

Symbol key: ＼ = no response ☐ = response
Circled numbers indicate the sequence of signal presentations.
Note: the presentation range is truncated at 70 dB to save space.

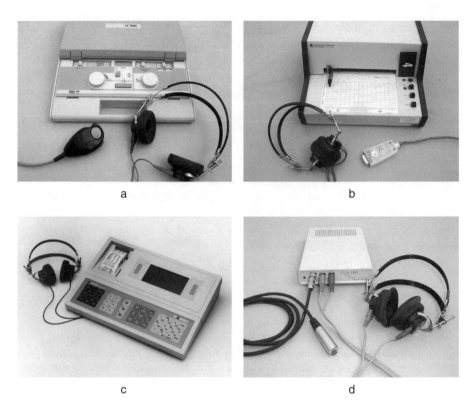

Figure 11.4—Representative audiometers of various types: manual (a), self-recording (b), microprocessor (c), and computer-controlled (d).

ation of a handswitch) create a pen tracing on a response card. In units manufactured for use in industry, the subject traced for 30 seconds per test frequency in each ear. The midpoints of the pen excursions for a given tracing can be interpreted as the hearing threshold at that frequency, as long as it is possible to draw a horizontal line which is crossed at least six times by the tracing. See Figure 11.5 for samples of acceptable and unacceptable tracings. Although self-recording audiometers are now largely obsolete, many current employees have past tests which were administered using them.

During the 1980s *microprocessor* audiometers became the preferred technology for automatic test administration. These audiometers contain a computer chip which controls the audiometer to present signal presentation trials following a paradigm similar to a manually administered audiogram, defines the subject's threshold based on an algorithm for interpreting the subject's response pattern to the series of trials, and prints out the results and/or transfers the retained information to a computer for electronic data storage. A variant of this concept is the *computer-controlled* audiometer, in which software in a personal computer

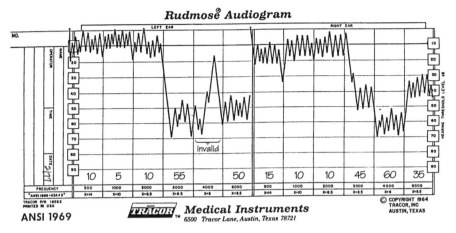

Figure 11.5—Sample tracings from a self-recording or Bekesy audiometer, with invalid tracing noted and thresholds interpreted for acceptable tracings.

drives the audiometer, which usually cannot be used separately from the computer. Independently operable microprocessor audiometers may also be connected to a computer in order to take advantage of software for recordkeeping and for immediate feedback of audiogram results to test subjects.

The type of audiometer used may influence measured thresholds, affecting comparisons among tests obtained with different test methods. Thresholds obtained using self-recording audiometers are generally about 2.5 dB more sensitive than thresholds obtained using manual audiometry with the standard 5-dB step size (Harris, 1980). Manual and computer-controlled audiometry yield identical thresholds if an identical method is employed by both the human examiner and the microprocessor algorithm (Jerlvall et al., 1983). However, it is difficult to obtain exact descriptions of the algorithms from microprocessor audiometer manufacturers, and different units employ different paradigms. Therefore, there may be small but systematic differences among thresholds obtained with various brands in comparison to each other or to those measured by a human examiner. Research detailing these differences is lacking, and a standards working group (Working Group 76 under ANSI Accredited Standards Committee S3, Bioacoustics) established to develop minimum reliability requirements across paradigms has not yet drafted a standard. Although small differences are insignificant in an absolute sense, HCP personnel should avoid unnecessary changes in test methods because inconsistencies obscure the longitudinal trend identification which is the goal of monitoring audiometry. Rates across a population of defined hearing threshold shifts (such as OSHA standard threshold shift, or STS) can be significantly affected by changes in audiometric methods (as well as other factors). In shopping for an audiometer, therefore, the buyer should inquire what data the manufacturer can provide regarding reliability and validity of threshold measurements. Manufacturers are likely to advertise rapid test speed; the buyer should ensure that speed is not achieved at the expense of reliability and validity. Another feature of microprocessor algorithms which can affect threshold validity is the timing of signal presentations; if presentations are too regular, the subject may respond according to the established rhythm even if s/he does not actually hear the tones, leading to spuriously sensitive apparent threshold measurements.

The advantages of automated audiometry (whether the older self-recording method or the modern microprocessor variant) include relieving the examiner of a repetitive task while freeing several minutes to carry out recordkeeping or other activities, maintaining absolutely consistent testing methods, and eliminating the possibility of human error in threshold documentation. However, some subjects will not respond well to automated audiometry and will require manual testing to obtain reliable thresholds. Therefore, the hearing examiner needs to be qualified to administer a manual audiogram (as through CAOHC certification) and should be alert to recognize subjects who would benefit from manual testing. When the validity of an automated test is in question, it is desirable for the audiometer to be capable of printing out a complete history of all responses and nonresponses at varying presentation levels as a permanent record. When a manually administered test is necessary, the microprocessor audiometer should be easy to use for this purpose.

Calibration of Audiometers

Hearing threshold measurements will be only as accurate as the calibration of the audiometer used to obtain them. Documentation of calibration checks is a critical aspect of HCP recordkeeping to support the validity of audiograms. Requirements for audiometers and the characteristics of the pure-tone signals they generate are detailed in ANSI standard S3.6 *Specification for Audiometers*, which is regularly updated (currently ANSI S3.6-1996). However, OSHA's HCA refers to the now obsolete version ANSI S3.6-1969. Older audiometers which cannot meet the tighter tolerances of the current ANSI standard are still acceptable for OSHA purposes if they meet the 1969 standard referenced in Appendix E of OSHA's HCA.

Daily Calibration Checks and Listening Checks

Before each day of use, the audiometer's operation must be verified as required by OSHA's HCA. The first step is for a human listener to perform a qualitative listening check to ascertain that the pure-tone signals are present and undistorted at each test frequency in each earphone, that the signal level increases and decreases appropriately as the presentation level control is changed, and that no extraneous sounds are heard (such as clicking, crackling, buzzing, or "crosstalk" sounds in the nontest earphone). If any of these problems is detected, then the audiometer needs repair and recalibration before it can be used. At least monthly the audiometer's physical condition should be inspected for signs of wear, including loss of tension in the headband for the earphones, hardening of the earphone cushions, fraying of electrical cords, loosening of electrical jacks, etc. Sample forms for listening check results and physical inspections are available (Suter, 1993).

The second step in the daily calibration check is to verify the earphone SPL outputs by testing a subject with known, stable thresholds in the normal range. If a human subject is used, then the process is known as a biological calibration check. However, it is also acceptable to use a commercial device called a bioacoustical simulator in place of a human subject. A bioacoustical simulator consists of receptacles for the audiometer earphones connected to a microphone circuit which detects the pure-tone signals at relatively high SPLs corresponding to around 60–70 dB hearing level. One such device is shown in Figure 11.6. Regardless whether the subject is human or simulated, if measured thresholds deviate by ± 10 dB from previously established stable values, then OSHA stipulates that the audiometer cannot be used until it has been recalibrated. If human subjects are used, it is wise to have multiple subjects so that testing can proceed even if one individual is absent or exhibits a temporary threshold elevation due to a cold. Each subject should be used on a regular basis, to ensure that his/her hearing levels remain stable.

Figure 11.6—A bioacoustical simulator with audiometric earphones in position for the daily output check, and microphone (front right) that can be used to monitor test booth noise levels.

Note that human subjects who have thresholds of 0 to 5 dB are not acceptable as biological calibration subjects because their measured thresholds are already near the lower limit of the audiometer's range. If the audiometer's SPL output increased, the thresholds of these subjects could not show 10 dB of improvement. Therefore, subjects with hearing levels of 15–25 dB are ideal as biological test subjects.

Each time the audiometer undergoes acoustic or exhaustive calibration, the reference hearing levels of the biological test subjects or bioacoustic simulators must be re-established separately for *each* audiometer the subjects will be matched with. It is a good idea to use the average thresholds for at least three tests per audiometer/subject pairing as the reference hearing levels to serve as targets for future daily calibration checks. A sample form for recording reference levels and daily calibration and listening checks is shown as Figure 11.7.

Acoustic and Exhaustive Calibrations

The OSHA HCA mandates that audiometers receive an acoustic calibration annually, entailing verification that the output SPLs at each test frequency are correct when the presentation level is set to 70 dB on the hearing level dial, and that the presentation level control accurately increases and decreases the output SPLs throughout the range of hearing levels. Every 2 years an exhaustive calibration is required, including additional measurements such as the accuracy of the frequency of test tones, signal rise time and fall time, harmonic distortion, etc., as specified in ANSI S3.6-1996. For explanatory descriptions of calibration procedures and rationale, see Wilber (1994). An illustration of an earphone mounted in one type of coupler for output measurements is shown as Figure 11.8.

Although it is possible to pack up an audiometer and ship it away for calibration, this is undesirable because the owner has no way of knowing whether rough handling during the return shipment may have affected the unit. In addition, when the audiometer is calibrated in place, then the measurements assess the entire system as actually used, including the electrical jacks connecting the audiometer to the earphones through the wall of the test booth.

HCP personnel should instruct the calibration service to measure and document output SPLs through each earphone before recommending any adjustments, and not to make adjustments without permission. If adjustments are necessary, each one should be documented, and then the output SPLs must be remeasured and rerecorded after calibration. The comparison of the pre- and postcalibration measurements and the adjustment notes document the size of any necessary changes to the audiometer's output SPLs. To avoid unnecessary seesaw effects in output SPLs, HCP personnel should not allow adjustments of the audiometer output if the deviations from target values are within the measurement accuracy of the calibration instrumentation and procedures used (that is, the range of values which would be expected if the measurement process were repeated multiple times in a row, repositioning the earphones in the coupler before each measurement).

AUDIOMETER BIOLOGICAL CALIBRATION LOG

AUDIOMETER: make: ________ model: ________ serial: ________ L phone: ________ R phone: ________

date of last acoustic calibration: ________ date of last exhaustive calibration: ________ (must be within 2 years)

TEST SUBJECT'S NAME: ________________ If subject is inexperienced, do 3 practice audiograms prior to reference tests.

REFERENCE HTLs:	LEFT EAR Thresholds by Frequency, Hz 500	1000	2000	3000	4000	6000	8000	RIGHT EAR Thresholds by Frequency, Hz 500	1000	2000	3000	4000	6000	8000
Ref Test 1 (/ /)														
Ref Test 2 (/ /)														
Ref Test 3 (/ /)														
Average of above =														
Reference HTLs: round average to nearest 5 dB														

ACCEPTABLE RANGE														
Low: Ref. HTL minus 5 dB														
High: Ref. HTL plus 5 dB														

On a daily calibration check, if the subject's measured threshold falls outside his or her acceptable range, then an acoustic calibration is needed unless a second subject's biological check is OK. If all subjects show a consistent trend toward one edge of their ranges, consider a recalibration.

CALIBRATION CHECKS:	LEFT EAR Thresholds by Frequency, Hz 500	1000	2000	3000	4000	6000	8000	RIGHT EAR Thresholds by Frequency, Hz 500	1000	2000	3000	4000	6000	8000	Listening check	OHC's initials
DATE / /																
DATE / /																
DATE / /																
DATE / /																
DATE / /																
DATE / /																
DATE / /																
DATE / /																
DATE / /																
DATE / /																
DATE / /																
DATE / /																
DATE / /																
DATE / /																
DATE / /																

Figure 11.7—Sample form for documenting the calibration check / listening check of the audiometer before each day of use.

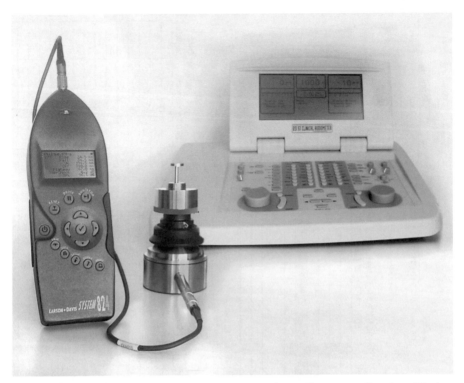

Figure 11.8—A supra-aural earphone in a coupler for audiometer calibration. (Photo courtesy of Larson•Davis.)

Choosing a reputable calibration service is important to achieve consistency in audiometer performance over time. HCP personnel should contract a calibration service only after inquiring about the training of the technicians (such as certification by the National Association of Special Equipment Distributors [NASED, 1999]), whether the calibration of the instrumentation used to measure audiometer performance is traceable and frequently reverified, and what the measurement error is for the calibration procedures and instrumentation used. The goal is to ensure that measurement results will be the same regardless of the particular technician who performs the calibration and the particular instruments in use that day. Switching among calibration services needlessly is likely to increase variability in audiometric threshold measurements, thereby reducing the ability of monitoring audiometry to identify real changes in hearing levels, as well as adding artifactual threshold shifts which then require follow-up actions.

Modern audiometers with solid state electronics are much more stable than older units (which contain potentiometers), and they allow separate calibration adjustments of the output SPLs in each earphone. Older audiometers depend upon a matched frequency response for the left and right earphones to maintain equivalent outputs in the two sides; calibration adjustments affect both earphones in this case. The earphones are the most vulnerable part of the audiometer; their frequency response can change if they are dropped or otherwise mistreated. Note that earphones cannot be switched between audiometers, as the calibration is correct only for the specific combination of audiometer plus earphones. If earphones need to be replaced, then the audiometer must be recalibrated with the new earphones. Annual calibration time is a convenient time to replace hardened earphone cushions and earphone headbands with weakened tension.

Whenever multiple audiometers are used in a HCP, special care is needed in calibration control to achieve equivalence in SPL outputs among all the audiometers which might be used to test an employee. In this case it is a good idea to obtain an acoustic calibration check if a persistent deviation of only ± 5 dB is found from the reference thresholds established for an audiometer as paired with one or more biological subjects or bioacoustical simulators. There will be increased variability in measured thresholds for employees if the audiometers are not equivalent, even if each audiometer falls within the ANSI S3.6-1996 tolerances for output SPLs (which are up to 10 dB wide at higher frequencies).

Process Factors in Obtaining Valid and Reliable Audiograms

The audiometric test results depend just as heavily upon the human beings involved as upon the technical aspects of the test environment and instrumentation.

Interaction Between Subject and Examiner

When the employee arrives for an audiogram, the results are strongly influenced by the amount of effort the subject decides to devote to the process. Threshold measurements can easily be 10 dB more sensitive if the subject concentrates on listening and responding, compared to the same subject just going through the motions. The interpersonal behavior of the examiner affects subject motivation, as does the perceived relevance of the test to the subject. In a good HCP, the OHC reminds the employee that the audiogram is needed to detect any significant hearing change from baseline thresholds and then agrees to share the results with him/her as soon as the test is completed. This immediate feedback stimulates the employee to participate actively, especially if accompanied by the sincere interest of the examiner in preventing hearing loss. In contrast, if the examiner views quick subject turnover as the main goal, taking insufficient time to instruct the employee and eliminating any feedback about the results, then the subject's motivation to listen and respond will be correspondingly reduced. Threshold measurements in such a situation will be more variable, probably resulting in false threshold shift indications requiring later retesting.

Employees correctly interpret the audiometric monitoring situation as reflecting the employer's degree of commitment to hearing conservation. If the employer's selected OHC provides a rushed, impersonal audiogram, then employees lose motivation not only for the audiometric phase, but also for the entire HCP. Such token audiometry is most likely to occur when the employer searches for the lowest-bid vendor in contract-testing situations, or chooses an uninterested individual as an in-house OHC.

Subject Factors

The subject's health and alertness at the time of the audiogram can affect the results. Before administering the audiogram, the examiner should record the employee's answers to a brief auditory history questionnaire (see later section on recordkeeping). This may be relevant to the validity of the test in the following situations.

Baseline audiograms and TTS: If the employee has had significant noise exposure without adequate hearing protection in the hours preceding the audiogram, temporary threshold shift (TTS) could be present, falsely elevating threshold measurements. It is important for the baseline audiogram to be uncontaminated by TTS because the baseline thresholds should represent the employee's true hearing ability at that date. Therefore, OSHA requires that baseline audiograms be preceded by 14 hours without any unprotected noise exposure. Employees should be instructed to avoid off-the-job noise exposure in the 14 hours prior to a baseline audiogram. If the baseline is administered at the beginning of the workshift, it will not be affected by workplace noise exposure. However, the baseline may also be scheduled for later during the workshift if the employee is fitted with HPDs which provide adequate attenuation and instructed to wear them properly and consistently until the audiogram appointment. If HPDs are relied upon to eliminate pre-baseline noise exposure, then types which yield a more reliable acoustic seal (such as earmuffs or foam earplugs, as discussed in Chapter 10) should be fitted to the individual by the OHC before the beginning of the workshift.

Annual audiograms and TTS: As discussed earlier, it is desirable to administer annual audiograms throughout the workshift in order to detect TTS if it is present, so that inadequate protection is identified before NIPTS occurs. An annual audiogram can be administered even if unprotected exposure is reported, but it will need to be repeated if an OSHA standard threshold shift (STS) is found. Detection of TTSs smaller than OSHA STS should trigger counseling and a recheck of HPD fit.

Current ear health: If the employee reports currently having a cold, allergy, or sinus problem, or other condition affecting his or her hearing, this should be documented. The OHC can proceed with the audiogram anyway, realizing that it may need to be repeated if apparent threshold shifts are found.

Earcanal status: The OHC should press on the subject's pinna and tragus prior to placing the earphone cushions to identify the small percentage of subjects whose pinnas and/or earcanals may collapse under the pressure of the audiometric headset. Such collapse is more prevalent in older subjects than in young people; for a summary and examples, see O Mahoney and Luxon (1996). Collapsing earcanals can be held open with tubing, or insert earphones can be used; testing the subject while his/her mouth is opened sufficiently to open the jaw joint may also yield accurate thresholds (Reiter and Silman, 1993). If collapsing earcanals are not identified before testing, then threshold measurements may be falsely elevated by 15–50 dB (Reiter and Silman, 1993), with the possibility of artifactual fluctuations across tests (O Mahoney and Luxon, 1996).

Likewise, if a cursory otoscopic check of the earcanals (using a pen light or otoscope) reveals that the subject's earcanals appear occluded with cerumen or have other obvious abnormalities, this should be noted. (For further details on the importance of otoscopic checks and auditory history information, see later section on recordkeeping.) Excess earwax will not start to affect thresholds unless the canal is more than 75% blocked (Roeser and Ballachanda, 1997), but it is sometimes hard to determine visually whether there is a sound path past the cerumen. If apparent threshold shifts are found, then the audiogram can be repeated later after the employee either uses a softening agent (nonprescription drops) to reduce the wax accumulation or visits a physician or audiologist to have the wax removed.

Cleanliness and infection control: A clean audiometric testing facility inspires confidence in employees, whereas a sloppy facility may reduce their motivation. Earphone cushions should be cleaned with disinfectant wipes on a regular basis, and especially after use by any subject with fluid drainage from the earcanals. Although earphone cushion covers are commercially marketed for infection control, they cause a small reduction of the signal SPL reaching the subject's earcanal. Therefore, such covers should not be used unless their effect is accounted for in the audiometer's calibration and they are used consistently, in order to avoid an artifactual contribution to threshold shifts. Otoscope tips must be disinfected after every use, or disposable tips can be purchased for one-time use. OHCs need training from the professional supervisor regarding proper procedures when a subject with signs of active ear infection is encountered; written guidance is also available (Ballachanda et al., 1996; Cohen and McCullough, 1996). Good housekeeping for audiometric test booths includes regular cleaning of chairs, response buttons, and other surfaces, as well as adequate ventilation to prevent stale odors in the booth. An extra fan placed in the open booth during testing breaks may be helpful.

State of alertness: If the employee is too fatigued to remain awake during the audiogram or is under the influence of substances (alcohol or drugs) which interfere with alertness, then the examiner can document the situation and reschedule the audiogram for another day. If thresholds have been measured prior to the examiner's making this decision, then the audiogram should be marked as invalid with the reason explained.

Instructions and Method

Details of audiometric method can significantly affect the reliability and validity of results. A good way to evaluate audiometry (whether by a contract testing service or by in-house personnel) is to ask to be given an audiogram, then observe the methods used by the OHC.

Instructions: Consistent instructions are important even if the employee has received audiograms in the past. The instructions should explain the listening task, define a response criterion which encourages response to faint signals, and motivate the listener by promising immediate discussion of the results. A sample set of instructions is shown in the accompanying text box. If employees are not proficient in English, tape-recorded instructions in other languages and written or pictorial instructions, as well as demonstration by the OHC, are needed to ensure that the subject understands the task. (It is best if the OHC can speak the subject's language, which also assists with explaining the results and teaching the subject how to fit HPDs.) For any subject, the OHC should stop the test to reinstruct if necessary. Talk-over capability on the audiometer (which allows the OHC to speak into a microphone circuit on the audiometer and be heard through the subject's earphones) is a desirable feature which saves time for reinstructions or clarifications.

Sample Instructions to the Subject

(These instructions not only define the listening task and the response criterion, but also motivate the subject to listen carefully.)

"OK John, let's see how good your hearing is this year. As soon as we finish the audiogram we can talk about the results. I won't know whether your hearing is changing or not unless you show me the best you can hear. I will place the earphones just over your earcanals; do not change the earphone position yourself.

Through the earphones you will hear tones at different pitches. At the beginning you will hear a few tones at louder levels, so that you can learn what you are listening for. The tones will go on and off in a beep-beep pattern. As soon as you hear the tones, press this button (or raise your finger). As soon as the beeps go away, release the button (or put down your finger).

Please pay close attention and concentrate on listening. When the tones are very faint, respond if you *think* you hear them, even if you aren't sure. Don't wait until the tones are louder before responding. We want to measure the faintest sounds you can hear.

We'll begin in your left ear. I can hear you outside the booth, so let me know if you have any problems. After the audiogram is complete we'll compare the results to your baseline to look for any hearing changes."

Earphone placement: The examiner needs to place the audiometric earphones carefully to align the center of the earphone's speaker with the subject's earcanal. The subject should not be allowed to reposition the earphones; the subject should never be told to put them on, as is the unfortunate practice in some testing facilities.

Pulsed tones: The test tones should be pulsed (typically half a second on, half a second off in a beep-beep-beep pattern) in order to minimize potential confusion of tinnitus with the signal and to facilitate responding near threshold.

Threshold criterion: The decision rule used by the examiner administering a manual audiogram or the microprocessor paradigm administering a computer-controlled test should correctly identify the hearing threshold as the lowest decibel level at which the subject responds appropriately in at least 50% of trials. Many testers (human or machine) use a 2-out-of-3 rule, which is adequate if the subject always responds at the same hearing level or responds twice at the lowest hearing level (one response at 20 dB and two responses at 15 dB). However, if the subject responds once at a lower level and twice at a higher level (for example, once at 15 dB and twice at 20 dB), then a fourth trial is needed to determine which hearing level qualifies as threshold, as illustrated in Table 11.4. If this fourth trial is omitted in cases where it is needed, then additional variability is introduced into threshold measurements across annual audiograms for the individual, thus increasing the incidence of measured shifts from baseline.

Identification of potentially invalid tests: The OHC should be alert for audiograms which are so different from the subject's past results (either better or worse) or so unusual in pattern that their validity is questionable. On-the-spot use of computer software enables to OHC to identify unusual results as soon as the test is over. The supervising audiologist or physician can provide guidance regarding whether to repeat a suspect audiogram immediately. Unless the subject's responses are deemed to be related to temporary health conditions (such as a cold) or temporary attention problems (such as the influence of drugs or alcohol), then an immediate retest is more efficient than returning for another appointment later.

Immediate Feedback to the Subject

The moment when the employee steps out of the test booth is the prime opportunity to influence that person to protect his or her hearing. The employee is interested at this point and often asks "Did I pass?" Even though the audiometric record will be reviewed later by a professional, it is critical for the OHC to make a quick comparison at this point between baseline and current HTLs to give the employee immediate feedback as to whether hearing is stable or changing. If thresholds have improved or remained stable, a word of praise is in order! If thresholds have worsened, then the technician can make a general statement such as "Your thresholds at several high frequencies are poorer today than in the past. The audiologist (or physician) will review your audiogram and give you more information later, but let's recheck your hearing protector fit now to see if you are getting adequate protection from noise." By expressing concern about the apparent changes, asking the employee about (and documenting) possible nonoccupational causes (both off-the-job noise and health-related factors), rechecking HPD fit, and counseling the employee about the importance of consistent HPD wear, the OHC motivates the employee to take responsibility for his/her hearing. This brief one-to-one interaction indicates the employer's commitment to the HCP and reinforces the causal connection between the employee's self-protective behaviors (or lack thereof) and personal hearing changes. If this motivational opportunity is lost, later actions cannot replace it.

Many microprocessor audiometers and all computer-controlled audiometers have the capacity to compare current thresholds to baseline, thereby aiding the OHC in giving immediate feedback about hearing trends. Software programs can quickly display audiogram graphs illustrating the difference between baseline and current thresholds, enabling the OHC to show trends to the employee visually. Computer-screen graphics can also illustrate to the employee how his/her current thresholds compare to typical results for others of the same age and gender. These graphic display capabilities are particularly effective motivational tools which should be *top priorities* in establishing or updating the audiometric protocol for the HCP, whether audiograms are administered by on-site personnel or by contract services. Sample graphs of these types are shown as Figures 11.9 and 11.10. Note that the software should identify significant shifts which are smaller than age-corrected OSHA STS for counseling purposes (in addition to identifying STSs for purposes of scheduling retest audiograms).

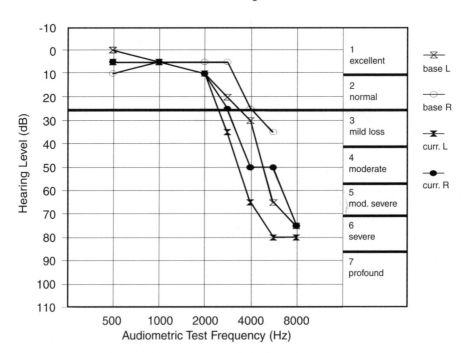

Figure 11.9—Sample graph intended for employee feedback, illustrating a comparison of current thresholds to baseline.

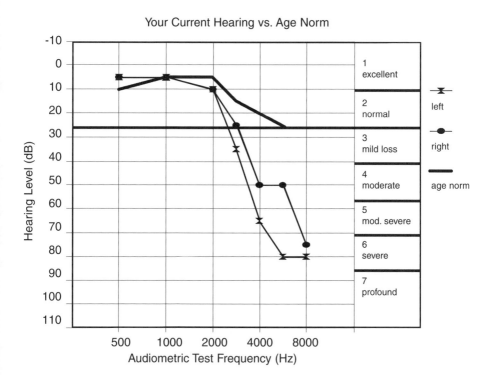

Figure 11.10—Sample graph intended for employee feedback, illustrating a comparison of the subject's current hearing thresholds to the expected age-effect hearing levels for the subject's age/sex/race.

Another extremely useful graphic way to demonstrate the effect of developing hearing loss on speech understanding is a "count-the-dots" representation of an audibility index related to the speech intelligibility index (see Chapter 14). Several versions of this tool have been published (Killion et al., 1993), and one version is illustrated as Figure 11.11 (after Lundeen, 1996). The shaded area at hearing levels of about 20–50 dB represents the spectral content of normal conversational speech heard in a quiet room at a typical listening distance of one meter. For presentation in audiogram format, the units have been converted from SPLs to the equivalent hearing level values. If all of an individual's hearing thresholds are more sensitive than the hearing level values of the shaded range, then all the sounds in conversational speech are audible. The 100 dots within the shaded area are spaced to correspond to the informational value of different frequencies to speech intelligibility; each dot represents 1 percent of the speech information. Note that the dots in the 1000–4000 Hz range are more closely spaced than those at other frequencies because this range is most critical for speech intelligibility. If an individual's audiogram is superimposed on the dot plot, then the number of dots at hearing level values less than his/her thresholds is the percent of speech infor-

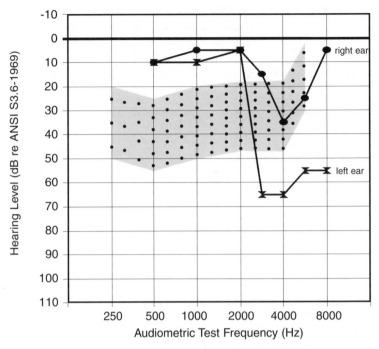

Figure 11.11—"Count-the-dots" audiogram for assessing the audibility of normal conversational speech at one meter (after Lundeen, 1996), with plotted thresholds showing 13% of speech information inaudible for the right ear and 35% inaudible for the left ear.

mation which is inaudible to that individual. In Figure 11.11 the illustrated audiogram (typical of an asymmetrical hearing loss from gunfire) shows 35 percent of speech information inaudible to the left ear, while the right ear shows 13 percent inaudible. This tool can demonstrate to employees that their speech understanding will worsen if their hearing loss progresses. Although the brain can fill in the gaps for a small percentage of inaudible information based on context (as we all do when listening over the limited band-pass of the telephone), a precipitous hearing loss above 2000 Hz will be much more difficult to cope with than a beginning notch affecting only 4000–6000 Hz. Even without actually counting the dots, a quick sketch of their own audiograms on a dot plot can convince employees of the need to protect their hearing.

Audiogram Review and Written Feedback

All audiograms must be reviewed to assess the need for follow-up actions (discussed in the next section). OSHA allows "routine" tests to be reviewed by the

OHC, while "problem" audiograms must be reviewed by the professional. Many HCPs choose to send all audiograms for professional review rather than having the OHC spend time to identify problem records following the professional's guidance. This choice depends on whether the technician's time is in demand for other duties, as well as cost. The benefit depends partially on whether the professional reviewer does actually provide the desirable extra detail in feedback to employees about their hearing trends. (Also see a later section regarding the advantage of entering all audiograms into a database, regardless of who reviews them.)

Routine Audiograms

Although OSHA does not require this, it is helpful for every employee to receive written feedback about the audiogram results, in addition to the immediate verbal feedback discussed earlier. Written feedback helps employees retain an understanding of their hearing status and hearing changes over time; having a permanent record of their results adds to the educational and motivational benefit of audiometric monitoring. (Unrelated to OSHA or to HCP effectiveness, see Chapter 18 regarding requirements in some workers' compensation jurisdictions to notify employees who exhibit hearing impairment.)

Use of commercially available software can automate the OHC's preparation of written feedback for routine audiograms, even allowing a feedback sheet with an audiogram graph to be printed out and given to the employee at the time of the test (which does not replace one-to-one verbal comments made by the hearing examiner). However, written feedback for routine tests can also be done by hand using check-box forms such as illustrated in Figure 11.12. (The professional supervisor must guide the OHC in identifying nonroutine audiogram results for which review by the supervisor is needed, as discussed below.)

The feedback information which is useful to give the employee includes: current hearing status, comparison of current thresholds to expected age-effect hearing levels, comparison of current results to the baseline, and general recommendations. The more fully the current hearing status is described, the more effective the feedback will be in warning the employee of progressive hearing change. Noise-exposed personnel can easily be educated to understand their audiograms and hearing trends; this fosters personal responsibility for protecting one's own hearing. However, feedback is not very helpful if results are classified into such broad categories that the status classification does not change in response to 15-dB shifts in hearing thresholds. Therefore, select a professional reviewer who provides useful feedback to each employee.

Nonroutine Problem Audiograms

OSHA has not exhaustively listed all those audiograms which constitute a "problem." Clearly, problem tests include those showing OSHA standard threshold shift (STS), further explained below. In an interpretation letter (OSHA, 1994), other examples of problem tests included "audiograms that show large differences in hearing thresholds between the two ears, audiograms that show

RESULTS OF YOUR ANNUAL AUDIOGRAM

Name: _____ ID number: _____ Date: _____

	LEFT EAR	RIGHT EAR

PAST HEARING STATUS

On your baseline audiogram your hearing status codes were: _____ _____

Last year your hearing status codes were: _____ _____

CURRENT HEARING STATUS

This year your hearing status codes are: _____ _____

The typical hearing status code for your age/gender/race is: _____ _____

HEARING CHANGE

The change in your hearing compared to past tests is:

normal: stable hearing or minor fluctuations

Congratulations on preserving your hearing! _____ _____

greater change than expected

Be sure to wear your hearing protectors during _____ _____

noise exposure on and off the job.

significantly greater change than expected

The audiologist will review your record. _____ _____

Wear your hearing protectors carefully and

consistently to avoid noise damage to your hearing.

HEARING STATUS CODES

Letter Codes:
The number code indicates your worst hearing level at low frequencies: .5, 1, and 2 kHz.

A: Low frequency hearing is within normal limits. No hearing levels are worse than 25 dB. Speech is usually easy to understand unless there is significant hearing loss at 3000 Hz.

B: Beginning hearing loss for low frequencies. Some hearing levels are worse than 25 dB. Speech may be difficult to understand, especially in a noisy situation.

C: Significant hearing loss for low frequencies. The average of hearing levels at 500, 1000, and 2000 Hz is worse than 25 dB. Speech is often difficult to understand, even in a quiet situation.

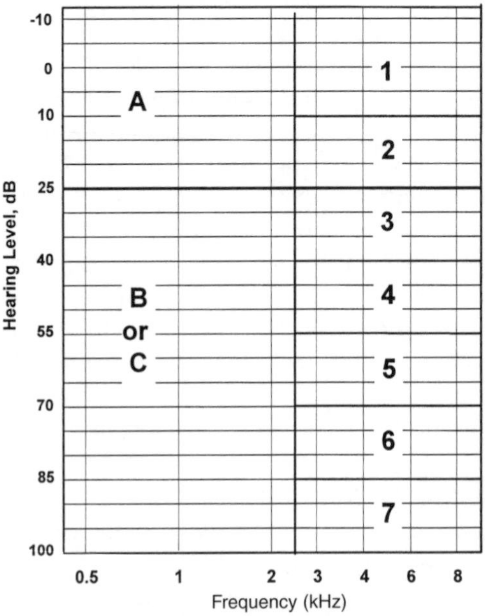

Number codes:
The number code indicates your worst hearing level at high frequencies: 3, 4, and 6 kHz. (8 kHz is not coded.)

1: Excellent hearing No hearing levels are worse than 10 dB.

2: Hearing within normal limits No hearing levels are worse than 25 dB.

3: Mild hearing loss No hearing levels are worse than 40 dB.

4: Moderate hearing loss No hearing levels are worse than 55 dB.

5: Moderately severe hearing loss. No hearing levels are worse than 70 dB.

6: Severe hearing loss No hearing levels are worse than 85 dB.

7: Profound hearing loss Some hearing levels are worse than 85 dB.

Figure 11.12—Sample of a checklist written feedback form which can be manually completed by the OHC under the professional supervisor's guidance.

unusual hearing loss configurations that are atypical of noise-induced hearing loss, and audiograms with thresholds that are not repeatable." The audiologist or physician supervisor must establish criteria for problem audiograms requiring professional review. NHCA is developing a position pamphlet on this topic which can be used for guidance once it is available (NHCA, in preparation A).

The professional's criteria for problem audiograms requiring review may include the types of tests shown in the accompanying text box.

Problem Audiograms Which Need Professional Review

1. Baseline audiograms showing significant pre-existing hearing loss,
2. OSHA STS,
3. Significant threshold shifts which do not constitute OSHA STS, including precursor shifts at STS frequencies as well as shifts at other audiometric frequencies,
4. Significant progression of hearing loss into the speech frequencies for employees with pre-existing high-frequency hearing loss on their baselines,
5. Audiograms with patterns suggesting potential medical ear problems (especially if accompanied by confirming auditory history and/or otoscopic inspection findings), such as some types described in guidelines from the American Academy of Otolaryngology—Head and Neck Surgery (AAO-HNS, 1997),
6. Audiograms with interaural threshold differences requiring masking to determine true thresholds in the poorer ear, and
7. Audiograms showing significant improvement from the current baseline, either at STS frequencies or at other frequencies.

The written feedback which the professional reviewer provides for employees with problem audiograms should be more detailed than the feedback provided for routine tests. In addition to the information mentioned above, the reviewer's feedback should include numerical and/or graphic baseline and current HTLs, the amount of shift at each test frequency, and the age norms relevant to the employee. Choices of age-effect reference data are discussed in Chapter 17. Note that the age norms used for purposes of counseling employees do not need to be the same as the presbycusis age-correction data allowed by OSHA in identifying STS. The sample written feedback sheet shown in Figure 11.13 uses data from Annex C of ANSI S3.44-1996 as age norms for feedback purposes. The professional should also make specific recommendations relevant to the employee's hearing trends and particular problems. The professional's comments can be more specific if auditory history and otoscopic check results are obtained and sent with the record for review (see further discussion in later section on recordkeeping); collection of this information is strongly recommended and will be demanded by most competent professionals (although OSHA does not require it). The reviewer's feedback has extra impact if the on-site HCP personnel go over the professional's comments with employees who showed problem audiograms, rather than simply mailing them the results. This is also a good opportunity to recheck HPD fit and reverify the employee's ability to achieve a good acoustic seal.

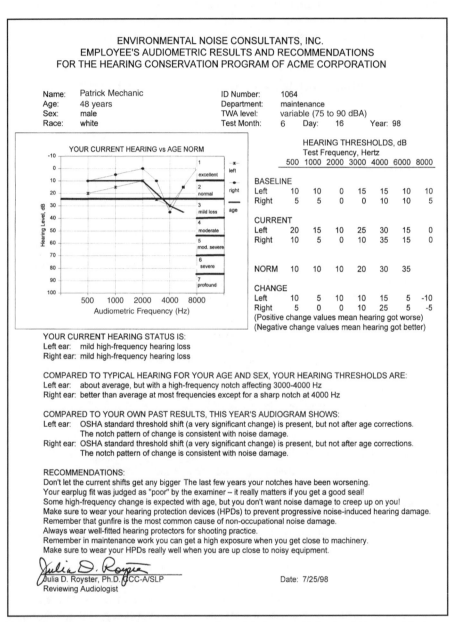

ENVIRONMENTAL NOISE CONSULTANTS, INC.
EMPLOYEE'S AUDIOMETRIC RESULTS AND RECOMMENDATIONS
FOR THE HEARING CONSERVATION PROGRAM OF ACME CORPORATION

Name:	Patrick Mechanic	ID Number:	1064
Age:	48 years	Department:	maintenance
Sex:	male	TWA level:	variable (75 to 90 dBA)
Race:	white	Test Month:	6 Day: 16 Year: 98

HEARING THRESHOLDS, dB
Test Frequency, Hertz

	500	1000	2000	3000	4000	6000	8000
BASELINE							
Left	10	10	0	15	15	10	10
Right	5	5	0	0	10	10	5
CURRENT							
Left	20	15	10	25	30	15	0
Right	10	5	0	10	35	15	0
NORM	10	10	10	20	30	35	
CHANGE							
Left	10	5	10	10	15	5	-10
Right	5	0	0	10	25	5	-5

(Positive change values mean hearing got worse)
(Negative change values mean hearing got better)

YOUR CURRENT HEARING STATUS IS:
Left ear: mild high-frequency hearing loss
Right ear: mild high-frequency hearing loss

COMPARED TO TYPICAL HEARING FOR YOUR AGE AND SEX, YOUR HEARING THRESHOLDS ARE:
Left ear: about average, but with a high-frequency notch affecting 3000-4000 Hz
Right ear: better than average at most frequencies except for a sharp notch at 4000 Hz

COMPARED TO YOUR OWN PAST RESULTS, THIS YEAR'S AUDIOGRAM SHOWS:
Left ear: OSHA standard threshold shift (a very significant change) is present, but not after age corrections.
          The notch pattern of change is consistent with noise damage.
Right ear: OSHA standard threshold shift (a very significant change) is present, but not after age corrections.
          The notch pattern of change is consistent with noise damage.

RECOMMENDATIONS:
Don't let the current shifts get any bigger The last few years your notches have been worsening.
Your earplug fit was judged as "poor" by the examiner – it really matters if you get a good seal!
Some high-frequency change is expected with age, but you don't want noise damage to creep up on you!
Make sure to wear your hearing protection devices (HPDs) to prevent progressive noise-induced hearing damage.
Remember that gunfire is the most common cause of non-occupational noise damage.
Always wear well-fitted hearing protectors for shooting practice.
Remember in maintenance work you can get a high exposure when you get close to machinery.
Make sure to wear your HPDs really well when you are up close to noisy equipment.

Julia D. Royster
Julia D. Royster, Ph.D. CCC-A/SLP   Date: 7/25/98
Reviewing Audiologist

Figure 11.13—Sample of the detailed written feedback which should be provided to the employee by the professional reviewer.

When employees are alerted and counseled about beginning threshold shifts, they are less likely to develop OSHA STSs through further progression of the hearing change due to inadequate protection from noise.

Follow-up Steps After Audiograms

Certain follow-up steps are mandated by OSHA or other regulations, while other types of follow-up are needed based on the judgment of the audiologist or physician who is supervising the audiometric phase of the HCP.

OSHA Standard Threshold Shift (STS)

OSHA's HCA defines an STS as a change of 10 dB or more in the average of hearing thresholds at the audiometric test frequencies of 2, 3, and 4 kHz in either ear when the current audiogram is compared to the baseline. Allowance for the contribution of age-related change may be made by employing age corrections as described in Appendix F of the HCA; these corrections are optional. Two equivalent ways of applying age corrections are illustrated as Table 11.5. (A third way,

TABLE 11.5

Two ways of applying OSHA age corrections (from Appendix F of OSHA's HCA) for a hypothetical STS in a male employee (only relevant thresholds for one ear are shown).

| | Row | | Hearing Threshold (dB) | | | |
| | | | Audiometric Frequency (Hz) | | | |
			2000	3000	4000	Average
Method 1:	A	baseline thresholds, age 30	10	20	25	18.3
	B	current thresholds, age 40	20	30	40	30.0
	C	OSHA Table F-1 values, age 30	4	6	9	6.3
	D	OSHA Table F-1 values, age 40	6	10	14	10.0
	E	Difference, Row D minus Row C	2	4	5	3.7
	F	Age-corrected current thresholds (Row B minus Row E)	18	26	35	26.3
	G	Age-corrected threshold shift (Row F minus Row A)	8	6	10	8.0*
Method 2:	H	baseline thresholds, age 30	10	20	25	18.3
	I	current thresholds, age 40	20	30	40	30.0
	J	threshold shift (Row I minus Row H)	10	10	15	11.7
	K	OSHA Table F-1 values, age 30	4	6	9	6.3
	L	OSHA Table F-1 values, age 40	6	10	14	10.0
	M	Difference, Row L minus Row K	2	4	5	3.7
	N	Age-corrected threshold shift (Row J minus Row M)	8	6	10	8.0*

*The STS is no longer present after age corrections because this value is less than 10 dB.

often used within computer software, involves subtraction of the age-related hearing level values from Appendix F from every audiogram, including the baseline, before calculating hearing change; for hand review the methods in Table 11.5 are more efficient.)

OSHA requires specific follow-up steps for employees who show STS unless one of two conditions applies:

1. A retest within 30 days of the initial annual test shows that the STS is not persistent, or
2. A physician determines that the STS is not work-related or aggravated by occupational noise exposure.

If neither of the above conditions is met, then the follow-up actions listed in the following text box are required by OSHA. These actions are intended to increase the degree of protection from noise for the affected individual (see paragraph [g][8] of the HCA [OSHA, 1983a]):

Follow-up Actions for OSHA STS

1. Inform the employee in writing of the STS within 21 days of the determination that it has occurred.
2. For an employee not already wearing HPDs (due to TWAs less than 90 dBA), fit the employee with HPDs, train the wearer in their proper use and care, and require HPD utilization.
3. For an employee already wearing HPDs, refit devices (providing greater attenuation if needed to achieve an assumed TWA exposure of 85 dBA or less under the HPD, as explained in Chapter 10), and retrain the wearer in their proper use and care.
4. Refer the employee for a clinical audiological evaluation and/or otological examination (at the employer's expense) if additional testing is needed or if a medical pathology of the ear may be caused or aggravated by wearing HPDs.
5. Inform the employee of the need for an evaluation (at his/her own expense) if a medical pathology is suspected which is unrelated to the use of HPDs.

It is highly desirable for the OHC to recognize that an STS has occurred at the time of the audiogram. Microprocessor or computer-controlled audiometers or software programs facilitate rapid identification of STS, but it can also be done by hand. Logistically, since it is often difficult to ensure that an employee will be retested within 30 days, it is expedient to do and document follow-up actions 2 and 3 while the individual is still present at the end of the annual audiogram. This ensures that the employee will benefit from the only follow-up steps which actually affect the worker's degree of protection from noise. It is wise to go ahead and perform these follow-up steps even though a later retest might show enough improvement that the STS would fail to persist; taking this action on the basis of potentially temporary threshold shift is important for motivating employees.

Likewise, it is expedient for the OHC to recognize the STS and schedule a retest within the 30-day window, rather than seeking the professional reviewer's authorization merely to do a retest. The retest can be scheduled as soon as practical within the 30-day time limit. If the OHC suspects that inattention by the subject or other transient factors affected the results, an immediate retest can be done. To avoid TTS contamination of delayed retests, it is prudent to schedule them at the beginning of a workshift. If the employee has a cold or other medical condition felt to be temporarily causing the STS, the retest should be delayed enough to let the condition clear (though not past the 30-day limit). Documentation of these types of auditory history details will assist the professional in evaluating the shift when the records are reviewed.

The records can be sent to the reviewing audiologist or physician as soon as the retest is obtained (or when the 30-day window expires without a retest). The professional will determine whether follow-up steps 4 or 5 are needed, and the written feedback from the professional can serve as step 1 if worded appropriately. Alternately, the employer may choose to provide the employee a separate written notice of STS occurrence. OSHA's interpretation of "within 30 days of the determination" for step 1 notification is that the determination should be made by the professional as quickly as practical (OSHA, 1983b) but without any fixed deadline. As discussed later in the section on recordkeeping, OSHA STS is recordable as an occupational hearing loss in some state OSHA jurisdictions, and different time limits apply for recordability purposes.

Follow-up for Other Significant Threshold Shifts

Due to the OSHA requirements for STS follow-up, HCP personnel sometimes become fixated on STS and mistakenly believe that it defines the smallest detectable hearing change which may be caused by noise. This misconception may go so far that OHCs tell employees they have shown "no change" in hearing unless an age-corrected STS has occurred! It is important to consider the weaknesses of STS as a shift criterion and the value of reacting to other significant threshold shifts (Royster, 1996a). Because STS averages together three test frequencies which differ in vulnerability to noise damage (2 kHz being less susceptible to NIPTS than either 3 or 4 kHz, as discussed in Chapter 17), it is not an early warning indicator of incipient NIHL. Individuals with normal baseline hearing will usually show shifts of 15–20 dB at 3 and 4 kHz before the STS average changes by 10 dB. At the other extreme, OSHA STS does not detect communicatively significant shifts for persons whose baselines show pre-existing high-frequency hearing loss; such individuals may develop a 20-dB shift at 2 kHz before the STS average changes by 10 dB. These two opposing deficiencies are illustrated by the examples in Figure 11.14.

Application of age corrections in calculating STSs compounds the difficulty of recognizing significant hearing change in persons with pre-existing hearing loss

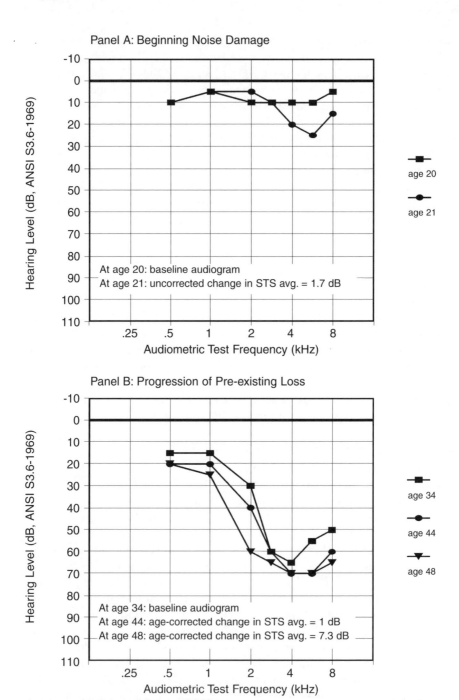

Figure 11.14—Sequential audiograms showing two different deficiencies of OSHA STS: it does not catch beginning NIHL at an early stage (panel A), and it does not catch progression of NIHL at 2000 Hz for employees with pre-existing high-frequency hearing loss on the baseline (panel B).

494

on baseline. In other words, a person with pre-existing NIHL would be expected to show less age-related hearing change from age 40 to age 60 than a person without such prior noise damage. The OSHA age corrections over-adjust for persons with pre-existing noise-induced hearing loss. These same individuals can least afford to lose additional hearing, especially as the loss begins to affect 2 kHz and lower frequencies. To alert such employees to communicatively significant hearing change, the professional reviewer should define significant shifts smaller than age-corrected STS for those workers with pre-existing hearing loss as "problem" tests requiring review (see next paragraph). Likewise, the OHC should give such employees immediate feedback concerning further encroachment of pre-existing loss deeper into the speech range.

Some audiogram reviewers mistakenly feel that individuals with normal hearing on their baselines do not need feedback about hearing changes until their hearing levels approach the 25-dB limit of the range of normal hearing (refer to the hearing level ranges at the right-hand side of Figure 11.9). Note that persons who actually hear better than 0 dB (for example, with thresholds of -10 dB, as exhibited by many black females) will have baseline threshold values measured as 0 dB if zero is the lower limit of the audiometer. The first 10 dB of noise-induced hearing change for such individuals will not even be detected using such a limited-range audiometer. Even if the true baseline hearing levels were actually 0 dB, changes within the normal range are important. A persistent 15-dB shift at 4000 Hz indicates the probability of early NIPTS even if the resulting threshold is still just 15 dB; the affected individual needs to be warned about the trend of developing hearing loss before it becomes a significant communicative problem.

In effective HCPs, STS rates are reduced by providing counseling, audiogram review by the supervising professional, and follow-up actions to increase protection for employees who show significant shifts other than OSHA STS, such as those described in the following text box. The supervising professional should include shifts such as these in the list of "problem audiogram" types which the OHC should send for professional review.

Significant Threshold Shifts Which Merit Review and Follow-up

1. A persistent 15-dB shift at any test frequency from 500 through 6000 Hz (advocated by NIOSH [1998] based on research by Royster [1992, 1996]),
2. A shift of 15 dB or more from baseline at 1, 2, 3, or 4 kHz; this shift criterion constitutes a mandatory review/referral requirement in the U.S. military (DoD, 1996),
3. An OSHA STS before age corrections,
4. Persistent 10-dB shifts at two adjacent test frequencies,
5. A persistent 10-dB shift at 1000 or 2000 Hz for employees who already show high-frequency hearing loss.

NIOSH (1998) has advised against applying age corrections, based on the rationale that they obscure hearing changes which are significant for the affected individuals, as well as the fact that median corrections necessarily over-correct for one tail of the presbycusis distribution (those persons who show less-than-median presbycusis). On the other hand, presbycusis is inevitable; therefore, not all hearing loss shown by noise-exposed persons is noise-induced. HCPs cannot prevent age-related hearing loss. Moreover, age corrections under-correct for those persons who show greater-than-median presbycusis. The professional reviewing the audiometric record should counsel the individual based on educated opinions regarding the cause(s) of the pattern of changes shown; familiarity with the characteristics of the development and progression of noise-induced hearing loss (ACOM, 1989; Dobie, 1993a) is essential for such decisions. It is counterproductive for the HCP if employees who are doing a good job of protecting their hearing from NIPTS are admonished for poor performance when their hearing shows changes consistent with expected age effects. This is just as unfortunate a mistake as to fail to warn employees with hearing changes consistent with NIPTS, just because OSHA-allowed age corrections reduce the degree of shift to less than STS. Therefore, the choice of an expert professional reviewer is critical to avoid the unjustified opinion that all hearing changes shown in noise-exposed persons are work-related.

Referral for Audiological and/or Medical Evaluation

The audiologist or physician reviewing the audiometric record will judge whether there is a need for a referral at employer expense, or for a recommendation to the employee to see an audiologist or physician at his/her own expense. The employer must pay for the referral if further testing is needed (as when an employee cannot provide a valid or reliable audiogram using the employer's OHC and audiometer, or when interaural differences in hearing require clinical audiometry with masking) or when an earcanal irritation or infection is caused by wearing HPDs, or a middle ear infection is potentially aggravated or hindered from clearing by wearing earplugs (OSHA, 1998). For example, in persons with chronic fungus in the earcanals or persons with earcanal drainage from a middle ear infection, earplugs may aggravate the problem or slow the healing process.

The findings of auditory history details and otoscopic inspection results (see later section on recordkeeping) are extremely helpful to the professional in judging whether a medical referral is appropriate. It is strongly recommended that such information be obtained during each HCP audiogram.

When the employer makes a medical or audiological referral, the employee's complete audiometric records should be sent in advance to the audiologist and/or physician who will see the employee, along with a letter stating clearly the reason for the referral and what information is needed back. If it is necessary due to geographic constraints to refer to an audiologist or physician who is not experienced in occupational hearing conservation, explicitly request that hearing thresholds be measured at 500–8000 Hz, including 3000 and 6000 Hz (which a clinician unfamiliar with hearing conservation may omit unless the thresholds at adjacent frequencies differ by 20 dB or more). In the case of ear irritation or infection, ask the physician specifically whether the employee can wear insert HPDs or whether earmuffs should be worn until the condition has resolved, explaining that HPD wear is a requirement of the employee's job. A sample referral form, with space for the reasons for referral to be detailed, is shown as Figure 11.15.

DATE: 6/15/98

TO: Tim Panic, M.D.
         Anytown Otology Associates, Box 789, Anytown, USA

FROM: Susie Que, Occupational Hearing Conservationist
         Acme Corporation, Box 123, Anytown, USA

RE: referral of our employee John Doe, SS# _123-45-6789_ d.o.b. _2/10/46_

This employee is included in the Acme hearing conservation program due to daily OSHA 8-hour time-weighted-average (TWA) noise exposures of 85 dB or higher.
Our sound surveys indicate the typical TWAs for _John Doe's_ job are _85-89_ dBA.

This employee has shown an OSHA Standard Threshold Shift (STS) after age corrections for the time elapsed since his baseline audiogram. The STS in the _right_ ear has been confirmed on a retest. A summary record of all audiograms is attached for your use in evaluating the time course of the hearing threshold shifts.

Acme's consulting audiologist, _Jane Smith, CCC-A_, suspects that a medical ear condition may be completely or partially responsible for this hearing shift. Her notes to you are shown below.

Please provide Acme Corporation with your written answer to the questions below, together with copies of your audiological evaluation results. Please mail these items to me at the address shown above. Thanks for your assistance.

Audiologist's comments and questions:

Mr. Doe showed stable thresholds on 3 audiograms between 10/95 (hire date) and 4/97, with normal hearing bilaterally. He has worn foam earplugs, and the OHC reports he achieved a good acoustic seal. His 6/1/98 audiogram shows shifts of 30 dB or more in the right ear only across most test frequencies. He reports having seen a physician about episodes of dizziness.

Mr. Doe's low workplace noise exposure is unlikely to have contributed to this sudden change affecting one ear only.

Please include the following in your evaluation:
1. A complete diagnostic audiometric evaluation performed by a licensed and certified audiologist, including threshold measurements at 3000 and 6000 Hz.
2. Advice to Mr. Doe if any treatment is suggested.
3. For the employer's fulfillment of OSHA responsibilities, a written report to Acme regarding:
    A. whether this OSHA STS is caused or aggravated by workplace noise exposure or the wearing of hearing protection devices,
    B. whether this OSHA STS is caused by a medical ear condition,
    C. whether this employee can safely wear insert hearing protectors, or if earmuffs are necessary to avoid aggravation of any medical ear condition.

Figure 11.15—Sample form letter for use when referring an employee for evaluation by a physician and an audiologist.

Another potential reason for referral at the employer's expense occurs when an employee shows a potentially recordable hearing shift which may be due to non-noise or nonoccupational causes, and the employer wants to obtain a medical opinion regarding whether the shift is caused or aggravated by workplace noise (requiring entry on the OSHA log as an occupational illness) or non–work-related (allowing a log entry to be lined out). Again, all audiometric and noise-exposure records for the employee should be sent to the examining physician, together with a letter very clearly stating the question about which an opinion is sought. Past research suggests that about half of referred cases (based strictly on audiometric results, not symptom complaints) may have otologic diagnoses other than NIHL (Dobie and Archer, 1981).

Referral rates may be approximately halved if confirmation by a retest is required prior to the referral (Simpson et al.,1995). Retesting for referrals does not have to fall within OSHA's 30-day time window for STS retests.

In addition to recommending referrals at the employer's expense in cases discussed above, the audiologist or physician who reviews problem audiograms may recommend that certain other employees seek medical and/or audiological evaluations at their own expense when the results of air-conduction audiograms (plus auditory history information about ear symptoms, if available) warrant such an assessment. This recommendation would appear on the reviewer's written feedback to the employee. The reviewer might recommend evaluation for reasons ranging from the possibility of a potentially serious medical pathology such as an acoustic nerve tumor, to the suspicion of a treatable conductive hearing loss, to the probable benefit the employee might receive from trying hearing aids.

Several sets of referral guidelines are available which address both audiological criteria (hearing status on baseline audiograms, and hearing changes from baseline) and medical criteria (symptoms and complaints, as well as direct observations by the hearing examiner). The AAO-HNS has published criteria for otologic referral (AAO-HNS, 1997). The U.S. Department of Defense has developed criteria for audiological and/or medical referral (unpublished but described in Schulz, 1999) which may also be useful for guidance in non-military settings. A committee of the NHCA is also preparing a set of guidelines (NHCA, in preparation B; described in Schulz, 1999) considering the above plus other recommendations (Health and Safety Executive, 1995).

The HCP gains credibility in employees' eyes when non-noise-related otological conditions of which the worker was unaware are identified as a result of annual audiometric monitoring. If the employee follows the recommendation to seek medical and/or audiological assessment, it is wise to ask the individual to sign a release form allowing the referral source to share the results with the employer. If this information is available in the employee's workplace medical file, the professional who reviews the annual audiograms can provide better advice to the employee in the future.

The audiometric evaluation report received in response to a referral will usually include hearing thresholds measured using a bone-conduction vibrator to determine whether the hearing loss is sensorineural (affecting the cochlea and/or neural pathways), in which case thresholds by air conduction (using earphones to present the signals) and by bone conduction will be similar. Alternately, an air-bone gap (a difference between thresholds measured by air versus bone conduction) indicates a conductive hearing loss component (affecting the outer ear or the middle ear) which might be medically treatable. The sample clinical audiometric evaluation form shown as Figure 11.16 shows a conductive hearing loss in one ear. The speech recognition threshold (also called speech reception threshold or SRT) values indicate the lowest hearing levels at which the subject could repeat two-syllable spondaic words (such as baseball, ice cream). The word recognition scores (sometimes called speech discrimination scores) illustrated indicate the percentage of a list of one-syllable phonemically balanced words which the subject could repeat at a given presentation level, such as 45 dB hearing level (for typical conversational speech) or 30–40 dB above the SRT (for subjects with hearing loss). Speech scores are typically normal for conductive loss but are often reduced for sensorineural hearing loss, which involves distortion from the damaged cochlea (see Chapter 5).

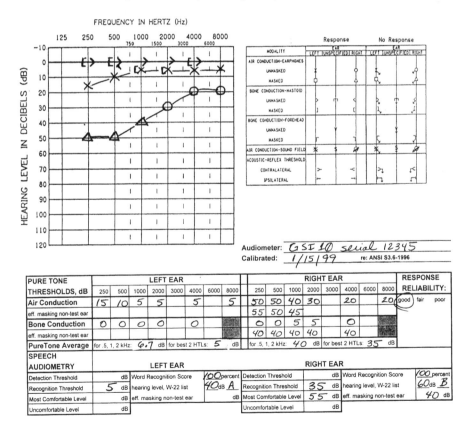

Figure 11.16—Sample clinical audiometric evaluation results showing both air- and bone-conduction thresholds (illustrating conductive hearing loss in the left ear) and speech testing results.

499

Baseline Revision for Improvement or for OSHA STS

OSHA allows the baseline or reference audiogram to be revised by the reviewing professional (not by computer software) either for significant improvement or for persistent STS. Improvements may occur when a temporary medical condition (such as a cold, ear infection, or earcanal blockage) was present at the time of the original baseline, or when the employee learns to take audiograms better, yielding more sensitive thresholds. Revision for improvements affords the employee more protection by allowing detection of STS from a baseline which is more representative of his/her true hearing ability. On the other hand, baseline revision for persistent STS prevents the same shift from being identified repeatedly; only additional hearing shifts resulting in STS from the revised baseline result in new follow-up actions.

Because OSHA did not provide explicit guidance on baseline revision procedures, different professionals have followed widely diverse policies (Royster and Stewart, 1990). To provide guidance based on a consensus of involved professionals, the NHCA established a committee to develop guidelines on this topic. After research to compare various suggestions (Royster, 1996b), the NHCA accepted the consensus recommendations for baseline revision summarized as Table 11.6 and applied and illustrated in Table 11.7 (also see Appendix IV).

TABLE 11.6
NHCA guidelines for audiometric baseline revision for the purpose of tracking OSHA STS (does not apply to baselines for other purposes).
See Appendix IV for additional details.

Revision Rules	
Improvement	Revise for improvement of 5 dB or more in the average of thresholds at 2, 3, and 4 kHz if the improvement is present on two consecutive audiograms.
	Revise to the test with the better average, or to the earlier test if both tests show equivalent averages.
OSHA STS	Revise for OSHA STS if the shift is persistent on two consecutive audiograms at least 6 months apart.
	Revise to the test with the better average, or to the earlier test if both tests show equivalent averages.

Additional Requirements	
	Each ear is considered independently, and the baselines for the two ears are revised separately. On revision, thresholds for all test frequencies in the affected ear are revised together.
	Age corrections do not apply to improvement. Age corrections are optional for OSHA STS.
	Revision must be made by an audiologist or physician; computer software may only identify records for professional consideration.
	Professional over-ride of the guidelines must be justified by documentation of the specific reasons for the override.

Note that the NHCA guidelines require persistence of an observed hearing change on two tests (usually two annual tests) before revision is made, to avoid premature baseline changes. The decision to handle baseline for each ear separately (allowing the left and right ears to have baselines from two dates if the ears' trends are dissimilar) simplifies decision making and allows computer algorithms to more easily identify audiograms which the professional reviewer should consider for potential baseline revision. If consensus guidelines are followed for baseline revision, then records may be easily transferred between reviewers or testing providers without discontinuity; otherwise it may be necessary to re-identify baselines (starting with the first test for each employee) when a new professional is selected to work with the HCP.

Baseline for Purposes Other Than OSHA STS Detection

Revision of the baseline from which OSHA STS is detected prevents repetitive identification of the same STS. However, the original baseline is still used for the identification of other types of threshold shifts. Some of these are voluntary, such as otologic referrals under AAO-HNS guidelines. One is mandatory: in jurisdictions where federal OSHA recordability criteria apply, the original baseline is used to detect 25-dB shifts in the average of thresholds at 2, 3, and 4 kHz. (This is discussed in a later section on logging occupational hearing loss.) To ensure timely detection of these various types of shifts, the audiometric recordkeeping system of the professional reviewer must allow separate baselines to be maintained simultaneously for STSs, medical referrals, and recordability. Generally the initial baseline serves all three purposes unless or until a persistent shift occurs, resulting in a revised baseline for detection of another later shift in that ear. Examples shown in Table 11.7 illustrate baseline revisions for persistent OSHA STS, recordable occupational hearing loss using federal OSHA criteria, voluntary medical referral, and persistent improvement.

Recordkeeping for the Audiometric Phase

Some records are needed to document compliance with OSHA's HCA, while others are needed to enhance the effectiveness of audiometric monitoring beyond mere compliance, or to assist the employer in potential worker's compensation claims or other legal actions. Consistency and completeness of records is important (Gasaway, 1985a, 1985b). Forms should be developed via careful planning and changed only when needed. OHCs and professionals involved in recordkeeping require a clear understanding of their respective responsibilities, so that uninterpretable "blank fields" do not occur.

Auditory History and Otoscopic Check

Although not required by OSHA, auditory history information should be collected and an otoscopic check performed at the time of every audiogram. The

TABLE 11.7
Two examples illustrating baseline revisions.

Serial audiometric records for two individuals, illustrating baseline revisions for different purposes.
See far left-hand columns for age and test type. See far right-hand column for test date.

Test Types: B = initial baseline audiogram   LSB = left STS baseline
                 A = annual audiogram   LRB = left recordability baseline
                 R = retest within 30 days   LMB = left medical baseline
                                              RSB = right STS baseline
                                            RRB = right recordability baseline
                                            RMB = right medical baseline

EXAMPLE 1: STS AND RECORDABILITY

Name: M.W.   Birthdate: 3/28/36   Gender: male   Race: white

Test Age	Type	.5	1	2	3	4	6	8	STS avg.	change from LSB	age-corr. change	Reviewer decision	Initial or revised BL	change from LRB	age-corr. change
46	B	15	15	15	20	30	35	20	21.7				LSB & LRB		
47	A	15	15	15	30	45	40	35	30.0	8.3				8.3	
48	A	20	15	15	30	50	40	30	31.7	10.0 *	9.3			10.0	
49	A	25	20	20	35	50	40	35	35.0	13.3 *	11.7 a		LSB	13.3	
50	A	20	20	25	40	55	40	35	40.0	18.3 *	16.0 a			18.3	
50	R	20	20	20	40	55	45	40	38.3	16.7 *	14.3 a	revise LSB		16.7	
51	A	20	25	25	45	60	35	40	43.3	8.3				21.7	
52	A	20	25	30	50	55	45	40	45.0	10.0 *	8.0			23.3	
53	A	20	20	30	55	55	45	35	46.7	11.7 *	9.0			25.0	20.7
54	A	30	20	35	65	65	55	35	55.0	20.0 *	17.0 a			33.3	28.3 b
54	R	25	20	30	55	50	40	40	46.7	11.7 *	8.7			25.0	20.3
55	A	25	25	35	60	55	45	40	50.0	15.0 *	11.0 a			28.3	22.6
55	R	20	20	35	60	55	40	40	50.0	15.0 *	11.0 a		LSB	28.3	22.6
56	A	25	25	35	60	55	35	50	50.0	15.0 *	10.3 a	revise LSB		28.3	22.0
57	A	25	25	40	60	60	55	50	53.3	3.3				31.7	24.7
58	A	20	30	60	65	65	50	55	63.3	13.3 *	10.7 a			41.7	33.4 b
58	R	25	30	50	65	60	50	60	58.3	8.3	5.7	revise LRB→	LRB	36.7	28.4 b
59	A	20	25	55	60	55	45	50	56.7	6.7				-1.6	
60	A	20	25	55	65	60	55	65	60.0	10.0 *	6.0			1.7	

Column groups: Left Ear Thresholds (dB) by frequency (kHz); STS avg.; Detection of STS or improvement (change from LSB, age-corr. change); Baseline Revisions (Reviewer decision, Initial or revised BL); Detection of 25-dB shift for 200 log (change from LRB, age-corr. change).

EXAMPLE 2: IMPROVEMENT AND MEDICAL REFERRAL

Name: L.G.   Birthdate: 9/14/42   Gender: male   Race: black

Test Age	Type	.5	1	2	3	4	6	STS avg.	change from LSB	age-corr. change	Reviewer decision	Initial or revised BL	History Information and Reviewer's Recommendations
31	B	25	25	30	25	30	25	28.3				LSB & LMB	
32	A	15	10	5	10	15	20	10.0	-18.3 i				
33	A	5	10	10	10	15	15	11.7	-16.7 i		rev. LSB, LMB		improvement persists
34	A	10	5	5	5	0	15	3.3	-6.7 i				
35	A	15	15	10	15	15	25	13.3	3.3				
36	A	10	15	5	15	10	25	10.0	0.0				
37	A	10	5	0	5	10	10	5.0	-5.0 i				
38	A	5	5	0	5	5	30	3.3	-6.7 i		rev. LSB, LMB	LSB & LMB	improvement persists

Column groups: Left Ear Thresholds (dB) by frequency (kHz); STS avg.; Detection of STS or improvement (change from LSB, age-corr. change); Reviewer Actions (Reviewer decision, Initial or revised BL); History Information and Reviewer's Recommendations.

Note A: Low-frequency shifts suggest potential medical ear condition.
               Recommend employee seek medical treatment unless already under physician's care.
Note B: Apparent conductive hearing loss probably related to eardrum perforation with cotton swab. Refer to otolaryngologist
               for clinical audiogram, diagnosis, and opinion regarding work-relatedness and safety of wearing HPDs.

TABLE 11.7 — continued

The federal OSHA recordability position is illustrated (age-corrected 25-dB average shift at 2,3,4 kHz). Some state OSHA programs use different criteria (such as STS) for recording occupational hearing loss.

Codes:
i = improvement of at least 5 dB in STS average (2,3,4 kHz)
* = OSHA STS before age corrections from current STS baseline
a = OSHA STS after age corrections from current STS baseline
# = 25-dB shift before age corrections from recordability baseline
b = 25-dB age-corrected shift from recordability baseline
m = medical/audiological referral (clinical exam) at employer's expense
r = recommend that employee seek medical ear treatment at own expense

.5	1	2	3	4	6	8	STS avg.	change from RSB	age-corr. change	Reviewer decision	Initial or revised BL	change from RRB	age-corr. change	Date
								Detection of STS or improvement		Baseline Revisions		Detection of 25-dB shift for 200 log		
15	10	15	20	35	40	25	23.3				RSB & RRB			04/26/82
15	15	15	25	40	40	80	26.7	3.3				3.3		03/03/83
20	20	15	25	45	55	70	28.3	5.0				5.0		04/04/84
20	15	15	30	50	50	60	33.3	10.0 *	8.3			10.0		03/12/85
20	20	20	30	55	55	65	35.0	11.7 *	9.4			11.7		03/24/86
20	15	20	30	55	50	60	35.0	11.7 *	9.4			11.7		04/05/86
15	15	20	35	65	50	60	40.0	16.7 *	14.0 a		RSB	16.7		03/23/87
20	10	25	35	70	65	70	43.3	20.0 *	16.3 a	revise RSB		20.0		03/14/88
20	10	30	40	65	65	75	45.0	5.0				21.7		03/21/89
20	20	30	35	75	60	75	46.7	6.7				23.3		05/22/90
20	15	25	35	70	60	70	43.3	3.3				20.0		06/10/90
20	20	35	40	65	60	80	46.7	6.7				23.3		05/07/91
20	20	30	40	65	65	75	45.0	5.0				21.7		05/12/91
20	20	35	50	70	65	80	51.7	11.7 *				28.3	22.0	04/30/92
20	25	40	65	70	75	85	58.3	18.3 *	15.0 a		RSB & RRB	35.0	28.0 b	07/29/93
25	45	70	75	80	100	110	75.0	35.0 *	29.3 a			51.7	43.3 b	06/14/94
25	35	45	75	75	75	85	65.0	25.0 *	19.3 a	revise both		41.7	33.3 b	07/14/94
20	30	35	60	65	65	85	53.3	-5.0				-5.0		06/20/95
25	30	40	65	65	75	105	56.7	-1.7				-1.7		06/18/96

.5	1	2	3	4	6	8	STS avg.	change from RSB	age-corr. change	Reviewer decision	Initial or revised BL	History Information and Reviewer's Recommendations	Date
								Detection of STS or improvement		Reviewer Actions			
5	15	15	35	25	15		25.0						09/16/73
10	20	15	30	20	25		21.7	-3.3					07/20/74
30	30	20	25	20	30		21.7	-3.3		r		See Note A.	06/25/75
5	20	5	20	10	25		11.7	-13.3 i				Improved at low freqs.	08/20/76
5	25	15	25	25	30		21.7	-3.3					09/15/77
10	0	10	30	10	20		16.7	-8.3					11/01/78
55	55	55	65	65	70		61.7	36.7 *	34.4 a	m		See Note B.	06/12/79
45	45	45	65	55	75		55.0	30.0 *	27.4 a	revise → RSB,RMB		See Note C.	07/03/80

Note C: Recommend that employee continue medical treatment for surgical repair of perforated eardrum.

most efficient approach is to take a comprehensive history when the baseline audiogram is performed, including the following general topic areas: past noisy jobs, past military noise exposure, past noisy hobbies, and use of firearms (all the preceding with dates and documentation of whether HPDs were worn); general health and medication history; ear health history and any otological treatment; any history of tinnitus, hearing aid use, or relatives with hearing loss. After the comprehensive history has been documented once, at subsequent annual audiograms an abbreviated annual update history can be taken, focusing on any problems experienced within the last year, as well as any ear symptoms currently present (such as tinnitus, or ear congestion due to a cold or allergies).

To save time, many HCPs ask the employee to fill out a written auditory history form. However, more information is likely to be solicited if the OHC interviews the employee. At the least the OHC should personally question the employee about any items which indicate a change or an abnormality, then document details about these areas. For example, if the employee checks "yes" regarding having seen a doctor about his ears in the preceding 12 months, then the OHC needs to document the reason, the name of the physician seen, the treatment and outcome, etc.

The purpose of the brief otoscopic check is to document any obvious abnormalities which would be helpful for the professional who reviews the record to know about, such as cerumen partially or completely blocking the earcanal, foreign objects in the earcanal, tenderness in the area of the pinna and earcanal, or any indications of possible pathology or injury involving the external ear. The OHC must be trained by the professional supervisor to perform such checks competently and safely, and to document the desired information. The OHC is not assisting in a medical diagnostic process by performing these checks, but rather is simply recording basic information which may aid the professional reviewer in deciding whether a referral is needed (Dobie, 1993b). Written guidance in otoscopy is available (Foltner, 1986; Wormald and Browning, 1996), but references are no substitute for hands-on training by the professional supervisor.

HPD Check

Since the adequacy of HPD fit and the employee's ability to achieve a good acoustic seal with these HPDs are critical to employee protection, these factors should be assessed and documented during each annual audiogram session. Employees should be required to bring their HPDs with them to the audiogram appointment. After asking the employee to put on the HPDs, the OHC can evaluate the seal and record the results as part of the auditory history update. The OHC can reinstruct the employee in proper earplug insertion if needed to achieve a good seal. If the HPDs need to be replaced due to signs of deterioration, or due to inadequate fit and/or incorrect size for the individual, then the OHC can issue new devices and teach the employee how to use them correctly for good attenuation. The OHC should document the type and size of any new HPDs issued and have the employee sign the record to verify that s/he has been taught about their use and care. This documentation is useful not only for OSHA purposes, but for the use of the professional reviewing audiometric records, who needs to consider the individual's hearing changes in light of HPD adequacy over time. Although replacement HPDs may be available from supervisors or dispenser boxes throughout the year, this annual HPD fit check and retraining by the OHC is the best way to reinforce the importance of proper HPD utilization and to emphasize how the employee's personal HPD use (on and off the job) can affect personal hearing status.

Test-Related Records

Records related to audiometric testing are needed to support the fact that the equipment and environment met all OSHA requirements (and hopefully stricter ANSI standards) and that the audiogram was administered by qualified personnel in such a way as to obtain valid and reliable results. OSHA's HCA specifies that certain items must be kept either on the audiogram itself or with the audiometric records; these items are starred in the documentation list shown in Table 11.8. The table also includes additional records needed to demonstrate OSHA compliance or carry out the provisions of the HCA, plus other documentation dictated by common sense. Although an OSHA inspector might not ask to see all these items, they are needed by a professional reviewing audiometric records, as well as to support test validity in workers' compensation claims or other legal contexts.

Audiograms

All audiograms must be retained, even those replaced by a later retest and those deemed invalid for some reason (from equipment malfunction to employee noncooperation). Invalid tests should be clearly marked as such, including the identified problem or an indication of the probable cause. OSHA's HCA specifies that audiometric records must be retained for the duration of the affected employee's tenure; however, for purposes of workers' compensation claims, it is recommended that they be retained for 30 years after termination. It is also prudent to administer exit audiograms at the time of reassignment out of a noisy job (or at termination, if the employee gives notice); though not required by OSHA, exit audiograms are useful for workers' compensation purposes.

OSHA stipulates that if there is a transfer of ownership of the employer's business, then all records must be transferred to the successor employer. Employees' baselines are retained as they were prior to the transfer of ownership (OSHA, 1994) and the record is evaluated for cumulative hearing change under both ownerships. That is, the new owner cannot start over with new baselines from the time of purchase.

A cumulative sheet listing the numerical thresholds of all audiograms in chronological order is a practical necessity for the professional reviewer; the reviewer will be frustrated and hampered if each audiogram is on a separate sheet in a medical record. The cumulative sheet also assists the OHC attempting to counsel the employee about hearing change, though the computer graphics

TABLE 11.8
Required and recommended documentation for the audiometric phase of the HCP.

Codes: * records required to be kept on or with the audiogram
c records needed to support HCA compliance and carry out its requirements
r additional desirable records to support the HCP and for legal purposes

Type of record		Information to document
Audiogram data and employee identification	*	employee's name, job classification, and TWA
	*	date of audiogram
	*	name of audiometric examiner
	*	audiometer identification and date of most recent calibration
	*	hearing thresholds in each ear at 500, 1000, 2000, 3000, 4000, 6000 Hz
	r	hearing thresholds in each ear at 8000 Hz
	r	employee's unique identification number and race
	c	employee's gender and date of birth
	c	make, model, and serial numbers for audiometer and earphones
	c	indication of the type of audiogram: baseline, annual, retest, exit, etc.
	c	for baseline audiograms: verification of 14 hours quiet prior to test
	r	credentials of audiometric examiner (indicating current certification)
	r	examiner's judgment of subject's response reliability
	r	notes explaining specific reasons for questionable test validity
Supporting data	c	audiometer daily calibration checks and listening checks
	c	audiometer annual acoustic and/or exhaustive calibration certificates
	c	ambient OB SPLs in the audiometric test booth during normal use
	r	audiometer pre- and post-cal output SPLs to track adjustments
	r	credentials of the professional supervisor of OHCs who perform audiograms
	r	otoscopic check results
	r	auditory history questionnaire responses
	r	HPD type and size issued for each ear
	r	HPD fit check and verification that employee can achieve an acoustic seal
HCP classification	c	whether employee is currently classified as included in the HCP
	r	date when employee entered the HCP (based on TWA above action level)
	r	date when employee exited the HCP (due to termination or TWA reduction)
Routine record review	r	examiner's notes about time-of-test employee counselling
	r	written feedback provided by OHC or professional reviewer
Problem audiogram review & follow-up	c	written notification of employees showing OSHA STS
	c	completion of follow-up actions required by OSHA for STS
	c	professional reviewer's designation of revised baselines (if any)
	r	credentials of professional reviewer
	r	written feedback and recommendations provided by professional reviewer
	r	documentation of actions taken to fulfill reviewer's recommendations

display is even better for this purpose. If the data are computerized, a new cumulative summary sheet can be printed each time a new test is added. However, it is simple to keep such a record by hand as well.

Another documentation need involves recordkeeping for employees who refuse to participate in audiometric monitoring. Although OSHA requires the employer to offer audiograms, OSHA has no authority to require an individual employee to take audiograms. Often this situation can be avoided by educating the employee about the purpose of the test and assuring the worker that the results will not be used to discriminate against him or her. However, the employer may elect to require participation in all safety and health programs as a condition of employment. Repeated refusal to participate would then become justification for dismissal.

In safety-critical jobs, minimum hearing requirements may be established to qualify to enter or retain the job assignment, such as for transport truck drivers, forklift drivers, air traffic controllers, aircraft pilots, police officers, etc. (LaCroix, 1996; MacLean, 1995). Audiograms for qualification purposes may be needed for employees who are not part of the OSHA HCP. The OHC should be aware that employees trying not to become disqualified on the basis of hearing loss may be motivated to indicate falsely sensitive threshold responses if the testing procedure facilitates this (as with an excessively rhythmic presentation pattern). Special testing by an audiologist (such as measurements of speech comprehension in a noisy background) may be necessary to demonstrate qualification for hearing-critical jobs; such testing is beyond the scope of the typical HCP. (Also see *Employment Criteria and the ADA* in Chapter 16.)

Employer's Access to Data Obtained by a Contractor

When the employer contracts with an external service to administer audiograms, the employer should make certain that all desired records will be maintained and available to the employer. External providers should clarify their own recordkeeping policies to the employer. Contracts should include the service's agreement to transfer to the employer relevant information regarding the testing program's facilities and equipment, as well as all the audiometric data obtained on behalf of the employer, either upon request or as a routine matter.

Documentation of Follow-up

For cases of OSHA STS, records must document the execution of the required follow-up steps, including the professional reviewer's disposition and recommendations, the OHC's fitting or refitting of adequately attenuating HPDs to the employee and training or re-training of the employee in proper HPD use and care, and the employer's written notification to the employee of the occurrence of the STS. If a referral to an audiologist and/or physician was made at employer expense, the written reports and results of the examinations should be retained in the employee's record. If a physician has provided a written opinion that an STS is not work-related, that statement must be retained.

If a revised baseline has been set by the audiologist or physician following a persistent shift, the professional must specify the new baseline for use by the OHC in OSHA STS identification (or for 25-dB recordability purposes, as applicable). Note that only the professional, not the OHC, can revise baselines.

These records are important not only for OSHA recordkeeping, but also for use by the professional reviewer when evaluating this employee's file in the future, or in workers' compensation or other legal actions. If follow-up actions taken to increase the employee's protection from noise are not documented, a future reviewer will be unable to know that they have occurred and may recommend redundant steps.

Use of Computerized Recordkeeping Systems

Modern computers can streamline recordkeeping, reduce the accumulation of paper files, and identify tasks which need completion (such as missed audiograms, overdue calibration checks, etc.). When the professional reviewer is in another location, electronic records can be sent via diskette or modem for review, avoiding mailing costs and eliminating the possibility that hard-copy records will be lost. However, electronic back-ups are mandatory to ensure that digital records are not lost through device malfunction. Password protection may be necessary to ensure that only authorized persons have the ability to change certain records (such as the authority to revise audiometric baselines).

In evaluating software, employers should look for the capacity of the software to accept open-ended remarks, since yes/no categorical information storage cannot reflect case-specific information about medical histories, reviewer comments, and similar individualized details. Selection or development of software should involve all those HCP team members who will use the system, including the professional supervisor.

Commercial software products may meet some employers' needs as sold. Other employers may choose to purchase customized versions of commercial software, or even to develop their own. Some recordkeeping systems include modules for other HCP elements besides audiometric monitoring, as well as modules for other occupational health and safety programs. Achieving compatibility and ease of data exchange across different information systems is critical to ensure that the information can be shared as needed (for example, between the personnel, medical, industrial hygiene, and safety departments). For example, it is critical that noise exposure monitoring data and audiometric data be compatible, sharing a common identification code (such as social security number or employee number) to allow match-ups. The employer should also ensure that any software selected has the capability to output the database in a format (such as standard ASCII) which would allow easy translation to a different format later if the employer chooses to change software.

Logging Occupational Hearing Loss for OSHA

Occupational hearing loss is recordable on the OSHA Form 200 log (Bureau of Labor Statistics, 1986) either as an occupational illness (for nontraumatic loss) or as an occupational injury (for immediate acoustic trauma). Traumatic injuries are recordable in column 2 or 6 (depending on whether lost work days are involved). Nontraumatic hearing loss is logged in column 7(f), unless wearing HPDs caused or aggravated a medical pathology of the ear, in which case the loss is logged in column 7(g).

The decibel amount of hearing change which requires logging has been a topic of controversy, with varying policies in force at different times in different OSHA jurisdictions. Federal OSHA currently requires (OSHA, 1991) that an age-corrected shift of 25 dB or more from the original baseline in the average of thresholds at 2, 3, and 4 kHz be logged as an occupational illness on the OSHA Form 200. In contrast, several state OSHA programs require recording of age-corrected 10-dB OSHA STSs at the time of this publication (California, Michigan, North Carolina, South Carolina, and Tennessee), and Oregon uses the 25-dB criterion without allowing age corrections (Megerson, 1995). Once the federal recordkeeping final rule is promulgated (see below), changes in many state programs will be likely.

In 1996 OSHA held hearings about proposed new recordkeeping rules (OSHA, 1996) and received oral testimony and written comment submissions for consideration before promulgating a final rule. The proposed rule would require recording of 15-dB shifts from baseline averaged across 2, 3, and 4 kHz. Numerous professional groups advocated recording of 10-dB OSHA STSs after the application of age corrections and the determination (rather than presumption) of work-relatedness; the groups' recommendations also included other technical comments (AIHA, 1996; Coalition to Preserve OSHA and NIOSH and Protect Workers' Hearing, 1996). At the time of this publication, OSHA had not yet issued the final recordkeeping rule.

HCP personnel should check periodically for changes in OSHA recordability policies in their own jurisdictions. Some corporations with facilities spread across jurisdictions with different recording rules have elected to adopt a uniform internal policy which complies with the most conservative jurisdiction. OSHA's website posts "interpretation and compliance letters" in response to requests for clarification regarding details of recordability issues as well as other aspects of hearing conservation regulations. HCP personnel may find these letters helpful.

Evaluating an Audiometric Testing Program

In assessing an in-house audiometric testing program, considering whether to change from in-house audiometric testing to an external service provider, or choosing among several providers, HCP personnel may use the recommendations in this chapter as a standard of good practice. Technical variables (equipment, background noise, procedures, etc.) as well as more qualitative factors (such as the dedication of personnel, interaction of OHCs with employees, individualization of professionals' written feedback and recommendations to employees) affect the reliability of the audiometric data and determine how effective the HCP's audiometric phase will be in identifying incipient hearing changes and preventing their progression. HCP personnel are urged to strive for excellence and consistency in the audiometric phase in order to reap its benefits.

Effects of Procedural and Equipment Changes on Audiometric Data

Many technical factors can systematically alter the absolute value of measured hearing thresholds in the absence of any real hearing change. The difference between self-recording audiometry and manual audiometry has already been mentioned. A similar impact could also occur from switching between micro-processor audiometers with different testing algorithms, from changing between two testing services, or from inconsistent calibration of different audiometer units of the same type. To illustrate the magnitude of the artifactual threshold shifts which may occur, mean HTL data for two sites over time are plotted as Figure 12.18 (in next chapter). Group 1 continued in an exemplary in-house audiometric monitoring program over all the years shown, whereas for Group 2 there were changes first to one mobile testing service, and then to another service. After the first change STS rates increased dramatically, and high data variability continued to be a problem (also see Figures 12.19 and 12.20 for more data on this same site). The implication is clear: consistency in audiometric monitoring is of paramount importance. Though changes are sometimes necessary, frequent changes without good reason should be avoided.

Audiometric Data as an Indicator of HCP Effectiveness

The audiometric database comprises the ultimate test of HCP success, since it provides evidence of whether occupational hearing loss is being prevented. However, the usefulness of the data for this purpose depends on the reliability of threshold measurements.

STS Rates

Most employers are anxious to reduce STS rates, which are sometimes used inappropriately as an indicator of HCP quality by managers and insurance carriers. Sometimes corporate personnel will wrongly compare STS rates for plants with very different characteristics regarding noise exposures, workforce age distribution and seniority, etc. HCP personnel should concentrate on ensuring that the HCP is effective (the only sensible strategy for minimizing STS rates) rather than supporting misguided rate-reduction attempts such as repeatedly retesting STS cases to see if a non-STS audiogram can possibly be obtained within the allowed 30 days.

Even in a perfectly effective program, non-noise-related STSs will occur; therefore, zero percent rates should not be expected. Factors which influence annual rates of new age-corrected STSs include the following (Royster and Stewart, 1997).

1. High turnover reduces rates. Persons who don't stay with the job long enough to receive annual tests cannot show STS even if NIPTS is occurring.

2. Low turnover increases rates. In a stable workforce, STS rates will climb as employees get older because median age corrections do not completely remove the influence of age-related hearing change for persons with above-median presbycusis.

3. A high proportion of older employees increases rates. OSHA age corrections do not go above age 60, although presbycusis accelerates at older ages.

4. Calibration and/or testing methodology differences between baselines and later audiograms can either increase or decrease rates, depending on the particular change involved. This influence underscores the importance of achieving consistent, reliable audiometry.

5. Significant pre-existing hearing loss on the baseline audiogram reduces the probability of later STS for that employee (see Chapter 17). In HCPs where a large proportion of employees worked for years and developed NIPTS before the initiation of the HCP, STS rates will be reduced until after those persons retire.

6. Previous STS incidence decreases current rates unless there is lots of employee turnover. Most employees will not develop more than one STS in a working lifetime due to on-the-job noise exposure.

Due to the above factors, it is unwise to use STS rates as an indicator of HCP quality. If rates are very high, that is clearly bad, but NIPTS is unlikely to be the only factor involved. Conversely, low rates may result from reasons other than adequate protection from noise. Royster and Stewart (1997) found that 2.5% of new confirmed STSs per year is an average rate across numerous datasets exhibiting the full range of factors outlined above which make STS rates difficult to interpret; consistent with the preceding discussion, this average should not be used as a target (also see Chapter 12).

Taking the above factors into account, it may still be informative to compare STS rates for different exposure groups or departments within the same facility. For example, HCP personnel may find that maintenance personnel show high STS rates although dosimetry shows their daily TWAs exceed 85 dBA in only half the samples. This might indicate that education for maintenance workers needs to emphasize the potential for acoustic trauma from brief durations when their ears are very close to operating machinery, the importance of keeping HPDs handy to wear during intermittent periods of exposure, and the extra risk of hearing loss from off-the-job noisy hobbies often pursued by individuals with mechanical aptitudes. In another scenario, discrepant STS rates for departments with similar TWAs might be associated with differences in HPD utilization.

Other Approaches Using the Audiometric Database

It is strongly urged that the audiometric results from HCP testing be entered into computer databases so that their potential can be tapped for purposes such as comparing hearing levels of different groups for epidemiological studies, detecting the effects of audiometer calibration changes on audiogram results, etc. Too often year after year of informative results are simply filed and forgotten. See Chapter 12 for other strategies for using audiometric data to evaluate HCP effectiveness.

Capturing the Full Potential of Audiometric Monitoring

Audiometry can be the best motivational tool in the HCP through the interaction between the OHC and the employee, the rechecking of HPD fit at audiogram time, the examiner's immediate verbal feedback about hearing changes, the professional reviewer's written feedback and recommendations, and the employer's follow-up actions to increase employee protection. In addition, the audiometric database proves the value of the HCP in preventing occupational hearing loss for the exposed population (see Chapter 12). This chapter provides guidance for implementing the audiometric phase in such a way that its enormous potential benefits to employee and employer will be realized. In contrast, managers at far too many worksites waste this potential either unwittingly (through ignorance of what constitutes excellence) or intentionally (through a decision to forego excellence in exchange for an immediate, small cost savings). HCP personnel must be vigilant in reminding management that the goal is *hearing loss prevention* rather than regulatory compliance, and that qualitative aspects of audiometric monitoring make all the difference.

References

AAO-HNS (1997). "Otologic Referral Criteria For Occupational Hearing Conservation Programs" (a pamphlet available by request from publisher), American Academy of Otolaryngology—Head and Neck Surgery, Alexandria, VA.

ACOM (1989). "Occupational Noise-Induced Hearing Loss" (a position statement developed by the Noise and Hearing Conservation Committee), *J. Occup. Med. 31*, 996.

AIHA (1996). "AIHA Position Statement: Recommended Criterion for Recording Occupational Hearing Loss on OSHA Form 300," *Am. Ind. Hyg. Assoc. J. 57*(7), 661–662.

ANSI (1978). "Method for Manual Pure-Tone Threshold Audiometry," ANSI S3.21-1978 (R 1997), Acoustical Society of America, New York, NY.

ANSI (1991). "Permissible Ambient Noise Levels for Audiometric Test Rooms," ANSI S3.1-1991, Acoustical Society of America, New York, NY.

ANSI (1996). "Specification for Audiometers," ANSI S3.6-1996, Acoustical Society of America, New York, NY.

ANSI (1999). "Maximum Permissible Ambient Noise Levels for Audiometric Test Rooms," ANSI S3.1-1999, Acoustical Society of America, New York, NY.

Ballachanda, B. B., Roeser, R. J., and Kemp, R. J. (1996). "Control and Prevention of Disease Transmission in Audiology Practice," *Am. J. of Audiology 5*(1), 74–82.

Bureau of Labor Statistics (1986). *Recordkeeping Guidelines for Occupational Illnesses and Injuries* (O.M.B. No. 1220-0029), U.S. Department of Labor, Washington, DC.

Coalition to Preserve OSHA and NIOSH and Protect Workers' Hearing (1996). "Comments In Response To OSHA's Notice of Proposed Rule, Occupational Injury and Illness Recording and Reporting Requirements," comments submitted to OSHA docket No. R-02, Washington, DC.

Cohen, M. R., and McCullough, T. D. (1996). "Infection Control Protocols for Audiologists," *Am. J. of Audiology* 5(1), 20 – 22.

Dobie, R. A. (1993a). *Medical-Legal Evaluation of Hearing Loss*. Van Nostrand Reinhold, New York.

Dobie, R. A. (1993b). "Otoscopy in Hearing Conservation Programs," *CAOHC Update 4*(3), 5 and 7.

Dobie, R. A., and Archer, R. J. (1981). "Otologic Referral Criteria In Industrial Hearing Conservation Programs," *J. Occup. Med. 23*, 755 – 761.

DoD (1996). "DODI 6055.12 Department of Defense Hearing Conservation Program," available via the internet at http://web7.whs.osd.mil/text/i605512p.txt

Foltner, K. A. (1986). "Visual Evaluation of the External Ear and Eardrum," in *Noise & Hearing Conservation Manual*, 4th edition, edited by E. H. Berger, W. D. Ward, J. C. Morrill, and L. H. Royster, American Industrial Hygiene Association, Fairfax, VA, 217 – 232.

Frank, T., Greer, A. C., and Magistro, D. M. (1997). "Hearing Thresholds, Threshold Repeatability, and Attenuation Values for Passive Noise-Reducing Earphone Enclosures," *Am. Ind. Hyg. Assoc. J. 58*(11), 772 – 778.

Frank, T., and Williams, D. L. (1993). "Effects of Background Noise on Earphone Thresholds," *J. Am. Acad. Audiology 4*, 201 – 212.

Frank, T., and Williams, D. L. (1994). "Ambient Noise Levels in Industrial Audiometric Test Rooms," *Am. Ind. Hyg. Assoc. J. 55*(5), 433 – 437.

Franks, J. R., Engle, D. P, and Themann, C. L. (1992). "Real Ear Attenuation at Threshold for Three Audiometric Headphone Devices: Implications for Maximum Permissible Ambient Noise Level Standards," *Ear & Hearing 13*(1), 2 – 10.

Franks, J. R., and Lankford, J. E. (1984). "Noise Reduction Characteristics of Prefabricated Sound-Isolating Enclosures: a Laboratory, Field Measurement, and Theoretical Analysis," *Ear & Hearing 5*(1), 2 – 12.

Franks, J. R., Merry, C. J., and Engel, D. P. (1989). "Noise Reducing Muffs for Audiometry," *Hearing Instruments 40*(11), 29 – 30,32 – 33,36.

Gasaway, D. C. (1985a). "Documentation: the Weak Link in Audiometric Monitoring Programs," *Occupational Health & Safety 54*(1), 28 – 33.

Gasaway, D. C. (1985b). *Hearing Conservation: A Practical Manual and Guide*, Prentice-Hall, Inc., Englewood Cliffs, NJ.

Gasaway, D. C. (1996). "To Prevent Noise-Induced Hearing Loss—Aim Between the Ears," paper in proceedings of the 21st Annual NHCA Hearing Conservation Conference, February 1996 in San Francisco, CA, National Hearing Conservation Association, Denver, CO.

Harris, J. D. (1980). "A Comparison of Computerized Audiometry by ANSI, Bekesy Fixed-Frequency, and Modified ISO Procedures in an Industrial Hearing Conservation Program," *J. Auditory Research 20*, 143 – 167.

Health and Safety Executive (1995). *A Guide to Audiometric Testing Programmes*, HSE Books, Sudbury, England.

Hirschorn, M., Kremenic, N., and Singer, E. E. (1982). "Field Evaluation of Audiometric Rooms," *Hearing Instruments 33*(7), 24 – 25.

Jerlvall, L., Dryselius, H., and Arlinger, S. (1983). "Comparison of Manual and Computer-Controlled Audiometry Using Identical Procedures," *Scand. Audiol. 12*, 209 – 213.

Killion, M. C., Mueller, H. G., Pavlovic, C. V., and Humes, L. E. (1993). "A Is for Audibility," *The Hearing J. 46*(4), 29.

LaCroix, P. (1996). "Hearing Standards and Employment Decisions: The Federal Model," *Spectrum 13*(3), 6 – 8.

Lankford, J. E., Franks, J. R., and Martin, C. L. (1981). "Ambient Noise Levels in Prefabricated Sound Rooms," *Hearing Instruments 32*(8), 14 – 15 and 43.

Lankford, J. E., Perrone, D. C., and Thunder, T. D. (1999). "Ambient Noise Levels In Mobile Audiometric Testing Facilities," *Am. Assoc. Occup. Health Nurses J. 47*, 163 – 167.

Lundeen, C. (1996). "Count-the-Dot Audiogram in Perspective," *Am. J. Audiology 5*(3), 57 – 58.

MacLean, S. (1995). "Employment Criteria in Hearing Critical Jobs," *Spectrum 12*(4), 20 – 23.

Megerson, S. C. (1995). "Noise in Washington over Hearing Loss Recordability," *CAOHC Update 6*(2), 4 – 5.

NASED (1999). "NASED Certification of Audiometric Technicians." Additional information and lists of certified technicians available from National Association of Special Equipment Distributors, P.O. Box 870923, Stone Mountain, GA 30087.

NHCA (1996). "A Practical Guide to Mobile Hearing Testing and Selecting a Service Provider" (Guide #3), National Hearing Conservation Association, Denver, CO.

NHCA (in preparation A). "A Practical Guide to Audiometric Review Criteria," National Hearing Conservation Association, Denver, CO.

NHCA (in preparation B). "A Professional Guide to Audiometric Referral Criteria," National Hearing Conservation Association, Denver, CO.

NIOSH (1998). *Criteria for a Recommended Standard: Occupational Noise Exposure Revised Criteria 1998*, DHHS (NIOSH) Publication No. 98-126, National Institute for Occupational Safety and Health, Cincinnati, OH.

O Mahoney, C. F., and Luxon, L. M. (1996). "Misdiagnosis of Hearing Loss Due to Ear Canal Collapse: A Report of Two Cases," *J. Laryngol. Otol. 110*, 561 – 566.

OSHA (1983a). "Occupational Noise Exposure: Hearing Conservation Amendment; Final Rule," Occupational Safety and Health Admin., 29 CFR 1910.95, *Fed. Regist. 48*(46), 9738 – 9785.

OSHA (1983b). Interpretation letter dated 4/6/83 from R. Leonard Vance to John E. Schroll. Accessible via the internet at http://www.osha-slc.gov/OshDoc/Interp_data/ I19830406A.html.

OSHA (1991). Memorandum to OSHA regional administrators dated 6/4/91 from Patricia Clark and Stephen Newell regarding recording of hearing loss and cumulative trauma disorders. Accessible via the internet at http://www.osha-slc.gov/OshDoc/Interp_data/ I19910604C.html.

OSHA (1993). Interpretation letter dated 8/31/93 from Roger A. Clark to Fredrik Lindgren. Accessible via the internet at http://www.osha-slc.gov/OshDoc/Interp_data/ I19930831B.html.

OSHA (1994). Interpretation letter dated 5/9/94 from Ruth E. McCully to J. Christopher Nutter. Accessible via the internet at http://www.osha-slc.gov/OshDoc/Interp_data/I19940509A.html.

OSHA (1996). "29 CFR Parts 1904 and 1952, Occupational Injury and Illness Recording and Reporting Requirements; Proposed Rule," *Fed. Regist. 61*(23), 4030 – 4067.

OSHA (1998). Interpretation letter dated 6/2/98 from John B. Miles, Jr. to Debra A. Rowland. Accessible on the internet at http://www.osha-slc.gov/OshDoc/Interp_data/I19980602.html.

Rebke, E. (1992). "The Benefits of Field Testing the Acoustic Performance of Sound Isolation Rooms," *Canadian Acoustics 20*(4), 11 – 16.

Reiter, L. A., and Silman, S. (1993). "Detecting and Remediating External Meatal Collapse During Audiologic Assessment," *J. Am. Acad. Audiol. 4*, 264 – 268.

Roeser, R. J., and Ballachanda, B. B. (1997). "Physiology, Pathophysiology and Anthropology/Epidemiology Of Human Earcanal Secretions," *J. Am. Acad. Audiol. 8*(6) 391 – 400.

Royster, J. D. (1985). "Audiometric Evaluations for Industrial Hearing Conservation," *Sound and Vibration 19*(4), 24 – 29.

Royster, J. D. (1992). "Evaluation of Different Criteria for Significant Threshold Shifts in Occupational Hearing Conservation Programs," Contract report submitted to National Institute for Occupational Safety and Health, Cincinnati, OH, December 1992. Available from National Technical Information Service as NTIS document PB93-159143.

Royster, J. D. (1996). "Evaluation of Additional Criteria for Significant Threshold Shift in Occupational Hearing Conservation Programs," Contract report submitted to National Institute for Occupational Safety and Health, Cincinnati, OH, August 1996. Available from National Technical Information Service as NTIS document PB97-104392.

Royster, J. D. (1996a). "Technical Comments in Response to OSHA's Proposed Rule on Occupational Injury and Illness Recording and Reporting Requirements With Respect to Occupational Hearing Loss," Appendix C of testimony to OSHA's docket No. R-02 concerning 29 CFR Parts 1904 and 1952, from the Coalition to Preserve OSHA and NIOSH and Protect Worker's Hearing.

Royster, J. D. (1996b). "New NHCA Guidelines for Baseline Revision," *Spectrum 13*(2), 1 and 4 – 9.

Royster, J. D., and Royster, L. H. (1990). *Hearing Conservation Programs: Practical Guidelines for Success,* Lewis Publishers, Inc., Chelsea, MI.

Royster, J. D., and Stewart, A. P. (1990). "Revising Baselines: Diversity Abounds!" *Spectrum 7*(4), 1 and 4 – 7.

Royster, J. D., and Stewart, A. P. (1997). "What Affects STS Rates?" paper in proceedings of the 22nd annual NHCA Hearing Conservation Conference, February 1997 in Orlando, FL, National Hearing Conservation Association, Denver, CO.

Schulz, T. Y. (1999). "Hearing Conservation Referral Criteria," paper in proceedings of the 24th Annual NHCA Hearing Conservation Conference, February 1999 in Atlanta, GA, National Hearing Conservation Association, Denver, CO.

Simpson, T. H., Stewart, M., and Blakley, B. W. (1995). "Audiometric Referral Criteria for Industrial Hearing Conservation Programs," *Arch. Otol. Head and Neck Surg. 121*, 407 – 411.

Singer, E. E. (1984). "Physical Environment for the Performance of Basic and Follow-up Audiometric Studies," in *Occupational Hearing Conservation*, edited by M.H. Miller and C.A. Silverman, Prentice-Hall, Inc., Englewood Cliffs, NJ, pp. 62 – 74.

Suter, A. H. (1993). *Hearing Conservation Manual*, Third Edition, Council for Accreditation in Occupational Hearing Conservation, Milwaukee, WI.

Wilber, L. A. (1994). "Calibration, Puretone, Speech and Noise Signals," in *Handbook of Clinical Audiology*, fourth edition, edited by J. Katz, Williams & Wilkins, Baltimore, MD, pp. 73 – 94.

Wormald, P. J., and Browning, G. G. (1996). *Otoscopy — A Structured Approach,* Singular Publishing Group, Inc., San Diego, CA.

The Noise Manual, revised 5th edition, edited by E.H. Berger,
L.H. Royster, J.D. Royster, D.P. Driscoll, and M. Layne
©2003 American Industrial Hygiene Association

12

Evaluating Hearing Conservation Program Effectiveness

Julia Doswell Royster and Larry H. Royster

Contents

Introduction

The goal of occupational hearing conservation programs (HCPs) is to prevent significant on-the-job noise-induced hearing loss; therefore, the ultimate outcome measure of HCP performance can be found in the audiometric database of annual audiograms for noise-exposed personnel. This chapter covers several methods of using audiometric database analysis (ADBA) to evaluate HCP success and gives examples of their application and interpretation. ADBA is a *performance-based* approach to HCP evaluation, which yields information not obtainable through compliance-oriented audits (J. D. Royster, 1992a). The HCP is judged as acceptable if the noise-exposed population does not exhibit significant development or progression of occupational noise-induced hearing loss (NIHL), and if audiometric variability is low.

Tabulation Assessment Approaches

Tallies of HCP completeness and tabulation of HCP statistics describe HCP status but do not provide outcome-oriented performance measures.

HCP Audits

One approach to evaluating HCPs is to assess whether they are *complete* by performing a checklist tally of whether the desired program elements are present, and whether they meet specifications (such as compliance with a regulation or adherence to more stringent standards). The difficulty with this approach is in judging the quality of HCP efforts: an element may be present but carried out in such a cursory manner as to be ineffective. Sample checklists are available that do address qualitative factors (Franks et al., 1996; J. D. Royster and Royster, 1990), but the auditor must have the expertise to judge quality, plus the time to delve deeply enough when assessing HCP implementation. A knowledgeable hearing conservationist can add items to evaluate achievement of facility- or company-specific goals.

Audits may also include annual performance data on HCP statistics such as the percent of personnel wearing HPDs correctly, percent of audiograms administered on time, percent of audiometric retests achieved within 30 days, and other similar measures. Indicators based on a broader view of HCP achievement include numbers of employees removed from the HCP due to reduction of OSHA time-weighted average noise exposures (TWAs) by engineering noise controls, decreasing (or increasing) average TWAs for departments, or overall reductions (or increases) in the percentage of the workforce exposed to hazardous noise.

Audit approaches may be useful, but they fail to get at the bottom-line indicator of HCP success: is occupational hearing loss being prevented in exposed workers?

Individual Hearing Shift Tabulations

Many practitioners rely on the annual rates of OSHA standard threshold shift (STS) or other significant hearing shifts to indicate program quality. These measures, which tally shifts from baseline audiometric thresholds, do attempt to identify the beginning or the progression of occupational NIHL in individual cases, but they are imperfect.

518

Such shifts may result from nonoccupational causes (including off-the-job noise, presbycusis, and pathology); therefore, a 0% rate is not expected even in a population without on-the-job noise exposure. It would be possible to develop normative data for expected OSHA STS rates (or other shifts) in nonexposed populations who participate in annual audiometric monitoring, but such data are not currently available.

Because the rate of noise-induced permanent threshold shift (NIPTS) progression decelerates over time (see Chapter 17), the probability of STS occurrence declines if there is a significant amount of pre-existing NIPTS in the individual's baseline audiogram. The application of OSHA age corrections further reduces the probability of shifts in ears with pre-existing NIPTS (see Chapter 11). In populations with large numbers of senior employees, a significant percentage will show age-corrected STSs even if perfectly protected from noise, simply because some individuals exhibit much greater age-related hearing change than assumed by the median values used for the OSHA age corrections. The validity of baseline audiograms, as well as policies for baseline revision, are obviously critical in any tabulations of hearing shifts from baseline. Baseline quality control can be achieved if guidelines for baseline revision are followed (see Chapter 11). Finally, most workers in the U.S. have TWAs that would not be expected to cause more than one STS in a working lifetime even without any protection from noise. A completely ineffective HCP might enjoy low STS rates after the passage of enough time for individuals to develop this first and only STS.

The factors mentioned above make it potentially misleading to compare shift rates for different plant sites or groups without first understanding the baseline hearing status, current age, and past STS history characteristics of the noise-exposed populations and then ensuring that the groups to be compared are similar regarding these variables. For further discussion, see Franks et al. (1989) and Simpson et al. (1993).

Based on rates of OSHA STS across many different industries, annual rates of new confirmed age-corrected OSHA STS typically should be less than 2.5% across the entire population (J. D. Royster and Stewart, 1997). However, unusual characteristics of the local population (such as high turnover or, in contrast, a predominance of long-term employees) may make this general benchmark inapplicable to a particular situation. Furthermore, this overall benchmark does not apply if the evaluator decides to investigate STS rates in groups of recently hired young employees, who generally have greater potential for NIPTS development (J. D. Royster, 1992b and 1996; J. D. Royster and Stewart, 1997).

In addition to population characteristics, audiometric testing factors can also play havoc with shift rates. Any change in audiometry that systematically affects measured thresholds in the same direction (toward better hearing or toward worse hearing) for one or more of the STS frequencies (2, 3, and 4 kHz) will also drive shift rates down or up. For example, in many HCPs the change from previously used automatic Békésy-tracing audiometers to currently favored microprocessor audiometers caused measured thresholds at every test frequency to show an artifactual worsening of 2.5 – 5 dB. In one HCP, the poorer hearing threshold levels (HTLs) measured by microprocessor resulted in 50% of the population showing an OSHA STS in the year the audiometer change occurred. Of course, if a systematic change had caused HTLs to improve, OSHA STSs would have declined, and neither the testing service nor the employer would have been motivated to account for the fortuitous yet artifactual rate drop.

A cause of STSs more common than systematic audiometric bias is unreliable audiometric data in which measured HTLs randomly vary up and down from one year to the next, while group mean HTLs over time show very little trend toward worsening hearing. Excess audiometric variability may be related to poor quality control of technician training, inadequate instructions to and motivation of test subjects, inconsistent audiometer calibration (especially across multiple audiometers), and noisy or distracting testing environments. As an example, one employer grew concerned when annual STS rates in non–noise-exposed headquarters personnel were consistently high (about 6% new confirmed age-corrected OSHA STSs annually) — just as high as for employees in noise-exposed occupations. This finding implied that factors other than occupational noise were responsible, but the simple shift rates themselves yielded no further clues.

Reasons for Using ADBA

In contrast to simple tabulation approaches, ADBA techniques allow the evaluator to understand the audiometric data and the reasons behind data patterns. For instance, in the STS example described in the preceding paragraph, ADBA techniques identified unreliable audiometry as the cause of high STS rates for two reasons: neither the noise-exposed nor the nonexposed group showed median HTLs significantly worse than expected for their age, and group mean HTLs were fairly stable over time in spite of large random shifts in individuals' HTLs between sequential pairs of audiograms. Explorations of the audiometric test provider's protocol revealed several contributing deficiencies: inadequate control of calibration across multiple audiometers, switching among audiometers from year to year, insufficient instruction to test subjects, inadequate control of earphone placement, and no immediate feedback to employees about their results at the time of the audiogram (reducing their motivation to listen and respond carefully).

This example introduces ADBA concepts: using group HTL data for populations with varying degrees of noise exposure, as well as comparing HTL variability between noise exposure groups, to determine whether occupational NIHL is being prevented and whether apparent shifts are related to occupational noise or to other factors.

ADBA results can provide concrete evidence to management regarding progress in the HCP (meriting praise for HCP personnel) or needs for improvement (justifying more management emphasis or greater resource allocations). If simple graphical presentations are prepared, ADBA results can be used to motivate supervisors or groups of employees toward better participation in the HCP. ADBA results can also support HCP decision-making, such as whether to require HPD use in a borderline-TWA area, or whether a certain style of earplug is adequate for a given noise environment. Finally, ADBA results provide HCP team members with a new form of monitoring data by which to identify those areas of the HCP that need more attention. Audiometric variability analyses provide fresh feedback to the HCP team members each year, and the variability statistics are sensitive enough to identify changes in data quality immediately, whether due to testing problems or due to changes in employees' degree of protection from noise.

ADBA procedures can detect the specific areas in which the HCP needs improvement so that appropriate corrective steps can be implemented. Certain departments within the plant may need special attention, or the results may show a more general deficiency in one of the phases of the HCP. Patterns of ADBA results help the reviewer differentiate between group threshold changes due to inconsistent audiometric testing, audiometer calibration shifts, inadequate hearing protector use, and other sources (L. H. Royster and Royster, 1984, 1988).

ADBA results provide tangible evidence of problems that can be used to demonstrate to managers the need for HCP improvement. With the support of the ADBA findings, the responsible professional is more likely to obtain approval for the resource allocations or policy changes necessary to correct problems. ADBA statistics can also show managers evidence of the benefits the company derives when the HCP is improved. Potential compensation costs for occupational hearing loss will decline after an effective HCP is established (L. H. Royster and Royster, 1984), and management is often impressed by the resulting savings.

The motivational benefits of ADBA are invaluable in gaining support for the HCP among employees and supervisors. Simple bar charts prepared from the ADBA statistics for various departments can convince supervisors that better enforcement of hearing protector utilization is necessary. The same information can be discussed with employees in safety meetings or presented in poster format to show that hearing trends are truly better for groups whose members consistently and properly wear their hearing protection. Employees can also learn to understand their own threshold changes better if they compare themselves to the normal age-effect reference data used in ADBA evaluations.

In short, ADBA results can be applied to give meaningful feedback to all levels of personnel involved in the HCP in order to stimulate interest and support for a stronger hearing conservation effort — a real need (Dobie, 1995; Melnick, 1984).

Information ADBA Provides

In contrast to the review of individuals' audiogram histories, ADBA evaluates hearing trends for employees as a group (L. H. Royster and Royster, 1990). ADBA detects early signs of occupational hearing loss in the population so that corrective actions can be taken *before* large threshold shifts occur for many individuals. In addition, the average hearing changes for groups of employees can serve as a reference comparison for interpretation of individual shifts. The top portion of the flow chart shown as Figure 12.1 illustrates the reciprocal relationship between ADBA and individual audiogram review. A review of each person's audiogram is necessary to identify significant hearing changes and trigger appropriate follow-up actions. However, the results from audiogram reviews cannot easily be used to assess how well the HCP is protecting the exposed population as a whole.

ADBA procedures should not necessarily be expected to identify the relative contributions of all the factors that affect the database, such as audiometric methods, test-

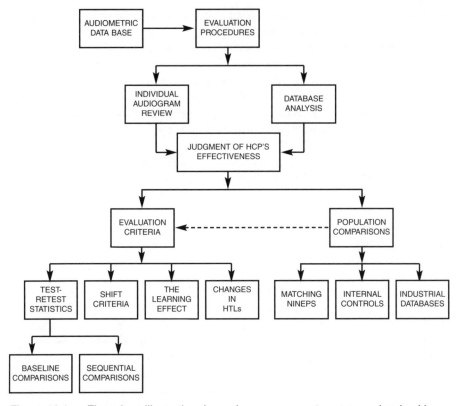

Figure 12.1 — Flow chart illustrating the various component processes involved in using audiometric database analysis to evaluate the effectiveness of the hearing conservation program.

ing environment, effectiveness of hearing protection devices, off-the-job noise exposures, and population characteristics. However, ADBA does give a warning if the net result of all these parameters indicates problems. ADBA techniques should be usable by personnel in general industry, who may have only limited computing and analytical abilities. Therefore, the methods developed are easy to apply and the results easy to understand. If an audiometric database has been computerized, the analysis is accomplished more quickly, but results for small samples can be tabulated by hand to give a reliable indication of HCP effectiveness.

ADBA Options

The ADBA techniques described in the following sections employ two types of comparisons: *population comparisons* of the amounts of hearing loss shown by employees included in the HCP versus that shown by control groups, and *criterion comparisons* of year-to-year audiometric variability versus criteria for reliable audiometric data. The criterion comparison methods are included in Draft ANSI S12.13-1991. Population comparisons offer an historical perspective; because this method is based on absolute HTL values, the results will change only slowly if hearing trends are followed over time. In contrast, because criterion comparisons are sensitive to annual changes in HCP implementation, fluctuations in annual results allow HCP personnel to monitor audiometric data quality on an ongoing basis. If annual ADBA results warn of developing problems, HCP personnel can take quick action to solve them.

Population Comparisons

The HTL data for groups of noise-exposed employees may be compared to reference population data either at a single point in time or longitudinally in terms of HTL changes. However, because NIPTS develops gradually, especially at moderate noise exposures, the passage of considerable time is required before HCP success or failure can be gauged by HTL comparisons for noise-exposed groups. If an inappropriate reference database is selected, the comparison can be misleading.

Comparisons at One Point in Time

In this simple ADBA method the group median HTLs for exposed populations are compared to the expected age-effect HTLs for appropriate reference populations without occupational noise exposure. If the reference population is comparable to the exposed population in every relevant respect except for occupational noise exposure, then the difference in HTLs between the two groups may be interpreted as the typical amount of occupational NIPTS present. (See Chapters 4 and 17 for discussions of HTL differences between screened presbycusis populations and unscreened age-effect populations.) This approach is useful in estimating the reduction in NIPTS for employees who were hired after the implementation of an HCP, compared to employees who worked for years without benefit of HCP efforts, and/or compared to the predicted hearing loss expected for the exposure history assuming no protection at all (see Chapter 17).

For example, consider Figure 12.2, illustrating data for employees of a textile plant's weaveroom, where the typical TWA has been 103 dBA for many years. Median HTL data are plotted for two groups of white male employees age 35–45 years old, with 15–25 years of service, compared to expected age-effect HTLs for the white male nonindustrial-noise-exposed population (NINEP) data from Annex C of ANSI S3.44-1996 (described in Chapter 17). Group A was taken from audiograms performed in 1977, 5 years after the initiation of the HCP at this facility; therefore, Group A would have worked unprotected for 10–20 years prior to HCP initiation. Because most of the NIPTS occurs in the first 10–15 years of exposure, inclusion in the HCP would have occurred too late to prevent very much of the NIPTS for Group A. In contrast, Group B was gathered from 1997 audiograms, and this group would have been protected by HCP efforts for their entire careers. Clearly Group B shows much less hearing loss than Group A, and their HTLs are close to those expected for their age. Assuming that audiometric testing procedures and test experience of the two groups were equivalent, and that daily noise exposures were similar in the two

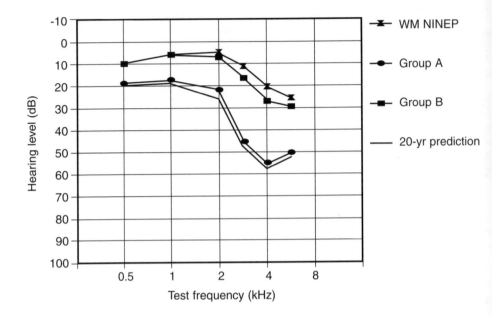

Figure 12.2 — Mean HTLs for Group A (unprotected for first 15 years of exposure) and Group B (protected by an HCP) compared to expected age-effect HTLs for the white male NINEP at age 40 and the predicted total hearing loss for white males at age 40 with 20 years of unprotected exposure at 103 dBA.

time periods, the difference in hearing between Groups A and B can be interpreted as the amount of NIHL prevented by hearing conservation efforts. The bold line on the graph shows the predicted HTLs at age 40 after 20 years of unprotected exposure at 103 dBA. The prediction was made using ANSI S3.44-1996 (see Chapter 17) and is a slight extrapolation beyond the 100-dB daily exposure range of the model.

For examples of this type of population comparison analysis, see Pelausa et al. (1995), Roberts (1997), Rosler (1994), Schulz (1996), and Wolgemuth et al. (1995).

Comparisons of HTL Trends Over Time

Another population comparison technique is to assess hearing change trends for groups of noise-exposed employees followed longitudinally over time in comparison to age-effect reference population trends. If the noise-exposed population shows hearing change no greater than that expected due to age effects, then the population is assumed to be adequately protected. In discussing population comparisons, we will use the model of hearing loss prediction standardized in ANSI S3.44-1996, which is discussed in Chapter 17. In a nutshell, this model predicts total hearing loss as the sum of NIPTS and age-related hearing threshold level (HTLA), minus a correction factor or compression term.

There are several difficulties with this seemingly simple population comparison method. First, the reference population chosen must be *appropriate* as a control group: it must be comparable to the exposed population in every way except for occupational noise exposure. Second, currently available reference data are based on cross-sectional studies (subjects of varying ages and exposure histories, each tested only once) rather than longitudinal studies (each subject retested after multiple time intervals). Because repeated experience with audiometric testing results in improvements in measured thresholds, referred to as the *learning effect*, the exposed population's HTL trends will include a learning-effect influence whereas those of the age-effect population will not. Finally, if substantial pre-existing NIPTS exists in an exposed population, then their future age-related hearing changes will be reduced due to the correction factor in the ANSI S3.44-1996 model.

Large discrepancies in predicted HTLs will result if different reference populations are used to estimate the age-related component of total hearing loss, as illustrated in Figure 12.3. The top curve represents the expected growth of NIPTS at 4 kHz for median susceptibility ears due to unprotected exposure to a daily TWA of 95 dBA. The lower curves show the resulting predicted total hearing loss at 4 kHz over time if this NIPTS is combined with the HTLA hearing levels for four different reference populations. Because the NIPTS component is the same in each case, the differences are due to the discrepancies in age-effect HTLs for the databases shown. The differences among the predictions become larger with advancing age because some reference populations show much larger age-effect hearing changes than others do.

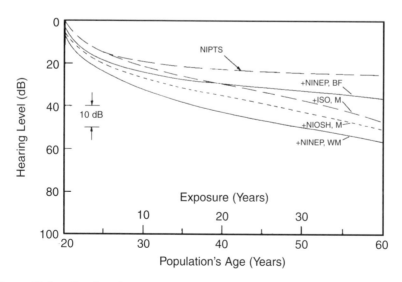

Figure 12.3 — Predicted population HTLs at 4 kHz as a function of age and years of exposure to a TWA of 95 dB for typical ears (0.5 fractile), calculated by combining the NIPTS component (top curve) with the age-related components for each of four HTLA populations to yield the total hearing losses shown by the four lower curves.

For comparisons to occupationally noise-exposed populations, we prefer the reference data offered in Annex C of ANSI S3.44-1996, which are NINEPs unscreened for any hearing influences except on-the-job noise exposure (L. H. Royster, Driscoll, Thomas, and Royster, 1980; L. H. Royster and Thomas, 1979). The only screening criterion used to eliminate subjects from the NINEPs was more than 2 weeks of exposure to significant on-the-job noise during the subject's prior work experience. Because gender and race are factors associated with hearing ability, separate NINEPs are available for black and white females and males. Unless a local database is developed to match an industrial population in terms of every age-effect hearing influence, we feel that the NINEPs provide the best available controls against which to compare industrial audiometric data in estimating the amount of on-the-job hearing change.

In Figure 12.3 the four HTLA populations used are the NIOSH male (NIOSH, M) reference population (NIOSH, 1972), which is also referenced in OSHA's Hearing Conservation Amendment; the ISO male (ISO, M) presbycusis population defined in Annex A of ANSI S3.44-1996; and NINEP populations of white males (NINEP, WM), and black females (NINEP, BF) from Annex C of ANSI S3.44-1996. If the exposed population were free of NIPTS at age 20 and experienced unprotected noise exposure at a TWA of 95 dB thereafter, then the expected changes in HTLs over time would be as shown in Figure 12.3, computed by combining the NIPTS data with the thresholds for the specified HTLA populations. The predicted thresholds differ significantly depending upon the HTLA population selected, due to factors including age effects and sex and race characteristics.

In contrast, if the population were properly protected from on-the-job noise since initial exposure at age 20, then the expected threshold levels over time would be as shown in Figure 12.4, depending upon the HTLA population selected. These curves

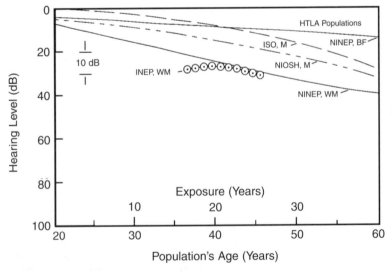

Figure 12.4 — Mean HTLs at 4 kHz over time for a white male INEP compared to the expected age-effect HTLs for four different HTLA populations.

represent the rates of change in thresholds with age in the absence of industrial noise exposure for these same reference populations. Also plotted in Figure 12.4 are the measured mean threshold levels at 4 kHz for a white male industrial-noise-exposed population (INEP) in which each employee has been given 10 consecutive annual audiometric evaluations. The typical TWA for this population has remained relatively constant at 87 dBA over the 10 years represented by the test results, and the wearing of hearing protection is a condition of employment. Extensive ADBA results have also confirmed adequate protection from on-the-job noise exposures for this low-noise reference INEP.

Figure 12.4 shows that the amount of pre-existing on-the-job NIPTS assumed for the white male INEP would differ significantly depending upon the male population selected as the HTLA reference. The amount of pre-existing NIPTS for this white male INEP would be overestimated if a screened presbycusis HTLA population such as the male ISO data were used, or if a mixed-race population such as the male NIOSH database were used. Because the learning effect results in improved thresholds over several years of testing for the INEP, and the NINEP age-effect curve does not include learning improvements, the best estimate of pre-existing NIPTS is the difference between the INEP mean threshold on the first test and the NINEP age-effect hearing level at the same age. (It is assumed that the white male INEP did not have significant prior audiometric test experience.)

A second observation from Figure 12.4 is that the predicted rate of change, or slope (in dB/year), of the HTLA curves could differ significantly depending on the mean age of the INEP and on the particular HTLA population selected. In this example the INEP HTL curve becomes parallel to the white male NINEP curve after the learning effect diminishes, because the INEP had little pre-existing hearing loss in excess of age-effects. In other cases where substantial pre-existing NIPTS exists, the INEP may actually show less change over time than predicted by age-effects (see Chapter 4).

As shown in Figure 12.1 there are at least three different types of control populations that can be utilized as comparison groups for evaluating the changes in an INEP: matching NINEPs or other similar HTLA populations (as discussed above), industrial audiometric databases that are known to be representative of effective HCPs, and appropriately selected internal control populations of non–noise-exposed employees. Internal control groups are especially useful in attempting to account for possible audiometer calibration shifts in the data and in establishing the minimum test-to-test variability achievable for the audiometric testing environment and techniques used in a particular plant facility.

Estimating the amount of on-the-job NIPTS depends not only upon selecting an appropriate HTLA population but also, as noted earlier, upon the relative audiometric experience of the two populations compared. Therefore, if we assume that the NINEP is an appropriate HTLA population, the best indicator of pre-existing on-the-job NIPTS would be a comparison between the initial mean thresholds of the INEP and the NINEP hearing level at the age corresponding to the mean age of the INEP at the time of the first test, as indicated in Figure 12.4. Comparisons of later tests to the NINEP curve would underestimate NIPTS because of the learning effect.

The preceding discussion has briefly pointed out the importance of considering several parameters when attempting to evaluate the effectiveness of an HCP by using population comparisons. These parameters include age, gender, race, the learning effect, the choice of an appropriate HTLA population, and the amount of pre-existing NIPTS at various test frequencies. If an INEP's initial hearing thresholds were already poor, less shifting would be anticipated with continued exposure to the same noise environment. In such a case the population can be divided by hearing level in order to evaluate trends for the groups with more sensitive hearing on baseline. Unfortunately, some authors (e.g., Lee-Feldstein, 1993) have failed to account for such factors and have erroneously concluded that INEPs with pre-existing hearing loss were properly protected because they showed no more hearing change over time than expected due to aging for populations without pre-existing hearing loss.

Criteria Comparisons

The difficulties posed by population comparisons, especially the necessity to wait for several years of data to become available before beginning to track rates of change in HTLs compared to expected age-effects, make these methods unsuitable for proactive, ongoing evaluations of HCP effectiveness. To overcome this deficiency, methods for criteria comparisons were developed.

Learning-Effect Indicators

Learning-related statistics can be used during the first few years of the HCP, when the affected employees are new to the program and have little audiometric test experience. They may also be used to analyze established HCPs if the analysis is restricted to the initial years of testing for recently hired employees. The learning-related criteria had their origin in two observations made while studying industrial audiometric databases (Berger, 1976; Lilley, 1980; L. H. Royster, Lilley, and Thomas 1980; L. H. Royster and Thomas, 1979). The first observation was that low-noise-exposed populations and populations subjectively judged to be properly wearing hearing protection always exhibited an improvement in indicated mean threshold levels as a function of repeated testing, if not already audiometrically experienced. The second observation was that the variability in the test–retest data, either with respect to the baseline audiogram or between sequential audiograms, seemed to be inversely related to the level of protection afforded the noise-exposed populations.

The learning effect, or observed improvement in the population's mean hearing levels over the first few audiograms, was illustrated in Figure 12.4 by the improvements in measured mean HTL data for the selected white male INEP. The INEP's mean HTLs approach the line representing the expected NINEP age-effect trend, rather than remaining parallel to the line.

A clearer understanding of the degree of learning expected is demonstrated in Figure 12.5. The mean threshold levels at 6 kHz are shown for a white male INEP restricted to those employees with five consecutive approximately annual audiometric tests. Also shown are the age-effect curve for the white male NINEP, and a parallel curve (line A) that passes through the value for the INEP's mean threshold for the first test. Intuitively

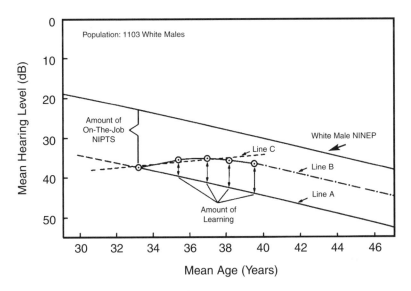

Figure 12.5 — Mean HTLs at 4 kHz as a function of population mean age on the first five audiograms for a white male INEP, shown with the age-effect curve for the white male NINEP. Line A shows the HTLs expected in the absence of learning. Line B shows the learning-effect improvement in HTLs predicted for a properly protected INEP population. Line C is the linear regression through mean HTLs for tests 1–4.

one would expect that, in the absence of significant on-the-job noise exposure, the INEP would exhibit age effects approximately equal to the NINEP. However, as shown, the white male INEP's mean thresholds improve relative to line A up to approximately the fifth or later test, after which they are projected to follow line B. The observed level of learning is indicated by the vertical difference between lines A and B, which is about 6 dB. The white male NINEP curve was obtained from a single test of a population without significant prior audiometric experience. It is assumed that the NINEP age-effect hearing curve also would have been improved by learning if this population had been given several audiometric evaluations (L. H. Royster, Driscoll, Thomas, and Royster, 1980; L. H. Royster and Thomas, 1979).

Typically, acceptable HCPs will exhibit slope magnitudes of slightly less than 0 dB/year to -1.0 dB/year over the first 4 years of annual audiometric evaluations (L. H. Royster and Royster, 1982). As shown in Figure 12.5, line A exhibits a positive slope because the line represents increasing threshold levels with increasing population age. However, line C represents decreasing threshold levels with increasing age and therefore exhibits a negative slope. If line C were horizontal (parallel to the mean age axis), then line C would exhibit a slope of 0 dB/year, indicating no change in threshold level with increasing age.

Previous observations for properly protected or low-noise-exposed populations indicate a typical maximum learning effect of 5–6 dB (Berger, 1976; Lilley, 1980;

L. H. Royster, Driscoll, Thomas, and Royster, 1980; L. H. Royster, Lilley, and Thomas, 1980; L. H. Royster and Royster, 1982). The degree of learning tends to increase for segments of the population with poorer thresholds at the higher test frequencies (J. D. Royster and Royster, 1981).

Obviously, learning can be partly or completely overshadowed by on-the-job NIPTS. As an example, if an audiometrically inexperienced INEP exhibited NIPTS exactly equal in magnitude over time to the degree of learning exhibited, then the mean thresholds would follow Line A in Figure 12.5 as age-effect hearing decline occurred. If the amount of NIPTS incurred was greater than the degree of learning, the mean thresholds would decline faster than line A, indicating an even less effective HCP. However, if the population already had enough audiometric experience to have exhausted the learning effect, then the interpretation would be altered: thresholds following line A would indicate adequate protection.

Since learning diminishes significantly after the first three to four audiometric tests, a simple approximation of the degree of learning can be derived by fitting a linear regression line through the measured mean thresholds at 4 kHz for the first four tests, as indicated by line C in Figure 12.5. The magnitude of the slope of line C, in dB/year, is then used in a learning-related criterion for program effectiveness. For protected populations without prior audiometric experience, this slope should be a negative value of up to about -1 dB/year over the first 4 years of testing. The slope is plotted on the horizontal axis of the criterion graph shown as Figure 12.6.

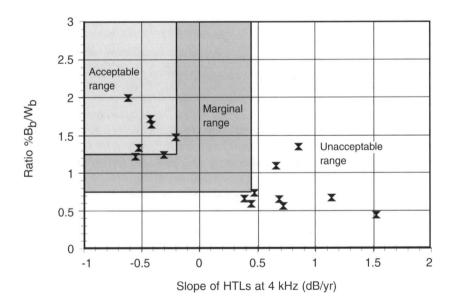

Figure 12.6 — Criterion ranges for rating HCP effectiveness based on learning-related ADBA indicators on tests 1–4. Plotted points show values for datasets available to S12/WG12.

To obtain the plotting coordinate for the vertical axis of Figure 12.6, a different statistic is calculated, as described next. The quantity plotted on the vertical axis is a ratio of percentages of employees showing defined shifts from test 1 to tests 2, 3, and 4. For each of these comparisons to baseline, the percent of employees showing a shift of 15 dB or more toward better hearing at any test frequency (0.5 – 6 kHz) in either ear is tabulated as the *Percent Better re: baseline* (%B_b), and the percent showing a shift of 15 dB or more toward worse hearing in either ear at any of the same frequencies is tabulated as the *Percent Worse re: baseline* (%W_b). Due to the learning effect, shifts toward better hearing should outnumber shifts toward worse hearing in early test years. The statistic plotted on the vertical axis of Figure 12.6 is the ratio formed from the maximum %B_b value and the maximum %W_b value from comparisons of tests 2 –4 back to test 1. The relationship between the slope at 4 kHz and the %B_b/%W_b ratio allows HCP effectiveness to be classified for the first 4 years of testing for a group, as shown in Figure 12.6 by the ranges for acceptable, marginal, and unacceptable values for these learning-related statistics. The data points plotted in Figure 12.6 are actual industrial databases analyzed by Working Group 12 of Accredited Standards Committee S12, Noise (S12/WG12), showing a range of HCP effectiveness.

Variability-Based Indicators

Analysis methods involving baseline comparisons or rates of change in HTLs over time share the disadvantage that the expected or possible changes are a function of the amount of pre-existing hearing loss shown on the baseline test, as well as the age of the population at baseline. The quality and validity of the baseline audiogram itself are also critical. These methods are useful in familiarizing HCP personnel (or professionals assisting them) with the hearing characteristics of a particular noise-exposed population. However, the dependence of the results on population-specific factors makes such methods less suitable for standardization in universally applicable HCP performance criteria.

To avoid the difficulties with longitudinal trend analyses, year-to-year variability-based methods were utilized by S12/WG12. Criteria were based on comparisons of sequential pairs of annual monitoring audiograms only about 1 year apart. This short elapsed time span allows age corrections to be ignored since aging causes very little change within such a short period, even for older persons. The members of S12/WG12 tested potential analysis methods on data from real HCPs, then included the most useful procedures as Draft ANSI S12.13-1991 "Evaluating the Effectiveness of Hearing Conservation Programs" (ANSI, 1991). This draft standard provides criterion ranges for HCP performance. The criterion ranges were determined by using the results for control databases of non-noise and low-noise-exposed employees, as well as employees protected by HCPs judged by professionals to be well-implemented. Criterion ranges are provided for the two primary statistics defined below.

Percent Worse sequential (%W_s) is the percent of the selected group showing a change of 15 dB or more toward worse hearing at any audiometric frequency (.5 – 6 kHz) in either ear between two sequential audiograms.

Percent Better or Worse sequential (%BW$_s$) is the percent of the selected group showing a change of 15 dB or more either toward better hearing or toward worse hearing at any audiometric frequency (.5–6 kHz) in either ear between two sequential audiograms.

The sample data for individuals presented in Table 12.1 illustrate how each employee is either counted or not counted in the %W$_s$ and %BW$_s$ statistics for each sequential test comparison. Each statistic is tabulated for comparisons of successive pairs of sequential audiograms: test 1 to test 2, test 2 to test 3, test 3 to test 4, and so forth.

By isolating the variability analysis to consideration of just two consecutive tests, age-related influences are virtually eliminated. If the remaining variability is high, it may indicate either noise-related TTS and NIPTS, or non–noise-related testing problems that make the audiometric data unreliable for detecting either individual shifts or population trends. In other words, high variability is a sign of trouble, but its presence does not necessarily mean that the population is inadequately protected from noise. The evaluator must determine whether the cause of high variability is noise related or testing related (or both). The presence of excess audiometric variability renders the HCP ineffective because, no matter why the data are too variable, the audiogram reviewer cannot aggressively take follow-up actions to halt the progression of NIPTS if that reviewer cannot have confidence that an indicated threshold shift is real rather than artifactual.

The criterion ranges for rating HCP performance according to Draft ANSI S12.13-1991 are shown in Table 12.2. Due to the influence of the learning effect during the first four tests for a population, the %BW$_s$ is not evaluated until test 5 and later; only the %W$_s$ is evaluated for early tests. Both statistics are considered in later years of testing.

Different shift magnitudes were considered in trial evaluations of audiometric databases from low-noise-exposed industrial control populations, and populations in HCPs judged as effective and as ineffective, in order to select an appropriate size shift to trigger the shift statistics (L. H. Royster, Lilley, and Thomas, 1980). If too small a shift is used, then very high percentages of employees will be included even for the low-noise-exposed populations. Conversely, if the level of shift required to flag an individual employee is too high, then the statistic would not be sensitive enough to

TABLE 12.1
Example of determining whether an individual employee will be counted in the Percent Worse sequential and Percent Better or Worse sequential statistics in each of three sequential test comparisons.

Test	Left Ear HTLs (dB) Frequency (kHz)						Right Ear HTLs (dB) Frequency (kHz)						Tests Compared	Employee Counted		
	.5	1	2	3	4	6	.5	1	2	3	4	6		%W$_s$	%B$_s$	%BW$_s$
1	20	5	5	10	30	50	20	10	0	20	35	50				
2	0	-5	0	0	25	35	0	0	-5	-5	25	40	1-2	no	yes	yes
3	0	-5	0	0	25	50	0	0	0	0	20	45	2-3	yes	no	yes
4	0	0	0	5	25	30	5	0	0	-5	35	40	3-4	yes	yes	yes

TABLE 12.2

Criterion ranges for rating HCP performance (adapted from Draft ANSI S12.13-1991) using the values of the Percent Worse sequential and the Percent Better or Worse sequential statistics.

HCP Rating	Sequential Comparisons of First Four Tests (1–2, 2–3, or 3–4)	Sequential Comparisons of Any Later Tests (4–5, 5–6, or Higher)	
	$\%W_s$	$\%W_s$	$\%BW_s$
Acceptable	<20	<17	<26
Marginal	20 to 30	17 to 27	26 to 40
Unacceptable	>30	>27	>40

distinguish between good and poor HCPs. The final shift magnitude selected, 15 dB, seemed to be a satisfactory compromise. Figure 12.7 illustrates $\%BW_s$ data for two sequential tests for one INEP, showing that most test-to-test variability in these audiograms is less than 15 dB.

The rationale for using overall audiometric variability as an indicator of HCP effectiveness is an assumption that it is not practical for industry to produce an industrial audiometric database that exhibits low test–retest variability without implementing program elements that are characteristic of effective HCPs (L. H. Royster et al., 1982).

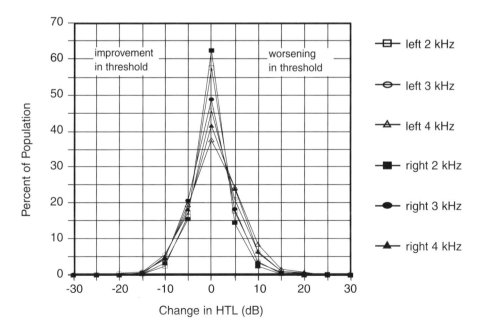

Figure 12.7 — Percentages of employees showing shifts of -30 to +30 dB between tests 7 and 8 at 2, 3, or 4 kHz in the left and right ears.

Examples include: (1) taking the time necessary to obtain an acceptable audiometric test, (2) providing effective feedback to employees at the time of the yearly audiometric evaluation, and (3) properly fitting hearing protectors and instructing employees in their use, checking for correct insertion, and replacing worn-out protectors.

Because the criterion comparison option is sensitive to annual changes in HCP implementation, changes in annual results allow HCP personnel to monitor audiometric data quality on an ongoing basis. If the $\%W_s$ and/or $\%BW_s$ rates for a new testing cycle are higher than before, HCP personnel can determine whether high variability is related to noise exposure (temporary threshold shift [TTS] and NIPTS) or to audiometric testing factors yielding poor quality data. If noise exposure is the cause, then the evaluators can identify the underlying problems in the HCP (for example, a decline in HPD use) as they develop, so that deficiencies can be corrected before they result in significant permanent hearing loss in many employees. On the other hand, if the variability is related to testing deficiencies, then steps can be taken to ensure more reliable audiometry (whether it is performed in-house or contracted to an external provider).

Application of ADBA Concepts

The examples in this section illustrate the steps in using ADBA and the types of information that the evaluator can gain from the analysis. Other sample analyses can be found in the literature (Reynolds et al., 1990; Thomas, 1995).

Selection of Groups for Analysis

Because ADBA criterion ranges differ for sequential comparisons of tests 1–2, 2–3, and 3–4 (affected by the learning effect) versus comparisons for tests 4–5 and higher, the groups whose HTLs are evaluated in ADBA should be separated in terms of whether the learning effect is an active influence (the first three comparisons per person) or not (all later comparisons). ADBA criteria assume that TTS can be detected if employees are inadequately protected; therefore, retest audiograms preceded by quiet periods (such as retests conducted to see whether OSHA STSs are persistent) are discarded from ADBA analyses. In order to benefit from the detection of TTS, S12/WG12 limited its criteria development process to audiometric results obtained *during* the workshift; testing done *prior* to daily noise exposure should yield lower variability.

In order to detect testing-related problems such as audiometer calibration shifts, audiograms should be evaluated on the basis of the annual testing cycles between calibrations. For simplicity, we will assume that a cycle corresponds to a calendar year. The simplest overall analysis of the entire database is to look at the mean HTL trends and variability-based indicators for the most recent N years of testing for all employees whose last N audiograms are test 4 or higher. For example, the evaluator might choose the most recent 5 years of testing (yielding four sequential test comparisons) for all employees whose most recent audiogram is test 10 or higher. This would eliminate newly hired employees (whose HTLs are still being influenced by the learning effect), while including the largest possible group of other employees. If the overall variability for this group is acceptable, then smaller subgroups (divided by noise exposure, for example) probably also show acceptable variability.

For more detailed analyses, groups can be separated out by any variable of interest: by department, by noise exposure, by the type of HPDs worn, or by HPD enforcement policy (mandatory versus optional), and so forth. To examine long-term HCP effectiveness, the evaluator might select the most recent 12 tests for all employees whose most recent audiogram was test 16 or higher. To examine HCP effectiveness for new hires, the evaluator might look at the first 4 years of data for employees hired 4 years ago. To determine the difference in audiometric reliability between a former testing service and the current service, ADBA methods can be applied to the last three audiograms before the switch and the first three audiograms after the switch for all employees present across this period.

Whatever the group selected, there must be constant group membership over the analysis period. That is, the same N people must be followed over time for the selected period of analysis. Otherwise, changing group membership would affect the mean HTLs, and the variability could not be properly compared from one year to the next. Several analysis strategies will be discussed in the following sections.

Determination of HCP Effectiveness

The first sample database to be examined, called "ANSI015," was contributed to S12/WG12. The data are for 356 males employed at a steel plant, whose audiograms were performed by a mobile van vendor from the initiation of the program. The mean HTLs on the first audiogram are shown in Figure 12.8 for decade age groups of employees. Although race was unknown for individuals within this population, most were reported to be Caucasian. Therefore, reference HTLA data for the white male NINEP age-effect database are shown for comparison at ages 25, 35, 45, and 55 years. The youngest group shows about 10 dB greater high-frequency hearing loss than expected, and the oldest group shows about 20 dB more than expected. Since TWAs were about 95 dBA, occupational noise is a likely contributor to this hearing loss.

The mean HTL trends for the noise-exposed employees over the first eight audiograms are plotted as Figure 12.9 to look for any aberrations. The 6-kHz data show several abrupt jumps that suggest problems. Furthermore, the 3-kHz data demonstrate a remarkable calibration anomaly, with mean HTLs improving by about 10 dB from test 1 to test 2, then later declining by about the same amount from test 7 to test 8.

It is unknown how this problem escaped detection by the testing service for so many years! On test 8, when the 3-kHz values were readjusted, OSHA STS rates rose sharply, commanding attention. However, if mean HTLs over time had been examined, the problem would have been detected on test 2. Interestingly, this same database has been described as indicating excellent protection of employees by researchers who looked only at tests 1 – 6, failed to detect the calibration aberration, and used only OSHA STS rates as a criterion (Adera and Gaydos, 1997). This example points out the importance of getting to know the characteristics of audiometric data before relying on them to make qualitative judgments, as well as the danger in depending completely on percentages of shifts from baseline as indicators of HCP success.

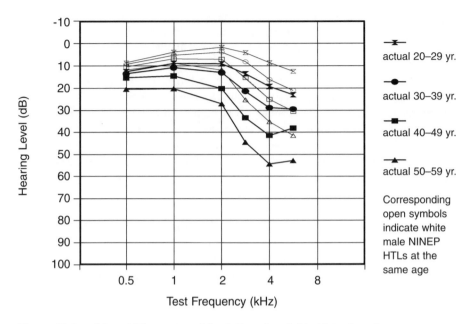

Figure 12.8 — Mean HTLs on test 1 for 356 males divided into decade age groups, compared to expected age-effect HTLs for the white male NINEP age-effect population at decade midpoint ages (open symbols).

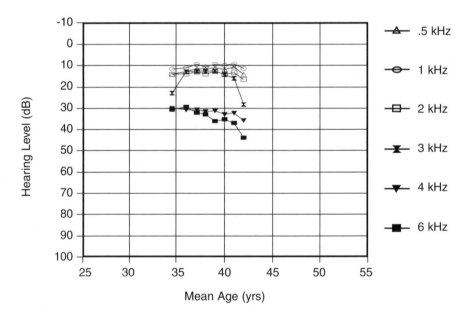

Figure 12.9 — Mean HTLs at each audiometric test frequency as a function of population mean age on tests 1–8 for 356 males.

Because no obvious calibration problems are apparent at 4 kHz, the slope data over the first 4 years can be analyzed using the learning-related criteria. The slope of mean HTLs at 4 kHz, calculated from the data plotted in Figure 12.9, is 0.4 dB/year. Apparently the employees are developing hearing loss at a rate fast enough to overshadow the learning effect and cause measured HTLs to worsen. The maximum $\%B_b/\%W_b$ ratio is calculated from the variability data shown in Table 12.3. Because these percentages include all test frequencies, even the aberrant 3-kHz HTLs, the $\%B_b$ percentages are quite high until baseline comparison of tests 1 to 8. The maximum $\%B_b/\%W_b$ ratio over the first four tests is 59.5%/34.3% = 1.7. The intersection of the slope and ratio values is plotted as shown in Figure 12.10 to obtain the HCP rating based on learning-related criteria. The point falls close to the borderline between mar-

TABLE 12.3
Percent Worse, Percent Better, and Percent Better or Worse values in baseline test comparisons and in sequential test comparisons across 8 tests for 356 males.

	Baseline Comparisons				Sequential Comparisons		
Tests	$\%W_b$	$\%B_b$	$\%BW_b$	Tests	$\%W_s$	$\%B_s$	$\%BW_s$
1–2	24.1	58.4	67.4	1–2	24.1	58.4	67.4
1–3	31.4	59.5	73.6	2–3	28.6	18.5	42.7
1–4	34.3	57.6	71.9	3–4	21.9	18.8	37.6
1–5	44.4	58.4	76.6	4–5	25.5	14.6	38.4
1–6	49.7	57.6	81.1	5–6	22.8	21.1	41.0
1–7	56.4	52.2	81.1	6–7	31.7	23.3	50.8
1–8	78.9	23.3	86.8	7–8	70.2	20.8	81.7

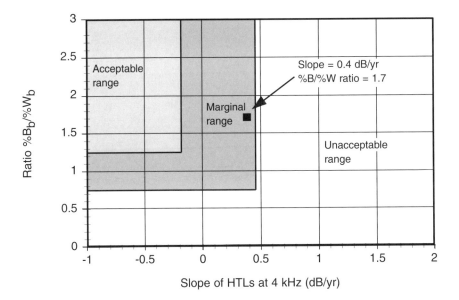

Figure 12.10 — Learning-related criteria over tests 1–4 for 356 males.

ginal and unacceptable HCP performance, indicating problems. Without the calibration aberration at 3 kHz, the $\%B_b/\%W_b$ ratio value would have been quite small and the HCP performance would have dropped into the unacceptable range.

The sequential variability data $\%W_s$ and $\%BW_s$, from Table 12.3, are plotted in Figures 12.11 and 12.12 respectively. The $\%W_s$ values fall generally within the marginal range until comparison of tests 7 and 8, when the 3-kHz calibration adjustment makes the last value increase markedly. The $\%BW_s$ values are also marginal to unacceptable for those tests that are classified (test comparisons 4 – 5 and later), with extremely high values for the comparisons of tests 1 – 2 and 7 – 8, which are impacted by the 3-kHz calibration adjustment.

In summary, the HCP represented by the ANSI015 data can be judged as ineffective over this time period. Even if the test comparisons affected by the remarkable calibration jumps are discounted, the variability statistics over tests 2 through 7 were undesirably high. Mean HTLs at 4 and 6 kHz were worsening steadily. Although it would be necessary to investigate actual HCP policies (for HPD use, etc.) to identify all the causes of these unacceptable ADBA findings, it is likely that both poor audiometric test quality and inadequate protection from noise are contributing to these results. The ADBA findings serve as a warning to HCP personnel to make improvements in the performance of the testing provider as well as in on-site employee protection.

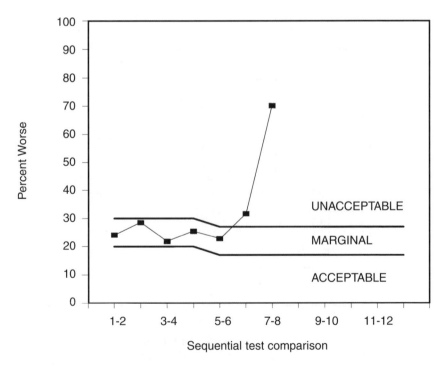

Figure 12.11 — Percent Worse sequential values over tests 1 –8 for 356 males.

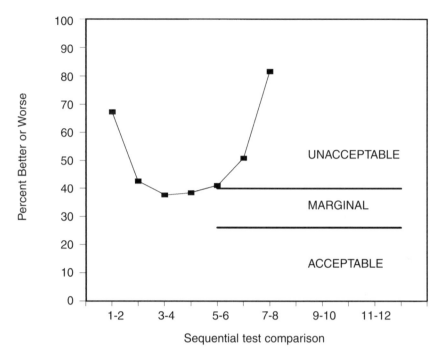

Figure 12.12 — Percent Better or Worse sequential values over tests 1 –8 for 356 males.

HCP Effectiveness for Later Years of Testing

The second sample database to be examined includes audiograms from 1973 to 1991 for a meat processing plant. Senior employees have accumulated many annual tests, but few share the same year for test 1. In order to obtain as large a group size as possible, senior employees were included in this analysis if their 1991 audiogram was test 9 or higher; then the most recent 5 years of data were evaluated (eliminating the learning-effect influence). This selection yielded 63 noise-exposed employees, plus 29 non–noise-exposed employees who also received audiograms (an internal control group).

The 1991 mean HTLs for whites and blacks within each noise exposure group are shown as Figure 12.13. The mean thresholds for blacks are quite similar by exposure group except at 8 kHz, but a difference at this frequency alone would not be expected due to noise exposure. The white exposure groups show a mean difference of about 6 dB at both 3 and 4 kHz, which is consistent with potential noise-induced damage. The presence of a notch in the audiogram pattern for the noise-exposed whites also suggests some NIHL.

The variability indicators $\%W_s$ and $\%BW_s$, tabulated separately for the nonexposed and noise-exposed groups over the most recent five tests, are plotted as Figures 12.14

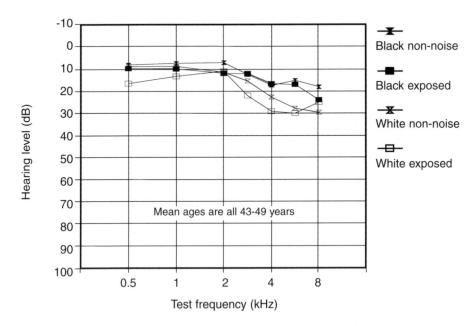

Figure 12.13 — Mean HTLs in 1991 for black and white employees grouped by noise exposure.

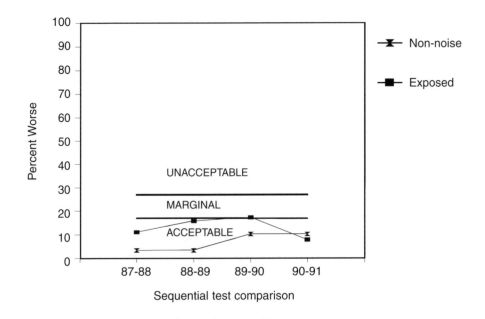

Figure 12.14 — Percent Worse sequential values over the most recent 5 years of testing for employees grouped by noise exposure.

and 12.15. The nonexposed group data fall well within the acceptable ranges for both indicators, while the noise-exposed group data are closer to the borderline of the marginal range (or just across the boundary) in several comparisons. These findings indicate that testing-related variability is low (since the data are good for the non-noise group). While HCP effectiveness is acceptable for the noise-exposed employees, the higher variability for exposed workers compared to the internal control group suggests that protection might still be improved. In fact, this HCP had not required HPD use in a department with TWAs of 85–89 dBA.

Evaluation of Hearing Protector Adequacy

ADBA procedures were used to evaluate the HCP for a textile facility where the daily TWA in the high-speed spinning department was 107 dBA. The values of the %BW$_s$ statistic and the learning-related statistic (the combination of %B$_b$[max]/%W$_b$[max] in relation to the slope of the mean threshold levels at 4 kHz) indicated that the HCP was unacceptable in terms of the level of protection being provided the noise-exposed workforce (L. H. Royster, Royster, and Cecich, 1984). Because the HCP appeared to exhibit the administrative characteristics of effective programs (L. H. Royster, Royster, and Berger, 1982), it was suspected that the available hearing protection devices were inadequate to protect the noise-exposed population at such a high TWA, causing excessive variability in the audiometric database as indicated by the ADBA findings. The three hearing protectors used by the employees over

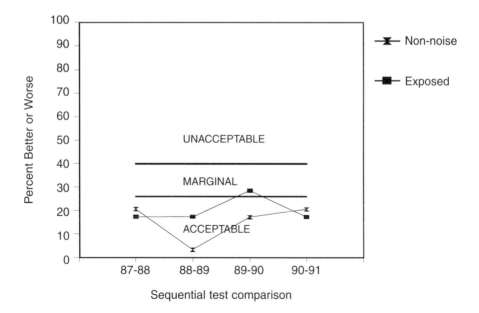

Figure 12.15 — Percent Better or Worse sequential values over the most recent 5 years of testing for employees grouped by noise exposure.

the most recent 4 years of the HCP were the North Com-fit® earplug, Flents Silenta® Model 080 earmuff, and the E-A-R® Classic foam earplug.

The real-world effectiveness of these three hearing protection devices was investigated in two ways. First, the prevention of TTS was assessed by measuring the employees' hearing levels at the beginning and end of a work shift. The mean threshold shifts (pretest minus posttest) across all test frequencies were: Com-fit earplug, - 2.5 dB; Silenta earmuff, - 0.5 dB; and E-A-R earplug, 0.4 dB (negative values indicate the presence of TTS). In summary, the measured mean TTS was greater for the Com-fit earplug, while the Silenta earmuff protector exhibited a smaller mean TTS, and the E-A-R foam earplug wearers actually showed a slight improvement in hearing from pretest to posttest, indicating the highest relative level of protection of the three devices compared.

The second method of investigating relative protection by the three devices was to apply the %BW$_s$ statistic to the most recent 3 years of audiometric data for the three wearer groups. The mean percentages of employees exceeding the %BW$_s$ statistic, averaged over the last three sequential audiogram comparisons, are presented in Figure 12.16. The relative ranking of real-world protection obtained from the %BW$_s$ findings is identical to that obtained from the TTS study. That is, the E-A-R foam plug wearers exhibited the lowest level of variability between annual audiometric tests as evaluated using the %BW$_s$ statistic, while the Silenta earmuff wearers exhib-

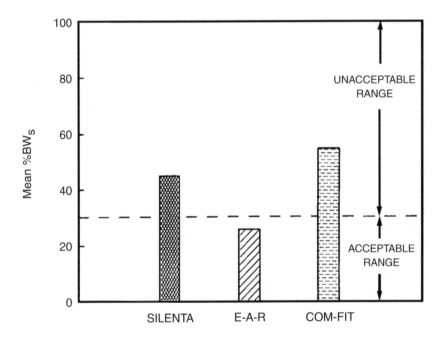

Figure 12.16 — Mean values of the Percent Better or Worse sequential statistic for each HPD-wearer group over the most recent three annual audiograms.

ited intermediate variability, and the Com-fit wearers exhibited the highest level of variability. A very important finding was that the %BW$_s$ statistic accurately evaluated the relative effectiveness of the two hearing protectors for every sequential comparison for the two groups that included at least 30 ears (E-A-R and Com-fit). The observation suggests that the %BW$_s$ statistic exhibits the potential to predict the relative effectiveness of different hearing protectors using the results of only two annual audiometric tests.

Adequacy of Protection from Noise

When testing factors are the same for departments or other groups, differences in ADBA results between the groups can be assumed to be due to different degrees of protection from noise. The %W$_s$ data for three separate plants that shared the same mobile testing service provider are presented in Figure 12.17. Plant A had TWAs below 85 dBA and performed annual testing as an employee wellness benefit; it is therefore a non-noise control group. The acceptable %W$_s$ data for Plant A confirm that the audiometric testing variability was low. In contrast, the %W$_s$ data are higher for Plants B and C. Because audiometric testing factors (shared with Plant A) are not a problem, the marginal to unacceptable %W$_s$ data for Plants B and C indicate inadequate employee protection from noise. (A similar analysis could be performed for noise-exposure groups within a single plant, as long as a non-noise control group received audiograms.)

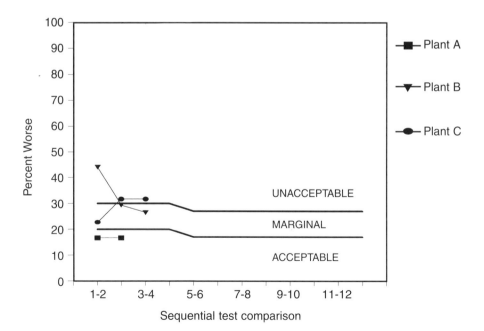

Figure 12.17 — Percent Worse sequential values over tests 1–4 for three plants whose audiograms were provided by the same mobile testing service.

Audiometric Testing Reliability

When a corporation split, audiometric testing for one division remained in-house, while the second division chose to contract with an external testing service. The data for long-term employees with tests before and after the split provide a study of audiometric variability across testing sources. Shown as Figure 12.18 are mean HTLs over time for 10 tests for two groups: 41 employees who received in-house testing for all 10 tests (Group 1), and 13 employees who received 4 tests in-house, followed by 4 tests from the initial testing contractor, and finally 2 tests from a second contractor (Group 2). Note that the groups were chosen to eliminate the true first 4 audiograms for each subject, so that learning-effect influences would not affect the data. The tests are renumbered as A to J, with test provider changes for Group 2 occurring on tests E and I, while Group 1 continued with in-house testing through all tests. Note the abrupt changes in measured mean HTLs for Group 2 on test E (first provider change). These testing-related changes caused an abrupt increase in OSHA STSs, since employees' baseline audiograms were not comparable to test E. Note also the later jumps in measured mean HTLs on tests I and J; apparently the second contractor's data were not comparable even from one year to the next. Reasons for the large shifts in measured HTLs between providers include numerous testing factors: calibration differences between audiometers, testing paradigm differences between types of microprocessor

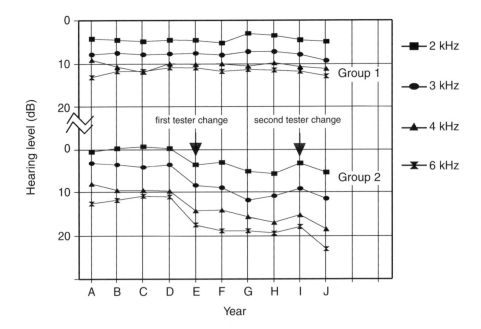

Figure 12.18 — Mean HTLs as a function of test year (A through J) for Group 1 (in-house audiometry in all years) and Group 2 (in-house audiometry in years A–D, one mobile test provider in years E–H, then a second mobile service provider in years I–J).

audiometers, background noise levels and distracting vibrations in the audiometric test booth, earphone placement, instructions by the tester, and motivation of the subjects to respond attentively. For some combination of these reasons (possibly with other reasons as well) the contract service providers obtained measured HTLs that were not as sensitive as those previously measured in-house for the same employees.

The sequential variability criteria also show corresponding changes, as shown in Figures 12.19 for %W$_s$ and 12.20 for %BW$_s$. Not only are absolute HTLs different as measured by the two contract testing services, but test-to-test thresholds are less reliable than for the in-house testing program, with values in the marginal to unacceptable range. The only exception is test comparison H to I for %W$_s$; the acceptable value obtained here results from the jump toward better hearing in mean HTLs on test I, when the contract service changed for the second time. This is essentially an artifact.

This example illustrates that it is very undesirable to change audiometric service providers frequently. Before any change is made, it is wise to examine evidence of the testing reliability achieved by the provider being considered. The consequences of high data variability are costly both in terms of individual employee follow-up effort for apparent shifts, and in terms of population statistics such as the OSHA STS rate.

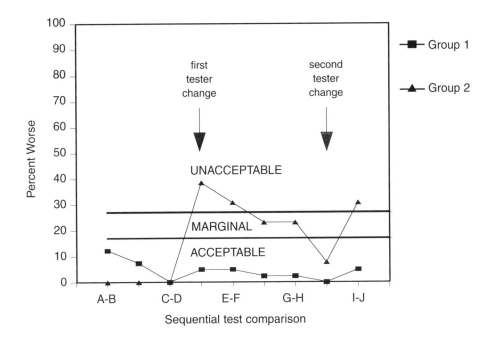

Figure 12.19 — Percent Worse sequential values over the most recent 10 years of testing (A–J) for Groups 1 and 2 (with changes in audiometric test provider indicated for Group 2).

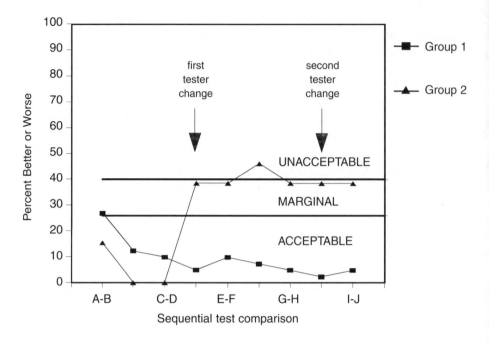

Figure 12.20 — Percent Better or Worse sequential values over the most recent 10 years of testing (A–J) for Groups 1 and 2 (with changes in audiometric test provider indicated for Group 2).

Concluding Remarks

This chapter has reviewed concepts for audiometric database analysis (ADBA) and demonstrated the application of several techniques. ADBA methods allow HCP personnel to assess whether HPD use is adequate to prevent TTS, whether contracted testing services are providing reliable audiograms, and whether the HCP overall is preventing occupational hearing loss. Annual ADBA results are a highly useful tool for assessing HCP performance changes from year to year. Don't waste the great potential of the audiometric data by failing to try ADBA!

References

Adera, T., and Gaydos, J. (1997). "Identifying Comparison Groups for Evaluating Occupational Hearing Loss: A Statistical Assessment of 22 Industrial Populations," *Am. J. Indust. Med. 31*, 243–249.

ANSI (1991). "Evaluating the Effectiveness of Hearing Conservation Programs," Acoustical Society of America, Draft ANSI S12.13-1991, New York, NY.

ANSI (1996). "Determination of Occupational Noise Exposure and Estimation of Noise-Induced Hearing Impairment," Acoustical Society of America, ANSI S3.44-1996, New York, NY.

Berger, E. H. (1976). "Analysis of the Hearing Levels of an Industrial and Nonindustrial Noise Exposed Population," M.S. thesis, North Carolina State University, Raleigh, NC.

Dobie, R. A. (1995). "Prevention of Noise-Induced Hearing Loss," *Arch. Otolaryng. Head Neck Surg. 121*, 385–391.

Franks, J. R., Davis, R. R., and Krieg, E. F. Jr. (1989). "Analysis of a Hearing Conservation Program Data Base: Factors Other than Workplace Noise," *Ear & Hearing 10*, 273–280.

Franks, J. R., Stephenson, M. R., and Merry, C. J. (1996). "Preventing Occupational Hearing Loss — A Practical Guide," U.S. Department of Health and Human Services (NIOSH) Publication No. 96-110, National Institute for Occupational Safety and Health, Cincinnati, OH.

Lee-Feldstein, A. (1993). "Five-Year Follow-Up Study of Hearing Loss at Several Locations Within a Large Automobile Company," *Am. J. Indust. Med. 24*, 41–54.

Lilley, D. T. Jr. (1980). "Analysis Techniques for Evaluating the Effectiveness of Industrial Hearing Conservation Programs," M.S. thesis, North Carolina State University, Raleigh, NC.

Melnick, W. (1984). "Evaluation of Industrial Hearing Conservation Programs: A Review and Analysis," *Am. Ind. Hyg. Assoc. J. 45*(7), 459–467.

NIOSH (1972). "Criteria for a Recommended Standard. Occupational Exposure to Noise," U.S. HEW, Report No. HSM 73-11001, Cincinnati, OH.

Pelausa, E. O., Abel, S. M., Simard, J., and Dempsey, I. (1995). "Prevention of Noise-Induced Hearing Loss in the Canadian Military," *J. Otolaryngol. 24*, 271–280.

Reynolds, J. L., Royster, L. H., and Pearson, R. G. (1990). "Hearing Conservation Programs (HCPs): The Effectiveness of One Company's HCP in a 12-hr Work Shift Environment," *Am. Ind. Hyg. Assoc. J. 51*(8), 437–446.

Roberts, M. (1997). "Has the Hearing Conservation Program Worked?," Proceedings, 22nd annual conference, National Hearing Conservation Association, Denver, CO.

Rosler, G. (1994). "Progression of Hearing Loss Caused by Occupational Noise," *Scand. Audiol. 23*(1), 13–37.

Royster, J. D. (1992a). "Using Draft ANSI S12.13-1991 to Evaluate Your Hearing Conservation Program," *Spectrum 9*(1), 1, 5.

Royster, J. D. (1992b). "Evaluation of Different Criteria for Significant Threshold Shifts in Occupational Hearing Conservation Programs," Contract report submitted to National Institute for Occupational Safety and Health, Cincinnati, OH, December 1992. Available from National Technical Information Service as NTIS document PB93-159143.

Royster, J. D. (1996). "Evaluation of Additional Criteria for Significant Threshold Shift in Occupational Hearing Conservation Programs," Contract report submitted to National Institute for Occupational Safety and Health, Cincinnati, OH, August 1996. Available from National Technical Information Service as NTIS document PB97-104392.

Royster, J. D., and Royster, L. H. (1981). "Judging the Effectiveness of Hearing Protection Devices by Analyzing the Audiometric Database," in *Proceedings of Noise-Con '81*, edited by L. H. Royster, F. D. Hart, and N. D. Stewart, Noise Control Foundation, Poughkeepsie, NY, 157–160.

Royster, J. D., and Royster, L. H. (1990). *Hearing Conservation Programs: Practical Guidelines for Success,* Lewis Publishers, Chelsea, MI.

Royster, J. D., and Stewart, A. P. (1997). "What Affects STS Rates?" Proceedings, 22nd annual conference, National Hearing Conservation Association, Denver, CO.

Royster, L. H., Driscoll, D. P., Thomas, W. G., and Royster, J. D. (1980). "Age Effect Hearing Levels for a Black Nonindustrial Noise Exposed Population," *Am. Ind. Hyg. Assoc. J. 41*(2), 113–119.

Royster, L. H., Lilley, D. T. Jr., and Thomas, W. G. (1980). "Recommended Criteria for Evaluating the Effectiveness of Hearing Conservation Programs," *Am. Ind. Hyg. Assoc. J. 41*(1), 40–48.

Royster, L. H., and Royster, J. D. (1982). "Methods of Evaluating Hearing Conservation Program Audiometric Data Bases," in *Personal Hearing Protection in Industry*, edited by P. W. Alberti, Raven Press, New York, NY, 511–540.

Royster, L. H., and Royster, J. D. (1984). "Making the Most out of the Audiometric Data Base," *Sound and Vibration 18*(5), 18–24.

Royster, L. H., and Royster, J. D. (1988). "Getting Started in Audiometric Data Base Analysis," *Seminars in Hearing 9*, 325–338.

Royster, L. H., and Royster, J. D. (1990). "Important Elements and Characteristics of Hearing Conservation Programs and Determination of Their Effectiveness," *Environ. Int. 16*, 339–352.

Royster, L. H., Royster, J. D., and Berger, E. H. (1982). "Guidelines for Developing an Effective Hearing Conservation Program," *Sound and Vibration 16*(5), 22–25.

Royster, L. H., Royster, J. D., and Cecich, T. F. (1984). "An Evaluation of the Effectiveness of Three Hearing Protection Devices at an Industrial Facility with a TWA of 107 dB," *J. Acoust. Soc. Am. 76*(2), 485–497.

Royster, L. H., and Thomas, W. G. (1979). "Age Effect Hearing Levels for A White Nonindustrial Noise Exposed Population (NINEP) and Their Use in Evaluating Industrial Hearing Conservation Programs," *Am. Ind. Hyg, Assoc. J. 40*(6), 504–511.

Schulz, T. Y. (1996). "Alternative Methods to Evaluate Hearing Conservation Program Effectiveness," Proceedings, 21st annual conference, National Hearing Conservation Association, Denver, CO.

Simpson, T. H., McDonald, D., and Stewart, M. (1993). "Factors Affecting Laterality of Standard Threshold Shift in Occupational Hearing Conservation Programs," *Ear & Hearing 14*(5), 322–331.

Thomas, J. (1995). "The Application of Audiometric Data Base Analysis to Selected Air Force Bases." M.S. technical report, University of North Carolina at Chapel Hill, Chapel Hill, NC.

Wolgemuth, K. S., Kamhi, A. G., Luttrell, W. E. and Wark, D. J. (1995). "The Effectiveness of the Navy's Hearing Conservation Program," *Military Med. 160*, 219–222.

The Noise Manual, revised 5th edition, edited by E.H. Berger,
L.H. Royster, J.D. Royster, D.P. Driscoll, and M. Layne
©2003 American Industrial Hygiene Association

13

Room Noise Criteria

Warren E. Blazier, Jr.

Contents

Introduction

The selection of appropriate and practical acoustical criteria for building spaces presents unusual challenges due to the extraordinarily wide variety of types of rooms and their corresponding uses. Similarly, there are a wide variety of environmental control and mechanical system noise sources serving those spaces. Although practicing hearing conservationists are more generally concerned with industrial environments, where the principal problem is with hazardous noise, they often have to deal with multiple criteria when visiting different plant facilities. For example, it may be necessary to determine if the background noise level in a training room near the production floor is acceptable, or if it is not, by how much the room noise level should be reduced in order to improve an existing situation.

This chapter is intended to acquaint the reader with some of the more common methods that are used to rate or describe background noise, and to contrast the advantages and disadvantages of each in application to a variety of environmental situations. Criteria for interpreting these ratings as a function of space-use, as well as a method of measuring the room noise that is present are provided. For the reader who wants immediate access to the room noise criteria provided in this chapter, see Table 13.3.

Sound Rating Methods

The purpose of this section is to present sufficient information to make a knowledgeable decision on which background noise rating method is most appropriate for the objectives of a given set of circumstances. There are several methods currently being used to express indoor sound ratings, and these include the traditional A-weighted sound level (dBA), tangent Noise Criteria curves (NC) as developed by Beranek (1957), the European NR (Kosten and Van Os, 1962), plus the more complex Room Criterion method (RC) from Blazier (1981), the Balanced Noise Criteria (NCB) from Beranek (1989), and the recent RC Mark II (Blazier, 1997). Each sound rating method was developed based on data for specific applications, hence *not all methods are equally suitable for the rating of noise in the wide variety of applications encountered in practice.*

For example, the simpler forms of noise rating, such as the A-weighted sound level, tangent-NC, and the European NR, are single-number metrics that are only *level-sensitive* and provide no information regarding the spectral content of the noise. The spectral content of a noise strongly influences a listener's subjective judgment of character or sound "quality." Therefore, where the concern is principally with only the level or perceived loudness of the noise, use of these simple methods is usually adequate. However if a noise problem is known to exist, not because it is perceived by the listener as too loud, but because its character is offensive, then more complex methods of diagnostic noise rating must be drawn upon to identify the basis for complaint and to provide guidance for remedial action.

Simple, Level-Sensitive Metrics

A-Weighted Sound Level, dBA

The use of A-weighted sound level (dBA) is most frequently encountered in the evaluation and rating of *outdoor* environmental noise, but it also has application indoors for assessing the potential for acceptable speech communication in the presence of background noise, and in specifying limits for human exposure to high-level noise. The A-weighted sound level is the simplest of the available rating metrics because it can be read directly on a typical sound level meter (SLM) with the frequency-weighting switch set on A.

Tangent-NC Noise Rating, NC-xx

The tangent-NC method of noise rating (Beranek, 1957) has been used extensively in the U.S. for over 30 years. The method is based on the use of a family of curves expressed in terms of octave-band sound pressure levels extending over the frequency range from 63 Hz to 8,000 Hz. Each curve in the family has a numerical designation, for example NC-40, which is keyed to a set of acceptability criteria expressed in terms of an identified space use. In conventional practice, the NC rating is determined by plotting the octave-band spectrum of the noise in question on the family of NC curves, and then assigning a rating based on the highest NC curve that is just *tangent* to the spectrum.

The NC curves were originally defined in 5-NC increments, but it has become common practice to interpolate between the curves in 1-NC increments, *despite the fact that people are not so precise in detecting level differences*. Furthermore, it is not a requirement of the procedure that the spectrum being rated follow or approximate the shape of the NC curve — *only that the rating number correspond to the highest curve tangent to the spectrum*. Because of this convention, a sound rating using the tangent-NC procedure means only that no point in the octave-band spectrum *exceeds* the stated NC curve. For example, it cannot be used to evaluate the full effect of the sound on speech communication, because it only signifies that the speech-masking potential is likely less than the rating number; how much less is not revealed.

The use of the tangent-NC method of noise rating is illustrated in Figure 13.1. In this example, a spectrum to be rated is plotted on the family of NC noise criterion curves. The highest NC curve that is tangent to this spectrum is the NC 35 curve at 2000 Hz. Therefore, the spectrum receives a rating of NC 35.

Note that this rating does not provide any information about the shape of the noise spectrum, only the fact that at some point in the frequency range it is tangent to an NC 35 curve.

European NR Noise Rating

The European NR noise rating (Kosten and Van Os, 1962) is similar in most respects to the NC noise rating used in the U.S. The principal difference is in the shape of the family of criterion curves, which are more permissive in the low-frequency region than the NC curves. The NR curves were originally developed for rat-

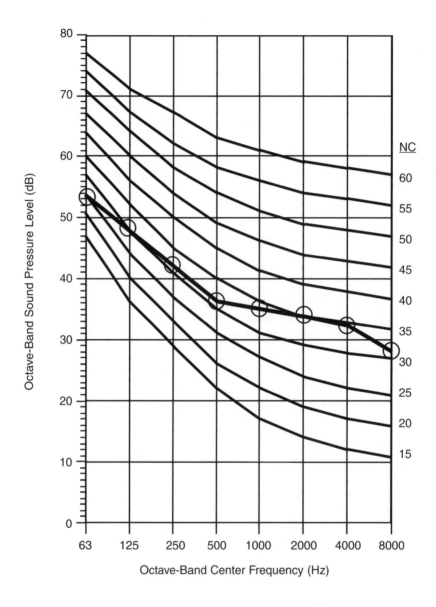

Figure 13.1 — NC noise criterion curves. The tangent-NC rating of the spectrum illustrated is NC 35.

ing *outdoor* environmental noise, but are used by some today to rate background noise in rooms. In most situations, where the noise spectrum of interest does not contain predominant low-frequency components, NR ratings are essentially equivalent to NC ratings, with the same attendant disadvantages discussed above.

Level-Sensitive Ratings Do Not Always Correlate With Subjective Opinion

Neither the A-weighted sound level, nor the tangent-NC/NR ratings, identifies the character or *quality* of background noise as perceived by the human ear. For example, it is possible for several noises having identical ratings expressed in dBA to rank-order differently when judged by a listener on a scale of relative acceptability. The same is true of ratings expressed as tangent-NC/NR values.

An example of this contradiction is illustrated in Figure 13.2. In this illustration, three distinctly different-sounding noise spectra are shown that have identical A-weighted sound level ratings of 42 dBA, and also, coincidentally, identical tangent-NC ratings of NC 35. These three spectra receive a rating of NC 35, because the highest tangent to an NC curve in each instance is the NC 35. The (H) spectrum is tangent at 2000 Hz, the (N) spectrum is tangent at 500 Hz, and the (R) spectrum is tangent at 250 Hz.

Although the character or "quality" of each of the above spectra will be significantly different as perceived by the human ear, neither type of rating reveals such subjective attributes. The spectrum labeled (H) will be usually judged as "hissy," the spectrum labeled (R) "rumbly," and the spectrum labeled (N) "neutral" — that is to say, no one part of the spectrum dominates the character of the noise. Although it is recognized that there is great demand for the simplicity of single-number noise ratings, it is not possible to describe the factors of level, frequency, and time variability, which are important to the subjective assessment. Therefore, the user of single-number noise ratings should be aware of these pitfalls.

Noise Ratings That Are Both Level and Quality Sensitive

Loudness vs. Annoyance

There is significant scientific evidence to support the premise that people are able to categorize noises of different spectral character on a scale of *relative loudness*, without regard to differences in the way the signals "sound" to the ear (Broner, 1998; Zwicker, 1989). For example, the classical equal-loudness contours for pure tones were determined by comparing tones at various frequencies with a fixed reference tone at 1000 Hz, and adjusting the level of each for equal loudness. The A-weighted frequency compensation network available on standard sound level meters was established from these data. In a similar way, the equal-loudness contours for bands of noise were developed, and these form the basis for the family of curves used in the NC rating method discussed above.

If on the other hand, people are asked to make a comparison of different noise spectra on the basis of *relative annoyance*, a different rank ordering is frequently obtained. For example, spectra that were judged equally loud in loudness comparison tests may be ranked as significantly different when people are asked to judge them on the basis

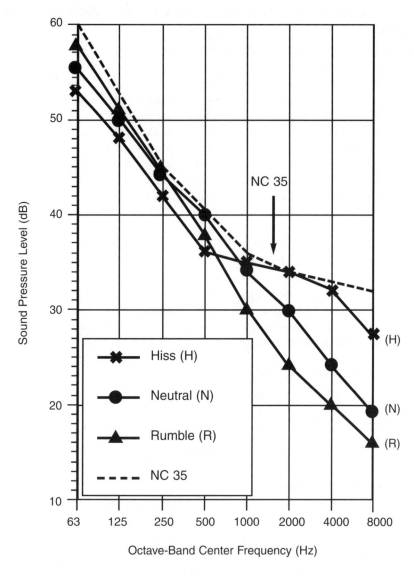

Figure 13.2 — Three noise spectra having identical A-weighted sound levels (42 dBA), and identical tangent-NC ratings (NC35).

of relative annoyance (Broner, 1998; Zwicker, 1989). Although the relationship between loudness and annoyance is complex, frequently cited reasons for this change in rank ordering are differences in the relative spectral content of the signals and the time-patterns of fluctuations in level. For example, a spectrum with significant low-frequency fluctuations in level (rumble) may rank relatively low in terms of subjective loudness, but relatively high in terms of subjective annoyance!

In regard to level fluctuations in the signal, it is important for the industrial hygienist to be aware that the common practice of recording noise data as the time-average over a given interval can obscure the presence of annoying short-term signal fluctuations. This can result in a situation where the room occupant is dissatisfied with the acoustical environment, although the appropriate *steady-state* room noise criterion may not have been exceeded (Zwicker, 1989). Problems of this nature are difficult to diagnose without the aid of sophisticated instrumentation not commonly available to the hygienist, but the human ear is often an adequate instrument to subjectively evaluate whether or not the time-pattern of the signal is likely responsible for the occupant complaint.

With regard to differences in spectral content, recent research indicates that people seem to prefer a neutral, balanced background noise spectrum, in which no one frequency range is more pronounced than another. When a spectrum is unbalanced, producing a rumble, roar, or hiss, people generally react negatively (Blazier, 1996; Broner, 1994). In fact, a recent study in Sweden suggests that at even moderate sound levels a spectrum that is significantly unbalanced at low frequencies leads to a feeling of oppressiveness and a reduction in work output when the exposure period is on the order of 1 hour or greater (Persson-Waye et al., 1997). This aspect should be considered by the industrial hygienist if situations are encountered in which tasks requiring concentration and attention to detail are suspected of being put at a disadvantage by the quality of the background noise in the work area.

Thus, it is becoming clear that peoples' reactions to background noises of differing spectral content are complex, and that relative loudness is only one of a series of factors that determine the rank ordering process on a scale of subjective acceptability (Zwicker, 1989). This justifies the need for more complex noise rating procedures that assess a noise spectrum not only in terms of relative loudness, but also with respect to sound quality attributes.

ANSI Standard S12.2-1995, "Criteria for Evaluating Room Noise"

The ANSI Standard S12.2-1995, *Criteria for Evaluating Room Noise*, represents the culmination of many years of effort to develop a standard for correlating objective measurements of room noise with criteria of acceptability as a function of space use. However, the standard as presently structured sets forth *two* different procedures for noise rating, which yield conflicting assessments of noise spectra that contain substantial low-frequency content in the frequency range below 100 Hz (rumble). Unfortunately, the standard provides no guidance to aid in the choice of methodology, so the user must decide between the two procedures based upon his or her own judgment and practical experience.

One rating method is the NCB (Beranek, 1989), which is intended for application to *occupied* spaces containing noise attributable not only to the heating, ventilating, and air-conditioning (HVAC) system, but also to space-activity noises produced by people, telephones, and office equipment. Because space-activity noises are virtually impossible to predict in advance, an NCB rating is inappropriate as the basis for specifying permissible levels of noise attributable to an HVAC system.

The alternative method of noise rating is the RC (Blazier, 1981). Although this procedure was originally intended for rating noise in rooms due to the HVAC system alone, it is also applicable to the rating of room noise that is the composite of several different sources. This procedure for noise rating has been recommended by the American Society of Heating, Refrigerating and Air-Conditioning Engineers (ASHRAE) since 1987. It is much less tolerant of low-frequency noise than the NCB procedure in the frequency region below 100 Hz.

RC Mark II Noise Rating, RC-xx(yy)

One shortcoming common to both the RC and NCB noise rating methods of ANSI 12.2-1995 is that no guidelines are provided to estimate the variation in subjective response as a function of the magnitude of spectrum-imbalance. However a recent new procedure, known as the RC Mark II (Blazier, 1997), has been introduced that incorporates a method for estimating listener response on a 3-point subjective scale (acceptable-marginal-objectionable), and includes a revised procedure for the assessment of sound quality that has improved reliability over that of the original RC methodology in ANSI 12.2-1995. This new procedure was developed primarily for rating the acceptability of HVAC system noise, but is also applicable to all sources of broadband noise in rooms. The procedure used to estimate the subjective response of a room occupant is based on the collective experience of the author and others in the diagnosis of field complaints. It is anticipated that future research funding will be available from ASHRAE to more scientifically evaluate its reliability.

RC Mark II Noise Rating and Diagnostic Procedures

There are three elements utilized in the RC Mark II methodology: (1) a family of RC (room criteria) curves, (2) a procedure for determining the RC rating number (level) and the noise spectral balance (quality), and (3) a procedure for estimation of occupant satisfaction when the design criterion is not exactly achieved (Blazier, 1995). The RC Mark II rating is a two-dimensional expression that takes the form of RC-xx(yy). The first term, "xx," is a *quantitative* descriptor, and is the value of the RC reference curve corresponding to the arithmetic average of the levels in the principal speech frequency region, namely the levels in the 500-, 1000-, and 2000-Hz octave bands. The second term, "yy," is a *qualitative* descriptor that identifies the perceived "character" of the sound: (N) for neutral, (LF) for low-frequency rumble, (MF) for mid-frequency roar, and (HF) for high-frequency hiss. There are also two subcategories of the low-frequency descriptor: (LF_B), denoting a moderate but perceptible degree of sound-induced ceiling/wall vibration, and (LF_A), denoting a noticeable degree of sound-induced vibration. These subcategories are assigned whenever the low-frequency region of the spectrum being rated extends into the B or A ranges of the RC curves illustrated in Figure 13.3.

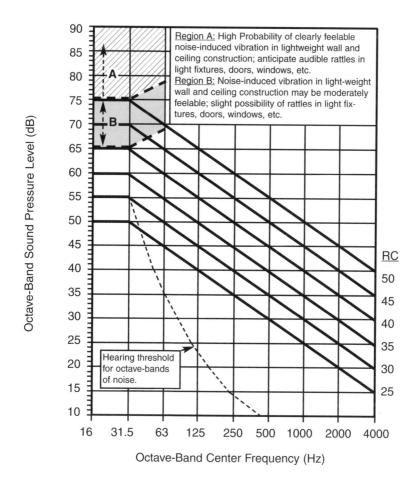

Figure 13.3 — Family of RC Mark II room criterion curves.

Room Criteria Curves (Mark II)

The family of Room Criterion (RC) curves shown in Figure 13.3 forms the basis of the RC Mark II noise rating procedure. Each reference curve identifies the shape of a neutral, bland-sounding spectrum, indexed to a curve number corresponding to the sound level in the 1000-Hz octave band. The shape of these reference curves is similar to those in the original RC procedure (Blazier, 1981), except for modifications to the levels at 16 Hz recommended as a result of recent ASHRAE research (Broner, 1994). Regions A and B denote levels at which sound can induce vibration in light wall and ceiling construction. In region A, the acoustically induced vibration may be clearly felt and is capable of causing audible rattles in light fixtures, furniture, doors, windows, and so forth. In region B similar effects are potentially possible, but to a lesser extent. The threshold of hearing for octave-bands of noise, as defined by ANSI 12.2-1995, is also illustrated. (No threshold value for the 16 Hz octave-band has been standardized.)

Procedure for Determining the RC Mark II Noise Rating

The first step is to determine the appropriate RC reference curve. This is done by obtaining the arithmetic average of the levels in the 500, 1000, and 2000 Hz octave bands. The RC reference curve is chosen to be that which has the same value at 1000 Hz as the calculated average value.

The second step is the assignment of subjective sound quality, and this is done by calculation of the *Quality Assessment Index*, QAI (Blazier, 1995). The QAI is a measure of the degree to which the shape of the spectrum under evaluation deviates from the shape of the RC reference curve. The procedure requires calculation of the energy-averaged spectral deviations from the RC reference curve in each of three groups of octave bands: LF (16–63 Hz), MF (125–500 Hz), and HF (1000–4000 Hz). The procedure for the LF region is given by Equation 13.1, and is repeated in the MF and HF regions by substituting the corresponding values at each frequency. Values greater than the corresponding RC curve are treated as positive Δs and those below the curve as negative.

$$LF = 10 \log [(10^{0.1\Delta L_{16}} + 10^{0.1\Delta L_{31.5}} + 10^{0.1\Delta L_{63}})/3] \qquad (13.1)$$

In this way, three specific spectral deviation factors are associated with the spectrum being rated, and these are expressed in dB with either positive or negative values. The QAI is the *range* in dB between the highest and lowest values of the spectral deviation factors rounded to the nearest 0.1 dB.

The final step is to assign a sound quality descriptor to the spectrum being rated. This depends upon (a) the magnitude of the QAI, and (b) the spectral deviation factor having the highest positive value. For example, if the value of the QAI ≤ 5 dB, the spectrum is assigned a neutral (N) rating, independent of the deviation factor having the highest positive value. However, if the QAI exceeds 5 dB, the sound quality descriptor of the rating corresponds to the deviation factor having the highest positive value.

Example of Determining the RC Mark II Rating of a Noise Spectrum

Figure 13.4 and Table 13.1 illustrate the method used to determine the RC Mark II rating of a background noise spectrum. The sample spectrum is plotted in Figure 13.4 together with other pertinent information used in the rating process. The tabular data shown in Table 13.1 are required for the calculation procedure. (The overall process can be simplified by setting up a spreadsheet to do the calculations.)

The first step in the rating process is to calculate the arithmetic average of the values in the 500, 1000, and 2000 octave bands. The values are 42, 35, and 28 dB as shown in Table 13.1; the arithmetic average is 35 dB. Thus, the reference curve upon

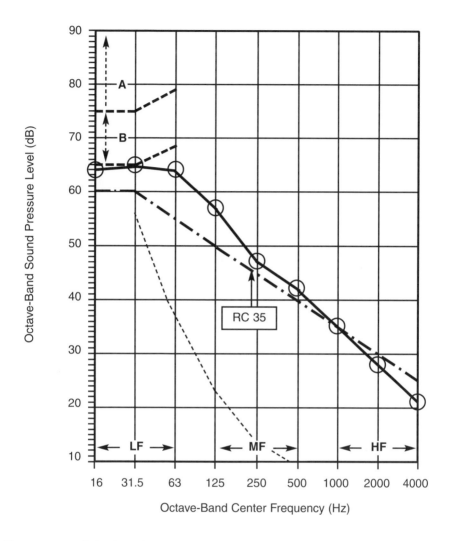

Figure 13.4 — Selection of the RC reference curve, based on the arithmetic average of the levels in the 500, 1000, and 2000 Hz octave bands of the spectrum illustrated. The appropriate curve is RC 35.

which the analysis will be based is RC 35. This curve is illustrated as a bold dashed line in Figure 13.4. The second step is to compute the three spectral deviation factors using Equation 13.1, and then calculate the QAI. The results are (LF) = 6.6, (MF) = 4.4, and (HF) = -1.7. The value of the QAI is the range between the maximum and minimum spectral deviation factors. As tabulated in Table 13.1, the QAI = 8.3 dB, [6.6 - (-1.7)]. The final step is to determine the sound quality rating descriptor. Since the QAI > 5 dB, the spectral deviation factor having the highest *positive* value is chosen

— in this case the (LF) descriptor. Thus, the rating of the noise spectrum in this example is RC 35 (LF). A listener should perceive this spectrum as "rumbly" in character.

TABLE 13.1
Tabular solution to the RC rating of the noise spectrum illustrated in Figure 13.4.

Input Data		Reference Curve	Deviation
OB Ctr. Freq. (Hz)	Level (dB)	RC 35 (dB)	(dB)
16	64	60	4
31.5	65	60	5
63	64	55	9
125	57	50	7
250	47	45	2
500	42	40	2
1000	35	35	0
2000	28	30	-2
4000	21	25	-4

Avg. 500,1000, 2000 Hz (dB)	Spectrum Deviation Factors (dB)	
35	(LF)	6.6
	16–63 Hz	
Reference Curve	(MF)	4.4
	125–500 Hz	
RC 35	(HF)	-1.7
	1000–4000 Hz	
Quality Rating	RC Rating	
(LF)	RC 35 (LF)	
	QAI = 8.3	

Estimating Occupant Satisfaction Using the RC Rating and the Quality Assessment Index (QAI)

In general, there are three principal factors that appear to influence the acceptability of room background noise to people:

1. The *relative loudness* of the noise, with respect to that produced by normal activities taking place in the surrounding area. If noise is clearly noticeable above normal activity sounds, then it is likely to be distracting and subject to complaint. The *numerical* part of the RC rating is a measure of this factor, relative to an assigned space criterion.

2. The potential for *task-interference*. For example, if the background noise level is high enough that the intelligibility of necessary speech communication is impaired, the frustration that results is frequently a cause for complaint. The *numerical* part of the RC rating is also a measure of the potential for acceptable speech communication.

3. The "*quality*" of the background noise and its subtle influence on the subjective judgment of an occupant. For example, if the character of the background noise is more or less neutral, a condition that exists whenever the spectrum is well-

balanced and no one frequency region is more perceptible than another, the subjective judgment is usually most influenced by the relative loudness of the noise. However, if the spectrum of the background sound is imbalanced, and its character is perceived as a "rumble," "roar," or "hiss," this can have a negative effect on subjective judgments. The sound-quality descriptor part of the RC rating addresses this issue.

The QAI, as a measure of the extent of spectrum imbalance, is useful in estimating the probable reaction of an occupant when the background noise is not of optimum sound quality. A QAI of 5 dB or less corresponds to a generally acceptable condition, *provided that the perceived level of the sound is in a range consistent with the given type of space-occupancy.* A QAI that exceeds 5 dB, but is less than or equal to 10 dB, represents a marginal situation in which the acceptance by an occupant is questionable. However, a QAI greater than 10 dB will likely be objectionable to the average occupant. As an aid in interpreting the meaning of the sound-quality descriptor, and the QAI, refer to Table 13.2.

TABLE 13.2
Interpreting the sound-quality descriptor and Quality Assessment Index (QAI).

Sound-Quality Descriptor	Subjective Perception	Magnitude of QAI	Estimated Occupant Evaluation
(N) Neutral (Bland)	Balanced, no one frequency range dominant	QAI ≤ 5 dB	Acceptable
(LF) Rumble	Low-frequency range dominant (16–63 Hz)	5 dB < QAI ≤ 10 dB	Marginal
		QAI > 10 dB	Objectionable
(LF$_A$) Rumble, Clearly Perceptible Surface Vibration	Low-frequency range dominant (16–63 Hz)	5 dB < QAI ≤ 10 dB	Marginal
		QAI > 10 dB	Objectionable
(LF$_B$) Rumble, Moderately Perceptible Surface Vibration	Low-frequency range dominant (16–63 Hz)	5 dB < QAI ≤ 10 dB	Marginal
		QAI > 10 dB	Objectionable
(MF) Roar	Mid-frequency range dominant (125–500 Hz)	5 dB < QAI ≤ 10 dB	Marginal
		QAI > 10 dB	Objectionable
(HF) Hiss	High-frequency range dominant (1000–4000 Hz)	5 dB < QAI ≤ 10 dB	Marginal
		QAI > 10 dB	Objectionable

Trade-Off Between Perceived Sound Level and Sound Quality

Undoubtedly situations will occur in the assessment of room background noise when the numerical part of the RC rating is *less than* the specified maximum for the space use, but the sound quality descriptor is other than the desirable (N). For example, a maximum of RC 40(N) is specified, but the actual noise environment turns out to be RC 35(MF). Is a problem indicated, or is this an acceptable trade-off?

Although there have been no controlled psychoacoustic studies that address this specific issue, intuitively one might suspect that there is some trade-off between level and quality, if the differences are moderate. For example, the marginal situation indicated by a QAI > 5 dB, but ≤ 10 dB may be offset by a perceived sound level 5 dB less than specified. However if the QAI is > 10 dB, it is doubtful that such a trade-off exists. There is probably more of a safety factor if the trade-off involves an (MF) or (HF) rating, rather than an (LF), (LF_B), or (LF_A) rating. Even at moderate levels, if the dominant portion of the background noise occurs in the very-low-frequency region, there are indications that some people experience a sense of oppressiveness or depression in the environment (Persson-Waye et al., 1997).

Limitations in Application of the RC Mark II Rating Methodology

The RC Mark II rating procedure is based on the application of data measured in octave bands. Because of this, it is principally applicable to the analysis of only broadband noises that do not contain audible tonal or narrow-band components. If only octave-band data are available, the bandwidth is generally too wide to identify and quantify the presence of tonal components with any degree of certainty.

The use of 1/3-octave, or narrower band analyzers is generally required to supplement the octave-band data when investigating complaints triggered by the presence of audible tonal components. However, at the present time there are no agreed-upon standards for assessing and rating noises containing such components. Although considerable research has been directed at the question of detectability of tones in noise, and procedures have been developed, such as those appearing in ANSI S12.10, Appendix B, for objectively quantifying this relationship, to date, there is no consensus about the penalty to be applied in rating broadband noise containing audible tonal components, however some form of penalty is usually justified.

Recommended Room Noise Criteria for Specific Applications

As discussed in the section, *Sound Rating Methods*, the selection of an appropriate room noise criterion for a given space-use depends upon the objectives of the application. For example, if the principal objective is reliable speech communication and to a lesser extent the "quality" of the sound, then the simpler level-sensitive noise rating metrics such as A-weighted sound level or tangent-NC/NR are usually adequate. On the other hand, if the objectives of the application include a concern for sound quality, the use of a more complex noise rating such as the RC Mark II is advisable.

Most of the various noise-rating metrics currently available have one commonality: speech-communication criteria are a consideration in their formulation. For example, the numerical part of an RC rating can be used to describe the quality of speech communication in a room as a function of vocal effort, or as an indicator of satisfactory telephone use. Somewhat less reliably, A-weighted sound levels and tangent-NC/NR ratings can also provide similar information, if their magnitude is not heavily biased by frequencies below 500 Hz. As a general rule, A-weighted sound levels will usually be on the order of 5–7 units higher than the corresponding tangent-NC/NR or RC values.

Table 13.3 lists room background noise criteria applicable to representative types of spaces. Note that when the *quality* of the background noise is a significant concern, the RC Mark II rating metric is recommended. These recommendations are based on the judgment of the author and other experienced individuals in the field. Although they represent normally acceptable conditions, it is recognized that in some situations economics and other practical considerations may preclude meeting the recommended criteria. A margin of about 5 rating units above the maximum values listed in each category is likely permissible before serious problems can be anticipated with the level or quality of the background sound.

TABLE 13.3
Recommended background noise criteria as a function of space-use.

Application	Communication Environment	Sound-Quality Also of Concern
Multipurpose auditoria (without sound amplification).	< NC 25; < 32 dBA	< RC 25 (N)
High-level management offices and conference rooms.	NC 25–30; 32–37 dBA	RC 25–30 (N)
Company reception rooms; greeting visitors, etc.	NC 30–35; 37–42 dBA	RC 30–35 (N)
Mid-level management offices, conference rooms, and multipurpose auditoria (w/sound amplification).	NC 30–35; 37–42 dBA	RC 30–35 (N)
Medical facilities: Nurses carrying out medical exams, etc.	NC 30–35; 37–42 dBA	RC 30–35 (N)
Lower-level management offices and conference rooms. Satisfactory for conferences at a 2-2.5 m table (normal voice). Telephone use satisfactory.	NC 35–40; 42–47 dBA	RC 35–40 (N)
Industrial training rooms, assembly spaces for announcements to lower-level management, employees, greeting visitors, etc.; normal to raised speaking voice.	NC 35–40; 42–47 dBA	RC 35–40 (N)
Cafeteria/recreational spaces.	NC 40–45; 47–52 dBA	RC 40–45 (N)
Control rooms in process units, meeting rooms adjacent to noisy work areas. Raised voice at 2-4 m. Telephone use slightly difficult.	NC 40–45; 47–52 dBA	RC 40–45 (N)
Production-area supervisory offices. Telephone use difficult without amplification.	NC 45–50; 52–57 dBA	RC 45–50 (N)
Guard stations, shipping and receiving rooms. Telephone use difficult without amplification.	NC 45–50; 52–57 dBA	RC 45–50 (N)

Recommended Field Measurement Methodology

In the commissioning of HVAC systems in buildings, the achievement of a specified room noise criterion must frequently be documented. Because measurement procedures are often not specified, considerable confusion ensues when interested parties each use different methodologies that achieve differing results. Complicating this is the fact that most real rooms encountered in typical field situations exhibit some point-to-point variation in sound pressure level. Thus measurements made at different locations in a room will not usually be the same.

Except where there are audible tonal components in the noise, the differences in measured sound pressure levels at several locations in a room are not usually large enough to be significant to the casual observer (± 2 dB). However, when audible tonal components are present, the variations due to standing waves may exceed 5 dB, and these are generally noticeable to average listeners, depending upon their location in the room.

Unfortunately in the commissioning process, where precise limits usually are established as the basis for demonstrating compliance, the outcome can be controversial, unless the measurement procedure to be followed has been specified in detail. At the present time, there is no general agreement regarding an acoustical measurement procedure for commissioning HVAC systems. However, the Air-Conditioning & Refrigeration Institute (ARI) has taken an initial step in this direction by incorporating a "suggested procedure for field verification of NC/RC levels" in ARI Standard 885-90.

ARI 885-90 recommends defining the octave-band *room* sound pressure level of the spectrum as the energy average of each octave-band level, measured at four (4) specified points in the space of concern. The recommended locations of these four points are illustrated in Figure 13.5. ARI 885-90 also provides a procedure for processing the data obtained at each measurement location to calculate the energy-aver-

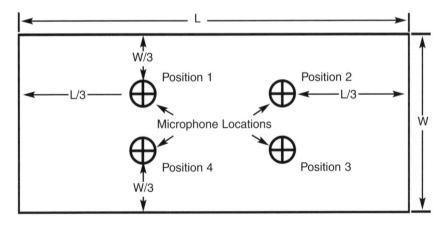

Figure 13.5 — Recommended microphone locations for measuring the average room sound pressure level. Each microphone location is 1.5 m above the floor.

aged level in each octave-band over the frequency range of interest. Thus, the average octave-band spectrum of the noise can be determined for comparison to a specified room noise criterion.

The author recognizes that situations may arise where the measurement of room noise levels, other than those attributed to the HVAC system, are of concern. However, the procedure discussed above should be applicable in most cases.

References

ARI (1990). "Procedure for Estimating Occupied Space Sound Levels in the Application of Air Terminals and Air Outlets," Air-Conditioning and Refrigeration Institute 885-90, Arlington, VA.

ANSI (1985). "Methods for the Measurement and Designation of Noise Emitted by Computer and Business Equipment," (See Appendix B, "Identification of Prominent Discrete Tones") ANSI S12.10-1985, Acoustical Society of America, New York, NY.

ANSI (1995). "Criteria for Evaluating Room Noise," ANSI S12.2-1995, Acoustical Society of America, New York, NY.

Beranek, L. L. (1957). "Revised Criteria for Noise in Buildings," *Noise Control 3*(1), 19–27.

Beranek, L. L. (1989). "Balanced Noise Criterion (NCB) Curves," *J. Acoust. Soc. Am. 86*, 650–654.

Blazier, W. E. Jr. (1981). "Revised Noise Criteria for Application in the Acoustical Design and Rating of HVAC Systems," *Noise Control Eng. 16*(2), 64–73.

Blazier, W. E. Jr. (1995). "Sound Quality Considerations in Rating Noise from Heating, Ventilating, and Air Conditioning (HVAC) Systems in Buildings," *Noise Control Eng. J. 43*(3), 53–63.

Blazier, W. E. Jr. (1996). "Room Noise Criteria: The Importance of Temporal Variations in Low-Frequency Sounds from HVAC Systems," in *Proceedings of Noise-Con 96*, edited by J. D. Chalupnik, S. E. Marshall, and R. C. Klein, Noise Control Foundation, Poughkeepsie, NY, 687–692.

Blazier, W. E. Jr. (1997). "RC Mark II; a Refined Procedure for Rating the Noise of Heating, Ventilating and Air-Conditioning (HVAC) Systems in Buildings," *Noise Control Eng. J. 45*(6), 243–250.

Broner, N. (1994). "Determination of Relationship Between Low-Frequency HVAC Noise and Comfort in Occupied Spaces — Objective Phase," ASHRAE Rept. 714-RP, Atlanta, GA.

Broner, N. (1998). "Low Frequency Noise Loudness vs. Annoyance," *Proceedings of Inter-Noise 98* (on CD), Noise Control Foundation, Poughkeepsie, NY.

Kosten, C. W., and Van Os, G. J. (1962). "Community Reaction Criteria for External Noises," National Physical Laboratory Symposium No. 12, *The Control of Noise*, London, Her Majesty's Stationary Office.

Persson-Waye, K., Benton, S., and Leventhall, H. G. (1997). "Effects on Performance and Work Quality Due to Low-Frequency Ventilation Noise," *J. Sound Vib. 205*(4), 467–474.

Zwicker, E. (1989). "On the Dependence of Unbiased Annoyance on Loudness," in *Proceedings of Inter-Noise 89*, edited by G. C. Maling, Jr., Noise Control Foundation, Poughkeepsie, NY, 809–814.

The Noise Manual, revised 5th edition, edited by E.H. Berger,
L.H. Royster, J.D. Royster, D.P. Driscoll, and M. Layne
©2003 American Industrial Hygiene Association

14   Speech Communications and Signal Detection in Noise

G. S. Robinson and J. G. Casali

Contents

Introduction

Individuals working in noisy environments must be able to hear and communicate while performing their jobs. A machinist must listen to the sounds being produced by the material being machined so as to be alerted to a potential problem before a possibly catastrophic accident occurs. An equipment operator must be able to hear the machine to be sure it is operating properly and in order to fix minor problems or make adjustments. Workers must be able to hear auditory alarms such as fire and evacuation signals or vehicle backup alarms and must also be able to communicate with one another face-to-face and over an intercom or public address system. Unfortunately, noise, hearing protectors, and an increased auditory threshold can make hearing speech and other auditory signals difficult.

In many circumstances, hearing protectors can improve an individual's ability to hear signals in noise compared to the unoccluded condition. However, this effect depends greatly on the attenuation characteristics of the HPD, the hearing threshold of the wearer, and the level and spectrum of the noise and the target signal. Even under the best conditions, auditory signals and speech will be less audible in noise than in quiet. The problem, then, is to determine whether or not speech will be intelligible or an auditory signal will be audible in a particular noise condition, and if not, at what level the speech or signal must be presented to be intelligible or audible. These are the questions that are addressed in this chapter.

We will begin our discussion by presenting a few basic principles of masking which apply to speech signals as well as nonspeech signals. Several methods for predicting the detectability of signals in noise will then be presented, followed by a discussion of relevant standards governing alarm and warning signals as well as guidelines for design and selection of such signals. Issues related specifically to speech intelligibility will be presented next, including discussions of real versus synthesized speech and message content. Several methods for estimating the intelligibility of speech in noise will be presented and various methods for improving communication will be discussed. This chapter will conclude with general recommendations concerning the design and selection of both verbal and nonverbal auditory communications systems as well as considerations relating to the hearing-impaired worker.

Hearing Interference and the Masking of Nonspeech Signals by Noise

Concept of the Masked Threshold

Masking is defined as the increase in the threshold of audibility of one sound (the *masked* sound) caused by the presence of another sound (the *masking* sound or *masker*). In a typical industrial setting, the masked sound might be an alarm or warning (e.g., a vehicle backup alarm or an evacuation siren), speech from a coworker or over an intercom, or a sound produced by a machine (e.g., the sound of a slipping drive belt, metal being machined on a lathe, or air escaping from a relief valve or ruptured pneumatic line). Masking becomes a problem when an intentional or incidental sound conveying useful or important information is rendered inaudible or, in the case of speech, unintelligible by another sound. Typical masking sounds include engine or motor noise, ventilation or fan noise, conveyor noise, and machinery or material forming noise to name a few. It is also possible for one signal to mask another signal if both are active at the same time.

In this chapter, various aspects of the masking phenomenon will be discussed and methods for calculating a masked threshold or, in the case of speech, an estimate of intelligibility, will be presented. In these discussions, it is important to remember that the masked threshold is, in fact, a *threshold*; it is not the level at which the signal is clearly audible. For the purposes of our discussion, a functional definition of an auditory threshold is the sound pressure level (SPL) at which the stimulus is *just audible* to an individual listening intently for it in the specified conditions. If the threshold is determined in quiet, as is the case during an audiometric examination, it is referred to as an *absolute* threshold. If, on the other hand, the threshold is determined in the presence of noise, it is referred to as a *masked* threshold.

Basic Principles and Problems

The degree to which one sound masks another and the resulting masked threshold depend on the physical characteristics of the sounds (levels, spectra, periods, etc.), the hearing threshold of the listener, the etiology of the listener's hearing loss (if present), the attenuation characteristics of the hearing protector (if used), and the manner in which sound is processed by the inner ear and brain. Obviously, not all of these factors can be controlled, but it is important to understand how they are related in order to understand how a change in one or more of these factors affects the audibility of a sound.

Amplitude and Frequency Considerations

In general, the greater the background noise level relative to the signal (hereafter, including speech), the more difficult it will be to hear the signal. Conversely, if the level of the background noise is reduced or the level of the signal is increased, the masked signal will be easier to hear. Manipulating the relative levels of the background noise and signal is perhaps the easiest, but not necessarily the best or most appropriate method for increasing the audibility of a sound.

If the background noise level can be reduced through engineering controls (Chapter 9), existing audible alarms, warnings, and displays should become easier to distinguish from the background noise. If reducing the background noise is not possible or is impractical, it may be possible to increase the level of the signals. Although most off-the-shelf alarms and warning devices have a preset output level, it is possible to increase the effective level of the devices by placing multiple alarms or warning devices throughout the work area instead of relying on a single centrally located device. This approach can also be used for variable-output devices such as public address (PA) systems since simply increasing the output of such systems often results in distortion of the amplified speech signal, thereby reducing intelligibility. Simply increasing the signal level without adding more sound sources can have the undesirable side effect of increasing the noise exposures of workers in the area of the signal if it is used too often. If the levels are extremely high, for example near 115 dB, employees could experience temporary threshold shifts or tinnitus if they are in the vicinity of the device when it is broadcasting.

One problem directly related to the level of the background noise is distortion within the inner ear. At very high noise levels, the cochlea becomes overloaded and cannot accurately transduce the acoustic energy reaching it, resulting in the phenomenon known as *cochlear distortion*. In order for a signal, including speech, to be audible at very high noise levels, it must be presented at a higher level, relative to the background noise, than would be necessary at lower noise levels. This is one reason why it is best to make reduction of the background noise a high priority in the workplace. Doing so not only helps preserve the workers' hearing, but also makes it much easier to hear auditory signals, often making the workplace safer.

In addition to manipulating the levels of the auditory displays, alarms, warnings, and background noise, it is also possible to increase the likelihood of detection of an auditory display or alarm by manipulating its spectrum so that it contrasts with the background noise and other common workplace sounds. In a series of experiments, Wilkins and Martin (1982, 1985) found that the contrast of a signal with both the background noise and irrelevant signals was an important parameter in determining the detectability of a signal. For example, in an environment characterized by high-frequency noise such as sawing and/or planing operations in a wood mill, it might be best to select a warning device with strong low-frequency components, perhaps in the 700- to 800-Hz range. On the other hand, for low-frequency noise such as might be encountered in the vicinity of large-capacity ventilation fans, an alarm with strong mid-frequency components in the 1000- to 1500-Hz range might be a better choice.

When considering masking of a tonal signal by a tonal noise or a narrow band of noise, masking is greatest in the immediate vicinity of the masking tone or, in the case of a band-limited noise, the center frequency of the band. (This is one reason why increasing the contrast in frequency between the signal and noise can increase the audibility of a signal.) However, the masking effect does spread out above and below this frequency, being greater at the frequencies above the frequency of the masking noise than at frequencies below the frequency of the masking noise (Egan and Hake, 1950; Wegel and Lane, 1924). This phenomenon is referred to as the *upward spread of masking*, and it becomes more pronounced as the level of the masking noise increases, probably due to cochlear distortion. In practical situations, masking by pure tones would seldom be a problem, except in instances where the noise contains strong tonal components, or if two warnings with similar frequencies were activated simultaneously. Although less pronounced, upward spread of masking does occur when either broadband or band-limited noise are used as maskers. This phenomenon is illustrated in Figure 14.1.

Perhaps the most common form of masking with respect to a typical industrial workplace occurs when a signal is masked by a broadband noise. In examining the masking of pure-tone stimuli by white noise (white noise, which had equal energy per Hz, and sounds like a TV tuned to a channel with no signal), Hawkins and Stevens (1950) found that masking was directly proportional to the level of the masking noise, irrespective of the frequency of the masked tone. In other words, if a given background white noise level increased the threshold of a 2500-Hz tone by 35 dB, then the

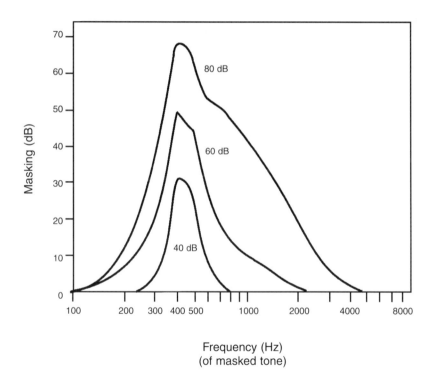

Frequency (Hz)
(of masked tone)

Figure 14.1 — Illustration of upward spread of masking of a pure tone by three levels (40, 60, and 80 dB) of a 90-Hz wide band of noise centered at 410 Hz. The ordinate (y-axis) is the amount (in dB) by which the absolute threshold of the masked tone is raised by the masking noise while the abscissa is the frequency of the masked tone. [Reprinted with permission from Egan, J. P,. and Hake, H. W. (1950). On the masking pattern of a simple auditory stimulus. *Journal of the Acoustical Society of America,* 22 (5), Fig. 6, 627.]

threshold of a 1000-Hz tone would also be increased by 35 dB. Furthermore, they found that for the noise levels investigated, masking increased linearly with the level of the white noise, meaning that if the level of the masking noise were increased 10 dB, the masked thresholds of the tones also increased by 10 dB. (This finding does not refute the existence of cochlear distortion discussed earlier; rather, the noise levels used were insufficient to produce such distortion. At sufficiently high noise levels, cochlear distortion does occur and masking becomes nonlinear.)

To explain observed masking phenomena, Fletcher (1940) developed what would become *critical band theory*. According to this theory, the ear behaves as if it contains a series of overlapping auditory filters, with the bandwidth of each filter being proportional to its center frequency. When masking of pure tones by broadband noise is considered, only a narrow "critical band" of the noise centered at the frequency of the tone is effective as a masker and the width of the band is dependent only on the frequency of the tone being masked. In other words, the masked threshold of a pure tone could be predicted simply by knowing the frequency of the tone and the spectrum level (dB per Hz) of the masking noise, assuming the noise spectrum is reasonably flat in the region around the tone. Thus, the masked threshold of a tone in white noise would simply be:

$$L_{mt} = L_{ps} + 10 \log (BW) \qquad (14.1)$$

where:

L_{mt} is the masked threshold (in dB),
L_{ps} is the spectrum level of the masking noise (in dB), and
BW is the width (in Hz) of the auditory filter centered around the tone.

Strictly speaking, this relationship applies only when the masking noise is flat (i.e., white noise with equal energy per Hz) and when the masked signal has a duration greater than 0.1 second. However, an acceptable approximation may be obtained for other noise conditions as long as the spectrum level in the critical band does not vary by more than 6 dB (Sorkin, 1987). In many industrial environments, the background noise is likely to be sufficiently constant and can often be presumed to be flat in the critical band for a given signal. The exception to this assumption is a situation where the noise has prominent tonal components and/or fluctuates a great deal. Example masked threshold calculations using this relationship appear later in this chapter.

Detection versus Attention

As defined in the preceding discussion, a masked threshold is the level at which a signal must be presented in order for it to be just audible under the specified noise conditions. Most laboratory-based experiments investigating issues related to the masked threshold are conducted such that the experimental subjects not only expect a specific signal, but are also listening intently for it. While such a paradigm is good for determining true thresholds, it is not representative of situations occurring in the real world.

In the real world, workers do not sit around waiting for an alarm; instead, they are doing their jobs, concentrating on the task at hand. Conceptually, it is not difficult to understand how someone engrossed in a job might fail to notice an auditory alarm or warning that might otherwise be audible. Unfortunately, attention is an extremely difficult variable to quantify and control in an empirical study. Most often, experimental subjects are simply asked to perform some secondary task (e.g., playing a video game or driving a vehicle simulator) while also performing the listening/detection task. Few such studies have been performed, and the results are contradictory; some researchers find no difference in detection thresholds (Wilkins and Martin, 1982, 1985) while others find significant differences (Fidell, 1978).

Field tests are even more difficult due to the nearly complete lack of control researchers have over the experimental conditions (noise and signal level, workload, etc.). In the absence of reliable, quantitative data, the best and most conservative course of action is to assume that workers may not notice an alarm or warning if they are engrossed in their jobs. A method of designing alarms that accounts for this lack of attention to auditory warnings will be discussed later.

Detection versus Recognition/Identification

Regardless of how well an auditory alarm or warning is designed or how well it contrasts with the background noise and other signals, humans have a limited ability to distinguish one item from a group of similar items. A basic principle in the field of ergonomics places this number between five and nine, depending on the stimulus (Miller, 1956). In a test of auditory alarms, Patterson (1982) found that subjects quickly learned the first 7 of a set of 10 different alarms and could learn the additional alarms after a 1-hour training session. A week later, the same subjects were able to immediately recognize, on average, 7 of the 10 alarms. Patterson suggested that the number of auditory alarms be limited to 7 or 8 to avoid confusion and eliminate the need for constant retraining.

Perceived Urgency of Signals

Perceived urgency is another factor important in the design and selection of auditory warnings. Problems arise when the reaction elicited by an alarm does not match the severity of the circumstances related to the alarm. Patterson (1982) describes the situation on aircraft flight decks where emergency alarms are often the result of normal inflight procedures. Even in industrial situations, disruptive alarms can draw an operator's attention away from the task at hand, and can also divert attention from the emergency while the operator attends to the alarm. Conversely, disaster could also result if an alarm warning of a critical situation were perceived to be routine or unimportant. To alleviate this problem, Patterson (1982) suggests that auditory alarms be differentiated by type: "advisory alarms" or "*attensons*," intended to draw the operator's attention to additional information explaining the situation, and "immediate action alarms." Furthermore, he suggests that alarms be carefully designed or selected so that the perceived urgency matches the urgency of the situation responsible for the alarm.

The problem remains, however, of which acoustic parameters can be manipulated (and how they should be manipulated) to achieve a desired level of perceived urgency.

Edworthy et al. (1991) found that the perceived urgency of a warning signal could indeed be varied in a predictable manner by manipulating such signal characteristics as fundamental frequency, amplitude envelope, harmonic delay, rhythm, speed, pitch range, and pitch contour. Haas and Casali (1995) determined that perceived urgency increased and reaction time decreased as warning signal pulse level increased and inter-pulse interval decreased.

Unfortunately, it is not yet possible to assign a quantitative value of urgency to a given situation and then specify signal parameters so that an appropriate alarm can be designed or an off-the-shelf unit selected; however, with continued research, this scenario might become a reality someday. The important thing to remember is that it is not always appropriate to use the most obnoxious alarm with the highest possible output available. Each situation must be carefully considered and the alarm that will elicit the appropriate response selected; good judgment, common sense, and a thorough understanding of the situation should guide this process.

Complex versus Tonal Signals

Complex signals (i.e., signals possessing multiple frequency components, periodic or aperiodic temporal patterns, and with modulating frequencies and/or amplitudes) are more easily detected in noisy environments than are simple tonal signals. There are several reasons for this. First, a complex signal with multiple fundamental components as well as strong harmonic and inharmonic elements is more likely to have at least one component that exceeds the masked threshold in nearly any noise. Furthermore, a complex signal contains more information than does a simple pure tone. The brain can use this additional information in processing the auditory stimulus and will be better able to distinguish the desired signal from the background noise. Manipulation of the various acoustic characteristics of a complex auditory signal also allows the perceived urgency of the signal, its contrast with both the background noise and other signals, and its attention-getting ability to be manipulated, not to mention allowing the listener to distinguish one alarm from another.

Signal Localization and Speed Judgments

The brain localizes sounds in space by utilizing subtle phase and intensity differences between the sounds received in each of the listener's ears, for judging direction, as well as the balance between the direct and reflected sound, for judging distance (McMurtry and Mershon, 1985). Speed and movement are judged by perceived changes in position, level, and pitch as the sound source approaches or recedes. The presence of noise can have a detrimental impact on an individual's ability to localize sound.

As background noise level increases, sound sources are judged to be closer to the observer than they actually are. This is because the noise masks the reflected component of the sound reaching the observer's ears (McMurtry and Mershon, 1985). Judging the direction of the sound is less affected by the presence of noise. However, if the noise source and signal are co-located or the signal level is not sufficiently above its masked threshold (by approximately 10 to 15 dB), then problems in directional judgments of high-frequency signals can occur (Small and Gales, 1991). An example is a backup alarm affixed to a gasoline-powered forklift.

HPD Effects

The intended purpose of hearing protectors is to reduce the noise level reaching the ear to prevent damage. However, hearing protectors generally affect the sound spectrum as well, which can influence masking. Because they also change the level of the signal present at the ear, they can improve (for normal-hearing listeners), or detract from, the ability to hear signals in noise. Hearing protectors can also affect the ability to localize sounds in space. For a discussion of these effects see Chapter 10.

Hearing Aid Effects

Individuals with a hearing loss sufficient to require the use of hearing aids are at a distinct disadvantage in the workplace when it comes to their ability to hear auditory alarms, warnings, or even speech. Activation of hearing aids in high levels of noise so as to hear speech or auditory alarms increases the risk of additional damage to hearing due to amplification of the ambient noise (Humes and Bess, 1981). Turning off the hearing aids increases the chance that the signals being listened for may be below threshold, and since it has been shown that vented hearing aid inserts do not function well as hearing protectors (Berger, 1987), there is still a risk of further hearing damage by doing so.

It has been suggested that some benefit may be obtained if a programmable in-the-ear (ITE) hearing aid is worn under an earmuff (Berger, 1987); however, doing so makes it impossible to quantify the wearer's noise exposure and offers no guarantee that the individual will indeed be able to hear the necessary alarms. Others (Royster et al., 1991) have suggested a compensatory visual display be located at the worker's location to augment the auditory warnings. It might even be possible to give such individuals devices that function as vibrating message pagers. These devices, however, would have to be integrated into the plant's alarm system and such a system could be quite expensive. With the lack of quantitative research in this area, no blanket recommendation can be made concerning such individuals. Each case should be evaluated individually, different candidate solutions assessed and tried, and a workable solution amenable to both the employer and employee ascertained. In some cases, it may be necessary to relocate the employee from a hearing-critical job.

Prediction of Signal Detectability in Noise

Several methods have been proposed for calculating masked thresholds in noise. We will present two such methods. Either method may be used to estimate unoccluded or occluded (wearing hearing protection) masked thresholds. All of the example calculations presented below, and those presented in the section on speech intelligibility, will use the same hypothetical noise spectrum shown in Figure 14.2. The octave band and 1/3 octave-band levels for this noise sample appear in Table 14.1. Use of a single noise spectrum for all of the example calculations allows direct comparisons of the various methods.

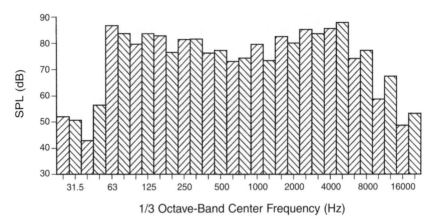

1/3 Octave-Band Center Frequency (Hz)

Figure 14.2 — Spectrum of the hypothetical noise sample (95.8 dB, 94.2 dBA) used in the example calculations for masked threshold and speech intelligibility presented in the text.

The warning signal we will consider for our calculations is an industry-standard backup alarm typically found on commercial trucks and construction equipment. It has strong tonal components at 1000 and 1250 Hz and strong harmonic components at 2000 and 2500 Hz. The alarm has a 1-s period and a 50% duty cycle (i.e., it is "on" for 50% of its period). The levels in all other 1/3 octave bands are sufficiently below those in the bands mentioned as to be inconsequential.

Critical Band Method

The first method for calculating the masked threshold of the signal components is based on critical band theory. As mentioned earlier, the masked threshold for a tonal signal can be approximated using critical band theory if the masking noise is sufficiently flat. In many cases, this is a valid assumption and we will make that assumption for our example. Therefore, the masked threshold (L_{mt}, in dB) is computed using Equation (14.1), repeated below for convenience:

$$L_{mt} = L_{ps} + 10 \log (BW) \qquad (14.1)$$

where:

L_{ps} is the spectrum level of the masking noise (in dB) in the vicinity of the signal component being considered, and

BW is the width (in Hz) of the auditory filter centered around the signal component in question.

The spectrum level of the noise in each of the 1/3 octave bands containing the signal components *is not* the same as the band level measured using an octave-band or 1/3 octave-band analyzer. Spectrum level refers to the level per Hz, or the level that would be measured if the noise were measured using a filter 1-Hz wide. If we assume

TABLE 14.1

Hypothetical 95.8 dB (94.2 dBA) noise spectrum used in the masked-threshold and speech intelligibility calculations presented in the text.

Center Frequency (Hz)*	1/3 Octave-Band Level (dB)	Octave-Band Level (dB)
25	52.0	
31.5	50.7	54.7
40	42.9	
50	56.4	
63	86.8	88.5
80	83.7	
100	79.7	
125	83.7	87.1
160	82.8	
200	76.5	
250	81.4	85.1
315	81.6	
400	76.3	
500	77.3	80.7
630	73.1	
800	74.4	
1000	79.6	81.5
1250	73.4	
1600	82.6	
2000	80.1	87.9
2500	85.3	
3150	83.7	
4000	85.7	90.9
5000	88.0	
6300	74.2	
8000	77.3	79.1
10000	58.7	
12500	67.4	
16000	48.7	67.6
20000	53.3	

* Frequencies in **boldface** type are octave-band center frequencies.

that the noise is flat within the bandwidth of the 1/3 octave-band filter, then the spectrum level can be estimated using the following equation.

$$L_{ps} = 10 \log \left(\frac{10^{L_{pb}/10}}{BW_{1/3}} \right) \tag{14.2}$$

where:

L_{ps} is the spectrum level of the noise (in dB) within the 1/3 octave band,

$BW_{1/3}$ is the bandwidth (in Hz) of the 1/3 octave band, calculated by multiplying the center frequency (f_c) of the band by 0.232, and

L_{pb} is the sound pressure level (in dB) measured in the 1/3 octave band in question.

577

TABLE 14.2
Calculation of masked threshold using critical band theory.

| | Frequency of Signal Component (Hz) | | | |
	1000	1250	2000	2500
Step 1: Determine the bandwidth of the 1/3 octave band filter centered at the frequency of each signal component.				
$BW_{1/3} = 0.232f_c$	232	290	464	580
Step 2: Calculate the spectrum level L_{ps} of the noise in the appropriate 1/3 octave band.				
$L_{ps} = 10 \, Log \left(\dfrac{10^{L_{pb}/10}}{BW_{1/3}} \right)$	55.9	48.8	53.4	57.7
Step 3: Determine the bandwidth of the auditory filters centered around each signal component.				
$BW = 0.15f_c$	150.0	187.5	300.0	375.0
Step 4: Calculate the masked threshold for the signal component.				
$L_{mt} = L_{ps} + 10 \, Log(BW)$	77.7	71.5	78.2	83.4

Note: The 1/3 octave-band levels of the masking noise (Figure 14.2, Table 14.1) at the center frequencies given above are 79.6, 73.4, 80.1, and 85.3 respectively. See the discussion in the text.

Finally, the bandwidth of the auditory filter can be approximated by multiplying the frequency of the masked signal/tone by 0.15 (Patterson, 1982; Sorkin, 1987). Since our alarm has strong tonal components at four frequencies, it is necessary to calculate the masked threshold at each of these frequencies. The step-by-step calculations appear in Table 14.2. If the signal levels measured in one or more of the 1/3 octave bands considered exceed these masked threshold levels, then the signal should be audible.

Note that the masked thresholds calculated for each signal component are slightly below the 1/3 octave-band levels centered at the same frequencies (Table 14.1). This makes sense since the bandwidths of the auditory filters are slightly less than the bandwidths of 1/3 octave-band filters with the same center frequencies. These same procedures can also be used for octave-band data by simply substituting the bandwidth of the appropriate octave-band filter ($BW_{1/1} = 0.707f_c$) for the $BW_{1/3}$ term in Step 2.

ISO 7731-1986(E) Method

International Standard 7731-1986(E), *Danger Signals for Work Places — Auditory Danger Signals* (ISO, 1986) presents a method for calculating the masked threshold for a signal in noise, based on either octave-band or 1/3 octave-band noise data. Both analyses are presented herein.

Step 1:   Starting at the lowest octave-band or 1/3 octave-band level available, the masked threshold (L_{mtl}, in dB) for a signal in that band is:

$$L_{mtl} = L_{pb1},\ (14.3)$$

where L_{pb1} is the sound pressure level (in dB) measured in the octave band or 1/3 octave band in question.

Step 2:   For each successive octave-band or 1/3 octave-band filter n, the masked threshold (L_{mtn}, in dB) is the noise level in that band (in dB) or the masked threshold in the preceding band, less a constant; whichever is *greater*:

$$L_{mtn} = max(L_{pbn} ; L_{mtn-1} - C),\ \ \ \ \ \ \ \ \ \ \ \ \ \ \ \ \ \ (14.4)$$

where $C = 7.5$ dB for octave-band data or 2.5 dB for 1/3 octave-band data.

This procedure, unlike the earlier critical band procedure, presumes the auditory filter is equal to the 1/3 octave band or to the octave band, and also takes upward spread of masking into account by comparing the level in the band in question to the level in the preceding band.

The masked thresholds for each octave band and 1/3 octave band of noise for our hypothetical sample are shown in Table 14.3. For the purposes of our example signal (the backup alarm), only the thresholds for the 1/3 octave bands centered at 1000, 1250, 2000, and 2500 Hz and the thresholds for the octave bands centered 1000 and 2000 Hz are relevant. As before, if the signal levels measured in one or more of these bands exceed the calculated masked threshold levels, then the signal should be audible. Comparing these masked thresholds to those obtained in the previous example, we can see that they are similar, with the masked thresholds obtained using the ISO procedure being slightly higher (more conservative estimates).

Both the critical band and ISO methods are based on critical band theory and may be used to calculate unoccluded or occluded masked thresholds. Calculating an occluded masked threshold for a particular signal requires: (1) subtracting the attenuation of the HPD from the noise spectrum to obtain the noise spectrum effective when the HPD is worn, (2) calculation of a masked threshold for each signal component using the procedures outlined in the preceding discussion, which results in the signal-component levels that would be *just audible* to the listener when the HPD is worn, and (3) adding the attenuation of the HPD to the signal-component thresholds to provide an estimate of the environmental (exterior to the HPD) signal-component levels that would be required to produce the under-HPD threshold levels calculated in Step 2. Although not difficult, this procedure does require a reasonably reliable estimate of the actual attenuation provided by the hearing protector. The manufacturer's data supplied with the HPD are unsuitable for this purpose because they overestimate the real-world performance of the hearing protector, as explained in Chapter 10. Furthermore, if a 1/3 octave-band masking computation is desired, the manufacturer's attenuation data, which are avail-

TABLE 14.3

Masked threshold calculations according to ISO 7731-1986(E) for octave-band and 1/3 octave-band methods.

Center Freq. (Hz)*	1/3 Octave-Band Level (dB)	Masked Threshold (dB)**	Octave-Band Level (dB)	Masked Threshold (dB)**
25	52.0	52.0		
31.5	50.7	50.7	54.7	54.7
40	42.9	48.2		
50	56.4	56.4		
63	86.8	86.8	88.5	88.5
80	83.7	84.3		
100	79.7	81.8		
125	83.7	83.7	87.1	87.1
160	82.8	82.8		
200	76.5	80.3		
250	81.4	81.4	85.1	85.1
315	81.6	81.6		
400	76.3	79.1		
500	77.3	77.3	80.7	80.7
630	73.1	74.8		
800	74.4	74.4		
1000	79.6	**79.6**	81.5	**81.5**
1250	73.4	**77.1**		
1600	82.6	82.6		
2000	80.1	**80.1**	87.9	**87.9**
2500	85.3	**85.3**		
3150	83.7	83.7		
4000	85.7	85.7	90.9	90.9
5000	88.0	88.0		
6300	74.2	85.5		
8000	77.3	83.0	79.1	83.4
10000	58.7	80.5		
12500	67.4	78.0		
16000	48.7	75.5	67.6	75.9
20000	53.3	73.0		

* Frequencies in **boldface** type are octave-band center frequencies.
** Thresholds in **boldface** type are the masked thresholds for the signal components of the backup alarm described in the text.

able for only nine selected 1/3 octave bands, are insufficient for the computation. Finally, both methods fail to take the listener's hearing level into account. It is simply assumed that if the calculated masked thresholds are above the listener's absolute thresholds the sounds should be audible.

On the Conservative Nature of Masked Threshold Prediction

It is important to remember that masked thresholds estimated using any prediction technique are not necessarily accurate or *exact*, nor are they intended to be. They do provide conservative masked threshold estimates for a large segment of the population representing a wide range of hearing levels for nonspecific noise environments and signals. Many factors affect the audibility of a signal in noise, including, but not limited to, the bandwidth of the signal and noise (pure tones versus narrow- or broadbands), the contrast of the signal(s) and the noise (e.g., tonal signals in broadband noise), and the temporal characteristics of the signals and noise. While many of these factors can be quantified, factors such as the contrast between the signals and noise are less quantifiable. In general, the more complex the signal and the greater the contrast with the background noise, the lower the masked threshold (and the better the audibility) will be. If any masked threshold prediction technique is designed to provide an acceptable prediction for a worst-case scenario in which there is little contrast between the noise and signal, it will overestimate, probably by a reasonably large margin, the masked threshold for a more ideal situation.

In psychophysics, the concept of the just noticeable difference (JND) is used to express the difference in magnitude required for an observer to distinguish between two stimuli or the change in magnitude of a stimulus necessary for the observer to notice a change in the stimulus (Gescheider, 1985). For auditory stimuli presented at levels greater than 20 dB above threshold (a condition met in all practical noise-related applications), this difference varies from about 2 dB for pulsed low-frequency tones to less than 0.5 dB for broadband sounds (Small and Gales, 1991). When two identical noise bands are presented simultaneously, the SPL of the two sounds together will be 3 dB greater than the SPL of either individual sound, which is well above the JND for any sound. In fact, the second sound will begin to have a calculable (if not a measurable) effect on the overall SPL when it is more than 15 dB below the SPL of the first sound. To achieve a 0.5 dB increase in the overall SPL of the two sounds together, the second sound need be only 9.1 dB below the SPL of the first sound. An increase in the overall SPL of the two sounds together of at least 1 dB would require only that the individual sound pressure levels differ by no more than 5.9 dB. Table 2.1 can be used to approximate the SPL resulting from the combination of two or more uncorrelated noise sources.

In applying the ISO-standard method to predict masked thresholds for octave-band or 1/3 octave-band analyses, the masked threshold in any band is the level in that band or the level in the preceding band less a constant, whichever is greater. By definition, the masked thresholds predicted using the ISO-standard procedures will be equal to or greater than the noise levels in the bands being considered. If the band levels of a particular signal are at or above the band levels of the noise, it stands to reason that the overall SPL in that band would increase sufficiently that the change in level would be noticeable to the listener, even if the signal itself were not clearly audible. Of course, an increase in the level of one or a few individual 1/3 octave bands does not necessarily equate to a significant increase in the overall broadband level of a noise; however, many incidental and intentional sounds contain energy in a fairly wide range of frequencies and can be considered to be broadband sounds. In these cases, if their band levels are somewhat close to the band levels of the noise, they could easily cause a sufficient increase in the overall noise level to allow that change in level to be noticed.

This discussion is not intended to represent an exhaustive treatise on the subject of masked threshold prediction; rather, it is intended only to discuss a few of the reasons that existing masked threshold prediction techniques provide conservative estimates. This is not a shortcoming of the methodology; to the contrary, it is a strength. The conservative overestimation of the masked threshold serves as a safety factor so that for most cases, if a sound is sufficiently above its masked threshold as determined by a suitable prediction technique, it should be audible to most people. Problems do arise, however, when individuals suffer from a severe hearing loss.

Design of Warning Signals and Alarms for Workplaces

The methods outlined in the preceding discussions are intended to determine the *threshold* of the signal, not the level at which it can be reliably heard 100% of the time. To ensure audibility and overcome any decrement due to inattention, the signal level must be increased above its threshold level. Exactly how much the signal level should be increased is open to debate, but many researchers suggest that 10 to 15 dB is adequate for most situations (Coleman et al., 1984; Sorkin, 1987; Wilkins and Martin, 1978). Specific guidelines are presented below.

Wilkins and Martin (1978) classified auditory alarm and warning devices into four broad categories based on their design. These categories were: siren, horn, bell, and electronic device. Regardless of type, the sound produced by any given alarm or warning device can be varied in an almost infinite variety of ways by manipulating only three parameters (level, spectrum, and period), although other parameters can also be varied. Level, and to a slightly less extent spectrum, are the primary determinants of signal detectability, while temporal patterning and spectrum aid discrimination and identification (Wilkins and Martin, 1978).

Of these three parameters, level and spectrum are the most likely to be specified in an alarm or warning standard. However, surprisingly few such standards exist. Perhaps the most comprehensive standard is the aforementioned International Standard 7731-1986(E), *Danger Signals for Work Places — Auditory Danger Signals* (ISO, 1986). This standard not only presents guidelines for calculation of the effective masked threshold of audibility, but also specifies the spectral content and minimum signal-to-noise ratios (S/N) of the signals, and requires manufacturers of such devices to consider individuals suffering from hearing loss or those wearing HPDs (although it gives no quantifiable recommendations or procedures for doing so). The major requirements of ISO 7731-1986(E) are summarized in Table 14.4.

A standard that attempts to ensure recognition in addition to detection is ANSI S3.41-1990, *Audible Emergency Evacuation Signal* (ANSI, 1990). [This is the U.S. version of ISO 8201-1987(E), *Acoustics — Audible Emergency Evacuation Signal* (ISO, 1987)]. In addition to requiring minimum sound pressure levels and establishing specific temporal characteristics for emergency evacuation alarms (see Table 14.5), this standard requires that such alarms be clearly distinguished from other alarms or warnings that might be present in the area. In establishing a specific temporal pattern for an alarm intended to warn of a specific condition in both Europe and North America, these standards are a first step in establishing consistent cross-cultural mean-

TABLE 14.4
Summary of the major requirements of ISO 7731-1986(E), Danger Signals for Work Places — Auditory Danger Signals.

Parameter	Requirement or Guideline
Level	**Broadband estimate:** A-weighted SPL of signal should exceed that of the background noise by 15 dB or more.
	Octave-band estimate: Signal shall exceed the masked threshold (calculation method specified) by at least 10 dB in one or more octave bands as specified (see spectral content requirement below).
	1/3 octave-band estimate: Signal shall exceed the masked threshold by at least 13 dB in one or more 1/3 octave bands as specified.
Spectral Content	The signal shall have its energy concentrated in the frequency range from 300 to 3000 Hz. Sufficient energy shall be present in the frequency range below 1500 Hz to satisfy the needs of individuals suffering from hearing loss or wearing HPDs.
Temporal Patterning	Pulsed signals are preferred. Pulse duration should be between 0.2 and 5 Hz. Pulse rate and duration are to be different from any periodically varying ambient noise in the work area.
Audibility and Discriminability	On-site listening tests are to be performed to ensure that the signal is both audible and discriminable. A minimum of 10 subjects are to be used and the test is to be repeated five times. Subjects are to be representative of the workers who will be working in the area (in terms of age and hearing levels) and shall wear hearing protection if appropriate. 100% detection/discrimination is required.

ings for important warnings. By presenting examples of how the temporal requirements can be met using different types of devices, this standard allows a great deal of flexibility in designing and selecting a suitable signaling device.

The requirements of a few United States standards (ANSI, 1990; DoD, 1981; NFPA, 1979; SAE, 1978; UL, 1981) for audible alarms and warnings are summarized in Table 14.5. For the most part, these standards specify only minimum levels and bandwidths; they do not consider individuals with hearing loss or those wearing HPDs. Although none of the standards are in complete agreement, two summary observations can be made. There appears to be a consensus that the signal level should be about 15 dB above the noise level; and, while wider bandwidths are allowed, all of the standards that contain specific bandwidth information include the frequency range from 700 to 2800 Hz. Frequencies higher than about 3000 Hz are unsuitable because individuals suffering from a noise-induced hearing loss and/or presbycusis may have significant high-frequency hearing loss and would thus be at a disadvantage.

In addition to the various standards summarized in Tables 14.4 and 14.5, several authors have developed alarm and warning signal design guidelines, which attempt to maximize signal detectability and/or discriminability. Based on a lengthy series of experiments, Coleman et al. (1984) recommend that signals should be at least 15 dB

TABLE 14.5
Summary of the requirements of U.S. standards for auditory alarms and warnings.

Standard	Requirements
ANSI S3.41-1990/ISO 8201-1987(E); Audible Emergency Evacuation Signal	**Level:** Minimum sound output of 65 dBA (FAST response measured during the "on" phase of the device). The minimum level is raised to 75 dBA if the device is intended to awaken sleeping individuals (measured at the individual's head position). The A-weighted output level of the alarm during its "on" phase must exceed the highest A-weighted SPL of the background noise over a continuous 60-s interval. **Temporal pattern:** Requires a specific pattern of three "on" phases with a period of 1 s followed by an "off" phase of 1.5 s.
SAE J994b-1978; Performance, Test and Application Criteria for Electronically Operated Backup Alarm Devices	**Spectral content:** Predominant frequencies are to be in the range from 700 to 2800 Hz. **Level:** Levels required vary by type. Five types specified: A - 112 dBA, B - 107 dBA, C - 97 dBA, D - 87 dBA, and E - 77 dBA. Measurement to be made at a distance of 4 ft. from the alarm. **Temporal pattern:** Recommends periods of 1 to 2 s with 50% duty cycles.
MIL-STD-1472C(1981); Human Engineering Design Criteria for Military Systems, Equipment and Facilities	**Spectral content:** Predominant frequencies are to be in the range from 200 to 5000 Hz, but preferably from 500 to 3000 Hz. When the distance to the alarm exceeds 300 m, only frequencies below 1000 Hz should be utilized. **Level:** Requires the signal to be 20 dB above the noise in at least one octave band in the operating frequency range.
NFPA 72A-1979; Local Protective Signaling Systems	**Level:** Recommends that the signal output should be 15 dB above the steady state background noise level. If the noise varies, recommendation made that the signal output be 5 dB higher than the *maximum* noise level. **Temporal pattern:** Recommended on-time of 0.5 to 1 s and off-time of 0.5 s.

— *continued on next page* —

TABLE 14.5 — continued

Standard	Requirements
UL 464-1981; Audible Signal Appliances	**Level:** Minimum sound output of 75 dBA. Measurement of signal output specified to be in accordance with ANSI S1.21-1972, *Methods for the Determination of Sound Power Levels of Small Sources in Reverberation Rooms*, which has been superseded by ANSI S1.31-1980, *Precision Methods for the Determination of Sound Power Levels of Broad-Band Noise Sources in Reverberation Rooms* and ANSI S1.32-1980, *Precision Methods for the Determination of Sound Power Levels of Discrete-Frequency and Narrow-Band Noise Sources in Reverberation Rooms.*
	Temporal pattern: Single-stroke devices shall operate at a rate of 60 impulses/min with a 50% duty cycle.

above masked threshold, across the entire spectrum whenever possible. To avoid a startle response, signals should be no more than 25 dB above threshold and have rise and fall times on the order of 20 ms. When signal levels exceed 90 dB, consideration should be given to the possibility of the signal contributing to the noise dose of the exposed individuals. Temporal patterns (inverse of the period — number of periods [i.e., on/off cycles] per unit time) should be on the order of 1 to 4 Hz while modulation (amplitude and/or frequency fluctuations within a single period of the alarm) should be 20 Hz or higher. Complex signals consisting of harmonically related components with a fundamental frequency below 1000 Hz should be used.

Sorkin (1987) suggests that signal levels 6 to 10 dB above masked threshold are adequate to ensure 100% detectability, while signals that are approximately 15 dB above their masked threshold will elicit rapid operator response. He also suggests that signals more than 30 dB above the masked threshold could result in an unwanted startle response and that no signal should exceed 115 dB. (This suggested upper limit on signal level is consistent with OSHA hearing conservation requirements [OSHA, 1983], which prohibit exposure to continuous noise levels greater than 115 dBA.) These recommendations are in line with those of other authors (Deatherage, 1972; Wilkins and Martin, 1982).

In addition to the need to hear an alarm or signal in noise, it is often also necessary to simultaneously communicate verbally with coworkers, either face-to-face or using an intercom or telecommunications system. If the alarm or warning is presented at too high a level, either in absolute terms or relative to the background noise, simultaneous speech communications can be extremely difficult if not impossible. This point was addressed by Patterson (1982) in the context of aircraft flight decks; however, similar situations are also common in industrial and office environments.

A major practical shortcoming in most of the auditory warning and alarm standards and literature cited above is the general assumption that listeners will possess normal hearing and that HPDs will not be used. The single exception is ISO 7731-1986. However, even that standard fails to provide quantitative guidelines concerning listeners with a hearing impairment or wearing HPDs, specifying only that the signal must be audible.

It cannot be overemphasized that the noise and signal levels cited in the preceding discussion (and, for that matter, in the entire chapter), refer to the levels measured *at the listener's location.* Measurement made at some central location or the specified output levels of the alarm or warning devices are not representative of the levels present at a given workstation and cannot be used for masked threshold calculations.

In an outdoor environment, the sound level of an alarm or warning will generally decrease as the distance from the source increases. Furthermore, buildings or other large structures in the source-receiver path can create "shadow zones" in which little or no sound is audible. It is for these reasons that MIL-STD-1472 (DoD, 1981) recommends that frequencies below 1000 Hz be used for outdoor alarms since low frequencies are less susceptible to atmospheric absorption and more readily diffract around barriers. Similar problems can be encountered indoors as well. Problems associated with the general decrease in SPL with increasing distance as well as shadow zones created by walls, partitions, screens, machinery, and even moving and stationary vehicles must be considered. Although the atmospheric absorption of high-frequency sound is not generally a problem indoors, many of the materials and structures used to protect workers from unwanted noise will also block or absorb the sound of an alarm or warning. Furthermore, since different construction materials reflect and absorb sound depending on frequency, not only do the sound levels change from position to position, but the spectra of both the noise and signals can change as well. Finally, since most interior spaces reverberate to some degree, we must also be concerned with phase differences between reflected sounds that can result in superposition effects of enhancement or cancellation of the sound from location to location.

It is for these reasons that it is necessary to know the SPL at the listener's location when considering masked thresholds. General noise level data (and possibly even noise spectra) should be available as a result of the noise survey (Chapter 7). With a little extra effort, the level and spectra of the various alarms can also be obtained and the audibility of each evaluated. If spectral data cannot be obtained, ISO 7731-1986 does present guidelines for using broadband A-weighted SPL measurements; however, this should be a last-resort option and can result in unnecessarily high signal levels.

Masking of Speech Signals by Noise

Many of the concepts presented in the earlier discussion of masking of nonspeech signals by noise apply equally well to the masking of speech. For speech, however, the concern is not simply audibility or detection, but rather intelligibility. The listener must understand *what* was said, not simply know that something was said. Furthermore, speech is a very complex broadband signal whose components are not only differentially susceptible to noise, but are also highly dependent on vocal effort, the gender of the speaker, and the content and context of the message. This is in addition to the effects of HPD use by the speaker and/or listener (see Chapter 10), hearing loss of the listener, or speech signal degradation occurring in a communications system.

Speech Characteristics and Noise Parameter Effects
Speech/Noise Ratio

As was the case with nonspeech signals, the signed difference between the speech level and the background noise level is referred to as the signal-to-noise ratio, or more appropriately, the speech-to-noise ratio (also abbreviated S/N). The speech level referred to is usually the long-term rms level measured in dB.

When background noise levels are between 35 and 110 dB SPL, an S/N of 12 dB is usually adequate to reach a normal-hearing individual's threshold of intelligibility (Sanders and McCormick, 1993); however, it is quite impossible for any individual to sustain the vocal efforts required in the higher noise levels without electronic amplification (i.e., a public address system). The *threshold of intelligibility* is defined as the "level at which the listener is just able to obtain without perceptible effort the meaning of almost every sentence and phrase of" continuous speech (Hawkins and Stevens, 1950, p. 11); essentially, this is 100% intelligibility. Intelligibility decreases as the S/N decreases, reaching 70 – 75% (as measured using phonetically balanced words) at an S/N of 5 dB, 45 – 50% at an S/N of 0 dB, and 25 – 30% at an S/N of -5 dB (Acton, 1970).

Forced Vocal Effort

In low-to-moderate noise levels, people automatically modulate their vocal effort to overcome the masking effects of the background noise and communicate with other individuals. This characteristic is referred to as the Lombard reflex (Lane and Tranel, 1971). However, there is an upper limit to this ability. Mean speech levels for male and female speakers in quiet are shown in Table 14.6. As can be seen, speech levels for females are 2 to 7 dB less than for males, depending on vocal effort. Although extreme efforts can produce levels greater than 90 dB (Kryter, 1985), such levels cannot be maintained for long periods. Since a relatively high positive S/N is necessary for reliable speech communications in the presence of background noise, it should be obvious that in high noise levels (greater than about 75–80 dB), unaided speech cannot be relied upon except for short durations over short distances.

TABLE 14.6
Mean speech levels for male and female speakers in quiet (at 1m) (adapted from Kryter, 1985).

Vocal Effort	Speech Level, dB	
	Males	Females
Casual	52.0	50.0
Normal	58.0	55.0
Raised	65.0	63.0
Loud	76.0	71.0
Shout	89.0	82.0

Speech sounds themselves differ in the vocal effort with which they are produced. Consonants are produced with much less intensity than are vowels. Sanders and McCormick (1993, p. 199) point out that the *a* in *talk* possesses 680 times the speech power of the *th* in *then*, a difference of 28 dB. Since consonants are far more critical to intelligibility than are vowels, any circumstance that compromises consonant recognition (e.g., excessive noise-induced hearing loss) can have an adverse effect on speech intelligibility, more so than if only the vowel sounds are affected (e.g., by peak-clipping the speech signal in a telecommunications system).

Speech Bandwidth Considerations

The speech bandwidth extends from 200 to 8000 Hz with male voices generally having more energy at the low frequencies than female voices (Kryter, 1974); however, the region between 600 and 4000 Hz is most critical to intelligibility (Sanders and McCormick, 1993). This also happens to be the frequency range at which most auditory alarms are presented, providing an opportunity for the direct masking of speech by an alarm or warning. Therefore, speech communications in the vicinity of an activated alarm can prove difficult.

The greater importance of consonants to intelligibility also makes speech differentially susceptible to masking by band-limited noise, depending of course on the level of the noise. At low levels, bands of noise in the mid- to high-frequency ranges directly mask consonant sounds, thus impairing speech intelligibility more so than would low-frequency sounds presented at similar levels. However, at high levels, low frequency bands of noise can also adversely affect intelligibility due to upward spread of masking into the critical speech bandwidth.

When electronic means are used to overcome problems associated with speech intelligibility, it is important to understand how the systems themselves may exacerbate the problem if they are not designed properly. Most industrial telecommunications systems (i.e., intercoms, telephones, PA systems) do not transmit the full speech bandwidth nor do they reproduce the entire dynamic range of the human voice. To reduce costs and simplify the electronics, such systems often filter the signal and pass (transmit) only a portion of the speech bandwidth (e.g., the telephone passband is generally 300 to 3600 Hz). If the frequencies above 4000 Hz or the frequencies below 600 Hz are filtered out (not transmitted), there is little negative impact on speech intelligibility. However, if the frequencies between 1000 and 3000 Hz are filtered out of the signal, intelligibility is severely impaired. To ensure adequate intelligibility, it is imperative that such systems pass the frequencies most critical to speech intelligibility.

In addition to filtering the speech signal, it is possible to clip the speech peaks so that the full dynamic range of the speaker's voice is not transmitted to the listener. This clipping may be intentional on the part of the designer in order to reduce the cost of the system, or it may be an artifact of the amplitude distortion caused by an overloaded amplifier. Either way, the effects on intelligibility are the same. Since the speech peaks contain primarily vowel sounds and intelligibility relies predominantly on the recognition of consonants, there is little loss in intelligibility due strictly to peak clipping. However, if the clipping is caused by distortion within the amplifier, there may be ancillary distortion of the speech signal in other ways that could adversely affect intelligibility.

Real versus Synthesized Speech

Speech synthesizers are becoming increasing popular as the necessary technology improves and becomes less expensive. Such systems are already commonplace in phone-based systems as well as in some consumer electronics, transportation systems, and computer applications. However, care must be exercised when such systems are considered for use in an industrial setting where speech intelligibility is critical.

Live continuous speech contains subtle nuances that make it more than just the sum of the words spoken. The pronunciation of the same speech sound differs not only from word to word, but also with differences in preceding and trailing words. Since speech synthesizers cannot reproduce all of the subtleties of live speech, their utility is limited.

In an investigation of synthetic speech displays for possible use in long-haul trucks, Morrison and Casali (1994) determined that even at an S/N of 15 dB, intelligibility of synthesized speech was just 72% in realistic levels of truck noise. When speech noise (i.e., as might be produced by a radio or CB) was added to the truck noise, intelligibility dropped to less than 60%. By comparison, the intelligibility for live speech calculated using the Articulation Index (AI, now referred to as the Speech Intelligibility Index, SII — ANSI, 1997) exceeded 90% in all cases. The authors caution against the use of synthetic speech displays for critical functions in high-noise environments.

Other Influences

Facial Cueing

One factor that can aid speech intelligibility in noise is face-to-face contact between the speaker and listener. At low speech-to-noise ratios, the ability of the listener to view or "read" the lips of the speaker can have a tremendous positive effect on the intelligibility of the speech signal. Erber (1969) found that at an S/N of -10 dB, intelligibility increased from less than 20% to almost 80% simply by allowing the listener to see the lips of the speaker. The benefit associated with the listener's ability to see the speaker's lips decreases as S/N increases, becoming negligible at positive S/N.

Message Content and Context

Only about 50% of the words spoken in conversational speech are intelligible when presented in isolation from other words (Pollack and Pickett, 1964). This fact implies the importance of context to speech intelligibility. Context supplies invaluable clues to the listener as to what words to expect and to the subject of the conversation (Sorkin and Kantowitz, 1987). In addition to context, other factors that affect intelligibility of sentences include their syntax (how words and other sentence elements combine to form grammatically correct sentences) and prosodic structure (how individuals naturally group words when speaking sentences). In general, intelligibility is higher for complete sentences than for isolated words and higher for whole words than for single letters (Sanders and McCormick, 1993). The greater intelligibility of words over individual letters is the reason that the military and police use phonetic alphabets (alpha, bravo, charlie, etc.) instead of individual letter names when communicating information such as map coordinates, call signs, or license plate numbers. Finally, intelligibility is also maximized when the vocabulary of possible words/phrases is restricted and well known to the listener (Miller et al., 1951). These statements hold for speech heard in noise or in quiet.

Acoustic Environment

The acoustic environment (barriers, reverberation, etc.) can also have a detrimental effect on speech intelligibility. It should be obvious that as the distance between the listener and the speech source (person or loudspeaker) increases, the ability to understand the speech can be adversely affected if the S/N decreases sufficiently. In the same vein, barriers in the source-receiver path can create shadow zones in which the S/N is insufficient to allow the listener to understand what is said. Finally, speech intelligibility decreases linearly as reverberation time increases. Each one second increase in reverberation time will result in a loss of approximately 5% in intelligibility (Sanders and McCormick, 1993).

Prediction and Empirical Testing of Speech Intelligibility in Noise

Indirect Techniques

Unlike the situation with nonspeech signals, there are numerous methods available for estimating speech intelligibility in noise or determining if intelligibility may be a problem in a particular environment. Each method has its advantages and disadvantages depending on the application. We will examine two such methods, the Preferred Speech Interference Level (PSIL) and the Speech Intelligibility Index (SII). The latter has the advantage of being standardized (ANSI, 1997), and can be used to predict intelligibility for both normal and hearing-impaired listeners under a variety of listening conditions, including the use of hearing protectors. However, this increased flexibility comes at the cost of simplicity.

The PSIL (Peterson and Gross, 1978) is the arithmetic average of the noise levels measured in the octave bands centered at 500, 1000, and 2000 Hz. Most useful when the spectrum of the background noise is flat, the PSIL is intended only as an indication of whether or not there is likely to be a communications problem, not as a predictor of intelligibility. Consider the hypothetical noise spectrum presented earlier (Table 14.1). The PSIL for this spectrum would be: $(80.7 + 81.5 + 87.9)/3 =$ 83. With this information we can use Figure 14.3 to determine how difficult verbal communication is likely to be in this level of noise. At a PSIL of 83, verbal communications will be "difficult" at any speaker–listener distance greater than about 18 inches. Even at closer distances, a "raised" or "very loud" voice must be used. If octave band levels are not available, the A-weighted sound level may also provide approximate guidance concerning the speech-interfering effects of background noise, also shown in Figure 14.3. The PSIL is a useful tool for estimating the degree of difficulty that can be expected when verbal communications are attempted in a steady, flat background noise. If the background noise is not flat, is predominated

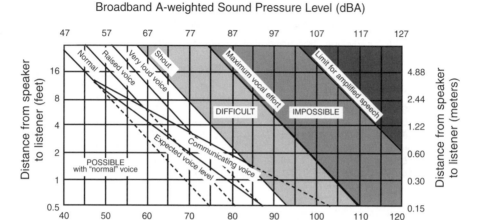

Figure 14.3. — Illustration showing the relationship between the PSIL, speech difficulty, vocal effort, and speaker–listener separation [from Sanders, M. S., and McCormick, E. J. (1993). *Human Factors in Engineering and Design* (7th ed.). Fig. 7–9, Pg. 209, New York: McGraw-Hill. Reprinted with Permission of McGraw-Hill Inc.].

by or contains strong tonal components, or fluctuates a great deal, the utility of the PSIL is lessened.

The SII (ANSI, 1997) is a flexible tool for estimating speech intelligibility in noise. Four calculation methods are available: the critical band method (most accurate), the 1/3 octave-band method, the equally contributing critical band method, and the octave-band method (least accurate). At a minimum, the calculations require knowledge of the spectrum level of the speech and noise as well as the listeners' hearing thresholds. Where speech spectrum level(s) are unavailable or unknown, the standard offers guidance in their estimation. The methodology accounts for the existence of external masking noise, differences in speaker vocal effort, reverberant speech, monaural and binaural listening, hearing loss, varying message content, and insertion loss (HPD use) or gain (amplified communications systems). Although quite flexible in the number and types of conditions to which it can be applied, application of the standard is limited to natural speech, otologically normal listeners with no linguistic or cognitive deficiencies, and situations that do not include sharply filtered bands of speech or noise. It can also be applied to those with a hearing loss or those wearing hearing protection, but it will only account for changes in the threshold levels, not for any distortions or upward spread of masking effects arising from those threshold changes.

The general steps used in calculating the SII and estimating intelligibility are presented below. Two example calculations of the SII are presented in Table 14.7. The first considers the general condition of a normal-hearing individual listening (binaurally) to speech in noise without a hearing protector (Case 1). The second example considers the use of an HPD (a high-attenuation earmuff) by an individual with a conductive hearing loss (Case 2). In each case, the noise considered is the same hypothetical noise

TABLE 14.7
Calculation of the speech intelligibility index (SII).

Case 1   Normal-hearing listener in noise without hearing protection.

Step 1	Step 2			Step 3	Step 4	Step 5	Step 6	Step 7		Step 8
i CF_i	E'_i	N'_i	T'_i	Z_i	X'_i	D_i	L_i	K_i	A_i	I_iA_i
1 250	36.4794	62.62	-1.7	62.62	-5.6	62.62	1.000	0.0000	0.0000	0.0000
2 500	43.2194	55.22	-1.7	55.22	-11.4	55.22	1.000	0.1000	0.1000	0.0167
3 1000	45.2894	53.02	-1.7	53.02	-14.4	53.02	0.9358	0.2423	0.2267	0.0538
4 2000	38.2994	56.42	-1.7	56.42	-19.4	56.42	0.9314	0.0000	0.0000	0.0000
5 4000	28.3894	56.42	-1.7	56.42	-27.6	56.42	0.9434	0.0000	0.0000	0.0000
6 8000	14.6994	41.62	-1.7	41.62	-8.8	41.62	0.9777	0.0000	0.0000	0.0000

SII = 0.0705

Case 2   Hearing-impaired listener wearing a high-attenuation earmuff.

Step 1	Step 2			Step 3	Step 4	Step 5	Step 6	Step 7		Step 8
i CF_i	E'_i	N'_i	T'_i	Z_i	X'_i	D_i	L_i	K_i	A_i	I_iA_i
1 250	10.7794	36.92	28.3	36.92	24.4	36.92	1.0	0.00	0.00	0.0000
2 500	4.4194	16.42	13.3	16.42	3.6	16.42	1.0	0.10	0.10	0.0167
3 1000	3.0894	10.82	33.3	10.82	20.8	20.8	1.0	0.00	0.00	0.0000
4 2000	-0.5006	17.62	28.3	17.62	10.6	17.62	1.0	0.00	0.00	0.0000
5 4000	-13.2106	14.82	23.3	14.82	-2.6	14.82	1.0	0.00	0.00	0.0000
6 8000	-28.0006	-1.08	63.3	-1.08	56.2	56.2	1.0	0.00	0.00	0.0000

SII = 0.0167

presented earlier (Table 14.1). The speaker–listener distance is assumed to be two meters (a little over 6 feet), and the vocal effort of the speaker is assumed to be a "shout." It is extremely important for the reader to realize that the procedures summarized below are in no way intended as a substitute for the much more detailed information presented in the standard. The reader is encouraged to obtain a copy of the standard and study it thoroughly before attempting to perform his/her own calculations.

Step 1:   Select the calculation method. The example calculations presented herein utilize the octave-band method. Although this is the least accurate method, it is the simplest of the four procedures and octave-band noise data are most likely to be the type of noise data available to hearing conservationists. Furthermore, the octave-band method allows the use of existing hearing threshold data and HPD attenuation data (supplied by the manufacturer on the product packaging).

Step 2:   Determine the equivalent speech spectrum level (E'_i), the equivalent noise spectrum level (N'_i), and the equivalent hearing threshold level (T'_i).

Based on the "standard speech spectrum level" for a shout presented in the standard, modified for a speaker–listener distance of more than 1 meter (and use of an HPD in Case 2), the equivalent speech spectrum (E'_i, in dB) is calculated as:

TABLE 14.8
**Values used in intermediate calculations to determine the values contained
in Table 14.7**

i	F_i	E_i	G_i	$BW_{adj.}$	HL	X_i	U_i	I_i
1	250	42.5	-25.7	22.48	30	-3.9	34.75	0.0617
2	500	49.24	-38.8	25.48	15	-9.7	34.27	0.1671
3	1000	51.31	-42.2	28.48	35	-12.5	25.01	0.2373
4	2000	44.32	-38.8	31.48	30	-17.7	17.32	0.2648
5	4000	34.41	-41.6	34.48	25	-25.9	9.33	0.2142
6	8000	20.72	-42.7	37.48	65	-7.1	1.13	0.0549

$$E'_i = E_i - 20 \log (d/d_0) + G_i \tag{14.5}$$

Where:

E_i is the standard speech spectrum level (in dB) given in the standard for each octave band i (Table 14.8).

d/d_0 is the ratio of the actual speaker–listener distance (d, in meters) to the reference distance (d_0 = 1 meter). This ratio is 2 in our example.

G_i is the insertion gain/loss of the hearing protector (see Case 2) or amplification system (in dB), if used (Table 14.8).

The equivalent noise spectrum level (N'_i, in dB) is the noise level that would be measured at the middle of the listener's head if the listener were not present, calculated as follows:

$$N'_i = BL_i - BW_{adj.} + G_i \tag{14.6}$$

Where:

BL_i is the measured noise level (in dB) in octave band i from Table 14.1.

$BW_{adj.}$ (in dB) is an adjustment given in the standard to convert the octave-band level to a band-limited spectrum level (Table 14.8).

G_i is the insertion gain/loss (in dB) of the hearing protector (see Case 2) or amplification system, if used (Table 14.8).

For the case of normal hearing and monaural listening, the equivalent hearing threshold level (T'_i, in dB) is 0. For thresholds other than 0 dBHL, the average hearing threshold level should be used, that is to say, T'_i = HL. For the case of normal hearing and binaural listening, T'_i = - 1.7 dB and for binaural listening with thresholds other than 0 dBHL, and assuming negligible binaural differences between the hearing levels of the two ears, T'_i = HL - 1.7 dB.

Step 3:   Determine the equivalent masking spectrum level (Z_i, in dB). For the octave-band method,

$$Z_i = N'_i \tag{14.7}$$

Step 4:   Using the reference internal noise spectrum level (X_i, see Table 14.8, in dB) presented in the standard, determine the equivalent internal noise spectrum level (X'_i, in dB).

$$X'_i = X_i + T'_i \tag{14.8}$$

Step 5:   Determine the equivalent disturbance spectrum level (D_i, in dB), the larger of the two noise spectrum levels calculated in Steps 3 (the equivalent masking spectrum level, Z_i) and 4 (equivalent internal noise spectrum level, X'_i).

$$D_i = Max(Z_i, X'_i) \tag{14.9}$$

Step 6:   Calculate the level distortion factor (L_i, in dB) using the standard speech spectrum level at a normal vocal effort as presented in the standard (U_i, see Table 14.8, in dB).

$$L_i = 1 - [(E'_i - U_i - 10)/160] \tag{14.10a}$$

When considering individuals with a hearing loss, this equation is modified to include the portion of the equivalent hearing threshold level (T'_i) due solely to the individual's hearing loss (J_i, in dB).

$$L_i = 1 - [(E'_i - U_i - 10 - J_i)/160] \tag{14.10b}$$

In either case, if L_i as calculated above is greater than 1.0 dB, use 1.0 instead.

Step 7:   First, calculate a temporary variable, K_i as

$$K_i = (E'_i - D_i + 15)/30 \tag{14.11}$$

such that K_i must fall in the range $0 \leq K_i \leq 1$.

Next, calculate the band audibility function (A_i).

$$A_i = L_i * K_i \tag{14.12}$$

Step 8:   Calculate the SII by summing the product of the band importance function (I_i) for each octave band as given in the standard and the band audibility function (A_i) calculated in Step 7, as

$$SII = \Sigma (I_i * A_i) \tag{14.13}$$

Note: Although components of equations 14.11 through 14.13 (E, D, K, L) have units of dB, the SII is dimensionless, so the dimensions were dropped after Step 6. Alternatively, one may think of the Band Importance Function (I_i) having units of dB^{-2}.

As stated in the standard, the SII resulting from the calculations described above and illustrated in Table 14.7 is an indication of the proportion of the speech cues that would be available to the listener for "average speech" under the speech/noise conditions for which the calculations were performed. (Intelligibility is greatest when SII = 1.0, indicating that all of the speech cues are reaching the listener and poorest when SII = 0.0, indicating that none of the speech cues are reaching the listener.) Unlike previous versions of the standard, no procedures are presented to allow conversion of the calculated SII score to a percentage that can be used to predict the proportion of the speech message that could be understood by the listener. The standard does, however, suggest that "good" communications can be achieved when the SII exceeds 0.75 and that communications will be "poor" when the SII is less than 0.45. When specific speech materials (e.g., nonsense syllables, phonetically balanced words, monosyllables, and others) are being considered, the standard provides alternative values for the Band Importance Function (I_i). As can be seen from the example calculations, neither listener is able to understand much of the speech directed toward him/her since the SII is less than 0.1 in both cases. For case 1, if the speaker-to-listener distance were reduced to 0.17 m (less than 1 foot), the SII would increase to 0.3179, a better than fourfold increase in the SII (but still inadequate for acceptable communications). The calculations are left as an exercise for the reader.

Direct Empirical Tests

When conditions are such that the PSIL and the SII (or any other predictive technique) are not suited to estimating intelligibility, then there is little choice but to conduct an empirical test to actually measure intelligibility. Fortunately, for such circumstances there exists a standard (ANSI S3.2-1989, *American National Standard Methods for Measuring the Intelligibility of Speech Over Communications Systems*), which not only provides guidance for conducting such tests but also provides sets of standard speech stimuli as well. The document, which is intended for designers and manufacturers of communications systems, provides valuable insight into the subject of speech intelligibility and how various factors associated with the speaker, transmission path/environment, and listener can affect it. Although space does not permit a detailed description of the procedures, the strategy involves presenting speech stimuli to a listener in an environment that replicates the conditions of concern and measuring how much of the speech message is understood. The speech stimuli may be produced by a trained talker speaking directly to the listener while in the same environment or via an intercom system. Alternatively, the materials may be recorded and presented electronically. Use of recorded stimuli and/or electronic presentation of the stimuli offers the greatest control over the speech levels presented to the listener.

Improving Communications and Signal Detection in Noise

Special-Purpose HPDs

Designers and manufacturers of hearing protectors have long recognized the need for improved communication in noisy environments and, as a result, have developed a number of electronic and mechanical devices intended to do just that. These devices are covered in detail in Chapter 10.

Final Recommendations for Nonverbal Signal Design

The detection of audible alarms and warnings will present a problem as long as excessive noise levels are prevalent in the workplace. However, based on the research and practitioner experience to date, it is possible to make some recommendations that can help reduce the severity of the problem.

- Use the most appropriate standard for guidance in selecting an alarm or warning device. Even though a standard may not exist that covers the specific application, those that do exist can still be used to guide the selection process.
- Limit the number of discrete signals to seven or eight. More than this could cause confusion as to the meanings of the signals.
- Ensure that the signals contrast well against the background noise and that the various signals sound sufficiently different to avoid confusion.
- Whenever possible, calculate the audibility of signals in the noise in which they are to be heard and verify the results using a listening test. Perhaps the most defensible calculation method is outlined in ISO 7731-1986(E), *Danger Signals for Work Places — Auditory Danger Signals* (ISO, 1986).
- Signal and noise levels *at the listener's location* should be used for all such calculations.
- Include material on the detection of auditory signals in the training provided to the workforce and stress the *positive* effects that proper use of HPDs can have on the detection of such signals.
- Encourage input and feedback from employees regarding the audibility of signals in their workplaces. Such information can identify problem areas before an accident occurs.
- Signals should be selected so that their perceived urgency matches the condition to which they call attention.
- Alarm and warning signals should have a fundamental frequency in the range between about 800 and 3000 Hz.
- Signals intended for use outdoors or that need to be heard over great distances should have fundamental frequencies below 1000 Hz so they will be less susceptible to atmospheric absorption and more easily diffracted around corners and barriers.
- Signals should be at least 6 to 10 dB (with 15 dB preferred) above their masked threshold, across their entire spectrum whenever possible.

- Signals that are more than 20 to 25 dB above their masked threshold may elicit an undesirable startle response from individuals in the area.
- Signal levels and frequency of occurrence should be closely monitored to ensure that they do not add to the noise exposure of the employees in the area. If the noise level is such that excessively high signal levels are required, then measures should be taken to reduce the background noise.

Final Recommendations for Speech Communications Design

As with the detection of nonspeech signals, verbal communication will always be problematic as long as excessive noise levels are so prevalent. Based on the preceding discussion, however, it is possible to make some recommendations that can help reduce the severity of the problem.

- Whenever possible, quantify the problem using the most appropriate method available (PSIL, SII, or other).
- Noise levels, and if possible, speech levels *at the listener's location* should be used for all such calculations.
- Whenever possible, decrease the distance between the speaker and the listener and encourage the use of hand and facial cues, although these are ancillary aids that should not take the place of verbal stimuli.
- Include material on speech intelligibility in noise in the training provided to the workforce and stress the *positive* effects that proper use of HPDs can have on the audibility of speech. Teach the tactics available to maximize intelligibility (see bullets relating to vocabulary construction).
- Encourage employees to speak more forcefully to overcome the tendency to lower the voice while wearing HPDs. However, increased vocal effort should not be continued for long periods because of the potential to irritate the speaker's vocal tract.
- Encourage input and feedback from employees regarding speech intelligibility in their workplaces. Such information can identify problem areas before an accident occurs.
- Select HPDs that are appropriate for the noise environment, and do not overprotect. Excessive HPD attenuation has the potential to reduce the level of the consonants so important to intelligibility to the point that these critical speech sounds become inaudible.
- Improve message content by encouraging and implementing consistent sentence construction for standard messages.
- Avoid the use of single letters, and use whole words (phonetic alphabet) or complete sentences whenever possible.

Special Considerations for Hearing-Impaired Workers

As mentioned previously, individuals suffering from a hearing loss pose a special problem in hearing auditory signals, including speech. Such individuals cannot be denied the opportunity to work, legally or ethically, simply because they have a hearing loss. On the other hand, if they cannot hear critical auditory signals, they could pose a risk to themselves and others. It is impossible to make any blanket recommendations for such individuals because of the lack of relevant research into the topic and because every situation is unique. Each situation must be examined separately and a solution acceptable to both the employer and employee ascertained. However, there are several avenues that can be pursued in an attempt to reach a solution.

- If the loss is not too severe, following the guidelines presented above may prove to be sufficient in many cases.
- Use of a special-purpose HPD or a communications headset (see Chapter 10) may also help the hearing-impaired employee hear the necessary auditory signals.
- Auditory alarms can be supplemented with visual displays. Color-coded lights can substitute for nonverbal auditory displays, and text-based displays can be used instead of verbal communications.
- In special cases, tactile (vibratory) alarm displays may be considered. The primary disadvantage is that these are personal displays that must be worn against the body.

References

Acton, W. I. (1970). "Speech Intelligibility in a Background Noise and Noise-Induced Hearing Loss," *Ergonomics 13*(5), 546–554.

ANSI (1989). "Methods for Measuring the Intelligibility of Speech Over Communications Systems," ANSI S3.2-1989, Acoustical Society of America, New York, NY.

ANSI (1997). "Methods for the Calculation of the Speech Intelligibility Index," ANSI S3.5-1997, Acoustical Society of America, New York, NY.

ANSI (1990). "Audible Emergency Evacuation Signal," ANSI S3.41-1990 (R1996), Acoustical Society of America, New York, NY.

Berger, E. H. (1987). "EARLog #18 — Can Hearing Aids Provide Hearing Protection?" *Am. Ind. Hyg. Assoc. J. 48*(1), A20–A21.

Coleman, G. J., Graves, R. J., Collier, S. G., Golding, D., Nicholl, A. G. McK., Simpson, G. C., Sweetland, K. F., and Talbot, C. F. (1984). "Communications in Noisy Environments," Institute of Occupational Medicine, Ergonomics Branch, Report Number TM/84/1, Edinburgh, UK.

Deatherage, B. H. (1972). "Auditory and Other Sensory Forms of Information Presentation," in *Human Engineering Guide to Equipment Design*, edited by H. P. Van Cott and R. G. Kincade, Wiley, New York, NY, 123–160.

DoD (1981). "Human Engineering Design Criteria for Military Systems, Equipment and Facilities," Department of Defense, MIL-STD-1472C, Washington, DC.

Edworthy, J., Loxley, S., and Dennis, I. (1991). "Improving Auditory Warning Design: Relationship Between Warning Sound Parameters and Perceived Urgency," *Human Factors 33*(2), 205–231.

Egan, J. P., and Hake, H. W. (1950). "On the Masking Pattern of a Simple Auditory Stimulus," *J. Acoust. Soc. Amer. 22*(5), 622–630.

Erber, N. P. (1969). "Interaction of Audition and Vision in the Recognition of Oral Speech Stimuli," *J. Speech and Hearing Res. 12*, 423–425.

Fidell, S. (1978). "Effectiveness of Audible Warning Signals for Emergency Vehicles," *Hum. Factors 20*(1), 19–26.

Fletcher, H. (1940). "Auditory Patterns," *Rev. Mod. Physics 12*, 47–65.

Gescheider, G. A. (1985). *Psychophysics-Method, Theory, and Application, Second Edition*, Lawrence Erlbaum Assoc., Hillsdale, NJ.

Haas, E. C., and Casali, J. G. (1995). "Perceived Urgency of and Response Time to Multi-Tone and Frequency-Modulated Warning Signals in Broadband Noise," *Ergonomics 38*(11), 2313–2326.

Hawkins, J. E., and Stevens, S. S. (1950). "The Masking of Pure Tones and of Speech by White Noise," *J. Acoust. Soc. Amer. 22*(1), 6–13.

Humes, L. E., and Bess, F. H. (1981). "Tutorial on the Potential Deterioration in Hearing Due to Hearing Aid Usage," *J. Speech and Hearing Res. 24*(1), 3–15.

ISO (1986). "Danger Signals for Work Places — Auditory Danger Signals," International Organization for Standardization, ISO 7731-1986(E), Geneva, Switzerland.

ISO (1987). "Acoustics — Audible Emergency Evacuation Signal," International Organization for Standardization, ISO 8201-1987(E), Geneva, Switzerland.

Kryter, K. D. (1974). "Speech Communication," in *Human Engineering Guide to Equipment Design*, edited by H. P. Van Cott and R. G. Kincade, Wiley, New York, NY, 161–226.

Kryter, K. D. (1985). *The Effects of Noise on Man*, Academic Press, Orlando FL.

Lane, H., and Tranel, B. (1971). "The Lombard Sign and the Role of Hearing in Speech," *J. Speech and Hearing Res. 14*, 677–709.

McMurtry, P. L., and Mershon, D. H. (1985). "Auditory Distance Judgments in Noise, With and Without Hearing Protection," in *Proceedings of the Human Factors Society 29th Annual Meeting*, Human Factors Society, Santa Monica, CA, 811–813.

Miller, G. A. (1956). "The Magical Number Seven, Plus or Minus Two: Some Limits on Our Capacity for Processing Information," *Psychological Rev. 63*, 81–97.

Miller, G. A., Heise, G. A., and Lighten, W. (1951). "The Intelligibility of Speech as a Function of the Context of the Test Materials," *J. Experimental Psychol. 41*, 329–335.

Morrison, H. B., and Casali, J. G. (1994). "Intelligibility of Synthesized Voice Messages in Commercial Truck Cab Noise for Normal-Hearing and Hearing-Impaired Listeners," in *Proceedings of the Human Factors and Ergonomics Society 38th Annual Meeting*, Human Factors and Ergonomics Society, Santa Monica, CA, 801–805.

NFPA (1979). "Local Protective Signaling Systems," National Fire Protection Association, ANSI/NFPA 72A-1979, Boston, MA.

OSHA (1983). "Occupational Noise Exposure; Hearing Conservation Amendment; Final Rule," Occupational Safety and Health Administration, 29CFR1910.95. *Fed. Regist. 48*(46), 9738–9785.

Patterson, R. D. (1982). "Guidelines for Auditory Warning Systems on Civil Aircraft," Civil Aviation Authority, Airworthiness Division, Paper 82017, Cheltanham, England.

Peterson, A., and Gross, E. Jr. (1978). *Handbook of Noise Measurement, Eighth Edition*, General Radio Co., New Concord, MA.

Pollack, I., and Pickett, J. (1964). "The Intelligibility of Excerpts from Conversational Speech," *Lang. and Speech 6*, 165–171.

Royster, J., Atack, R., Cook, G., Dobie, R., Glaser, R., and Miller, M. (1991). "The Hearing-Impaired Worker: Hearing Aids, Hearing Protectors, and Employment Criteria," *Spectrum* Suppl. 1, 8, p. 19.

SAE (1978). "Performance, Test, and Application Criteria for Electronically Operated Backup Alarm Devices," Society of Automotive Engineers, Inc. ANSI/SAE J994b-1978. Warrendale, PA.

Sanders, M. S., and McCormick, E. J. (1993). *Human Factors in Engineering and Design*, McGraw-Hill, New York, NY.

Small, A. M., and Gales, R. S. (1991). "Hearing Characteristics," in *Handbook of Acoustical Measurement and Noise Control, Third Edition*, edited by C. M. Harris, McGraw-Hill, New York, NY, 17.1–17.25.

Sorkin, R. D. (1987). "Design of Auditory and Tactile Displays," in *Handbook of Human Factors*, edited by G. Salvendy, Wiley, New York, NY, 549–576.

Sorkin, R. D., and Kantowitz, B. H. (1987). "Speech Communication," in *Handbook of Human Factors*, edited by G. Salvendy, Wiley, New York, NY, 294–309.

UL (1981). "Audible Signal Appliances," Underwriters Laboratories Inc., ANSI/UL 464-1981, Melville, NY.

Wegel, R. L., and Lane, C. E. (1924). "The Auditory Masking of One Pure Tone by Another and its Probable Relation to the Dynamics of the Inner Ear," *Physiological Rev. 23*, 266.

Wilkins, P. A., and Martin, A. M. (1978). "The Effect of Hearing Protectors on the Perception of Warning and Indicator Sounds — A General Review." Technical Report No. 98, Institute of Sound and Vibration Research, Univ. Southampton, England.

Wilkins, P. A., and Martin, A. M. (1982). "The Effects of Hearing Protection on the Perception of Warning Sounds," in *Personal Hearing Protection in Industry*, edited by P. W. Alberti, Raven Press, New York, NY, 339–369.

Wilkins, P.A., and Martin, A. M. (1985). "The Role of Acoustical Characteristics in the Perception of Warning Sounds and the Effects of Wearing Hearing Protection," *J. Sound and Vibration 100*(2), 181–190.

The Noise Manual, revised 5th edition, edited by E.H. Berger,
L.H. Royster, J.D. Royster, D.P. Driscoll, and M. Layne
©2003 American Industrial Hygiene Association

15

Community Noise

Dennis P. Driscoll, Noral D. Stewart, and Robert R. Anderson

Contents

Concepts in Community Noise

Introduction

There are historic references to noise being a problem in cities. In the 1920s, noise sources such as new modes of transportation, ventilation systems, industrial plants, and loudspeakers were becoming more common. The coming of jet aircraft renewed interest in environmental noise in the 1960s. In fact, transportation noise as a whole is a major source of community annoyance. However, this chapter will focus on sources the industrial hygienist can control, namely industrial noise.

In 1972 the United States Congress affirmed the growing danger that noise presents to the health and welfare of the nation's population, particularly in urban areas, through its Congressional finding and statement of policy (Anon., 1972). Over 25 years later this statement is being echoed by the recent formation of public action and awareness groups,[1] and the increased attention in the media to noise in our society. This attention has led to the call for rational environmental noise standards in local communities (Erdreich, 1998).

The primary reasons for limiting noise in the community are to reduce speech and/or sleep interference, and to limit annoyance. People are not usually annoyed if the sound is of the level and quality they expect in their community, and does not interfere with speech or sleep. A side effect of annoyance is stress that can affect some health conditions. Besides the physical effect on people, increased noise in a previously quiet community can change the value of property.

The quality of the sound and a community's characteristics also must be considered. Much depends on the existing conditions and expectations of the community. In densely populated areas, the emphasis is on controlling the overall growth of noise. However, in quieter, less densely populated areas, a new noise that might go undetected in a noisier community can become very noticeable and cause complaints. Often, in these quieter areas, the quality of the sound is as

---

[1] A clearinghouse of related information is available from the public awareness group at Noise Pollution Clearinghouse, Montpelier, VT, and through their website at www.nonoise.org.

important as the quantity. Unusual sounds such as discrete tones and impulsive sounds need more attention. Sometimes tones are masked near a source, but clearly audible in quieter areas farther away. The frequency content of sound changes with distance. A source with an acceptable spectrum nearby can sound like a rumble at greater distances. Sounds with strong low-frequency content require special attention (Berglund and Lindvall, 1995; Berglund et al., 1996). Most criteria for community noise based on overall sound levels measured outdoors assume a balanced sound spectrum. When there is strong low-frequency dominance, the sound can more easily penetrate homes. Thus, such sounds are more annoying indoors than a sound of similar overall level but balanced spectrum.

Congress intended that states and cities retain primary responsibility for control of community noise when it passed the Noise Control Act of 1972. This has resulted today in a diversity of noise regulations among local communities and states, as well as in many locations that lack any noise ordinances at all. The widely varying approaches to regulating noise in communities pose a significant challenge to companies that operate multiple facilities, and to the people charged with the responsibility to assess compliance with those regulations.

An industrial hygienist may need to evaluate community noise for several reasons:
- Compliance of noise produced by facilities operating in regions with local ordinances,
- Determination of acceptable noise levels and noise characteristics for new equipment,
- Evaluation of site suitability for a new facility,
- Resolution of complaints from neighbors.

Research on community noise has concentrated on sources related to transportation (airports, trains, highway and street traffic, etc.), military (aircraft low-level fly-overs, heavy vehicles maneuvering, firing ranges, etc.), and ventilation systems (outside air conditioners and blowers, noise from ventilation stacks, etc.). These sources are widespread, affect large areas, and there are readily available mechanisms to fund the research. This research has emphasized establishing acceptable quantities of sound for typical areas that are affected, and reducing sound accordingly. Less research is available on isolated and unique noise sources in quieter communities where the noise is unexpected. An industrial hygienist is most likely to be faced with noise from an industrial plant disturbing a few local neighbors. However, in some cases, distinctive or new sounds can annoy neighbors several kilometers away. In some circumstances people farther from the source can be more annoyed than those near it.

Measures of Noise in the Community

The basic noise measures or descriptors used in community noise are discussed in Chapter 3. These include the sound level, the *equivalent continuous sound level* ($L_{eq,T}$) (now called time-average sound level in most standards), and the sound exposure level (SEL). Overall sound levels for community noise are usually A-weighted. The C-weighted sound level is used in special circumstances related to impulsive noise. A 3-dB (equal-energy) exchange rate is always used for time-average sound levels. Octave-band or 1/3 octave-band levels are sometimes used to evaluate sound quality.

A long-term average sound level over a 24-hour period can be used to describe community noise. The *day–night average sound level* (DNL), symbolized as L_{dn}, has a 10-dBA night-time penalty added to all sound between 10:00 p.m. and 7:00 a.m. (Equation 15.1a). A variation of this adds an evening penalty of 5 dBA from 7:00 p.m. until 10:00 p.m. It is used primarily in California, where it is called the *community noise equivalent level* (CNEL). Communities with very different noise characteristics can have the same DNL. Without a strong local noise source, such as an airport, freeway, or industrial plant, the expected DNL in communities of at least 200 people per km^2 can be estimated using Equation 15.1b (EPA, 1974).

$$L_{dn} = 10 \log 1/24[15 \times 10^{(L_d/10)} + 9 \times 10^{(L_n+10)/10}] \text{ dBA} \qquad \textbf{(15.1a)}$$

where, L_d is the equivalent-continuous sound level from 7 a.m. until 10 p.m.

L_n is the equivalent-continuous sound level from 10 p.m. until 7 a.m.

$$L_{dn} = 26 + 10 \log (\text{number of people/km}^2) \text{ dBA} \qquad \textbf{(15.1b)}$$

Community sound levels are also sometimes analyzed using statistical measures. The sound level is sampled using a fast or slow time response, or sometimes very short samples of equivalent (time-average) sound level. The levels exceeded various percentages of time are calculated, with the results, which are called percentile levels, used to give an indication of the variation in the sound. The level exceeded 90% of the time is often used as a measure of the background sound present without transient or intermittent sounds. Many early regulations, before the widespread availability of averaging meters, were based on the sound level exceeded 10% of a measurement period. The number of samples measured should be at least 10 times the difference in decibels between the highest and lowest level.

United States Federal Government Guidelines and Regulations

Most United States federal guidelines for community noise are based on the DNL (EPA, 1974). The Environmental Protection Agency (EPA) recommended that DNL should be kept below 55 dBA in residential areas "to protect public health and welfare with an adequate margin of safety" (EPA, 1974). This level corresponds to that normally present in a typical suburban community of about 770 people per km^2. This goal did not consider economic or technological feasibility and was not intended as a regulation. The study recognized that many people lived in both quieter and noisier areas, including densely populated urban areas. It provided methods to evaluate problems and the potential for noise complaints based on DNL. These involved adjusting or normalizing the DNL for specific circumstances before comparing the DNL to criteria based primarily on expectations in densely populated urban areas.

The United States Department of Housing and Urban Development (HUD) has noise criteria for areas where it funds or finances housing (HUD, 1979). These recognize the need to build housing in densely populated areas where the desirable noise levels of DNL 55 cannot be achieved. They are based on surveys of the percentage of people highly annoyed by existing noise in areas where they live. Sound levels up to DNL 65 dBA are considered normally acceptable by HUD. Sound levels between DNL 65 and DNL 75 are normally unacceptable. However, housing can be funded when steps are taken to reduce the noise reaching the interior of homes. For single-family homes, there is often a requirement for barriers to reduce outside noise over DNL 70. The Department of Defense and Federal Aviation Administration also use DNL 65 as their regulatory goal. They do not recognize significant noise impacts from aircraft or military activities below this level. The Federal Highway Administration (FHA) uses a 1-hour equivalent (time-average) sound level criteria of 67 dBA to determine when to consider noise barriers for new highway projects. Before actually building barriers, the project has to further qualify based on the cost and benefit of the barrier per protected home.

DNL and normalized DNL work best to characterize the long-term acoustical character of a community as influenced by noise sources that are continually present as steady-state sounds or frequently occurring events over most of the day every day. DNL does not work well for infrequently occurring loud sounds that may be disturbing to a community without strongly affecting the long-term average sound level. Even the normalized DNL for continuous sounds may not always properly account for unique characteristics of the sound. For instance, the correction for discrete tone sounds may be insufficient (see *Assessment for Prediction of Community Response*). DNL also is not a practical measure for enforcement use by communities because of the long-term evaluations needed to establish it.

Local Noise Ordinances

Noise from industry and business in North America is regulated, if at all, primarily by local governments. There are state noise regulations in approximately 13 states; however, enforcement is often tenuous at best. Community ordinances can be classified as general nuisance ordinances or as a combination of nuisance and quantitative components. A nuisance ordinance is typically a prohibition of making or allowing to be made any unreasonable or excessive noise. Because this type of ordinance does not specify a sound level limit, compliance is a matter of satisfying subjective response by typically two or more listeners. Quantitative ordinances specify sound level limits and usually provide stronger legal control over undesirable sound levels than is attainable with an ordinance containing only nuisance provisions. However, these ordinances can vary greatly in the measurements required. They can range from a single not-to-exceed A-weighted sound level at a nonspecified location, to a matrix of source and receiver land-use categories with different limits for day and night and requirements for averaging or sampling over specified periods. Some also can contain octave or 1/3 octave-band criteria, or criteria to evaluate discrete-tone noises.

Quantitative ordinances usually require measurements over periods of less than an hour. The measurement method may be a simple A-weighted sound level, an equivalent (time-average) sound level, and/or a level exceeded 10% or 50% of the measurement period. If the measurement does not involve sampling or averaging, the regulation may have different limits depending on the duration of the noise. If the primary limit is based on levels exceeded 10% or 50% of the time, there is often a higher limit never to be exceeded. Sometimes the ordinance will only mention a level not to be exceeded using slow response. The limits in such cases are often too low for a sound of short duration or too high for continuous sounds.

The primary limits for sound entering residential areas are usually 55 to 60 dBA in the daytime, and 50 to 55 dBA at night as measured at the boundary or property line of the complainant. It is worth noting that some local ordinances impose limits on noise at the boundary of the source property. Sometimes, nighttime limits are as low as 45 dBA especially in rural areas or less densely populated cities, and daytime limits are as high as 65 dBA especially in densely populated areas (EPA, 1975). Ordinances will usually allow higher levels for sound entering commercial or industrial properties. Sometimes, ordinances allow more noise entering residential areas from industrial properties than from other residential properties. The definition of daytime and night-time varies, but night is most commonly 10:00 p.m. until 7:00 a.m. Without access to expert advice, local governments sometimes set limits unreasonably high or low, or require instruments no longer available. Because conditions and expectations vary within different parts of most local jurisdictions, and the ordinances must usually be kept simple, they cannot prevent all problems. Sound levels that comply with the ordinance can still be objectionable to a portion of the population. It is particularly difficult to prevent problems from distinctive sounds like discrete tones without some complexity in the ordinance.

Voluntary Noise Measurement and Assessment Standards

Where there is no regulatory requirement, or when there are complaints in spite of regulatory compliance, the investigator must determine the best way to evaluate the noise. Sometimes, a voluntary standard developed by a national or international group such as the International Standards Organization (ISO) can help. Many countries (but not the United States) have adopted a three-part international standard for description and measurement of environmental noise which addresses (1) basic quantities and procedures (ISO, 1982), (2) acquisition of data pertinent to land use (ISO, 1987a, 1998), and (3) application to noise limits (ISO, 1987b).

In North America, the Acoustical Society of America develops American National Standards Institute (ANSI) standards related to community noise. Additional standards are also provided by the American Society for Testing and Materials (ASTM). The ANSI standards concentrate primarily on measurement and evaluation methods rather than setting specific criteria for acceptability based on those methods.

ANSI S12.9, *American National Standard Quantities and Procedures for Description and Measurement of Environmental Sound,* is a five-part standard which (in separate documents) addresses (1) descriptors for noise (ANSI, 1988), (2) measurement of long-term, wide-area sound (ANSI, 1992), (3) short-term measurements with an observer present (ANSI, 1993), (4) noise assessment and prediction of long-term community response (ANSI, 1996), and (5) sound level descriptors for determination of compatible land use (ANSI, 1998). Part Four provides adjustments to measured sound levels for certain sound characteristics such as tonality and impulsiveness. Note that long-term community response and land-use compatibility are best used as indicators of acceptance of existing noise by people who choose to live with it. They may not indicate the reaction of an existing community to a new noise. The land-use compatibility and community-response criteria assume noises without characteristics such as tonality, impulsiveness, low-frequency dominance, or clearly heard speech or music.

ASTM E1686 *Standard Guide for Selection of Environmental Noise Measurements and Criteria* (ASTM, 1996a) discusses additional methods to measure and evaluate community noise which are not covered in this chapter. Other ASTM standards include the guide E1014 for measuring sound levels using simple instruments (ASTM, 1984), guide E1780 for measuring outdoor sound received from a nearby fixed source (ASTM, 1996c), and guide E1779 for preparing a measurement plan for conducting outdoor sound measurements (ASTM, 1996b). ASTM E1503 *Standard Test Method for Conducting Outdoor Sound Measurements Using a Digital Statistical Analysis System* (ASTM, 1992) provides a detailed method for using sophisticated instruments in major studies.

Sometimes the sound emitted by a source must be established to allow calculation of the sound expected at a distant location. Standards for individual sources include ANSI/ASTM PTC 36 (ASME, 1985), ANSI S12.34 (ANSI, 1997a), and ANSI S12.36 (ANSI, 1997b). ISO 8297 (ISO, 1994) provides a method to determine the sound emission of multi-source industrial plants.

Factors Other Than Absolute Sound Level Influencing Community Reaction to Noise

Most noise regulations are based on sound level, possibly with lower limits at night or penalties for sounds with tonal or impulsive characteristics. However, research indicates many important factors influence community reaction and annoyance produced by noise. Those identified by the EPA (1974) were:

- Frequency content of the noise,
- Duration of the noise,
- Time of day the noise occurs,
- Time of year the noise occurs,
- History of prior exposure to the noise source,
- Perceived attitude of the noise source owner,
- Special characteristics of the noise that make it especially irritating,
- Ratio of intruding noise level to normal background noise level.

Other studies have identified additional factors that are very much related to community reaction and annoyance. These include whether the complainant believes s/he is being ignored or treated unfairly, or perceives the noise as:

- Unnecessary, or unnecessarily loud,
- A threat to personal health or safety,
- A threat to economic investment (property value),
- Beyond his or her control.

A most important factor is the difference in sound level between a new noise and other expected and existing noise in the neighborhood. The most significant finding of the EPA community reaction studies (EPA, 1974) was that widespread complaints and legal actions are likely when the average level of nondistinctive noise from a single source is regularly more than 5 dB above the average level of other existing sounds in the community. Vigorous community action results for differences of 20 dB. Some noises such as discrete tones are more irritating or difficult to ignore because of the way they sound. People expect not only quiet, but a pleasant sound quality if sound is audible. These unpleasant and distinctive sounds often cause complaints if they are detectable at any level. The acoustical designers of vehicles, appliances, and other products today spend much of their effort on "sound quality." Some common industrial sources such as high-pressure or material-handling fans or positive-displacement blowers produce strong discrete tones. Power presses can produce repetitive impulsive sounds. Speech and music have information content that makes them difficult to ignore. These factors affect the quality of the sound in the community even at otherwise acceptable levels.

Factors and Conditions Affecting
Sound Propagation Outdoors

As sound propagates outdoors it generally decreases in magnitude with increasing distance from the source; however, the attenuation is not totally a function of spherical divergence. There are several meteorological and physical conditions that affect the rate of attenuation. The meteorological conditions include variations in air temperature with increased elevation, relative humidity, wind speed and direction, and atmospheric factors such as cloud coverage. The physical effects include topography, natural and artificial barriers, and vegetation.

Often a primary question one needs to answer is what will be the effect on community noise when an industrial plant is built, expands, or adds new equipment outside the building, or a residential subdivision encroaches upon the facility's property line? To answer this question it is important to know what factors affect outdoor sound propagation, and how to estimate attenuation to select locations. ANSI S12.18, *American National Standard for Outdoor Measurement of Sound Pressure Level (SPL)* describes procedures for outdoor sound measurement, including a discussion of the attenuation effects due to the various elements mentioned above (ANSI, 1994). This standard is useful, not only for measurement procedures, but also for estimating SPLs at different locations from the source. For sound radiating from a point source in a free field (directivity factor, Q=1), the SPL per octave band at a given distance may be calculated from:

$$L_p = L_W - A_{total} - 10.9, \text{ dB} \tag{15.2}$$

where,

$L_p =$ the octave-band sound pressure level, in dB, at the location of interest,

$L_W =$ the octave-band sound power level (PWL) of the source, in dB, and

$A_{total} =$ the total attenuation at each octave band, in dB.

The total attenuation (A_{total}) for each octave band in Equation 15.2 is calculated by:

$$A_{total} = A_{div} + A_{air} + A_{env} + A_{misc}, \text{ dB} \tag{15.3}$$

where,

A_{div} is the attenuation due to geometrical divergence,

A_{air} is the air absorption,

A_{env} is the sound reduction due to the effects of the environment, and

A_{misc} is the attenuation resulting from all other factors, such as foliage, barriers, etc.

Because high-frequency sounds have relatively short wavelengths their sound energy will decrease rapidly with increasing distance due to atmospheric absorption. Conversely, low-frequency sounds with much longer wavelengths will often carry several kilometers from the source and are usually the cause of complaints from citizens. This variation by frequency must be accounted for when calculating the total attenuation. Once the individual attenuation values are known for each octave band, they can be logarithmically added together using Equation 2.11, and the resultant value may be used in Equation 15.2 along with the known PWL to estimate the SPL (see example problem presented later in this chapter).

Geometrical Divergence (A_{div})

Geometrical divergence, often termed spreading losses, occurs as sound waves propagate and expand from a source, and in turn become less intense as they dissipate over larger spherical areas. The divergence is not a function of frequency, and attenuation is estimated by:

$$A_{div} = 20 \log (r/r_0), \text{ dB} \tag{15.4}$$

Where,

$r =$ distance from the point source in meters (m), and

$r_0 =$ reference distance of 1 m.

For distances far from the source, the geometrical divergence results in a 6-dB decrease per doubling of distance from a point source, which equates to a 20-dB decrease for each tenfold increase of distance. For a line source, such as a busy highway or long runs of noisy pipelines stretching perpendicular to the measurement location (e.g., a petrochemical plant), the geometrical divergence will be 3-dB decrease per doubling of distance.

Air Attenuation or Atmospheric Absorption (A_{air})

Sound energy decreases in a quiet calm atmosphere by two mechanisms: (1) heat conduction and viscosity in the air, and (2) relaxation of air molecules as they vibrate (Kurze and Beranek, 1988). The atmospheric absorption losses depend on frequency, temperature, and relative humidity. Of these three factors, relative humidity is the dominant variable, followed by the frequency and then the temperature.

For various temperatures the attenuation due to air absorption may be determined by (Piercy and Daigle, 1991):

$$A_{air} = \alpha'r/1000 \text{ dB} \tag{15.5}$$

where,

$\alpha' =$ the air attenuation coefficient, dB/km, and

$r =$ distance from source to receiver, m.

The air attenuation coefficient values are presented in Table 15.1 for various temperatures and relative humidity, as a function of frequency (ANSI, 1994). Should temperature and humidity values differ from those presented in Table 15.1, interpolation may be used to estimate the air attenuation coefficients. Calculations employing Equation 15.5 reveal that air attenuation becomes significant at distances over 300 m and frequencies above 1000 Hz. For example, at 20°C and relative humidity of 70%, the attenuation at 1000 Hz is 5.0 dB/km. At 200 meters this amounts to an attenuation of 1.0 dB. However, at 2 km the attenuation is a significant 10 dB. For dry air with a relative humidity of 10%, these attenuation values are 2.8 dB and 28 dB for 200 m and 2 km, respectively. For the same 10% relative humidity at 20°C, at a distance of 2 km using the absorption coefficients at 250 Hz and 2000 Hz, these attenuation values are 3.2 dB and 90 dB, respectively. Clearly, as distance from the source increases, there is a significant increase in sound attenuation at the higher frequencies with a relatively small increase at the lower frequencies (see Table 15.1).

TABLE 15.1
Air attenuation coefficients ∝′, at 1 atmosphere for sound propagation in open air (db/km).*

Temperature	Relative Humidity (Percent)	Octave-Band Frequency (Hz)					
		125	250	500	1000	2000	4000
	10	0.96	1.8	3.4	8.7	29	96
	20	0.73	1.9	3.4	6.0	15	47
30°C	30	0.54	1.7	3.7	6.2	12	33
(86°F)	50	0.35	1.3	3.6	7.0	12	25
	70	0.26	0.96	3.1	7.4	13	23
	90	0.20	0.78	2.7	7.3	14	24
	10	0.78	1.6	4.3	14	45	109
	20	0.71	1.4	2.6	6.5	22	74
20°C	30	0.62	1.4	2.5	5.0	14	49
(68°F)	50	0.45	1.3	2.7	4.7	9.9	29
	70	0.34	1.1	2.8	5.0	9.0	23
	90	0.27	0.97	2.7	5.3	9.1	20
	10	0.79	2.3	7.5	22	42	57
	20	0.58	1.2	3.3	11	36	92
10°C	30	0.55	1.1	2.3	6.8	24	77
(50°F)	50	0.49	1.1	1.9	4.3	13	47
	70	0.41	1.0	1.9	3.7	9.7	33
	90	0.35	1.0	2.0	3.5	8.1	26
	10	1.3	4.0	9.3	14	17	19
	20	0.61	1.9	6.2	18	35	47
0°C	30	0.47	1.2	3.7	13	36	69
(32°F)	50	0.41	0.82	2.1	6.8	24	71
	70	0.39	0.76	1.6	4.6	16	56
	90	0.38	0.76	1.5	3.7	12	43

*Note: Air attenuation coefficient values of temperature and relative humidity (or frequency) intermediate to those shown in the table may be obtained by interpolation.
Source: From ANSI S12.18-1994: "Outdoor Measurement of Sound Pressure Level," with permission.

Attenuation Due to Environmental Effects (A_{env})

In addition to divergence and air absorption, sound propagating from a source is also attenuated by the environment, such as the ground, wind, and temperature gradients. Figure 15.1 illustrates the propagation path from source to receiver. The magnitude of the reflected sound will depend upon the type of ground surface, the angle of incidence (ψ), and frequency (Piercy and Daigle, 1991). ANSI S12.18 classifies ground surfaces for grazing angles less than 20° as follows (ANSI, 1994):

- *Hard Ground*—Open water, asphalt, or concrete pavement, and other ground surfaces having very low porosity tend to be highly reflective, absorbing very little acoustic energy upon reflection. Tamped ground, for example, as often occurs around industrial sites, can be considered as hard ground.

Attenuation by the Environment (A_{env})

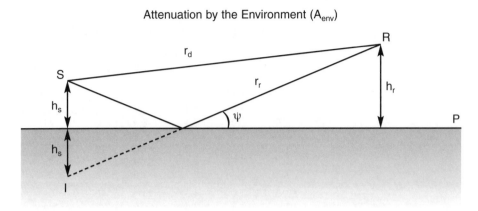

Figure 15.1 — Paths for propagation from source S to receiver R. The direct ray is r_d, and the ray reflected from the plane P (which effectively comes from image source I) is r_r, whose length is measured from plane P to R. Source: From Piercy and Daigle (1991), with permission.

- *Soft Ground*—Ground covered by grass, shrubs, or other vegetation, and all other porous grounds suitable for the growth of vegetation such as farming land.
- *Very Soft Ground*—New-fallen snow is even more absorptive at low frequencies than grass-covered ground, as is ground covered in pine needles or similarly loose material. It is recommended by ANSI that measurements above snow-covered ground be avoided unless operation of the sound source is intimately tied with the ground condition.
- *Mixed Ground*—A ground surface which includes both hard and soft areas.
- *At angles off the ground greater than 20°*, which can commonly occur at short ranges or in the case of elevated sources, soft ground becomes a good reflector of sound and can be considered hard ground.

Sound outdoors reaches a receiver by both direct and reflected paths. For distances of approximately 100 m or less, termed short-range propagation, the attenuation values are primarily due to ground effects and the presence of any barriers. Table 15.2 presents the attenuation values at each octave band from 125 – 4000 Hz for hard, soft, and very soft ground surfaces. For mixed ground conditions the attenuation values will need to be calculated for both hard and soft surface areas. A_{env} then becomes the value interpolated between these two results based on the proportion of soft to hard ground.

For distances over 100 m, termed long-range propagation, the wind and temperature conditions will play an important role, while barriers and ground effects have minimal influence. The effects of wind and temperature on sound transmission are described later in this chapter; however, for purposes of determining the long-range attenuation of sound these conditions should be assumed to be

TABLE 15.2
Values of environmental attenuation A_{env} in decibels for short-range propagation [r < 100 m (300 ft)].*

Hard ground (asphalt, concrete)	
$(r_r - r_d) \ll$ all λ	$(r_r - r_d) \gg$ all λ
– 6.0	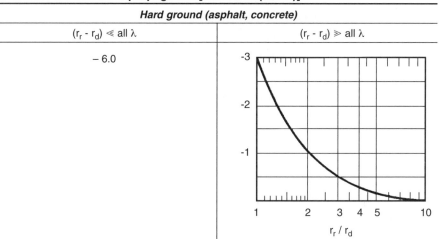

Soft ground (grass, vegetation), h_r = 1.8 m

Source Height (m)	Distance (m)	Frequency (Hz)					
		125	250	500	1000	2000	4000
0.01	10	- 5.7	- 5.0	- 3.6	- 1.4	1.1	4.1
	20	- 5.6	- 4.6	- 1.8	1.9	5.1	8.5
	40	- 5.5	- 3.9	- 1.4	6.7	10.1	13.7
	60	- 5.4	- 3.3	4.2	9.8	13.2	16.9
	80	- 5.4	- 2.7	6.8	12.2	15.5	19.3
	100	- 5.3	- 2.2	9.2	14.0	17.4	21.1
0.3	10	-5.4	- 4.3	- 0.9	5.9	- 2.5	- 1.9
	20	- 5.4	- 4.0	- 0.1	6.3	- 0.1	- 3.0
	40	- 5.4	- 3.4	2.9	10.2	4.1	- 2.9
	60	- 5.3	- 2.8	5.8	13.1	7.1	- 0.4
	80	- 5.2	- 2.2	8.4	15.3	9.3	1.7
	100	- 5.2	- 1.7	10.8	17.1	11.1	3.4
1.2	10	- 4.0	2.0	0.1	- 3.0	- 3.0	- 3.0
	20	- 4.8	- 1.9	7.5	- 2.7	- 3.0	- 3.0
	40	- 4.9	- 2.1	6.9	0.5	- 3.0	- 3.0
	60	- 4.9	- 1.6	9.1	2.9	- 3.0	- 3.0
	80	- 4.8	-1.0	11.6	4.8	- 2.8	- 3.0
	100	- 4.8	- 0.5	13.8	6.4	- 1.5	- 3.0

— continued on next page —

*Note: Refer to Figure 15.1 for illustration of r_d and r_r, which are the paths for sound wave propagation from source to reviewer.
Source: From Piercy and Daigle (1991), with permission.

TABLE 15.2 — continued
Values of environmental attenuation A_{env} in decibels for short-range propagation [r < 100 m (300 ft)].*

		Very soft ground (snow, pine forest), $h_r = 1.8$ m					
Source	*Distance*	*Frequency (Hz)*					
Height (m)	*(m)*	*125*	*250*	*500*	*1000*	*2000*	*4000*
0.01	10	- 3.1	0.8	3.9	6.0	7.3	7.0
	20	- 1.5	5.2	8.6	10.9	12.3	11.9
	40	1.4	11.1	14.0	16.3	17.7	17.3
	60	3.9	14.8	17.3	19.6	21.0	20.7
	80	6.2	17.3	19.7	22.0	23.4	23.1
	100	8.4	19.3	21.6	23.8	25.3	24.9
0.3	10	- 2.3	2.8	5.0	- 0.8	- 3.0	- 3.0
	20	- 0.8	7.0	9.1	2.9	- 2.9	- 3.0
	40	2.0	12.8	14.2	7.9	1.4	- 3.0
	60	4.6	16.5	17.5	11.2	4.5	- 1.3
	80	6.9	19.0	18.2	13.5	6.8	0.8
	100	9.1	21.0	21.7	15.4	8.6	2.6
1.2	10	0.1	4.5	- 2.5	- 2.5	- 2.5	- 2.5
	20	0.9	7.0	- 0.7	- 3.0	- 3.0	- 3.0
	40	3.6	11.6	3.3	- 3.0	- 3.0	- 3.0
	60	6.3	14.8	6.3	- 0.6	- 3.0	- 3.0
	80	8.7	17.1	8.5	- 1.5	- 3.0	- 3.0
	100	10.9	18.9	10.3	3.2	- 2.6	- 3.0

*Note: Refer to Figure 15.1 for illustration of r_d and r_r, which are the paths for sound wave propagation from source to reviewer.
Source: From Piercy and Daigle (1991), with permission.

advantageous to sound propagation. Toward long-range propagation, the distance between source and receiver is divided into three zones, as depicted in Figure 15.2. The environmental factor for each zone is as follows (Piercy and Daigle, 1991):

1. The *source zone* covers a distance of $30h_s$ between the source and receiver (see Figure 15.2), with a maximum of r, where h_s is the source height and r is the distance from the source S to receiver R.

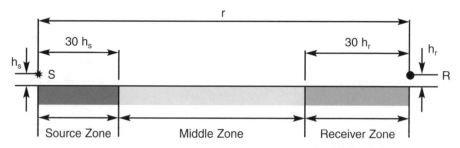

Figure 15.2 — Three zones between a source *S* and receiver *R* separated by distance *r*, used in determining the ground attentuation A_{env} at long ranges.
Source: From Piercy and Daigle (1991), with permission.

2. The *receiver zone* starts at the receiver and stretches back a distance of 30 h$_r$, with a maximum of r, where h$_r$ is the receiver height.
3. The *middle zone* covers the region between the source and receiver zones.

The surface area around each zone has the following *ground factor* G:
Hard ground: G = 0,
Soft ground: G = 1,
Mixed ground: G equals the fraction of the ground that is soft.
Note: For very soft ground there is no available value. However, it is suggested a value of 1 be used. The user is cautioned that using a factor of 1 for very soft ground will underestimate the actual ground attenuation, particularly in the lower frequency range from 100 – 500 Hz.

For the octave-band environmental attenuation values at long-range, Table 15.3 is utilized as follows:

TABLE 15.3
Expressions to be used in calculating the octave-band environmental attenuation (A$_{env}$) in decibels at long range.*

Octave-Band Frequency (Hz)	A$_s$ and A$_r$ (dB)	A$_m$ (dB)
63	- 1.5	- 3e
125	(a)(G) - 1.5	- 3e(1 - G)
250	(b)(G) - 1.5	- 3e(1 - G)
500	(c)(G) - 1.5	- 3e(1 - G)
1000	(d)(G) - 1.5	- 3e(1 - G)
2000	(1 - G)(- 1.5)	- 3e(1 - G)
4000	(1 - G)(- 1.5)	- 3e(1 - G)
8000	(1 - G)(- 1.5)	- 3e(1 - G)

Distance r(m)	Source or Receiver Height (m)				
	0.5	1.5	3.0	6.0	> 10.0
	Factor a				
50	1.7	2.0	2.7	3.2	1.6
100	1.9	2.2	3.2	3.8	1.6
200	2.3	2.7	3.6	4.1	1.6
500	4.6	4.5	4.6	4.3	1.6
> 1000	7.0	6.6	5.7	4.4	1.7
	Factor b				
50	6.8	5.9	3.9	1.7	1.5
100	8.8	7.6	4.8	1.8	1.5
> 200	9.8	8.4	5.3	1.8	1.5

— continued on next page —

*G is the ground factor, h is height, and r is distance from source to receiver. The subscripts s, r, and m indicate source, receiver, and middle zones, respectively. (See Figure 15.2.) The factor e is equal to {1 - [30(h$_s$ + h$_r$)/r]}.
Source: From Piercy and Daigle (1991), with permission.

TABLE 15.3 — continued
Expressions to be used in calculating the octave-band environmental attenuation (A_{env}) in decibels at long range.*

Distance (m)	Source or Receiver Height (m)				
	0.5	*1.5*	*3.0*	*6.0*	*> 10.0*
Factor c					
50	9.4	4.6	1.6	1.5	1.5
100	12.3	5.8	1.7	1.5	1.5
> 200	13.8	6.5	1.7	1.5	1.5
Factor d					
50	4.0	1.9	1.5	1.5	1.5
> 100	5.0	2.1	1.5	1.5	1.5

*G is the ground factor, h is height, and r is distance from source to receiver. The subscripts s, r, and m indicate source, receiver, and middle zones, respectively. (See Figure 15.2.) The factor e is equal to $\{1 - [30(h_s + h_r)/r]\}$.
Source: From Piercy and Daigle (1991), with permission.

Step 1: Determine A_s, which is the source zone attenuation portion of A_{env}, using the appropriate ground factors,

Step 2: Determine A_r, which is the receiver zone attenuation portion of A_{env},

Step 3: Calculate A_m, which is the middle zone attenuation portion of A_{env}. Note: for the middle zone to exist, $r > 30(h_s + h_r)$ must be satisfied,

Step 4: The total A_{env} in any octave band will be:

$$A_{env} = A_s + A_r + A_m \qquad (15.6)$$

EXAMPLE 15.1, Predicting Sound Levels at the Property Line

Consider the following example:

Management of a manufacturing plant plans an expansion that will include a large gas turbine located in the center of a 20 m × 20 m concrete skid or pad outside the new building structure. It is anticipated the turbine's exhaust will be a dominant source of noise and could significantly impact a residential area located at the facility's property line 1450 meters away. The point of the turbine discharge is 3 m above grade and the receiver height is 1.5 m. The ground surface area around the concrete skid and at the receiver is grass, while the ground cover between the source and receiver zones is 75% grass and 25% asphalt parking lot. Finally, there is no foliage or trees between the source and receiver locations. To investigate whether a potential community noise problem will result, it is necessary to estimate the overall A-weighted sound level at the property line. The turbine manufacturer reports the following exhaust sound power levels per octave band:

Octave-band center frequency (Hz):	125	250	500	1000	2000	4000
Exhaust L_W (dB):	144	145	144	138	137	134

Equation 15.2 is used to calculate the SPL at the location of interest, however, Equation 15.3 is needed to first determine the total attenuation (Recall $A_{total} = A_{div} + A_{air} + A_{env} + A_{misc}$). Note: many of the attenuation factors are frequency-dependent. For purposes of this example and to demonstrate use of the equations and tables, *all values will be estimated for 250 Hz.*

Step 1: Use Equation 15.4 to predict A_{div}, the attenuation due to divergence

$A_{div} = 20 \log r/r_0$
$= 20 \log (1450/1) = 63.2$ dB

Step 2: Calculate the A_{air} value using Equation 15.5 and Table 15.1. For calculation purposes assume the temperature is 20°C with a relative humidity of 70%. From Table 15.1 at 250 Hz for the given temperature and relative humidity, the attenuation coefficient is 1.1 dB/km. Therefore, the A_{air} at this frequency is:
$A_{air} = \alpha'r/1000 = (1.1)(1450)/1000 = 1.6$ dB

Step 3: Calculate the environmental attenuation using Equation 15.6 and Table 15.3. Recall that Equation 15.6 is:
$A_{env} = A_s + A_r + A_m$ dB
The first term to determine is A_s:
For the source zone: $30h_s = (30)(3) = 90$ m,
Next, from Table 15.3 at 250 Hz:
$A_s = (b)(G) - 1.5$ dB
Note: Since the proposed turbine is to be located in the center of a 20 m × 20 m concrete skid, 10 m of the source zone is classified as "hard," and the remaining 80 m is grass or "soft." Thus, the ground factor G is:
$G = (90 - 10)/90 = 0.89$
Therefore, using Table 15.3:
$A_s = (b)(G) - 1.5 = (5.3)(0.89) - 1.5 = 3.2$ dB
Note: b = 5.3 at 250 Hz, which is given in the table.
The second term to calculate is A_r:
Here for the receiver zone: $30h_r = (30)(1.5) = 45$ m,
From Table 15.3 at 250 Hz:
$A_r = (b)(G) - 1.5$ dB
Since the receiver is located on grass, the ground is considered "soft" and G = 1. Therefore,
$A_r = (b)(G) - 1.5 = (8.4)(1) - 1.5 = 6.9$ dB

Note: b = 8.4 at 250 Hz, which is given in the table.

The final component to determine is the middle zone. Recall for A_m to exist the expression

$r > 30(h_s + h_r)$ must be satisfied. In this example, $r = 1450$, and $1450 > 30(3 + 1.5) = 135$ is satisfied.

Therefore, from Table 15.3 at 250 Hz:

$A_m = -3e(1 - G)$ dB, where $e = \{1 - [30(h_s + h_r)/r]\}$

Now, $e = \{1 - [30(3 + 1.5)/1450]\} = 0.91$

and,

$A_m = -3(0.91)(1 - 0.75) = -0.7$ dB

Note: G = 0.75 since 75% of the ground cover in the middle is grass.

Finally, sum up each term to get A_{env}:

$A_{env} = A_s + A_r + A_m$ dB

$= 3.2 + 6.9 - 0.7 = 9.4$ dB

Step 4: Since there is no interfering foliage or trees to provide additional attenuation, A_{misc} is zero.

Step 5: Determine the total attenuation at 250 Hz from Equation 15.3:

$A_{total} = A_{div} + A_{air} + A_{env} + A_{misc} = 63.2 + 1.6 + 9.4 = 74.2$ dB

Step 6: Use Equation 15.2 to calculate the L_p at this frequency:

$L_p = L_W - A_{total} - 10.9$ dB

$= 145 - 74.2 - 10.9 = 59.9$ dB

Step 7: Find the A-weighted sound level for the 250-Hz octave band: The sound level for the 250-Hz band is 59.9-8.6 = 51.3 dBA. Note: the -8.6 value is the conversion factor at 250 Hz when going from linear SPL to A-weighting (see Table 3.1).

Step 8: Repeat steps 1–7 for all other frequencies of concern, then use Equation 2.11 to logarithmically add all A-weighted octave-band values to calculate the overall A-weighted sound level at the property line. Completing steps 1–7 for 125, 500, 1000, 2000, and 4000 Hz, yields A-weighted octave-band values of 45.3, 58.3, 56.9, 52.0, and 28.4 dBA, respectively. Then inputting these data into Equation 2.11, including 51.3 dBA at 250 Hz, results in an estimated overall sound level of 62 dBA. As discussed previously, many local noise ordinances limit sound entering residential areas to 55–60 dBA during daytime hours and 50–55 dBA at night; therefore, it is likely that a sound level of approximately 62 dBA will be unacceptable according to the local noise ordinance, as well as in the perception of the neighbors.

Effects of Wind and Temperature

Sound wave propagation follows a predictable model in a still environment. However, sound will not conform to any predictable pattern in windy conditions. As temperature changes occur, there is a corresponding change in the speed of sound as follows:

$$c = c_o \sqrt{\frac{T}{T_o}}$$

(15.7)

Where,

c = speed of sound
T = temperature (K° or R°)
c_o = speed of sound in air at reference temperature T_o

It is a natural phenomenon that temperature usually decreases with increasing elevation during daytime hours, and increases with elevation at night. Under normal daytime conditions, the velocity of sound is greatest at lower elevations, and sound waves bend or refract upward as depicted in Figure 15.3. This often results in a shadow zone near the ground, and the attenuation significantly increases with distance. This additional sound reduction will typically be 10–20 dB or more above the expected attenuation due to ground effects.

Figure 15.4 exhibits the sound spreading pattern that occurs during temperature inversions when the temperature increases with elevation. This condition is more common at night due to radiation cooling of the ground, and during sunrise and sunset. Since the speed of sound is faster in warmer upper layers of air,

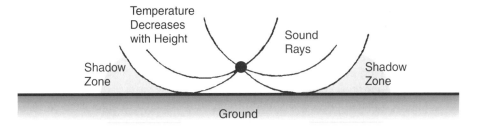

Figure 15.3 — Wave propagation during daytime.

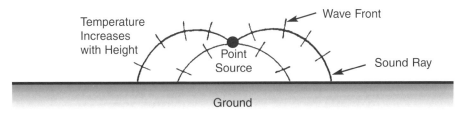

Figure 15.4 — Wave propagation during inversion.

sound waves will actually bend downward as they propagate from the source. This condition results in little to no attenuation due to the environment for several hundred meters, and produces a favorable condition for sound propagation.

Figure 15.5 illustrates how sound wave propagation behaves with wind gradients. As sound extends upwind, the spreading waves refract upward and create a shadow zone with excess attenuation near the ground. Because of this condition, it is not recommended that sound level measurements be conducted upwind of the source. On the other hand, as sound radiates downwind, the waves bend downward resulting in a condition advantageous to propagation. This explains why sound levels downwind of a noise source are more easily detected or heard as compared to the listening conditions upwind. Consequently, it is recommended that measurements be conducted downwind of the source.

One other phenomenon that often occurs is sound traversing large distances. Since spreading patterns for sound will vary or fluctuate with increased elevation, wind, and temperature, it is common to hear or detect sound as a warble or intermittent event several kilometers away. This is especially true for low-frequency sounds, such as a locomotive horn, or an outside warning alarm at an industrial facility.

Miscellaneous Attenuation Effects (A_{misc})

Attenuation of sound resulting from rain, dense fog, and falling snow is practically zero. Therefore, these conditions may be ignored, with the possible exception of snow-covered ground that may change the classification of the ground-surface rating as described previously. For the most part, these conditions affect other environmental factors such as altering the wind and temperature gradients, which are accounted for when calculating the air and environmental attenuation values.

A common misconception is that a few rows of trees can be planted along the property line to help reduce community noise. While it is true that trees often block the visual line of sight to the source, and as a result provide a psychological noise-reduction benefit, in reality a series of trees a few meters deep is acoustically transparent and provides no measurable attenuation. Table 15.4 presents the attenuation due to sound propagation through foliage, such as trees

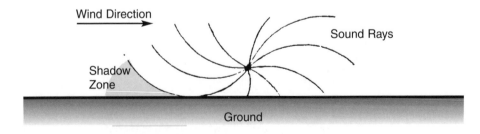

Figure 15.5 — Wave propagation with wind.

TABLE 15.4
The attenuation due to propagation through foliage, such as trees and bushes.

	Octave-Band Center Frequency (Hz)								
	31.5	63	125	250	500	1000	2000	4000	8000
A_{misc} (dB/m)	0.02	0.02	0.03	0.04	0.04	0.05	0.06	0.08	0.12

Source: From Piercy and Daigle (1991), with permission.

and bushes. The type of tree, density of planting, and noise source characteristics are the controlling factors toward their acoustical benefit. A good rule of thumb is that for the first 100 m of dense forest, the average attenuation will be approximately 4–8 dBA provided both the source and receiver are within, or relatively close to, the trees. For distances greater than 100 m, no rule of thumb applies, however, a more detailed discussion of this issue may be found in Piercy and Daigle (1991).

Measuring Community Noise

A person measuring community noise must often comply with the requirements of appropriate ordinances and standards. The referenced standards provide technical guidelines, some of which are discussed briefly in this section. The measurement guidelines should match the goal of the sound survey. Some standards require that measurements be conducted under the most favorable weather and physical conditions for sound propagation. This requirement ensures that data are collected during sound propagation conditions that typically correspond to a majority of complaints from neighbors. However, the goal of many community noise measurements is to document noise in the community for various propagation conditions.

Factors that Influence Community Noise Measurement

Seasonal factors, weather, measurement locations, and source operating parameters are all conditions that will affect community noise measurement results. These factors should be identified during the planning of the measurement process, and should be accounted for to the greatest degree practical.

Seasonal Factors Affecting Sound Present in a Community

Seasonal variations in plant, insect, and wildlife conditions can influence the sound present. Suppose the distance between the source and receiver is less than 100 m and heavily forested with deciduous trees. Sound levels from the source reaching the receiver could be much less in the summer than the winter. Insects and wildlife produce sound that will significantly influence sound measurements. In some cases, almost steady sound from insects, tree frogs, or large flocks of wild birds can dominate the overall A-weighted sound level.

Intermittent bird sounds can be eliminated by measuring percentile levels. Frequency analysis is essential when the overall level is dominated by steady, high-frequency insect sounds. It is not unusual in some places for insect sounds to exceed the limits of local ordinances for several hours, especially at night. However, this high-frequency insect noise does not mask annoying noises with lower spectral content.

Monitoring and Documenting Meteorological Conditions
Weather has a major effect on the propagation of sound as described previously. Therefore, weather conditions must be monitored and documented for community noise measurements.

- Wind speed and direction should be monitored directly at the measurement site and documented. Measurements should be avoided when wind speeds approach 19 kph (5 meters per second, 12 mph). For low SPLs or low-frequency sound, even winds more than 10 kph can cause problems. One problem is wind interaction with the microphone. Therefore, a windscreen should always be used to minimize this problem. Sound levels radiated from sources at considerable distances in the presence of high wind speeds may be highly variable and not representative of conditions with lesser wind.
- Ambient air temperature, relative humidity, barometric pressure and cloud cover corresponding to time of measurement should be recorded. These data are typically available via radio or the Internet from a nearby meteorological station, usually located at an airport.
- Measurements should not be conducted during measurable precipitation or thunder, since these conditions will artificially raise the background sound level, as well as potentially affect the performance of the acoustical instrumentation.
- Recognize that snow cover or water-saturated ground can influence results (see *Attenuation Due to Environmental Effects*).

Measurement Location
Measurement location factors directly influence measurement results. They include distance from the source, topography, ground surface cover, and reflective surfaces. Locations of measurement sites should be documented on a scaled map to permit estimation of distance from the source as well as to facilitate repeat measurements. The following factors should be noted and considered in the selection of the measurement sites.

- Topography and elevation changes affecting line-of-sight to the source are factors to consider in selection of measurement location(s).
- Measurements over large paved areas should be avoided unless the goal is specifically to document the sound level at such areas.
- Large reflecting surfaces such as buildings will influence sound levels. The locations of such surfaces should be carefully documented. If the goal is to obtain data easily related to the output of a source, measurements should be avoided near such surfaces. In such cases, measure at least

7.5 m and preferably 15 m from such surfaces. However, the goal of community noise measurements is often to measure sound at and near a home. In those cases, measurements are appropriately made at locations near the home with documentation of reflecting surfaces. It is a good survey procedure to locate the microphone at least 1.5 m from smaller objects such as trees, posts, bushes, etc., if possible.

Source Conditions

The operating conditions of the source also influence measurement results. The operating conditions desired for testing should be selected and documented. This may include particular production or process conditions correlated to the time of measurement. When measurements are made far from a source, simultaneous measurements near the source are advisable, especially if source output is variable.

Measurement Protocol

Site Selection

Selecting a measurement location will depend upon the purpose of the sample. If the goal is to assess the sound reaching a specific location at a specific time, then the terrain must be accepted as is. However, if the primary purpose is to document the sound output of a specific source, it is best to optimize the conditions. The site may be specified by standard or ordinance. Otherwise, measurement sites should be selected to allow for description of the acoustic environment and to be able to assess its impact on the surrounding community. The most common location to start with is the source property line near potentially affected neighbors. This site will allow for initial assessment without intrusion. Sometimes it may not be possible to measure at the boundary line. That location may not be feasible or representative because of extreme elevation differences, obstructions to the source, etc. In this case, select a location closer to and within line of sight of the source in question.

Microphone Height and Orientation

The microphone position above the ground should usually be between 1.2 and 1.8 m. This may be specified by ordinance or standard. Higher microphone locations may be needed if the line of sight between source and receiver is high above the ground. The microphone orientation should provide a sound incidence angle for the primary source according to manufacturer's instructions. Brief measurements can be made with a hand-held sound level meter being careful to hold it away from the body. For longer measurement periods, the meter or preferably the microphone should be mounted on a tripod. This allows the operator to stay away from the microphone during measurements.

Measurement and Observation

The sensitivity calibration of the measurement system should be checked before and after the survey period. While measurements are occurring, the person conducting the tests should note environmental conditions and events, logging them with observations of levels. Background sound levels with the source under study shut down should be measured where possible. If this is not possible, try to estimate the background level with a measurement at a similar site removed from the source.

The measurement period will often be specified by a standard, regulation, or local ordinance. Otherwise, professional judgment is required by the surveyor to determine the appropriate amount of sampling time needed to satisfy the goals of the survey. The purpose of the measurements and the characteristics of the sound must then be considered. If the measurement only needs to demonstrate levels above a given criterion, and steady sound is clearly above that criterion, a very short period (less than 1 minute) can be acceptable. However, very long periods may be necessary to document statistically reliable indications of long-term sound levels. For DNL measurements it may be necessary to sample the noise over several days, even weeks.

Instrumentation

The quantities to be measured and required instrumentation will vary depending on the goals of the measurement and the procedure specified by standard, regulation, or ordinance.

Conventional Sound Level Meter

For simple ordinances specifying sound levels not to be exceeded, and for steady sound near a source, a conventional sound level meter can be used. Sampled data with a conventional meter also can be used to estimate a time-average sound level or percentile sound levels. This method is not advised if the data are part of a litigation record, unless the method is specified by the governing ordinance. The period of observation should be established based on the operating characteristics of the source. If the noise is comparatively steady, less time is needed (e.g., 5 minutes). If the noise fluctuates, more sampling time (e.g., 20 minutes) is recommended. Set the instrument for slow response and log the sound level at 10-second intervals. See both ANSI S12.9 Part 3 and ASTM 1014 for additional information and guidance on measurement procedures (ANSI, 1993; ASTM, 1984).

Integrating Sound Level Meter

An integrating sound level meter can be used to measure time-average sound level, maximum sound level, and peak sound pressure levels. The measurement period should be established based on the nature of the source and local ordinance requirements. Measurement periods typically range from 10 minutes to 1 hour. During the measurement, log events and conditions that may influence the measurement. The log will serve as the record to explain the measurement. An example log sheet is shown below (see Figure 15.6).

Time	L_{AS}	L_{ASmax}	Local Noise/ Traffic	Train	Plane	Tone?	Description of Events	Wind Dir	Wind Speed (mph)	Temp (Deg F)	Cloud Cover
10:00	67	84			X		Plane overhead, intermittent traffic	WSW	<10	67	Cloudy
10:10	64	76					Intermittent traffic	WSW	<10	67	Cloudy
10:20	63	73					Intermittent traffic	WSW	<10	67	Cloudy
10:30	64	76					Intermittent traffic	WSW	<10	67	Cloudy
10:40	67	82	X				Intermittent traffic, lawn mowing	WSW	<10	67	Cloudy
10:50	68	83	X				Intermittent traffic, lawn mowing	WSW	<10	67	Cloudy
11:00	64	75					Intermittent traffic	WSW	<10	67	Cloudy
11:10	62	72					Intermittent traffic	WSW	<10	67	Cloudy
11:20	61	73					Intermittent traffic	WSW	<10	67	Cloudy
11:30	65	79			X		Plane overhead, intermittent traffic	WSW	<10	67	Cloudy
11:40	64	71					Intermittent traffic	WSW	<10	67	Cloudy
11:50	65	81		X			Distant train horn	WSW	<10	67	Cloudy
12:00	64	79					Intermittent traffic	WSW	<10	67	Cloudy

Figure 15.6 — Example data log sheet.

Data-Logging Devices

There are a variety of microprocessor-based data-logging devices that may be used to maintain descriptive statistics of the data sampled. These systems range from the more sophisticated integrating sound level meters to environmental monitoring stations. Industrial dosimeters also can be used. However, make sure they are set for a 3-dB exchange rate and an adequately low threshold level (not the default threshold of 80 dB). Data logged by the instrument are stored in memory for later retrieval and analysis. These devices are typically programmable and can include valuable statistics such as percentile levels and DNL. Measurement periods are typically designed to be longer in these instruments with sampling rates corresponding to sample length (limited by memory). These instruments can be left unattended. However, it is advisable to have an observer, especially if the data are to be used in litigation. The most useful percentile levels are the time-average sound levels that are exceeded 10% and 90% of the time. The level exceeded 10% of the time is a criterion used in some ordinances. The 90th percentile level can help define the steady noise level in the absence of intermittent noises. The level exceeded 1% of the time can be a useful indication of normal maximum sound levels due to short events when the actual maximum varies among events. Note that the percentile levels and maximum levels will be influenced by the selection of fast or slow response, or sample duration for instruments using short samples of time-average sound level.

Frequency Analyzers

The frequency spectrum of the community sounds can be measured and recorded using octave-band or 1/3 octave-band filters or fast Fourier transform (FFT) analyzers. Octave-band and 1/3 octave-band filters may allow measurement of all frequencies simultaneously or require serial measurement of each band. The data can be compared to criteria specified in an ordinance or regulation. Some ordinances specify a method of evaluating the presence of a discrete tone using 1/3 octave-band data. For the tone to be considered present, the 1/3 octave band of concern must exceed the arithmetic average for the two adjacent bands by some specified amount. Annex C of ANSI S12.9-1996 Part 4 gives guidance defining these differences, as shown in Table 15.5 (ANSI, 1996). This method will not always properly identify a discrete-tone problem. The user of Table 15.5

TABLE 15.5
Guidance for determining the existence of a pure tone.

Range of 1/3 Octave-Band Center Frequencies (Hz)	Difference Between Arithmetic Average of SPLs in Two Adjacent Bands (dB)
25-125	15
160-400	8
500-10,000	5

Note: Obtain the arithmetic average of the SPLs in the 1/3 octave bands immediately above and below the frequency of concern. Subtract this average value from the SPL in the 1/3 octave band containing the suspected pure tone. If the difference equals or exceeds the value indicated for the respective frequency range listed in Table 15.5, a discrete or pure tone may be assumed to exist.
Source: From ANSI S12.9-1996 Part 4, Annex C, with permission.

is cautioned that a tone at or near the boundary between 1/3 octave bands will share the sound energy between the two bands giving a false indication of no tone. Also, nontonal sound covering most of a band, but with little content in adjacent bands, will falsely indicate that a tone is present. An FFT analyzer is used to measure narrow-band frequencies, with the frequency resolution determined by the surveyor. A method using FFT analysis over octave-band and 1/3 octave-band measurements to more clearly identify the presence of a pure tone is described by Lilly (1994).

Interpreting Results

After collection, data must be organized and analyzed. Similar techniques can be applied to project the effect of new noise sources and to evaluate the need for noise control. There are numerous methods for describing and classifying community noise. This section will discuss interpreting the data for compliance with existing or potential regulations, and for community reaction.

Compliance with Existing or Potential Regulations

Depending on the jurisdiction (local, state, or provincial), rules limiting noise in the community may be found in general ordinances, zoning codes, or health regulations. However, compliance with these regulations does not assure community satisfaction. Most businesses want to be perceived as good neighbors. Regulatory compliance also is not always a satisfactory defense in legal proceedings. Many local ordinances contain specific clauses preserving the rights of plaintiffs to bring legal action against noise sources that comply with the ordinance. The plaintiff faces a heavy burden in that case, to prove the noise either is a nuisance or reduces property value. In some communities there also may be multiple applicable regulations. If there are no regulations, it is advisable to search for regulations in nearby jurisdictions. This could suggest the type of regulation the community might adopt in the future. Realize that simplified ordinances can sometimes be very restrictive. A 55-dBA limit is more stringent for unsteady sound if interpreted as a maximum or instantaneous level rather than an average level over a reasonable time. Lacking local guidance, typical regulation limits can be considered as references.

Assessment for Prediction of Community Response

A procedure for evaluating community reaction based on DNL was proposed by the EPA (1974) and updated by two of the original authors (von Gierke and Eldred, 1993). This procedure works best when the sound is broad-band in content, and present most days for much of the day. It normalizes the sound for various factors including existing sound levels. The expected DNL of the new source alone is first determined and adjusted by the factors shown in Table 15.6. These factors correct for seasonal variation, previous exposure and community attitudes, and the presence of tones or impulses. Larger correction factors than the EPA-proposed values, taken from ANSI S12.9 Part 4, have been added to this

TABLE 15.6
Corrections added to the measured noise level to obtain normalized level.

Type of Correction	Description	Amount Added to Measured Level in dB
Seasonal correction	Summer (or year-round operation).	0
	Winter only (or windows always closed).	- 5
Correction for previous exposure & community attitudes	No prior experience with intruding noise.	+5
	Community has had some previous exposure to intruding noise, but little effort is being made to control the noise. This correction may also be applied in a situation where the community has not been exposed to the noise previously, but the people are aware that bona fide efforts are being made to control the noise.	0
	Community has had considerable previous exposure to the intruding noise, and the noise maker's relations with the community are good.	- 5
	Community is aware that operation causing noise is very necessary and it will not continue indefinitely. This correction can be applied for an operation of limited duration and under emergency circumstances.	-10
Pure tone or impulse	No pure tone or impulsive character.	0
	Pure tone or impulsive character present.	+5
	Highly impulsive sounds, gunfire, hammering, drop hammering, pile driving, drop forging, pneumatic hammering, pavement breaking, metal impacts during rail-yard shunting operation, and riveting.	+12

Source: From EPA (1974).

table for highly impulsive sounds. The existing DNL without the new source is then arithmetically subtracted from the DNL expected for the new source alone. The DNL existing in the community can be estimated from sound measurements using Equation 15.1a, Equation 15.1b or from Table 15.7. The resulting difference is then compared to Table 15.8 to predict response. Notice that even the addition of a new sound, equal in level to the existing sound, will produce an increase in the overall level. Thus, some sporadic complaints are expected even when the normalized change is zero. A clearly dominant sound from a single new source will produce widespread complaints.

TABLE 15.7
Typical community noise levels.

Community Description	DNL (dBA)
Rural and sparsely populated areas	35–50
Quiet suburban (260 people/km^2, remote from large cities and from industrial activity and trucking)	50
Normal suburban community (770 people/km^2 not located near industrial activity)	55
Urban residential community (2600 people/km^2 not immediately adjacent to heavily traveled roads and industrial areas)	60
Noisy urban residential community (near relatively busy road or industry or 7700 people/km^2)	65
Very noisy urban residential community (26,000 people/km^2)	70

Source: Adapted from EPA (1974).

TABLE 15.8
Expected community reaction for normalized DNL difference.

Normalized Change in DNL (dBA)	Reaction
- 5	None
0	Sporadic complaints
+ 5	Widespread complaints
+ 14	Threats of legal action
+ 21	Vigorous action

Source: From EPA (1974).

EXAMPLE 15.2, Expected Community Reaction to a New Noise Source

For example, suppose a new industrial plant is to be built in a suburban area. It is not near other industry but there are two existing residential communities nearby. Noise controls can eliminate tones and impulsive sounds, and the sound produced will be steady 24 hours a day. Atmospheric effects will produce some variation in sound level reaching the communities. The DNL reaching the communities will be 52 – 55 dBA for the closer community and 45 – 46 dBA for the other. The population densities are about 500 people per km^2 for the closer community and 1000 people per km^2 for the other. What is the expected reaction in the two communities?

Table 15.7 indicates the existing DNL in the communities will be close to 55 dBA. Actual DNL for the two communities can be estimated to be 53 and 56 dBA using Equation 15.1b. Since the communities have no prior experience with or expectation of the noise, Table 15.6 indicates 5 dBA should be added to the source noise level or DNL reaching each community (52 – 55 dBA and 45 – 46 dBA) as described above. This gives a normalized DNL for the source of 57 – 60 dBA in the closer community and 50 – 51 dBA in the other. Next, we subtract the estimated existing DNL in the communities from the normalized DNL due to the source. For the closer community we subtract 53 from 57 – 60 and get a difference of 4 to 7 dBA. For the other community, we subtract 56 from 50 – 51 and get -4 to -5 dBA. From Table 15.8, we see there will probably be no reaction in the more distant and densely populated community. However, we can expect widespread complaints from the closer and more sparsely populated area.

In some cases the use of DNL will underestimate community reaction. This is most likely when the sound occurs only occasionally (once a day or less) in short periods of loud sound not typical for the community. These short periods of noise could be loud when they occur but not significantly change the DNL. This can be a particular problem if the noise occurs during evening or weekend periods when people are home and possibly trying to enjoy the outdoors. It is better in these cases to use actual sound levels during the events, rather than the DNL, for both the new noise and the existing noise. Using actual sound levels may overestimate community reaction but will be more reliable when the normalized change is large with them and small using DNL.

DNL or any measure based on overall A-weighted sound levels will not work well for distinctive sounds, such as speech, music, or discrete tones. The A-weighted sound level also can be misleading for strong low-frequency sounds where the C-weighted sound level is more than 10 dB greater than the A-weighted sound level. This includes high-energy impulsive sounds such as quarry and mining explosions, demolition and industrial processes using high explosives, explosive industrial circuit breakers, and other explosive sources where the equivalent mass of dynamite exceeds 25 g. Other sources of disturbing low-frequency noise include industrial exhaust stacks, outside blowers or fans, vacuum trucks used to clean parking lots, heavy vehicles (e.g, 18-wheel trucks) traveling on highways and over bridges, wind turbines, etc. It is worth noting that when the SPLs are less than 65 dB and relatively steady at the octave-band frequencies of 16, 31.5, and 63 Hz, it is unlikely that an annoyance problem exists. Residents may be annoyed, however, when sound less than 65 dB in these same frequencies fluctuates rapidly. See Annexes B and D of ANSI S12.9 Part 4 (1996) for guidance.

Often the problems due to strong low-frequency noise are evident only inside homes. The long wavelengths of low-frequency sounds can easily penetrate a building's structure and excite room resonances. The results include audible sound and possibly rattles due to vibration induced by the noise. Such rattles make the annoyance equivalent to a noise at least 10-dB higher. Resonant tones will often be amplified leaving the sound inside the home even more dominated by low frequencies.

Adjustments to Account for Background Sound Levels

When the difference between the level due to the source of concern and the background level is less than 10 dB, it is sometimes desirable to determine the level due solely to the primary source. This can be done by using Equation 2.13. The result can also be approximated using Table 15.9.

TABLE 15.9
Adjustments to account for background sound levels.
The contribution of the background sound level (without source under study operating) may be accounted for under the following conditions.

Condition	Comment	Action
The sound pressure level increases over the background sound pressure level by 10 dB or more.	The measured operating sound pressure level is due to the source.	No adjustment necessary.
The sound pressure level increases over the background sound pressure level between 4 and 10 dB.	The measured operating sound pressure level consists of elements of both source and background.	Apply adjustment to measured level using Table.
The sound pressure level increases over the background sound pressure level by 3 dB or less.	The sound pressure level due to the source is equal to or less than the background sound pressure level.	The two contributions cannot be separated.

NOTE: Where the difference is 3 dB or less, report the unadjusted source level and identify it as being "masked" by the background level.

Adjustment of measured level to account for the effect of background sound.

Difference Between Measured Level and Background Level (dB)	Adjustment to be Made to Measured Level (dB) to Obtain Corrected Source Level
4	- 2.2
5	- 1.7
6	- 1.3
7	- 1.0
8	- 0.8
9	- 0.6
10	- 0.4
Greater than 10	0

Source: From ANSI S12.18-1994, with permission.

Report and Documentation

The report of noise measurements taken in the community should reflect the purpose of the study. The report must adequately describe the conditions of the measurements so that the findings are taken in context. Furthermore, in the event that measurements need to be repeated, the report should be sufficient to serve as a reference for future measurements.

Elements that should be considered for a report include:

1. A clear statement of the purpose of the measurements (e.g., cursory check of conditions, documentation of a source output, evaluation of ordinance compliance, evaluation of land use compatibility, prediction of community response, etc.).

2. Description of methodology for obtaining measurements—including rationale for choice (e.g., ordinance specification, satisfaction of purpose, limitations in source operation, etc.).

3. Description of the setting—including the surrounding area, terrain, land use classifications, etc.

4. Description of noise source(s) within the environment—including temporal characteristics, tonal qualities, operation/process relationship of major sources. The description should also include background and transient sources.

5. Description of measurement site(s)—including specific location of site, rationale for selection, position relative to source(s), description of terrain and objects near the site.

6. Plan view of site—a topographic map including source locations, measurement sites, significant objects such as buildings, major vegetation, and other locations of interest (including nearby residences etc.). Significant ground slopes should also be indicated. An example of a plan view is shown in Figure 15.7.

7. The sound descriptors (e.g., maximum sound level, equivalent sound level, percentile sound level, day–night average sound level, etc.) used to describe/evaluate the source(s)—including rationale for use of such descriptors.

8. Documentation of instrumentation—including manufacturer, model, and serial numbers of all meters, microphones, calibrators, and other instrumentation used in the study. Sampling rates and settings should also be included, as well as the pre- and postsurvey calibration readings.

9. Description of meteorological conditions—including typical wind speed and direction, temperature, relative humidity, and cloud cover, supplemented with a brief description of weather conditions during time of measurements. Wind speed and direction corresponding to time intervals should also be documented in a separate log appended to the report.

10. Exceptions to standard procedures—including deviations due to ordinance requirements, site limitations, or purpose.

11. Other observations—including description of occurrences during the measurement periods that could have an effect on the data collected.

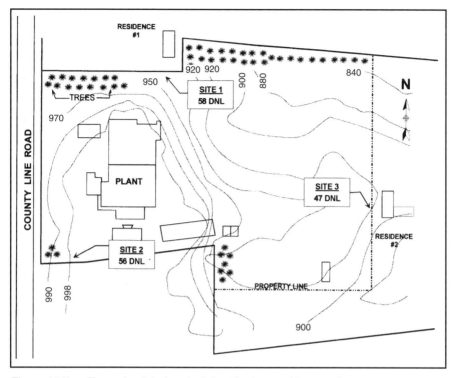

Figure 15.7 — Example plot plan depicting the two residences, three measurement sites, principal geographic features, elevation contours, and measured sound descriptors.

12. Acoustical data—including measurement data (background and source test data), results of comparisons to criteria or ordinances, and other conclusions. Measurement data should be presented with the knowledge that the readers may not have any technical understanding of sound or the evaluation criteria. Methods to simplify understanding should be used where possible. A time-history of the measurement can be illustrative of the conditions as shown in Figure 15.8.
13. Executive summary—it is recommended that a summary of the study including purpose and findings be included at the beginning of the report.

Summary

Most industrial companies will face the potential of a community noise problem. Each surrounding community is different and will tolerate varying levels of noise. Factors influencing community tolerance include:
• Visibility of noise source. Some members of the community may be more concerned with "visual" noise sources (e.g., stacks, vents, etc.).

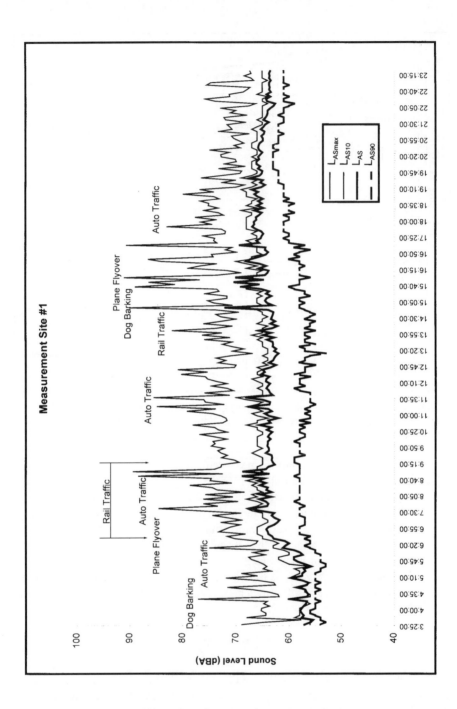

Figure 15.8 — Example of time-history log.

- Noise sources that cannot be associated with the operation of the facility or seem foreign to the community. Some members of the community may interpret these sources as potentially dangerous.
- Noise centered within a narrow frequency band (pure tones).
- Noises that can startle the community (impulsive noise).
- Noise that is random in occurrence and duration (may be related to lack of control).
- Low-frequency noise that may cause vibrations and/or resonances within residential structures.
- A very low pre-existing background noise level.

If a community noise problem is suspected, the following information should be considered:

- Review current local noise control ordinance. If there is none, refer to any state guidelines for information on what is expected for monitoring and compliance.
- Conduct perimeter (property line) sound level measurements. Compare to limits specified in the local ordinance. Check for pure tones. Many ordinances have definitions and special restrictions for tone generation.
- Be aware of the time of the noise complaint. Certain sounds may be noticed at greater distances in the evening or early morning due to meteorological effects, as well as lower background noise, and may not be discernable during the day.

Additional follow-up steps may include the following:

- Meet with the community/complainant. This shows that the company is concerned about being a good neighbor. Sometimes the noise complaint is related to another issue and noise is being used to get attention and response.
- Open communications. Consider creating a "noise hot-line" that the community can call 24 hours a day. Avoidance or quick resolution of a noise issue is always in the plant's best interest. In addition, a well-documented list of complaint calls can be cross-referenced with plant operating conditions to track down possible problems.
- Inform the community of any unusual noise emissions prior to noise generation. Typically, complaints will come when a "normal" noise environment changes. In addition, a noise generated between 7 p.m. and 7 a.m. is generally more likely to cause complaints than an identical noise occurring during daytime hours.
- Elimination of noise sources may also cause complaints—if the old noise source masked a dominant tone or other "offensive" noise.

References

Anon. (1972). The United States Code, Title 42 The Public Health and Welfare, Section 4901: Congressional findings and statement of policy, 42USC4913 (1972).

ANSI (1988). "Quantities and Procedures for Description and Measurement of Environmental Sound, Part 1," ANSI S12.9-1988/Part 1, Acoustical Society of America, New York, NY.

ANSI (1992). "Quantities and Procedures for Description and Measurement of Outdoor Environmental Sound, Part 2: Measurement of Long-Term, Wide Area Sound," ANSI S12.9-1992/Part 2, Acoustical Society of America, New York, NY.

ANSI (1993). "Quantities and Procedures for Description and Measurement of Environmental Sound, Part 3: Short-Term Measurements with an Observer Present," ANSI S12.9-1988/Part 3, Acoustical Society of America, New York, NY.

ANSI (1994). "Outdoor Measurement of Sound Pressure Level," ANSI S12.18-1994, Acoustical Society of America, New York, NY.

ANSI (1996). "Quantities and Procedures for Description and Measurement of Environmental Sound, Part 4: Measurement of Long-Term, Wide Area Sound," ANSI S12.9-1996/Part 4, Acoustical Society of America, New York, NY.

ANSI (1997a). "Engineering Methods for the Determination of Sound Power Levels of Noise Sources for Essentially Free-Field Conditions over a Reflecting Plane," ANSI S12.34-1988 (R 1997), Acoustical Society of America, New York, NY.

ANSI (1997b). "Survey Methods for Determination of Sound Power Levels of Noise Sources," ANSI S12.36-1990 (R 1997), Acoustical Society of America, New York, NY.

ANSI (1998). "Quantities and Procedures for Description and Measurement of Environmental Sound, Part 5: Sound Level Descriptors for Determination of Compatible Land Use," ANSI S12.9-1998/Part 5, Acoustical Society of America, New York, NY.

ASME (1985). "Measurement of Industrial Sound," American Society of Mechanical Engineers ANSI/ASME PTC 35-1985, New York, NY.

ASTM (1984). "Guide for Making Measurements of Outdoor A-Weighted Sound Levels," American Society for Testing and Materials E1014, West Conshohocken, PA.

ASTM (1992). "Test Method for Conducting Outdoor Sound Measurements Using a Digital Statistical Analysis System," American Society for Testing and Materials E1503, West Conshohocken, PA.

ASTM (1996a). "Guide for Selection of Environmental Noise Measurements and Criteria," American Society for Testing and Materials E1686, West Conshohocken, PA.

ASTM (1996b). "Guide for Preparing a Measurement Plan for Conducting Outdoor Sound Measurements," American Society for Testing and Materials E1779, West Conshohocken, PA.

ASTM (1996c). "Guide for Measuring Outdoor Sound Received from a Nearby Fixed Source," American Society for Testing and Materials E1780, West Conshohocken, PA.

Berglund, B., Hassmen, P., and Job, R. F. S. (1996). "Sources and Effects of Low-Frequency Noise," J. Acoust. Soc. Am. 99, 2985–3002.

Berglund, B., and Lindvall, T. (eds.). (1995). Community Noise [Archives of the Center for Sensory Research Vol. 2(1)], Stockholm Univ. and Karolinska Inst., Stockholm, Sweden.

EPA (1974). "Information on Levels of Environmental Noise Requisite to Protect Public Health and Welfare with an Adequate Margin of Safety," U.S. Environmental Protection Agency, Report #550/9-74-004, Washington, DC.

EPA (1975). "Model Community Noise Ordinance," U.S. Environmental Protection Agency, Washington, DC.

Erdreich, J. (1998). "Developing Rational Environmental Noise Standards," *Sound and Vibration 32*(12), 26–28.

HUD (1979). "Environmental Criteria and Standards," Housing and Urban Development, *Fed. Regist. 44*(135), 24CFR Part 51, 40860-40866.

ISO (1982). "Acoustics—Description and Measurement of Environmental Noise—Part 1: Basic Quantities and Procedures," International Organization for Standardization, 1996-1:1982, Switzerland.

ISO (1987a). "Acoustics—Description and Measurement of Environmental Noise—Part 2: Acquisition of Data Pertinent to Land Use," International Organization for Standardization, 1996-2:1987, Switzerland.

ISO (1987b). "Acoustics—Description and Measurement of Environmental Noise—Part 3: Application to Noise Limits," International Organization for Standardization, 1996-3:1987, Switzerland.

ISO (1994). "Acoustics—Determination of Sound Power Levels of Multisource Industrial Plants for Evaluation of Sound Pressure Levels in the Environment—Engineering Method," International Organization for Standardization, 8297:1994, Switzerland.

ISO (1998). "Amendment 1 to ISO 1996-2:1987," International Organization for Standardization, 1996-2/AMD 1:1998, Switzerland.

Kurze, U., and Beranek, L. L. (1988). "Sound Propagation Outdoors," in *Noise and Vibration Control*, edited by L. L. Beranek, Institute of Noise Control Engineering, Poughkeepsie, NY, pp. 164–193.

Lilly, J. G. (1994). "Environmental Noise Criteria for Pure Tone Industrial Noise Sources," in *Proceedings of Noise-Con 94 Progress in Noise Control for Industry*, edited by J. M. Cuschieri, S. A. L. Glegg, and D. M. Yeager, Noise Control Foundation, Poughkeepsie, NY, 757–762.

Piercy, L. E., and Daigle, G. A. (1991). "Sound Propagation in the Open Air," in *Handbook of Acoustical Measurements and Noise Control, Third Edition*, edited by C.M. Harris, McGraw-Hill, New York, NY, reprinted by Acoustical Society of America, 1998, pp. 3.1–3.26.

von Gierke, H. E., and Eldred, K. McK. (1993). "Effects of Noise on People," *Noise News International 1*(2), 67–89.

The Noise Manual, revised 5th edition, edited by E.H. Berger,
L.H. Royster, J.D. Royster, D.P. Driscoll, and M. Layne
©2003 American Industrial Hygiene Association

16 Standards and Regulations

Alice H. Suter

Contents

The Need for Standards and Regulations

Millions of people throughout the world are exposed to levels of noise that are potentially hazardous. Because of lack of knowledge in some cases and lack of concern in others, a substantial number of those exposed are losing their hearing. Occupational noise, as with many other occupational hazards, cannot be adequately dealt with by marketplace incentives. Workers' compensation for hearing loss, at least in the U.S., is relatively insignificant as a motivation for noise control. If workers' hearing is to be conserved, it is almost always up to government bodies to inform and regulate employers.

An estimated 9 million U.S. workers are exposed to time-weighted average (TWA) sound levels of 85 dBA and above. The number of noise-exposed workers in various types of occupations is described in Table 16.1.

TABLE 16.1
Number of American workers exposed above 85 dB L_{A8hn}.*

Agriculture	323,000
Mining	353,100
Construction	513,000
Manufacturing and Utilities	5,124,000
Transportation	1,934,000
Military	976,000
Total	9,223,100

*From Simpson and Bruce (1981).

In the U.S., the advent of federal regulations for noise control was preceded by a series of consensus standards, issued by such organizations as the American Standards Association (the predecessor of ANSI), the National Research Council Committee on Hearing and Bioacoustics (CHABA), and the "Intersociety Committee." Although these bodies reflected an ever-increasing level of knowledge by the professional community, there was a need for uniformity. The federal government satisfied that need with the promulgation of a noise regulation in 1969 under the authority of the Walsh-Healey Public Contracts Act of 1935. The same is happening currently in Europe under the direction of the Council of the European Communities (CEC).

While uniformity is desirable, it is sometimes achieved through a process of simplification at the expense of precision. An example of this is the evolution of several consensus standards from the 1960s into the Department of Labor's 1969 noise regulation (DOL, 1969). These standards consisted of complex criteria, conveying allowable on-times, off-times, and numbers of exposure cycles, all of which were simplified into the 5-dB trading rule (now referred to as the exchange rate) (Suter, 1992).

The Need to Be Informed

Safety and health professionals who are well informed about the regulations that apply to their companies will be doing employers as well as employees a service. Those who are employed by international corporations will have to deal with the regulations of each country in which their company manufactures, and these professionals must be aware of how the regulations differ from those in the U.S. Even if the employer simply wants to sell goods abroad, safety and health professionals need to advise management about noise emission and labeling requirements where they exist. This chapter will briefly discuss some of these requirements, but the reader needs to obtain actual copies of the pertinent regulations or directives to be adequately informed.

Terminology

In the field of occupational noise, the terms "standard," "regulation," and "legislation," are often used interchangeably, leading to confusion. They do, however, have different meanings. A standard is a codified set of rules or guidelines that is often an acknowledged measure of comparison for quantitative or qualitative values. Standards may be developed under the auspices of a consensus group, such as the American National Standards Institute (ANSI), and they are usually not obligatory. By contrast, in the U.S., and in many other countries as well, a regulation is a rule or order prescribed by a government authority, usually the executive or administrative branch, which has the force of law. Regulations may incorporate standards to precisely define particular actions required by such laws. Like regulations, legislation also represents a type of law, but it is developed by legislating authorities or by local governing bodies (elected representatives, such as members of Congress). Legislation is usually broader in scope and more general than regulations, and in fact is often the underlying law that instructs a branch of government to create regulations. For example, the U.S. Congress enacted the Occupational Safety and Health Act of 1970 that created the Occupational Safety and Health Administration (OSHA) and empowered it to write regulations concerning safety and health in the workplace.

Even with the move toward uniformity, the terms describing standards and regulations may differ among nations and even within the same nation. There are regulations to control the noise exposure of workers over a period of time, such as a working day (e.g., OSHA regulations), and there are regulations to control the noise emitted by products, such as those issued by the CEC regarding machinery noise (CEC, 1989).

The CEC issues "directives," which may be considered orders or instructions. All members of the European Community were directed to "harmonize" their "laws, regulations, and administrative procedures" with the 1986 CEC directive on occupational noise exposure by January 1, 1990 (CEC, 1986). This means that the noise regulations of the member countries had to be at least as protective as the CEC directive.

Some nations use a "code of practice," which is somewhat less formal. For example, the Australian National Standard for Occupational Noise consists of two short paragraphs on exposure levels, followed by a 35-page code of practice, providing practical guidance on how the standard should be implemented (Worksafe Australia, 1993). Codes of practice generally do not have the legal force of regulations or legislation. However, according to Zusho and Miyakita (1996), the Japanese standard for occupational noise uses the term "administrative guidance," and yet the standard includes the word "shall" throughout, making this guidance obligatory. Another term that occasionally appears is "recommendation," which is more like a guideline than a regulation and is not enforceable.

In this chapter, the terms "regulation," "legislation," "directive," "standard," "code of practice," and "recommendation" will be used as the different nations, agencies, or groups use them, but an effort will be made to clarify which are compulsory and which are not.

Consensus Standards

Certain consensus groups are currently quite active in the process of developing consensus noise standards. These groups can be national to the U.S., such as the American National Standards Institute (ANSI) and the American Society for Testing Materials (ASTM). Others are international, like the International Organization for Standardization (ISO) and the International Electrotechnical Commission (IEC).

ISO has long been active in the area of noise standardization through its Technical Committee ISO/TC 43, *Acoustics*, made up of members from some 27 countries. There is no doubt that the pressure by the European Community to harmonize standards, thereby eliminating technical barriers to trade, has accelerated the work of ISO standards committees. ISO Subcommittee SC1, *Noise* is primarily concerned with noise exposure, emission, and measurement, as well as hearing conservation and machinery noise. Examples of published standards are

ISO 4871:1984, "Acoustics: Noise Labelling of Machinery and Equipment" and ISO 1999:1990, "Acoustics: Determination of Occupational Noise Exposure and Estimation of Noise-Induced Hearing Impairment," the latter of which (and its U.S. version, ANSI S3.44-1996) will be discussed in detail in Chapter 17. In 1993, von Gierke reported that ISO TC43/SC1 had issued 58 standards and that 63 were in a state of revision or preparation (von Gierke, 1993).

IEC also develops and publishes standards pertaining to instrumentation related to noise measurement and hearing conservation. Examples are: IEC 651:1979, "Sound Level Meters," IEC 373:1990, "Mechanical Coupler for Measurements of Bone Vibrators," and IEC 654-1:1992, "Audiometers — Part 1: Pure-Tone Audiometers." These standards are developed primarily by IEC Technical Committee 29: *Electroacoustics*.

Most ANSI standards related to noise are coordinated by the Acoustical Society of America, which provides the secretariat for the four committees: S1 *Acoustics*, S2 *Mechanical Shock and Vibration*, S3 *Bioacoustics*, and S12 *Noise*. Working groups under each S committee actually develop the standards. For example, S1 has overseen the development of standards dealing with the performance characteristics of sound level meters and dosimeters, and standards for calibrating and making measurements with these instruments. Performance specifications for audiometers and the conduct of audiometric tests are under the auspices of S3, and S12 has published standards on such topics as measuring impulse noise and determining the insertion loss of noise barriers. At this time there are more than 100 ANSI working groups and 65 ANSI standards in place in the field of acoustics.

Damage-Risk Criteria

The term *damage-risk criteria* refers to criteria specifying the risk of hearing impairment from various levels of noise. Many factors enter into the development of criteria in addition to the data showing the amount of hearing loss that results from a certain amount of noise exposure. There are both technical and policy considerations.

Policy considerations include the thorny questions of what proportion of the noise-exposed population should be protected and how much hearing loss constitutes an acceptable risk. Should we protect even the most susceptible members of the exposed population against any loss of hearing, or should we protect only against a compensable hearing handicap? These questions have a bearing on which hearing impairment formula to use, and different governmental and consensus bodies have varied widely in their selections.

Hearing Impairment Formulas

In earlier years, regulatory decisions were made that allowed substantial amounts of hearing loss as an acceptable risk. The most common definition used to be an average hearing threshold level (or "low fence") of 25 dB or greater at the audiometric frequencies 500, 1000, and 2000 Hz. Since that time, the definitions of "hearing impairment" or "hearing handicap" have become more conservative, with various organizations advocating different definitions. For example, OSHA and the Mine Safety and Health Administration (MSHA) now use an average of 25 dB at 1000, 2000, and 3000 Hz (MSHA, 1999; OSHA, 1981), the most recent NIOSH criterion uses 1000, 2000, 3000, and 4000 Hz (NIOSH, 1998), and the American Council of Governmental Industrial Hygienists (ACGIH) bases its limits on average hearing levels at 500, 1000, 2000, and 3000 Hz, although the threshold limit value (TLV®) is designed to protect 4000 Hz as well (ACGIH, 1994). Other definitions may incorporate fences lower or higher than 25 dB, or a broader range of frequencies.

In general, as definitions include higher frequencies and lower fences, the allowable exposures become more stringent and a higher percentage of the exposed population will appear to be at risk from given levels of noise. For there to be no risk of any hearing loss from noise exposure, even in the more susceptible members of the exposed population, the permissible exposure limit would have to be as low as 75 dB L_{A8hn} (see EPA, 1974). In fact, the CEC has established a level of 75 dB L_{A8hn} as the level at which the risk is negligible, and this level has also been put forward as a goal for Swedish production facilities (Kihlman, 1993).

Overall, the prevailing thought on this subject is that it is acceptable for a noise-exposed workforce to lose some hearing, but not too much. As for how much is too much, there is no consensus at this time. In all probability, most nations draft standards and regulations that attempt to keep the risk at a reasonably low level while taking technical and economic feasibility into account, but without coming to consensus on such matters as the frequencies, fence, or percentage of the population to be protected.

NIPTS and Risk

Criteria for noise-induced hearing loss may be presented in either of two ways: noise-induced permanent threshold shift (NIPTS) or percentage risk. NIPTS is the amount of permanent threshold shift remaining in a population after subtracting the threshold shift that would occur "normally" from causes other than occupational noise. The percentage risk is the percentage of a population with a certain amount of noise-induced hearing loss remaining after subtracting the percentage of a similar population not exposed to occupational noise that would exceed that same hearing loss due to aging and other causes. This concept is sometimes called *excess risk*. Unfortunately, neither method is without problems.

The problem with using NIPTS alone is that it makes it difficult to summarize the effects of noise on hearing. The data are usually set out in a large table showing noise-induced threshold shift for each audiometric frequency as a function of noise level, years of exposure, and population centile (see Chapter 17). The concept of percent risk is more attractive because it uses single numbers and appears easy to understand. But the trouble with percent risk is that it can vary enormously depending on a number of factors, particularly the selection of the reference population, the location of the fence, and the frequencies used to define hearing impairment (or handicap).

With both methods, the user needs to be sure that the exposed and nonexposed populations are carefully matched for such factors as age and nonoccupational noise exposure. Some would argue that populations should also be matched for sex and race, although this is not always practicable. Especially when using the percent risk method, it is very important to choose an appropriately matched population since small differences in population hearing threshold levels can have large effects on the percentage of people exceeding the selected fence.

Table 16.2 shows an example of the percentage risk method taken from ISO 1999, Annex D (ISO, 1990). (The same graph and information are reprinted as Annex E in ANSI S3.44-1996.) According to the ISO and ANSI data and methods, the risk in a male population due to noise at 90 dBA after 30 years of exposure would be 13 percent, using the frequencies 1000, 2000, and 4000 Hz and a fence of 27 dB (see left column of data). Using the same data and method, but changing just a few parameters, the risk becomes 19.5 percent (right column). By examining hearing impairment at age 60 instead of age 50, with 40 years of exposure instead of 30, the risk has increased considerably.[1]

TABLE 16.2
Example of two populations at risk from noise exposure
(adapted from ISO 1999, Annex D and ANSI S3.44, Annex E).

Noise exposure level	90 dB L_{A8hn}	90 dB L_{A8hn}
Sex	Male	Male
Age	50	60
Years of exposure	30	40
Non-noise population	Database A	Database A
Frequencies	1,2,4 kHz	1,2,4 kHz
Fence	27 dB	27 dB
Risk from aging	3%	15.5%
Risk from noise plus aging	16%	35%
Percent risk	13%	19.5%

ACGIH Threshold Limit Values®

The ACGIH sets TLVs® and biological exposure indices for various chemical and physical agents in the workplace. Although its recommendations do not directly affect current government regulations, such as those of OSHA and MSHA, they carry considerable weight in the scientific and technical communities. Some large companies and municipalities make a policy of adopting ACGIH recommendations.

---

[1] These figures are slightly different from those that would be obtained using the example in ISO Annex D (ANSI Annex E). In the ISO/ANSI examples, the user makes an adjustment to the NIPTS values, whereas it is more correct to make this adjustment to the hearing loss from aging, as has been done here.

In 1994 the ACGIH revised its TLV® for noise by changing from the 5-dB to the 3-dB exchange rate (ACGIH, 1994). The current TLV® recommends a permissible exposure limit (PEL) of 85 dBA and specifies TLVs® for 24 hours at 80 dBA down to 0.11 seconds at 139 dBA. No exposure to any type of noise is recommended in excess of a peak C-weighted sound level of 140 dB, and a hearing conservation program "with all its elements" is necessary when the TLVs® are exceeded. When the ACGIH issues a revised TLV®, as it did in 1994, documentation is also published to explain and support the Conference's decision (Sliney, 1993). This documentation is available, along with TLV® booklets, from the ACGIH headquarters in Cincinnati.

NIOSH Criteria Document

The National Institute for Occupational Safety and Health (NIOSH) has the responsibility for conducting research and establishing criteria that are then considered by other federal agencies and may be implemented in future regulations. Recently NIOSH (1998) issued "Criteria for a Recommended Standard: Occupational Noise Exposure, Revised Criteria 1998," which is an update of an earlier document (NIOSH, 1972). This document provides a review of scientific and technical information concerning prevention of occupational hearing loss, plus NIOSH's recommendations for a workplace noise standard. Unless the recommendations are adopted by regulatory agencies, they are not binding. However, hearing conservation professionals may want to review the document because it may influence what is considered "best" practice and hence future litigation in the courts, and it can provide suggestions that would enhance their local hearing conservation programs (HCPs).

Selected highlights of the revised criteria document include a recommended exposure limit (REL) of 85 dBA, an exchange rate of 3 dB, a new definition of standard threshold shift (STS) as a shift for the worse of 15 dB at any test frequency confirmed to be persistent by a follow-up audiogram, and a recognition of the need for derating hearing protector labeled attenuation values, or preferably testing those products in conformance with Method B of the new standard (ANSI, 1997) so that no derating is necessary. Key features of the NIOSH recommendations as compared to the current OSHA and MSHA regulations (see next section) are provided in Table 16.3. Note that unlike the OSHA and MSHA regulations discussed below, which have an action level and a permissable exposure limit, NIOSH uses a single REL above which action is required.

Federal Government Regulations (U.S.)

U.S. Department of Labor

Most federal agencies within the U.S., with the exception of the Department of Defense (DOD), follow the lead of the Labor Department and OSHA in the development of health and safety regulations. Another agency in the Labor Department, the Mine Safety and Health Administration, is also concerned with safety and health.

TABLE 16.3

Comparison of U.S. hearing conservation regulations, interpretations, and recommendations.

The table is provided to permit a quick comparison of the hearing conservation requirements of U.S. general industry (OSHA, 1983a), mining (MSHA, 1999), and recommendations of the NIOSH (1998) *Criteria for a Recommended Standard: Occupational Noise Exposure*. The OSHA regulation appears in its entirety in Appendix I of this text. The MSHA regulation is not included because it is so new (as of this printing) that it may yet be challenged in court. Please note following conditions for use of this table.

1. The MSHA regulation was published September 13, 1999 with an effective date of September 13, 2000. This table is current as of spring 2000 but litigation could cause changes before implementation. Check with MSHA (see web address below) for latest status.
2. The *Criteria Document* is a NIOSH recommendation, not a compliance document, but can be construed as a "best practices" guide.
3. Recordable or reportable hearing loss is addressed under OSHA in 29 CFR 1904, and directly in the MSHA rule.

This analysis is not intended to be all-inclusive; please check with the applicable agency for updates and current status. OSHA information is available at <http://www.osha.gov>; MSHA at <http://www.msha.gov>; and NIOSH at <http://www.cdc.gov/niosh>.

Issue	Description and Definition	OSHA 29 CFR 1910.95	MSHA 30 CFR Part 62	NIOSH Pub. No. 98-126
Action Level (AL)	The time-weighted average (TWA) exposure which requires program inclusion, hearing tests, training, and optional hearing protection.	AL = 85 dBA TWA. AL is exceeded when TWA ≥ 85 dBA, integrating all sounds from 80–130 dBA.	Similar to OSHA, except integration is for all sounds from 80 *to at least* 130 dBA.	Does not have AL; rather has a single Recommended Exposure Limit (REL, see next row) for hearing loss prevention, noise controls and HPDs.
Permissible Exposure Limit (PEL)	The TWA, which when exceeded, requires feasible engineering and (MSHA)/or (OSHA) administrative controls, and mandatory hearing protection.	PEL = 90 dBA TWA. PEL is exceeded when TWA > 90 dBA, integrating all sounds from 90–140 dBA, as inferred from Table G-16 of 1910.95(b).	Similar to OSHA, except integration range is explicit in the reg. (62.101, Definitions), and is for all sounds from 90 *to at least* 140 dBA.	REL = 85 dBA TWA. REL is exceeded when TWA ≥ 85 dBA, integrating all sounds from 80–140 dBA
Exchange Rate	The rate at which exposure accumulates; the change in dB TWA for halving/doubling of allowable exposure time.	5 dB.	Same as OSHA.	3 dB.

— continued on next page —

TABLE 16.3 — continued

Comparison of U.S. hearing conservation regulations, interpretations, and recommendations.

Issue	Description and Definition	OSHA 29 CFR 1910.95	MSHA 30 CFR Part 62	NIOSH Pub. No. 98-126
Ceiling Level	The limiting sound level above which employees cannot be exposed.	No exposures > 115 dBA; there is evidence that this ceiling level is not being enforced.	"P" code violation issued for any protected or unprotected exposures > 115 dBA.	No protected or unprotected exposure to continuous, varying, intermittent, or impulsive noise > 140 dBA.
Impulse Noise	Noise with sharp rise and rapid decay in level, ≤ 1 sec. in duration, and if repeated, occurring at intervals > 1 sec.	To be integrated with measurements of all other noise, but *should* not exceed 140-dB peak SPL.	To be integrated with measurements of all other noise.	To be integrated with measurements of all other noise, but not to exceed 140 dBA.
Monitoring	Assessment of noise exposure.	Once to determine risk and HCP inclusion; from there as conditions change resulting in potential for more exposure.	Mine operator must establish system to evaluate each miner's exposure sufficiently to determine continuing compliance with rule.	Every 2 years if any exposure ≥ 85 dBA TWA.
Noise Control	Investigation and implementation of feasible engineering and administrative control measures.	Feasible controls required where TWA > 90 dBA; subsequent compliance policy (which may be changed/revoked by OSHA at any time) permits proven effective HCP in lieu of engineering where TWA < 100 dBA.	Feasible engineering and administrative controls required for TWA > 90 dBA; even if controls do not reduce exposure to the PEL, they are required if feasible (i.e., ≥ 3-dBA reduction). Administrative controls must be provided to the miner in writing and posted.	Feasible controls to 85 dBA TWA. Administrative controls must not expose more workers to noise.
Hearing Protection	Exposure requirements and conditions for use of hearing protection devices (HPDs).	Optional for ≥ 85 dBA TWA; mandatory for > 90 dBA TWA, and for ≥ 85 dBA TWA for workers with STS. Protect to 90 or to 85 with STS. Choices must include a "variety" which is interpreted as at least 1 type of plug and 1 type of muff.	Use requirements same as OSHA, but amount of protection not specified, and choices must include 2 plugs and 2 muffs. Double hearing protection (muff plus plug) required at exposures >105 dBA TWA.	Mandatory for ≥ 85 dBA TWA; must protect to 85. Double hearing protection (muff plus plug) recommended at exposures > 100 dBA TWA.

Evaluation of Hearing Protector Effectiveness	Method of assessing adequacy of HPDs.	Use manufacturer's labeled NRRs to assess adequacy, but subsequent compliance policy stipulates 50% derating of NRRs to compare relative effectiveness of HPDs and engineering controls	No method included in standard. Preamble to regulation indicates that compliance guide will follow with suggested procedures.	Labeled NRRs must be derated by 25% for muffs, 50% for foam plugs, and 70% for other earplugs unless data available from ANSI S12.6-1997 Method B.
Supervisor of Audiometric Testing	The person who conducts or who is responsible for the conduct of audiometric testing and review.	Licensed or certified audiologist, otolaryngologist, or other physician.	Licensed or certified audiologist, or physician.	Audiologist or physician.
Audiometric Technician	The person who conducts audiometric testing and routine review under guidance of a professional supervisor.	Must be responsible to supervisor (see above). CAOHC certified, or has demonstrated competence to supervisor. When microprocessor audiometers used, certification not required.	Must be under direction of supervisor (see above). Must be certified by CAOHC or equivalent certification organization.	Must be under direction of supervisor (see above). Must be certified by CAOHC or equivalent certification organization.
Audiometry	Initial and ongoing hearing tests used to assess the efficacy of hearing conservation measures.	Required annually for all workers exposed ≥ 85 dBA TWA. Baseline test within 6 months of exposure; 12 months if using mobile testing service, with HPDs in the interim.	Same as OSHA, but choice of whether or not to take an audiogram is at miner's discretion.	Required for all workers exposed ≥ 85 dBA TWA. Baseline test preplacement or within 30 days of exposure Best practice is to test workers exposed > 100 dBA TWA twice per year.
Quiet Period Prior to Baseline Audiogram	Period of nonexposure to workplace noise required prior to baseline audiogram.	14 hrs.; use of HPDs acceptable as alternative.	Same as OSHA.	No exposure to noise ≥ 85 dBA for 12 hrs.; HPDs can not be used as alternative.

— continued on next page —

TABLE 16.3 — continued
Comparison of U.S. hearing conservation regulations, interpretations, and recommendations.

Issue	Description and Definition	OSHA 29 CFR 1910.95	MSHA 30 CFR Part 62	NIOSH Pub. No. 98-126
Background Noise	Permissible noise in audio-metric test chamber during testing.	Levels specified as 40 dB @ 500 and 1000 Hz, 47 dB @ 2000 Hz, 57 dB @ 4000 Hz, and 62 dB @ 8000 Hz.	According to scientifically validated procedures.	Per ANSI S3.1-1999 or latest revision; 19 dB more stringent than OSHA at 500 Hz, and 13 to 25 dB more stringent at other frequencies.
Audiogram Review and Employee Notification	Required actions following audiograms.	Not specified unless STS is detected; see STS follow-up.	Audiograms must be reviewed within 30 days and feedback provided in writing to each miner within 10 days thereafter.	Not specified unless STS is detected; see STS follow-up.
STS (OSHA/MSHA– Standard Threshold Shift; NIOSH–Significant Threshold Shift;	A change in hearing compared to an earlier (baseline) hearing test that requires follow-up action.	≥ 10-dB average shift from baseline hearing levels at 2000, 3000 and 4000 Hz in either ear.	Same as OSHA.	≥ 15-dB shift for the worse from baseline at any test frequency, in either ear, confirmed with follow-up test for same ear/frequency.
STS Retests	Follow-up audiogram that is permitted or required when initial STS is detected.	May obtain retest within 30 days and substitute for annual audiogram.	Same as OSHA.	Must provide confirmation audiogram within 30 days.
STS Follow-up	Required actions when an STS is detected.	Notify worker within 21 days; unless STS is not work-related, must fit or re-fit employee with HPDs and select higher attenuation if necessary, refer for audio/otological exam if more testing needed or problem due to HPDs, and inform employee of need for exam if problem unrelated to HPD usage is suspected.	Within 30 days of receiving evidence or confirmation of STS, unless STS is not work-related, must retrain the miner, provide miner an HPD or a different HPD, and review effectiveness of any engineering and administrative controls to correct deficiencies.	Notify worker within 30 days; must take action such as explain effects of noise, reinstruct and refit with HPDs provide additional training in hearing loss prevention, or reassign to quieter area.

Baseline Revision	Procedures for revising the baseline audiogram to reflect changes in hearing.	Annual audio substituted for baseline when STS is *persistent* or thresholds show significant improvement.	Annual audio substituted for baseline when STS is *permanent* or thresholds show significant improvement.	Annual audio substituted for baseline when confirming audiogram validates an STS.
Presbycusis or Age Correction	Adjustments for hearing levels for anticipated effects of age.	Allowed.	Same as OSHA.	Not allowed.
Recordable or Reportable Hearing Loss	Amount of hearing loss triggering reporting requirements on workplace injury/illness logs.	By OSHA directive, ≥ 25-dB average shift from original baseline at 2000, 3000, and 4000 Hz, in either ear, w/age correction; rule change pending.	≥ 25-dB average shift from baseline, or revised baseline, at 2000, 3000, and 4000 Hz in either ear.	Not indicated.
Training and Education	Description of the annual training and educational component of the hearing conservation program.	Annual for all employees exposed ≥ 85 dB TWA; include effects of noise, HPDs, and purpose and explanation of audiometry.	Same as OSHA, except must begin within 30 days of enrollment in HCP, and include description of mine operator and miner's responsibilities for maintaining noise controls.	Same as OSHA, but must also include psychological effects of noise, and roles and responsibilities of both employers and workers in program.
Warning Signs and Postings	Requirements to post signs for noisy areas or to post regulations.	Hearing conservation amendment shall be posted in workplace.	No requirements for posting reg., but when administrative controls are utilized the procedures must be posted.	Signs must be posted at entrance to areas with TWAs routinely ≥ 85 dBA.
Record Retention	Specification on retention of data, and transfer requirements if employer goes out of business.	Noise surveys for at least 2 yrs., hearing tests for duration of employment, with requirement to transfer records to successor if employer goes out of business.	Employee noise exposure notices and training records for duration of enrollment in HCP + 6 months, and hearing tests for duration of employment + 6 months, with requirement to transfer records to successor mine operator.	Noise surveys for 30 yrs., hearing tests for duration of employment + 30 yrs., calibration records for 5 yrs., with record transfer per 29CFR1910.20(h).

OSHA

In 1970, the U.S. Congress enacted the Occupational Safety and Health Act (Public Law 91-596), which gave employers the duty to protect their workers from all kinds of chemical and physical hazards, including noise. Congress gave the responsibilities for research and criteria development to NIOSH, located in the U.S. Department of Health and Human Services, presumably to separate these functions from the more political and pragmatic rule-making process. The promulgation and enforcement of regulations were given to OSHA.

Even before Congress passed the Occupational Safety and Health Act, the Department of Labor had issued a noise regulation, as mentioned above, and in 1971 this regulation became applicable to all industry engaging in interstate commerce. It called for a PEL of 90 dBA with a 5-dB exchange rate (the relationship between noise level and duration), reduction of noise levels to the PEL by engineering or administrative controls whenever feasible, the provision and wearing of hearing protection devices (HPDs) above the PEL, and, in section (c), the conduct of a "continuing, effective hearing conservation program" for employees exposed above the PEL.

In 1981 OSHA amended its existing noise regulation, making specific requirements for noise measurement, audiometric testing, the use and care of HPDs, employee training and education, and recordkeeping (OSHA, 1981). The agency promulgated a revision of the hearing conservation amendment in 1983, which is still in effect at this time (OSHA, 1983a). Although the hearing conservation amendment replaces section (c) of the existing noise standard, sections (a) and (b) detailing the PEL and the requirement for feasible engineering or administrative controls are still in effect. The key features of the regulation are summarized in Table 16.3 and the complete regulation (29 CFR[2] 1910.95) appears in Appendix I in this text. See Table G-16a in Appendix I for an explicit listing of PEL values.

The 1970 Occupational Safety and Health Act also allows state agencies to issue and enforce their own regulations, as long as these regulations are at least as protective as those promulgated by the federal OSHA. About half of the states have chosen to do this while the rest choose to have federal enforcement.

In addition to the regulation for general industry (which includes utilities and maritime), OSHA has a separate noise regulation for the construction industry (29 CFR 1926.52). It is very similar to the original noise regulation promulgated under the authority of the Walsh-Healey Contracts Act, and has not yet been amended for specific requirements for hearing conservation programs, although in late 1999 OSHA announced its intention to issue a notice of proposed rulemaking to update and expand this regulation. To date it has not been rigorously enforced.

---

[2] CFR is the abbreviation for the *Code of Federal Regulations*, which is a compilation of regulations published annually by the federal government. Regulations, such as those issued by OSHA or MSHA will be reprinted identically in the CFR each year until such time as they are officially changed. Announcements of changes will always appear first in the *Federal Register*, which is analogous to the government's newspaper.

There are other occupational settings under OSHA's jurisdiction which, however, are not covered by OSHA's noise regulation. Examples are agriculture and oil and gas well drilling and servicing. These and other such nonindustrial occupations are presumably covered by the Occupational Safety and Health Act's "General Duty" clause, Section 5 (a), which states: "Each employer . . . shall furnish to each of his employees employment and a place of employment which are free from recognized hazards that are causing or are likely to cause death or serious physical harm to his employees. . . ." This clause, however, is rarely invoked for noise exposure.

Enforcement of OSHA's regulations is carried out by compliance officers (inspectors) located in OSHA's regional, area, and district offices, and, of course, in state OSHA agencies. Inspections are supposed to be unannounced, but if voluntary entry is denied, compliance officers must obtain a warrant prior to entering the workplace. Upon arrival, OSHA representatives need to confer with representatives of management and employees. They also have the duty to issue a citation for any hazard that is not properly controlled, in addition to the hazard for which the inspection was initiated.

The general requirement for hearing conservation programs has been enforced fairly seriously over the past 2 decades. Although the agency's preference for engineering controls has continued to prevail for all other hazards, these controls have not been actively enforced for noise. In 1983, OSHA instructed its federal compliance officers to issue citations for the absence of engineering controls only when such controls were economically as well as technically feasible *and* when employee noise exposures exceed an average TWA of 100 dBA (OSHA, 1983b). Engineering noise control is not considered necessary when effective hearing conservation programs are in place, but OSHA has never defined the word "effective." Even when exposures exceed a TWA of 100 dBA, federal OSHA personnel are to use their discretion when deciding whether or not to issue a citation. As a result, the number of citations for the absence of feasible engineering controls has declined dramatically, along with the impetus for industry to use them.

But two facts are important to remember in this regard. First, because this instruction is a policy and was not put forward after "public notice and comment," it does not have the force of regulation and can be rescinded. In fact, as of late 1999, there were indications that this policy may be withdrawn. Second, some of the state OSHA programs do not subscribe to the federal policy and still enforce the use of engineering controls to some extent. Industrial personnel as well as safety and health consultants in "state program" states should check with the authorities as to whether or not the 100-dBA policy holds in their state.

MSHA

Shortly before Congress passed the Occupational Safety and Health Act, it enacted the Federal Coal Mine Health and Safety Act of 1969 and later the Federal Mine Safety and Health Amendments Act of 1977. Under the authority of these acts MSHA promulgated noise regulations for surface and underground coal mines, and different regulations for surface and underground metal and nonmetal mines. All of these regulations were similar to the Labor Department's original noise regulation of 1969. In 1999 MSHA issued a single comprehensive noise regulation that applies uniformly to all kinds of mining (MSHA, 1999). The new regulation is similar to the 1983 OSHA noise regulation in that it retains the 90-dBA PEL and the 5-dB exchange rate, and calls for an 85-dBA action level. Many of the hearing conservation requirements are similar, except that MSHA's noise monitoring provisions are very general and the audiometric testing requirements are also somewhat less specific. Two additions are the requirements for dual hearing protection at TWAs greater than or equal to 105 dBA, and for mine operators to provide miners with a choice of at least 2 plugs and 2 muffs. Table 16.3 compares the OSHA and MSHA regulations as well as the NIOSH Criteria Document.

Other Federal Agencies

Most federal agencies promulgating noise regulations take their lead from OSHA, as mentioned above. There are some variations, however. For example, the U.S. Department of Transportation regulates the noise exposure of truck and bus drivers through the Federal Highway Administration's Bureau of Motor Carrier Safety (DOT, 1973). The Bureau issued a noise standard in 1973 calling for a maximum noise level of 90 dBA at the driver's position, along with requirements for measuring this level. A 2-dB tolerance is permitted, so the effective enforcement level is 92 dBA. The standard does not specify a time-weighted average level, but the maximum permissible driving time is 10 hours. Although the standard has been unchanged to date (as 49 CFR 393.94), it could be revised at any time.

Another example of a federal noise program is the U.S. Coast Guard guidelines for crews on board U.S. commercial vessels (Coast Guard, 1982). The guidelines recommend the evaluation of crew members' 24-hour exposures, a criterion level, $L_{OSHA,24}$ of 82 dBA, the wearing of HPDs at an $L_{OSHA,24}$ of 85 dBA, and an action level of $L_{OSHA,24}$ of 77 dBA, where HCPs should be initiated. This is equivalent to OSHA's criteria on a 24-hr basis.

Readers should keep in mind that any federal noise recommendations or regulations may be changed at any time, but the latter must follow public notice and comment procedures.

Department of Defense

DOD Instructions

The Department of Defense (DOD) issues "Instructions" periodically on various matters, including safety and health, to be used by all three branches of the U.S. Armed Services. A revised Instruction was issued on April 22, 1996 for the conduct of hearing conservation programs (DOD, 1996). Like the preceding instruction, (DOD, 1991), it requires heads of all departments to establish and maintain hearing conservation programs when equivalent daily exposures equal or exceed 85 dBA and requires an exchange rate at least as protective as 4 dB. Unlike its predecessor, it "strongly recommends" the use of a 3-dB exchange rate.

Because the primary means of reducing exposure to hazardous noise is engineering controls, the 85-dBA requirement (using either the 3-dB or 4-dB exchange rate) may be considered a PEL. The specific aspects of the hearing conservation program are:

- Noise measurement and analysis
- Safety signs and labels, including labeling of tools, vehicles, or pieces of equipment emitting hazardous levels of noise (unless an entire area is designated noise hazardous)
- Noise abatement when feasible, including the establishment of priorities for noise control resources according to a Risk Assessment Code, and purchase requirements for new equipment
- Fitting, training in, and use of HPDs
- Education of noise-exposed personnel
- Audiometric testing
- Personnel assignments, which may include criteria for hearing sensitivity
- Access to information, training materials, and records
- Recordkeeping
- Program performance evaluation.

Each branch of the service may promulgate its own rules, but they must be at least as stringent as the current DOD Instruction. At this time the Army, Air Force, and Navy are standardizing some 175 reports so there will be a uniform approach to gathering and reporting data among all three services. The program should be fully developed and in use by 1999 (Ohlin, 1998).

Army

The U.S. Army's instruction (Army, 1994a) now requires enrollment in a hearing conservation program when equivalent daily exposures equal or exceed 85 dBA, whereas the previous Army instructions specified sound levels of 85 dBA or above. Furthermore, the Army has now adopted the 3-dB exchange rate. Whereas the old system based on levels may have been easier to administer, the new system based on exposures includes fewer soldiers in the HCP, but the ones who are involved in the program are those who actually need to be.

The Army policy references the U.S. Army Environmental Hygiene Agency Technical Guide 181 (Army, 1994b), which contains detailed recommendations for sampling strategies, dosimetry protocol, advice for measuring work shifts longer or shorter than 8 hours, and other kinds of guidance which could be useful to professionals outside the military as well.

Navy

The current U.S. Navy instructions are for shore-based personnel (Navy, 1994a). The program requirements for shipboard personnel are basically the same, with the exception that these personnel must have a medical exam to confirm readiness for certain kinds of duty if the total of hearing levels at 3000,

4000, and 6000 Hz in both ears exceeds 270 dB (Navy, 1994b). The Navy's HCP conforms to the DOD Instruction, but unlike the Army and the Air Force, its PEL is 84 dBA with a 4-dB exchange rate (Navy, 1994a and 1994b).

Air Force

Requirements of the current U.S. Air Force regulation are consistent with the DOD instruction and specify an 85-dB PEL using a 3-dB exchange rate (Air Force, 1993). Also included are limits for whole-body effects (no octave- or 1/3-octave band levels above 145 dB for frequencies from 1 Hz to 40 kHz) and for exposure to ultrasound (1/3-octave band levels from 10 kHz to 50 kHz ranging from 80 dB to 115 dB respectively).

International Standards and Regulations for Occupational Exposure to Noise

With the advent of jet travel and multinational corporations the world seemed to become smaller. Now with the communications revolution, involving faxes, modems, and the Internet, the world is smaller still. Industrial health and safety professionals have new responsibilities for knowledge of and compliance with the regulations and practices of other nations.

Keeping abreast of occupational noise standards around the world is a challenging task, since nations may amend or revise their standards at any time. Table 16.4 represents an attempt to summarize key elements of various nations' standards, but the reader needs to be aware that the information may change in the coming months and years. Although the standards of many nations are outlined here, the list is necessarily incomplete. Anyone doing business in a foreign country should *always* consult the newest versions of that country's noise standards and existing enforcement policies. The information presented here is meant to provide guidance only and should not be used for compliance purposes.

Recommendations of the I-INCE Working Party on Upper Noise Limits in the Workplace

The General Assembly of the International Institute of Noise Control Engineering (I-INCE) approved a Working Party in 1992 to review current knowledge and practice on upper noise limits in the workplace. Made up of 11 members from 8 different countries, the Working Party was convened by Tony F.W. Embleton (Embleton, 1994). After several drafts and much deliberation, the recommendations were approved by the I-INCE Member Societies (Embleton, 1997). The recommendations are intended especially to help countries that have no current regulations. They are summarized below:

- An exposure limit, L_{A8hn} of 85 dBA, which incorporates the contribution from all sounds (including short-term, high-intensity sounds). This limit should be

TABLE 16.4
Permissible exposure limits (PEL), exchange rates, and other
requirements for noise exposure according to nation.

Nation Date	PEL Equivalent 8-hour dBA	Exchange Rate (dBA)	L_{ASmax} rms L_{pk} SPL	Level* (dBA) Engineering Control	Level* (dBA) Audiometric Test	Comments
Argentina	90	3	115 dBA**	85	85	
Australia, 1993	85	3	140 dB peak	85	85	Note (1)
Austria	85		110 dBA	90		
Brazil, 1992	85	5	115 dBA 130 dB peak	90	85	
Canada, 1990	87	3		87	84	Note (2)
CEC, 1986	85	3	140 dBC peak	90	85	Notes (3), (4)
Chile	85	5	115 dBA 140 dB peak			
China, 1985	70-90	3	115 dBA			Note (5)
Finland, 1982	85	3		90		
France, 1990	85	3	135 dBC peak	90	85	
Germany, 1990	85 70, 55	3	140 dBC peak	90	85	Note (3) Note (6)
Hungary	85	3	125 dBA 140 dBC peak	90		
India, 1989	90		115 dBA 140 dBA peak			Note (7)
Israel, 1984	85	5	115 dBA 140 dBC peak			
Italy, 1990	85	3	140 dBC peak	90	85	
Japan, 1992	90	3		85	85	Note (8)
Netherlands,	85	3	140 dBC peak	90	80	
New Zealand, 1995	85	3	140 dB peak	85	85	Note (9)
Norway, 1982	85 70, 55	3	110 dBA		80	Note (10)
Poland	85	3	115 dBA 135 dBC peak			
Spain, 1989	85	3	140 dBC peak	90	80	
Sweden, 1992	85	3	115 dBA fast 140 dBC peak	85	85	
Switzerland	85 or 87	3	125 dBA 140 dBC peak	85	85	
U.K., 1989	85, 90	3	140 dBC peak	90	90	Note (11)
U.S., 1983	90	5	115 dBA 140 dB peak	90	85	Note (12)

— continued on next page —

TABLE 16.4 — continued
Permissible exposure limits (PEL), exchange rates, and other requirements for noise exposure according to nation.

Nation Date	PEL Equivalent 8-hour dBA	Exchange Rate (dBA)	L_{ASmax} rms L_{pk} SPL	Level* (dBA) Engineering Control	Level* (dBA) Audiometric Test	Comments
Uruguay	90	3	110 dBA			
I-INCE, 1996	85	3	140 dBC peak	85	85	Note (13)

Sources for Table 16.4:

Jorge P. Arenas, Institute of Acoustics, Universidad Austral de Chile, Valdivia, Chile (for South American data; paper presented at the 129th meeting of the Acoustical Society of America, 1995.)

Pamela Gunn, Dept. Occup. Health, Safety & Welfare, Perth, Western Australia (personal communication).

Tony F.W. Embleton, "Technical assessment of upper limits on noise in the workplace," *Noise/News International*, I-INCE, Poughkeepsie, NY, 1997.

ILO, Noise Regulations and Standards, CIS data base, International Labour Office, Geneva, Switzerland (summaries), 1994.

Published standards of various nations and papers as referenced.

Notes and Comments for Table 16.4:

* Like the PEL, the levels initiating the requirements for engineering controls and audiometric testing also, presumably, are average sound levels normalized to 8 hours.

** Unless otherwise stated, the L_{max} is specified in terms of the slow meter response.

Note (1) The data listed are for the Australian national standard (Worksafe Australia, 1993); the various states issue their own standards. The national standard is supplemented by a code of practice, which includes goals and specific procedures for engineering controls, hearing tests, and other elements of the hearing conservation program. There is a recent standard developed jointly by noise and hearing conservation professionals from Australia and New Zealand (SA/SNZ, 1998). It refers to "occupational noise management" and provides guidance on applying noise control measures to existing equipment and purchasing quieter equipment in the future.

Note (2) There is some variation among the individual Canadian provinces: Ontario, Quebec, and New Brunswick use 90 dBA with a 5-dB exchange rate; Alberta, Nova Scotia, and Newfoundland use 85 dBA with a 5-dB exchange rate; and British Columbia uses 85 dBA with a 3-dB exchange rate. All require engineering controls to the level of the PEL. Manitoba requires certain hearing conservation practices above 80 dBA, hearing protectors and training on request above 85 dBA, and engineering controls above 90 dBA.

Note (3) The Council of the European Communities (CEC, 1986) and Germany (UVV Laerm, 1990) state that it is not possible to give a precise limit for the *elimination* of hearing hazard and the risk of other health impairments from noise.

Note (4) Those countries comprising the European Community were required to have standards that conformed to the CEC Directive, or were at least as stringent, by January 1, 1990. This means that they must require the employer to reduce the noise level as far as possible, taking technical progress and the availability of control measures into account. They must also require the declaration of sound power levels of machinery to be purchased, and the use of reduced noise reflection in buildings, regardless of sound pressure or exposure levels.

Note (5) China requires different levels for different activities: e.g., 70 dBA for precision assembly lines, processing workshops, and computer rooms; 75 dBA for duty, observation, and rest areas; 85 dBA for new workshops; and 90 dBA for existing workshops.

— continued on next page —

TABLE 16.4 — continued

Note (6) Germany also has noise standards of 55 dBA for mentally stressful tasks and 70 dBA for mechanized office work.

Note (7) India — recommendation only.

Note (8) The Japanese standard makes the wearing of hearing protection mandatory at 90 dBA. It also states that efforts shall be made to reduce the noise level to 85 dBA, and employers whose workplaces exceed 90 dBA shall "forth with [sic] tackle the noise level" to bring it below 90 dBA (Zusho and Miyakita, 1996).

Note (9) New Zealand has joined with Australia in a coordinated occupational noise management standard (SA/SNZ, 1998).

Note (10) Norway requires a PEL of 55 dBA for work requiring a large amount of mental concentration, 70 dBA for work requiring verbal communication or great accuracy and attention, and 85 dBA for other noisy work settings. Recommended limits are 10 dB lower. Workers exposed to noise levels greater than 85 dBA should wear hearing protectors.

Note (11) The United Kingdom "Noise at Work" regulation requires employers to reduce the risk to the lowest level reasonably practical. Two "action levels" are specified: 85 dBA is required for noise exposure assessment, training, the provision of hearing protection devices to workers who request them, and the maintenance and use of equipment; and 90 dBA is required for noise reduction by "means other than ear protectors" and mandatory provision and wearing of HPDs. Audiometric testing is not required but recommended at an L_{A8hn} of 90 dBA under separate regulations, the "Management of Health and Safety at Work Regulations, 1992."

Note (12) These levels apply to the OSHA noise standard, covering workers in general industry and maritime. As explained above, the U.S. military services require standards that are somewhat more stringent.

Note (13) Reflects a consensus of the members of the International Institute of Noise Control Engineering (see details below).

achieved as soon as possible given pertinent economic and "sociological" factors. If impulse noise is further addressed, the limit should be 140 dBC_{pk}.

- An exchange rate of 3 dB regardless of time-variation.
- Engineering controls are the preferred method of reducing noise exposure. A long-term noise control program should be implemented when L_{A8hn} exceeds 85 dBA. Noise should be reduced to the lowest level that is economically and technologically reasonable, even when there may be no risk of long-term hearing damage. This is to minimize other negative effects, such as reduced productivity, stress, and disruption of speech communication. Consideration should be given to sound and vibration isolation between noisier and quieter areas at the design stage of any new installation. Maximum sound power level and sound pressure level (at the operator's position) should be included in purchase specifications for new and replacement machinery.
- Hearing protection devices should be encouraged when noise control measures are unable to reduce exposures to an L_{A8hn} of 85 dBA and should be mandatory over 90 dBA.

- Employers should conduct audiometric testing of workers exposed to an L_{A8hn} of 85 dBA at least every 3 years or at shorter intervals depending upon exposure level and past history of the individual worker. Records of these tests should be preserved in the worker's permanent file.

Summary of Nations' Noise Exposure Standards and Regulations

At this time, most nations use 85 dBA as the PEL. A few use 90 dBA and some require varying PELs depending upon the nature of the work (China, Germany, and Norway). Even though most nations have placed the PEL at 85 dBA, more than half still use 90 dBA for compliance with requirements for engineering control, as allowed by the CEC. In these cases the PEL may be considered an acceptable "upper limit" (see Embleton, 1997), but nations may have chosen to require 90 dBA for engineering controls for economic reasons. The 85-dBA "upper limit" is not to be confused with the "action level" used in the U.S. As stated above, the CEC requires its member countries to urge employers to control noise to the lowest level practicable.

Nearly every nation listed above has adopted the 3-dB exchange rate, with the exception of Brazil, Chile, Israel, and civilian U.S., which use the 5-dB rule.

Most nations limit impulsive noise exposure to a peak sound pressure level of 140 dB, (or dBC as recommended by the CEC), while a few use slightly lower limits (Brazil, France, Poland). Quite a few limit continuous noise to 115 dBA, while some use lower limits (e.g., Norway at 110 dBA) and some higher limits (e.g., Switzerland at 125 dBA).

Standards and Regulations Concerning Effects Other Than Hearing Loss

In addition to protecting workers against hearing loss, several nations include requirements to protect against other adverse effects of noise. Both the CEC directive and the German standard acknowledge that workplace noise involves a risk for the health and safety of workers beyond hearing loss, but that current scientific knowledge of the nonauditory effects does not enable precise safe levels to be set. They assume that reduction of noise to the lowest practicable level will lower the risks of these effects.

The nonauditory health effects are not the only ones addressed. For example, the Norwegians include a requirement that noise levels must not exceed 70 dBA in work settings where speech communication is necessary. The Germans advocate noise reduction for the prevention of accident risks, and both Norway and Germany require a maximum noise level of 55 dBA to enhance concentration and prevent stress during mental tasks.

Special Features of Certain Nations' Standards and Regulations

Some countries have special noise standards or regulations for different kinds of workplaces. In Norway, for example, auto repair shops, fish oil and meal factories, and foundries have their own noise regulations. Office noise is regulated in Germany, Japan, and the Netherlands, and steel mills and daycare centers have their own noise regulations in Sweden. Other nations treat noise as one of many regulated hazards in a particular process. Examples would be computer work and construction excavation in Sweden, cutting and welding in Denmark, and precision assembly work in China. Some nations use innovative approaches to attack the occupational noise problem. For example, the Netherlands has a separate regulation for newly constructed workplaces, and Australia and Norway give information to employers for instructing manufacturers to provide quieter equipment.

In 1989 the CEC established a directive requiring manufacturers to include instructions on noise level when any machinery exceeded a continuous level of 70 dBA or a peak level of 130 dBC, or when sound power levels exceeded 85 dBA (CEC, 1989). In response, EC member nations have regulated certain types of equipment for noise. For example, chain saw noise is regulated by France, Japan, Norway, and Sweden, construction equipment by Finland and France, and tractors by Finland, Hungary, and the Netherlands.

In addition to the above, some nations have promulgated separate standards or regulations for hearing protection devices (such as the CEC, Netherlands, and Norway) and for hearing conservation programs (such as France, Norway, Spain, Sweden, and the U.S.).

Enforcement and Compliance

There is no indication from the above material of the degree to which these standards and regulations are enforced. Some specify that employers "should" take certain actions (as in codes of practice or guidelines), while most specify that employers "shall." Those that use "shall" are more apt to be mandatory, but individual nations vary widely in their ability and inclination to enforce. Even within the same nation, such as the U.S., enforcement of occupational noise regulations may vary considerably among jurisdictions and particularly with the administration in power.

There is relatively little written information about the degree to which employers and government agencies comply with their nations' noise standards. Informal contact with professionals in various nations, however, indicates that most large companies tend to comply, some mid-sized companies comply, and few small companies do. In some nations, such as Sweden and the U.K., it appears that large companies make conscientious efforts to comply with noise regulations by reducing noise and protecting workers, whether or not standards are vigorously enforced by governmental bodies. In developing nations, such as China, the regulations are likely to be disregarded by most employers.

Use of HPDs

The use of HPDs has been documented by several researchers and tabulated by Berger (1996), who found that the percentage of HPD use in the manufacturing industries ranges from 3 percent for a study in Thailand to 84 percent in a Canadian study. Use among other noise-exposed individuals varies greatly according to several parameters, such as occupation, activity, nation, and definition of "use." It appears that HPDs are used more consistently in industrialized nations, such as the U.S., Canada, U.K., and Finland, than they are in developing nations, such as Thailand, Malaysia, and Egypt. Even in industrialized nations, however, HPD use can be erratic.

Implementation of HCP Elements

An informal survey of members of ISO TC43 WG5 (Royster and Royster, 1996) received responses from representatives of Austria, Canada, Italy, New Zealand, South Africa, Sweden, and Switzerland. Although this information is limited, it sheds some light on hearing conservation practices in some of the industrialized nations. The author found that all of the countries, with the possible exception of Italy and one Canadian province, require hearing conservation programs. Most of the countries whose representatives were surveyed require audiometric testing of noise-exposed personnel in government, industry, and the military. Enforcement of the various aspects of hearing conservation programs was defined in the survey as "hearing conservation program implementers actually receive penalties for failure to implement the phase." Engineering and administrative controls are always included, but often not enforced, according to the survey's definition of enforcement. The same is true for sound surveys and educational/motivational programs. The use of HPDs is included in all of the surveyed programs and appears to be the one element that is consistently enforced.

Enforcement of the OSHA Noise Regulation

The approach in the U.S. is rather different from that of other countries in that it is federal and state OSHA policy to issue citations for noncompliance whenever it is encountered. Enforcement by OSHA, however, has decreased over recent years (Suter, 1989), both for engineering controls and for other elements of the hearing conservation program. The number of inspections specifically for noise has declined and the penalties for noncompliance are minimal in most cases. As mentioned above, federal and most state OSHA inspectors do not issue citations for the absence of feasible engineering controls until TWAs reach 100 dBA (OSHA, 1983b), and even then rarely. However, this policy is subject to change. In this respect, incentive for noise control in the U.S. has not differed significantly from that in other nations.

In all probability, some European nations surpass the U.S. in the implementation of engineering noise control. This is particularly true of the U.K., where compliance with the "Noise at Work Regulations" (HSE, 1994) has been unusually successful. Perhaps the U.K.'s progress in achieving noise reduction could serve as a model for other nations.

Employment Criteria and the ADA

Two areas of standards and regulation have emerged relatively recently and should be of interest to professionals in occupational hearing conservation. One is the 1990 enactment of the Americans with Disabilities Act (ADA) and subsequent regulations. The other is the development of requirements for hearing sensitivity for employment in certain "hearing critical" jobs. These two areas are necessarily linked. The result is a complex set of issues, as those who formulate the regulations and those who hire persons with hearing impairments perform a delicate balancing act. They must try to protect the rights of hearing-impaired individuals on the one hand and the safety of coworkers and the public on the other, while making sure that the individual can adequately perform the job in question.

The Americans with Disabilities Act (ADA)

Regulations issued under the authority of the ADA require that employers may not discriminate in hiring or promotion against a person with a disability if he or she is otherwise qualified for the job. Employers may, however, ask all applicants about their ability to perform a job and even have them demonstrate this ability before being hired. Employers may also require a medical examination. If the person is qualified, the employer is then obligated to make "reasonable accommodation" to the person's limitations, as long as these accommodations do not impose an "undue hardship" on the operation of the business (defined as a significant difficulty and/or expense). Employers may determine that a person poses a significant risk of harm to the health or safety of that individual or others, and this risk cannot be eliminated or reduced by reasonable accommodation. This determination must be based on an individual assessment of that particular person's abilities and the specific functions expected of anyone in that job (Tri-Alliance, 1992).

Reasonable accommodations for a hearing-impaired individual may involve a variety of adjustments, some of which are already commonly used, like fax machines, personal computers, e-mail capabilities, and the addition of lighting. Others, such as assistive listening systems, telecommunication devices for the deaf, captioned videotapes, and visual warning signals can be installed at relatively modest expense. In some cases interpreters using sign language may be necessary. The type and extent of these accommodations depend upon the abilities and preferences of the hearing-impaired employee and the technical skills and responsibilities of the job. A Technical Assistance Manual for Title I of the ADA is available from the Equal Employment Opportunity Commission in Washington, DC, and free technical assistance to employers may be available from organizations such as the Job Accommodation Network (Jones, 1995).

Hearing Sensitivity Requirements for Employment
Federal Government

The Office of Personnel Management (OPM) issues physical or medical requirements for the performance of certain jobs, which apply to civilian employees of the government. LaCroix has summarized the role of the OPM in directing federal agencies' policies about interviewing and hiring persons with disabilities, including hearing impairment (LaCroix, 1996a and 1996b). Federal agencies must state the physical requirements of a job before a determination can be made about an applicant's qualifications. For jobs that require "normal hearing," however, there is no definition of "normal." The result, according to LaCroix, is a "patchwork" of requirements for federal employees, which sometimes appears to be arbitrary. In some instances a prosthesis may be allowed in the demonstration of adequate functioning.

The OPM maintains that medical standards and physical requirements apply equally to applicants and to individuals who are already employed (LaCroix, 1996b). Federal agencies may require an audiological evaluation when good hearing is necessary for the safe and efficient performance of a job, and a person may be disqualified when the job is in an environment that presents a substantial risk to his or her hearing. Yet, according to LaCroix, applicants for jobs in the federal service and in many states may not be denied employment solely because the employee's condition may become aggravated and result in a claim for worker's compensation at a later date (LaCroix, 1996a). Hence, the delicate balance.

Some of the federal positions with requirements for hearing sensitivity approved by OPM include customs inspectors, border patrol agents, criminal investigators, and motor vehicle operators, each with a slightly different requirement (LaCroix, 1996a). Federal agencies with varying requirements for hearing abilities include the ATF, FBI, IRS, and FAA (for air traffic controllers and pilots) (MacLean et al., 1993).

States and Cities

Most of the available information from states and cities concerns law enforcement officials. A survey of audiometric criteria in state and local agencies was conducted by MacLean et al. (1993). The authors found at that time that 28 states had criteria for law enforcement officials and 16 were planning new or revising existing criteria. The specificity of these requirements ranged from no requirements at all to very specific ones. For example, Alaska and Louisiana had no requirements for hearing and no plans to formulate any. Alabama and Maine had no requirements but were working on them, Arizona and New Mexico had only the requirement for "normal hearing" in new employees, and some states (e.g., North Carolina, Tennessee, and Arkansas) still required only whispered-voice tests at the time of the survey. Examples of states with very specific criteria would be Oregon, which required hearing sensitivity for new hires of at least 25 dB at 500, 1000, 2000, and 3000 Hz, aided speech reception threshold of 35 dB and scores of at least 90 percent on speech discrimination tests. Pennsylvania's requirements appear to be similar except for a maximum speech reception threshold of 25 dB.

Of approximately 40 city governments surveyed, 15 had audiometric criteria for law enforcement officials in place, 7 were in the process of developing them, and 10 had no plans at the time.

Both the ADA and the OPM policies stress the importance of evaluating the performance abilities of individuals in specific jobs on a case-by-case basis. An example of this is the case of a paramedic who was refused a job in the Nashville Fire Department because of a profound hearing loss in one ear, even though he had successfully completed the requirements of the job selection process and had performed as a paramedic in other jobs. He sued and the court found in his favor, stating that employers must not use personnel standards that require any categorical exclusions based on an applicant's physical conditions and that all hiring decisions must be based on individualized assessment (MacLean, 1998).

Summary

More than 9 million American workers and additional millions of workers in other nations are exposed to levels of noise that are capable of causing permanent hearing loss. Many of these workers are covered by noise regulations, directives, and recommendations issued by government bodies. In addition, there are numerous consensus standards designed to provide guidance and uniformity, some of which are incorporated into government regulations. These standards and regulations provide a foundation on which professionals can base their programs to prevent noise-induced hearing loss. Thus it is critical for professionals to be knowledgeable about the particulars of the standards and regulations that apply in their own circumstances. Professionals should also keep in mind that the various standards and regulations are subject to change, that in some cases they provide only minimal protection, and that they can, therefore, be improved upon to the benefit of employers and employees alike.

References

ACGIH (1994). "1994–1995 Threshold Limit Values® for Chemical Substances and Physical Agents, and Biological Exposure Indices," American Conference of Governmental Industrial Hygienists, Cincinnati, OH.

Air Force (1993). "Hazardous Noise Program," U.S. Air Force Occupational Safety and Health Standard 48-19.

ANSI (1996). "Determination of Occupational Noise Exposure and Estimation of Noise-Induced Hearing Impairment," American National Standards Institute, S3.44-1996, New York, NY.

ANSI (1997). "Methods for Measuring the Real-Ear Attenuation of Hearing Protectors," American National Standards Institute, S12.6-1997, New York, NY.

Army (1994a). "Medical Services: Hearing Conservation." DA PAM 40-501, as amended by cover memo from Col. Frederick J. Erdtmann, SGPS-PSP (40-5), 24 June 1994.

Army (1994b). "Noise Dosimetry and Risk Assessment," U.S. Army Environmental Hygiene Agency, Bio-Acoustics Division and Industrial Hygiene Field Services Program, Technical Guide 181, Aberdeen Proving Ground, MD.

Berger, E. H. (1996). "International Activities in the Use, Standardization, and Regulation of Hearing Protection." *J. Acoust. Soc. Am. 99*(4) pt.2, 2463.

CEC (1986). "Council Directive of 12 May 1986 on the Protection of Workers from the Risks Related to Exposure to Noise at Work," Council of the European Communities, 86/188/EEC.

CEC (1989). "Council Directive of 14 June 1989 on the Approximation of the Laws for the Member States Relating to Machinery," Council of the European Communities, 89/392/EEC.

Coast Guard (1982). "Recommendations on Control of Excessive Noise." Dept. of Transportation, U.S. Coast Guard, Navigation and Vessel Inspection Circular No. 12–82.

DOD (1991, 1996). "DOD Hearing Conservation Program," Department of Defense Instruction, No. 6055.12.

DOL (1969). "Occupational Noise Exposure," Bureau of Labor Standards, 34 *Fed. Reg.*, 7948–7949.

DOT (1973). "Vehicle Interior Noise Levels," Department of Transportation, Federal Highway Administration, Bureau of Motor Carrier Safety. 38 *Fed. Reg.*, 30880–30882.

Embleton, T. F. W. (1994). "Report by I-INCE Working Party on 'Upper Noise Limits in the Workplace,'" in *Proceedings of INTER-NOISE 94*, edited by S. Kuwano, Noise Control Foundation, Poughkeepsie, NY, 59–66.

Embleton, T. F. W. (1997). "Technical Assessment of Upper Limits on Noise in the Workplace: Final Report." *Noise News International 4*, 203–216.

EPA (1974). "Information on Levels of Environmental Noise Requisite to Protect Public Health and Welfare with an Adequate Margin of Safety," Environmental Protection Agency, 550/9-74-004, Washington, DC.

HSE (1994). "Assessment of Compliance with the Noise at Work Regulations 1989," *Health & Safety Executive*, Research Paper 36, HSE Books, Sudbury, UK.

ISO (1984). "Acoustics: Noise Labelling of Machinery and Equipment," International Organization for Standardization, ISO 4871:1984, Switzerland.

ISO (1990). "Acoustics: Determination of Occupational Noise Exposure and Estimate of Noise-Induced Hearing Impairment," International Organization for Standardization, ISO 1999:1990, Switzerland.

Jones, A. L. (1995). "ADA and Accommodation Issues for Individuals with Hearing Loss," *Spectrum* Suppl. 1(12), 36.

Kihlman, T. (1993). "Sweden's Action Plan Against Noise," *Noise/News International 1*(4), 194–208.

LaCroix, P. (1996a). "Hearing Standards and Employment Decisions: The Federal Model (Part One)," *Spectrum 13*(3), 6–8.

LaCroix, P. (1996b). "Hearing Standards and Employment Decisions: The Federal Model (Part Two)," *Spectrum 13*(4), 1, 12–13.

MacLean, S. (1998). "Hearing Ability and ADA: A Case Study," *Spectrum 15*(1), 8.

MacLean, S., Karlsen, E. A., Sostarich, M. E., and Ator, G. (1993). "Current Audiometric Criteria in Law Enforcement: Where Do We Go from Here?" poster paper presented at the annual convention of the American Speech-Language-Hearing Assoc., Anaheim, CA.

MSHA (1999). "Health Standards for Occupational Noise Exposure; Final Rule," Mine Safety and Health Administration, 30 CFR Part 62, 64 *Fed. Reg.*, 49548–49634, 49636–49637.

Navy (1994a). "Hearing Conservation and Noise Abatement," U.S. Navy, OPNAVINST 5100.23D, Chapter 18.

Navy (1994b). "Navy Occupational Safety and Health (NAVOSH) Program Manual for Forces Afloat," U.S. Navy, OPNAVINST 5001.19C, Chapter 14.

NIOSH (1972). "Criteria for a Recommended Standard — Occupational Exposure to Noise," National Institute for Occupational Safety and Health, U.S. Dept. of HEW, Report No. HSM 73–11001, Cincinnati, OH.

NIOSH (1998). "Criteria for a Recommended Standard: Occupational Noise Exposure: Revised Criteria," National Institute for Occupational Safety and Health, U.S. Dept. HHS, Report DHHS (NIOSH) 98–126, Cincinnati, OH.

Occupational Safety and Health Act (1970). PL 91-596.

Ohlin, D. (1998). Personal communication.

OSHA (1981). "Occupational Noise Exposure: Hearing Conservation Amendment." Occupational Safety and Health Administration, 46 *Fed. Reg.*, 4078–4179.

OSHA (1983a). "Occupational Noise Exposure: Hearing Conservation Amendment; Final Rule," Occupational Safety and Health Administration, 29 CFR 1910.95; 48 *Fed. Reg.*, 9738–9785.

OSHA (1983b). "OSHA Instruction CPL 2-2.35, Nov. 9, 1983," in *Guidelines for Noise Enforcement*, Occupational Safety and Health Administration, Washington DC.

Royster, L. H., and Royster, J. D. (1996). "The Audiometric Phase of the Hearing Conservation Program: Summary of Activities from a Global Standpoint," *J. Acoust. Soc. Am. 99*(4) Pt. 2, 2463.

SA/SNZ (1998). "Occupational Noise Management, Parts 0 – 4." Standards Australia and Standards New Zealand, AS/NZS 1269.0-1998 through 1269.4-1998, Homebush, Australia.

Simpson, M., and Bruce, R. (1981). "Noise in America: The Extent of the Noise Problem," U.S. Environmental Protection Agency, EPA Report No. 550/9-81-101.

Sliney, D. H. (1993). "Report from the ACGIH Physical Agents TLV Committee: Review of the Threshold Limit Value for Noise," *Appl. Occup. Environ. Hyg. 8*(7), 618–623.

Suter, A. H. (1989). "Noise Wars," *Technology Rev. 92*(8), 42–49.

Suter, A. H. (1992). "The Relationship of the Exchange Rate to Noise – Induced Hearing Loss," Prepared under contract to the National Institute for Occupational Safety and Health, NTIS No. PB-93-118610, Cincinnati, OH.

Tri-Alliance (1992). "Americans with Disabilities Act: Facts Sheet on Employment," in *Putting the ADA to Work For You,* [videotape]. (American Occupational Therapy Association, American Physical Therapy Association, American Speech-Language-Hearing Association). American Speech-Language-Hearing Association, Rockville, MD.

UVV Laerm (1990). "UVV Laerm," Arbeitsstaettenverordnung, Federal Minister of Labour and Social Affairs, Noise Safety Regulation, Germany.

von Gierke, H. E. (1993). "Noise Regulations and Standards: Progress, Experiences, and Challenges," in *Noise & Man '93: Noise as a Public Health Problem, 3*, edited by M. Vallet, Institut National de Recherche sur les Transports et leur Securite, Arcueil Cedex, France, pp. 547 – 554.

Worksafe Australia (1993). "National Standard for Occupational Noise" [NOHSC:1007(1993)], "National Code of Practice for Noise Management and Protection of Hearing at Work" [NOHSC:2009(1993)], Sydney, Australia.

Zusho, H., and Miyakita, T. (1996). "Conservation and Compensation for Noise-Induced Hearing Loss in Japan," *J. Acoust. Soc. Am. 100* (4, Pt. 2), 2673.

The Noise Manual, revised 5th edition, edited by E.H. Berger,
L.H. Royster, J.D. Royster, D.P. Driscoll, and M. Layne
©2003 American Industrial Hygiene Association

17

Prediction and Analysis of the Hearing Characteristics of Noise-Exposed Populations or Individuals

Larry H. Royster, Julia Doswell Royster, and Robert A. Dobie

Contents

Background

This chapter describes the use of ANSI S3.44-1996, an American National Standard that provides a model for estimating the hearing threshold levels resulting from the combination of aging and noise exposure. There are numerous benefits from having available an empirically based scientific model for demonstrating the development of the hearing threshold levels for populations or, from a statistical standpoint, for individuals. Several examples are as follows:

- educational benefits to professionals and employees from knowing how noise-induced permanent threshold shifts (NIPTS) develop,
- availability of reference *age-effect* hearing threshold level (HTL) trends over time by frequency, gender, and race for comparison to employee population trends. (Age-effect norms are assumed to include all contributors to hearing loss except for occupational noise exposure [Royster et al., 1980; Royster and Thomas, 1979].),
- information to support decisions regarding the causes of measured OSHA standard threshold shifts,
- allows the comparison of age-effect hearing threshold levels to existing population and/or individual thresholds in order to estimate the NIPTS component,
- availability of reference population thresholds by audiometric frequency (normal audiogram format) for comparison to actual audiograms in order to support decisions regarding the potential cause of observed audiogram shifts,
- provides the basis for allocating the occupational and nonoccupational components of observed population and/or individual hearing thresholds,
- allows the prediction of the effects of nonoccupational noise exposures and the creation of related graphs for educational purposes, and finally,
- brings together the often dramatically different opinions expressed by consultants in legal cases and gives the courts one yardstick by which to judge the expressed expert opinions.

The ANSI S3.44 model is a very useful resource for professionals involved in hearing conservation. Professionals can use model predictions to demonstrate to management the amount of hearing loss expected in the noise-exposed worker population in the absence of an effective hearing conservation program (HCP), as well as the minimal loss expected for an adequately protected population. It can be used to demonstrate the reduction in potential workers' compensation liability resulting from establishing an effective HCP, thereby motivating managers to fund the HCP adequately. In terms of counseling individual noise-exposed employees, professionals can use the ANSI S3.44 model to show the worker the range of hearing loss s/he might develop if s/he does not properly wear HPDs on and off the job. The model can also be used to demonstrate that it is nearly pointless for a worker to say s/he will start wearing hearing protectors when s/he is older, as after only 10 years s/he will have accumulated most of the NIPTS at the higher audiometric test frequencies that s/he would ever get (assuming the noise exposure does not significantly increase).

The audiologist or physician can also use the ANSI S3.44 model in evaluating hearing loss compensation claims. Although the model predicts HTLs for populations, *not for individuals*, ANSI S3.44 states that the standard can be applied to individuals in terms of statistical probabilities. That is, the measured HTLs for the individual claimant can be compared to the population predictions for a similar noise exposure history to determine whether the claimant's HTLs are within the range of loss to be expected for selected population fractiles. If so, then on-the-job noise exposure can be accepted as a probable contributing cause of the loss (together with any other documented medical causes and nonoccupational noise contributions), assuming no evidence to the contrary.

If the individual shows greater loss than predicted for the most susceptible allowed fractile (the 0.05 fractile), then either (1) s/he may be one of the 5 people in 100 who is very susceptible to noise, or (2) s/he may have additional noise exposure from prior jobs and/or off-the-job activities, or (3) non–noise-related causes are present, or (4) errors have been made during data measurement and/or collection, or (5) a combination thereof.

If several annual hearing tests are available, then the model can be used to evaluate the likelihood that some NIPTS predated the employee's date of hire. This is achieved by comparing the ANSI S3.44 predicted rate of change in hearing at different audiometric test frequencies to the employee's actual change in hearing over time.

Introduction

ANSI S3.44-1996, *Determination of Occupational Noise Exposure and Estimation of Noise-Induced Hearing Impairment*, is an adaptation of international standard ISO 1999 (ISO, 1990). ANSI S3.44 provides a statistical model for predicting the hearing threshold levels for a population resulting from the NIPTS due to a specified noise exposure combined with an assumed age-related hearing loss for a user-defined set of population characteristics. From a statistical standpoint, such as the legal criterion of "more likely than not," predictions may also be made for individuals within the population.

Because the model was developed using industrial data that represented certain types of noise environments, the standard recommends that the application of the model be restricted. It is appropriate to situations where the noise is steady, intermittent, fluctuating, irregular, or impulsive, but only up to instantaneous peak sound pressure levels of 140 dB. Predictions are based on an assumed 40-hour week including five daily 8-hour exposures to equivalent continuous A-frequency weighted sound pressure levels of 75–100 dBA for up to 40 years of exposure. Up to 60 hours of actual noise exposure per week may be normalized to an equivalent 40-hour weekly exposure; for example, this would accommodate five 12-hour workdays. However, the model assumes that the variation in noise exposures for different days is limited. The standard states that the worst-day daily equivalent noise exposure should not be more than 10 dB greater than the average daily value over time and no more than 12 hours of noise exposure are allowed on any one day. Use of the model for noise exposure conditions that do not meet these limitations would be an extrapolation of the model.

It makes little difference how a math model is developed. When it is applied in real-world environments, if it is found to give reasonable and reliable predictions (Henderson and Saunders, 1998), then in fact it becomes an acceptable model. The authors have applied the present ANSI S3.44 and earlier ISO 1999 models (as well as the forerunner Burns and Robinson [1970] model) to many different industrial audiometric databases and found these models to be very reliable in their predictions if the proper age-effect population is utilized. If it works, it is usable. If it works, it is an acceptable model regardless of its origin. If it works, make use of it!

The ANSI S3.44 Model

The math model defined in the standard is presented below as Equation 17.1.

$$\text{HTLAN} = \text{NIPTS} + \text{HTLA} - ((\text{NIPTS} \times \text{HTLA})/120), \tag{17.1}$$

where HTLAN is total hearing threshold level associated with age and noise, NIPTS is the noise-induced permanent threshold shift component, and HTLA is the age-related threshold component.

The last term in the model, $((\text{NIPTS} \times \text{HTLA})/120)$, is a correction factor that reaches a significant value when the predicted level for either component, or their sum, becomes larger than about 40 dB. It is intended to prevent predicting thresholds significantly exceeding 100 dB HL, which is normally the upper limit of audiometers used to give hearing evaluations in industry. This correction term recognizes the fact that if an individual obtains a large amount of NIPTS early in life, then this individual's hearing will not be able to change due to age effects in the same manner as would have occurred had his/her hearing not been damaged by noise earlier on. The reverse situation also is assumed to hold; if a worker enters a high noise exposure environment at age 60 after experiencing significant age-effect hearing loss, then the new occupational noise exposure will not have the same impact on hearing as it would have had when s/he was 20 years old and had normal hearing.

NIPTS Component

The model predicts NIPTS based on the L_{A8hn}, the 8-hour level that is equivalent in acoustic energy to the population's total daily noise exposure using the 3-dB exchange rate. If the noise exposure was constant and 8 hours in length, then L_{A8hn} would be equivalent to the OSHA TWA (time-weighted average level over 8 hours that would produce the same noise dose as was measured for the actual noise exposure using a 5-dB exchange rate).

The NIPTS predicted by the model for selected noise exposure levels is shown in Table 17.1. A plot of the data for an L_{A8hn} of 95 dBA for the median susceptibility fractile is shown in Figure 17.1. Note how NIPTS develops first at the audiometric test frequencies of 3, 4, and 6 kHz, and more than half of the loss occurs within the first 10 years of noise exposure. Note also that even at this fair-

TABLE 17.1

NIPTS predicted by the ANSI S3.44-1996 model for different susceptibility fractiles of the population after 10 to 40 years of exposure to different L_{A8hn} values.

Data are shown for three fractiles:
        the 0.9 fractile (tough ears — 90% of ears are worse than this)
        the 0.5 fractile (typical ears — 50% are worse, 50% are better)
        the 0.1 fractile (tender ears — only 10% are worse than this)

L_{A8hn} (dBA)	Freq. (Hz)	10 Years of Exposure			20 Years of Exposure			30 Years of Exposure			40 Years of Exposure		
		0.9	0.5	0.1	0.9	0.5	0.1	0.9	0.5	0.1	0.9	0.5	0.1
85	500	0	0	0	0	0	0	0	0	0	0	0	0
	1000	0	0	0	0	0	0	0	0	0	0	0	0
	2000	0	1	1	1	1	2	1	1	2	1	2	2
	3000	2	3	5	3	4	6	3	4	7	3	5	7
	4000	3	5	7	4	6	8	5	6	9	5	7	9
	6000	1	3	4	2	3	5	2	3	6	2	4	6
90	500	0	0	0	0	0	0	0	0	0	0	0	0
	1000	0	0	0	0	0	0	0	0	0	0	0	0
	2000	0	2	6	2	4	8	3	5	9	4	6	10
	3000	4	8	13	7	10	16	8	11	18	9	12	19
	4000	7	11	15	9	13	18	10	14	19	11	15	20
	6000	3	7	12	4	8	14	5	9	15	6	10	15
95	500	0	0	1	0	0	1	0	1	1	0	1	1
	1000	1	2	4	2	3	5	2	3	5	2	3	6
	2000	0	5	13	5	9	17	7	12	20	9	14	22
	3000	8	16	25	13	19	31	16	22	34	18	23	37
	4000	13	20	27	16	23	32	18	25	34	19	26	36
	6000	5	14	23	8	16	26	10	18	28	12	19	29
100	500	2	4	8	3	5	9	4	6	11	5	7	11
	1000	3	6	12	6	9	15	7	10	17	8	11	19
	2000	0	8	23	8	16	31	13	21	35	16	24	39
	3000	13	26	41	21	32	51	26	35	56	29	38	60
	4000	20	31	42	25	36	49	28	39	53	30	41	56
	6000	9	23	37	14	27	42	17	29	46	19	30	48

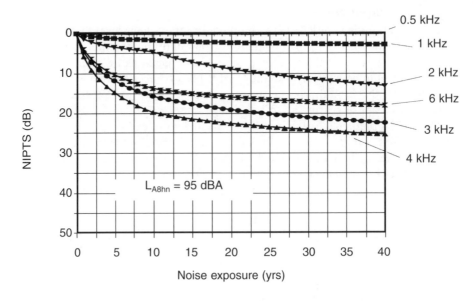

Figure 17.1 — ANSI S3.44 predicted median NIPTS at different audiometric test frequencies for an L_{A8hn} of 95 dBA vs. noise exposure in years.

ly high level of noise exposure, the NIPTS developed at the 0.5 and 1 kHz frequencies is insignificant.

The development of NIPTS is presented in audiometric format in Figure 17.2 for 10-year noise exposure increments, for two different noise exposure levels (L_{A8hn}, 85 and 95 dBA). This allows the reader to more fully appreciate the relative impact that the NIPTS component has on the population's or individual's total hearing threshold levels. If one had plotted the data in Figure 17.2 using a 0 – 30 dB NIPTS vertical axis, then a misleading impression would have been projected of the effects of different noise exposures on the affected population. Note the almost insignificant predicted NIPTS for the median susceptibility fractile for an L_{A8hn} of 85 dBA.

The previous figures have been for a population or individual assumed to exhibit median susceptibility. However, we must recognize, as the model does, that significant individual differences exist in susceptibility to NIPTS. Therefore the model predicts differing rates of progression of NIPTS (and different absolute amounts of NIPTS after a given exposure duration) for populations or individuals in different susceptibility fractiles, from the 0.05 fractile (very susceptible ears — only 5% of ears are more susceptible) through the 0.50 fractile (median ears) to the 0.95 fractile (very tough ears — 95% of ears are more susceptible). The NIPTS for persons more vulnerable than the 0.05 fractile or less vulnerable than the 0.95 fractile cannot be predicted. The extreme tail areas of the susceptibility distribution curve are simply mathematical oddities. If such

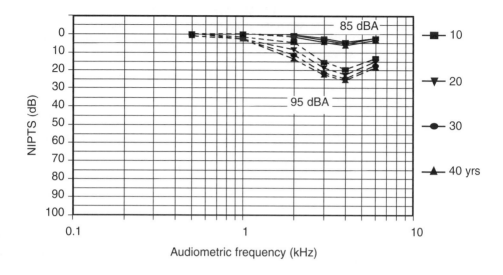

Figure 17.2 — ANSI S3.44 predicted median NIPTS (audiometric format) vs. audiometric test frequency for two different noise exposure levels (L_{A8hn} of 85 and 95 dBA) for different years of noise exposure.

was not the case then in large industrial databases one should find individuals on a fairly regular basis who develop significant hearing loss within a few weeks of employment. Also one should find a significant number of individuals who exhibit significant hearing loss after a few noise exposures from attending noisy sporting events.

HTLA Component

ANSI S3.44 provides three different reference populations (Annexes A, B, and C) that may be selected to represent the HTLA component of the model. It also allows the user of the standard to select an alternative reference population if justification for such a choice can be established. As an example, if in the future it is determined that to adequately predict the hearing thresholds of Hispanic populations none of the existing age-effect databases in the present standard is sufficient, then an additional reference HTLA population should be developed.

Presented as Tables 17.2 and 17.3 are the hearing threshold level data included in the standard as Annexes A and C. Annex A represents a composite of reference populations that are highly screened with respect to pathology and nonoccupational noise exposures and therefore represent the presbycusis component of hearing, that is, hearing loss due only to aging.

Annex C represents reference populations unscreened except to eliminate individuals with more than 2 weeks of occupational noise exposure over their working lifetimes (Royster and Thomas, 1979; Royster et al., 1980). The data in

675

TABLE 17.2

Age-related hearing levels expected from presbycusis in a highly screened, otologically normal population without any noise exposure from ISO 7029, as reprinted in ANSI S3.44-1996, Annex A.

Data are shown for three fractiles:

the 0.9 fractile (tough ears — 90% of ears are worse than this)
the 0.5 fractile (typical ears — 50% are worse, 50% are better)
the 0.1 fractile (tender ears — only 10% are worse than this)

Gender	Freq. (Hz)	Age 30 Years			Age 40 Years			Age 50 Years			Age 60 Years		
		0.9	0.5	0.1	0.9	0.5	0.1	0.9	0.5	0.1	0.9	0.5	0.1
Males	500	-6	1	9	-5	2	11	-4	4	14	-3	6	18
	1000	-6	1	9	-5	2	11	-4	4	14	-2	7	19
	2000	-7	1	11	-6	3	15	-4	7	21	-1	12	29
	3000	-7	2	13	-5	6	19	-2	12	29	3	20	42
	4000	-7	2	14	-4	8	23	0	16	37	7	28	55
	6000	-8	3	16	-5	9	26	0	18	41	8	32	62
Females	500	-6	1	9	-5	2	11	-4	4	14	-3	6	18
	1000	-6	1	9	-5	2	11	-4	4	14	-2	7	19
	2000	-6	1	10	-5	3	13	-3	6	18	-1	11	25
	3000	-7	1	11	-5	4	15	-3	8	21	0	13	30
	4000	-7	1	12	-6	4	17	-3	9	25	1	16	35
	6000	-8	2	14	-6	6	21	-2	12	31	2	21	45

Annex C, which are binaural average data, include the influences of gunfire, nonoccupational noise exposures, otological pathology, etc.: all of the contaminations seen in industrial populations except significant on-the-job noise exposure. Such populations, screened only for occupational noise exposure, are referred to as nonindustrial noise-exposed populations (NINEPs) representing the age-effect component of hearing loss.

The standard also includes Annex B, which is an unscreened population collected by the U.S. Public Health Service (National Center for Health Statistics, 1965). However, since the population in Annex B only includes data for different gender groups (not for race groups) and the hearing levels presented are very similar to those in Annex C, this database is not discussed herein.

Unlike Annexes A and B, which show data only by gender, Annex C includes data for black and white race groups. Years of analyzing age-effect and industrial data have clearly shown that black populations typically have significantly better hearing than whites, for whatever reasons.

The user of ANSI S3.44 must appreciate the critical nature of choosing the appropriate reference population when using the model. Figures 17.3 and 17.4 present data that clearly demonstrate the extreme importance of selecting the proper data (Annex C) for the HTLA component. Presented in Figure 17.3 are the observed population thresholds for a large industrial database that is predominately male and white. If the ANSI S3.44 model is used to predict the hear-

TABLE 17.3

Age-effect hearing levels at decade age intervals for preliminary fractiles of the unscreened nonindustrial noise-exposed populations (NINEPs) separated by race and sex as shown in Annex C of ANSI S3.44-1996.

Data are shown for three fractiles:

the 0.9 fractile (tough ears — 90% of ears are worse than this)
the 0.5 fractile (typical ears — 50% are worse, 50% are better)
the 0.1 fractile (tender ears — only 10% are worse than this)

Group	Freq. (Hz)	Age 30 Years 0.9	0.5	0.1	Age 40 Years 0.9	0.5	0.1	Age 50 Years 0.9	0.5	0.1	Age 60 Years 0.9	0.5	0.1
White	500	3	9	17	4	10	19	5	11	21	6	13	24
Males	1000	-1	5	13	0	6	17	1	8	20	2	10	24
	2000	-4	3	14	-1	6	20	1	10	29	3	15	41
	3000	-1	6	27	3	12	38	6	20	48	9	31	56
	4000	1	12	37	6	21	50	11	30	58	16	41	63
	6000	4	17	43	10	26	58	15	36	67	20	47	71
Black	500	-1	6	14	-2	5	13	-4	6	16	-6	9	24
Males	1000	-4	1	7	-4	2	9	-6	3	16	-9	6	28
	2000	-6	0	5	-5	1	8	-6	4	16	-7	7	30
	3000	-5	3	13	-4	5	19	-2	9	28	0	16	41
	4000	-3	3	15	-3	7	22	-2	14	34	-3	25	50
	6000	-4	5	17	-3	9	25	1	14	36	7	21	51
White	500	2	9	16	3	11	19	5	13	24	7	16	33
Females	1000	-1	5	11	0	7	16	0	10	23	1	14	32
	2000	-5	2	9	-3	5	14	0	8	25	3	13	39
	3000	-5	3	10	-3	6	16	0	12	27	3	20	42
	4000	-2	5	14	1	10	22	3	17	34	6	26	50
	6000	-1	7	15	2	13	25	7	22	39	13	33	58
Black	500	4	10	18	4	10	18	2	10	20	0	11	26
Females	1000	-1	4	14	-1	5	17	-1	6	18	-1	7	19
	2000	-5	2	13	-4	3	16	-3	5	18	-3	8	20
	3000	-4	1	12	-3	2	15	-2	6	19	-2	11	24
	4000	-3	4	14	-1	6	17	-1	8	20	-1	11	24
	6000	-1	6	19	2	9	24	3	15	29	2	21	35

ing thresholds for this population based on the best noise exposure data available, an L_{A8hn} of 86 dBA (abbreviated as PRE86 in the figures), and if the highly screened reference HTLA population from Annex A is selected (ISOA), then the predicted HTLs are uniformly better than what has been measured for this population. If the predicted thresholds and the measured thresholds are uniformly different across test frequency, as is the case here, then some factor other than NIPTS should be expected.

Shown in Figure 17.4 are similar predictions using the reference age effect (NINEP) white male population provided in Annex C. Now note how much closer the agreement is between the measured and predicted thresholds. Clearly, the

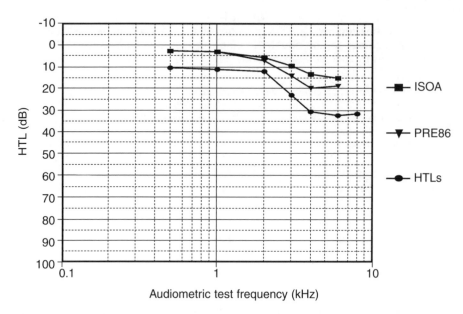

Figure 17.3 — Measured and predicted (PRE86) median population HTLs for one population using ISO A controls (presbycusis, a highly screened population) for the age effect (HTLA) population.

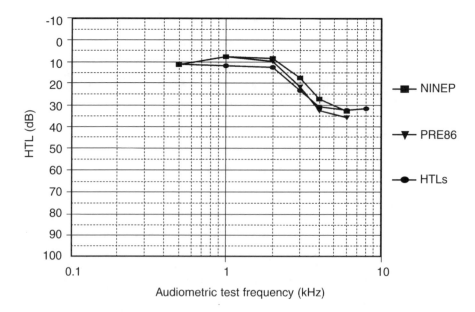

Figure 17.4 — Measured and predicted (PRE86) median population HTLS for one population using NINEP controls (screened only for occupational noise exposure) for the age effect (HTLA) population.

data in Figures 17.3 and 17.4 (as well as multiple other applications of the ANSI S3.44 model by the authors) support the selection of Annex C as the reference population of choice when predicting thresholds for industrial populations.

Related Considerations

With respect to the ANSI S3.44 model in general, three very important observations can be made:

1. In industrial populations, as well as the armed services, 90–95% of worker TWAs are less than 95 dBA. Comparing the data presented in Tables 17.1 and 17.3, it is obvious that for the vast majority of the noise-exposed populations the dominant factor affecting the hearing threshold levels of the population is the age-effect component.

2. The damage done by noise occurs relatively quickly after the onset of exposure. Unless the individual's noise exposure increases significantly over time, very little can be done later on in the noise exposure career to correct the impact of an inadequately protected early noise exposure period.

3. The damage caused by typical industrial noise is very frequency dependent, with the audiometric test frequencies of 3, 4, and 6 kHz being the most susceptible to NIPTS.

Determination of Noise Exposure

The measure of daily noise exposure preferred for use in ANSI S3.44 is indicated by the abbreviation L_{A8hn}. This stands for the equivalent continuous A-frequency weighted sound pressure level normalized to an assumed 8-hour workday. The L_{A8hn} is conceptually similar to the OSHA 8-hour TWA except that it is calculated using the 3-dB exchange rate (approximating an equal-energy trading rule), not OSHA's 5-dB exchange rate.

For employees who work 8-hour shifts in steady-state noise, the L_{A8hn} and the OSHA TWA will be equivalent. However, if the noise environment exhibits significant variability, then L_{A8hn} will generally be a higher value than the OSHA TWA, typically by 2 to 3 dBA (Petrick et al., 1996; Royster and Royster, 1994). If employees' actual exposure durations are longer or shorter than 8 hours and the same criterion level is used, such as 90 dBA, the differences between the daily exposures calculated using the 3-dB and 5-dB exchange rates can vary. For 12-hour workshifts, for example, the OSHA TWA is often higher than the L_{A8hn}.

ANSI S3.44 allows the use of an alternative exchange rate other than 3 dB to determine daily noise exposures for intermittent types of noise exposures if justified by the user of the standard. If the user elects to use an alternative exchange rate (such as OSHA's 5-B exchange rate), the standard suggests that certain conditions be met:

- sound levels above 105 dBA should contribute less than 10% daily OSHA noise dose, and

- there should be at least seven off-times (with sound pressure levels below 80 dBA) spaced evenly through the day.

If the user opts to employ an alternative exchange rate, then the daily equivalent continuous sound pressure level is referred to as an *equivalent effective level* since it is not the same as the L_{A8hn}.

An appropriate Equation to use in combining different sound levels and corresponding exposure periods is given below as Equation 17.2.

$$L_{x,T} = A \log [(1/T) \times (t_1 \times 10^{(L_1/A)} + t_2 \times 10^{(L_2/A)} + ...$$
$$... + t_n \times 10^{(L_n/A)})], \qquad (17.2)$$

where n is the number of levels to be combined, x is the exchange rate deemed appropriate, T is the time duration over which the equivalent level is being determined, and A is a constant depending on the exchange rate selected (A = 10 for the 3-dB exchange rate, or A = 16.61 for the 5-dB exchange rate). Note: When x is equal to 3 dB and T is equal to 8 hours, then $L_{x,T}$ is equivalent to L_{A8hn}.

When combining daily equivalent exposures over time periods longer than a day, then the 3-dB exchange rate is appropriate. As an example of the calculation of a long-term noise exposure level, consider the noise exposure profile shown in Figure 17.5. If an equivalent yearly noise exposure is required over the first 20 years of exposure, ages 20 to 39, then the above Equation would take on the following form:

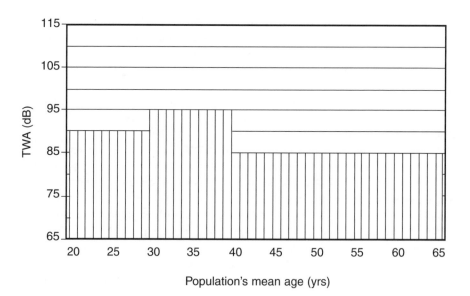

Figure 17.5 — An assumed noise exposure profile that presents yearly equivalent TWAs as a function of the population's mean age.

$$L_{eq,20yr} = 10 \log [(1/20) \times (10 \times 10^{(90/10)} + 10 \times 10^{(95/10)})], \quad \textbf{(17.3)}$$
$$= 93.2 \text{ dB}$$

On the other hand, if the equivalent noise exposure level for the total 45-year work history of this population or individual is required, then:

$$L_{eq,45yr} = 10 \log [(1/45) \times (10 \times 10^{(90/10)} + 10 \times 10^{(95/10)}$$
$$+ 25 \times 10^{(85/10)})], \quad \textbf{(17.4)}$$
$$= 90.4 \text{ dBA}$$

If the last 25 years of noise exposure at 85 dBA were not considered, then the equivalent level of the first two noise exposures at 90 and 95 spread over the total 45 years of exposure would have resulted in an $L_{(eq,45yr)}$ of 89.7 dBA. In other words, the contribution of the population's last 25 years of exposure to a yearly equivalent level of 85 dBA contributed only 90.4 - 89.7 = 0.7 dB. This example clearly demonstrates that it is the high noise exposure years of a population's or individual's exposure history that dominate the overall equivalent exposure level.

Applications of the Model

The following examples illustrate ways in which the ANSI S3.44 model can be applied.

Approach A

The ANSI S3.44 model provides mathematical expressions that can be used to predict, for any set of metrics within the limits of the model, the hearing threshold levels for a population or any individual from a statistical standpoint. Therefore, one approach to making predictions based on the model is to program the Equations so that the user can input whatever set of constraints is desired. One example in using the ANSI S3.44 model is now demonstrated.

The assumed noise exposure profile is shown in Figure 17.5. The population and/or individual is assumed to begin noise exposure at age 20 and is exposed to an L_{A8hn} of 90 dBA for 10 years. Then the noise exposure increases to 95 dBA for the next 10 years, and finally the exposure drops to 85 dBA for 25 years.

Next it is assumed that this population and/or individual is appropriately represented by the median susceptibility fractile, or 50th percentile group. That is, 50% of the population would exhibit equal or worse hearing and 50% of the population would exhibit equal or better hearing. Although predictions can be made using ANSI S3.44 separately for each audiometric frequency, predictions are shown here for the average of the hearing levels at 0.5, 1, 2, and 3 kHz (for the combination of frequencies used today in most compensation formulas, reference Chapter 18).

Because an industrial population is being modeled, the appropriate age-effect population to use is the one included in Annex C of the standard, or Table 17.3 of this chapter. In this case the white male reference data are illustrated. As required by the model, the same level of susceptibility (median) is assumed for the age-effect component as for the NIPTS component.

Next the equivalent equal-energy noise exposure level after each successive year of noise exposure is calculated. The resulting predicted values and changes for the NIPTS, HTLA, and HTLAN components (Equation 17.1) of the model are shown in Figure 17.6. First, the bottom curve indicated by the squares represents the predicted NIPTS for the noise exposure profile defined. Note that the growth of NIPTS follows a smooth growth curve up to the point where the assumed higher noise exposure starts (age 29). Then a new growth curve starts and continues at an ever slower rate up to the point that the noise exposure profile drops back to 85 dBA (age 39). Since the accumulated noise exposure at the reduced level of 85 dBA never equals the equivalent previous cumulated noise exposure, no more NIPTS is predicted for the population or the individual.

The curve indicated by the triangles represents the NINEP age-effect (HTLA) component of the hearing loss of the population or individual. Note that it starts off at a given value and begins to accelerate in magnitude as age increases.

Finally, the population's or individual's total hearing, represented by the circles, is the sum of the NIPTS and HTLA components minus the contribution of the correction factor. As noise exposure proceeds, Approach A subtracts the cor-

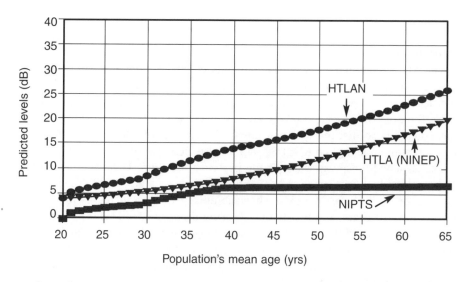

Figure 17.6 — HTL components (HTLAN, HTLA, and NIPTS) predicted using ANSI S3.44 for the assumed noise exposure profile of Figure 17.5 averaged across the audiometric test frequencies of 0.5, 1, 2, and 3 kHz (Approach A and an assumed white male age effect population).

rection factor from the HTLA component (assuming that noise exposure begins at a young age). It seems logical that if hair cells are destroyed in the early years of noise exposure, they cannot be returned to use as would be the case if the correction factor were later on in the noise exposure cycle applied partially or totally to the NIPTS component. Therefore in this approach the correction factor is associated with the HTLA component.

The ability to allocate a population's or individual's hearing between occupational noise exposure and aging is often beneficial. Therefore, assuming that the population's or individual's hearing is composed of the two components indicated in Figure 17.6, NIPTS and HTLA, the percentage contribution curves shown in Figure 17.7 result. When the population starts off its noise exposure cycle, its hearing is composed entirely of the HTLA component. However, as NIPTS develops, a smaller and smaller percentage of the total hearing threshold is due to the HTLA component. At around 18 years of noise exposure, or around 37 years of age, when the dominant noise exposure is over, the breakdown in allocation begins to reverse, until finally at age 65, the allocation of total hearing loss would be approximately 76% to HTLA (age effects) and 24% to NIPTS.

Approach B

The previous discussion (Approach A) assumes that, when exposure levels vary over the years, the best estimate of NIPTS is found by first calculating an equivalent exposure level (L_{eq}) for the entire period under consideration (on a year-to-year basis), then finding the expected NIPTS for L_{eq} and the assumed duration. Approach A also assumes that L_{eq} is to be calculated with a 3-dB exchange rate, giving a level that, if constantly present in the workplace, would

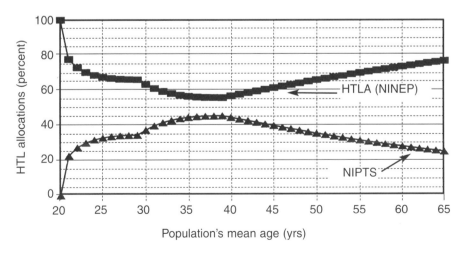

Figure 17.7 — Allocation of the HTLAN components (HTLA and NIPTS) shown in Figure 17.6 as a function of the population's mean age.

have delivered the same amount of acoustic energy to the worker's ear as the varying noise exposure being considered.

ANSI S3.44 is actually somewhat unclear regarding the combination of levels over long periods of time. Section 3.3 states that the period of summation is "a whole day . . . or a longer period that is to be specified, for example, a working week." Section 3.6 gives an Equation for calculating an equivalent daily exposure level from several daily exposures, again in the context of variations within a week. The only reference to the combination of levels over longer periods is in section 4.4.2, where the applicability of the standard is called into question in cases where L_{eq} on the worst day is more than 10 dB greater than L_{eq} "averaged over a longer period (not exceeding 1 year)." To some readers, all of this suggests that varying levels should not be combined into equivalent levels for periods exceeding 1 year.

If a 3-dB equal energy combination were valid over years of exposure, one would have to conclude that x years at y dBA would yield the same NIPTS as x/2 years at (y + 3) dBA, or x/4 years at (y + 6) dBA. Referring to Table 17.1, one immediately finds some limited exceptions to this rule. For example, 10-year exposures at 95 dB are predicted to cause NIPTS of 4 dB at 1 kHz for the most susceptible 10% of the population. Using a 3-dB L_{eq} rule, we'd expect the same NIPTS for 40 years at 89 dBA, yet 40 years at 90 dBA yields zero NIPTS. The formulas that were used to generate these tables were created to fit pooled data from empirical studies and do not conform exactly to an equal-energy model.

An alternative approach (B) was suggested by Robinson (1991), and is most easily understood graphically using the same hypothetical case as in Approach A (see Figure 17.8 and Table 17.4). The first exposure (10 years at 90 dBA) is expected to produce 2.6 dB NIPTS for the pure-tone average of 0.5, 1, 2, and 3 kHz. The key assumption of Approach B is that this first exposure was equivalent (in hazard, if not in energy) to a briefer exposure at a higher level (in this case, 2 years at 95 dBA), because the two exposures predict the same amount of NIPTS. This is illustrated by the horizontal dashed arrow to the left in Figure 17.8, or the line drawn between 2.6 dB in the L_{A8hn} 90 and 95 dB columns in Table 17.4. Now adding an additional 10 years of noise exposure along the 95-dBA growth curve, which is equivalent to adding 10 years of exposure on top of the initial 10 years of exposure on the 90-dBA curve, results in a prediction of 6.3 dB of NIPTS after 20 years of exposure. The effect of the third exposure (25 years at 85 dBA) would be estimated in the same way, but as the dashed arrow to the right suggests, no amount of exposure at 85 dBA would ever be expected to produce 6.3 dB of NIPTS, and we conclude that the third exposure period is unlikely to cause further harm.

In this example, Approaches A and B yield similar results. Both predict median NIPTS of 2.6 dB after 10 years, about 6 dB after 20 years, and no further NIPTS after that. In cases of stable exposure over the years, their results will always be identical. Nevertheless, the two approaches can give slightly different results in some cases. For example, assume an exposure history of 30 years at 80 dBA followed by 10 years at 100 dBA. Approach A finds the equivalent level

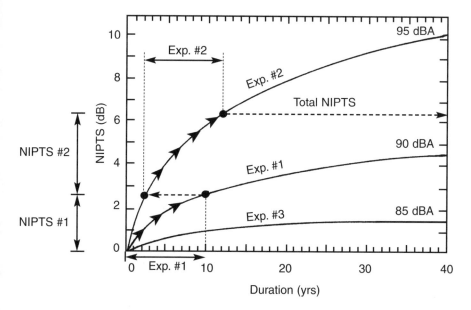

Figure 17.8 — Predicted NIPTS using the curve walking procedure, Approach B, for the average NIPTS across the audiometric test frequencies of 0.5, 1, 2, and 3 kHz.

to be 94.1 dBA using Equation 17.2. Estimated NIPTS after 40 years would be about 9 dB using Approach A, compared to about 11 dB with Approach B. Note that 10 years at 100 dBA alone predicts about 11 dB NIPTS (Table 17.1); the result from Approach A seems to imply that the 30 years at 80 dBA had somehow prevented some of the NIPTS from the later noise exposure.

Both approaches have been illustrated using the pure-tone average of 0.5, 1, 2, and 3 kHz. At least for Approach B (also referred to as "curve-walking" [Dobie, 1993]), results may vary slightly if predictions are first made frequency-by-frequency, then averaged. Similarly, if a percentile other than the median is used, perhaps because an individual's hearing levels suggest an unusual degree of susceptibility to NIHL, results will vary slightly for both approaches. These differences are usually inconsequential.

Probably the most important caveat in applying either approach (A or B) to an individual case is that data specific to that person should be given greater weight (if of good quality) than these epidemiological predictions. For example, if a good series of audiograms showed that an individual exposed as in Figure 17.5 came to the workplace at age 20 with normal hearing, developed substantial hearing loss over the next 10 years, then displayed stable hearing levels until age 65, it would be unreasonable to conclude (as both Approaches A and B would predict) that the second 10-year exposure at 95 dBA contributed more to his hearing loss than the first 10-year exposure at 90 dBA. There could be several reasons for such discrepancies between individual-specific data and epidemiological predictions: errors in noise exposure measurements or extrapolations, inaccurate history of hearing protection device usage, and contributions of nonoccupational noise, to name a few. Allocation estimates for individuals often require judgmental consideration of conflicting lines of evidence.

TABLE 17.4
Predicted NIPTS averaged for the four audiometric test frequencies of 0.5, 1, 2, and 3 kHz.

Exposure Years	Daily Equivalent Exposure Level (dB)			
	85.0	90.0	95.0	100.0
0	0.0	0.0	0.0	0.0
1	0.3	0.8	1.6	3.2
2	0.4	1.2	2.6	5.1
3	0.5	1.5	3.3	6.4
4	0.6	1.8	3.8	7.4
5	0.7	1.9	4.2	8.2
6	0.7	2.1	4.6	9.0
7	0.8	2.3	4.9	9.6
8	0.8	2.4	5.2	10.1
9	0.9	2.5	5.5	10.6
10	0.9	2.6	5.7	11.0
11	1.0	2.7	6.0	11.7
12	1.0	2.9	6.3	12.2
13	1.0	3.0	6.5	12.7
14	1.1	3.1	6.8	13.2
15	1.1	3.2	7.0	13.7
16	1.1	3.3	7.2	14.1
17	1.1	3.4	7.4	14.5
18	1.2	3.4	7.6	14.9
19	1.2	3.5	7.8	15.2
20	1.2	3.6	7.9	15.5
21	1.2	3.6	8.1	15.9
22	1.3	3.7	8.2	16.2
23	1.3	3.8	8.4	16.4
24	1.3	3.8	8.5	16.7
25	1.3	3.9	8.6	17.0
26	1.3	3.9	8.8	17.2
27	1.3	4.0	8.9	17.5
28	1.4	4.1	9.0	17.7
29	1.4	4.1	9.1	18.0
30	1.4	4.1	9.2	18.2
31	1.4	4.2	9.3	18.4
32	1.4	4.2	9.4	18.6
33	1.4	4.3	9.5	18.8
34	1.4	4.3	9.6	19.0
35	1.5	4.4	9.7	19.2
36	1.5	4.4	9.8	19.4
37	1.5	4.4	9.9	19.5
38	1.5	4.5	10.0	19.7
39	1.5	4.5	10.1	19.9
40	1.5	4.6	10.2	20.0

Annotations in the 90.0 column indicate a "10 years" span from years 2 to 10; annotations in the 95.0 column indicate a "10 years" span from years 2 to 12.

Conclusions and Recommendations

The ANSI S3.44 model for predicting the hearing thresholds or changes in hearing thresholds of populations or individuals from a statistical standpoint is a very powerful tool that is available to all professionals with an interest in the effects of noise on humans. As with any math model, its validity lies in its ability to predict what is actually observed in real-world situations. Two of the authors have had extensive experience in applying the model to population databases representative of different gender, race, and ethnic groups from around the world (Royster et al., 1984; Royster and Royster, 1986). The model has in fact predicted rather consistently the hearing thresholds for these populations. When one realizes the difficulties and opportunities for error in obtaining sound survey noise exposure data and industrial audiometric data for large populations, and the inherent population differences, it is truly remarkable that consistency in the predictions has been achieved. Such findings suggest that the ANSI S3.44 model is worth further utilization by fellow professionals.

Availablility of ANSI S3.44

Major research libraries may have copies of standards in the reference section. Copies of ANSI S3.44-1996 may be purchased from the Standards Secretariat of the Acoustical Society of America (Appendix II). Note that if a hand calculator is used, making predictions is quite time-consuming. A computer program that allows rapid predictions is also available for a tax-deductible contribution to a student scholarship fund managed by the Acoustical Society of America (Royster et al., 1983).

References

ANSI (1996). "Determination of Noise Exposure and Estimation of Noise-Induced Hearing Impairment," ANSI S3.44-1996, Acoustical Society of America, New York, NY.

Burns, W., and Robinson, D. W. (1970). *Hearing and Noise in Industry*, Her Majesty's Stationery Office, London, United Kingdom.

Dobie, R. A. (1993). *Medical-Legal Evaluation of Hearing Loss*, Van Nostrand Reinhold, New York.

Henderson, D., and Saunders, S. S. (1998). "Acquisition of Noise-Induced Hearing Loss by Railway Workers," *Ear & Hearing* 19(2), 120–130.

ISO (1990). "Acoustics — Determination of Noise Exposure and Estimation of Noise-Induced Hearing Impairment," International Standard ISO 1999, second edition, Geneva, Switzerland.

National Center for Health Statistics (1965). "Hearing Levels of Adults by Age and Sex, United States, 1960–62." *Vital and Health Statistics*, Public Health Service Publication No. 1000, Series 11, No. 11. U.S. Government Printing Office, Washington, DC.

Petrick, M. E., Royster, L. H., Royster, J. D., and Reist, P. (1996). "Comparison of Daily Noise Exposures in One Workplace Based on Noise Criteria Recommended by ACGIH and OSHA," *Am. Ind. Hyg. Assoc. J. 57*, 924–928.

Robinson, D. W. (1991). "Relation Between Hearing Threshold and Its Component Parts," *British J. of Audiology 25*, 93–103.

Royster, J. D., and Royster, L. H. (1986). "Documentation of Computer Tape of Audiometric Data," Final report submitted to DHHS PHS CDC NIOSH to fulfill order # 85-35350, June, DHHS, Washington, DC.

Royster, L. H., Driscoll, D. P., Thomas, W. G., and Royster, J. D. (1980). "Age Effect Hearing Levels for a Black Nonindustrial Noise-Exposed Population (NINEP)," *Am. Ind. Hyg. Assoc. J. 41*, 113–119.

Royster, L. H., and Royster, J. D. (1994). "The Impact of Using a 3-dB vs. 5-dB Exchange Rate on the Predicted Employee's 8-h Equivalent A-weighted Sound-Pressure Level in Three Different Types of Industrial Work Environments," *J. Acoust. Soc. Am. 96*, 3273.

Royster, L. H., Royster, J. D., and Cecich, T. F. (1984). "An Evaluation of the Effectiveness of Three Hearing Protection Devices at an Industrial Facility With a TWA of 107 dB," *J. Acoust. Soc. Am. 76*, 485–497.

Royster, L. H., Royster, W. K., and Royster, J. D. (1983). "The N.A.N. Computer Program for Estimating a Population's Hearing Threshold Characteristics," version 3.0, Environmental Noise Consultants, Inc., P.O. Box 30698, Raleigh, NC 27622-0698, EFFECTIVE_HCPS@compuserve.com.

Royster, L. H., and Thomas, W. G. (1979). "Age Effect Hearing Levels for a White Nonindustrial Noise Exposed Population (NINEP) and Their Use in Evaluating Industrial Hearing Conservation Programs," *Am. Ind. Hyg. Assoc. J. 40*, 504–511.

The Noise Manual, revised 5th edition, edited by E.H. Berger,
L.H. Royster, J.D. Royster, D.P. Driscoll, and M. Layne
©2003 American Industrial Hygiene Association

18

Workers' Compensation[*]

Robert A. Dobie and Susan C. Megerson

Contents

---

[*] Portions of this chapter are adapted from Allen Cudworth's chapter of the same name that appeared in the prior (4th) edition of this same text.

Introduction

There are many reasons for preventing hearing loss. One reason that is sometimes underestimated is the cost of paying workers for their noise-induced hearing loss (NIHL), an occupational disease now scheduled (or otherwise covered) in workers' compensation (WC) laws of most states. Payment is provided via WC insurance coverage, which is required of employers in all states. Outside of WC laws, hearing loss, monetary loss, or injury resulting from negligent failure to control noise may result in civil liability to the employer or to the noise-source manufacturer. Both WC claims and civil litigations are completely independent of the regulations of the Occupational Safety and Health Administration (OSHA). Compliance with OSHA regulations will not absolve employers of WC or civil liability, nor will OSHA citations necessarily lead to WC or civil awards. Persons having responsibilities in hearing conservation programs should become acquainted with at least the basic legal aspects of the industrial noise problem.

Workers' Compensation History

Prior to the 20th century, occupational injury and disease were rarely compensated. An injured worker who sued the employer for damages was unlikely to win for several reasons — the cost of litigation, the possibility of the worker's contributory negligence, and the belief that a worker in a dangerous job voluntarily accepts the risks thereof. In the early 1900s, a concept emerged whereby it became the employer's obligation to compensate employees for any injury incurred during, and resulting from, the course of their employment. In this country, the first WC law was the Federal Employees Compensation Act of 1908. Now, compensation acts are in effect in all states. In addition, federal employees can be compensated for NIHL by the Veterans Administration (for military personnel) and by the Department of Labor (for civilians). WC laws usually provide a fairly streamlined process in which a worker does not have to prove negligence on the employer's part, and will receive a payment based on the severity of the injury and the worker's income. In return, the worker gives up the right to sue and therefore to receive whatever a judge or jury might find reasonable.

Originally, compensation was to be paid only when a job-related injury resulted in income loss to the employee. However, in recent years, almost all states recognize partial or total loss of function in activities of daily living, with or without wage loss, as the basis for an award.

Although the loss-of-function concept appeared earlier, it was not until 1948 that it was tested in a case involving occupational hearing loss. In that year, a precedent was set (*Slawinski v. J. H. Williams,* 1948) — the Supreme Court of New York decided that Slawinski was entitled to an award for partial loss of hearing even though he had not been disabled (i.e., prevented from earning full wages at his regular employment). Similarly, in 1953, the Wisconsin Supreme

Court (*Green Bay Drop Forge v. Industrial Commission*, 1953) upheld a Wisconsin Industrial Commission ruling that an employee was entitled to compensation for partial hearing loss even though the Wisconsin law, like the New York law, required disability (defined in terms of wage loss) before NIHL could be compensated. Further confirmation of this concept was given in a 1955 decision under the United States Longshoremen's and Harbor Workers' Compensation Act (1927).

In 1959 and 1961, the Maryland courts discounted the above precedents and ruled that no compensation would be paid in the absence of disability with wage loss. Ultimately on April 14, 1967, the Maryland legislature made occupational hearing loss compensable without consideration of wage loss.

NIHL is usually interpreted as an occupational disease as opposed to a traumatic injury (hearing impairment from trauma is usually associated with explosions, head injuries, etc., and has always been covered by WC). The "gradual injury" concept was formulated in an attempt to win compensation in a state where injury was compensable, but disease was not. This concept was tested on an NIHL case in Georgia in 1962 (*Shipman v. Employers Mutual Liability Insurance*, 1962). The Georgia court ruled that an employee's NIHL resulted from the cumulative effect of a succession of injuries caused by each daily noise insult to the ear. By this reasoning, it held that the loss was compensable under the accidental injury provisions of the Georgia law. In general, WC laws now cover occupational diseases as well as injuries. Thus, the distinction is no longer relevant in most cases.

Finally, railroad workers involved in interstate commerce and merchant seamen are not covered by state or federal WC programs. For them, damages may be awarded in civil litigation. In such cases, the employee must prove negligence on the part of the employer.

As previously mentioned, a worker who has received a WC award generally cannot subsequently sue the employer for hearing loss. Yet there remains the possibility of suing third parties, such as the companies that make noise-generating tools and machinery that the employer used. Several large class-action lawsuits of this type (not governed by WC laws) have been filed in recent years.

Workers' Compensation for Hearing Loss

Table 18.1 provides a summary of WC practices and awards throughout the United States and Canada. Data were obtained in late 1998 and early 1999 by a written survey of WC officials. In some cases, the state or province supplied a copy of relevant statutes *in lieu* of completing the written survey. In these instances, the authors extracted the appropriate data from the materials provided. In cases where the written survey appeared to conflict with the statutes, the authors utilized information from the official published document.

In general, almost all states and all Canadian provinces indicated that hearing loss is compensable in their jurisdictions. The following four states reported that

TABLE 18.1

Hearing loss statutes in the United States and Canada.

Jurisdiction	1. Is occupational hearing loss compensable?	2. Is minimum noise exposure required for filing?	3. Schedule in weeks (one ear)	4. Schedule in weeks (both ears)	5. Maximum compensation (one ear)	6. Maximum compensation (both ears)	7. Hearing impairment formula	8. Waiting period	9. Is deduction made for presbycusis?	10. Is award made for tinnitus?	11. Provision for hearing aid?	12. Credit for improvement with hearing aid?	13. Is hearing loss prior to employment considered in compensation claim?	14. Statute of limitations for hearing-loss claim	15. Penalty for not wearing hearing protection devices?	16. Self-assessment of hearing impairment considered in rating/award?	Comments
Alabama	Yes	No	53	163	$11,660	$35,860	ME	No	No	Yes-I	Yes	No	Yes	2 yrs.	Yes-D	Poss	**3–6:** awards based on temporary disability and permanent partial impairment according to AMA guidelines; **12:** unless hearing aid enables worker to return to work; **13:** as long as there has been substantial aggravation at work.
Alaska	Yes	No	*	*	*	*	AAO-79	No	No	Yes-I	Yes	No	No*	2 yrs.	No	No	
Arizona	Yes	No	86	260	$23,100	$69,300	ME	No	No	Yes-I	Yes	No	Yes	1 yr.	No	No	
Arkansas	Yes	No	42	158	$11,296	$42,502	AAO-79	No	Poss	No	Poss	No*	No*	Yes*	No	No	**14:** statute of limitations and other hearing loss issues currently before Board of Appeals.
California	Yes	No	50*	311*	$8,040*	$58,863*	AAO-79	No	No	Yes	Yes	Yes	Yes		Yes-P	Yes	**3–6:** awards modified by age and occupation at time of injury.
Colorado	Yes	No	35	139	$5,250	$20,850	AAO-79	No	Yes	Yes	Yes	No	Yes	1 yr.	Yes-P	No	
Connecticut	Yes	No	35	104	*	*	ME*	3 days	Poss	Poss	Poss	Poss	Poss	1 yr.	Yes-P	Poss	**5–6:** no maximum reported—award is number of weeks scheduled benefit at claimant's compensation rate; **7:** case law has supported AAO-79.
Delaware	Yes	No	75	175	$30,833	$71,944	ME	No	No	No	Yes	Yes	No	2 yrs.	No	No	
District of Columbia	Yes	No	39	150	$34,880	$134,170	AAO-79	6 mo.	Poss	Poss	Poss	Poss	Poss	1 yr.	Poss	Poss	
Florida	Yes	No	18	105	$8,892	$51,870	AAO-79	6 mos.	No	Yes-I	NR	No	Yes	2 yrs.	Yes-P	No	
Georgia	Yes*	Yes*	NA	150	NA	NR	AAOO-59	6 mos.	NR	No	NR	NR	Yes	NR	Yes-D	NR	**1:** no awards granted for monaural hearing loss unless pre-existing deafness in other ear; **2:** 90 dBA for 90 days.
Hawaii	Yes	Yes	52	200	$26,416	$101,600	AAOO-59	No	No	Yes	Yes	No	No	2 yrs.	No	No	

State																	Comments
Idaho	Yes*	No	NR	175	NR	$42,639	ME	No	No	Yes-I	No	No	Yes	1 yr.	No	No	**1:** only hearing loss due to work-related trauma/injury is considered.
Illinois	Yes	Yes*	50-100	200	$43,989	$87,978	Other*	No	No	Yes-I	Yes	No	Yes	2-3 yrs.	No	No	**2:** 90 dBA TWA; **7:** avg > 30 dB at 1000, 2000, and 3000 Hz, 1.82% per dB.
Indiana	Yes*	No	*	*	$12,500	$39,500	ME	Yes	Yes	NR	No	NR	Yes	2 yrs.	Yes-D	Yes	**1:** only hearing loss due to work-related trauma/injury is compensable, but case law evolving—likely to consider NIHL in future; **3-4:** awards paid for temp. disability (up to 500 wks) or permanent partial impairment based on % of max. awards.
Iowa	Yes	Yes	50	175	$43,600	$152,600	AAO-79	Yes	Yes	Yes	Yes	No	Yes	2 yrs.	Yes-D	No	See p. 701 this chap., adj. for age.
Kansas	Yes	No	30	110	$10,980	$40,260	AAO-79	Yes	Yes	Yes-I	Yes	Yes	Yes	200 days	Yes-D	No	
Kentucky	Yes	No	520	Lifetime	*	*	ME	No	No	No	No	No	No	3-5 yrs.	No	No	**5-6:** award based on % impairment and average weekly wage.
Louisiana	Yes*	No	100	100	NR	$36,700	ME	No	No	Yes-I	No	No	Yes	1 yr.	Yes-D	No	**1:** only hearing loss due to work-related trauma/injury is considered.
Maine	Yes	Yes	50	200	$20,100	$80,400	AAOO-59	Yes*	Yes*	Yes-I	Yes	Yes	Yes	Yes	Yes-D	No	**9:** 1/2 dB for each year over 40 yrs. of age.
Maryland	Yes	Yes*	125	250	NR	NR	AAOO-59	Poss	Poss	Poss	Yes	No	Yes	Yes	NR	Poss	**2:** exposure to harmful noise for 90 days or more.
Massachusetts	Yes	No	*	*	*	*	ME	Yes	Yes	Yes	Yes	NR	Yes	Yes	Yes	No	**3-6:** based on statewide average wage — maximum $700/week.
Michigan	No*																**1:** no compensation for occupational hearing loss — employees are compensated only if an injury to the ear (that may or may not result in hearing loss) causes a loss of wages.
Minnesota	Yes	No	*	*		*	AAO-79	No	No	No	Yes	No	Yes	3 yrs.	No	No	**3-6:** hearing loss compensation is based on a percentage of whole body impairment (max for one ear, 6% whole body; for both ears, 35% whole body).
Mississippi	Yes	No	40	150	$11,191	$41,967	ME	No	No	Yes	Yes	No	Yes	Yes	No	Yes	**9:** 1/2 dB per year over age 40 yrs.; **15:** results in 15% penalty.
Missouri	Yes	No	49	180	$14,442	$53,051	AAOO-59	Yes*	Yes*	Yes	Yes	No	Yes	No	Yes-P*	No	
Montana	Yes	Yes*	40	200	NR	NR	AAOO-59	Yes*	Yes*	No	No	No	Yes	No	No	No	**2:** 90 dB daily for 90 days or more; **9:** 1/2 dB per year over age 40 years.
Nebraska	Yes	No	50	*	$22,200	*	ME	No	Poss	Poss	Yes	Yes	Yes	2 yrs.	No	No	**4 and 6:** average weekly wage x life expectancy.
Nevada	Yes	No	*	*	*	*	AAO-79	No	Poss	Yes	Yes	No	Yes	No	Yes-P	No	**3-6:** awards determined according to AMA guide and state statute.

table continued on next page.

Data compiled in 1998/1999 by Susan Megerson with assistance from Cyd Kladden.

KEY: Poss - possible; **NA** - non-applicable; **NR** - no response; ***** - see comments; **AAOO-59** - avg. > 25 dB at 500, 1000, and 2000 Hz; **AAO-79** - avg. > 25 dB at 500, 1000, 2000, and 3000 Hz; **ME** - medial evidence, hearing loss formula at discretion of consulting physician; **I** - only if impairment also present; **D** - claim denied; **P** - penalty applied.

TABLE 18.1 — continued
Hearing loss statutes in the United States and Canada.

Jurisdiction	1. Is occupational hearing loss compensable?	2. Is minimum noise exposure required for filing?	3. Schedule in weeks (one ear).	4. Schedule in weeks (both ears).	5. Maximum compensation (one ear).	6. Maximum compensation (both ears).	7. Hearing impairment formula.	8. Waiting period.	9. Is deduction made for presbycusis?	10. Is award made for tinnitus?	11. Provision for hearing aid?	12. Credit for improvement with hearing aid?	13. Is hearing loss prior to employment considered in compensation claim?	14. Statute of limitations for hearing-loss claim.	15. Penalty for not wearing hearing protection devices?	16. Self-assessment of hearing impairment considered in rating/award?	Comments
New Hampshire	Yes	No	30	123	$25,200	$103,320	ME	No	No	No	Yes	No	Yes	2 yrs.	No	No	2: 90 dB TWA; 7: avg. > 30 dB at 1000, 2000, and 3000 Hz, 1.5% per dB.
New Jersey	Yes	Yes*	60	200	$8,280	$48,200	Other*	4 wks.	No	Yes	Yes	Yes	Yes	2 yrs.	Yes-D	Yes	2: exposure to harmful noise for 90 days or more.
New Mexico	Yes	No	40	150	$15,039	$56,397	ME	7 days	No	Yes	No	No	Yes	1 yr.	Poss-P	No	
New York	Yes	Yes*	60	150	$24,000	$60,000	AAO-79	3 mos.	No	No	No	No	Yes	90 days	No	No	2: 90 dBA for at least 90 days; 7: if hearing loss due to injury, then "medical evidence" is utilized.
North Carolina	Yes	Yes*	70	150	$37,240	$79,800	AAO-79*	6 mos.	No	No	Yes	No	Yes	2 yrs.	Yes-D	No	
North Dakota	Yes	No	5	100	$695	$13,900	AAO-79	No	Poss	Yes-I	Yes	No	Yes	No	No	No	1: only permanent total hearing loss in one or both ears is compensable.
Ohio	Yes*	No	25	125	$13,525	$67,625	ME	No	No	No	Yes	No	No	2 yrs.	No	No	10: up to 5% for tinnitus in cases of unilateral hearing loss.
Oklahoma	Yes	No	104	312	$22,194	$66,583	AAO-79	No	No	Yes*	Yes	No	Yes	2 yrs.	Yes-D	No	
Oregon	Yes	No	*	*	$27,240	$87,168	Other*	No	Yes	Yes	Yes	No	Yes*	1 yr.	No	No	3-4: one time permanent partial disability award based on impairment; 7: avg > 25 dB at 500, 1000, 2000, 3000, 4000, and 6000 Hz, 1.5% per dB; 13: if baseline completed within 180 days of hire.
Pennsylvania	Yes	No*	60	260	*	*	AAO-79	No	No	No	No	No	Yes*	3 yrs.	No	No	2: case law has adopted OSHA standards as employer defense; 5 and 6: # weeks x 2/3 average wage; 13: only if pre-employment testing was performed at employer's expense.

Jurisdiction																	Comments
Rhode Island	Yes	Yes	17*	100*	$1,530*	$9,000*	AAO-79	6 mos.	Yes	Poss-I	Poss	No	No*	2 yrs.	No	No	3: 60 weeks if loss due to trauma; 4: 200 weeks for trauma; 5: $5400 for trauma; 6: $18,000 for trauma;13: current employer is solely responsible for occupational loss.
South Carolina	Yes	No	80	165	$38,678	$79,773	AAO-79	No	No	No	No	No	Yes	No	No	No	
South Dakota	Yes	Yes*	50	150	$20,400	$61,200	AAO-79	No	Yes*	No	Yes	No	Yes	2 yrs.	Yes-D	No	2: 90 dBA TWA; 9: 1/2 dB for each year over 45 yrs.
Tennessee	Yes	No	75	150	$38,625	$77,250	ME	No	Yes	Yes-I	Yes	Poss	Yes	1 yr.	Poss-D	Yes	3-6: no maximum scheduled awards.
Texas	Yes	No	*	*	*	*	AAO-79	No	No	No	Yes	No	Yes	30 days	No	No	
Utah	Yes	Yes*	NR	109	NR	$35,425	AAO-79	6 wks.	NR	NR	NR	NR	Yes	180 days	NR	NR	2: 90 dBA TWA or impact/impulsive noise 140 dB or greater.
Vermont	Yes	No	24	142	$17,448	$103,052	ME	No	No	Yes-I	Yes	No	No	Yes	Yes-D	No	
Virginia	Yes*	No	50	100	*	*	AAO-79	No	No	No	Poss	No	Yes	2-5 yrs.	No	No	1: only hearing loss due to work-related trauma or injury is considered; 5-6: # weeks x average weekly wage.
Washington	Yes	No	N/A	N/A	$10,837	$65,023	AAO-79	No	No	Yes-I	Yes	No	Yes	2 yrs.	No	No	
West Virginia	Yes	No	*	*	*	*	AAO-79	2 mos.	Yes	No	Yes	No	Yes	3 yrs.	No	No	3-6: hearing loss compensation is based on a percentage of whole body impairment (max for both ears, 22.5% whole body).
Wisconsin	Yes	No	36	216	*	*	Other*	7 days	No	No*	Yes	No	Yes	No	No	No	5-6: depends on year of retirement; 7: avg. ≥ 30 dB at 500, 1000, 2000, and 3000 Hz; 10: not compensable since 1/1/92.
Wyoming	Yes	No	*	*	*	*	ME	No	NR	NR	No	No	Yes	Yes	Yes-D	NR	3-6: based upon rating of impairment.
U.S. DOL – FECA	Yes	Yes	52	200	NR	NR	AAO-79	No	NR	NR	No	No	No	3 yrs.	No	No	FECA - federal employees compensation act
U.S. DOL – Longshoremen	Yes	No	52	200	*	*	AAO-79	No	No	No	Yes	No	No	1 yr.	No	No	5-6: based upon compensation rate, increases annually.
Guam	Yes	No	52	200	$13,000	$50,000	ME	No	Yes	Yes-I	Yes	No	Yes	1 yr.	No	No	
Alberta	Yes	Yes*	NA	NA	$3,184	$19,105	NR	No	No	Yes-I	Yes	No	Yes	5 yrs.	No	No	2: 85 dBA for 8 hrs.
British Columbia	Yes	Yes*	NA	NA	*	*	Other*	No	No	No	Yes	No	Yes	No	No	No	2: 85 dB LEX,8h for 2 yrs.; 5-6: based on % of annual wage; 7: avg > 28 dB at 500, 1000, and 2000 Hz; 2.5% per dB
Manitoba	Yes	Yes*	NA	NA	*	*	Other*	No	Yes*	No	Yes	No	Yes	No	Poss-D	No	2: 85 dB for 2 yrs.; 5-6: lump sum paid based on %; 7: avg ≥ 35 dB at 500, 1000, 2000, and 3000 Hz; 9: 1/2 dB per year over 60 yrs. of age.

table continued on next page.

Data compiled in 1998/1999 by Susan Megerson with assistance from Cyd Kladden.

KEY: Poss - possible; NA - non-applicable; NR - no response; * - see comments; AAO-59 - avg. > 25 dB at 500, 1000, and 2000 Hz; AAO-79 - avg. > 25 dB at 500, 1000, 2000, and 3000 Hz; I - only if impairment also present; D - claim denied; P - penalty applied.
ME - medial evidence, hearing loss formula at discretion of consulting physician;

TABLE 18.1 — continued
Hearing loss statutes in the United States and Canada.

Jurisdiction	1. Is occupational hearing loss compensable?	2. Is minimum noise exposure required for filing?	3. Schedule in weeks (one ear).	4. Schedule in weeks (both ears).	5. Maximum compensation (one ear).	6. Maximum compensation (both ears).	7. Hearing impairment formula.	8. Waiting period.	9. Is deduction made for presbycusis?	10. Is award made for tinnitus?	11. Provision for hearing aid?	12. Credit for improvement with hearing aid?	13. Is hearing loss prior to employment considered in compensation claim?	14. Statute of limitations for hearing-loss claim.	15. Penalty for not wearing hearing protection devices?	16. Self-assessment of hearing impairment considered in rating/award?	Comments
New Brunswick	Yes	No	NA	NA	*	*	ME	No	No	Yes-I	Yes	No	Yes	Yes*	No	No	**5-6**: depends on percent of disability; **14**: prior to retirement.
NW Territories	Yes	Yes*	*	*	$147/month	$887/month	Other*	Yes*	No	No	Yes	No	Yes	1 yr.	No	No	**2**: 90 dB for 8 hrs/day for 2 yrs.; **3-4**: no maximum time period; **7**: avg ≥ 30 dB at 500, 1000, 2000, and 3000 Hz; **8**: when removed from exposure.
Nova Scotia	Yes	Yes*	*	*	*	*	Other*	No	Yes*	Yes*	Yes	No	Yes	5 yrs.	No	No	**2**: 85 dB for 8 hrs/day for 5 yrs.; **3-6**: awards based on pre-injury wages and % impairment with no maximums; **7**: avg ≥ 35 at 500, 1000, 2000 and 3000 Hz; **9**: 1/2 dB per year over age 60 yrs.; **10**: 2-5% awarded if specific criteria are met.
Ontario	Yes	Yes*	NA	NA	*	*	AAO-79	No	No	Yes	Yes	No	Yes	6 mo.	No	No	**2**: 90 dB for 5 yrs.; **5-6**: lump sum awards based on % impairment, age at date of accident and maximum medical recovery; **9**: 1/2 dB per year over age 60 yrs.
Prince Edward Island	Yes	Yes*	*	*	*	*	ME	No	Yes*	Yes*	Yes	No	Yes	No	No	No	**2**: 2 yrs. minimum; **3-6**: lump sum awards based on % impairment; **9**: 1/2 dB per year over age 60 yrs.; **10**: 2% maximum.
Quebec	Yes	Yes*	*	*	*	*	ME	No	Yes	No	Yes	No	Yes	No	No	NR	**2**: 90 dB for 8 hrs/day for 2 yrs.; **3-6**: lump sum award based on % impairment and age at time of injury.
Saskatchewan	Yes	No	*	*	$1,130	$13,560	Other *	No	No	Yes-I	Yes	No	Yes	No	No	No	**3-4**: lump sum pension and hearing aid costs; **7**: avg ≥ 30 dB at 500, 1000, 2000, and 3000 Hz.
Yukon	Yes	Yes*	*	*	*	*	AAO-79	2-5 yrs.	Yes*	Yes-I	Yes	No	Yes	No	No	No	**2**: 85 dB for 8 hrs/day; **3-6**: awards based on % impairment and ability to return to work; **9**: maximum 2% for each year over age 45 yrs.

KEY: **Poss** - possible; **NA** - non-applicable; **NR** - no response; *. - see comments; **AAO-59** - avg. > 25 dB at 500, 1000, and 2000 Hz; **AAO-79** - avg. > 25 dB at 500, 1000, 2000, and 3000 Hz; **I** - only if impairment also present; **D** - claim denied; **P** - penalty applied. **ME** - medial evidence, hearing loss formula at discretion of consulting physician;

only work-related hearing loss caused by injury (single event) is compensable: Idaho, Indiana, Louisiana, and Virginia. Michigan responded that compensation is considered only if an injury to the ear (single event), which may or may not result in hearing loss, causes a loss of wages. Georgia reported that in cases of monaural loss, awards are provided only in the presence of pre-existing deafness in the other ear. Ohio stated that only total hearing loss in one or both ears is compensable.

All states, provinces, territories, and agencies were surveyed regarding specific statutes and practices for assessing hearing-loss claims and awarding compensation within their jurisdictions. A number of states compensate all work-related hearing loss but follow different procedures for hearing loss resulting from injury and resulting from long-term noise exposure (considered a disease process). In many states and provinces, the written statutes provide little information specific to hearing loss. Actual WC practices in these jurisdictions are heavily dependent upon interpretation of general WC standards and case law (which is flexible in application and can evolve over time). A number of states reported that statutes have been revised over the years, and individual claims are evaluated using different criteria depending upon date of injury or date of claim.

Due to the many special circumstances and qualifiers surrounding processing of a WC claim for hearing loss, it is important for those involved to become completely familiar with regulations and procedures within their jurisdiction. In practice, WC claim evaluations are subject to many nuances that require an in-depth understanding of applicable regulations and practices. Readers are cautioned not to rely on Table 18.1 as a definitive single source of information. Professionals should contact the appropriate WC office to request the most current and comprehensive information available.

Impairment, Handicap, and Disability

The American Academy of Otolaryngology-Head and Neck Surgery (AAO-HNS), the American Medical Association (AMA), and the World Health Organization (WHO) all define the terms *impairment, handicap,* and *disability* slightly differently. All agree that impairment is the reduction of function outside the normal range. However, handicap and disability are defined differently by each organization. AAO-HNS defines handicap as a "disadvantage imposed by an impairment sufficient to affect the individual's efficiency in activities of daily living." AMA uses nearly identical language to define handicap in the introductory chapter of the *Guides to the Evaluation of Permanent Impairment* (1993). However, in the hearing loss chapter of the *Guides*, the term "binaural hearing impairment" is used to denote interference with activities of daily living (with the adjective "binaural" indicating that both ears are being considered). It is important to note that AAO-HNS's hearing handicap (HH) and AMA's binaural impairment (BI) refer to exactly the same concept—the same formula (to be described) is used to calculate both HH and BI. This formula is intended to estimate an individual's degree of difficulty with understanding speech under everyday circumstances. Most states use BI as a measure of determining the amount of compensation to be paid.

697

"Disability" means different things to AAO-HNS ("inability to be employed at full wages"), to AMA ("alteration of an individual's capacity to meet . . . demands . . . or . . . requirements because of an impairment"), and to WHO ("inability to perform an activity in a manner or within a range considered normal"). In this chapter, we will limit our discussion to the term BI as defined by the AMA; it is essentially synonymous with what AAO-HNS calls HH and with what WHO calls disability.

BI can be estimated by interviewing a hearing-impaired person (self-report), by measurements of ability to understand speech, or by calculations based on pure-tone audiometry. Self-report is rarely advocated for use in evaluating individuals in medical-legal settings because most clinicians consider it likely that exaggeration cannot be controlled. On the other hand, comparing self-report to audiometric data for groups of hearing-impaired people who are not seeking compensation provides the best validation of methods of estimation of BI from pure-tone or speech tests (Schow and Gatehouse, 1990). In other words, people with high BI scores calculated from the audiogram would be expected to report more difficulties in everyday life than people with low BI scores. Results of the survey summarized in Table 18.1 confirmed that most WC systems do not consider self-assessment in the determination of awards. Only five states (California, Indiana, Mississippi, New Jersey, and Tennessee) responded definitively that self-assessment is taken into account.

Arguably, since speech understanding is the most important thing we do with our ears, it is perhaps surprising that speech tests are not widely used to estimate BI. There are several reasons why pure-tone tests are preferred:

(1) they are less susceptible to exaggeration (ASHA, 1981),

(2) they are faster than speech tests that would have acceptable reliability, and

(3) an attempt to simulate speech conditions of "everyday life" would require testing under multiple conditions (varying speech-to-noise ratios, degrees of redundancy, etc.), and would require a way to weight the scores achieved under the different conditions.

Pure-tone thresholds generally correlate better with self-report than do clinical speech test performance scores (Hardick et al., 1980; King et al., 1992) and thus form the basis of BI estimation in all U.S. jurisdictions (except the Veterans Administration, which combines pure-tone and speech scores to determine awards).

Calculation of Binaural Impairment

Many methods have been used over the years to estimate BI (for a historical discussion, see Dobie, 1993). The Subcommittee on Noise of the American Academy of Ophthalmology and Otolaryngology (AAOO, which was a predecessor of AAO-HNS) developed the first widely-used method in 1959 (AAOO, 1959). In 1961, the method was approved by the American Medical Association. The 1959 AAOO formula is still included in some WC laws.

The AAOO formula considered only those frequencies that were believed to be important to the hearing of speech in quiet environments. Specifically, hearing threshold levels were to be determined by pure-tone air-conduction audiometry at frequencies of 500, 1000, and 2000 Hz. If the arithmetic average of the hearing levels (HLs) at these three frequencies for an ear was 25 dB HL or less, no impairment was said to exist in that ear. This value of beginning impairment, referred to as the "low fence," reflected the fact that the ability to understand everyday speech was not appreciably affected in persons having hearing thresholds above 0 but less than 25 dB HL. Every decibel above 25 dB HL constituted a 1.5 percent monaural impairment for that ear. Thus, an average hearing level of 92 dB represented total monaural impairment ($1.5 \times [92 - 25] = 100\%$). Binaural impairment was determined by a weighted average in which the percentage of impairment in the better ear was multiplied by a factor of five and added to the worse-ear impairment, with the resultant sum divided by six. Although there appears to have been no research basis for the 5-to-1 weighting of the better ear, it is clear that equal weighting is not appropriate. Total loss of hearing in one ear would not cause a 50 percent impairment in the abilty to understand speech under everyday conditions.

In 1979, the American Academy of Otolaryngology (AAO, which was a successor to AAOO and a predecessor of AAO-HNS) revised its guide for the evaluation of BI in order to recognize the importance of higher frequencies for speech understanding in noisy backgrounds (AAO, 1979). The AMA endorsed this change in 1979. The newer definition of BI is based on the average loss over 4 frequencies (500, 1000, 2000, and 3000 Hz) while the low fence of 25 dB HL was retained for the point of beginning impairment. The high fence at 92 dB HL and the better ear weighting were also unchanged. Because noise exposure and aging cause more change at 3000 Hz than at the lower frequencies, the effect of this change is usually to increase the estimated degree of impairment for a given amount of hearing loss.

Results of the recent survey (Table 18.1) revealed that the most popular method for calculating binaural impairment is the AAO-79 method. It is notable that a full third of jurisdictions surveyed stated that they do not specify a formula to be used (impairment ratings are based on "medical evidence"). Of those jurisdictions that do require a certain method for determining impairment, 68 percent of U.S. states/territories and 25 percent of Canadian provinces responded that they utilize the AAO-79 formula by specific reference or by virtue of a requirement to follow the most recent AMA guideline for determining impairment (which specifies the AAO formula). Six states are still utilizing the AAOO-59 formula and several states and provinces have adopted other variations (see Table 18.1, Column 7).

Table 18.2 compares the percentage BI for one audiogram that was calculated using four different formulas. It can be seen that the addition of 3000 Hz in the 1979 AAO equation increases the BI by a significant amount.

Awards

The process for determining awards is typically based on applying the impairment rating to some sort of schedule for lump sum payments or extended payments based

TABLE 18.2
Sample hearing impairment formulas and example calculations.

Formula	Audiometric Frequencies Used (Hz)	Low Fence (dB HL)	High Fence (dB HL)	% Per Decibel Loss Above Low fence	Better Ear/ Worse Ear Weighting
AAOO-1959	500, 1000, 2000	25 dB	92 dB	1.5	5/1
AAO-1979	500, 1000, 2000, 3000	25 dB	92 dB	1.5	5/1
New Jersey	1000, 2000, 3000	30 dB	97 dB	1.5	5/1
Illinois	1000, 2000, 3000	30 dB	85 dB	1.82	1/1

Example calculation from audiogram below, using AAO-1979:

Frequency (Hz)	500	1000	2000	3000	4000	8000
Right Ear (dB)	15	25	45	55	65	45
Left Ear (dB)	30	45	60	85	90	70

For each ear average threshold level = $\dfrac{\text{thresholds at } 500, 1000, 2000, 3000 \text{ Hz}}{4}$

Right Ear $\dfrac{(15 + 25 + 45 + 55)}{4} = 35\ dB$

Left Ear $\dfrac{(30 + 45 + 60 + 85)}{4} = 55\ dB$

To obtain % impairment, subtract 25 dB and multiply by 1.5:

Right Ear = $(35 - 25) \times 1.5 = 15\%$
Left Ear = $(55 - 25) \times 1.5 = 45\%$

Binaural Impairment (BI)% $= \dfrac{(5 \times \text{Better Ear}) + (1 \times \text{Poorer Ear})}{6}$

$= \dfrac{(5 \times 15) + (1 \times 45)}{6} = \dfrac{75 + 45}{6} = 20.0\%$

Comparison of BI for four formulas using example audiogram:

Formula	AAOO-1959	AAO-1979	New Jersey	Illinois
BI	9.2%	20.0%	22.9%	41.0%

on a percentage of the individual's wages. There is a great deal of variation across states and provinces in the amount of awards available. Depending on the state, a totally deaf worker could receive as little as $9000 or as much as $150,000 over and above replacement of wage losses (see Table 18.1). Once again, the reader is cautioned to contact local WC offices for the most definitive and up-to-date information available.

Some employers find benefit in tracking total "potential liability" for hearing loss claims for a given plant, division, or company. An example of the computation of potential claims is provided by Berger (1985).

Other Considerations

Beyond the key issues of the BI formula that a state uses and the maximum dollar payments available, a number of additional factors can effect the potential compensation. They are discussed in the following sections.

Adjustments for Age

It is generally recognized that hearing deteriorates with the passage of time, even without noise exposure. This age effect, often termed presbycusis (but which also includes sociocusis and nosoacusis — see Chapter 4) is characterized by a gradual, yet significant loss of hearing at all frequencies — usually, the higher the frequency, the greater the degree of loss. For most WC laws, an individual's total impairment is compensable as long as an occupational noise exposure substantially contributed to the overall impairment. However, many states do consider aging effects.

Over 40 percent of states and provinces (Table 18.1) indicated that some type of deduction in impairment/award may be made for presbycusis. A number of jurisdictions deduct 0.5 dB from the person's average hearing level for every year of the person's age beyond 40, prior to the calculation of BI. This is called "age correction." It can have the effect of making an otherwise compensable loss noncompensable. For this reason, age correction prior to BI calculation is often considered unfair (AAO-HNS, 1998). In contrast, allocation or apportionment refers to a process in which BI is calculated first, followed by an estimation or calculation of the relative contributions of occupational noise, aging, and other factors. For example, Iowa requires that the initially calculated BI score be reduced by a factor equal to the ratio of median expected age-related threshold shift (for the 0.5, 1, 2, and 3 kHz pure-tone average) during the period of employment to the final pure-tone average. Pennsylvania permits expert testimony regarding the contribution of aging and other causes, and awards are frequently adjusted or denied based on such testimony. AAO-HNS (1998) has published a monograph on the evaluation of noise-exposed workers that includes a discussion of allocation from the medical point of view.

Waiting Period

WC laws require that hearing impairment be both occupationally related and *permanent*. In the process associated with sensory damage from noise exposure, there is a temporary threshold shift that disappears upon termination of the exposure. Although there is considerable variation in recovery time depending on the level and duration of the noise insult, most of the temporary effect is gone within 48 hours in a quiet environment. For economic reasons, longer recovery periods have been assumed in some states, and an employee may not be able to pursue a claim until six months of nonexposure have passed. Seventy percent of U.S. states and provinces surveyed indicated that no waiting period is necessary for filing a WC claim. For those jurisdictions that do impose a waiting period, reported time-frames range from 3 days to 6 months. Ironically, such provisions usually do not apply to traumatically-induced hearing loss, which often *can* require three to six months for full stabilization (Segal et al., 1988).

Use of Hearing Protection Devices (HPDs)

Most WC laws do not address the question of personal hearing protection. It is not generally clear whether the wearing of adequately-fitted hearing protection constitutes removal from exposure, although in at least one state, the law specifically adopts such a provision. Over 40 percent of the U.S. jurisdictions surveyed indicated that claims would be denied or an award penalty assessed if an individual was found to have willfully disregarded a requirement to wear HPDs. Canadian provinces reported no such provisions.

Statute of Limitations

Most WC laws were originally established to compensate employees for traumatic injury, and contain a provision encouraging timely filing of claims. The provision, called a statute of limitations, also applies to NIHL claims and requires that a date of injury be established. In some states, the date of injury is "the last day exposed," while in others it is the date the employee became aware of hearing loss or its work relatedness. The statute of limitations varies from state to state, and can be as short as 30 days to as long as 5 years.

Apportionment or Allocation Among Employers

The WC system was originally established on the principle that at the time of the employee's injury, the employer should pay for the loss of earnings of the employee. However, in a case of NIHL, where several employers might be involved, a new approach was necessary. This led to the inclusion of an apportionment clause in the acts of many states. For example, if the last employer at the date of injury can establish that the employee began employment with hearing loss, prior employers (or a special fund) may be tapped for any NIHL present at the time of employment.

Allocation among employers does have the effect of encouraging preplacement audiometry, and can be a costly potential liability in the purchase of manufacturing operations. The survey summarized in Table 18.1 reported that only 17 percent of WC jurisdictions reported that they have no provision for apportionment of pre-existing hearing loss.

Hearing Aids

In most states, hearing impairment is evaluated without consideration of the effect a hearing aid or other prosthesis might have on the claimant's ability to understand speech. Most states and all Canadian provinces now include some provision for hearing aids as part of the compensation. However, the award for NIHL is generally based on a pure-tone audiogram in the absence of such a device.

Tinnitus

Although tinnitus (ringing or head noises) typically accompanies NIHL, it often occurs in the absence of noise exposure. As shown in Table 18.1, tinnitus is recognized in a number of states, and the award for compensation may be modified by its presence. Over half the states and provinces responding to the WC survey indicated that awards may be made for tinnitus. Many of these jurisdictions qualified that the award is available only in combination with hearing impairment, not for tinnitus alone. There are no objective measurements for tinnitus, and the award (if any) is based on the claimant's symptoms and on the physician's opinion.

Duration and Level of Exposure

Many states include a provision that excludes an NIHL claim where the occupational noise exposure is below a specified level. In the U.S., the most commonly specified minimum exposure is 90 dB (or 90 dBA time-weighted average [TWA]). In Canada, provinces require a minimum of from 85 to 90 dB exposures for claims to be considered. Other states require that the noise exposure duration must exceed a minimum number of days in order to file a valid claim. Other characteristics, such as A-weighting or TWA, may also be addressed in the WC law.

Considerations for Handling Claims

In the broadest sense, a valid NIHL claim exists when it can be shown that there is permanent impairment, and that the impairment arose out of and in the course of employment. A number of considerations go into validating claims when they arise in industry. In establishing the validity of a claim, testimony and records become extremely important.

The Validation of a Claim

The following is a check list of some of the considerations to be examined as part of clinical evaluation of a claim:

(1) Has a hearing loss been established? If so, has its severity been established? The most recent audiometric data should be complete and professionally obtained by a clinical audiologist certified by the American Speech-Language-Hearing Association (ASHA) or licensed by the state. All audiometric records should be compared to determine test-retest consistency and detect possible fluctuating hearing levels — such records may be invalid or may indicate nonnoise-related causes of hearing loss. Inconsistencies during the pure-tone testing on a given date (usually marked as unreliable on the audiogram), or inconsistencies between pure-tone thresholds and speech thresholds suggest invalid data. (See Chapter 11.)

(2) Is the type of hearing loss consistent with excessive exposure to noise? Monaural (one ear) losses do not usually result from noise exposure — conductive (middle ear) losses are never caused by continuous noise, although explosions can sometimes rupture the eardrum and cause a conductive loss. If the loss is sensorineural and bilateral, is the audiometric configuration consistent with noise exposure (greatest loss in 3 to 6 kHz range, especially in the early years of exposure)? Is the magnitude of the hearing impairment consistent with the extent of noise exposure? (See Chapter 17.)

(3) Are adequate noise monitoring records available to demonstrate that excessive noise exposure existed in sufficient degree and duration to result in hearing loss (i.e., consistent exposure for years to 8-hour TWAs of 85 dBA or greater)?

(4) Is the time course of the hearing loss consistent with the alleged cause? A "trajectory analysis" is useful when there are multiple hearing tests for the claimant under review. Generally, occupational hearing loss will be characterized by gradual worsening of hearing over the series of audiograms, with the largest changes in the early years (i.e., a decelerating trajectory as opposed to the accelerating changes typical of age-related hearing loss).

(5) Could the hearing loss be attributed to nonoccupational causes? Noise exposures resulting from a variety of hobbies or spare time activities (especially shooting) may have caused a portion of the hearing loss. A detailed personal case history should be obtained by the audiologist or physician, but often the most helpful documentation of nonoccupational exposure is found in hearing conservation program questionnaires completed by the claimant over the years.

(6) If exposed, was the employee provided with properly fitted protection and instructed in its use? In some states, the willful misuse of protection invalidates a claim for NIHL.

(7) Are the symptoms claimed plausible? Inappropriate complaints for NIHL include: (a) pain; (b) dizziness, vertigo, unsteadiness; and (c) a sense of "fullness" in the ear that extends beyond the time of noise exposure. One or more of the preceding complaints suggests alternative etiology, multiple etiology, or an attempt to falsely magnify the problem. Typical complaints for NIHL (or any other type of high-frequency hearing loss, such as age-related hearing loss) include: (a) difficulty hearing female or children's voices, (b) tendency to have radio and television volume at a high sound level, (c) a sense of embarrassment due to giving inappropriate responses to inquiries, (d) tinnitus, and (e) difficulty hearing speech in a noisy background.

The answers to these questions should assist the clinician in determining whether or not the hearing impairment is occupationally related, and whether the impairment resulted from or was aggravated by conditions at work. Clinical evaluation of WC claims is discussed in more detail by Dobie (1993) and by AAO-HNS (1998). The American College of Occupational Medicine (ACOM) has published criteria for diagnosis of occupational NIHL that are similar to those described above (ACOM, 1989). Specifically, occupational NIHL should be sensorineural, bilateral and symmetrical, greatest in the 3 to 6 kHz range, and should demonstrate a decelerating growth over time. In addition, once noise exposure ceases, further progression of hearing loss must be due to other causes. Both Dobie (1993) and the National Hearing Conservation Association (Tolley, 1990) have noted that some of the details of the ACOM criteria are debatable. For example, ACOM says that NIHL is never profound, but thresholds at specific frequencies can sometimes reach the profound range due to the combination of noise-induced and age-related changes, based on the ANSI S3.44

model. Similarly, that model predicts small additional threshold shifts at 3 to 6 kHz even after 20 years of exposure — ACOM states that by this time, an asymptote will usually have been reached.

Testimony and Records

Although most WC claims will be handled without controversy, either party has the right to a hearing in order to establish the validity of the claim and the amount of the award. In such a hearing, the records that establish hearing loss and noise exposure become important evidence.

The hearing is usually an adversary process, with the normal rules of evidence submission prevailing. The weight of the evidence is strongly affected by how it is obtained, its completeness, its adherence to accepted standards for accuracy, and its professional backup (Gasaway, 1985). A valid hearing conservation program (HCP) audiogram should include test results at all frequencies used in the evaluation of impairment plus those recommended for hearing conservation purposes. The baseline audiogram and the audiogram used to determine BI at career-end should each represent the permanent hearing thresholds (i.e., excluding temporary threshold shifts), and should therefore be obtained after at least 14 hours of quiet. In contrast, many hearing conservation professionals advise that annual audiograms be completed during the workday without special precautions to exclude temporary threshold shift (see Chapter 11 for discussion of annual audiometric testing). Indeed, detection of temporary shifts is a good thing if it leads to corrective action before hearing loss becomes permanent. A certified audiometric technician under the supervision of a physician or audiologist should obtain the audiometric data. Test results should be reviewed by the supervising professional. Records of audiometer calibration and background noise conditions should also be complete and of sufficient frequency to prove that accurate testing was done.

The establishment of noise exposure (or lack thereof) requires similar record-keeping and professional supervision (see Chapters 6 and 7). In addition to valid records of noise exposure and hearing status, a good hearing conservation program includes records of medical history and employee activities. These records should also be complete and professionally supervised.

From a legal point of view, the validity of records is established through testimony. Any member of the hearing conservation program team can be called upon to testify, so it is useful to review the role of the expert witness.

The Expert Witness

A fact witness knows something about a case that usually could not be provided by others who weren't there. If only ten people saw and heard the explosion, only those ten could testify as fact witnesses about how the explosion looked and sounded. In contrast, an expert witness is not usually at the scene, yet s/he possesses special knowledge, training, and/or experience that can assist a court in interpreting facts. Experts might be able to show that the explosion was sufficiently intense to damage the hearing of workers who were within 100 yards, based on facts provided to them regarding the nature and quantity of the explosive material. Physicians, audiologists, industrial hygienists, and others involved in hearing conservation may be qualified to testify as experts in NIHL cases. Sometimes the roles overlap — an expert witness may also be a fact witness. Readers interested in a more detailed discussion of the role of the expert witness may wish to consult the AAO-HNS monograph (1998).

Any witness (fact or expert) should remember that the attorney requesting testimony is *not* the witness's attorney. Each attorney represents the interests of a client, not the interests of the witness. Conflicts may arise — for example, an expert whose opinions are generally favorable to one side may find the attorney for that side pushing for even more favorable opinions. If the attorney's arguments are not persuasive, the expert should continue to offer opinions that are genuine and will preserve the expert's integrity and reputation.

The attorney will usually meet with a prospective witness to discuss the scope of testimony, the important issues, and even specific questions to be asked. It is the attorney's responsibility to frame questions that will elicit facts and opinions favorable to the client's case. The witness' duty is to answer the questions honestly, but with no more elaboration than is required or requested. The attorney will also indicate what should be brought to court (i.e., documents, exhibits, etc.)

Experts in most jurisdictions are asked to provide opinions that meet a standard of "reasonable scientific (or medical) certainty." This usually means "more probably true than false." This standard bothers some scientifically-trained people, who are inclined to say they can't be sure that X caused Y unless it's been proven to a 95 percent (or 99 percent) level of certainty. It is the attorney's duty to explain this to the expert, and to ask questions that permit the expert to answer both truthfully and (if possible) helpfully. The word "probably" solves the problem most of the time: "Based on the different dosimetry readings, Mr. Smith's typical daily exposure was probably between an 85- and 90-dBA TWA" (the 95 percent confidence interval may have been much wider).

The expert should review the facts of the case prior to the testimony. The attorney may have requested a written report. If so, this should be reviewed along with the reports of others. Often a report is not requested because it may limit subsequent testimony more than the attorney wishes.

Direct examination by the attorney requesting testimony usually begins with an exploration of qualifications (i.e., education, training, experience, etc.) Subsequently, there will be a discussion of the facts and the background material necessary for a lay jury or judge to understand the expert's opinion. This part of the expert's testimony often involves charts, diagrams, or models that can be used to present a brief introduction to the scientific issue in question. Finally, the expert's opinions are elicited.

Cross-examination by the opposing attorney attempts to discredit the expert's opinions, based on attacks on qualifications, inadequacy of facts considered, bias, conflicts with prior statements/writings by the expert, or conflicts with "mainstream" opinions in one's own field, etc.

Because prevention is the best medicine, experts should:

- stay out of areas where they are unqualified,
- review and carefully consider facts provided to them,
- avoid conflict of interest (the expert should not financially gain or lose based on the outcome of the case),
- maintain consistent opinions over time
- be prepared to explain how opinions have evolved based on experience, and
- know what others in the field have written and currently believe.

Experts will often be asked about "authoritative" texts and journal articles, and should be able to identify the most useful references. However, they should remember that even the best books contain controversial or even erroneous information. If there are unavoidable weaknesses in the expert's testimony, these are often best dealt with candidly on *direct* examination. However, it is the requesting attorney's responsibility to elicit these problems and to decide whether or not to bring them out early.

One's own words (i.e., reports, articles, prior testimony) or the writings of others may be read out loud during cross-examination in an attempt to show conflict with the direct testimony. It is usually wise to request the document being read from before responding to the question in order to ensure that the brief quotation hasn't been taken out of context.

In cross-examination, as in direct examination, it's important to pay close attention to the question and answer it without unnecessary elaboration. Some questions contain debatable or even false assumptions and should not be answered "yes" or "no." ("Have you stopped stealing from the company?")

The above discussion describes the order of events in courtroom testimony or in a *perpetuation* deposition, in which a witness's testimony is taken as a substitute for actually appearing at trial. In a *discovery* deposition, the opposing attorney may compel an expert's testimony in order to be better prepared for what the expert will say at trial. The opposing attorney may be the only one asking the questions, although the other attorney (the one who initially requested the expert's help in the first place) may ask followup questions at the end.

Because expert witnesses provide valuable services, they are usually paid for their time. It is unethical to base expert witness compensation on the outcome of the case. Rather, compensation should be based on hourly charges, on a lump sum, or on some other basis agreed to in advance of any services rendered.

Handling Claims

Although the purpose of this chapter is to provide guidance in the legal aspects of hearing-loss claims, the actual process of filing a claim or preparing a defense should be referred to persons skilled in such activities. Often, such matters are carried out by attorneys or by the representatives of insurance carriers. There are, however, steps that management can take that will eliminate some of the uncertainties and expedite the handling of valid claims.

In establishing the audiometric monitoring phase of the hearing conservation program (HCP), plan for the disposition of those persons showing abnormal hearing levels.

- In preplacement programs, a decision should be made as to when an employee is referred for a diagnostic work-up and how the results will be acted upon. The AAO-HNS has published recommended criteria for otologic referral in HCPs (AAO-HNS, 1997). If any hearing impairment exists at the time of preplacement testing, a diagnostic audiological test will more accurately define the level of impairment. Future audiograms will be used to determine if any further loss has occurred.
- In followup testing, a decision should be made as to the followup procedures for employees showing large or repeated threshold shifts. Will they be referred to their personal physicians or provided access to a qualified consultant? The insurance carrier and professional supervisor should be consulted in this decision.

Select consultants carefully. Attorneys, audiologists, and physicians vary considerably in their levels of expertise in dealing with WC claims and other cases involving NIHL.

Implications for HCPs

Requirements and standard practices for an HCP are thoroughly reviewed throughout this manual. As discussed in this chapter, although WC claims and civil litigation are independent of OSHA regulations, compensation implications should be considered in the design and implementation of the HCP.

Noise Surveys and Analysis

Many U.S. states and Canadian provinces specify minimum levels of noise exposure for individuals seeking compensation for hearing loss. Under basic WC law, it is the employer's responsibility to provide proof if an employee did not have sufficient exposure to cause the injury/illness. Therefore, proper documentation and retention of noise surveys and individual employee exposure assessments is essential for the later evaluation and defense of hearing loss claims. Many companies underestimate the importance of simple documentation of work areas and jobs typically considered to be safe (i.e., under 80 to 85 dBA TWA). (See Chapter 7.)

Audiometric Monitoring

Chapter 11 provides detailed guidance for the implementation of the audiometric monitoring portion of an HCP. Employers should be aware of the importance of adequate baseline/preplacement audiometric testing and exit audiograms to the defense of WC claims. In particular, baseline audiograms are essential if apportionment of NIHL is allowed in the applicable jurisdiction. For example, in New York, the last employer is responsible for total hearing loss unless a preplacement audiogram documented pre-existing hearing loss, and unless the previous employer was notified of such in writing within 90 days. Likewise, documentation of aural case history is also helpful in the evaluation of a claim. Earlier documentation of noisy hobbies and/or medical complications will help the professional reviewer weigh the possibility of nonoccupational causes of a hearing loss. Employers should also be aware of any special reporting requirements of their state or provincial WC regulations. As an example, Iowa requires that employees be provided written notification of hearing levels if an audiometric examination reveals hearing loss in a potentially compensable range — obviously not an OSHA requirement. Finally, employers should be aware that some states deny or reduce claims to employees who willfully failed to submit to reasonable periodic audiometric examinations. Each of these factors should be considered during the routine administration of the HCP.

Personal Protection

As mentioned earlier in this chapter, many U.S. states now reduce awards or invalidate claims if the employer can prove that the individual failed to utilize the provided and required hearing protection devices. See Chapter 10 for information on hearing protection and recommended documentation for fitting and review of such devices. In addition, Chapter 6 provides a general discussion of documentation protocols in HCPs. Any and all HCP records may be useful during later evaluations and defense of WC claims.

Summary

Hearing impairment resulting from occupational noise exposure or acoustic trauma is covered under each state's WC law, but is defined and accommodated in a variety of ways. Each state's jurisdiction is independent in its interpretation and handling of NIHL claims, but certain general rules apply. It is important that hearing conservationists, industrial hygienists, audiologists, and physicians be aware of these rules so that any hearing impairment found in exposed employees will be properly documented and appropriate steps taken. This chapter describes some of the features of NIHL and suggests the conditions that should be addressed in establishing the validity of a claim. Hearing impairment is quite prevalent in working populations, even in those with no known noise exposure. Furthermore, with liberalization of WC benefits, it becomes increasingly important to identify the true sources of hearing loss so that appropriate steps may be taken to reduce exposures and limit compensation to valid claims.

References

American Academy of Ophthalmology and Otolaryngology (1959). "Committee on Conservation of Hearing: Guide for the Evaluation of Hearing Impairment," *Trans. Am. Academy Ophthalmol. Otolaryngol. 63*, 236–238.

American Academy of Otolaryngology (1979). "Committee on Hearing and Equilibrium and the American Council of Otolaryngology Committee on the Medical Aspects of Noise: Guide for the Evaluation of Hearing Handicap," *J. Am. Med. Assoc. 241(19)*, 2055–2059.

American Academy of Otolaryngology — Head and Neck Surgery (1997). "Otologic Referral Criteria for Occupational Hearing Conservation Programs," Alexandria, VA.

American Academy of Otolaryngology — Head and Neck Surgery (1998). "Evaluation of People Reporting Occupational Hearing Loss," Alexandria, VA.

American College of Occupational Medicine (1989). "Occupational Noise-Induced Hearing Loss," *J. Occup. Med. 31(12)*, 996.

American Medical Association (1993). "Guides to the Evaluation of Permanent Impairment: Fourth Edition," Committee on Rating Mental and Physical Impairment, Chicago, IL.

American Speech-Language-Hearing Association (1981). "Task Force on the Definition of Hearing Handicap." *ASHA 23(4)*, 293–297.

ANSI (1996). "Determination of Noise Exposure and Estimation of Noise-Induced Hearing Impairment," ANSI S3.44-1996, Acoustical Society of America, New York, NY.

Berger, E. H. (1985). "EARLog #15 — Workers' Compensation for Occupational Hearing Loss," *Sound and Vibration 19(2)*, 16–18.

Dobie, R. A. (1993). *Medical-Legal Evaluation of Hearing Loss*, Van Nostrand Reinhold, New York, NY.

Federal Employees Compensation Act (1908).

Gasaway, D. C. (1985). "Documentation: The Weak Link in Audiometric Monitoring Programs," *Occup. Health Saf. 54(1)*, 28–33.

Green Bay Drop Forge v. Industrial Commission (1953). 265 W.S. 38, 60 N. W. 2d 409.

Hardick, E. J., et al. (1980). "Compensation for Hearing Loss for Employees Under Jurisdiction of the U.S. Department of Labor: Benefit Formula and Assessment Procedures (Contract No. J-9-E-9-0205)," Ohio State University, Columbus, OH.

King, P. F., et al. (1992). "Assessment of Hearing Disability: Guidelines for Medicolegal Practice," Whurr Publishers, London, England.

Longshoremen's and Harbor Workers' Compensation Act (1927). 33 U.S.C. 1901 *et seq.*

Schow, R. F. and S. Gatehouse (1990). "Fundamental Issues in Self-Assessment of Hearing," *Ear Hear. Suppl. 11(5)*, 6S–16S.

Segal, S., et. al. (1988). Acute Acoustic Trauma: Dynamics of Hearing Loss Following Cessation of Exposure," *Am. J. Otol. 9(4)*, 293–298.

Shipman v. Employers Mutual Liability Ins. and Lockheed Corp. (1962). 125 SE(2d) 72.

Slawinski v. J. H. Williams (1948). 208 N.Y. 546, 81 N.E. 2d 93 Aff g 273 Appellate Division, S825, 75, N.Y.S. 2d 888.

Tolley, C. (1990). "ACOM Publishes Position Paper on Noise-Induced Hearing Loss; NHCA Responds," *Spectrum 7*, 1–4.

Department of Labor
Occupational Noise Exposure
Standard

Appendix

I

[Code of Federal Regulations, Title 29, Chapter XVII,
Part 1910, Subpart G, 36 FR 10466, May 29, 1971;
Amended 48 FR 9776-9785, March 8,1983]

Occupational Noise Exposure

† 1910.95

(a) Protection against the effects of noise exposure shall be provided when the sound levels exceed those shown in Table G-16 when measured on the A scale of a standard sound level meter at slow response. When noise levels are determined by octave band analysis, the equivalent A-weighted sound level may be determined as follows:

Figure G-9

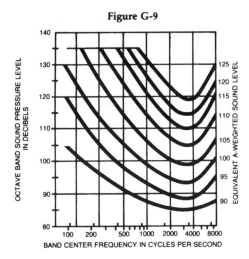

Figure G-9—Equivalent sound level contours. Octave band sound pressure levels may be converted to the equivalent A-weighted sound level by plotting them on this graph and noting the A-weighted sound level corresponding to the point of highest penetration into the sound level contours. This equivalent A-weighted sound level, which may differ from the actual A-weighted sound level of the noise, is used to determine exposure limits from Table 1.G-16.

(b) (1) When employees are subjected to sound exceeding those listed in Table G-16, feasible administrative or engineering controls shall be utilized. If such controls fail to reduce sound levels within the levels of Table G-16, personal protective equipment shall be provided and used to reduce sound levels within the levels of the table.

(2) If the variations in noise level involve maxima at intervals of 1 second or less, it is to be considered continuous.

TABLE G-16—PERMISSIBLE NOISE EXPOSURES

Duration per day, hours	Sound level dBA slow response
8	90
6	92
4	95
3	97
2	100
1½	102
1	105
½	110
¼ or less	115

Table G-16—When the daily noise exposure is composed of two or more periods of noise exposure of different levels, their combined effect should be considered, rather than the individual effect of each. If the sum of the following fractions: $C_1/T_1 + C_2/T_2 + \ldots + C_n/T_n$ exceeds unity, then, the mixed exposure should be considered to exceed the limit value. C_n indicates the total time of exposure at a specified noise level, and T_n indicates the total time of exposure permitted at that level.

Exposure to impulsive or impact noise should not exceed 140 dB peak sound pressure level.

Hearing Conservation Amendment*

(c) **Hearing conservation program.** (1) The employer shall administer a continuing, effective hearing conservation program, as described in paragraphs (c) through (o) of this section, whenever employee noise exposures equal or exceed

*In a unanimous eight-judge decision in *Forging Industry Association vs. Secretary of Labor* (No. 83-1420), the U.S. Court of Appeals for the Fourth Circuit Court, upheld the Secretary of Labor's promulgation of the hearing conservation amendment to the occupational noise exposure standard, and reversed the November 7, 1984, ruling by a three-judge panel which had found the amendment to be invalid. The appeal was argued June 3, 1985, and decided September 23 of the same year.

an 8-hour time-weighted average sound level (TWA) of 85 decibels measured on the A scale (slow response) or, equivalently, a dose of fifty percent. For purposes of the hearing conservation program, employee noise exposures shall be computed in accordance with Appendix A and Table G-16a, and without regard to any attenuation provided by the use of personal protective equipment.

(2) For purposes of paragraphs (c) through (n) of this section, an 8-hour time-weighted average of 85 decibels or a dose of fifty percent shall also be referred to as the action level.

(d) Monitoring. (1) When information indicates that any employee's exposure may equal or exceed an 8-hour time-weighted average of 85 decibels, the employer shall develop and implement a monitoring program.

(i) The sampling strategy shall be designed to identify employees for inclusion in the hearing conservation program and to enable the proper selection of hearing protectors.

(ii) Where circumstances such as high worker mobility, significant variations in sound level, or a significant component of impulse noise make area monitoring generally inappropriate, the employer shall use representative personal sampling to comply with the monitoring requirements of this paragraph unless the employer can show that area sampling produces equivalent results.

(2) (i) All continuous, intermittent and impulsive sound levels from 80 decibels to 130 decibels shall be integrated into the noise measurements.

(ii) Instruments used to measure employee noise exposure shall be calibrated to ensure measurement accuracy.

(3) Monitoring shall be repeated whenever a change in production process, equipment or controls increases noise exposures to the extent that:

(i) Additional employees may be exposed at or above the action level; or

(ii) The attenuation provided by hearing protectors being used by employees may be rendered inadequate to meet the requirements of paragraph (j) of this section.

(e) Employee notification. The employer shall notify each employee exposed at or above an 8-hour time-weighted average of 85 decibels of the results of the monitoring.

(f) Observation of monitoring. The employer shall provide affected employees or their representatives with an opportunity to observe any noise measurements conducted pursuant to this section.

(g) Audiometric testing program. (1) The employer shall establish and maintain an audiometric testing program as provided in this paragraph by making audiometric testing available to all employees whose exposures equal or exceed an 8-hour time-weighted average of 85 decibels.

(2) The program shall be provided at no cost to employees.

(3) Audiometric tests shall be performed by a licensed or certified audiologist,

otolaryngologist, or other physician, or by a technician who is certified by the Council for Accreditation in Occupational Hearing Conservation, or who has satisfactorily demonstrated competence in administering audiometric examinations, obtaining valid audiograms, and properly using, maintaining and checking calibration and proper functioning of the audiometers being used. A technician who operates microprocessor audiometers does not need to be certified. A technician who performs audiometric tests must be responsible to an audiologist, otolaryngologist or physician.

(4) All audiograms obtained pursuant to this section shall meet the requirements of Appendix C: *Audiometric Measuring Instruments.*

(5) *Baseline audiogram.* (i) Within 6 months of an employee's first exposure at or above the action level, the employer shall establish a valid baseline audiogram against which subsequent audiograms can be compared.

(ii) *Mobile test van exception.* Where mobile test vans are used to meet the audiometric testing obligation, the employer shall obtain a valid baseline audiogram within 1 year of an employee's first exposure at or above the action level. Where baseline audiograms are obtained more than 6 months after the employee's first exposure at or above action level, employees shall wearing [sic] hearing protectors for any period exceeding six months after first exposure until the baseline audiogram is obtained.

(iii) Testing to establish a baseline audiogram shall be preceded by at least 14 hours without exposure to workplace noise. Hearing protectors may be used as a substitute for the requirement that baseline audiograms be preceded by 14 hours without exposure to workplace noise.

(iv) The employer shall notify employees of the need to avoid high levels of non-occupational noise exposure during the 14-hour period immediately preceding the audiometric examination.

(6) *Annual audiogram.* At least annually after obtaining the baseline audiogram, the employer shall obtain a new audiogram for each employee exposed at or above an 8-hour time-weighted average of 85 decibels.

(7) *Evaluation of audiogram.* (i) Each employee's annual audiogram shall be compared to that employee's baseline audiogram to determine if the audiogram is valid and if a standard threshold shift as defined in paragraph (g) (10) of this section has occurred. This comparison may be done by a technician.

(ii) If the annual audiogram shows that an employee has suffered a standard threshold shift, the employer may obtain a retest within 30 days and consider the results of the retest as the annual audiogram.

(iii) The audiologist, otolaryngologist, or physician shall review problem audiograms and shall determine whether there is a need for further evaluation. The employer shall provide to the person performing this evaluation the following information:

(A) A copy of the requirements for hearing conservation as set forth in paragraphs (c) through (n) of this section.

(B) The baseline audiogram and most recent audiogram of the employee to be evaluated;

(C) Measurements of background sound pressure levels in the audiometric test room as required in Appendix D: *Audiometric Test Rooms.*

(D) Records of audiometer calibrations required by paragraph (h) (5) of this section.

(8) *Follow-up procedures.* (i) If a comparison of the annual audiogram to the baseline audiogram indicates a standard threshold shift as defined in paragraph (g) (10) of this section has occurred, the employee shall be informed of this fact in writing, within 21 days of the determination.

(ii) Unless a physician determines that the standard threshold shift is not work related or aggravated by occupational noise exposure, the employer shall ensure that the following steps are taken when a standard threshold shift occurs:

(A) Employees not using hearing protectors shall be fitted with hearing protectors, trained in their use and care, and required to use them.

(B) Employees already using hearing protectors shall be refitted and retrained in the use of hearing protectors and provided with hearing protectors offering greater attenuation if necessary.

(C) The employee shall be referred for a clinical audiological evaluation or an otological examination, as appropriate, if additional testing is necessary or if the employer suspects that a medical pathology of the ear is caused or aggravated by the wearing of hearing protectors.

(D) The employee is informed of the need for an otological examination if a medical pathology of the ear that is unrelated to the use of hearing protectors is suspected.

(iii) If subsequent audiometric testing of an employee whose exposure to noise is less than an 8-hour TWA of 90 decibels indicates that a standard threshold shift is not persistent, the employer:

(A) Shall inform the employee of the new audiometric interpretation; and

(B) May discontinue the required use of hearing protectors for that employee.

(9) *Revised baseline.* An annual audiogram may be substituted for the baseline audiogram when, in the judgment of the audiologist, otolaryngologist or physician who is evaluating the audiogram:

(i) The standard threshold shift revealed by the audiogram is persistent; or

(ii) The hearing threshold shown in the annual audiogram indicates significant improvement over the baseline audiogram.

(10) *Standard threshold shift.* (1) As used in this section, a standard threshold shift is a change in hearing threshold relative to the baseline audiogram of an average of 10 dB or more at 2000, 3000, and 4000 Hz in either ear.

(ii) In determining whether a standard threshold shift has occurred, allowance may be made for the contribution of aging (presbycusis) to the change in hearing level by correcting the annual audiogram according to the procedure described in Appendix F: *Calculation and Application of Age Correction to Audiograms.*

(h) Audiometric test requirements. (1) Audiometric tests shall be pure tone, air conduction, hearing threshold examinations, with test frequencies including as a minimum 500, 1000, 2000, 3000, 4000, and 6000 Hz. Tests at each frequency shall be taken separately for each ear.

(2) Audiometric tests shall be conducted with audiometers (including microprocessor audiometers) that meet the specifications of, and are maintained and used in accordance with, American National Standard Specification for Audiometers, S3.6 -1969.

(3) Pulsed-tone and self-recording audiometers, if used, shall meet the requirements specified in Appendix C: *Audiometric Measuring Instruments.*

(4) Audiometric examinations shall be administered in a room meeting the requirements listed in Appendix D: *Audiometric Test Rooms.*

(5) *Audiometer calibration.* (i) The functional operation of the audiometer shall be checked before each day's use by testing a person with known, stable hearing thresholds, and by listening to the audiometer's output to make sure that the output is free from distorted or unwanted sounds. Deviations of 10 decibels or greater require an acoustic calibration.

(ii) Audiometer calibration shall be checked acoustically at least annually in accordance with Appendix E: *Acoustic Calibration of Audiometers.* Test frequencies below 500 Hz and above 6000 Hz may be omitted from this check. Deviations of 15 decibels or greater require an exhaustive calibration.

(iii) An exhaustive calibration shall be performed at least every two years in accordance with sections 4.1.2; 4.1.3; 4.1.4.3; 4.2; 4.4.1; 4.4.2; 4.4.3; and 4.5 of the American National Standard Specification for Audiometers, S3.6-1969. Test frequencies below 500 Hz and above 6000 Hz may be omitted from this calibration.

(i) Hearing protectors. (1) Employers shall make hearing protectors available to all employees exposed to an 8-hour time-weighted average of 85 decibels or greater at no cost to the employees. Hearing protectors shall be replaced as necessary.

(2) Employers shall ensure that hearing protectors are worn:

(i) By an employee who is required by paragraph (b)(1) of this section to wear personal protective equipment; and

(ii) By any employee who is exposed to an 8-hour time-weighted average of 85 decibels or greater, and who:

(A) Has not yet had a baseline audiogram established pursuant to paragraph (g) (5) (ii); or

(B) Has experienced a standard threshold shift.

(3) Employees shall be given the opportunity to select their hearing protectors from a variety of suitable hearing protectors provided by the employer.

(4) The employer shall provide training in the use and care of all hearing protectors provided to employees.

(5) The employer shall ensure proper initial fitting and supervise the correct use of all hearing protectors.

(j) Hearing protector attenuation. (1) The employer shall evaluate hearing protector attenuation for the specific noise environments in which the protector will be used. The employer shall use one of the evaluation methods described in Appendix B: *Methods for Estimating the Adequacy of Hearing Protection Attenuation.*

(2) Hearing protectors must attenuate employee exposure at least to an 8-hour time-weighted average of 90 decibels as required by paragraph (b) of this section.

(3) For employees who have experienced a standard threshold shift, hearing protectors must attenuate employee exposure to an 8-hour time-weighted average of 85 decibels or below.

(4) The adequacy of hearing protector attenuation shall be reevaluated whenever employee noise exposures increase to the extent that the hearing protectors provided may no longer provide adequate attenuation. The employee [sic] shall provide more effective hearing protectors where necessary.

(k) Training program. (1) The employer shall institute a training program for all employees who are exposed to noise at or above an 8-hour time-weighted average of 85 decibels, and shall ensure employee participation in such a program.

(2) The training program shall be repeated annually for each employee included in the hearing conservation program. Information provided in the training program shall be updated to be consistent with changes in protective equipment and work processes.

(3) The employer shall ensure that each employee is informed of the following:

(i) The effects of noise on hearing;

(ii) The purpose of hearing protectors, the advantages, disadvantages, and attenuation of various types, and instructions on selection, fitting, use, and care; and

(iii) The purpose of audiometric testing, and an explanation of the test procedures.

(l) Access to information and training materials. (1) The employer shall make available to affected employees or their representatives copies of this standard and shall also post a copy in the workplace.

(2) The employer shall provide to affected employees any informational materials pertaining to the standard that are supplied to the employer by the Assistant Secretary.

(3) The employer shall provide, upon request, all materials related to the employer's training and education program pertaining to this standard to the Assistant Secretary and the Director.

(m) Recordkeeping. (1) *Exposure measurements.* The employer shall maintain an accurate record of all employee exposure measurements required by paragraph (d) of this section.

(2) *Audiometric tests.* (i) The employer shall retain all employee audiometric test records obtained pursuant to paragraph (g) of this section.

(ii) This record shall include:

(A) Name and job classification of the employee;

(B) Date of the audiogram;

(C) The examiner's name;

(D) Date of the last acoustic or exhaustive calibration of the audiometer; and

(E) Employee's most recent noise exposure assessment.

(F) The employer shall maintain accurate records of the measurements of the background sound pressure levels in audiometric test rooms.

(3) *Record retention.* The employer shall retain records required in this paragraph (m) for at least the following periods.

(i) Noise exposure measurement records shall be retained for two years.

(ii) Audiometric test records shall be retained for the duration of the affected employee's employment.

(4) *Access to records.* All records required by this section shall be provided upon request to employees, former employees, representatives designated by the individual employee, and the Assistant Secretary. The provisions of 29 CFR 1910.20 (a)-(e) and (g)-(i) apply to access to records under this section.

(5) *Transfer of records.* If the employer ceases to do business, the employer shall transfer to the successor employer all records required to be maintained by this section, and the successor employer shall retain them for the remainder of the period prescribed in paragraph (m) (3) of this section.

(n) **Appendices.** (1) Appendices A, B, C, D and E to this section are incorporated as part of this section and the contents of these Appendices are mandatory.

(2) Appendices F and G to this section are informational and are not intended to create any additional obligations not otherwise imposed or to detract from any existing obligations.

(o) **Exemptions.** Paragraphs (c) through (n) of this section shall not apply to employers engaged in oil and gas well drilling and servicing operations.

(p) **Startup date.** Baseline audiograms required by paragraph (g) of this section shall be completed by March 1, 1984.

APPENDIX A: NOISE EXPOSURE COMPUTATION
This Appendix Is Mandatory

I. Computation of Employee Noise Exposure

(1) Noise dose is computed using Table G-16a as follows:

(i) When the sound level, L, is constant over the entire work shift, the noise dose, D, in percent, is given by: $D = 100\ C/T$ where C is the total length of the work day, in hours, and T is the reference duration corresponding to the measured sound level, L, as given in Table G-16a or by the formula shown as a footnote to that table.

(ii) When the workshift noise exposure is composed of two or more periods of noise at different levels, the total noise dose over the work day is given by:

$$D = 100 \ (C_1/T_1 + C_2/T_2 + \ldots + C_n/T_n),$$

where C_n indicates the total time of exposure at a specific noise level, and T_n indicates the reference duration for that level as given by Table G-16a.

(2) The eight-hour time-weighted average sound level (TWA), in decibels, may be computed from the dose, in percent, by means of the formula: $TWA = 16.61 \ log_{10}(D/100) + 90$. For an eight-hour work shift with the noise level constant over the entire shift, the TWA is equal to the measured sound level.

(3) A table relating dose and TWA is given in Section II.

TABLE G-16a

A-weighted sound level, L (decibel)	Reference duration, T (hour)	A-weighted sound level, L (decibel)	Reference duration, T (hour)
80	32	108	0.66
81	27.9	109	0.57
82	24.3	110	0.5
83	21.1	111	0.44
84	18.4	112	0.38
85	16	113	0.33
86	13.9	114	0.29
87	12.1	115	0.25
88	10.6	116	0.22
89	9.2	117	0.19
90	8	118	0.16
91	7.0	119	0.14
92	6.1	120	0.125
93	5.3	121	0.11
94	4.6	122	0.095
95	4	123	0.082
96	3.5	124	0.072
97	3.0	125	0.063
98	2.6	126	0.054
99	2.3	127	0.047
100	2	128	0.041
101	1.7	129	0.036
102	1.5	130	0.031
103	1.3		
104	1.1		
105	1		
106	0.87		
107	0.76		

In the above table the reference duration, T, is computed by

$$T = \frac{8}{2^{(L-90)/5}}$$

where L is the measured A-weighted sound level.

II. Conversion Between "Dose" and "8-Hour Time-Weighted Average" Sound Level

Compliance with paragraphs (c)-(r) of this regulation is determined by the amount of exposure to noise in the workplace. The amount of such exposure is usually measured with an audiodosimeter which gives a readout in terms of "dose." In order to better understand the requirements of the amendment, dosimeter readings can be converted to an "8-hour time-weighted average sound level" (TWA).

In order to convert the reading of a dosimeter into TWA, see Table A-l, below. This table applies to dosimeters that are set by the manufacturer to calculate dose or percent exposure according to the relationships in Table G-l6a. So, for example, a dose of 91 percent over an eight hour day results in a TWA of 89.3 dB, and, a dose of 50 percent corresponds to a TWA of 85 dB.

If the dose as read on the dosimeter is less than or greater than the values found in Table A-l, the TWA may be calculated by using the formula:

TWA =   16.61 $\log_{10}$ (D/100) + 90 where
TWA =   8-hour time-weighted average sound level and
D =   accumulated dose in percent exposure.

APPENDIX B: METHODS FOR ESTIMATING THE ADEQUACY OF HEARING PROTECTOR ATTENUATION

This Appendix Is Mandatory

For employees who have experienced a significant threshold shift [sic], hearing protector attenuation must be sufficient to reduce employee exposure to a TWA of 85 dB. Employers must select one of the following methods by which to estimate the adequacy of hearing protector attenuation.

The most convenient method is the Noise Reduction Rating (NRR) developed by the Environmental Protection Agency (EPA). According to EPA regulation, the NRR must be shown on the hearing protector package. The NRR is then related to an individual worker's noise environment in order to assess the adequacy of the attenuation of a given hearing protector. This Appendix describes four methods of using the NRR to determine whether a particular hearing protector provides adequate protection within a given exposure environment. Selection among the four procedures is dependent upon the employer's noise measuring instruments.

TABLE A-1
Conversion from "Percent Noise Exposure" or "Dose" to "8-Hour Time-Weighted Average Sound Level" (TWA)

Dose or percent noise exposure	TWA	Dose or percent noise exposure	TWA
10	73.4	109	90.6
15	76.3	110	90.7
20	78.4	111	90.8
25	80.0	112	90.8
30	81.3	113	90.9
35	82.4	114	90.9
40	83.4	115	91.1
45	84.2	116	91.1
50	85.0	117	91.1
55	85.7	118	91.2
60	86.3	119	91.3
65	86.9	120	91.3
70	87.4	125	91.6
75	87.9	130	91.9
80	88.4	135	92.2
81	88.5	140	92.4
82	88.6	145	92.7
83	88.7	150	92.9
84	88.7	155	93.2
85	88.8	160	93.4
86	88.9	165	93.6
87	89.0	170	93.8
88	89.1	175	94.0
89	89.2	180	94.2
90	89.2	185	94.4
91	89.3	190	94.6
92	89.4	195	94.8
93	89.5	200	95.0
94	89.6	210	95.4
95	89.6	220	95.7
96	89.7	230	96.0
97	89.8	240	96.3
98	89.9	250	96.6
99	89.9	260	96.9
100	90.0	270	97.2
101	90.1	280	97.4
102	90.1	290	97.7
103	90.2	300	97.9
104	90.3	310	98.2
105	90.4	320	98.4
106	90.4	330	98.6
107	90.5	340	98.8
108	90.6	350	99.0

TABLE A-1 — continued
Conversion from "Percent Noise Exposure" or "Dose" to 8-Hour Time-Weighted Average Sound Level" (TWA).

Dose or percent noise exposure	TWA	Dose or percent noise exposure	TWA
360	99.2	690	103.9
370	99.4	700	104.0
380	99.6	710	104.1
390	99.8	720	104.2
400	100.0	730	104.3
410	100.2	740	104.4
420	100.4	750	104.5
430	100.5	760	104.6
440	100.7	770	104.7
450	100.8	780	104.8
460	101.0	790	104.9
470	101.2	800	105.0
480	101.3	810	105.1
490	101.5	820	105.2
500	101.6	830	105.3
510	101.8	840	105.4
520	101.9	850	105.4
530	102.0	860	105.5
540	102.2	870	105.6
550	102.3	880	105.7
560	102.4	890	105.8
570	102.6	900	105.8
580	102.7	910	105.9
590	102.8	920	106.0
600	102.9	930	106.1
610	103.0	940	106.2
620	103.2	950	106.2
630	103.3	960	106.3
640	103.4	970	106.4
650	103.5	980	106.5
660	103.6	990	106.5
670	103.7	999	106.6
680	103.8		

Instead of using the NRR, employers may evaluate the adequacy of hearing protector attenuation by using one of the three methods developed by the National Institute for Occupational Safety and Health (NIOSH), which are described in the "List of Personal Hearing Protectors and Attenuation Data," HEW Publication No. 76-120, 1975, pages 21-37. These methods are known as NIOSH methods #1, #2 and #3. The NRR described below is a simplification of NIOSH method #2. The most complex method is NIOSH method #1, which is probably the most accurate method since it uses the largest amount of spectral information from the individual employee's noise environment. As in the case of the NRR method described below, if one of the NIOSH methods is used, the selected method must be applied to an individual's noise environment to assess the adequacy of the attenuation. Employers should be careful to take a sufficient number of measurements in order to achieve a representative sample for each time segment.

Note.—*The employer must remember that calculated attenuation values reflect realistic values only to the extent that the protectors are properly fitted and worn.*

When using the NRR to assess hearing protector adequacy, one of the following methods must be used:

(i) When using a dosimeter that is capable of C-weighted measurements:

(A) Obtain the employee's C-weighted dose for the entire workshift, and convert to TWA (see Appendix A, II).

(B) Subtract the NRR from the C-weighted TWA to obtain the estimated A-weighted TWA under the ear protector.

(ii) When using a dosimeter that is not capable of C-weighted measurements, the following method may be used:

(A) Convert the A-weighted dose to TWA (see Appendix A).

(B) Subtract 7 dB from the NRR.

(C) Subtract the remainder from the A-weighted TWA to obtain the estimated A-weighted TWA under the ear protector.

(iii) When using a sound level meter set to the A-weighting network:

(A) Obtain the employee's A-weighted TWA.

(B) Subtract 7 dB from the NRR, and subtract the remainder from the A-weighted TWA to obtain the estimated A-weighted TWA under the ear protector.

(iv) When using a sound level meter set on the C-weighting network:

(A) Obtain a representative sample of the C-weighted sound levels in the employee's environment.

(B) Subtract the NRR from the C-weighted average sound level to obtain the estimated A-weighted TWA under the ear protector.

(v) When using area monitoring procedures and a sound level meter set to the A-weighing [sic] network:

(A) Obtain a representative sound level for the area in question.

(B) Subtract 7 dB from the NRR and subtract the remainder from the A-weighted sound level for that area.

(vi) When using area monitoring procedures and a sound level meter set to the C-weighting network:

(A) Obtain a representative sound level for the area in question.

(B) Subtract the NRR from the C-weighted sound level for that area.

APPENDIX C: AUDIOMETRIC MEASURING INSTRUMENTS
This Appendix Is Mandatory

1. In the event that pulsed-tone audiometers are used, they shall have a tone on-time of at least 200 milliseconds.

2. Self-recording audiometers shall comply with the following requirements:

(A) The chart upon which the audiogram is traced shall have lines at positions corresponding to all multiples of 10 dB hearing level within the intensity range spanned by the audiometer. The lines shall be equally spaced and shall be separated by at least 1/4 inch. Additional increments are optional. The audiogram pen tracings shall not exceed 2 dB in width.

(B) It shall be possible to set the stylus manually at the 10-dB increment lines for calibration purposes.

(C) The slewing rate for the audiometer attenuator shall not be more than 6 dB/sec except that an initial slewing rate greater than 6 dB/sec is permitted at the beginning of each new test frequency, but only until the second subject response.

(D) The audiometer shall remain at each required test frequency for 30 seconds (±3 seconds). The audiogram shall be clearly marked at each change of frequency and the actual frequency change of the audiometer shall not deviate from the frequency boundaries marked on the audiogram by more than ± 3 seconds.

(E) It must be possible at each test frequency to place a horizontal line segment parallel to the time axis on the audiogram, such that the audiometric tracing crosses the line segment at least six times at that test frequency. At each test frequency the threshold shall be the average of the midpoints of the tracing excursions.

APPENDIX D: AUDIOMETRIC TEST ROOMS

This Appendix Is Mandatory

Rooms used for audiometric testing shall not have background sound pressure levels exceeding those in Table D-l when measured by equipment conforming at least to the Type 2 requirements of American National Standard Specification for Sound Level Meters, S1.4-1971 (R1976), and to the Class II requirements of American National Standard Specification for Octave, Half-Octave, and Third-Octave Band Filter Sets, Sl.11-1971 (R1976).

TABLE D-1
Maximum Allowable Octave-Band Sound Pressure
Levels for Audiometric Test Rooms

Octave-band center frequency (Hz)	500	1000	2000	4000	8000
Sound pressure level (dB)	40	40	47	57	62

APPENDIX E: ACOUSTIC CALIBRATION OF AUDIOMETERS

This Appendix Is Mandatory

Audiometer calibration shall be checked acoustically, at least annually, according to the procedures described in this Appendix. The equipment necessary to perform these measurements is a sound level meter, octave-band filter set, and a National Bureau of Standards 9A coupler. In making these measurements, the accuracy of the calibration equipment shall be sufficient to determine that the audiometer is within the tolerances permitted by American Standard Specification for Audiometers, S3.6-1969.

(1) Sound Pressure Output Check

A. Place the earphone coupler over the microphone of the sound level meter and place the earphone on the coupler.

B. Set the audiometer's hearing threshold level (HTL) dial to 70 dB.

C. Measure the sound pressure level of the tones that [sic] each test frequency from 500 Hz through 6000 Hz for each earphone.

D. At each frequency the readout on the sound level meter should correspond to the levels in Table E-1 or Table E-2, as appropriate, for the type of earphone, in the column entitled "sound level meter reading."

(2) Linearity Check

A. With the earphone in place, set the frequency to 1000 Hz and the HTL dial on the audiometer to 70 dB.

B. Measure the sound levels in the coupler at each 10-dB decrement from 70 dB to 10 dB, noting the sound level meter reading at each setting.

C. For each 10-dB decrement on the audiometer the sound level meter should indicate a corresponding 10 dB decrease.

D. This measurement may be made electrically with a voltmeter connected to the earphone terminals.

(3) Tolerances

When any of the measured sound levels deviate from the levels in Table E-1 or Table E-2 by ±3 dB at any test frequency between 500 and 3000 Hz, 4 dB at 4000 Hz, or 5 dB at 6000 Hz, an exhaustive calibration is advised. An exhaustive calibration is required if the deviations are 15 dB or greater at any test frequency.[1]

TABLE E-1 Reference Threshold Levels for Telephonics—TDH-39 Earphones			TABLE E-2 Reference Threshold Levels for Telephonics—TDH-49 Earphones		
Frequency, Hz	Reference threshold level for TDH-39 earphones, dB	Sound level meter reading, dB	Frequency, Hz	Reference threshold level for TDH-49 earphones, dB	Sound level meter reading, dB
500	11.5	81.5	500	13.5	83.5
1000	7	77	1000	7.5	77.5
2000	9	79	2000	11	81.0
3000	10	80	3000	9.5	79.5
4000	9.5	79.5	4000	10.5	80.5
6000	15.5	85.5	6000	13.5	83.5

[1]As amended June 28, 1983 in CFR 48 (125), p. 29687.

APPENDIX F: CALCULATIONS AND APPLICATION OF AGE CORRECTIONS TO AUDIOGRAMS

This Appendix Is Non-Mandatory

In determining whether a standard threshold shift has occurred, allowance may be made for the contribution of aging to the change in hearing level by adjusting the most recent audiogram. If the employer chooses to adjust the audiogram, the employer shall follow the procedure described below. This procedure and the age correction tables were developed by the National Institute for Occupational Safety and Health in the criteria document entitled "Criteria for a Recommended Standard . . . Occupational Exposure to Noise," [(HSM)-11001].

For each audiometric test frequency:

(i) Determine from Tables F-1 or F-2 the age correction values for the employee by:

(A) Finding the age at which the most recent audiogram was taken and recording the corresponding values of age corrections at 1000 Hz through 6000 Hz;

(B) Finding the age at which the baseline audiogram was taken and recording the corresponding values of age corrections at 1000 Hz through 6000 Hz.

(ii) Subtract the values found in step (i)(B) from the value found in step (i)(A).[1]

(iii) The differences calculated in step (ii) represented that portion of the change in hearing that may be due to aging.

Example: Employee is a 32-year-old male. The audiometric history for this right ear is shown in decibels below.

Employee's age	Audiometric test frequency (Hz)				
	1000	2000	3000	4000	6000
26	10	5	5	10	5
*27	0	0	0	5	5
28	0	0	0	10	5
29	5	0	5	15	5
30	0	5	10	20	10
31	5	10	20	15	15
*32	5	10	10	25	20

The audiogram at age 27 is considered the baseline since it shows the best hearing threshold levels. Asterisks have been used to identify the baseline and most recent audiogram. A threshold shift of 20 dB exists at 4000 Hz between the audiograms taken at ages 27 and 32.

(The threshold shift is computed by subtracting the hearing threshold at age 27, which was 5, from the hearing threshold at age 32, which is 25.) A retest

---

[1]As amended June 28, 1983 in CFR 48 (125), p. 29687.

audiogram has confirmed this shift. The contribution of aging to this change in hearing may be estimated in the following manner:

Go to Table F-1 and find the age correction values (in dB) for 4000 Hz at age 27 and age 32.

	Frequency (Hz)				
	1000	2000	3000	4000	6000
Age 32	6	5	7	10	14
Age 27	5	4	6	7	11
Difference ...	1	1	1	3	3

The difference represents the amount of hearing loss that may be attributed to aging in the time period between the baseline audiogram and the most recent audiogram. In this example, the difference at 4000 Hz is 3 dB. This value is subtracted from the hearing level at 4000 Hz, which in the most recent audiogram is 25, yielding 22 after adjustment. Then the hearing threshold in the baseline audiogram at 4000 Hz (5) is subtracted from the adjusted annual audiogram hearing threshold at 4000 Hz (22). Thus the age-corrected threshold shift would be 17 dB (as opposed to a threshold shift of 20 dB without age correction).[2]

APPENDIX G: MONITORING NOISE LEVELS
NON-MANDATORY INFORMATIONAL APPENDIX

This appendix provides information to help employers comply with the noise monitoring obligations that are part of the hearing conservation amendment.

What is the purpose of noise monitoring?

This revised amendment requires that employees be placed in a hearing conservation program if they are exposed to average noise levels of 85 dB or greater during an 8 hour workday. In order to determine if exposures are at or above this level, it may be necessary to measure or monitor the actual noise levels in the workplace and to estimate the noise exposure or "dose" received by employees during the workday.

When is it necessary to implement a noise monitoring program?

It is not necessary for every employer to measure workplace noise. Noise monitoring or measuring must be conducted only when exposures are at or above 85 dB. Factors which suggest that noise exposures in the workplace may be at this level include employee complaints about the loudness of noise, indications that employees are losing their hearing, or noisy conditions which make normal conversation difficult. The employer should also consider any information available

---

[2]Editor's Note: For additional discussion and examples of age correction, see pp. 491–492 of this text, as well as in the index.

TABLE F-1
Age Correction Values in
Decibels for Males

Years	Audiometric Test Frequencies (Hz)				
	1000	2000	3000	4000	6000
20 or younger ...	5	3	4	5	8
21	5	3	4	5	8
22	5	3	4	5	8
23	5	3	4	6	9
24	5	3	5	6	9
25	5	3	5	7	10
26	5	4	5	7	10
27	5	4	6	7	11
28	6	4	6	8	11
29	6	4	6	8	12
30	6	4	6	9	12
31	6	4	7	9	13
32	6	5	7	10	14
33	6	5	7	10	14
34	6	5	8	11	15
35	7	5	8	11	15
36	7	5	9	12	16
37	7	6	9	12	17
38	7	6	9	13	17
39	7	6	10	14	18
40	7	6	10	14	19
41	7	6	10	14	20
42	8	7	11	16	20
43	8	7	12	16	21
44	8	7	12	17	22
45	8	7	13	18	23
46	8	8	13	19	24
47	8	8	14	19	24
48	9	8	14	20	25
49	9	9	15	21	26
50	9	9	16	22	27
51	9	9	16	23	28
52	9	10	17	24	29
53	9	10	18	25	30
54	10	10	18	26	31
55	10	11	19	27	32
56	10	11	20	28	34
57	10	11	21	29	35
58	10	12	22	31	36
59	11	12	22	32	37
60 or older	11	13	23	33	38

TABLE F-2
Age Correction Values in
Decibels for Females

Years	Audiometric Test Frequencies (Hz)				
	1000	2000	3000	4000	6000
20 or younger ...	7	4	3	3	6
21	7	4	4	3	6
22	7	4	4	4	6
23	7	5	4	4	7
24	7	5	4	4	7
25	8	5	4	4	7
26	8	5	5	4	8
27	8	5	5	5	8
28	8	5	5	5	8
29	8	5	5	5	9
30	8	6	5	5	9
31	8	6	6	5	9
32	9	6	6	6	10
33	9	6	6	6	10
34	9	6	6	6	10
35	9	6	7	7	11
36	9	7	7	7	11
37	9	7	7	7	12
38	10	7	7	7	12
39	10	7	8	8	12
40	10	7	8	8	13
41	10	8	8	8	13
42	10	8	9	9	13
43	11	8	9	9	14
44	11	8	9	9	14
45	11	8	10	10	15
46	11	9	10	10	15
47	11	9	10	11	16
48	12	9	11	11	16
49	12	9	11	11	16
50	12	10	11	12	17
51	12	10	12	12	17
52	12	10	12	13	18
53	13	10	13	13	18
54	13	11	13	14	19
55	13	11	14	14	19
56	13	11	14	15	20
57	13	11	15	15	20
58	14	12	15	16	21
59	14	12	16	16	21
60 or older	14	12	16	17	22

regarding noise emitted from specific machines. In addition, actual workplace noise measurements can suggest whether or not a monitoring program should be initiated.

How is noise measured?

Basically, there are two different instruments to measure noise exposures: the sound level meter and the dosimeter. A sound level meter is a device that measures the intensity of sound at a given moment. Since sound level meters provide a measure of sound intensity at only one point in time, it is generally necessary to take a number of measurements at different times during the day to estimate noise exposure over a workday. If noise levels fluctuate, the amount of time noise remains at each of the various measured levels must be determined.

To estimate employee noise exposures with a sound level meter it is also generally necessary to take several measurements at different locations within the workplace. After appropriate sound level meter readings are obtained, people sometimes draw "maps" of the sound levels within different areas of the workplace. By using a sound level "map" and information on employee locations throughout the day, estimates of individual exposure levels can be developed. This measurement method is generally referred to as *area* noise monitoring.

A dosimeter is like a sound level meter except that it stores sound level measurements and integrates these measurements over time, providing an average noise exposure reading for a given period of time, such as an 8-hour workday. With a dosimeter, a microphone is attached to the employee's clothing and the exposure measurement is simply read at the end of the desired time period. A reader may be used to read-out the dosimeter's measurements. Since the dosimeter is worn by the employee, it measures noise levels in those locations in which the employee travels. A sound level meter can also be positioned within the immediate vicinity of the exposed worker to obtain an individual exposure estimate. Such procedures are generally referred to as *personal* noise monitoring.

Area monitoring can be used to estimate noise exposure when the noise levels are relatively constant and employees are not mobile. In workplaces where employees move about in different areas or where the noise intensity tends to fluctuate over time, noise exposure is generally more accurately estimated by the personal monitoring approach.

In situations where personal monitoring is appropriate, proper positioning of the microphone is necessary to obtain accurate measurements. With a dosimeter, the microphone is generally located on the shoulder and remains in that position for the entire workday. With a sound level meter, the microphone is stationed near the employee's head, and the instrument is usually held by an individual who follows the employee as he or she moves about.

Manufacturer's instructions, contained in dosimeter and sound level meter operating manuals, should be followed for calibration and maintenance. To ensure accurate results, it is considered good professional practice to calibrate instruments before and after each use.

How often is it necessary to monitor noise levels?

The amendment requires that when there are significant changes in machinery or production processes that may result in increased noise levels, remonitoring must be conducted to determine whether additional employees need to be included in the hearing conservation program. Many companies choose to remonitor periodically (once every year or two) to ensure that all exposed employees are included in their hearing conservation programs.

Where can equipment and technical advice be obtained?

Noise monitoring equipment may be either purchased or rented. Sound level meters cost about $500 to $1,000, while dosimeters range in price from about $750 to $1,500. Smaller companies may find it more economical to rent equipment rather than to purchase it. Names of equipment supplier may be found in the telephone book (Yellow Pages) under headings such as: "Safety Equipment," "Industrial Hygiene," or "Engineers-Acoustical." In addition to providing information on obtaining noise monitoring equipment, many companies and individuals included under such listings can provide professional advice on how to conduct a valid noise monitoring program. Some audiological testing firms and industrial hygiene firms also provide noise monitoring services. Universities with audiology, industrial hygiene, or acoustical engineering departments may also provide information or may be able to help employers meet their obligations under this amendment

Free, on-site assistance may be obtained from OSHA-supported state and private consultation organizations. These safety and health consultative entities generally give priority to the needs of small businesses. Write OSHA for a listing of organizations to contact for aid.

APPENDIX H: AVAILABILITY OF REFERENCED DOCUMENTS

Paragraphs (c) through (o) of 29 CFR 1910.95 and the accompanying appendices contain provisions which incorporate publications by reference. Generally, the publications provide criteria for instruments to be used in monitoring and audiometric testing. These criteria are intended to be mandatory when so indicated in the applicable paragraphs of Section 1910.95 and appendices.

It should be noted that OSHA does not require that employers purchase a copy of the referenced publications. Employers, however, may desire to obtain a copy of the referenced publications for their own information.

The designation of the paragraph of the standard in which the referenced publications appear, the titles of the publications, and the availability of the publications are as follows:

Paragraph designation	Referenced publication	Available from—
Appendix B	"List of Personal Hearing Protectors and Attenuation Data," HEW Pub. No. 76-120, 1975. NTIS-PB267461.	National Technical Information Service, Port Royal Road, Springfield, VA 22161.
Appendix D	"Specification for Sound Level Meters," S1.4-1971 (R1976).	American National Standards Institute, Inc., 1430 Broadway, New York, NY 10018.
§ 1910.95(k)(2), appendix E	"Specifications for Audiometers," S3.6-1969.	American National Standards Institute, Inc., 1430 Broadway, New York, NY 10018.
Appendix D	"Specification for Octave, Half-Octave and Third-Octave Band Filter Sets," S1.11-1971 (R1976).	Back Numbers Department, Dept. STD, American Institute of Physics, 333 E. 45th St., New York, NY 10017; American National Standards Institute, Inc., 1430 Broadway, New York, NY 10018.

The referenced publications (or a microfiche of the publications) are available for review at many universities and public libraries throughout the country. These publications may also be examined at the OSHA Technical Data Center, Room N2439, United States Department of Labor, 200 Constitution Avenue, NW., Washington, D.C. 20210, (202) 523-9700 or at any OSHA Regional Office (see telephone directories under United States Government-Labor Department).

APPENDIX I: DEFINITIONS

These definitions apply to the following terms as used in paragraphs (c) through (n) of 29 CFR 1910.95.

Action level—An 8-hour time-weighted average of 85 decibels measured on the A-scale, slow response, or equivalently, a dose of fifty percent.

Audiogram—A chart, graph, or table resulting from an audiometric test showing an individual's hearing threshold levels as a function of frequency.

Audiologist—A professional, specializing in the study and rehabilitation of hearing, who is certified by the American Speech-Language-Hearing Association or licensed by a state board of examiners.

Baseline audiogram—The audiogram against which future audiograms are compared.

Criterion sound level—A sound level of 90 decibels.

Decibel (dB)—Unit of measurement of sound level.

Hertz (Hz)—Unit of measurement of frequency, numerically equal to cycles per second.

Medical pathology—A disorder or disease. For purposes of this regulation, a condition or disease affecting the ear, which should be treated by a physician specialist.

Noise dose—The ratio, expressed as a percentage, of (1) the time integral, over a stated time or event, of the 0.6 power of the measured SLOW exponential time-averaged, squared A-weighted sound pressure and (2) the product of the criterion duration (8 hours) and the 0.6 power of the squared sound pressure corresponding to the criterion sound level (90 dB).

Noise dosimeter—An instrument that integrates a function of sound pressure over a period of time in such a manner that it directly indicates a noise dose.

Otolaryngologist—A physician specializing in diagnosis and treatment of disorders of the ear, nose and throat.

Representative exposure—Measurements of an employee's noise dose or 8-hour time-weighted average sound level that the employers deem to be representative of the exposures of other employees in the workplace.

Sound level—Ten times the common logarithm of the ratio of the square of the measured A-weighted sound pressure to the square of the standard reference pressure of 20 micropascals. Unit: decibels (dB). For use with this regulation, SLOW time response, in accordance with ANSI SI.4-1971 (R1976), is required.

Sound level meter—An instrument for the measurement of sound level.

Time-weighted average sound level—That sound level, which if constant over an 8-hour exposure, would result in the same noise dose as is measured.

The Noise Manual, revised 5th edition, edited by E.H. Berger,
L.H. Royster, J.D. Royster, D.P. Driscoll, and M. Layne
©2003 American Industrial Hygiene Association

Appendix II References for Good Practice

Contents

Introductory Remarks

This appendix is based upon a document originally prepared by the American Industrial Hygiene Association (AIHA) Noise Committee in conjunction with the National Hearing Conservation Association (NHCA). Its purpose is to provide a comprehensive (but not exhaustive) guide to standards, regulations, and recommended references pertaining to noise and hearing loss. Out-of-date standards, regulations, and historical references have not been included unless they bear directly on current practice. Not all textbooks have been referenced; rather, choices are limited to those which are the most recognized and respected by the hygiene and hearing conservation community. The appendix contains approximately 200 references.

The organization of this appendix is primarily based upon the type of information resource, but for ease of use and for those who wish to limit the depth and extent of their literature review, the first three sections list key references that provide an overview and synopsis of the issues.

Overview

A good overview of noise and hearing loss is provided by the following document that resulted from a consensus conference sponsored by the U.S. National Institutes of Health (NIH).

National Institutes of Health (1990). "Noise and Hearing Loss—Consensus Conference," *J. Am. Med. Assoc. 263*(23), 3185–3190.

Hearing Conservation Program Administration Guidelines

Several texts have been produced to aid in development and administration of hearing conservation programs. For additional titles see later section, "Handbooks, Textbooks, Guidelines, and Review Articles."

Berger, E. H., Royster, L. H., Royster, J. D., Driscoll, D. P., and Layne, M. (eds.). (2000). *The Noise Manual, 5th Edition*, American Industrial Hygiene Association, Fairfax, VA.

Franks, J. R., Stephenson, M. R., and Merry, C. J. (1996). "Preventing Occupational Hearing Loss—A Practical Guide," U.S. Department of Health and Human Services (NIOSH) Report 96–110.

Royster, J. D., and Royster, L.H. (1990). *Hearing Conservation Programs: Practical Guidelines for Success*, Lewis Pub., Chelsea, MI.

Suter, A. H. (1993). *Hearing Conservation Manual, 3rd Edition*, Council for Accreditation in Occupational Hearing Conservation, Milwaukee, WI.

Quick References

These publications and pamphlets provide concise and to-the-point information on specific issues.

ASHA (~1990). "Noise in Your Workplace," American Speech-Language-Hearing Association, Rockville, MD.

Berger, E. H. (1979-1999). E·A·R*Log Series* of Technical Monographs on Hearing and Hearing Protection, Nos. 1–21, E·A·R, Indianapolis, IN.

NHCA (1996). "A Practical Guide to: Selecting Hearing Protection," National Hearing Conservation Association, Guide #1, Denver, CO.

NHCA (1996). "A Practical Guide to: Fitting Hearing Protection," National Hearing Conservation Association, Guide #2, Denver, CO.

NHCA (1996). "A Practical Guide to: Mobile Hearing Testing and Selecting a Provider," National Hearing Conservation Association, Guide #3, Denver, CO.

NHCA (1998). "A Practical Guide to: Complying with OSHA's Hearing Conservation Amendment CFR 1910.95," National Hearing Conservation Association, Guide #4, Denver, CO.

NHCA (1996). "A Practical Guide to: Noise and Hearing in the Farming Community," National Hearing Conservation Association, Guide #5, Denver, CO.

NHCA (1997). "A Practical Guide to: Hearing Loss Prevention for Musicians," National Hearing Conservation Association, Guide #6, Denver, CO.

NHCA (1998). "A Practical Guide to: Firearm Use and Hearing," National Hearing Conservation Association, Guide #7, Denver, CO.

OSHA (1980). "Noise Control: A Guide for Workers and Employers," OSHA 3048, a highly recommended booklet adapted from a publication of the Swedish Work Environment Fund, Washington, DC. [also published as: Stepkin, R. L., and Mosely, R. E. (1984). *Noise Control: A Guide for Workers and Employers*, American Society of Safety Engineers, Des Plaines, IL.]

OSHA (1992, revised). "Hearing Conservation," OSHA 3074, synopsis of the Hearing Conservation Amendment, Washington, DC.

Professional Organizations, Publications, and Home Pages

Several organizations exist which address the issues of noise, vibration, and noise-induced hearing loss, and which produce regular publications with research and articles of interest. Telephone numbers and web page addresses are also listed.

Acoustical Society of America (ASA), New York, NY (516-576-2360).
Journal of the Acoust. Soc. of Am. (monthly); [http://asa.aip.org].

American Industrial Hygiene Association (AIHA), Fairfax, VA (703-849-8888).
Am. Ind. Hyg. Assoc. Journal (bimonthly); [http://www.aiha.org]

737

American Speech-Language-Hearing Association (ASHA), Rockville, MD (301-897-5700).

   ASHA (quarterly); [http://www.asha.org]

Council for Accreditation in Occupational Hearing Conservation, Milwaukee, WI (414-276-5338).

   CAOHC Update (quarterly); [http:// www.caohc.org]

Institute of Noise Control Engineering, Poughkeepsie, NY (914-462-4006).

   Noise Control Engineering Journal (bimonthly), Noise/News International (quarterly); [http://ince.org]

National Hearing Conservation Association (NHCA), Denver, CO (303-224-9022).

   Spectrum (quarterly); [http://www.hearingconservation.org]

TLVs® (Threshold Limit Values)

TLVs® are oft-cited guidelines produced by the American Conference of Governmental Industrial Hygienists (ACGIH) that influence corporate good practice as well as judicial and regulatory activities.

ACGIH (1999). "Supplements to the 6th Edition—Documentation of the Threshold Limit Values and Biological Exposure Indices: Infrasound and Low-Frequency Sound," American Conference of Governmental Industrial Hygienists, Cincinnati.

ACGIH (1996). "Supplements to the 6th Edition—Documentation of the Threshold Limit Values and Biological Exposure Indices: Noise," American Conference of Governmental Industrial Hygienists, Cincinnati.

ACGIH (1998). "Supplements to the 6th Edition—Documentation of the Threshold Limit Values and Biological Exposure Indices: Ultrasound," American Conference of Governmental Industrial Hygienists, Cincinnati.

Sliney, D. H. (1993). "Report from the ACGIH Physical Agents TLV Committee—Review of the Threshold Limit Value for Noise," *Appl. Occup. Environ. Hyg. 8*(7), 618–623.

Guidelines, Position Statements, and Criteria Documents

The following documents, produced by various professional organizations, represent guidelines and/or formal positions on issues pertaining to regulations and good practice.

AAO-HNS (1997). "Otologic Referral Criteria for Occupational Hearing Conservation Programs," Medical Aspects of Noise Subcommittee of Am. Academy of Otolaryngol.-Head and Neck Surgery, Washington, DC.

AAO-HNS (1988). "Guide for Conservation of Hearing in Noise," edited by J. D. Osguthorpe, Subcommittee on the Medical Aspects of Noise of the Committee on Hearing and Equilibrium, Am. Academy of Otolaryngol.-Head and Neck Surgery Foundation, Inc., Washington, DC.

AAO-HNS (1998). "Evaluation of People Reporting Occupational Hearing Loss," Subcommittee on the Medical Aspects of Noise, Am. Academy of Otolaryngol.-Head and Neck Surgery Foundation, Inc., Alexandria, VA.

ACOM (1989). "Occupational Noise-Induced Hearing Loss," American College of Occupational Medicine, *J. Occup. Med. 31*(12), 996.

AIHA (1996). "Recommended Criterion for Recording Occupational Hearing Loss on OSHA Form 300," *Am. Ind. Hyg. Assoc. J. 57*(7), 661–662.

AMA (1984). "Guides to the Evaluation of Permanent Impairment," American Medical Association, Chicago, IL.

AOMA (1987). "Guidelines for the Conduct of an Occupational Hearing Conservation Program," American Occupational Medicine Association Noise and Hearing Conservation Committee, *J. Occup. Med. 29*(12), 981–982.

ASHA (1996). "Guidelines on the Audiologist's Role in Occupational and Environmental Hearing Conservation," *ASHA 38*(Suppl. 16), 45–52 [update expected in late 2000].

ASHA (1991). "Combatting Noise in the '90s: A National Strategy for the United States," edited by E. Cherow, Rockville, MD.

CAOHC (1999). "Course Outline for Course Leading to Accreditation as an Occupational Hearing Conservationist," Council for Accreditation in Occupational Hearing Conservation, Milwaukee, WI.

CHABA (1993). "Hazardous Exposure to Impulse Noise," Committee on Hearing, Bioacoustics, and Biomechanics, National Research Council, National Academy Press, Washington, DC.

Driscoll, D. P., and Morrill, J. C. (1987). "A Position Paper on a Recommended Criterion for Recording Occupational Hearing Loss on the OSHA Form 200," *Am. Ind. Hyg. Assoc. J. 48*(11), A-714–A-716.

Embleton, T.F.W., Gibson-Wilde, B., Slama, J. G., Shaw, E.A.G., Gamba, R., Lazarus, H., Timar, P. L., Bellhouse, G., Ward, W. D., Roth, S. I., and Suter, A. H. (1997). "Technical Assessment of Upper Limits on Noise in the Workplace," *Noise/News Int. 5*(4), 203–216.

Garrett, B.R.B., and Royster, L. H. (1995). "NHCA Urges EPA Action on NRR: The Recommendations of the NHCA Task Force on Hearing Protector Effectiveness," *Spectrum 12*(3), 23–25.

MacLean, S. (1995). "Employment Criteria in Hearing Critical Jobs," *Spectrum 12*(4), 20–23.

MacLean, S., and Danielson, R. (1996). "Hearing Standards in the US Military," *Spectrum 13*(1), 16–17.

NHCA (1987). "Occupational Hearing Conservationist Training Guidelines," National Hearing Conservation Association, Denver, CO.

NHCA (1990). "ACOM Publishes Position Paper on Noise-Induced Hearing Loss; NHCA Responds," *Spectrum 7*(1), 1–4.

NMTBA (1976). "Noise Measurement Techniques," National Machine Tool Builders Association, McLean, VA.

Royster, J. D. (1996). "New NHCA Guidelines for Baseline Revision," *Spectrum 13*(2), 1 and 4–9.

Royster, J. D., and Royster, L. H. (1998). "A Guide to Developing and Maintaining an Effective Hearing Conservation Program," NC-OSHA Industry Guide #15, North Carolina Dept. of Labor, Raleigh, NC (also available at www.nonoise.org).

Schaffer, M. E. (1991). "A Practical Guide to Noise and Vibration Control for HVAC Systems," American Society of Heating, Refrigerating, and Air-Conditioning Engineers, Atlanta, GA.

WHO (1980). "Environmental Health Criteria 12—Noise," World Health Organization, Geneva, Switzerland.

National and International Standards

Consensus standards provide the basis for instrument and equipment design and calibration, good practice, and regulations in the U.S. and worldwide. Those with greatest relevance to hearing conservation, vibration exposure, and the practicing industrial hygienist are listed below.

ANSI (American National Standards Institute), New York, NY
[order S1, S2, S3, and S12 standards from 412–741-1979; on-line searches at http://www.ansi.org/public/std_info.html]

Committee S1—Acoustics
S1.1-1994. "Acoustical Terminology."

S1.4-1983 (R1997). "Specification for Sound Level Meters."

S1.11-1986 (R1998). "Specification for Octave-Band and Fractional-Octave-Band Analog and Digital Filters."

S1.13-1995. "Measurement of Sound Pressure Levels in Air."

S1.25-1991. "Specification for Personal Noise Dosimeters."

S1.40-1984 (R1997). "Specification for Acoustical Calibrators."

Committee S2—Mechanical Shock and Vibration
S2.40-1984 (R1990). "Mechanical Vibration of Rotating and Reciprocating Machinery— Requirements for Instruments for Measuring Vibration Severity."

S2.41-1985 (R1997). "Mechanical Vibration of Large Rotating Machines with Speed Range from 10 to 200 rev/s—Measurement and Evaluation of Vibration Severity *in situ*."

S2.60-1987 (R1997). "Balancing Machines—Enclosures and Other Safety Measures."

S2.61-1989 (R1997). "Guide to the Mechanical Mounting of Accelerometers."

Committee S3—Bioacoustics
S3.1-1999. "Maximum Permissible Ambient Noise Levels for Audiometric Test Rooms."

S3.2-1989 (R1995). "Method for Measuring the Intelligibility of Speech over Communication Systems."

S3.4-1980 (R1997). "Procedure for the Computation of Loudness of Noise."

S3.5-1997. "Methods for the Calculation of the Speech Intelligibility Index."

S3.6-1996. "Specification for Audiometers."

S3.14-1977 (R1997). "Rating Noise with Respect to Speech Interference."

S3.18-1979 (R1993). "Guide for the Evaluation of Human Exposure to Whole-Body Vibration."

S3.20-1995. "Bioacoustical Terminology."

S3.21-1978 (R1997). "Standard Method for Manual Pure-Tone Threshold Audiometry."

S3.29-1983 (R1996). "Guide to the Evaluation of Human Exposure to Vibration in Buildings."

S3.34-1986 (R1997). "Guide for the Measurement and Evaluation of Human Exposure to Vibration Transmitted to the Hand."

S3.40-1989 (R1995). "Guide for the Measurement and Evaluation of Gloves Which are Used to Reduce Exposure to Vibration Transmitted to the Hand."

S3.41-1990 (R1996). "Audible Emergency Evacuation Signal."

S3.44-1996. "Determination of Occupational Noise Exposure and Estimation of Noise-Induced Hearing Impairment."

Committee S12—Noise

S12.2-1995. "Criteria for Evaluating Room Noise."

S12.6-1997. "Methods for Measuring the Real-Ear Attenuation of Hearing Protectors."

S12.8-1998. "Method for Determination of Insertion Loss of Outdoor Noise Barriers."

S12.9-1988 Part 1 (R1993). "Quantities and Procedures for Description and Measurement of Environmental Sound, Part 1."

S12.9-1992 Part 2. "Quantities and Procedures for Description and Measurement of Outdoor Environmental Sound, Part 2: Measurement of Long-Term, Wide-Area Sound."

S12.9-1993 Part 3. "Quantities and Procedures for Description and Measurement of Environmental Sound, Part 3: Short-Term Measurements with an Observer Present."

S12.9-1996 Part 4. "Quantities and Procedures for Description and Measurement of Environmental Sound, Part 4: Noise Assessment and Prediction of Long-Term Community Response."

S12.9-1998 Part 5. "Quantities and Procedures for Description and Measurement of Environmental Sound, Part 5: Sound Level Descriptors for Determination of Compatible Land Use."

S12.13-1991. "Draft American National Standard: Evaluating the Effectiveness of Hearing Conservation Programs."

S12.14-1992 (R1997). "Methods for the Field Measurement of the Sound Output of Audible Public Warning Devices Installed at Fixed Locations Outdoors."

S12.16-1992 (R1997). "Guidelines for the Specification of Noise of New Machinery."

S12.19-1996. "Measurement of Occupational Noise Exposure."

S12.36-1990. "Survey Methods for the Determination of Sound Power Levels of Noise Sources."

S12.42-1995. "Microphone-in-Real-Ear and Acoustic Test Fixture Methods for the Measurement of Insertion Loss of Circumaural Hearing Protection Devices."

IEEE (Institute of Electrical and Electronics Engineers, Inc., New York, NY)

[order IEEE standards at 800-678-4333 or on-line at http://stdsbbs.ieee.org/]

ANSI/IEEE Std 260.4-1996 (Revision and redesignation of ANSI/ASME Std Y10.11-1984). "American National Standard Letter Symbols and Abbreviations for Quantities Used in Acoustics."

ISO (International Standards Organization, Geneva, Switzerland)

[order ISO standards from ANSI at 212-642-4900; on-line searches at http://www.iso.ch/]

ISO 226-1987. "Acoustics—Normal Equal-Loudness Level Contours."

ISO 1999:1990. "Acoustics—Determination of Occupational Noise Exposure and Estimation of Noise-Induced Hearing Impairment."

ISO 2631-1:1997. "Mechanical Vibration and Shock—Evaluation of Human Exposure to Whole-Body Vibration—Part 1: General Requirements."

ISO 2631-2:1989. "Evaluation of Human Exposure to Whole-Body Vibration—Part 2: Continuous and Shock-Induced Vibrations in Buildings (1 to 80 Hz)."

ISO 4869-1:1990. "Acoustics—Hearing Protectors—Part 1: Subjective Method for the Measurement of Sound Attenuation."

ISO 4869-2:1994. "Acoustics—Hearing Protectors—Part 2: Estimation of Effective A-Weighted Sound Pressure Levels When Hearing Protectors are Worn."

ISO/TR 4869-3:1989. "Acoustics—Hearing Protectors—Part 3: Simplified Method for the Measurement of Insertion Loss of Ear-Muff Type Protectors for Quality Inspection Purposes."

ISO/TR 4869-4:1998. "Acoustics—Hearing Protectors—Part 4: Measurement of Effective Sound Pressure Levels for Level-Dependent Sound-Restoration Ear-Muffs."

ISO 5349-1986. "Mechanical Vibration—Guidelines for the Measurement and the Assessment of Human Exposure to Hand-Transmitted Vibration."

ISO 7029-2000. "Acoustics—Statistical Distribution of Hearing Thresholds as a Function of Age."

ISO/DIS 9612.2-1995. "Acoustics—Guidelines for the Measurement and Assessment of Exposure to Noise in a Working Environment."

ISO/DIS 10843-1994. "Acoustics—Methods for the Physical Measurement of Single Impulses or Bursts of Noise."

Databases and Programs

Various individuals and organizations have compiled bibliographies, and data on noise levels, hearing protector performance, and hearing levels and make those materials available to the public for examination and related research.

Berger, E. H. (1999). "Bibliography on Hearing Protection: 1831-1999 (contains approximately 2500 entries)," E·A·R 82-6/HP, E·A·R/Aearo Company, Indianapolis, IN.

Berger, E. H., and Kladden, C. A. (1998). "Noise and Hearing Conservation Films and Videotapes: Reviews and Availability," E·A·R 82-10/HP, E·A·R/Aearo Company, Indianapolis, IN.

CCOHS (1993). "Noise Levels: A Database of Measured Industrial Noise Levels (1953–1993)," Canadian Center for Occupational Health and Safety, Hamilton, Ontario.

Franks, J. R., Theman, C. L., and Sherris, C. (1994). "The NIOSH Compendium of Hearing Protection Devices," U.S. Department of Health and Human Services, Pub. No. 95-105, Cincinnati, OH.

Green, C. (1987). *Traffic Noise: A Bibliography on Surface Transportation Noise, 1979–1988*, Letchworth, Herts, UK (Updated edition of "Traffic Noise: A Review and Bibliography on Surface Transportation Noise," 1964–1978).

Miller, T. W. Jr. (1995). "An Analysis of Octave Band Sound Pressure Level Data Bases and Various Metrics Based on the Data Bases Studied," MS Thesis, North Carolina State University, Raleigh, NC.

National Acoustic Laboratories (1998). "Attenuation and Use of Hearing Protectors, Eighth Edition," National Acoustic Laboratories, NSW, Australia.

Royster, J. D., and Royster, L. H. (1986). "Final Report: Documentation of Computer Tape of Audiometric Data," NTIS Rept. PB88-117916/KHX, Springfield, VA.

Royster, L. H., Royster, J. D., and Royster, W. K. (1996). "The N.A.N. Computer Program for Estimating a Population's Hearing Threshold Characteristics (Based on ANSI S3.44-1996)," Environmental Noise Consultants, Inc., Raleigh, NC.

Handbooks, Textbooks, and Review Articles

For those wishing to delve further into the issues of noise, vibration, and noise-induced hearing loss, a selection of useful and recognized references follows. The articles that are listed are limited to those which are reviews, as opposed to the research of an individual author or group of authors.

Acoustics and Noise Control

Barber, A. (1992). *Handbook of Noise and Vibration Control, 6th Edition*, Elsevier Science Pub. Ltd., Oxford, UK.

Beranek, L. L. (1986). *Acoustics*, Acoustical Society of America, New York, NY.

Beranek, L. L. (ed.). (1988). *Noise and Vibration Control*, Institute of Noise Control Engineering, Poughkeepsie, NY.

Beranek, L. L., and Ver, I. L. (eds.). (1992). *Noise and Vibration Control Engineering: Principles and Applications*, John Wiley & Sons, Inc., New York, NY.

Crocker, M. J. (ed.). (1997). *Encyclopedia of Acoustics* (4 Volumes), John Wiley & Sons, Inc., New York, NY.

Foreman, J.E.K. (1990). *Sound Analysis and Noise Control,* Van Nostrand Reinhold, New York, NY.

Harris, C. M. (ed.). (1991). *Handbook of Acoustical Measurements and Noise Control, Third Edition*, Acoustical Society of America, New York, NY (originally published by McGraw-Hill, Inc.).

Hassall, J. R., and Zaveri, K. (1988). *Acoustic Noise Measurements, 5th Edition*, Brüel & Kjær, Denmark.

Irwin, J. D., and Graf, E. R. (1979). *Industrial Noise and Vibration Control*, Prentice-Hall, Inc. Englewood Cliffs, NJ.

Pelton, H. K. (1993). *Noise Control Management*, Van Nostrand Reinhold, New York, NY.

Peterson, A.P.G. (1980). *Handbook of Noise Measurement, 9th Edition*, GenRad, Concord, MA.

Seto, W. W. (1971). *Schaum's Outline Series, Theory and Problems of Acoustics*, McGraw-Hill Book Co., New York, NY.

Community Noise

Cowan, J. P. (1994). *Handbook of Environmental Acoustics*, Int. Thompson Pub., New York, NY.

Nelson, P. M. (1987). *Transportation Noise Reference Book*, Butterworth, London, UK.

Transportation Research Board (1983). *Transportation Noise: Prediction and Analysis*, National Research Council, Washington, DC.

Transportation Research Board (1984). *Issues in Transportation Noise Mitigation: Highway and Railway Studies*, National Research Council, Washington, DC.

Transportation Research Board (1987). *Environmental Issues: Noise, Rail Noise, and High-Speed Rail*, National Research Council, Washington, DC.

Transportation Research Board (1990). *Energy and Environment 1990: Transportation-Induced Noise and Air Pollution*, National Research Council, Washington, DC.

WHO (1995). "Community Noise," Archives of the Center for Sensory Research 2(1), edited by B. Berglund and T. Lindvall, prepared for World Health Organization by Stockholm University, Sweden (available from www.who.int).

Effects of Noise: Auditory

Burns, W. (1968). *Noise and Man*, JB Lippincott, Philadelphia, PA.

Burns, W., and Robinson, D. W. (1970). *Hearing and Noise in Industry*, Her Majesty's Stationery Office, London, UK.

Johnson, D. L. (1978). "Derivation of Presbycusis and Noise Induced Permanent Threshold Shift (NIPTS) to be Used for the Basis of a Standard on the Effects of Noise on Hearing," Aerospace Medical Research Laboratory Report, AMRL-TR-78-128, Wright-Patterson AFB, OH.

Kryter, K. D. (1985). *The Effects of Noise on Man, Second Edition*, Academic Press, New York, NY.

Kryter, K. D. (1994). *Handbook of Hearing and the Effects of Noise: Physiology, Psychology and Public Health*, Academic Press, New York, NY.

Robinson, D. W. (1971). *Occupational Hearing Loss*, Academic Press, New York, NY.

Royster, J. D. (1996). "Noise-Induced Hearing Loss," in *Hearing Disorders (3rd edition)*, edited by J. L. Northern, Allyn & Bacon, Boston, MA, 177–188.

Shaw, E. A. G. (1985). "Occupational Noise Exposure and Noise-Induced Hearing Loss: Scientific Issues, Technical Arguments and Practical Recommendations," National Research Council Rept. NRCC/CNRC No. 25051, Ottawa, Canada.

Suter, A. H. (1993). "The Relationship of the Exchange Rate to Noise-Induced Hearing Loss," *Noise/News Int. 1*(3), 131–151.

Effects of Noise: Non-Auditory

Fay, T. H. (1991). *Noise & Health*, New York Academy of Medicine, New York, NY.

Greenberg, G. N., and Cohen, B. A. (1998). "Reproductive Hazards of the Workplace," in *Noise, Ultrasound and Vibration*, edited by L. M. Frazier and M. L. Hage, John Wiley & Son, Philadelphia, PA.

Loeb, M. (1986). *Noise and Human Efficiency*, John Wiley & Sons, Chichester, UK.

Smoorenburg, G. F., Axelsson, A., Babisch, W., Diamond, I. G., Ising, H., Marth, E., Miedema, H.M.E., Öhrström, E., Rice, C., Roscam Abbing, E. W., van de Wiel, J.A.G., and Passchier-Vermeer, W. (1996). "Effects of Noise on Health," Chapter 3 of a Report on *Noise and Health* prepared by a Committee of the Health Council of the Netherlands, *Noise/News Int. 4*(3), 137–150.

Suter, A. H. (1992). "Communication and Job Performance in Noise: A Review," American Speech-Language-Hearing Association, ASHA Monograph No. 28, Rockville, MD.

Hearing Conservation

Alberti, P. W. (ed.). (1982). *Personal Hearing Protection in Industry*, Raven Press, New York, NY.

Berger, E. H., Royster, L. H., Royster, J. D., Driscoll, D. P., and Layne, M. (eds.). (2000). *The Noise Manual, 5th Edition*, American Industrial Hygiene Association, Fairfax, VA.

Chasin, M. (1996). *Musicians and the Prevention of Hearing Loss*, Singular Publishing Group, Inc., San Diego, CA.

Feldman, A. S., and Grimes, C. T. (eds.). (1985). *Hearing Conservation in Industry*, Williams & Wilkins, Baltimore, MD.

Gasaway, D. C. (1985). *Hearing Conservation, A Practical Manual and Guide*, Prentice-Hall, Inc. Englewood Cliffs, NJ.

Lipscomb, D. M. (ed.). (1988). *Hearing Conservation in Industry, Schools, and the Military*, College-Hill Press, Boston, MA.

Miller, M. H., and Silverman, C. A. (1984). *Occupational Hearing Conservation*, Prentice-Hall, Englewood Cliffs, NJ.

Olishifski, J. B., and Harford, E. R. (1975). *Industrial Noise and Hearing Conservation*, National Safety Council, Chicago, IL.

Royster, J. D., and Royster, L. H. (1990). *Hearing Conservation Programs: Practical Guidelines for Success*, Lewis Pub., Chelsea, MI.

Suter, A. H. (1993). *Hearing Conservation Manual, 3rd Edition*, Council for Accreditation in Occupational Hearing Conservation, Milwaukee, WI.

Suter, A. H. (ed.). (1997). "Noise," Chapter 47 in *Encyclopedia of Occupational Health and Safety (4th edition)*, edited by J. Stellman, International Labor Office, Geneva, 47.1–47.18.

International Conferences/Multiple Topics

AIHA (1981). "Monograph Series — Noise and Hearing Conservation," American Industrial Hygiene Association, Fairfax, VA, a compendium of 62 selected articles from *AIHAJ*.

Axelsson, A., Borchgrevink, H. B., Hamernik, R. P., Hellstrom, P. A., Henderson, D., Salvi, R. J. (eds.). (1996). *Scientific Basis of Noise-Induced Hearing Loss*, Thieme, New York, NY.

Berglund, B., Berglund, U., Karlsson, J., and Lindvall, T. (eds.). (1988). *Proceedings of the 5th International Congress on Noise as a Public Health Problem*, Swedish Council for Building Research, Stockholm, Sweden.

Dancer, A. L., Henderson, D., Salvi, R. J., and Hamernik, R. P. (eds.). (1992). *Noise-Induced Hearing Loss*, Mosby Year Book, Inc., St. Louis, MO.

Hamernik, R. P., Henderson, D., and Salvi, R. (eds.). (1982). *New Perspectives on Noise-Induced Hearing Loss*, Raven Press, New York, NY.

Henderson, D., Hamernik, R., Dosanjh, D., and Mills, J. H. (eds.). (1976). *Effects of Noise on Hearing*, Raven Press, New York, NY.

Morata, T. C., and Dunn, D. E. (eds.). (1995). *Occupational Hearing Loss* [Occupational Medicine: State of the Art Reviews, 10(3)], Hanley & Belfus, Inc., Philadelphia, PA, 495–689.

Rossi, G. (ed.). (1983). *Proceedings of the Fourth International Congress on Noise as a Public Health Problem*, Centro Ricerche e Studi Amplifon, Milan, Italy.

Salvi, R. J., Henderson, D., Hamernik, R., and Coletti, V. (eds.). (1985). *Basic and Applied Aspects of Noise-Induced Hearing Loss*, Plenum Press, New York, NY.

Tobias, J. V., Jansen, G., and Ward, W. D. (eds.). (1980). *Proceedings of the Third International Congress on Noise as a Public Health Problem*, ASHA Reports 10, American Speech-Language-Hearing Association, Rockville, MD.

Vallet, M. (ed.). (1993). *Noise & Man '93: Noise as Public Health Problem*, Institut National de Recherche sur les Transports et Leur Sécurité, Arcueil Cedex, France.

Ward, W. D., and Fricke, J. E. (eds.). (1969). *Proceedings of the Conference, Noise as a Public Health Hazard*, ASHA Reports 4, American Speech and Hearing Association, Rockville, MD.

Medical/Legal

Dobie, R. A. (1993). *Medical–Legal Evaluation of Hearing Loss*, Van Nostrand Reinhold, New York, NY.

Hawke, M., Keene, M., and Alberti, P. W. (1984). *Clinical Otoscopy—A Text and Colour Atlas*, Churchill Livingstone, Edinburgh, UK.

Kramer, M. B., and Armbruster, J. M. (1982). *Forensic Audiology*, University Park Press, Baltimore, MD.

Royster, J. D. (1996). "Occupational and Forensic Audiology," *Current Opinion in Otolaryngol. & Head and Neck Surg. 4*, 345–352.

Senturia, B. H., Marcus, M. D., and Lucente, F. E. (1980). *Diseases of the External Ear, An Otologic-Dermatologic Manual, Second Edition*, Grune & Stratton, Inc., New York, NY.

Wormald, P. J., and Browning, G. G. (1996). *Otoscopy— A Structured Approach*, Singular Publishing Group, Inc., San Diego, CA.

Ototoxicity

Cary, R., Clarke, S., and Delic, J. (1997). "Effects of Combined Exposure to Noise and Toxic Substances: Critical Review of the Literature," *Annals of Occup. Hyg. 41*, 455–465.

Fechter, L. D. (1995). "Combined Effects of Noise and Chemicals," *Occup. Med. State of the Art Rev. 10*, 609–622.

Franks J. R., and Morata T. C. (1996). "Ototoxic Effects of Chemicals Alone or in Concert with Noise: A Review of Human Studies," in *Scientific Basis of Noise-Induced Hearing Loss*, edited by A. Axelsson, H. M. Borchgrevink, R. P. Hamernik, P. A. Hellstrom, D. Henderson, and R. J. Salvi, Thieme, New York, NY, 437–446.

Johnson, A. C., and Nylen, P. (1995). "Effects of Industrial Solvents on Hearing," *Occup. Med. State of the Art Rev. 10*, 623–640.

Vibration

Pelmear, P. L., Taylor, W., and Wasserman, D. E. (1992). *Hand–Arm Vibration: A Comprehensive Guide for Occupational Health Professionals*, Van Nostrand Reinhold, New York, NY.

Seto, W. W. (1964). *Schaum's Outline Series, Theory and Problems of Mechanical Vibrations*, McGraw-Hill Book Co., New York, NY.

Tempest, W. (1976). *Infrasound and Low Frequency Vibration*, Academic Press, New York, NY.

Wasserman, D. E. (1987). *Human Aspects of Occupational Vibration*, Elsevier Science Pub. B. V., Amsterdam, The Netherlands.

U. S. Government Regulations and Reports

Many government agencies are involved in developing regulations pertaining to noise, and others provide research and position statements for the scientific and regulatory communities.

DOD (Dept. of Defense; http://www.dtic.dla.mil/adm/direct_instruct.html)

"Occupational Noise Control and Hearing Conservation," NAVMEDCOM Instruction 6260.5, April 26, 1984.

"Design Criteria Standard, Noise Limits," MILSTD 1474D, February 12, 1997.

"Hearing Conservation Program," Air Force Occupational Safety and Health Standard 161-20, October 15, 1991 (under revision).

"Hearing Conservation," Department of the Army Pamphlet (DA PAM) 40-501, December 10, 1998.

"Hazardous Noise Program," Air Force Occupational Safety and Health Standard 48-19, March 1994.

"Navy Occupational Safety and Health Program Manual for Forces Afloat," Chapter B4, OPNAV Instruction 5100.19C, January 19, 1994.

"Navy Occupational Safety and Health Program Manual," Chapter 18, OPNAV Instruction 5100.23D, October 11, 1994.

"Noise Dosimetry and Risk Assessment," Army Technical Guide TG181, May 1994.

"DOD Hearing Conservation Program," Instruction 6055.12, April 22, 1996 [download from http://www.acq.osd.mil/ens/sh/NOISE.HTML].

DOE (Dept. of Energy; http://www.doe.gov/)

Worker Protection Management for DOE Federal and Contractor Employees, Order 440.1, September 1995.

Dept. of the Interior (http://www.doi.gov/indexj.html)

Bartholomae, R. C., and Parker, R. P. (1983). "Mining Machinery Noise Control Guidelines," A Bureau of Mines Handbook.

Department of the Interior Safety and Health Handbook (485 DM), June 10, 1991.

DOT (Dept. of Transportation, Washington, DC)

"Navigation and Vessel Inspection Circular No. 12-82, Recommendation on Control of Excessive Noise," U.S. Coast Guard Report, 1982.

"Qualification of Drivers; Hearing Deficiencies; Waivers; Proposed Rule and Notice," 49CFR Part 391 *Fed. Regist. 58*(239), 65634–65643, 1993.

"Vehicle Interior Noise Levels," 49CFR Ch. 111 Part 393.94, 1995.

"Revision of Airman Medical Standards and Certification Procedures and Duration of Medical Certificates; Final rule," 14CFR Parts 61 and 67, *Fed. Regist. 61*(54), 11238–11263, 1996.

EPA (Environmental Protection Agency, Washington, DC; http://www.epa.gov/access/ index.html)

EPA (1973). "Proceedings of the International Congress on Noise as a Public Health Problem," EPA Rept. 550/9-73-008.

EPA (1974). "Information on Levels of Environmental Noise Requisite to Protect Public Health and Welfare with an Adequate Margin of Safety," EPA Rept. 550/9-74-004.

EPA (1975). "Model Community Noise Control Ordinance," EPA Rept. 550/9-76-003.

EPA (1979). "Noise Labeling Requirements for Hearing Protectors," *Fed. Regist. 40* CFR Part 211, 44(190), 56130–56147.

Hattis, D., Ashford, N. A., Heaton, G. R., and Katz, J. I. (1976). "Some Considerations in Choosing an Occupational Noise Exposure Regulation," EPA Rept. 550/9-76-007.

Johnson, D. L. (1973). "Prediction of NIPTS Due to Continuous Noise Exposure," EPA Rept. 550/9-73-001-B.

Simpson, M., and Bruce, R. (1981). "Noise in America: Extent of the Noise Problem," EPA Rept. 550/9-81-101.

MSHA (Mine Safety and Health Administration, Pittsburgh, PA; http://www.msha.gov)

"Health Standards for Occupational Noise Exposure; Final Rule," 30 CFR Part 62, 64, *Fed. Reg.* 49458–49634, 49636–49637, 1999.

Valoski, M. P. (1994). "The Magnitude of the Noise-Induced Hearing Loss Problem in the Mining Industries," Informational Rept. No. 1220.

NIOSH (National Institute for Occupational Safety and Health, Cincinnati, OH; http://www.cdc.gov/niosh/homepage.html)

"Criteria for a Recommended Standard ... Occupational Exposure to Noise," U.S. Dept. HEW (NIOSH), Rept. HSM 73-11001 (1972).

"Criteria for a Recommended Standard ... Occupational Exposure to Hand–Arm Vibration," U.S. Dept. HHS (NIOSH), Pub. No. 89-106 (1989).

"Criteria for a Recommended Standard—Occupational Noise Exposure, Revised Criteria 1998," U.S. Dept. HHS (NIOSH) Pub. No. 98-126.

Franks, J. R., Stephenson, M. R., and Merry, C. J. (1996). "Preventing Occupational Hearing Loss—A Practical Guide," U.S. Dept. of HHS (NIOSH) Rept. 96-110.

Jensen, P., Jokel, C. R., and Miller, L. N. (1978). "Industrial Noise Control Manual (Revised Edition)," U.S. Department of Health, Education and Welfare (NIOSH), Contract No. 210-76-0149.

"NIOSH Publications on Noise and Hearing," U.S. Department of Health and Human Services, Division of Standards Development and Technology Transfer (1991).

OSHA (Occupational Safety and Health Administration; http://www.osha.gov)

"Occupational Noise Exposure," 29CFR1910.95 *Fed. Regist. 36*(105), 10518 (1971).

"Occupational Noise Exposure; Hearing Conservation Amendment," 29CFR1910.95 *Fed. Regist. 46*(11) 4078–4181 (1981).

"Occupational Noise Exposure; Hearing Conservation Amendment; Final Rule," 29CFR1910.95 *Fed. Regist. 46*(162), 42622–42639 (1983).

"OSHA Instruction CPL 2.45B CH-3, Violations of the Noise Standard, June 15," in *Field Operations Manual*, Government Institutes, Inc., Rockville, MD, IV-33-IV-36 (1992).

"OSHA Instruction TED 1.15, Noise Measurement, September 22, 1995," in *OSHA Technical Manual*, Government Institutes, Inc. Rockville, MD, II:5-1–II:5-18 (1996).

"Construction Industry Standards—Occupational Noise Exposure," 29CFR1926.52 (1983)

"Construction Industry Standards—Hearing Protection," 29CFR1926.101 (1983).

"Hearing Conservation Program Manual for Federal Agencies," prepared by Subcommittee on Occupational Noise, Federal Advisory Council on Occupational Safety and Health, OSHA 3089 (1984).

"Technical Manual, Section III, Chap. 5, Noise Measurement," on the web at http://www.osha-slc.gov/dts/osta/otm/otm_iii/otm_iii_5.html.

Other

"Environmental Noise Abatement," 32CFR Ch. V, Subpart G, 650.161–650.175, 1999.

Other Regulations and Reports

Australia

Mitchell, T., and Else, D. (1993). "Noise Control in Mining: Seventy-Five Noise Control Solutions," Worksafe Australia—Victorian Institute of Occupational Safety and Health, Ballarat University College.

Canada

"National Guidelines for Environmental Noise Control, Procedures and Concepts for the Drafting of Environmental Noise Regulations/By-Laws in Canada," Working Group on Environmental Noise of Federal/Provincial Advisory Committee on Environmental and Occupational Health, Ministry of National Health and Welfare, Environmental Health Directorate, Health Protection Branch, Ottawa, Canada (1989).

European Community

EEC (1986). "Council Directive of 12 May, 1986 on the Protection of Workers from the Risks Related to Exposure to Noise at Work," Doc. 86/188/EEC, Official Journal of the European Communities No. L137, 28–34.

United Kingdom

"Health and Safety—The Noise at Work Regulations," Document No. 1790, HMSO, United Kingdom, 1989.

Appendix III

Properties of Materials and Engineering Conversions

Properties of Solids, Liquids, and Gasses[*]

Solids

	Density kg/m³	Young's Modulus N/m²×10¹⁰	Shear Modulus N/m²×10¹⁰	Bulk Modulus N/m²×10¹⁰	Poisson's Ratio	Wave Velocity m/s		Characteristic Impedance MKS rayls ×10⁶	
						Bar	Bulk	Bar	Bulk
Aluminum	2700	7.1	2.4	7.5	0.33	5150	6300	13.9	17.0
Brass	8500	10.4	3.8	13.6	0.37	3500	4700	29.8	40.0
Copper	8900	12.2	4.4	16.0	0.35	3700	5000	33.0	44.5
Iron (cast)	7700	10.5	4.4	8.6	0.28	3700	4350	28.5	33.5
Lead	11300	1.65	0.55	4.2	0.44	1200	2050	13.6	23.2
Nickel	8800	21.0	8.0	19.0	0.31	4900	5850	43.0	51.5
Silver	10500	7.8	2.8	10.5	0.37	2700	3700	28.4	39.0
Steel	7700	19.5	8.3	17.0	0.28	5050	6100	39.0	47.0
Glass (Pyrex)	2300	6.2	2.5	3.9	0.24	5200	5600	12.0	12.9
Quartz (X-cut)	2650	7.9	3.9	3.3	0.33	5450	5750	14.5	15.3
Lucite	1200	0.4	0.14	0.65	0.4	1800	2650	2.15	3.2
Concrete	2600	-	-	-	-	-	3100	-	8.0
Ice	920	-	-	-	-	-	3200	-	2.95
Cork	240	-	-	-	-	-	500	-	0.12
Oak	720	-	-	-	-	-	4000	-	2.9
Pine	450	-	-	-	-	-	3500	-	1.57
Rubber (hard)	1100	0.23	0.1	0.5	0.4	1450	2400	1.6	2.64
Rubber (soft)	950	0.0005	-	0.1	0.5	70	1050	0.065	1.0
Rubber (rho-c)	1000	-	-	0.24	-	-	1550	-	1.55

[*]Kinsler, L. E., Frey, A. R., Coppens, A. B., and Sanders, J. V. "Tables of Physical Properties of Matter," *Fundamentals of Acoustics, Third Edition*, New York: John Wiley & Sons, Inc., 1982, pp. 461–463. Reprinted by permission John Wiley & Sons, Inc.

Liquids

	Temperature °C	Density kg/m³	Bulk Modulus N/m² ×10⁹	Ratio of Specific Heats	Speed of Sound m/s	Characteristic Impedance MKS rayls 10⁶	Coefficient of Viscosity N-s/m²
Water (fresh)	20	998	2.18	1.004	1481	1.48	0.001
Water (sea)	13	1026	2.28	1.01	1500	1.54	0.001
Alcohol (ethyl)	20	790	-	-	1150	0.91	0.0012
Castor oil	20	950	-	-	1540	1.45	0.96
Mercury	20	13600	25.3	1.13	1450	19.7	0.0016
Turpentine	20	870	1.07	1.27	1250	1.11	0.0015
Glycerin	20	1260	-	-	1980	2.5	1.2

Gasses (at a pressure of 1.013×10^5 N/m²)

	Temperature °C	Density kg/m³	Ratio of Specific Heats	Speed of Sound m/s	Characteristic Impedance MKS rayls 10⁶	Coefficient of Viscosity N-s/m²
Air	0	1.29	1.402	331.6	428	0.000017
Air	20	1.21	1.402	343	415	0.0000181
Oxygen	0	1.43	1.40	317.2	453	0.00002
CO_2 (low freq.)	0	1.98	1.304	258	512	0.0000145
CO_2 (high freq.)	0	1.98	1.40	268.6	532	0.0000145
Hydrogen	0	0.09	1.41	1269.5	114	0.0000088
Steam	100	0.6	1.324	404.8	242	0.000013

Conversion and Equivalence of Units

Quantity	SI Equivalence		English Equivalence	
Mass	1 lb_f - s^2/ft (slug)	= 14.594 kg	1 kg	= 2.2046 lb_m
	1 lb_f - s^2/ft (slug)	= 32.174 lb_m		
	1 lbm	= 0.45359 kg	1 kg	= 0.068522 slug (lb_f - s^2/ft)
Length	1 in	= 0.0254 m	1 m	= 39.370 in
	1 ft	= 0.3048 m	1 m	= 3.2808 ft
	1 mile (5280 ft)	= 1.6093 km	1 km	= 0.62137 mile
Area	1 in²	= 6.4516×10^{-4} m²	1 m²	= 1550.0 in²
	1 ft²	= 9.2903×10^{-2} m²		= 10.763 ft²
Volume	1 in³	= 1.6387×10^{-5} m³	1 m³	= 6.1023×10^4 in³
	1 ft³	= 2.8317×10^{-2} m³	1 m³	= 35.315 ft³
	1 US gallon	= 3.7854 liter	1 liter	= 0.26417 US gallon
	1 US gallon	= 3.7854×10^{-3} m³		

Conversion and Equivalence of Units
(continued)

Quantity	SI Equivalence		English Equivalence	
Force or weight	1 lb$_f$	= 4.4482 N	1 N	= 0.22481 lb$_f$
Torque or moment	1 lb$_f$ - in 1 lb$_f$ - ft	= 0.11298 N-m = 1.3558 N-m	1 N - m 1 N - m	= 8.8507 lb$_f$ - in = 0.73756 lb$_f$ - ft
Stress, pressure, or elastic modulus	1 lb$_f$/in^2 (psi) 1 lb$_f$/ft^2	= 6894.8 Pa = 47.880 Pa	1 Pa (N/m^2) 1 Pa	= 1.4504 × 10^{-4} lb$_f$/in^2 (psi) = 208.85 × 10^{-4} lb$_f$/ft^2
Mass density	1 lb$_m$/in^3 1 lb$_m$/ft^3	= 27.680 × 10^3 kg/m^3 = 16.018 kg/m^3	1 kg/m^3 1 kg/m^3	= 36.127 × 10^{-6} lb$_m$/in^3 = 62.428 × 10^{-3} lb$_m$/ft^3
Work or energy	1 lb$_f$ - in 1 lb$_f$ - ft 1 Btu 1 kWh	= 0.11298 J = 1.3558 J = 1055.350 J = 3.6 × 10^6 J	1 J 1 J 1 J 1 J	= 8.8507 in - lb$_f$ = 0.73756 ft - lb$_f$ = 0.94782 × 10^{-3} Btu = 0.27778 × 10^{-6} kWh
Power	1 lb$_f$ /s - in 1 lb$_f$ /s - in 1 lb$_f$ /s - in 1 hp	= 0.11298 W = 1.3558 W = 0.0018182 hp = 745.7 W	1 W 1 W 1 hp 1 W	= 8.8507 lb$_f$ /s - in = 0.73756 lb$_f$ /s - ft = 550 lb$_f$ - ft/s = 1.3410 × 10^{-3} hp
Area moment of inertia or second moment of area	1 in^4 1 ft^4	= 41.623 × 10^{-8} m^4 = 86.310 × 10^{-4} m^4	1 m^4 1 m^4	= 240.25 × 10^4 in^4 = 115.86 ft^4
Mass moment of inertia	1 lb$_f$-in-s^2	= 0.11298 kg-m^2	1 m^2 - kg	= 8.8507 lbf - in - s^2
Spring constant: Translational Torsional	 1 lb$_f$ / in 1 lb$_f$ / ft 1 lb$_f$ - in /rad 1 lb$_f$ - ft/rad	 = 175.13 N/m = 14.594 N/m = 0.11298 N-m/rad = 1.3558 N-m/rad	 1 N/m 1 N/m 1 N - m/rad 1 N - m/rad	 = 5.7102 × 10^{-3} lb$_f$ /in = 68.522 × 10^{-3} lb$_f$ /ft = 8.8507 lb$_f$ - in/rad = 0.73756 lb$_f$ - ft/rad
Damping constant: Translational Torsional	 1 lb$_f$ - s/in 1 in - lb$_f$ - s/rad	 = 175.13 N - s/m = 0.11298 N - s/rad	 1 N - s / m 1 N - m - s/rad	 = 5.7102 × 10^{-3} lb$_f$ - s/in = 8.8507 × lb$_f$ - in - s/rad

Angular Equivalents

1 rad	=	57.296 degrees
1 degree	=	0.017453 rad
1 rpm	=	0.016667 rev/s = 0.10472 rad/s
1 rad/s	=	9.5493 rpm
π rad	=	180°
2 π rad/cycle × 1 cycle/s	=	2 π rad/s

Pertinent Equivalents

0 dB SPL	=	20 µPa = 2 × 10^{-5} Pa
1 Pa	=	1 N/m^2 = 94 dB
1 atmosphere	=	1.013 × 10^5 N/m^2 = 194 dB
L_{A8hn} of 85	=	3640 Pa^2 s ~ 1.0 Pa^2 h
1 Joule	=	1 N - m
10 dyne/cm^2	=	1 N/m^2
10^5 dyne	=	1 N

Reference Sources

Beranek, L. L. (1988). *Noise and Vibration Control*, Inst. Of Noise Control Eng., Washington, DC.

IEEE/ASTM (1997). "Standard for Use of the International System of Units (SI): The Modern Metric System," SI 10-1997, New York, NY.

Irwin, J. D., and Graf, E. R. (1979). *Industrial Noise and Vibration Control*, Prentice Hall, Englewood Cliffs, NJ.

Kinsler, L. E., Frey, A. R., Coppens, A. B., and Sanders, J. V. (1982). *Fundamentals of Acoustics, Third Edition*, John Wiley & Sons, New York, NY.

Mechtly, E. A. (1973). *The International System of Units (2nd Revision)*, NASA SP-7012.

NIOSH (1975). "Compendium of Materials for Noise Control," Contract No. HSM 99-72-99, National Institute for Occupational Safety and Health, Cincinnati, OH.

Rao, S. S. (1995). *Mechanical Vibrations, 3rd Edition*, Addison-Wesley, Reading, MA.

Ver, I. L., and Holmer, C. I. (1988). "Interaction of Sound Waves with Solid Structures," in *Noise and Vibration Control*, edited by L. L. Beranek, McGraw-Hill, New York, NY.

Wandmacher, C. (1978), *Metric Units in Engineering—Going SI*, Industrial Press, New York, NY.

The Noise Manual, revised 5th edition, edited by E.H. Berger,
L.H. Royster, J.D. Royster, D.P. Driscoll, and M. Layne
©2003 American Industrial Hygiene Association

NHCA Professional Guide for Audiometric Baseline Revision

Appendix

IV

(reprinted with permission of the National Hearing Conservation Association)

What Is Baseline Revision?

In a hearing conservation program (HCP), each employee's baseline audiogram gives the reference hearing thresholds for that individual. The results of later monitoring audiograms are compared to the baseline to detect significant changes in hearing thresholds. When significant shifts for the worse are identified, follow-up actions are taken to improve employee protection from noise.

As specified in the Hearing Conservation Amendment (CFR 1910.95) promulgated by the Occupational Safety and Health Administration (OSHA), the baseline may be revised by the reviewing audiologist or physician either for significant improvement in measured thresholds or for persistent standard threshold shift (STS).

Because the baseline audiogram is so important for detecting hearing change and reacting to prevent additional change, NHCA assigned a special committee to develop guidelines for revising audiometric baselines. The 16-member committee conducted research and evaluated various strategies over several years. The guidelines given here, which were approved by the board of NHCA in March 1996, represent the consensus of the committee. Following these guidelines will provide consistency across professional reviewers and audiometric testing service providers, thereby increasing the degree of protection for noise-exposed workers.

Note: although the guidelines require persistence of hearing changes before the baseline is revised, protective follow-up actions for the employee are needed as soon as significant changes for the worse are first shown.

Definitions

OSHA STS: OSHA defines a standard threshold shift (STS) as a change for the worse in either ear of 10 dB or more in the average of thresholds at 2, 3, and 4 kHz, relative to the baseline.

Significant Improvement: OSHA does not specify a definition of significant improvement. However, an example in Appendix F of the Hearing Conservation Amendment illustrates revision of the baseline after an improvement of 5 dB in the average of hearing thresholds at 2, 3, and 4 kHz.

Baseline Audiogram: Initially the baseline is the latest valid audiogram obtained before entry into the HCP. If no appropriate pre-entry audiogram exists, baseline is the first valid audiogram obtained within 6 months of entry into the HCP (12 months for mobile testing). OSHA requires 14 hours of quiet prior to the original baseline.

Monitoring Audiograms: Subsequent to the baseline audiogram, new audiograms are obtained at least annually. To increase the preventive function of audiometry, many professionals suggest performing annual audiograms during the workshift in order to detect any noise-related temporary threshold shifts which may occur.

Age Corrections: OSHA permits optional application of age correction values (from Appendix F) to annual audiograms when comparing them to baseline for detection of STS, in order to account for median values of age change. Note: many professionals feel that if intervention for threshold shifts is delayed until after age-corrected STS has occurred, then significant hearing changes will not receive needed follow-up attention.

How to Use NHCA's Guidelines

Professional Review

These guidelines are meant to be employed only by a professional reviewer (audiologist or physician). Although the guidelines can be programmed by computer to identify records for potential revision, the final decision for revision rests with a human being. Because the goal of the guidelines is to foster consistency among professional reviewers, human over-ride of the guidelines must be justified by specific concrete reasons.

Separate Consideration of Each Ear

Each monitoring audiogram is compared to the baseline to detect improvement or OSHA STS (or other significant shifts). The two ears are examined separately and independently. If one ear meets the criteria for revision of baseline, then the baseline is revised for that ear only. Therefore, if the two ears show different hearing trends, the baseline for the left ear may be from one test date, while the baseline for the right ear may be from a different test date.

Use of Age Corrections

Age corrections do not apply in considering revisions for improvement. The audiologist or physician may choose whether to apply OSHA-allowed age corrections in evaluating baseline revision for persistent OSHA STS. Rule 2 operates in the same way whether optional age corrections are used or not.

Application Exceptions

These guidelines for baseline revision do not apply to the calculation of the 25-dB average shifts which in many states are recordable on the OSHA log for occupational illness and injury. The original baseline is the appropriate reference for that purpose. Neither do the guidelines apply to identification of other (non-STS) significant threshold shifts for the worse, which may be communicatively or medically important.

The Guidelines

Rule 1: Revision for Improvement

If the average of thresholds for 2, 3 and 4 kHz for either ear shows an improvement of 5 dB or more from the baseline value, and the improvement is present on one test and persistent on the next test, then the record should be identified for review by the audiologist or physician for potential revision of the baseline. The baseline for that ear should be revised to the improved test which shows the lower (more sensitive) value for the average of thresholds at 2, 3, and 4 kHz, unless the audiologist or physician determines and documents specific reasons for not revising. If the values of the three-frequency average are identical for the two tests, then the earlier test becomes the revised baseline.

Rule 2: Revision for Persistent OSHA Standard Threshold Shift

If the average of thresholds for 2, 3 and 4 kHz for either ear shows a worsening of 10 dB or more from the baseline value (OSHA STS), and the STS persists on the next annual test (or the next test given at least 6 months later), then the record should be identified for review by the audiologist or physician for potential revision of the baseline for persistent worsening. Unless the audiologist or physician determines and documents specific reasons for not revising, the baseline for that ear should be revised to the STS test which shows the lower (more sensitive) value for the average of thresholds at 2, 3, and 4 kHz. If both STS tests show the same numerical value for the average of 2, 3, and 4 kHz, then the audiologist or physician should revise the baseline to the earlier of the two tests, unless the later test shows better (more sensitive) thresholds for other test frequencies.

Following an STS, a retest within 30 days of the annual test may be substituted for the annual test if the retest shows better (more sensitive) results for the average threshold at 2, 3, and 4 kHz. If the retest is used in place of the annual test, then the annual test is retained in the record, but it is marked in such a way that it is no longer considered in baseline revision evaluations.

EXAMPLE

Male employee "L.M." born 10/05/63   See far right column for test date.

Test Age	Type	Left Ear Thresholds (dB) by frequency (kHz)							No Age Correction STS avg.	Change	Reviewer decision	Baseline status	With Age Correction (does not apply to improvement) Corr. avg.	Corrected change	Reviewer decision	Baseline status
		.5	1	2	3	4	6	8								
22	initial	10	5	5	10	25	40	15	13.3			B				B
23	annual	5	5	5	5	25	45	20	11.7	-1.7						
24	annual	0	-5	0	0	25	35	10	8.3	-5.0 impr		RB ↗				RB
25	annual	0	-5	0	0	25	40	20	8.3	-5.0 impr	revise					
26	annual	0	-5	-5	0	20	40	20	5.0	-3.3						
27	annual	0	0	0	5	25	30	10	10.0	1.7			9.0	0.7		
28	annual	5	5	5	5	35	40	15	15.0	6.7			13.7	5.3		
29	annual	5	0	5	10	40	40	15	18.3	10.0 STS		RB ↗	17.0	8.7		
30	annual	10	0	5	10	45	45	20	20.0	11.7 STS			18.3	10.0 STS		
30	retest	10	0	5	10	40	40	15	18.3	10.0 STS	revise		16.7	8.3		
31	annual	5	0	5	15	50	50	30	23.3	5.0			21.3	13.0 STS		
31	retest	5	5	5	10	50	50	35	21.7	3.3			19.7	11.3 STS		RB ↗
32	annual	5	0	5	15	55	55	40	25.0	6.7			22.3	14.0 STS		
32	retest	5	0	5	15	50	55	35	23.3	5.0			20.7	12.3 STS	revise	
33	annual	5	5	5	20	55	55	35	26.7	8.3			26.0	4.3		

If a retest within 30 days of an annual test confirms an OSHA STS shown on the annual test, the baseline will not be revised at that point because the required six-month interval between tests showing STS persistence has not been met. The purpose of the six-month requirement is to prevent baseline revision when STS is the result of temporary medical conditions affecting hearing. Although a special retest after six months could be given if desired to assess whether the STS is persistent, in most cases the next annual audiogram would be used to evaluate persistence of the STS.

Example Description

The example above illustrates how the baseline revision guidelines apply to one audiometric record. The abbreviations used are: B for baseline, RB for revised baseline, STS for OSHA STS, and impr for improvement. Revisions are shown both without use of age corrections, as well as with use of OSHA age corrections (with the choice being up to the professional in charge of revision). In the left ear, baseline is revised in 1988 for persistent improvement, to the test of 11/12/87. Subsequently the left ear shows persistent STS, with revision after the 1993 retest to the test of 5/21/92 (without using age corrections). With age corrections, the left ear shows persistent STS in 1995, with baseline revised to the test of 6/25/94. In the right ear, baseline revision for persistent STS without age corrections occurs in 1994 to the test of 5/28/93. With age corrections, the right baseline is revised in 1996 to the test of 06/01/95.

Note that the table shows values rounded to one tenth of a decibel, resulting in some apparent errors of one-tenth in the columns showing change from baseline. For example, one comparison indicates in the table values that $19.7 - 8.3 = 11.3$ because the underlying values are really $19.67 - 8.33 = 11.34$.

Right Ear Thresholds (dB) by frequency (kHz)							No Age Correction				With Age Correction (does not apply to improvement)				Date of Test
.5	1	2	3	4	6	8	STS avg.	Change	Reviewer decision	Baseline status	Corr. avg.	Corrected change	Reviewer decision	Baseline status	
15	10	5	5	20	40	15	10.0			B				B	08/03/85
10	5	0	5	25	35	10	10.0	0							11/04/86
0	-5	0	5	25	40	20	10.0	0							11/12/87
0	0	0	0	30	35	15	10.0	0							10/15/88
0	-5	0	5	35	40	10	13.3	3.3			12.0	2.0			12/12/89
5	0	0	0	40	40	15	13.3	3.3			11.7	1.7			11/23/90
0	0	0	5	40	40	15	15.0	5.0			13.0	3.0			04/14/91
5	0	0	5	45	50	20	16.7	6.7			14.7	4.7			05/21/92
10	5	5	10	55	45	20	23.3	13.3 STS			21.0	11.0 STS			05/22/93
5	0	0	10	50	45	15	20.0	10.0 STS		RB	17.7	7.7			05/28/93
10	5	0	10	50	55	20	20.0	10.0 STS			17.3	7.3			06/10/94
5	5	0	10	50	55	20	20.0	10.0 STS	revise		17.3	7.3			06/25/94
5	5	5	15	60	60	25	26.7	6.7			23.3	13.3 STS			05/07/95
5	0	5	15	55	65	30	25.0	5.0			21.7	11.7 STS		RB	06/01/95
10	0	5	10	60	60	25	25.0	5.0			21.7	11.7 STS	revise		05/02/96

Also recall that age corrections are not applied to baseline tests, but only to annual tests. Therefore, in the sections showing calculations with age corrections, the "corrected change" column shows change from the STS average without age corrections for the currently applicable baseline compared to the STS average with age corrections on the current annual test.

Names Index

This is one of two indexes for this text. It contains a complete listing of all names found throughout the text including those in end-of-chapter reference lists. A subject index can be found subsequent to this names index. Locators followed by the letter 'f' indicate a figure credited, and those followed by a 't' indicate a table credited.

Subject Index

This is one of two indexes for this text. A names index can be found on the preceding pages. Locators followed by an 'eq' indicate equations, those followed by 'f' indicate figures, those followed by 'ff' indicate the occasional mention of a term over the page range, those using an 'n' indicate a note, and those followed by 't' indicate a table. Abbreviations and acronyms are found under the abbreviation/acronym or the spelled-out term, depending upon the commonality of the term. See *Symbols and Abbreviations* (xiii – xx) for a complete list of abbreviations and acronyms.